SUBSURFACE RESTORATION

Edited by

C.H. Ward
National Center for Ground Water Research
Rice University

J.A. Cherry
Waterloo Centre for Groundwater Research
University of Waterloo

M.R. Scalf
Robert S. Kerr Environmental Research Laboratory
U.S. Environmental Protection Agency

ANN ARBOR PRESS, INC.
Chelsea, Michigan

Library of Congress Cataloging-in-Publication Data on file.

ISBN 1-57504-060-3

ANN ARBOR PRESS, INC.
121 South Main Street, Chelsea, Michigan 48118
Ann Arbor Press is an imprint of Sleeping Bear Press

PRINTED IN THE UNITED STATES OF AMERICA
1 2 3 4 5 6 7 8 9 0

About the Editors

C.H. (Herb) Ward

Dr. Ward is the Foyt Family Chair of Engineering in the George R. Brown School of Engineering at Rice University. He is also Professor of Environmental Science and Engineering and Ecology and Evolutionary Biology.

Dr. Ward has undergraduate (B.S.) and graduate (M.S. and Ph.D.) degrees from New Mexico State University and Cornell University, respectively. He also earned the M.P.H. in environmental health from the University of Texas.

Following 22 years as Chair of the Department of Environmental Science & Engineering at Rice University, Dr. Ward is now Director of the Energy and Environmental Systems Institute (EESI), a university-wide program designed to mobilize industry, government, and academia to focus on problems related to energy production and environmental protection.

Dr. Ward is also Director of the Department of Defense Advanced Applied Technology Demonstration Facility (AATDF), a distinguished consortium of university-based environmental research centers supported by consulting environmental engineering firms to guide selection, development, demonstration, and commercialization of advanced applied environmental restoration technologies for the DOD. For the past 16 years he has directed the activities of the National Center for Ground Water Research (NCGWR), a consortium of universities charged with conducting long-range exploratory research to help anticipate and solve the nation's emerging groundwater problems. He is also Co-Director of the EPA-sponsored Hazardous Substances Research Center/South & Southwest (HSRC/S&SW), whose research focus is on contaminated sediments and dredged materials.

Dr. Ward has served as President of both the American Institute of Biological Sciences and the Society for Industrial Microbiology, and is currently Vice President of the U.S. National Committee of the International Water Resources Association. He is the founding and current and Editor-in-Chief of the international journal *Environmental Toxicology and Chemistry*.

John A. Cherry

John A. Cherry received his B.S. degree in geological engineering from the University of Saskatchewan in 1962, his M.S. degree from the University of California at Berkeley in 1964 and a Ph.D. in hydrogeology from the University of Illinois in 1966. His research for the past thirty years has focused on field studies of groundwater contamination, including development of monitoring techniques, assessment of hydrogeologic factors in selection, design, and performance of waste disposal facilities and the integration of field information with laboratory and modeling studies for assessment of site contamination and remedial options.

Dr. Cherry has been a professor since 1971 in the Department of Earth Sciences at the University of Waterloo, where he teaches graduate students in contaminant hydrogeology and directs a university-consortium research program on subsurface contamination by common industrial organic chemicals (DNAPLs) under the sponsorship of major multi-national corporations and Canadian government agencies. From 1981–1987 he was Director of Waterloo's Institute for Groundwater Research. He presently holds a Chair in Contaminant Hydrogeology and is an Associate of Waterloo Centre for Groundwater Research. As a consultant to corporations and government agencies, he has assessed various problems associated with subsurface environmental problems pertaining to hazardous waste and industrial-site management. For his contributions to knowledge on groundwater contamination, Dr. Cherry has received awards from the American Geophysical Union, the Geological Society of America, the National Water Well Association (U.S.A.) and the Ontario government. He is a co-author of a textbook on hydrogeology, *Groundwater,* published by Prentice-Hall in 1979, and is also co-editor of a new reference book titled *Dense Chlorinated Solvents and other DNAPLs in Groundwater,* published by Waterloo Press in 1995.

M.R. (Dick) Scalf

Marion R. Scalf was born and grew up in McKinney, Texas. He earned a B.S. in Civil Engineering in 1962 and a M.S. in Environmental Engineering in 1963 from the University of Texas at Austin. He began his career with the U.S. Public Health Service in the United States nuclear testing program in Nevada. From 1965 to 1993, he held a succession of positions with EPA's Robert S. Kerr Environmental Research Laboratory in Ada, Oklahoma. In 1967, he was one of a small group of scientists and engineers that initiated the first major environmental research program directed at the protection and restoration of groundwater quality. This program was based on the development of scientific information on the processes that control contaminant movement in the subsurface environment and encompassed cooperative efforts with over one hundred different university and other research organizations throughout the United States and several foreign countries. In 1987, he was instrumental in initiating and developing EPA's highly successful Superfund technical support program, directed at applying the latest scientific information to remediation decision making.

Preface

Hindsight and a large body of evidence suggest that no other area of environmental management has been more misunderstood, complicated, expensive, wasteful, and controversial than subsurface remediation. Although groundwater quantity issues have been studied for many decades, groundwater quality science was almost totally ignored until the mid-1970s. This lack of attention and understanding resulted not only in extensive contamination and ignorance of the magnitude of the problem, but in unrealistic expectations for remediation. Closing the knowledge gap has been complicated by the heterogeneity of the subsurface and its relative inaccessibility for examination. And, although over two decades of research has made great strides in our understanding of subsurface science, that collective knowledge has not often been focused on remediation decisions that have proved to be both effective and efficient.

In the United States, implementation of the Comprehensive Environmental Response, Compensation, and Liability Act (CERCLA/"Superfund") of 1980 and the Resource Conservation and Recovery Act (RCRA) of 1976 led to the discovery of massive subsurface contamination—coming largely from leaking underground storage tanks at gasoline stations and fuel tank farms (depots) and industries and federal agencies where chlorinated solvents were widely used in large quantities for degreasing operations. From the early 1960s to the late 1980s, chlorinated solvents were used extensively in food and cosmetics production, metal fabrication, manufacture and repair of heat exchangers, electric motors, transformers, and electronics, and in most areas of computer development.

Hindsight also tells us that the loss of these products (fuels and solvents) to the environment resulted, for the most part, from their wide-scale use in commerce and from inadequate knowledge on which to base materials (chemicals) handling practices protective of human health and the environment. Hence, massive contamination of the subsurface exists, and our best current efforts at pollution prevention will not restore groundwater already contaminated to levels of quality required for beneficial use.

Subsurface and groundwater remediation efforts to date have largely focused, with generally poor results, on groundwater extraction and above ground treatment, commonly referred to as pump-and-treat technology. The presence and behavior of non-aqueous phase liquids (LNAPLs and DNAPLs) in the subsurface are now known to be the primary factors limiting the effectiveness of pump and treat approaches for groundwater cleanup (*Alternatives for Ground Water Cleanup,* National Academy Press, Washington, DC, 315 pp, 1994).

Studies by the U.S. Environmental Protection Agency and the National Academy of Science have concluded that the alternatives to pump-and-treat technology are few. The question then is, "Where do we go from here?" To explore this question and help assess the state-of-knowledge of subsurface science believed requisite to finding new solutions, we organized a multidisciplinary international conference and published a proceedings (*Subsurface Restoration Conference*, Rice University, Houston Texas, 342 pp, 1992). The conference helped to set the research agenda, and this book, long in the making, has further developed the concepts that guided the structure and organization of the conference.

We hope this treatise will be a useful reference for those making decisions on environmental remediation and to those involved in development of advanced subsurface remediation technology.

C.H. Ward
J.A. Cherry
M.R. Scalf

Contributing Authors

H.J. Albrechtsen
Technical University of Denmark
Department of Environmental Engineering
Building 115 D.T.H.
Lyngby DK-2800
Denmark

B.C. Alleman
Senior Engineer
Battelle Memorial Institute
505 King Avenue
Columbus, OH 43214

Mary P. Anderson
Professor of Geology and Geophysics
University of Wisconsin–Madison
Department of Geology and Geophysics
1215 W. Dayton Street
Madison, WI 53706

Michael D. Annable
Assistant Professor
Environmental Engineering Science
Department
University of Florida
Gainesville, FL 32611

Denie C.M. Augustijn
Assistant Professor
School of Management Studies
University of Twente
P.O. Box 217
7500 AE Enschede
The Netherlands

William P. Ball
Assistant Professor
Department of Geography and
Environmental Engineering
Johns Hopkins University
313 Ames Hall
Baltimore, MD 21218-2686

J.E. Barbash
Water Resources Division
U.S. Geological Survey
345 Middlefield Road
Menlo Park, CA 94025

James F. Barker
Professor
University of Waterloo
Centre for Groundwater Research
Waterloo, Ontario N2L 3G1
Canada

Philip B. Bedient
Chairman and Professor
Department of Environmental Science
and Engineering
Rice University
6100 Main Street, Mail Stop 317
Houston, TX 77005

Richard A. Brown
Director
Remediation Technology
Groundwater Technology, Inc.
310 Horizon Center Drive
Trenton, NJ 08691

John A. Cherry
Professor
Centre for Groundwater Research
University of Waterloo
200 University Avenue West
Waterloo, Ontario N2L 3G1
Canada

Thomas H. Christensen
Professor of Environmental
Engineering
Technical University of Denmark
Department of Environmental Science and
Engineering
Building 115 D.T.H.
Lyngby DK-2800
Denmark

G.P. Curtis
Environmental Engineer
Water Resources Division
U.S. Geological Survey
345 Middlefield Road
Menlo Park, CA 94025

Carl G. Enfield
Director, Processes and Systems Research Division
U.S. EPA Robert S. Kerr Environmental Research Laboratory
P.O. Box 1198
Ada, OK 74820

R.A. Ettinger
Shell Development
P.O. Box 1380
Houston, TX 77251

S. Feenstra
Applied Groundwater Research Ltd.
The Pentagon Building, Suite 207
2550 Argentia Road
Mississauga, Ontario L5N 5R1
Canada

W. Fish
Associate Professor
Oregon Graduate Institute
Department of Environmental Science and Engineering
P.O. Box 91000
Portland, OR 97291-1000

John C. Fountain
Professor
SUNY at Buffalo
Department of Geology
876 Natural Sciences & Mathematics Complex
Buffalo, NY 14260

R. Allan Freeze
Consultant
R. Allan Freeze Engineering, Inc.
3755 Nico Wynd Drive
White Rock, BC V4P 1J1
Canada

Robert W. Gillham
Director, Centre for Groundwater Research
University of Waterloo
Waterloo, Ontario N2L 3G1
Canada

G. Heron
Technical University of Denmark
Department of Environmental Engineering
Building 115 D.T.H.
Lyngby DK-2800
Denmark

Robert E. Hinchee
Senior Technical Manager
Parsons Engineering Science, Inc.
406 West South Jordan Parkway, Suite 300
South Jordan, UT 84095

Ronald E. Jessup
Scientific Programmer
Soil and Water Science Department
University of Florida
Gainesville, FL 32611-0290

Paul C. Johnson
Associate Professor of Engineering
Arizona State University
Department of Civil and Environmental Engineering
Tempe, AZ 85287-5306

B.M. Kurtyka, Esq.
Program Manager
BDM International, Inc.
1501 BDM Way
McLean, VA 20874

Linda S. Lee
Assistant Professor
Agronomy Department
Purdue University
West Lafayette, IN 47907

David N. Lerner
Professor of Environmental Management
University of Bradford
Civil and Environmental Engineering
Bradford BD7 1DP
United Kingdom

Stephen C.T. Lien
Director
Office of Environmental Management
U.S. Department of Energy
EM-54, Office of Technology Integration
20400 Century Blvd., Cloverleaf Bldg.
Germantown, MD 20874

John Lyngkilde
Technical University of Denmark
Department of Environmental Engineering
Building 115 D.T.H.
Lyngby DK-2800
Denmark

D.M. Mackay
Waterloo Centre for Groundwater Research
University of Waterloo
Waterloo, Ontario N2L 3G1
Canada

Ghislain de Marsily
Professor
University of Paris VI
Laboratoire de Géologie Appliquée
4 Place Jussieu, B123
Paris 75252 Cedex 05
France

Perry L. McCarty
Professor of Civil Engineering
Stanford University
Department of Civil Engineering
Stanford, CA 94305-4020

James W. Mercer
President
GeoTrans, Inc.
46050 Manekin Plaza, Suite 100
Sterling, VA 22170

James K. Mitchell
University Distinguished Professor
Virginia Polytechnic Institute and State University
Department of Civil Engineering
Blacksburg, VA 24061-0105

S.C. Mravik
Research Soil Scientist
U.S. EPA NRMRL/SPRD
P.O. Box 1198
Ada, OK 74820

Per H. Nielsen
Technical University of Denmark
Department of Environmental Engineering
Building 115 D.T.H.
Lyngby DK-2800
Denmark

Carl D. Palmer
Assistant Professor
Oregon Graduate Institute
Environmental Science and Engineering
19600 NW Von Neumann Drive
Beaverton, OR 97006-1999

R.M. Parker
Senior Engineer
Formerly, GeoTrans, Inc.

P. Suresh C. Rao
Graduate Research Professor
Soil and Water Science Department
University of Florida
Gainesville, FL 32611-0290

Martin Reinhard
Professor, Research
Stanford University
Department of Civil Engineering
Stanford, CA 94305-4020

Hanadi S. Rifai
Executive Director
Energy and Environmental Systems Institute
Rice University
6100 Main Street, Mail Stop 316
Houston, TX 77005

P.V. Roberts
Professor of Environmental Engineering
Department of Civil Engineering
Terman Engineering Center
Stanford University
Stanford, CA 94305-4020

Lewis Semprini
Associate Professor
Department of Civil, Construction, and Environmental Engineering
Oregon State University
202 Apperson Hall
Corvallis, OR 97331-2302

T.E. Short
Research Chemical Engineer
U.S. EPA NRMRL/SPRD
P.O. Box 1198
Ada, OK 74820

C.P. Spalding
Senior Hydrogeologist
GeoTrans, Inc.
46050 Manekin Plaza, Suite 100
Sterling, VA 20166

J.M. Thomas
7307 Churchill Road
McLean, VA 22101

Kent S. Udell
Associate Professor
University of California–Berkeley
Department of Mechanical Engineering
6165 Etcheverry Hall
Berkeley, CA 94720

W.A. Narin van Court
Geotechnical Engineer
Dames & Moore
140 Ohio Hill Road
Fairfield, ME 04937

C.H. Ward
Foyt Family Chair of Engineering
Director, Energy and Environmental Systems Institute
Professor, Department of Environmental Science and Engineering
Rice University
6100 Main Street, Mail Stop 316
Houston, TX 77005

John L. Wilson
Professor of Hydrology
New Mexico Institute of Mining and Technology
Building MSEC, Room 216
Socorro, NM 87801

John T. Wilson
Senior Research Microbiologist
U.S. EPA Robert S. Kerr Environmental Research Laboratory
P.O. Box 1198
Ada, OK 74820

A. Lynn Wood
Soil Scientist
National Risk Management Research Lab
U.S. Environmental Protection Agency
P.O. Box 1198
Ada, OK 74820

Contents

Section VII. Overview and Research Directions

Section I

Regulatory Strategy

Chapter 1

Regulatory Compliance Strategies for Subsurface Restoration

E.C. Jonlin, B.M. Kurtyka, and S.C.T. Lien

INTRODUCTION

The Department of Energy (DOE), the agency which formulates the national energy strategy, implements federal energy policy, and develops national energy resources, is also responsible for the design, development, production, and maintenance of nuclear weapons. DOE's industrial complex associated with nuclear weapons production (commonly known as the Nuclear Weapons Complex) was first operated by the Atomic Energy Commission (AEC), and then by the Energy, Research, and Development Administration (ERDA), which evolved into the present Department of Energy. Although the Nuclear Weapons Complex was initially composed of only a few laboratories, today the Complex is composed of 17 major facilities located on thousands of square miles of federal land in 13 states. During the past five decades, these facilities have together produced tens of thousands of warheads for the Department of Defense.

OVERVIEW OF THE NUCLEAR WEAPONS COMPLEX

The facilities in the Nuclear Weapons Complex can be classified into several categories according to function, namely: (1) weapons laboratories, which design and develop the weapons and test the various components and devices; (2) materials production facilities, which produce nuclear (radioactive) materials, and; (3) weapons production facilities, which fabricate the nuclear components, provide the various non-nuclear components, and assemble the weapons.

The development and manufacture of nuclear weapons is a complex process, involving the use and production of many hazardous chemicals, such as nitric acid, mercury, organic solvents, and highly radioactive chemicals and elements. Nuclear materials used in the manufacture of nuclear weapons include isotopes of uranium, hydrogen, and plutonium, and in the production/isolation of these radioactive elements, other radioactive elements (fission products) are generated.

CONTAMINATION ISSUES

During the operation of the Nuclear Weapons Complex over the past 40–50 years, while tremendous effort was placed on expediting the production of high-quality weapons, relatively little emphasis was placed on developing and implementing a systematic strategy and approach for the disposal of the various waste streams.[1] In the past, DOE utilized disposal practices that were similar to those in use by other industries, which included practices such as placing wastes in unlined trenches and ponds. In recent years, however, it has become apparent that these waste management practices were inadequate for disposing of the type and magnitude of waste that DOE generated, were inconsistent with sound environmental management practices, and were not in compliance with current environmental regulations.

DOE has come to realize that the radioactive and hazardous wastes generated during the production of nuclear weapons are far more dangerous to the environment and hazardous to human health than was previously believed. Former disposal practices have resulted in extensive contamination of soil, sediments, surface water, and groundwater, as well as vegetation and wildlife, throughout the Nuclear Weapons Complex.[2]

Examples of the types of environmental contamination which DOE must address are:

- **Volatile organic compounds (VOCs)**
  By far the most pervasive VOCs are trichloroethylene (TCE) and perchloroethylene (PCE), solvents used primarily as degreasers. VOC contamination of soil and groundwater is found at most of the sites throughout the Complex. Soil contamination by another VOC, carbon tetrachloride, is a significant problem at the Hanford site.
- **Heavy metals**
  Heavy metals such as mercury, lead, cadmium, and chromium are contaminating the soil and groundwater at many of the sites. Mercury contamination is a particular problem at the Savannah River Site.
- **Radioactive waste**
  Uranium, plutonium, radioactive strontium, thorium, and transuranics (radioactive elements which are heavier than uranium, such as neptunium and americium), are contaminants in groundwater and soil throughout the Complex. The Fernald Environmental Management Project in Ohio, where low-enriched uranium was reconstituted into uranium metal ingots, has extensive uranium contamination of soil.
- **Nitrate waste**
  The use of nitric acid in the plutonium-uranium extraction process (the "PUREX" process), employed at Hanford and the Savannah River Site, has resulted in significant nitrate contamination at these sites.
- **Tritium**
  Tritium is found in groundwater at many sites of the Complex. Because of its central role in the functioning of nuclear weapons and because it has a relatively short half-life (12.3 years) and had to be replaced often, huge quantities of tritium were generated and disposed of. Tritium contamination has been identified at the Idaho National Engineering Laboratory, the Oak Ridge National Laboratory, and the Hanford and Savannah River Sites.
- **Other contaminants**
  Arsenic, polychlorinated biphenyls (PCBs), and many other contaminants are also found at significant levels in the Complex.

At many of the sites the contamination is in the form of "mixed waste," meaning that it has both chemically hazardous and radioactive components. A substantial quantity of mixed waste is found in underground storage tanks, at buried waste sites, and in mixed waste landfills, all of which are presently leaking contaminants into surrounding soil, groundwater, surface water, and sediments due to improper or inadequate containment practices. The treatment and disposal of mixed waste pose special challenges not only from a technical standpoint, but also from a regulatory standpoint, in that the separate components of the mixed waste are regulated by different statutes and regulations.

DOE Is Required to Comply with Environmental Regulations

When the predecessor to DOE, the Atomic Energy Commission, was first established, it was designed to be self-regulating in terms of issues concerning human health and environmental protection. Because national defense required that the mission of the AEC (and later DOE) be carried out in the utmost secrecy, it seemed appropriate that other agencies not have oversight over DOE's activities, including its waste management practices.

However, in 1984 DOE's self-regulating role was challenged by the lawsuit *LEAF (Legal Environmental Assistance Foundation, Inc.) vs. Hodel.*[3] In this case, the court found that DOE was in violation of the Clean Water Act (CWA) and the Resource Conservation and Recovery Act (RCRA) at the Y-12 Plant in Oak Ridge, Tennessee. But most significantly, the court also determined that to the extent that the CWA and the RCRA were not in conflict with the Atomic Energy Act (which regulated the AEC), DOE was required to comply with the same environmental regulations as were all other waste generators, notwithstanding the AEC's "self-regulating" stance. This determination led to further scrutiny of the Nuclear Weapons Complex and of DOE in general, and it was revealed that the waste/contamination problem at DOE was extensive, that DOE was severely out of compliance with federal environmental regulations, and that environmental restoration activities and new waste management practices had to be put into effect.

In response to the new responsibilities placed upon DOE to conform with environmental regulatory requirements and cleanup standards, the Secretary of Energy, Admiral James D. Watkins, announced in June 1989 that environmental compliance and cleanup would immediately become a top priority for DOE.[4] As an indication of DOE's new commitment to environmental restoration and waste management, DOE releases an annually updated Five-Year Plan that describes its goals, strategies and specific programs for the assessment and cleanup of contaminated sites and facilities in order to meet standards prescribed in federal and state laws.[5]

Office of Environmental Restoration and Waste Management

The cornerstone of the Secretary's actions was the establishment of the Office of the Assistant Secretary for Environmental Restoration and Waste Management (EM). This Office is responsible for: (1) implementing the actual remediation of the environmental damage caused by past activities of the Nuclear Defense Complex, and (2) managing waste generated by current and future operations. EM seeks to remediate contaminated DOE sites and manage its inventory of regulated wastes within specific time frames and within regulatory constraints. EM's goal is to ensure that risks to human health and to the environment from past, present, and future operations are either eliminated or reduced to prescribed levels.

EM is made up of three programmatic offices, the Office of Waste Management (EM-30), the Office of Environmental Restoration (EM-40), and the Office of Technology Development (EM-50). EM-30 and EM-40 are responsible for accomplishing the actual tasks of waste management and environmental cleanup and are actively engaged in the establishment of numerous formal agreements with state and federal regulators describing EM's waste disposal and environmental cleanup strategies and the deadlines by which they will achieve their goals. The drivers for these compliance agreements are a number of regulatory acts applicable to the federal government (described below).

Because currently available technologies are insufficient for achieving waste disposal and for remediating environmental contamination in the time frame and to the extent to which the regulations specify,[6] the Office of Technology Development, EM-50, was created to coordinate the research, development, demonstration, testing, and evaluation of safe, cost-effective, and innovative new technologies for use by EM-30 and EM-40. As discussed below, EM-50 incorporates regulatory compliance considerations in the development of innovative technologies.

EM OFFICE OF TECHNOLOGY DEVELOPMENT REGULATORY COMPLIANCE STRATEGY

Over twenty environmental statutes and implementing regulations drive EM programs. The principal environmental statutes affecting EM activities include the Resource Conservation and Recovery Act (RCRA)[7] as amended by the Hazardous and Solid Waste Amendments (HSWA),[8] the Comprehensive Environmental Response, Compensation, and Liability Act (CERCLA)[9] as amended by the Superfund Amendments and Reauthorization Act (SARA),[10] and the National Environmental Policy Act (NEPA).[11]

These regulatory statutes are not specifically aimed at the development of technology, in that they do not dictate how or what technologies should be developed. However, by stipulating specific cleanup standards, the statutes do affect technology development. Only those technologies that make it possible to achieve environmental compliance can actually be implemented by EM-30 and EM-40, and thus the technologies undergoing development are designed so that they will meet regulatory requirements.

The overall DOE Environmental Compliance Strategy which lays out how EM-30 and EM-40 will comply with various regulatory requirements is a four-stage process designed to provide a systematic and consistent approach for completing site remediations throughout the DOE complex. It includes: (1) identification of compliance deficiencies, (2) notification of regulators, (3) negotiation of compliance agreements, and (4) performance of activities to achieve compliance. EM relies on the technologies developed by EM-50 to be able to meet the goals of its compliance agreements and remediate DOE sites.

SUBSURFACE CONTAMINATION AND CERCLA

Agreements concerning remediation of subsurface contamination are regulated by CERCLA.[12] CERCLA requires consideration of innovative technologies during the feasibility study stage, when information concerning performance, implementability, environmental impact, and cost is used to determine whether an innovative technology has comparable or superior performance to demonstrated technologies.[13] If the innovative technology is superior, further evaluation of the technology can continue, and additional criteria, such as overall protection of human health and environment, state and public acceptance, and reduction of toxicity, mobility, and volume are used to select the candidate technologies.[14] EM-50 works with the DOE parties to CERCLA compliance agreements to ensure that technologies under development receive appropriate consideration during this process.

Subsurface restoration of DOE sites frequently involves the need for treatment, storage, and/or disposal of hazardous wastes. CERCLA employs a concept known as "applicable or relevant and appropriate requirements" (ARARs) to ensure compliance with other regulatory schemes.[15] The management of hazardous waste is regulated by RCRA; therefore RCRA applies to the management of hazardous wastes encountered during subsurface restoration under CERCLA.

HAZARDOUS WASTE AND RCRA

RCRA has a more indirect, but equally significant effect on technology development. Among other types of wastes, RCRA regulates toxic waste,[16] of which DOE has large amounts. Many of the wastes generated, treated, and disposed of during subsurface restorations involve RCRA hazardous wastes subject to Land Disposal Restriction (LDR) requirements.[17] LDRs prohibit disposal of most RCRA hazardous wastes in or on the land (e.g., landfill, surface impoundment, injection well) unless the wastes meet certain Best Demonstrated Available Technology (BDAT) treatment standards.[18] Those wastes not subject to LDR need not meet these standards to be land disposed, but must still be managed according to certain requirements (e.g., in a landfill meeting design criteria).[19]

Most BDAT standards are concentration-based, meaning that the hazardous wastes must be reduced to certain concentrations (e.g., trichloroethylene concentration must be less than or equal to 0.062 mg/L in wastewaters).[20] Other BDAT standards are technology-based and involve the use of specific technologies which the EPA has accepted for treatment (e.g., incineration for certain mercury-containing wastes).[21]

EM-50 uses these standards as a benchmark to develop treatment technologies which, at a minimum, can achieve or surpass the concentration standards. EM-50 also seeks to obtain regulatory approval for treatment technologies that are shown to be equivalent or superior to the prescribed technologies, the BDATs.[22] Indeed, EM-50 is committed to developing technology solutions that are "better, safer, faster, cheaper" than those available. Because of the vast quantities of DOE waste, the complexity of DOE's waste problems, and the enormous costs which cleanup will incur, it is imperative that DOE develop alternative technologies for fulfilling RCRA requirements.[23]

Residues or secondary waste streams (e.g., ash, sludge, leachate, wastewater) that result from the treatment of RCRA hazardous wastes are also subject to RCRA waste management regulations.[24] This is known as the "derived-from" rule. The management of such residues is expensive and time-consuming, considering the low or non-existent concentration of hazardous constituents in them.[25] Therefore, EM-50 strives to develop innovative technologies that eliminate or reduce the volume of such secondary waste streams.

OTHER REGULATIONS

Non-solid secondary hazardous waste streams also must fulfill the CERCLA ARAR's requirements. Hazardous air emissions are regulated by the Clean Air Act (CAA)[26] according to the National Emission Standards for Hazardous Air Pollutants (NESHAPs).[27] Likewise, hazardous effluent may be subject to the provisions of the Clean Water Act (CWA)[28] and/or the Safe Drinking Water Act (SDWA).[29] The CAA and CWA both have permitting requirements and strict monitoring and control technology requirements that are expensive to implement and maintain. Injection of wastewaters directly into the ground requires permits and monitoring under SDWA.[30] Because the use of waste treatment technologies that subject facilities to additional regulatory and technical burdens is undesirable, EM-50 is pursuing the development of technologies that avoid or reduce exposure to these liabilities.

Having DOE sites in numerous states exposes EM-50 to a variety of EPA-authorized state regulatory programs. These programs are frequently patterned after their federal counterparts and they must be at least as strict, although they may be more stringent.[31] For example, before a technology designed for implementation at one DOE site can be transferred for use at a site in another state, it may need to be modified to comply with the other state's regulations. EM-50 plans its technology development to ensure that technologies intended for use in more than one state can meet applicable requirements.

RADIOACTIVE WASTE

A type of waste that is of special concern to EM is radioactive waste, including transuranic, high-level, and low-level radioactive wastes. Radioactive wastes represent a significant subsurface contamination problem across the Complex. Unlike the other waste forms, whose management is governed by regulations from outside agencies, radioactive waste is managed under the Atomic Energy Act (AEA)[32] which grants DOE the authority to regulate itself. DOE accomplishes this self-regulation through several DOE Orders, the most comprehensive being DOE Order 5820.2A, "Radioactive Waste Management."[33] DOE has committed itself to a goal of managing its radioactive and mixed wastes "in a manner that assures protection of the health and safety of the public, DOE, and contractor employees, and the environment."[34] To meet this goal, DOE Order 5820.2A makes the cleanup and management of radioactive waste subject to the requirements of RCRA, CERCLA, and other environmental laws, where applicable.[35] Therefore, subsurface restoration of radioactive waste contamination may be regulated in a manner similar to that for non-radioactive contamination.

Regulation of radioactive waste by this combination of regulations poses unique challenges for EM-50. As mentioned above, DOE has large quantities of mixed waste—radioactive waste that also contains a hazardous component. DOE and EPA have determined that the hazardous component of mixed waste is subject to RCRA (to the extent that such regulation does not conflict with the AEA),[36] but the management of the hazardous waste portion of mixed waste is complicated by the hazards presented by the radioactive portion. EM-50 is attempting to develop technologies that will enable safe, cost-effective, and correct characterization, separation, treatment, storage, and disposal of mixed waste.

PUBLIC PARTICIPATION

In developing its approach to regulatory issues, EM-50 has made the assumption that strictly following regulatory requirements is not enough to ensure that a new technology will be allowed to be implemented. Technology solutions must in addition be acceptable to the regulatory bodies and to the public. Thus, the performance of procedural compliance activities, such as requirements identification, compliance audits, and corrective activities, are augmented by other substantive activities such as interaction with regulators, public participation, integration with planning, and prioritization. These latter activities improve the likelihood of simultaneous completion of technology development and acceptance of the technology by the regulators and the public so that implementation can proceed without delay.

Requirements for public acceptability of remedial decisions are found in CERCLA and NEPA.[37] Early notification and continuous involvement in all phases of technology development should satisfy the substance of these requirements, complemented by timely and complete dissemination of information in the required formats.

ENVIRONMENTAL IMPACT AND NEPA

As part of a federal agency, EM-50 must follow NEPA by evaluating and documenting all of its activities which have potential for significant impact on the environment.[38] Recently, DOE promulgated a rule for implementing NEPA requirements.[39] This rule allows most EM-50 projects to be excluded (through a "Categorical Exclusion," or "CX") from extensive NEPA evaluation due to the knowledge that these projects will not have a significant environmental impact.[40]

However, full-scale implementation of the technologies by the end users (i.e., EM-30 and EM-40) may require more extensive evaluation, as part of an Environmental Assessment (EA) or an even more comprehensive Environmental Impact Statement (EIS).[41] In addition, some EM-50 projects may not qualify for a CX, and an EA or EIS will be necessary.[42] EM-50 submits such evaluations to public representatives for comment and resolution.

OFFICE OF TECHNOLOGY DEVELOPMENT STRATEGY: INTEGRATED PROGRAM AND INTEGRATED DEMONSTRATION APPROACH

EM-50's approach to technology development is designed to ensure their timely implementation. Accordingly, the technology development effort is organized into needs-driven, multi-faceted Research and Development, Demonstration, Testing, and Evaluation projects which expedite technology development. Through laboratory and bench-scale demonstrations, applied Research and Development (R&D) activities evaluate feasibility, applicability, and acceptability of new technologies, and identify critical engineering issues for full-scale testing and evaluation. Results from R&D projects provide the technical, regulatory, and economic basis for decisions to transfer technologies to the demonstration, testing, and evaluation stage. Demonstration, Testing, and Evaluation (DT&E) projects involve the evaluation of promising technologies through full-scale or near full-scale field demonstrations to validate their effectiveness, their usefulness, and their regulatory acceptability, as well as to provide the operational parameters for full-scale implementation and technology transfer.

EM-50's R&D and DT&E efforts are further divided into program elements called Integrated Programs (IPs) and Integrated Demonstrations (IDs), respectively. An Integrated Program is a set of research and development activities dedicated to providing technology solutions for a single problem category (e.g., minimization, characterization, treatment, storage, and disposal) common to environmental management activities across the DOE complex. An Integrated Demonstration provides a system for solving an environmental management problem end-to-end at a particular site. IDs address specific parts of environmental issues by applying multiple technologies and comparing them at various phases of the remedial investigation/feasibility study process associated with a given remediation or disposal problem. (See Figure 1.1.)

For both the IPs and the IDs considerable effort is extended toward rapidly transferring new technologies from the development stage to the implementation stage so that regulatory requirements can be met as quickly as possible. Activities are in fact prioritized to provide technology solutions to EM-30 and EM-40 so that they can meet the requirements of compliance agreements which have been established for the various sites.

The major program areas of EM-50, into which the IDs and IPs fall, include the following:

Groundwater and Soils Cleanup. Technologies and processes are being developed to remove or reduce hazardous and/or radioactive materials from contaminated soils and groundwater.

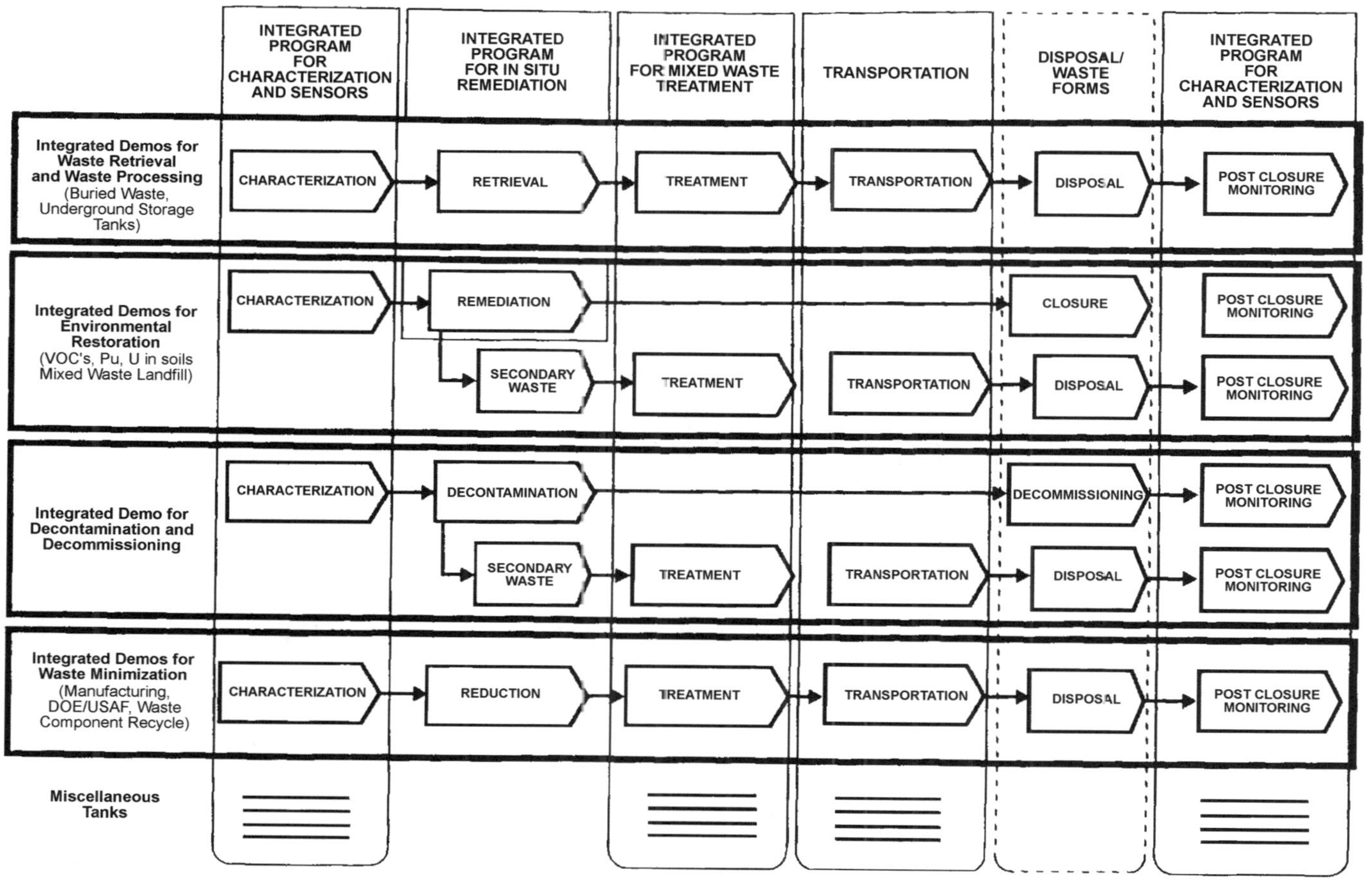

Figure 1.1. Integrated demonstrations provide beginning-to-end solutions. Integrated programs provide a coordinated technology base.

Waste Retrieval and Waste Processing. This program fills the need to develop more effective technologies for (1) the excavation or removal of hazardous and/or radioactively contaminated material from old disposal sites and/or storage facilities and (2) the treatment of contaminated material for transportation, storage, and disposal.

Waste Minimization and Waste Avoidance. Under development are technologies and processes for reducing, at the source, the toxicity or amount of hazardous materials generated by DOE activities such as the dismantlement of nuclear weapons, the decommissioning of facilities, and the management of existing waste.

SUBSURFACE RESTORATION: SOILS AND GROUNDWATER CLEANUP

Soils and groundwater cleanup addresses the amelioration of potential health and environmental risks from soils and groundwater contaminated with radioactive and/or hazardous materials. The development of new technologies for these problems is imperative because, according to an analysis conducted by the Office of Technology Assessment, "it may be impossible with current technology to remove contaminants from certain groundwater plumes and deeply buried soil, or even if possible, it may be extremely expensive or require prolonged periods of operation."[43] In addition to exploring new methods to remediate soils and groundwater, this research area also encompasses activities involving the evaluation of interim or temporary measures to retard migration or to contain radioactive and hazardous material until a suitable technology can be developed. Furthermore, mechanisms are being developed to increase the potential for U.S. firms to export technologies and systems to other countries needing remediation of contaminated soils and groundwater.

Specific soils and groundwater technology development focuses on remediation technologies for organic and inorganic hazardous, radioactive, and mixed waste contamination. New technologies need to be applicable to clean up a wide variety of sites because most DOE sites have contaminated soil and groundwater. Improved methods of performing site characterization, rapid and cost effective remediation, performance assessment, and post-remediation monitoring are all under development. Particular emphasis is on the demonstration of safe, cost-effective technologies for *in situ* remediation of contaminated soils and groundwater because *in situ* remediation technologies reduce the risk to workers and the community by minimizing disturbance of contaminated environmental media, and reduce costs by eliminating excavation and transportation costs. Systems which will ensure waste minimization are also being developed and optimized in order to reduce the amount of waste generated by the cleanup process itself. Methods of soil stabilization and revegetation are being developed and demonstrated as well.

Regulatory drivers for the development of technologies to clean up groundwater and soils are those mentioned above as well as the consent orders and interagency agreements that have been signed at DOE sites. The new technologies are also designed to meet the terms of compliance agreements and Records of Decision (RODs) for the specific sites.

Examples of EM's soils and groundwater technology development projects and how they respond to regulatory requirements include the following:

- ***In Situ* Remediation IP**
  The objectives of this IP are: 1) to develop and manage *in situ* remediation technology R&D activities, 2) to develop *in situ* remediation technologies to the point of field demonstration and to transfer the technologies to the users, and 3) to support the assessment of innovative technologies. R&D efforts are being undertaken in the areas of *in situ* physical, chemical, and biological treatment and containment for the cleanup of hazardous wastes, radioactive wastes, and mixed wastes in soils, groundwater, and storage tanks. Novel technologies such as electrokinetics and subsurface manipulation are funded under the current program. A particular benefit of the technologies being developed under this IP is that they reduce remediation costs by eliminating the need for waste excavation, transport, and disposal, and in addition they are potentially more effective. Thus, this IP answers the need to develop treatment technologies that are better than those already accepted by the EPA. *In situ* remediation technologies also have implications in the area of worker safety and health. By making possible reduced contact

exposure, *in situ* technologies minimize the health effects to workers and the public. Finally, on-site, *in situ* remediation also meets requirements of formal DOE agreements such as waste inventory reduction and helps DOE to avoid the Land Disposal Restriction requirements since contaminated soil and other waste do not need to be moved to another location.

- **Characterization, Monitoring, and Sensor Technology (CMST) IP**
  The purpose of this IP is to provide cost-effective and efficient CMST options for use in the Integrated Demonstrations, as well as in DOE environmental restoration and waste management operations. Technologies developed under this IP represent the critical first step in site and waste remediation as they underpin every step, from initial assessment and verification of the effectiveness of remedial actions to post-closure monitoring. Focus is placed on the development or improvement of technologies such as field-deployable instrumentation and noninvasive methods which address an identified need at a specific DOE site, and which are transferable to other waste streams and sites in the DOE complex. R&D emphasis has been placed on remote and minimally intrusive characterization and *in situ* measurements.
- **Volatile Organic Compounds (VOCs) in Non-Arid Soils ID**
  This ID focuses on the development of *in situ* bioremediation and offgas treatment technologies to remove/degrade VOC (primarily TCE and PCE) contamination at the Savannah River Site. TCE and PCE are hazardous wastes and, as such, their cleanup is regulated by RCRA. Other regulatory drivers for this ID include consent orders and interagency agreements at all of the major DOE sites for cleanup of volatile organic compounds.
- **Plutonium-Contaminated Soils ID**
  This ID is developing and demonstrating innovative waste minimization technologies for the cleanup of plutonium, other transuranics, and heavy metals in surface soils. In addition, various methods of soil stabilization and vegetation at arid sites will be explored and demonstrated. Various compliance agreements mandate that soils at the respective facilities be decontaminated. An Interagency Agreement (IAG) between the EPA, DOE, and the State of Nevada requires that plutonium-contaminated soils at the Nevada Test Site be cleaned up.
- **Uranium-Contaminated Soils ID**
  Research in this ID, to be demonstrated at the Fernald facility in Ohio, is aimed at remediating low levels of uranium contamination over large areas. Technologies developed in the ID will facilitate DOE's efforts to meet the terms of the Fernald compliance agreement to clean up the uranium contamination at a particular site at Fernald, and they will be applicable to cleanup of uranium at other sites.
- **VOCs in Arid Soils ID**
  This ID will involve the development and demonstration of cost-effective methods to characterize and remediate arid soils and groundwater contaminated with VOCs, such as carbon tetrachloride (CCl_4) and chloroform ($CHCl_3$), as well as other contaminants, such as plutonium, americium, and nitrate (NO_3^-). *In situ* characterization and remediation methods are a focus of this ID.
- **Mixed Waste Landfill ID**
  The purpose of this ID is to demonstrate, test, evaluate, and integrate *in situ* technologies for the characterization and remediation of heavy metals and mixed waste in landfills in arid environments. Landfill containment barriers and barrier emplacement technology, electrokinetic remediation of metal ions, and cryogenic containment technology will be evaluated in this ID.

SUMMARY AND CONCLUSION

As a reflection of DOE's commitment to doing what is right for the environment and for human health and safety, DOE established the Office of Environmental Restoration and Waste Management (EM) in 1989. This Office is devoted to ensuring that the Nuclear Weapons Complex will meet regulatory requirements for hazardous and radioactive waste cleanup and waste management, and within the required time frame. The EM Office of Technology Development (EM-50) responds to the compliance process by developing new technology solutions which are more effective, less costly, and faster to implement than currently available technologies for remediation.

For EM-50's programs to be effective, technical solutions to DOE's environmental problems must be implemented using a systems approach. Thus, EM-50 has adapted an accelerated technology development and implementation strategy through a combination of Integrated Programs and Integrated Demonstrations which solve problems "end-to-end" with minimum production of hazardous secondary waste streams. Technologies are researched, developed, demonstrated, tested, and evaluated within the context of real DOE problems. Research efforts specifically addressing DOE's subsurface environmental issues include: (1) development of interim or temporary containment and stabilization technologies to retard migration of contaminants, and (2) development of effective permanent solutions to resolve the subsurface problems existing in the DOE complex.

The ultimate step in the development of technology is implementation by the end user. For successful implementation, it is necessary that the technology comply with regulatory requirements. EM-50's regulatory compliance activities facilitate the implementation of its developed technologies.

REFERENCES

1. *The Nuclear Weapons Complex: Management for Health, Safety, and the Environment*, National Research Council (NRC), National Academy Press, Washington, DC (1990).
2. *Complex Cleanup: The Environmental Legacy of Nuclear Weapons Production*, Congress of the United States, Office of Technology Assessment (OTA), S/N 052-003-01222-7, Washington, DC (1991) (*OTA Report*).
3. 586 F. Supp. 1163 (E.D. Tenn. 1984).
4. *Environmental Restoration and Waste Management: Five Year Plan, Fiscal Years 1993–1997*, United States Dept. of Energy (DOE), National Technical Information Service, Springfield VA (1991), *Five Year Plan 1993–1997*.
5. *Id.*
6. *See* note 2, *OTA Report*.
7. 42 United States Code (U.S.C.) §§ 6901 to 6992k ("Solid Waste Disposal Act").
8. Pub. L. 98-616 (Nov. 8, 1984).
9. 42 U.S.C. §§ 9601 to 9675.
10. Pub. L. 99-499 (Oct. 17, 1986).
11. 42 U.S.C. §§ 4321 to 4370a.
12. 42 U.S.C. § 120(e).
13. 40 Code of Federal Regulations (C.F.R.) § 300.430(e)(5) (1991).
14. 42 U.S.C. § 9621(b); 40 C.F.R. § 300.430(e)(9)(iii) (1991).
15. 42 U.S.C. § 9621(d)(2)(A).
16. *See* e.g., 40 C.F.R. §§ 261.24 and 261.33(f) (1991) (toxicity "characteristic" wastes and "listed" toxic wastes).
17. 42 U.S.C. § 6924(d)-(m); 40 C.F.R. Part 268 (1991).
18. 42 U.S.C. § 6924(m); 40 C.F.R. Part 268, Subpart D (1991).
19. 42 U.S.C. § 6924(o); 40 C.F.R. Part 264 (1991).
20. 40 C.F.R. §§ 268.41 and 268.43 (1991).
21. 40 C.F.R. § 268.42 (1991).
22. *See* 40 C.F.R. § 268.42(b) (1991) (application requirements for approval of alternative BDAT treatment methods).
23. *See* note 4, *Five Year Plan 1993–1997* at 110–11.
24. 40 C.F.R. § 261.3(c)(2)(i) (1991).
25. The U.S. Environmental Protection Agency (EPA) is attempting to amend the derived-from rule to exempt certain residues from continued treatment and management regulation. However, a recent proposal was withdrawn in light of significant negative comments from the regulated community and the states. EPA has stated that it hopes to have a new proposal in 12–24 months.
26. 42 U.S.C. §§ 7401 to 7642.
27. 40 C.F.R. Part 61 (1991).
28. 33 U.S.C. §§ 1251 to 1387 ("Federal Water Pollution Control Act").
29. 42 U.S.C. §§ 300f to 300j-26 ("Public Health Service Act" (Title XIV)).
30. *Id.* at § 300h to 300h-7; 40 C.F.R. Parts 144-148 (1991).
31. 40 C.F.R. § 271.1(i) (1991).
32. 42 U.S.C. §§ 2011 to 2286(i).

33. Atomic Energy Act, Dated Sept. 26, 1988.
34. DOE Order 5820.2A, "Radioactive Waste Management," p. 1 (Sept. 28, 1988).
35. *Id.*
36. 51 Federal Register (Fed. Reg.) 24,504 (July 3, 1986); 53 Fed. Reg. 47,045 (Sept. 23, 1988).
37. *See* note 11; 40 C.F.R. Parts 1500-1517 (1991).
38. 42 U.S.C. § 4332.
39. 57 Fed. Reg. 15,122 (Apr. 24, 1992); *See also* Secretary of Energy Notice 15-90, "National Environmental Policy Act," (Feb. 5, 1990).
40. 57 Fed. Reg. 15,151-15,157 (Apr. 24, 1992) (to be codified at 10 C.F.R. § 1021.410 and Appendices A and B to Subpart D).
41. *Id.* at 15,146-15,151 and 15,157-15,158 (to be codified at 10 C.F.R. Part 1021, Subparts B and C and Appendices C and D to Subpart D).
42. *Id.*
43. *See* note 2, *OTA Report.*

Section II

Basic Science Required for Decision Making

Chapter 2

Geological Setting

M.P. Anderson

INTRODUCTION

The geological setting provides the framework for understanding hydrological, chemical and microbial processes. The geology affects the nature of these processes and the way in which they are monitored and measured. Unfortunately, there are significant difficulties in characterizing subsurface geology for contaminant investigations. The chief difficulty relates to the heterogeneity of geological materials (Anderson, 1996).

To say that porous materials in the subsurface are not homogeneous is both trite and profound. It is trite because both scientists and engineers would agree that the subsurface is heterogeneous. Yet groundwater hydrologists have readily accepted the assumption that the subsurface can be treated as an equivalent homogeneous porous medium ever since Theis (1935) demonstrated that this concept works when used to analyze flow in an aquifer stressed by pumping. The realization that the paradigm of an equivalent homogeneous porous medium does not work for contaminant transport in the subsurface was recognized early by Theis (1967), but was slow to gain acceptance. To admit that the subsurface is heterogeneous and cannot be simplified to an equivalent homogeneous medium is a profound concession.

BACKGROUND

Ever since 1856 when Darcy's law was published, groundwater hydrologists have characterized the subsurface using the representative elementary volume or REV. The concept of the REV recognizes that it is impractical or even impossible to describe hydrogeological processes and parameters at a micro scale, but in most situations this level of detail is unnecessary. The scale of operational interest is one that encompasses enough of the porous medium to yield representative values for its parameters. For example, consider porosity, which at the scale of a pore space is equal to one, and at the scale of a sand grain is equal to zero. A representative porosity, of course, lies somewhere in between these two extremes. When using the concept of the REV, one attempts to find the size of sample for which a representative value of the parameter is obtained (Figure 2.1).

The idea that a porous medium is homogeneous at the scale of the REV is a convenient construct used in derivation of the general governing equations for groundwater flow and contaminant transport and for measuring aquifer characteristics. The problem, of course, is that aquifers are not homogeneous at the scale of interest in most field investigations. For nearly 40 years after Theis introduced the idea of the equivalent homogeneous porous medium, groundwater hydrologists were interested in water supply problems, and it was possible to "solve" the heterogeneity problem by using the paradigm of the equivalent homogeneous porous medium for which we could measure effective aquifer properties by means of the pumping test. In an aquifer test, an aquifer is sampled at the scale influenced by pumping, and bulk average aquifer properties are calculated at that scale. In this way, a heterogeneous aquifer can be represented as an equivalent homogeneous porous medium described by effective parameters.

Applications of contaminant transport models in the early 1970s typically used the assumption that a heterogeneous aquifer could be represented by effective parameter values, thereby employing the assumption of the equivalent homogeneous porous medium. These early simulations were less than satisfactory (Anderson, 1979). Failure to simulate contaminant plumes correctly led eventually to the conclusion that the equivalent homogeneous porous medium approach does not work for contaminant investigations because contaminant transport is highly dependent on the nature and distribu-

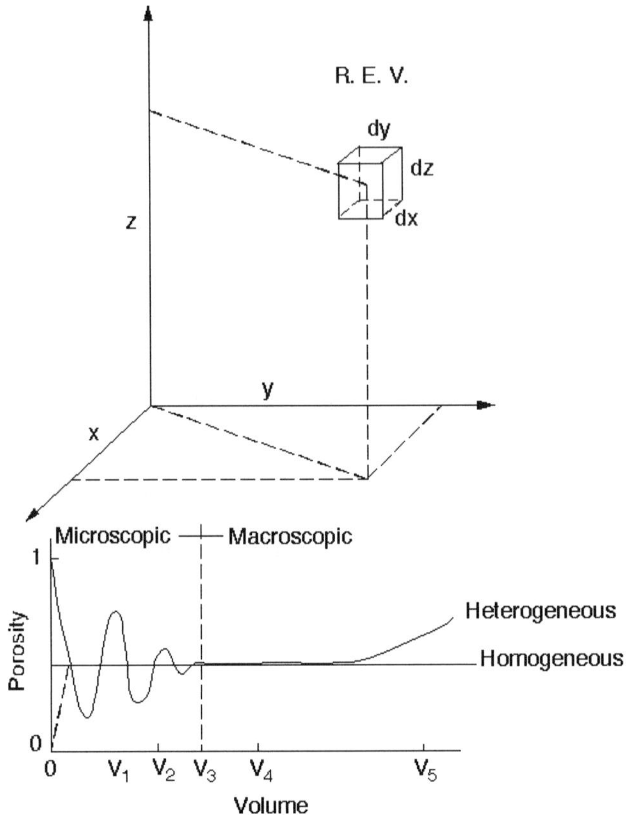

Figure 2.1. Definition of the representative elementary volume (REV) as it relates to porosity.

tion of heterogeneities. Therefore, a correct description of heterogeneity is critical in contaminant transport analysis.

Recent experiences with pump-and-treat systems for aquifer remediation demonstrate that these systems frequently do not operate as well as expected (NRC, 1994). The failure of groundwater models to predict remediation accurately is at least partly due to incomplete description of aquifer heterogeneity. Delineation of high-permeability layers that form preferential flow paths is important, as is the location and geometry of low-permeability units, which may temporarily trap contaminants and make aquifer remediation difficult to accomplish within a reasonable time frame.

An important consideration in assessing heterogeneity is measurement scale. Geologists traditionally work at the scale of the outcrop or at a much larger regional scale. Petroleum geologists additionally work at a smaller scale using cores. Hydrogeologists need information at a scale intermediate between the outcrop and the regional scale and also at the core scale when chemical and microbial processes are important. Description of the geological setting for purposes of aquifer remediation requires information on the three-dimensional distribution of geological units and values of hydraulic conductivities, effective porosities and retardation factors for each unit. Such information is not easily obtained. Complete geological characterization requires field and laboratory investigations on a regional scale larger than the site itself, on an intermediate scale the size of the contaminant plume, and on a microscopic scale.

Consider just three important hydrogeological parameters: hydraulic conductivity, effective porosity, and retardation factor. Hydraulic conductivity is a difficult parameter to quantify because of its spatial variability and because it can range over 13 orders of magnitude. Moreover, it can be measured on at least four scales. Aquifer tests measure hydraulic conductivity at a macro scale of 10s to 1000s of meters; slug tests are performed in the field at a scale of meters, permeameter tests measure hydraulic conductivity of packed sediment columns and grain size analyses are done at a micro scale. The scale of measurement should match the scale of analysis. For example, aquifer test data are appropriate for use in a numerical groundwater flow model with 100-m grid spacing. For analyses at the scale of meters, slug tests may be more appropriate.

Effective porosity is also different for each geological unit. This parameter is important in contaminant investigations because it is used to calculate groundwater velocity and travel times (Zheng and Bennett, 1995). Yet, it is almost impossible to measure. Total porosity is a function of the specific yield, plus water held by adsorption, and water in dead-end pores. Effective porosity is a function of specific yield and water held by adsorption. Effective porosity for sandy aquifers may be approximated by specific yield. However, for clayey material it is better to equate effective porosity to total porosity (Fetter, 1994). Measurements of specific yield and effective porosity can be made in the field by means of aquifer tests or tracer studies. Specific yield and total porosity can be measured in the laboratory on sediment columns or cores.

The retardation factor is dependent on the bulk mass density, the porosity, and the percent organic matter. It also is difficult to estimate in the field. There may be clues from an analysis of the environment of deposition that indicate the likelihood of organic material in the deposits. But it is necessary to analyze the materials themselves for percent organic material in order to calculate the retardation factor. Mineralogy may control retardation in the absence of organic material. Hence, it will be necessary to describe the mineralogy of deposits with low organic content.

DISCUSSION

Hydrogeological site characterization requires quantitative descriptions of geological heterogeneities at a scale smaller than the contaminant plume. Yet, geological data are traditionally qualitative rather than quantitative, and typically are reported at scales that are not directly relevant to hydrogeological problems.

The Nature of Geological Data

In order to construct a three-dimensional picture of the geology beneath a site, a geologist must first reconstruct the environment in which the deposits were formed. Examples of environments of deposition include fluvial, glaciofluvial, lacustrine, eolian, deltaic, and shallow marine. Traditionally, geologists have used conceptual models to illustrate the shapes and relative relations of deposits in a particular geological environment. These conceptual models typically take the form of block diagrams and/or stratigraphic columns that show the spatial relations of units of similar characteristics, know as facies. Hence, the models are termed facies models. The fundamental assumption inherent in a facies model is that facies transitions occur more commonly than would be expected if the process of deposition were random. The facies model captures these trends and is thought to represent the ideal example of a particular sedimentary environment. Facies models are constructed on at least two scales: a regional scale and a local scale. Geologists use a facies model as a norm for purposes of comparison, a framework for future observations and to help interpret processes operating in a particular depositional system.

Regional facies models represent large-scale trends in depositional environments and delineate facies assemblages. For example, in a facies model of an alluvial fan (Figure 2.2), there is a trend from coarse to fine sediment progressing away from the mountain front. In the facies model of an outwash deposit there are three facies assemblages: the coarse proximal deposits near the ice margin and the finer medial and distal sediment deposited by streams flowing out of the ice margin onto the outwash plain.

Outcrops are mapped in detail within each facies assemblage to produce site-specific local facies models. In fact, much of the sedimentological literature consists of detailed descriptions of facies at specific locations. Sedimentologists compare their detailed site-specific information against the trends predicted by a local generic facies model—such as the one shown in Figure 2.3—and in this way identify and/or verify the kinds of processes that formed a specific deposit, and deduce its geological history.

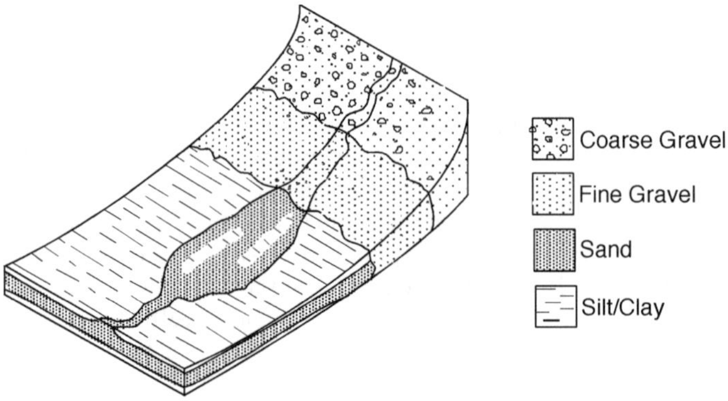

Figure 2.2. Facies model of an alluvial fan showing trend of coarse to fine sediment proceeding away from the mountain front.

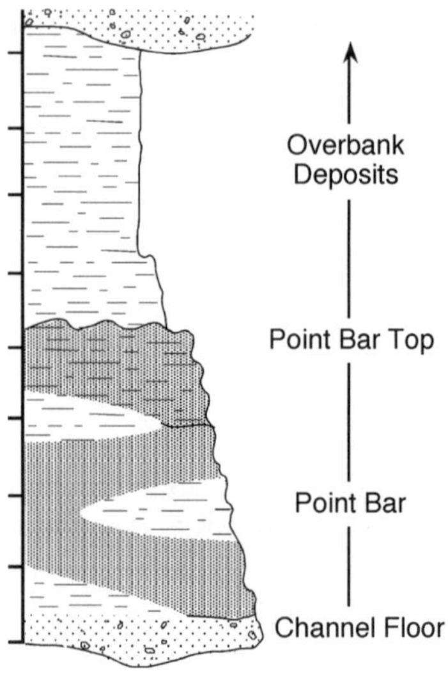

Figure 2.3. Stratigraphic column of a generic meandering stream deposit showing a fining upward sequence of deposits. Symbols are defined in Figure 2.2.

Hydrogeologists, however, would like to use facies models to help predict facies occurrences at sites where field information is limited. But using facies models for prediction is difficult because the ideal summary of a geological environment is unlikely to be found in any given field situation. Hence, a facies model is not very helpful in predicting the precise location of units of high permeability. For example, a facies model may indicate only the likelihood that such units exist somewhere in the deposit. Furthermore, hydrogeological site characterization requires information at an intermediate scale between the regional scale and the scale of an outcrop.

Among the few geological studies to address heterogeneity at a scale directly relevant to hydrogeological site characterization is the work of Miall (1985). He described eight geometrically and lithologically distinct fluvial deposits or elements, where an element consists of a suite of facies that are associated with a specific depositional process in the fluvial environment. For example, an alluvial environment would have a channel element with a characteristic sequence of deposits, as well as other elements for overbank deposits and gravel bars. Davis et al. (1993) attempted to use Miall's system to identify hydrogeological units in an alluvial system in New Mexico. Aiken (1993) applied Miall's system to an outwash deposit in Wisconsin.

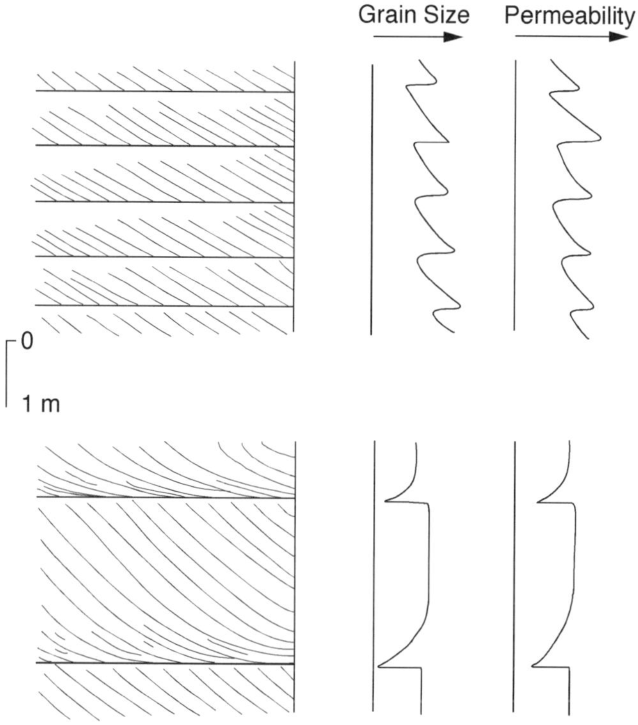

Figure 2.4. Cross-bedding with inferred variations in grain size and permeability (Selley, 1988, p. 69).

In the petroleum geological literature there are more applied contributions and another scale of investigation—the core scale. Petroleum geologists have studied bedding structures in sandstone cores, for example, and found that cross bedding (Figure 2.4) in water-deposited sands produces different trends in permeability when compared to eolian sands. In water-laid sediments permeability increases downward within a crossbed, whereas in eolian sediment, permeability decreases downward. From petroleum geologists we also find motivation to look at systems of dual permeability, namely sand/clay (sandstone/shale) sequences (Desbarats, 1990), since the distribution of shale lenses is important in oil recovery. The interconnection of permeable material is important in contaminant investigations as well (Anderson, 1991; Fogg, 1990).

In addition to the problems caused by different measurement scales, another difficulty with applying traditional geological techniques to hydrogeological investigations is that geological information is qualitative, while hydrogeological characterization requires quantitative descriptions. The possible variations of a general rock type, e.g., sandstone, are so large that cumbersome qualifiers are needed to give a precise description of a specific rock. Facies classification schemes also require qualifiers such as "sand, medium to very coarse, may be pebbly with solitary (alpha) or grouped (omicron) planar crossbeds, indicative of linguoid, transverse bars, sand waves (lower flow regime)" (Miall, 1985). While these sorts of descriptions are helpful to sedimentologists, hydrogeologists need quantitative information on grain size and sorting, for example, in order to translate geological descriptions into hydrogeological parameter values.

Other geological features of interest to hydrogeologists include faults and fractures. Faults can act as conduits if they consist of broken pieces of rock, or barriers to flow if they are smeared with clay. Bredehoeft et al. (1992) showed that faults in the Big Horn Basin of Wyoming conduct water vertically but act as horizontal barriers to flow. Information on faults may be gleaned from regional geological summaries and geological maps. But the smaller-scale and more numerous fractures that occur in both consolidated and unconsolidated materials typically are not the focus of purely geological investigations. Because of the difficulty in quantifying fracture patterns, the description and prediction of contaminant movement in fractured rock is a major challenge for hydrogeologists (Wang, 1991).

In summary, we find that the geological literature offers evidence of predictable large-scale regional trends in the distribution of geological units. But at the scale of an outcrop each deposit is unique in the detailed arrangement of heterogeneities. Architectural element analysis may provide some help in deciphering heterogeneities at scales in between the regional setting and the outcrop. We also find that a key feature of heterogeneity for oil recovery is the interconnection of permeable units that form preferential flow paths. Preferential flow paths are also important in hydrogeological investigations.

Hydrogeological Site Characterization

How does the type of geological information discussed above help in hydrogeological site characterization? Ideally we would like to have a three-dimensional picture of the subsurface geology beneath a site. A common way of representing subsurface geology is the fence diagram, which is an interpretation of the continuity of units based on information from boreholes. The challenge comes in using good geological judgment to do the correlation. Information about the regional geological setting and the type of facies or facies assemblages provides general guidelines about probable trends and features in the deposits.

In an outwash deposit, for example, proximal outwash is more heterogeneous (poorly sorted) and coarser grained than distal outwash. By identifying the facies assemblage, it should be possible to deduce some general information about the degree of heterogeneity. Similarly, knowing that a site consists of material deposited by a meandering river gives some indication of the types of facies or architectural elements present. However, generic geological facies models give no guidance regarding the precise location or arrangement or even the density of permeable deposits in a geological setting. This type of information can only be determined during site investigations with the help of boreholes and possibly with geophysics.

Typically, hydrogeological site characterizations rely simply on descriptions of geological units from borehole data without attempts to construct the units in three dimensions from facies analysis. However, there are at least two recent examples in the literature that demonstrate the application of facies analysis to hydrogeological site characterization (Jackson et al., 1985; Herweijer and Young, 1990). Both examples involved the use of numerous boreholes as well as conceptual facies models.

Jackson et al. (1985) identified seven hydrostratigraphic units ranging from stratified silt and clay to sand and gravel. The depositional environment was reconstructed following a conceptual facies model of subaqueous outwash deposition. Herweijer and Young (1990) reconstructed the depositional environment for a site located in an alluvial valley of the Tombigee River in Mississippi. They used conceptual facies models for deposition on a point bar of a meandering river and for deposition in a braided stream system.

Attempts at aquifer remediation frequently involve pump-and-treat systems. Design of a pump-and-treat system is facilitated by the use of a contaminant transport model, a basic element of which is a model of the flow system. The way in which borehole information is transferred to give nodal values of aquifer parameters at model grid points is critical to the accuracy of the flow simulation. Ideally geological insight into the environment of deposition at the site could help guide this effort. Especially critical is the identification of zones of high hydraulic conductivity that may form preferential flow paths for contaminant movement (Anderson, 1991; Desbarats, 1990; Fogg, 1986).

Several researchers are addressing the problem of defining the distribution of hydrogeological parameters from discrete point measurements. Philosophically these researchers are in two main camps. One camp believes that heterogeneity can be described stochastically by using random fields to represent hydraulic conductivity. The other group advocates incorporating geological structures directly into the representation of the conductivity field. Proponents of the stochastic approach generate random hydraulic conductivity fields with specific statistical properties in the hope of capturing the relevant features in the subsurface that govern contaminant movement. There are four main approaches, involving reliance on effective parameters, geostatistics, Monte Carlo simulation, and conditional simulation (Yeh, 1992). In addition, Ababou et al. (1989) use what they call a single realization approach.

The use of effective parameters is an extension of the equivalent homogeneous porous medium approach, whereby effective parameter values are defined statistically. It seems that even relatively homogeneous aquifers like the one that has been studied for many years now at the Borden site (Sudicky, 1986) may not be characterized adequately statistically using effective parameters (Burr et al., 1994). In this approach variograms are used to describe the hydraulic conductivity distribution in terms of mean, variance, and correlation length. But it is less likely that heterogeneous aquifers such as at the MADE site in Columbus, Mississippi (Boggs et al., 1992) can be adequately characterized in this way, although the effective parameter approach has been applied to this site as well (Adams and Gelhar, 1992).

The geostatistical approach relies on interpolation techniques such as kriging to fill in missing data. Monte Carlo simulations utilize multiple realizations of random conductivity fields to help define average transport behavior. Conditional simulations additionally use field measurements to fix values at known points. Conditional simulation models that rely on facies analysis to provide input data for the simulation are used by engineers to simulate heterogeneity in petroleum reservoirs (Haldorsen and Damsleth, 1990; Ravenne and Beucher, 1988). The single realization approach (Ababou et al., 1989) relies on small grid spacing and definition of a set of length scales to generate one statistically meaningful realization. This approach requires large numbers of nodes (of the order of 10^6) and solution using a supercomputer.

Other researchers advocate the use of geological information to help define the conductivity field (McKenna and Poeter, 1995; Anderson, 1989, 1991; Phillips and Wilson, 1989; Phillips et al., 1989; Fogg, 1990). The logical extension of a geologically based approach is the use of geological process simulation models, which allow the direct simulation of geological structures (Koltermann and Gorelick, 1996). An early paper by Price (1974), who used a random walk model to simulate alluvial fan deposition, anticipated the interest hydrogeologists would later show in process simulation models. More recently, Koltermann and Gorelick (1992) used a much more sophisticated process simulation model to reproduce the geological evolution of an alluvial fan in northern California over the past 600,000 years.

Webb (1994) developed a geological simulation model based on superposition of geomorphological surfaces to form sediment packages in a braided stream system. Results show the presence of high-conductivity preferential flow paths that funnel contaminants (Webb and Anderson, 1996). When a random distribution of hydraulic conductivities is used to represent variation within each facies unit, the high-conductivity units are not so obvious but still form preferential flow paths. Hence, even in seemingly random distributions, geological structures control flow. Funneling along preferential flow paths is a key feature of contaminant transport. Such funneling cannot be described by effective parameters or by a Fickian model of dispersion. Preferential flow paths will be difficult and perhaps impossible to identify in the field. Detailed field measurements are needed to estimate the probability of occurrence of interconnected paths for different types of deposits.

It is clear that hydrogeological analysis requires a quantitative description of the geological setting. This means taking geological information, which is traditionally qualitative, and developing ways of using it to produce realistic distributions of hydraulic conductivity. Groundwater hydrologists are also finding that the information and field methods in the traditional geological literature are not directly transferable to hydrogeological investigations owing to discrepancies in scale between traditional geological investigations and contaminant studies. Given that geological structures are an important part of heterogeneity and that uncertainty will always dictate a stochastic approach, it seems that some combination of methods is required. Conditional simulations may provide the best way to combine stochastic simulations with geological input.

CONCLUSION

There is no doubt that hydrogeological site characterization for aquifer remediation presents a challenge. Groundwater hydrologists have become much more concerned about geological characterization with the realization that geological heterogeneities control contaminant movement. The concept of the equivalent homogeneous porous medium and even the use of statistically defined effective parameters will have limited application in geologically complex heterogeneous field settings.

Let us assume we have a serious groundwater contamination problem for which we require a numerical flow and transport model in order to design a pump-and-treat system for aquifer remediation. Given a complex geological setting, we are faced with the daunting prospect of defining the nature and distribution of heterogeneities beneath the site. We have seen that this requires general geological information at a regional scale larger than the site as well as detailed site-specific borehole data and informed correlation of the stratigraphy by a trained geologist. Then we need to estimate values of the relevant hydrogeological parameters at appropriate scales for each hydrogeological unit.

The task of correlation of geological units between boreholes is aided by geological intuition and conceptual facies models. But facies models cannot predict the location of high conductivity preferential flow paths and units of low permeability, both of which are critical features of the hydrogeological setting for purposes of aquifer remediation. Some type of statistical analysis such as conditional simulation using geological information as input or conditional simulation with a geological simulation model is needed and will likely become an essential part of the protocol for hydrogeological site characterization in the future.

REFERENCES

Ababou, R., D. McLaughlin, L.W. Gelhar, and A.F.B. Tompson. 1989. Numerical simulation of three-dimensional saturated flow in randomly heterogeneous porous media. *Transport in Porous Media* **4**:549–565.

Adams, E.E. and L.W. Gelhar. 1992. Field study of dispersion in a heterogeneous aquifer, 2, Spatial moments analysis. *Water Resources Research* **28**:3293–3308.

Aiken, J.S. 1993. A three-dimensional characterization of a coarse glacial outwash deposit used for contaminant modeling. M.S. Thesis, Dept. of Geology and Geophysics, Univ. of Wisc.-Madison, 136 p.

Anderson, M.P. 1979. Using models to simulate the movement of contaminants through groundwater flow systems. *CRC Critical Reviews in Environmental Control*, **9**:97–156.

Anderson, M.P. 1989. Hydrogeologic facies models to delineate large-scale spatial trends in glacial and glaciofluvial sediments. *Geological Society of America Bulletin* **101**:501–511.

Anderson, M.P. 1991. Aquifer heterogeneity—A geological perspective. *In*: B. Hitchon (Ed.), *Parameter Identification and Estimation for Aquifer and Reservoir Characterization*, Assoc. of Ground-Water Scientists and Engineers, Columbus, OH, pp. 3–22.

Anderson, M.P. 1996. Characterization of Geological Heterogeneity. *In: Stochastic Subsurface Hydrology,* Cambridge University Press, in press.

Boggs, J.M., S.C. Young, L.M. Beard, L.W. Gelhar, K.R. Rehfeldt, and E.E. Adams. 1992. Field study of dispersion in a heterogeneous aquifer, 1, Overview and site description. *Water Resources Research* **28**:3281–3292.

Bredehoeft, J.D., K. Belitz, and S. Sharp-Hansen. 1992. The hydrodynamics of the Big Horn Basin: A study of the role of faults. *Amer. Assoc. of Petrol. Geol. Bull.* **76**:530–546.

Burr, D.T., E.A. Sudicky, and R.L. Naff. 1994. Nonreactive and reactive solute transport in three-dimensional heterogeneous porous media: Mean displacement, plume spreading, and uncertainty. *Water Resources Research* **30**:791–815.

Dagan, G. 1984. Solute transport in heterogeneous porous formations. *J. Fluid Mech.* **145**:151–177.

Davis, J.M., R.C. Lohmann, S.J. Colarullo, and R.M. Phillips. 1990. Alluvial aquifer heterogeneities in the Rio Grande Valley: Implications for ground water contamination. *Technical Completion Report*, New Mexico Water Resources Research Institute.

Davis, J.M., R.C. Lohmann, R.M. Phillips, J.L. Wilson, and D.W. Love. 1993. Architecture of the Sierra Ladrones Formation, central New Mexico; depositional controls on the permeability correlation structure. *Geological Society of America Bulletin,* **105**:998–1007.

Desbarats, A.J. 1990. Macrodispersion in sand-shale sequences. *Water Resources Research* **26**:153–164.

Fetter, C.W. 1994. *Applied Hydrogeology, 3rd Edition.* Merrill Publishing Co., Columbus, Ohio, 691 p.

Fogg, G.E. 1986. Groundwater flow and sand body interconnectedness in a thick, multiple-aquifer system. *Water Resources Research* **22**:679–694.

Fogg, G.E. 1990. Architecture of low-permeability geologic media and its influence on pathways for fluid flow. In: S.P. Neuman and I. Neretnieks (Eds.), *Hydrogeology of Low Permeability Environments*, International Assoc. of Hydrogeologists, pp. 19–40.

Haldorsen, H.H. and E. Damsleth. 1990. Stochastic Modeling. Society of Petroleum Engineers. Distinguished Author Series. SPE **20321**:404–412.

Herweijer, J.C. and S.C. Young. 1990. Use of detailed sedimentological information for the assessment of pumping and tracer tests in a shallow fluvial aquifer. *In:* B. Hitchon (Ed.), *Parameter Identification and Estimation for Aquifer and Reservoir Characterization*, Assoc. of Ground-Water Scientists and Engineers, Columbus, Ohio, pp. 101–115.

Jackson, R.E., R.J. Patterson, B.W. Graham, J. Bahr, D. Belanger, J. Lockwood, and M. Priddle. 1985. Contaminant hydrogeology of toxic organic chemicals at a disposal site, Gloucester, Ontario, 1. Chemical concepts and site assessment. NHRI paper No. 23, IWD Scientific Series No. 141, National Hydrol. Res. Instit., Ottawa, Canada.

Koltermann, C.E. and S.M. Gorelick. 1992. Paleoclimatic signature in terrestrial flood deposits, *Science*, **256**:1775–1782.

Koltermann, C.E. and S.M. Gorelick. 1996. Heterogeneity in sedimentary deposits: A review of structure-imitating, process-imitating, and descriptive approaches. *Water Resources Research* **32**:2617–2658.

McKenna, S.A. and E.P. Poeter. 1995. Field example of data fusion in site characterization. *Water Resources Research* **31**:3229–3241.

Miall, A.D. 1985. Architectural-element analysis: A new method of facies analysis applied to fluvial deposits. *Earth Sci. Rev.* **22**:261–308.

National Research Council (NRC). 1994. *Alternatives for Ground Water Cleanup,* National Academy Press, Washington, DC, 315 p.

Phillips, F.M. and J.L. Wilson. 1989. An approach to estimating hydraulic conductivity spatial correlation scales using geological characteristics. *Water Resources Research* **25**:141–143.

Phillips, F.M., J.L. Wilson and J.M. Davis. 1989. Statistical analysis of hydraulic conductivity distributions: A quantitative geological approach. *In: Proceedings of new field techniques for quantifying the physical and chemical properties of heterogeneous aquifers.* National Ground Water Assoc., Columbus, OH.

Price, W.E. 1974. Simulation of alluvial fan deposition by a random walk model. *Water Resources Research* **10**:263–274.

Raveme, C. and H. Beucher. 1988. Recent development in description of sedimentary bodies in a fluvio deltaic reservoir and their 3D conditional simulations. Society of Petroleum Engineers. SPE 18310:463–476.

Selley, R.C. 1988. *Applied Sedimentology*, Academic Press, San Diego CA, 446 p.

Sudicky, E.A. 1986. A natural gradient experiment on solute transport in a sand aquifer: Spatial variability of hydraulic conductivity and its role in the dispersion process. *Water Resources Research* **22**:2069–2082.

Theis, C.V. 1935. The relation between the lowering of the piezometric surface and the rate and duration of discharge of a well using groundwater storage. *Trans. Amer. Geophys. Union* **2**:519–524.

Theis, C.V. 1967. Aquifers and models. *In:* M.A. Marino (Ed.), *Proceedings of the Symp. on Ground-Water Hydrology*. Amer. Water Resour. Assoc. pp. 138–148.

Wang, J.S.Y. 1991. Flow and Transport in Fractured Rocks. *In: Contributions in Hydrology, U.S. National Report 1987–1990*, American Geophysical Union, pp. 254–262.

Webb, E.K. 1994. Simulating the three-dimensional distribution of sediment units in braided-stream deposits. *Journal of Sedimentary Research,* **B64**(2):219–231.

Webb, E.K. and M.P. Anderson. 1996. Simulation of preferential flow in three-dimensional, heterogeneous, random fields with realistic internal architecture. *Water Resources Research* **32**:533–545.

Yeh, T.-C. Jim. 1992. Stochastic modelling of groundwater flow and solute transport in aquifers. *Hydrological Processes* **6**:369–395.

Zheng, C. and G.D. Bennett. 1995. *Applied Contaminant Transport Modeling.* Van Nostrand Reinhold. 440 p.

Chapter 3

Physical/Chemical Processes Affecting the Subsurface Fate and Transport of Synthetic Organic Materials

W.P. Ball, G.P. Curtis, and P.V. Roberts

INTRODUCTION

Physical/chemical processes play critically important roles in controlling the fate of contaminants (and thus of remediation efforts) in subsurface environments. For many of the contaminants commonly found in groundwater, problems of predicting fate and transport are confounded not only by the heterogeneity of the subsurface environment but also by the diversity of potential chemical reactions that may occur (either abiotically or biologically mediated) and by the diversity of physical/chemical interactions among the various gaseous, aqueous, organic liquid, and solid phases present. Exacerbating the problem further is the fact that solid phase composition and morphology commonly are very poorly understood, even for comparatively homogeneous subsurface regimes.

Thus, there is a broad array of potentially important physical/chemical phenomena to be considered, and subsurface complexity makes each a challenge to understand and predict. In addition to phase movement (gas, water, or organic liquid advection and dispersion in the subsurface), other physical/chemical processes of special concern include (but are not limited to) the following: organic liquid dissolution and vaporization, chemical partitioning between gaseous, aqueous, and organic phases in the subsurface, surface complexation and adsorption (from vapor or water), complexation and solubilization of chemical species (ionic or nonionic) by chelating agents or other dissolved or colloidal material, chemical transformations and chemical reactions with aqueous or solid phase species (possibly catalyzed by natural substances or surfaces), and solid precipitation and dissolution. In each of these areas issues of physical mass transfer and diffusion may be critically important to understanding overall rates of chemical movement or reaction.

The role of various processes in subsurface transport and remediation and associated site characterization requirements has been the subject of EPA technology transfer efforts and related seminar publications (EPA, 1989; EPA, 1991) which provide a good introduction to contaminant transport and remediation. In this chapter we provide further review of physical/chemical phenomena affecting organic chemical fate and attempt to put them into relative perspective. Such perspective is provided in terms of: (1) a conceptual and mathematical framework for simulating contaminant fate and transport, and (2) a moment framework for applying the simulation models to the interpretation of field-scale plume movement (as reflected in plume mass, velocity and spreading).

An overview of physical/chemical processes especially relevant to organic chemical fate and transport is provided in this chapter, which is organized into the following three subsections: conceptual framework and mass conservation, phase equilibria and mass transport, and transport and fate of dissolved contaminants. The solute fate and transport section concludes with a description of one-dimensional transport modeling for the cases of equilibrium and nonequilibrium linear sorption with and without pseudo-first-order reaction kinetics. The effects of physical/chemical processes on field-scale transport are discussed in the context of moment analysis of contaminant plumes. Following a brief overview of the moment framework, we discuss case study results from the Borden field site, where extensive laboratory work and computer simulation have been coupled with field data. Example solutes are discussed, including a non-reacting tracer (chloride) a conservative sorbing solute (tetrachloroethene), and a species transformed by reaction (hexachloroethane).

PHYSICAL/CHEMICAL PROCESS MODELING IN THE SUBSURFACE

In saturated groundwater systems the movement of dissolved organic chemicals is controlled by advective transport and mixing, diffusion within the aqueous phase, interaction with aquifer solids, and in some cases, chemical transformation (which may or may not be biologically induced). More generally, contamination may also involve the unsaturated zone and/or the presence of organic liquids (often referred to as nonaqueous phase liquids or NAPLs), such that flow of nonaqueous phases and chemical transfer among phases are important. Where NAPLs are present, these phases are likely to represent the majority of contaminant mass and will often act as lingering sources of contamination to the other phases. Nonetheless, contamination of groundwater and associated aquifer solids often extends over a much greater volume than do nonaqueous pollution sources, and groundwater remediation presents a significant problem in its own right.

For groundwater remediation, it is critically important to be able to understand and simulate (if not predict) contaminant transport and fate. Such simulation relies on fundamental process understanding coupled with computer modeling, at a complexity level appropriate to the problem at hand (Committee on Ground Water Modeling Assessment, 1990). In this chapter, we provide brief overviews of important processes of phase equilibria and transfer, as well as more detailed discussion of physical/chemical processes relevant to the fate and transport of dissolved contaminants. First, however, we describe the basic conceptual and mathematical framework which underlies most numerical simulation models. Aside from its clear importance to modeling *per se*, this framework is also useful toward more general understanding in that it helps put individual processes in perspective with regard to their relative roles in the fate and transport of contaminants.

Conceptual Framework and Mass Conservation

Mathematical simulation of solute transport and attenuation requires the application of transport equations that maintain mass balance throughout the system and account for all major processes affecting solute movement and transformation. Several authors (e.g., Abriola and Pinder, 1985; and Corapcioglu and Panday, 1991) have summarized the set of macroscopic equations that maintains contaminant mass balance in a porous medium and incorporates advective transport, dispersion, reactivity within phases, and interphase transport. Essentially, the following equation is written for each substance within each phase (after Chen et al., 1992):

$$\partial(\rho^\alpha \varepsilon^\alpha \omega_i^\alpha)/\partial t + \nabla \bullet (\rho^\alpha \varepsilon^\alpha \mathbf{v}^\alpha \omega_i^\alpha) - \nabla \bullet (\rho^\alpha \varepsilon^\alpha \mathbf{D}_{Hi}^\alpha \bullet \nabla\omega_i^\alpha) = E_i^\alpha + G_i^\alpha \tag{3.1}$$

where

t = time [T];
ρ^α = mass density of phase α $[M_\alpha/L^3]$;
ε^α = volume fraction of phase α [-];
ω_i^α = mass fraction of substance i in phase α $[M_i^\alpha/M_\alpha]$;
$\mathbf{v}^\alpha$ = vector of mass average velocity for phase α [L/T];
$\mathbf{D}_{Hi}^\alpha$ = second rank tensor for the hydrodynamic dispersion coefficient of substance *i* with respect to phase α $[L^2/T]$;
E_i^α = external mass supply of substance *i* to the α phase due to interphase exchange $[M_i^\alpha/L^3\text{-}T]$; and
G_i^α = internal (reactive) source of substance *i* in phase α $[M_i^\alpha/L^3\text{-}T]$.

Note: Abbreviations shown in brackets represent the dimensionality of the variables being defined in consistent dimensional units, e.g., $[M_\alpha]$ = phase mass, $[M_i^\alpha]$ = substance mass in phase α, [L] = length, and [T] = time. Note that the generation term (G_i^α) and exchange term (E_i^α) are defined on a per unit volume of porous medium basis.

Overall mass conservation requires modeling of each phase, with chemically balanced reactions (appropriate G_i^α within a phase) as well as balanced transfer among phases (balanced E_i^α terms among phases). Rubin (1983) has comprehensively examined the issue of balancing reactive species sources and sinks in an article describing mathematical transport formulations for six broad classes of reaction. Other authors (Yeh and Tripathi, 1989; Mangold and Tsang, 1991) have reviewed numerical transport codes which have been used to couple field-scale physical flow and transport with chemical reaction. Some of the hydrogeochemical transport codes cited in those reviews include both chemical equilibria and kinetics of reaction, resulting in considerable model complexity. Explicit consideration of diffusional limitations in interphase exchange is, however, beyond the scope of all of the cited models. As evident from subsequent discussion in this chapter, such mass transfer limitations are potentially critical to remediation efforts in heterogeneous media, and especially so for sorbing organic solutes. Transfer of contaminants to and from gaseous or organic phases is briefly reviewed in the following section, while the subsequent section (Transport and Fate of Dissolved Contaminants) highlights the potential role of diffusional mass transfer limitations within the saturated zone.

Multiphase Equilibria, Transport and Mass Transfer

Many sites of contamination are characterized by significant contamination of the unsaturated zone, and organic liquids may be present in either saturated or unsaturated regions. In the unsaturated zone, gas phase advection and diffusion can be important means of transport for volatile substances. Where an immiscible organic phase is present, significant contaminant mass will be present in that phase, and movement, dissolution, and volatilization of the organic liquid will be important to understand. Finally, the gas and nonaqueous-liquid phases represent important sources and sinks of contaminant to the groundwater.

Vapor Phase Equilibria and Transport

Partitioning to the gas phase can be an important consideration for the fate and transport of many volatile synthetic organic compounds, especially in or near the unsaturated (vadose) zone. In this section, we briefly review aspects of vapor phase partitioning and mass transport relevant to subsurface contaminant fate.

Vadose Zone Transport

In terms of vadose zone transport of volatile organic chemicals, equations similar to Equation 3.1 are needed for both the aqueous and gas phases. Water movement in the unsaturated zone is itself a complex subject, beyond the scope of this chapter but well-reviewed elsewhere (e.g., Nielsen et al., 1986; van Genuchten and Jury, 1987). For the special case of steady-state water flow in homogeneous soils, the equations of aqueous phase solute transport in the unsaturated zone are very similar to those described subsequently under "Simulation of Dissolved Contaminant Transport," with the exception that interphase contaminant exchange with soil vapor needs to be considered.

In such unsaturated systems, gas phase transport may also be important to consider, either in terms of gas phase diffusion (e.g., Jury et al., 1983; Shoemaker et al., 1990; McCarthy and Johnson, 1993; Cho et al., 1993) or gas phase advection and dispersion (e.g., Sleep and Sykes, 1989; Gierke et al., 1990; Rasmuson et al., 1990; Brusseau, 1991b). Under steady-flow conditions at constant water content, gas phase advective-dispersion equations are of similar form to those of saturated aqueous flow (e.g., Brusseau et al., 1991b). However, the pressure gradient driving gas flow is often poorly quantified, is usually transient (owing to diurnal temperature fluctuations and barometric variations), and is generally more difficult to model than the liquid phase hydraulic head. Mechanisms of gas transport and equations for simulation have been reviewed by Thorstenson and Pollock (1989a,b) and by Massmann and Farrier (1992). In addition, although many modeling applications have assumed that the gas phase will be in equilibrium with the aqueous phase (e.g., Jury et al, 1983; Jury et al., 1990; Shoemaker et al., 1990), this assumption has not been rigorously tested in the field and may not be valid under many conditions. This latter issue is further addressed under "Rate of Volatilization" below.

Vapor Phase Partitioning

Understanding vapor phase partitioning is clearly important to vadose zone transport modeling, both when vapor-liquid equilibrium is assumed and when mass transfer is explicitly considered. In the latter case, the deviation from vapor-liquid equilibrium will be the driving force for mass transfer. In the presence of organic liquids, the equilibrium gas phase concentration of a contaminant will be dictated by its pure phase vapor pressure and its mole fraction and activity coefficient in the organic liquid. For organic liquid solutions with constituents of similar molecular structure, activity coefficients are close to unity. For benzene and toluene in gasoline or jet fuel, for example, activity coefficients can be safely assumed not to exceed 1.5 (Burris and MacIntyre, 1986; Burris and MacIntyre, 1987; Cline et al., 1991). For sufficiently unlike solvent/solute pairs, activity coefficients in the organic liquid may be an important consideration in determining which evaluation is possible using UNIFAC (Prausnitz et al., 1980) or other estimation techniques.

In the absence of organic liquids, equilibrium gas phase concentrations of a solute will depend on the solute's relative partitioning among liquid, vapor, and soil phases, as dictated by its volatility (Henry's constant) and extent of sorption to the soil solids. Pennell et al. (1992) describe several potential mechanisms of sorption enhancement under unsaturated conditions. However, these and other investigators (e.g., Chiou et al., 1985; Peterson et al., 1988) have observed that extents of sorption closely approximate those observed under saturated conditions when relative humidity in the soil gas exceeds 95%. In this context, it is useful to note that lower relative humidities will exist only when the vapor pressure of all water present is reduced below its normal bulk value. Because bulk water behavior is approached at short distances from solid surfaces (i.e., at adsorbed water thickness on the order of 5 nm; Etzler and Fagundus, 1987), high humidity conditions should be expected even at very low soil moisture contents. Nonetheless, for sufficiently dry soils (e.g., the first few centimeters of soil or at deeper depth in arid regions) and with solids of high specific surface area, enhanced adsorption may be important to consider (Shoemaker et al., 1990).

Rate of Volatilization

The rate of partitioning between the aqueous and gas phases can have an important effect on transport predictions (e.g., Sleep and Sykes, 1989; Zaidel and Russo, 1993); however, because of the complexities of understanding interfacial surface areas and gas phase flow, transfer rates are difficult to predict. Rates of volatilization of contaminants in the vadose zone, either from organic liquids or from aqueous solution, will depend on the exposed interfacial surface area, the hydrodynamics of gas and liquid flow, and the driving force for transfer (i.e., the difference between the actual and the equilibrium gas phase concentrations). Mass transfer correlations developed for trickle bed reactors (e.g., Turek and Lange, 1981) have been applied to predicting air-water mass transfer rates of volatile chemicals during groundwater infiltration (Gierke et al., 1990; Cho et al., 1993); however, Cho et al. (1993) found that modeling of VOC gas profiles during lower (natural)-velocity column-scale infiltration experiments required transfer rates over 50 times lower than the trickle bed predictions. Field-scale calibrated mass transfer values were another two orders of magnitude lower.

Deviations from equilibrium models are especially likely for conditions of active soil venting, where air velocities are typically high, or in spatially variable porous media, where slow mass transfer from impermeable zones might be expected (Sleep and Sykes, 1989). In terms of the localized mass transfer across the air-water interface, higher-volume-averaged mass transfer coefficients are expected in regions of high permeability, due both to the higher fluid velocity and to the lower water content (and higher air-water interfacial area), relative to the less permeable regions. For fully saturated impermeable regions, slow aqueous diffusion will be an important aspect of mass transfer. In these regards, the complexities of air-water mass transfer in the subsurface are somewhat similar to those encountered in the context of organic dissolution and sorption nonequilibrium (discussed subsequently).

Organic Phase Equilibria and Transport

As previously noted, organic liquids may constitute most of contaminant mass at many sites of remediation. Aside from the potential concern for direct contact or air contamination, pure or mixed-constituent organic liquids can represent continuing source terms for aqueous contamination. Examples of single-liquid contaminants include trichloroethene, tetrachloroethene, carbon tetrachloride, and other solvents, many of which are denser than water (DNAPL) and are present in the subsurface as both pools and ganglia (blobs) suspended by capillary forces in inter- or intra-particle pore spaces. For lighter-than-water liquids (LNAPLs), water table rise and fall may lead to dispersed ganglia in both the unsaturated and saturated zones, where volatilization or dissolution can occur. Mixed-constituent organic liquid contamination is also common. Examples of NAPL mixtures include creosote liquids below the water surface or floating layers of petroleum hydrocarbon mixtures (e.g., gasoline). Dissolution of either DNAPL or LNAPL in the saturated zone is an important continuing source of groundwater contamination at many (if not most) sites of organic chemical contamination.

Organic Phase Transport

Predicting the location and movement of NAPL phases through saturated and unsaturated porous media has been the subject of considerable research. Mathematical descriptions of multiphase flow and transport in the context of environmental remediation have been provided by several investigators (e.g., Allen, 1985; Abriola and Pinder, 1985; Pinder and Abriola, 1986; Parker, 1989; Parker et al., 1991). However, quantitative formulations of NAPL source functions at real contamination sites are generally not available (Mackay and Cherry, 1989; Johnson and Pankow, 1992). Moreover, implementation of multiphase transport equations is computationally difficult and information-intensive, even for well-defined experimental systems (Parker et al., 1991). Poulsen and Kueper (1992) provide valuable field evidence of PCE penetration into unsaturated porous media which suggests that soil, water, and gas contamination measurements typically taken at remediation sites are ineffective in detecting and accurately assessing the location and sizes of pools or ganglia of pure phase solute. These authors further imply that the size-scale of NAPL fingering and pools may be such that modeling will not be practical at sufficiently small scale to capture the microscopic chemistry and physics. Rather, modeling of average behavior will be required without explicit description of subscale contaminant distributions. Essaid et al. (1993) conducted a series of numerical simulations to identify significant processes that could control the distribution and migration of oil observed at the location of a crude oil spill site (as reported by Hess et al., 1992). The general features of the oil body could be reproduced only when hysteresis, oil entrapment, and spatial variability of the hydraulic properties were incorporated into the numerical simulations. The small-scale details of the observed oil distribution were not reproduced by the simulations.

Organic Phase Dissolution

When organic chemicals dissolve into groundwater from NAPL phases, volume-averaged rates of mass transfer will be dependent on numerous complex factors, including the concentration gradient for dissolution (defined as the difference between actual aqueous concentrations and those at equilibrium with the organic liquid), the organic/water surface area per unit volume of aquifer, and the mass transfer coefficient associated with the given hydrodynamic conditions. As with vapor equilibria, concentrations in equilibrium with mixed constituent liquids will depend not only on the contaminant's aqueous solubility, but also upon mole fraction and activity coefficients in the organic phase (e.g., Burris and MacIntyre, 1986; Burris and MacIntyre, 1987; Cline et al., 1991; Lee et al., 1992; Geller and Hunt, 1993).

Volume-averaged mass transfer coefficients have been well-studied in the context of flow through uniform packed beds, where they have been shown to be a function of contaminant diffusivity, fluid viscosity, pore sizes (as reflected in grain diameter), bulk fluid velocity, and porosity of the fluid medium (e.g., Hunt et al., 1988; Miller et al., 1990; Powers et al., 1991; Powers et al., 1992; Geller and Hunt, 1993). Powers et al. (1991) have provided a summary of some of the published mass transfer correlations for packed beds and flow around single spheres. Given our typically poor characterization of subsurface environments, specific simulation of dissolution rates at any given remediation site is a difficult proposition at best, even without considering the added difficulty associated with lack of knowledge about initial NAPL sources and movement. Moreover, large-scale heterogeneities in NAPL distribution will lead to important additional mass transfer effects that cannot be simply modeled by small-scale packed bed correlations.

Remediation in the Presence of NAPL

The above discussion highlights the many challenges that remain in understanding NAPL movement and dissolution. As noted by Sudicky and Huyakorn (1991), uncertainties related to multiphase flow and dissolution are still substantial and have yet to be systematically addressed. While mathematical analyses in this area are useful tools in the development of appropriate technologies for remediation, accurate *predictive* NAPL-related modeling at any given field site is currently beyond reach. Remediation in NAPL-contaminated zones therefore must rely on empirically evaluated application of appropriate technology, with crude prior predictive capability. In this context, it is useful to recognize that NAPL contamination often constitutes a much smaller part of the contaminated aquifer zone than does the dissolved contaminant plume, and that it is often advisable to physically or hydraulically isolate the source zone from the dissolved plume if possible (see Chapter 5, this volume).

In the following section, we address issues related to understanding and simulating the movement and remediation of dissolved contaminant plumes. Although much of the discussion is also relevant to scenarios where gas and immiscible phases are present, transport in and transfer to these phases are not explicitly considered. In the context of the aqueous phase transport discussed subsequently, gas- or NAPL phase sources and sinks would be modeled through their own transport equations, coupled to aqueous transport by additional $E_i^{\alpha q}$ terms not explicitly discussed below.

Transport and Fate of Dissolved Contaminants

Solutes in groundwater can be influenced by many different chemical processes including sorption and/or chemical reaction with (1) the aquifer solids, (2) transported (colloidal) mineral solids, (3) other solutes that are also transported, (4) other sorbed species, and (5) bacterial colonies or cells, either sorbed or in solution. In this chapter, we deal only with abiotic sorption and reaction, and we do not explicitly consider the potential for facilitated transport by colloidal phases. (Readers are referred to McCarthy and DeGueldre [1993] for a review of some recent developments in the latter area.) Modeling the transport of multiple solutes influenced by *equilibrium*-controlled reactions has received considerable attention and this work has been summarized previously (Yeh and Tripathi, 1989; Mangold and Tsang, 1991). Approaches to deal with multiple, simultaneous equilibrium and/or rate-controlled processes have only recently been proposed (Friedly, 1991; Friedly and Rubin, 1992). Treatment of these complex reaction networks, however, is beyond the scope of this chapter. Instead, we consider the mutual interactions of solutes that are subject to single first-order (or pseudo-first-order) chemical reaction kinetics in either the dissolved or sorbed phase and sorption that is either at equilibrium or rate-controlled. While such cases may appear to be chemically unrealistic, their consideration is still of value. Under the usually ill-defined circumstances of subsurface contamination, the use of more complex equilibrium or rate-controlled transport models may not be justified.

A transport model that accounts for non-equilibrium sorption and/or reactions with simple first-order kinetics may provide valuable insight into more complex transport processes and thus help guide remediation efforts. In addition, there are numerous cases where a given solute in the subsurface is transformed only by first-order reaction kinetics and is impacted by sorption processes that may not be at equilibrium. For example, transport of many aromatic hydrocarbons and chlorinated solvents may be controlled by both nonequilibrium sorption and reactions that can sometimes be approximated by a first-order decay such as hydrolosis, reactions with solids and even microbially mediated reactions. These processes are discussed in greater detail subsequently.

Advection and Dispersion of Dissolved Species

Groundwater movement is the principal means of transport for dissolved contaminants. It is therefore self evident that good hydrogeologic characterization and understanding is central to any remediation process. In the context of a contaminant plume, groundwater movement is important not only in regard to bulk advection, but also because of the important role of flow phenomena in solute mixing and dispersion. More specifically, a dispersion term (such as that indicated in Equation 3.1) is commonly used to simulate effects due not only to molecular diffusion of aqueous contaminants, but also to account for contaminant spreading related to flow variations within pores, mechanical mixing of pore water, and velocity variations in the porous medium. The effective hydrodynamic dispersion coefficient has been shown to depend upon the scale of application (Anderson, 1979; Pickens and Grisak, 1981; Molz et al., 1983; Gelhar et al., 1992).

There is a growing body of research aiming to describe large-scale dispersion as the product of smaller-scale variability of aquifer properties by applying transport equations (such as Equation 3.1) over heterogeneous fields with layered permeabilities (e.g., Molz et al., 1983, 1986) or randomly generated permeability fields (e.g., Schwartz, 1977; Smith and Schwartz, 1980; Tompson and Gelhar, 1990). Statistical approaches have also been developed for calculating effective macrodispersivities as the result of assumed distributions of the small-scale velocities (e.g., Dagan, 1982, 1986, 1990; Gelhar and Axness, 1983; Neuman et al., 1987). Sudicky and Huyakorn (1991) provide an excellent review of methodologies for describing solute migration and dispersion in heterogeneous aquifer systems. Gelhar et al. (1992) provide a thorough and critical review of field-scale dispersivity observations.

Even in one-dimensional column experiments, researchers have found that solute spreading is not always well-described by the diffusive Fickian mechanism reflected in the $\mathbf{D}_{\mathrm{H}i}^{\alpha}$ term of Equation 3.1. Such non-Fickian dispersion is an expected result of flow channeling, and may occur even in relatively homogeneous media due to particle size separation during column packing (Ripple et al., 1974). A number of investigators have used a "dead-zone" approximation to simulate the spreading effects associated with diffusion of solute into and out of soil aggregates or other regions of comparatively immobile water. "Two-site" models have been used to describe conservative tracer experiments through soil columns (e.g., Coats and Smith, 1964; van Genuchten and Wierenga, 1977; Nkedi-Kizza et al., 1983), where it was found that observed tracer behavior could be modeled by assuming separate compartments of mobile and immobile pore fluid. Such models have been expanded to include sorbing solutes, as described in greater detail subsequently.

Multidimensional transport models which incorporate heterogeneity of flow should also be capable of simulating mass-transfer related spreading effects associated with sorption in impermeable zones; in this case, transport through truly impermeable zones must be modeled through the molecular diffusion component of the transverse dispersion tensor, and proper modeling will require fine discretization and large computational effort. On the other hand, multidimensional models have the advantage of being able to explicitly model solute spreading effects associated with flow through regions of low (as well as zero) permeability. In practice, however, the configuration, geometry, sorption characteristics, permeabilities and diffusional characteristics of the subsurface are poorly understood, and particularly so with regard to their heterogeneity, such that simpler mobile-immobile transport codes are often appropriate approximations for modeling diffusional mass transfer resistance. For both types of models, stochastic distributions of parameters represent a useful means of evaluating modeling uncertainty (Dagan, 1986; Gelhar, 1986).

For organic chemicals, understanding dispersion can be further complicated by the processes of sorption, reaction, and related mass transfer. In addition to the direct effects of sorption on the aqueous concentrations relevant to advection and dispersion processes, sorption can significantly enhance the dispersive effect related to mass transfer of contaminants to and from immobile regions, especially if regions of lower permeability sorb more strongly (Valocchi, 1989; Chrysikopoulos et al.,

1990). In addition, the normally inconsequential impact of microscopic intraparticle regions on the dispersive effect can be greatly enhanced if there is strong sorption in the intraparticle regions. Finally, organic chemicals are often subject to aqueous or sorbed phase chemical reactions that will clearly alter mass balances, and chemical reactions are themselves affected by sorption and mass transfer. Sorption and reaction processes are reviewed in greater detail in the sections immediately following. Transport models incorporating these effects are discussed subsequently.

Sorption Equilibrium and Mass Transfer

Natural soil materials are complex, heterogeneous mixtures of rock fragments, clay minerals, inorganic precipitates, and soil organic matter. For most commonly encountered organic solutes, interactions with the natural geologic material will be important to consider. In this section, we briefly review issues of equilibrium sorption and mass transfer for organic solutes.

Sorption Equilibrium

From a phenomenological perspective, the sorbed concentration of a solute in equilibrium with any given aqueous phase concentration (for a specified temperature and set of solution conditions) is given by the experimental sorption isotherm, forms of which have been extensively reviewed elsewhere (e.g., Weber et al., 1991). Sorption of hydrophobic organic chemicals by natural solids has been extensively reviewed (e.g., Karickhoff, 1984; Curtis et al., 1986a; Chiou, 1989; Ball, 1989; Weber et al., 1991). Batch isotherm data are often empirically fitted to Freundlich isotherms:

$$q_e = K_f \, c_e^{1/n} \tag{3.2}$$

where

q_e = equilibrium aqueous concentration $[M_x/L^3]$;
K_F = Freundlich capacity coefficient $[(M_x/M_s)/(M_x/L^3)^{1/n}]$;
c_e = equilibrium sorbed concentration $[M_x/M_s]$; and
$1/n$ = Freundlich exponent (in this unit notation, M_x = mass of contaminant and M_s = mass of sorbent).

For nonpolar chemicals and soils of high organic carbon content, linear isotherms (i.e., $1/n = 1$) are an expected consequence of solute partitioning into a natural organic phase from dilute solution (Chiou, 1989). Empirical models have been successfully applied to the prediction of organic-matter partitioning (K_{oc}) on the basis of aqueous solubility and/or octanol-water partitioning coefficients (e.g., Schwarzenbach and Westall, 1981; Karickhoff, 1984; Curtis et al., 1986a), but such estimates are only approximate and there is a growing body of literature investigating differences attributable to the origin and type of natural organic matter (e.g., Chiou et al., 1986; Garbarini and Lion, 1986; Gauthier et al., 1987; Grathwohl, 1990; Rutherford et al., 1992).

There are also situations where surface adsorption to mineral surfaces may be equally as important as partitioning, either because very little organic matter is present (e.g., Stauffer and MacIntyre, 1986; Piwoni and Banerjee, 1989; Ball and Roberts, 1991a), because of an especially strongly *ad*sorbing phase such as kerogen or shale (Weber et al., 1992), or because the organic chemical has polar or ionic structural groups capable of engaging in surface complexation or ion exchange reactions with the mineral surface (e.g., Stumm et al., 1980; Means et al., 1982; Davis and Kent, 1990; Schindler, 1990; Zachara et al., 1990; Kung and McBride, 1991; Stone et al., 1993). Adsorption in the latter case can be especially complex to model, as detailed understanding of surface and solution chemistry is required (Stone et al., 1993).

Even for comparatively nonpolar chemicals, nonlinear sorption by aquifer solids has been demonstrated. For low-molecular-weight organic chemicals and natural aquifer solids, Freundlich exponents less than one are often observed (e.g., Miller and Weber, 1986; Piwoni and Banerjee, 1989;

Ball and Roberts, 1991a; Weber et al., 1992; Young and Ball, 1994), corresponding to higher sorbed:aqueous equilibrium ratios at low concentration. Weber et al. (1991) described how this condition leads to greater asymmetry or "tailing" of an eluting solute pulse or plume. Weber et al. (1992) provide extensive discussion of isotherm nonlinearity in the context of sorption by soils and sediments, and point out the importance of surface heterogeneity in understanding the phenomenon. By contrast, Backhus (1990) has suggested that very hydrophobic compounds may partition in similar linear manner to a variety of mineral surface phases.

Because the success of groundwater remediation depends so heavily on sorption/desorption at low concentration, issues of sorption mechanisms and modeling are important areas of ongoing research. At present, it is clear that truly accurate prediction of sorption is not possible without isotherms conducted under the specific conditions of interest—*a priori* estimation based on K_{ow} and organic carbon content are useful only as order-of-magnitude approximations, limited to nonpolar organic chemicals in soils of sufficiently high organic carbon content (on the order of 0.2 to 1% or higher). Moreover, the accurate measurement of soil organic carbon content is not always straightforward to obtain (Powell et al., 1989; Ball et al., 1990a; Weber et al., 1992). For best understanding, sorption isotherms need to be conducted with the chemical of interest (or an appropriate analogue) over the full aqueous concentration range of interest and with careful experimental procedure. Roy et al. (1991) have published recommended procedures for the conduct of batch-type sorption isotherms. Additional considerations relevant to the accurate measurement of volatile organic chemical uptake by low-carbon aquifer solids are discussed by Ball and Roberts (Ball, 1989; Ball and Roberts, 1991a).

For very hydrophobic compounds, issues of transport and sorption are further complicated by the possibility of facilitated transport by suspended colloids in the mobile groundwater, such as organic macromolecules, very fine metal oxides, or other suspended particulates. The evidence for and implications of facilitated transport have been reviewed by McCarthy and Zachara (1989), Backhus and Gschwend (1990), and Chin and Gschwend (1992). For the case of linear partitioning to aquifer solids and colloids, parameterization of equilibrium distributions has been described by Pankow and McKenzie (1991).

Sorption Related Mass Transfer

Sorption rate limitations are commonly recognized to be of importance in solute transport and, as described subsequently in this chapter, nonequilibrium transport models are often applied to experimental data. Nonetheless, prediction of contaminant transfer rates in the field is still elusive. At any given field site, the nature of mass transfer limitations will depend extensively upon the hydrogeologic setting and the types of aquifer solids present. At the field scale, mass transfer among regions of the porous medium will be important if there is significant spatial variability in hydraulic conductivity, and such mass transfer will become even more important to overall plume spreading if the sorption characteristics of the aquifer solids also vary spatially. Even with particle-scale local equilibrium, spatial variability in sorption will enhance solute dispersion over that of non-sorbing tracers, and the effect will be especially pronounced for the case where there is an inverse correlation between hydraulic permeability and strength of sorption (Garabedian et al., 1988; Valocchi, 1989; Chrysikopoulos et al., 1992), as might occur between a coarse gravel having weak sorption affinity and clay layers of enhanced carbon content and surface area.

Mass transfer limitations at the particle scale have also been observed, as reviewed elsewhere (Ball and Roberts, 1991c). Although very slow intragranular diffusion is difficult to detect in short-term experiments (hours to days of equilibration), slow particle-scale sorption and desorption are observed under conditions where sorbents and chemicals are allowed to remain in contact for an extremely long time (weeks to years; Ball and Roberts, 1991b; Pignatello, 1990a; Grathwohl et al., 1990; Pedit and Miller, 1992; Harmon et al., 1991; Young and Ball, 1994). On the other hand, carefully conducted batch and column studies with some quartz-based aquifer sands have shown comparatively rapid rates of sorption and desorption (MacIntyre et al., 1991; Larsen et al., 1992a,b).

In general, it is likely that rates of sorption and desorption will be highly dependent upon mineralogy and intragranular pore morphology (Ball and Roberts, 1991c; Pignatello, 1990b).

When present, particle-scale mass transfer limitations tend to augment and complicate the effect of field-scale mass transfer. Although diffusion length scales at the level of individual soil grains are small in comparison with field-scale aggregation or layering, diffusion rates may also be comparatively low owing to pore constrictivity (Wood et al., 1990; Ball and Roberts, 1991b,c), steric hindrance in an organic phase (Pignatello, 1989; Brusseau et al., 1991a), or both (Ball and Roberts, 1991b,c; Pignatello, 1990b). Irrespective of mechanism, transient intraparticle diffusion rates can be significantly slower than those predicted using the unretarded aqueous diffusivity, greatly slowing the approach to sorption or desorption equilibrium. Such effects can be important in the context of transformation or bioremediation of hydrophobic compounds (e.g., Rijnaarts et al., 1990; Criddle et al., 1991).

With regard to solute transport and retarded pore diffusion, the advection to diffusion ratios relevant to calculating the validity of the local equilibrium assumption (e.g., Valocchi, 1985; Bahr and Rubin, 1987; Roberts et al., 1987) are the same for both sorbing and non-sorbing solutes, since the advection of sorbing solutes is also retarded (Ball et al., 1991). If the immobile aqueous volume is low, however, nonequilibrium effects may be imperceptible in the transport of nonsorbing solutes. On the other hand, if significant sorption capacity exists in the immobile region (regardless of its aqueous volume), nonequilibrium effects can be a very significant aspect of sorbing solute transport. This capacity effect may also apply to intraparticle diffusion or diffusive effects related to larger zones of low permeability.

In a subsequent section of this chapter, we explore a case study where sorption-related mass transfer is believed to have influenced the transport of two sorbing solutes—one chemically conservative and one chemically reactive. First, however, we briefly review some aspects of organic chemical transformations in the subsurface.

Reactions of Organic Solutes in the Subsurface

Synthetic organic chemicals can be transformed in the subsurface environment by many different processes, including nucleophilic substitution reactions, oxidation reactions, and reduction reactions. Many of these reactions can be microbially mediated and may involve either intracellular reaction (requiring active microbial involvement) or reaction with enzymes external to the microbial cells. In addition, many aqueous reductants and other reactive species may be products of microbial metabolism (e.g., Bouwer and Wright, 1988; Bouwer, 1992). Microbiological transformations are discussed elsewhere in this volume and will not be specifically discussed here. Nonetheless, it is important to recognize that many subsurface reactions may depend directly or indirectly on microbial activity as, for example, in the microbiological mediation of the subsurface redox state.

Chemical reactions and mass transfer can be closely coupled processes: reactions can increase the rate of mass transfer by steepening concentration gradients, but at the same time, mass transfer can limit the overall rate of reaction by limiting the supply of reacting species. This is particularly true if the reaction occurs preferentially at mineral/water interfaces or in certain spatial domains. For example, sorbed species may not be available for reaction with large aqueous molecules or microbial cells (Rijnaarts et al., 1990; Criddle et al., 1991; Miller and Alexander, 1991; Mihelcic and Luthy, 1991), whereas reactions involving surface species or immobilized organic matter may be enhanced by sorption. Depending on the situation, sorption can either decrease or increase the overall reaction rate. In addition to these effects of equilibrium contaminant distribution, mass transfer limitations (nonequilibrium sorption) can limit the overall rate of reaction for either case. Thus, in sediment/water systems, it is important to account not only for the chemical reactions but also for the reaction site and the local conditions prevailing there. In the subsections below, we provide a brief overview of those abiotic transformation reactions currently believed to be most pertinent to organic chemical fate in the subsurface, with emphasis on current understanding with regard to the effect of sorption on such transformations. As highlighted below, the effects of sorption on transformation are still an area of considerable uncertainty and ongoing research.

Nucleophilic Substitution Reactions

Hydrolysis reactions have been studied for many years. Consequently, many structure-activity relations have been developed and comprehensive reviews of rate constants are available (Mabey and Mill, 1978; Mill and Mabey, 1988). In general, it has been found that hydrolysis reactions can occur by three parallel processes: an acid-catalyzed reaction, a base-catalyzed reaction, and a neutral process, the latter of which determines the minimum reaction rate (Mabey and Mill, 1978). Differences in hydrolysis rates in the sorbed and dissolved phase could result if the surfaces alter the concentrations of the reactants, alter the conformation of the reactants on the surface, or alter the immediate environment in which the reaction takes place (Stone, 1989a). Neutral hydrolysis reactions have been found to occur at the same rate in the sorbed phase as they do in the dissolved phase for both chlorinated hydrocarbons (Haag and Mill, 1988a) and several organophosphoro-thioate esters (Macalady and Wolfe, 1985). The alkaline hydrolysis rates of other esters decreased by approximately a factor of 10 in the presence of sediment (Macalady and Wolfe, 1985), which is consistent with the electrostatic repulsion of the negatively charged nucleophile (OH^-) from the predominantly negatively charged humic acids. Alkaline hydrolysis of specifically adsorbed monophenyl-terephthalate in aluminum oxide aqueous suspensions was increased by approximately a factor of 10 relative to particle-free water (Stone, 1989a). This observed increase is again in agreement with electrostatic considerations: aluminum oxides are positively charged and therefore would be expected to enhance the hydroxide concentration adjacent to the mineral surface. The addition of natural organic matter to the aluminum oxide suspension decreased rate enhancement probably because the organic matter both out-competed the monophenyl-terephthalate for the surface and decreased the positive charge at the surface (Stone, 1989b). Adsorption of phenyl picolinate onto TiO_2 and FeOOH surfaces was observed to catalyze the hydrolysis rate of phenyl picolinate, although similar effects were not observed for the case of SiO_2 or Al_2O_3 suspensions (Torrents and Stone, 1991). Because it is difficult to generalize on the effects of solid surfaces on hydrolysis reactions, reaction rates need to be considered on a case by case basis, accounting for the types of solid surfaces as well as the aqueous chemistry and the nature of the organic compound.

The hydrogen bisulfide anion (HS^-) is a powerful nucleophile encountered in anoxic waters, and its reactions with haloalkanes have drawn increasing interest (Schwarzenbach et al., 1985; Haag and Mill, 1988b). Assuming equal molar concentrations of reactant, reactions with HS^- as the nucleophile are 10^5 to 10^6 times faster than reactions with water or OH^- as the nucleophile, but owing to the high concentration of water, hydrolysis reactions often remain important in aquatic systems (Barbash and Reinhard, 1989). The reactions with HS^- in sediment/water systems are likely to be further complicated because HS^- can react with sediments to reduce amorphous iron oxide coatings and then form iron sulfide precipitates. The influence of sediments on the rate of hydrosulfide substitution reactions with alkyl halides has not been reported in the literature.

Oxidation-Reduction Reactions

Certain organic chemicals, particularly phenolic compounds and aniline derivatives, can undergo oxidation in the presence of transition metal oxides containing elements such as Fe and Mn (Stone and Morgan, 1984). Details of the reaction mechanisms have been summarized previously (Stone, 1991). These oxidation reactions have typically been studied in the presence of freshly precipitated solids, whereas comparatively little attention has been given to the reactions with subsurface sediments where natural organic matter coatings may block surface sites. The overall reaction involves the oxidation of the organic compound and the reduction and dissolution of the solid. In abiotic systems, the oxidation process typically stops at fairly complex organic compounds—rarely does the reaction result in aromatic ring cleavage, for example. This is in contrast to microbially mediated oxidations, which often mineralize the substrate to CO_2.

Organic chemicals containing strong electron-withdrawing substituents such as halogens or nitro groups are more susceptible to reductive transformation reactions in sediment/water systems than the corresponding compound with electron-donating groups (Macalady et al., 1986). Transformations involving other electron-withdrawing substituents include the reduction of nitro compounds to the corresponding amino compound, formation of aniline derivatives from azobenzenes, and the deoxygenation of sulfoxides to the corresponding thioether (Macalady et al., 1986).

Wolfe and Macalady (1992) recently summarized the current understanding and open issues concerning the reduction of organic compounds in the subsurface. Perhaps the most significant unknown at this time is with regard to the characterization of the environmentally relevant electron donors. Attempts to address this issue have involved the use of model compounds such as hydroquinones for the reduction of nitroaromatic compounds (Tratnyek and Macalady, 1989; Schwarzenbach et al., 1990) and for halogenated hydrocarbons (Curtis, 1991). These studies provide indirect evidence that hydroquinone moieties in natural organic matter may play an important role in the observed sediment-enhanced reduction of nitroaromatic compounds (Dunnivant et al., 1992) and alkyl halides (Curtis, 1991). The data strongly suggest that the natural organic matter can act as a mediator in the reduction reaction in that it promotes the transfer of an electron from an electron donor such as HS^- or Fe^{2+} to the organic compound. However, the reduction of carbon tetrachloride by HS^- is also promoted by sheet silicates (Kriegman-King and Reinhard, 1992) suggesting that many aquifer constituents may be capable of promoting reduction reactions. Because the rates of redox reactions may depend strongly on the reducing conditions in the aquifer, it may be inappropriate to model the transport of a solute that undergoes oxidation or reduction reactions (either abiotically or microbially mediated) by a simple, one-component transport model. Adequate description of the reducing conditions in an aquifer can depend on many other solutes that are also subject to transport.

Several reports have demonstrated that in a given system, reduction rates for a set of organic compounds can be correlated with a physical property such as the standard reduction potential (Schwarzenbach et al., 1990) or bond dissociation energy (Jafvert and Wolfe, 1987). However, similar success has not been reported in comparing reduction rates among a set of sediments. For example, the reduction of hexachloroethane in the presence of pond sediments (Jafvert and Wolfe, 1987) was found to be approximately 5 to 6 orders of magnitude faster than in the presence of a low organic carbon sand from a predominantly aerobic aquifer (Curtis, 1991). Although this trend agrees qualitatively with the anticipated faster reduction rate in anoxic solutions, there are no quantitative methods currently available to predict the observed differences.

Thus, at present, accurate characterization of surface-mediated organic chemical reaction requires independent laboratory estimates with the subsurface solids of interest. Research is ongoing to help identify those environments where such processes are most important to consider and how best to include their effects.

Simulation of Dissolved Contaminant Fate and Transport

In this section, a simple first-order kinetic modeling approach is described to illustrate how sorption, mass transfer, and transformation kinetics can be incorporated into the simulation of dissolved contaminant fate and transport. The one-dimensional linear case is used for simplicity of presentation; approaches involving other dimensions and geometries are conceptually analogous, although more complex in detail.

In considering aqueous phase transport in a fully saturated aquifer system and in the absence of organic liquids, the governing equation for the steady-state one-dimensional transport of a sorbing and reacting solute can be written as follows:

$$\theta_m \, \partial c_{i,m} / \partial t = -\theta_m \, v_m \, (\partial c_{i,m} / \partial x) + \theta_m \, D_h \, (\partial^2 c_{i,m} / \partial x^2) + E^m_{i,im} + E^m_{i,s} + G^m_i \qquad (3.3)$$

where

θ_m	=	mobile porosity = volume of mobile water/volume of porous medium $[L^3L^{-3}]$;
$c_{i,m}$	=	concentration of solute i in mobile pore water $[M_iL^{-3}]$;
v_m	=	average linear velocity of mobile pore water in the x direction $[LT^{-1}]$;
x	=	distance in the direction of transport from some specified datum [L];

D_h = coefficient of longitudinal hydrodynamic dispersion $[L^2T^{-1}]$;
$E^m_{i,im}$ = the source of solute i to the mobile liquid from the immobile pore water $[M_iT^{-1}L^{-3}]$;
$E^m_{i,s}$ = the source of solute i to the mobile liquid phase from the solid phase $[M_iT^{-1}L^{-3}]$; and
G^m_i = the rate of production of solute i in the mobile liquid phase $[M_iT^{-1}L^{-3}]$.

In Equation 3.3, an assumption of constant density for dilute aqueous solutions allows substitution of $c_{i,m}$ for the product $\rho_i^m\ \omega_i^m$ found in Equation 3.1. The mobile phase porosity, velocity and dispersion coefficient have also been assumed to be spatially and temporally constant. These assumptions are appropriate for dilute solute concentration in steady-state groundwater flow through a homogeneous porous medium. As in Equation 3.1, all mass flux terms are expressed on a per unit volume of porous medium basis.

The average linear velocity in the above equation is equal to the Darcy velocity (also known as specific discharge) divided by the effective porosity (volume mobile water/volume porous medium), appropriate to the representative elemental volume (REV) of interest (Bear, 1979). The average linear velocity can be directly estimated from monitoring of tracer movement or it can be derived using estimated values of hydraulic conductivity, potential gradient, and effective porosity. Note that the effective porosity differs conceptually from the total aquifer porosity because a part of the aquifer's porosity has been assumed to reside in extremely small pores in which advective and dispersive fluxes can be assumed negligible. Water in these pores is consequently called immobile water (van Genuchten and Wierenga, 1976a).

The terms $E^m_{i,im}$ and $E^m_{i,s}$ in Equation 3.3 represent the net flux of solute to the mobile pore water from the immobile pore water and from the sorbed phase. These transfers can be considered in terms of average immobile- and sorbed phase concentrations by writing separate mass balances for the sorbed and immobile phases. These mass balance equations and other terms of Equations 3.1 and 3.3 are explored in greater detail in subsections below. Note that in future developments the subscript i will be dropped, with the understanding that all concentrations and source rates refer to a single chemical constituent.

The Local Equilibrium Assumption

The most widely applied simplification to Equation 3.3 has become known as the Local Equilibrium Assumption (LEA). Its applications and limitations have been well documented (Rubin, 1983; Valocchi, 1985; Bahr and Rubin, 1987; Grove and Stollenwerk, 1987; Brusseau and Rao, 1989; Harmon et al., 1989; Weber et al., 1991). Under the LEA, immobile water does not need to be explicitly considered and mass transfer between mobile and sorbed solute is assumed to be quite rapid relative to groundwater advection. In this model, solute associated with immobile fluid is considered "sorbed" and reactions in the immobile fluid are considered as part of the "sorbed phase" mass balance.

Under the above conditions, the E^m_{im} term of Equation 3.3 is dropped and the E^m_s term can be derived from a mass balance in the sorbed region. Note that mass conservation across the sorbed phase/water interface requires that $E^s_m = -E^m_s$, where E^s_m is defined as a source term for solute to the sorbed phase from the mobile liquid $[M_iT^{-1}L^{-3}]$. Applying Equation 3.1 to the sorbed phase, where the dispersive and advective fluxes are zero, we obtain the following expression:

$$\rho_b\ \partial q/\partial t = E^s_m + G^s + G^{im} \qquad (3.4)$$

where

ρ_b = bulk density (mass of aquifer solids/volume of porous medium $[M_s/L^3]$);
θ = mass of sorbed solute per unit mass of solids $[M_i/M_s]$;
G^s = the rate of production of solute in the sorbed phase $[M_iT^{-1}L^{-3}]$; and
G^{im} = the rate of production of solute in immobile fluid $[M_iT^{-1}L^{-3}]$.

Invoking the LEA, we can combine Equations 3.3 and 3.4 and apply the chain rule to obtain the following transport equation:

$$[\theta_m + \rho_b(\partial q/\partial c_m)]\,(\partial c_m/\partial t) = -\theta_m\, v_m(\partial c_m/\partial x) + \theta_m\, D_H(\partial^2 c_m/\partial x^2) + G^m + G^{im} + G^s \qquad (3.5)$$

where $\partial q/\partial c_m$ is the change in sorbed concentration per unit change in aqueous concentration, as defined by the equilibrium sorption isotherm. Note that the process of combining Equations 3.3 and 3.4 to eliminate the E^s_m term also introduces two new reaction terms, G^{im} and G^s, into the mobile phase transport equation. In the simple case of the linear isotherm, the sorption isotherm is described by

$$q = K_d\, c_m \qquad (3.6)$$

where K_d is the linear distribution coefficient $[L^3/M_s]$. In this case,

$$[\theta_m + \rho_b K_d]\,(\partial c_m/\partial t) = -\theta_m v_m(\partial c_m/\partial x) + \theta_m D_H(\partial^2 c_m/\partial x^2) + G^m + G^{im} + G^s \qquad (3.7)$$

Largely because of its mathematical simplicity, this form of the transport equation has been broadly applied in practice. In the absence of the last three reaction terms and for a homogeneous porous medium, solving Equation 3.7 is identical to the case of a nonreactive, nonsorbing tracer, except that pore water velocity and dispersivity are reduced by a constant factor ($= [\theta_m + \rho_b K_d]/\theta_m$), known as the retardation factor, R. For example, most algorithms applied to date toward the optimization of well placement and pumping conditions for groundwater remediation have relied on LEA simplifications.

In contrast to the assumptions of simple homogeneous linear LEA models, sorption is likely to be far from uniform in its effect on transport. In particular, nonlinear sorption isotherms, rate limitations in the sorption process, large sorptive capacities in zones of low flow, and heterogeneities in the sorptive properties of the soil may all lead to enhanced spreading (greater effective dispersion) for sorbing solutes, an effect which is perhaps most commonly seen in the form of contaminant "tailing"—i.e., continuing low concentrations over greatly extended time periods when attempting to remediate by flushing with clean water. Nonequilibrium transport models are described subsequently in this chapter. First, however, it is useful to consider expressions for rates of reaction and to develop a transport and reaction framework for the simpler LEA case.

Rate Expressions and Pseudo-First-Order Reaction

The term G^m in Equations 3.3 and 3.7 accounts for the rate of formation or disappearance of solute in the aqueous phase. The actual expression that accounts for the rate of formation can exhibit various degrees of complexity. For the case of a single reaction between two species c_{1m} and c_{2m}, the rate of production of species "1" in mobile solution is given by:

$$G_1^m = -\,\theta_m\, k_m\, c_{1m}^{\ \nu_1}\, c_{2m}^{\ \nu_2} \qquad (3.8)$$

where

k_m = reaction rate constant $(M_iL^{-3})^{(1-1/(\nu_1+\nu_2)}(T^{-1})$ in the mobile phase;
c_{1m} = concentration of solute 1 (M_iL^{-3}); and
ν_1 = the reaction order with respect to solute 1. The negative sign indicates that the expression is for the assumed case of loss of solute 1 from solution and the unit notation for k_m assumes that species 1 and 2 are reported in like concentration units (e.g., "M_i/L^3" = moles/liter for both species).

A transport equation that includes a bimolecular rate expression such as that in Equation 3.8 must be solved simultaneously with a transport equation for c_{2m}. If c_{2m} is constant throughout the porous medium, $c_{2m} >> c_{1m}$, and the reaction order with respect to c_{1m} is unity, then the rate of disappearance of c_{1m} is given by the following pseudo-first-order expression:

$$G_1^m = -\,\theta_m\, k'_m\, c_{1m} \qquad (3.9)$$

where

k'_m = the pseudo-first-order rate constant for reaction of species 1 in the mobile phase.

Under certain conditions, the rate expression for the transformation of a solute by microorganisms can also be simplified to a first-order form. For example, when substrate consumption follows Monod kinetics (rate of change of substrate concentration S due to reaction = $k_M XS/[K_s + S]$), cells are not growing (cell concentration, X ≈ a constant), and the Monod half velocity (K_s) is sufficiently high ($K_s >> c_{1m}$), then the rate of substrate utilization can be approximated by an equation identical to Equation 3.9. k'_m is still a pseudo-first-order rate constant with units of inverse time, but is now equal to $(k_M X)/K_s$, where k_M is the maximum specific rate of substrate utilization. Alexander and Scow (1989) note that the conditions of the pseudo-first-order approximation for microbially mediated transformation reactions are probably rarely met, but the approximation is often used for mathematical simplicity. A more accurate description requires solving the complete coupled system for microbial cell concentration, concentrations of all important substrates, and the concentration of the electron acceptor (Molz et al., 1986; Baveye and Valocchi, 1989; Celia et al., 1989; Chen et al., 1992). However, in remediation of contaminated field sites, the pseudo-first-order approach may be adequate for simulation, given the typically equal uncertain initial conditions (i.e., initial quantity and distribution of the chemical) and the heterogeneous distribution of hydraulic, chemical, and microbiological properties in typical aquifers.

As reviewed previously, many types of reactions can occur in the sorbed phase. The term G^s in Equation 3.7 represents the rate of reaction in the sorbed phase. As in the case of reactions in solution, the general rate expression for a reaction in the sorbed phase can be strongly nonlinear. However, for the simplified case of a pseudo-first-order process, G^s is given by

$$G^s = -\, k_s\, \rho_b\, q \qquad (3.10)$$

where

k_s = the pseudo-first-order reaction constant for the sorbed phase $[T^{-1}]$ and the minus sign again indicates an assumed loss of solute due to a degradation reaction.

Equilibrium-Controlled Sorption and Pseudo-First-Order Reaction

Combining the pseudo-first-order rate laws for reactions in aqueous solution and in the sorbed phase (Equations 3.9 and 3.10) with the one-dimensional LEA transport equation for a linear sorption isotherm (Equation 3.7) gives

$$[\theta_m + \rho_b K_d]\,(\partial c_m/\partial t) = -\,\theta_m v_m(\partial c_m/\partial x) + \theta_m D_H(\partial^2 c_m/\partial x^2) - \theta_m k'_m c_m - k_s \rho_b K_d c_m \qquad (3.11)$$

As before, the substance subscript has been dropped ($c_{1m} = c_m$). Equation 3.11 can be rearranged to give

$$\partial c_m/\partial t = -\,(v_m/R)\,(\partial c_m/\partial x) + (D_H/R)(\partial^2 c_m/\partial x^2) - (k'_m/R)\,c_m(R-1)(k_s/R)\,c_m \qquad (3.12)$$

where R is the retardation factor $[= 1 + (\rho_b K_d/\theta_m)]$. Equation 3.13 describes the transport of a sorbing and reacting solute in a porous medium under conditions of linear local equilibrium. Equation 3.12 indicates that if $k_s = k'_m$, as would be expected for the case for a radioactive decay, for example, then the overall degradation rate is unchanged in the presence of the sorbing solids. If, on the other hand, $k_s = 0$, then the rate of degradation is reduced by a factor equal to R. This case could represent the important case of microbially mediated degradation, since it is often assumed that solute that is sorbed to solids is not available for biodegradation (Steen et al., 1980; Ogram et al., 1985; Criddle et al., 1991; Rijnaarts et al., 1990; Scow and Alexander, 1992). Therefore, in modeling the microbially mediated degradation in remediation efforts, it is important to account for the effect of sorption on the degradation rate, particularly for strongly sorbing solutes. Sorption may have other effects as well. In particular, if transformation is favored in only one of the two phases (sorbed or aqueous), slow uptake or release may represent a rate-limiting step for the overall loss rate. Modeling in this case requires explicit consideration of nonequilibrium sorption (described subsequently). As a point of interest, Equation 3.11 also indicates that slow reactions in the sorbed state can become more important for strongly sorbing solutes.

Nonequilibrium Transport

As previously noted and reviewed elsewhere (e.g., Harmon et al., 1989; Brusseau and Rao, 1989; Ball, 1989; Weber, 1991; Ball and Roberts, 1991b,c), the simple case of linear local equilibrium has often been found to be inadequate for describing transport in laboratory columns and in field studies. This lack of agreement is most often interpreted as evidence for mass transfer limitations in the sorption process. One approach to the incorporation of nonequilibrium sorption in modeling is to explicitly consider that portion of the porosity which exists as extremely small pores where advection and hydrodynamic dispersion fluxes are negligible (but where molecular diffusion may still be active), and to associate an appropriate fraction of the solid sorption affinity with these immobile regions. We now reintroduce the $E_{i,m}^{im}$ term of Equation 3.3 and provide a new mass balance equation for the immobile aqueous phase. In addition, we must now divide the sorbed phase mass balance (Equation 3.4) into separate equations for the mobile and immobile sorbed regions as well as the immobile fluid. For the immobile regions, unlike for the more rapidly dispersed mobile phase, local concentrations and gradients are sometimes explicitly modeled in order to more accurately consider diffusive fluxes. (See subsequent discussion.) With this in mind, we define average macroscopic immobile concentrations using the "ima" subscript to differentiate these average concentrations from more localized values. With this convention, the three mass balance equations are written as:

$$\theta_{im}\ \partial c_{ima}/\partial t = E_{m}^{im} + E_{sim}^{im} + G^{im} \tag{3.13a}$$

$$\rho_b\ \partial q_m/\partial t = E_{m}^{sm} + G^{sm} \tag{3.13b}$$

$$\rho_b\ \partial q_{ima}/\partial t = E_{im}^{sim} + G^{sim} \tag{3.13c}$$

where

θ_{im} = immobile porosity (volume of immobile water/volume of porous medium);
c_{ima} = average concentration of solute in the immobile pore water $[M_i L^{-3}]$;
E_m^{im} = the source of solute to the immobile liquid from the mobile liquid $[M_i T^{-1} L^{-3}]$;
E_{sim}^{im} = the source of solute to the immobile liquid from the immobile sorbed phase $[M_i T^{-1} L^{-3}]$;
G^{im} = the rate of production of solute i in the immobile liquid $[M_i T^{-1} L^{-3}]$;
q_m = mobile sorbed phase concentration $[M_i M_s^{-1}]$;
E_m^{sm} = the source of solute to the mobile liquid from the mobile sorbed phase $[M_i T^{-1} L^{-3}]$;
G^{sm} = the rate of production of solute in the sorbed phase in equilibrium with mobile liquid $[M_i T^{-1} L^{-3}]$;
q_{ima} = average immobile sorbed phase concentration $[M_i M_s^{-1}]$;
E_{im}^{sim} = the source of solute to the immobile sorbed phase from the immobile liquid $[M_i T^{-1} L^{-3}]$; and
G^{sim} = the rate of production of solute in the sorbed phase in equilibrium with the immobile liquid $[M_i T^{-1} L^{-3}]$.

In Equations 3.13a,b, and c, as elsewhere, the exchange terms (E) account for diffusive or other mass transfer exchanges of mass to or from the phase, and the generation (G) terms account for internal production or loss through transformation. Dependent upon the situation being modeled, investigators have formulated exchange terms both in terms of first-order transfer coefficients and in terms of Fickian diffusive flux into specific geometries. These formulations for exchange in sorption and desorption have been extensively reviewed elsewhere (e.g., Harmon et al., 1989; Ball et al., 1991c), but are very briefly reviewed in the text below.

If we assume linear local equilibrium *within* both the mobile and immobile regions, we can write equations similar to Equation 3.6 for both regions. Thus, we can define distribution coefficients for each region:

$$K_{d-m} = q_m / c_m \tag{3.14a}$$

and

$$K_{d-im} = q_{ima} / c_{ima} \tag{3.14b}$$

which are coupled by

$$q = q_m + q_{ima} \tag{3.14c}$$

Note that each sorbed phase concentration is defined in terms of the entire sorbent mass.

Equations 3.8 through 3.14 can be used together with Equation 3.3 to describe solute fate and transport in the mobile groundwater in terms of concentrations in each phase. The solution of the resulting equation, however, requires that the c_{ima} be related to c_m, or in other words, that a physical model be assumed to describe the mass transfer between the mobile and immobile liquids. Two approaches have been advanced to model the transfer of solute from the immobile to the mobile phase.

One commonly applied approach is to approximate the transfer as a first-order process (e.g., Coats and Smith, 1964; van Genuchten and Wierenga, 1976; Rao et al., 1979; Nkedi-Kizza et al., 1983; Lee et al., 1988; Brusseau et al., 1991a,b,c; Larsen et al., 1992a) or by some combination of two or more first-order processes (Brusseau et al., 1989; Brusseau, 1991a; Brusseau, 1992; Brusseau et al., 1992). Assumption of first-order transfer rates into (out of) zones of immobile water and sorbent is equivalent to assuming transfer between two well-mixed zones across a thin boundary layer. While not accurate in a mechanistic sense, the approach may be justified for case-specific simulation if calibration of the model can be conducted on the basis of transport data collected under time scales similar to those of the model application. For example, Brusseau and co-workers have suggested that rate coefficients can be predicted on the basis of strength of sorption and, toward this end, have proposed several inverse correlations between fitted first-order sorbed phase transfer coefficients and K_d (Brusseau and Rao, 1989; Brusseau et al., 1990; Brusseau et al., 1991a,b,c). However, such correlations should be applied with caution. For example, for the case of tetrachlorobenzene and Borden sand, Brusseau et al. (1991c) suggest that a column-based correlation agrees with diffusion rate data obtained from long-term batch experiments conducted by Ball and Roberts (1991b); however, the batch rates compared were those based on pulverized sample. Batch-measured sorption rates with an unpulverized coarse fraction were 40 times slower.

For a more mechanistic exploration of the nonequilibrium process in comparatively well-defined systems, it is often useful to apply more physically realistic models. As an alternative to the assumption of a well-mixed immobile region, several researchers have assumed that the concentration in the immobile phase is not uniform but is governed by diffusion in the immobile region (e.g., Rao et al., 1980; Nkedi-Kizza et al., 1984; Valocchi, 1985; Crittenden et al., 1986; Hutzler et al., 1986; Wu and Gschwend, 1986; Roberts et al., 1987; Miller and Weber, 1988; Harmon et al., 1989; Young and Ball, 1994). These authors have previously discussed the effects of diffusion-limited sorption on the one-dimensional transport of a sorbing solute. Young and Ball (1994) have also shown how independent (batch) estimates of diffusion rate can be used to model solute transport in a column. In the following section, we describe a mathematical framework that also incorporates first-order reaction kinetics, also independently measured through batch experiment.

Nonequilibrium Transport with Pseudo-First-Order Reaction

Equations 3.13a,b,c and 3.14a,b can be combined with Equation 3.3 and the pseudo-first-order rate equations (Equations 3.9 and 3.10) to obtain a transport equation for the mobile phase. The following equation results:

$$[\theta_m + \rho_b K_{d-m}](\partial c_m/\partial t) + [\theta_{im} + \rho_b K_{d-im}](\partial c_{ima}/\partial t) = -\theta_m v_m(\partial c_m/\partial x) + \theta_m D_H(\partial^2 c_m/\partial x^2) - \theta_{im} k'_{im} c_{ima} - \theta_m k'_m c_m - k_{s-im}\rho_b K_{d-im} c_{ima} - k_{s-m}\rho_b K_{d-m} c_m \tag{3.15}$$

In obtaining Equation 3.15, we have used a pseudo-first-order rate coefficient for reaction in immobile pore-fluid (k'_{im}). We have also used the following reciprocity relations: $E_{im}^{m} = -E_{m}^{im}$, $E_{sm}^{m} = -E_{m}^{sm}$, and $E_{im}^{sim} = -E_{sim}^{im}$. Equation 3.15, when combined with proper initial and boundary conditions and appropriate equations for linking c_{ima} and c_m (see below), describes the fate and transport of a linearly sorbing solute that reacts by pseudo-first-order reactions in a porous medium with a pore structure that can be segregated into mobile and immobile regions.

If immobile zones are comprised of soil aggregates or rock fragments containing intraparticle porosity, the immobile zones can be approximated as porous spheres. For these cases, the concentration in the immobile zone is governed by Fick's Second Law of diffusion:

$$[1 + (\rho_b/\theta_{im})K_{d-im}](\partial c_{im}/\partial t) = (D_{mol}/\chi_e)\,[(\partial^2 c_{im}/\partial r^2) + (2/r)(\partial c_{im}/\partial r)] - (k_{s-im}\rho_b K_{d-im} c_{im})/\theta_{im} - k'_{im} c_{im} \tag{3.16}$$

where c_{im} is the radially dependent concentration of solute in the immobile fluid (e.g., intraparticle pores), D_{mol} is the molecular diffusion coefficient, and χ_e is the effective tortuosity factor (Ball and Roberts, 1991b,c). The volume-averaged aqueous concentration in the immobile region (c_{ima}) is

$$c_{ima}(x,t) = (3/a^3)\int_0^a c_{im}(x,r,t)r^2 dr \tag{3.17}$$

where a is the radius of the sphere. The concentrations in the mobile zone and the immobile zone are coupled by assuming continuity at the boundary of the spherical zone (particle):

$$c_m(x,t) = c_{im}(x,a,t) \tag{3.18}$$

where x refers to all positions along the flow path. The simultaneous solution of Equations 3.16 through 3.18 under appropriate initial and boundary conditions represents the complete solution to the problem of diffusion-limited sorption into a spherical immobile region coupled with pseudo-first-order reaction in the solution and sorbed phases.

Analytical solutions to the full set of one-dimensional transport equations are available for certain special cases. These include first-order mass transfer without transformation reactions (van Genuchten and Wierenga, 1976) and more recently with transformation reactions (van Genuchten and Wagenet, 1989). An analytical solution for transport affected by radial diffusion with a linear isotherm without transformation reactions was derived by Rasmuson and Neretnieks (1980). In more general cases, the transport equations must be solved numerically (e.g., Rao et al., 1980; Goltz and Roberts, 1986; Crittenden et al., 1986; Curtis, 1991; Miller and Rabideau, 1993).

FIELD-SCALE SIMULATION AND ANALYSIS

Moment Analysis Framework for Transport Observations

Moment analysis offers an effective general approach for understanding solute behavior during groundwater transport. Theoretically expected moments can be calculated on the basis of the transport equations previously described (Aris, 1958; Valocchi, 1985; Goltz and Roberts, 1986), and these results can be used in connection with detailed sampling to provide useful information concerning conservation of mass, transport velocity and retardation, and plume spreading. For reasons detailed below, it is useful to distinguish between spatial and temporal moments.

Spatial moments describe the distribution of solute measured at various monitoring points, in a set of samples at a particular point in time. The moments are obtained by integrating, over the entire relevant spatial domain, the measured solution concentrations from a set of samples taken at a given point in time (Freyberg, 1986). If the solute is known to partition onto the stationary solids of the aquifer, the zeroth moment obtained from the direct integration must be multiplied by the partitioning ratio to obtain the total mass (Roberts et al., 1986). The first moment, normalized by the dissolved mass, gives the location of the aqueous plume's center of mass, and the similarly normalized second moment provides information on the spread of the solute about the center of mass (Freyberg, 1986). The time rate of change of the normalized first moment indicates the plume velocity, and the time rate of change of the normalized second moment yields insight into plume dispersion.

Temporal moments describe the solute behavior in a time series of samples taken at a particular monitoring point. Where the measured concentration values represent the response to a well-defined pulse, the temporal moments have a clear physical meaning. The zeroth moment again represents the solute mass, the first moment gives the average solute residence time, and the second moment represents the spreading of the concentration wave. In a one-dimensional situation for a conservative solute, the zeroth temporal moment requires no adjustment for solute partitioning, because ultimately all of the solute will appear at the monitoring point, but in a multidimensional scenario the zeroth moment decreases with distance from the source, because of transverse plume spreading. For a reacting solute or a reaction product, solute loss or production will also affect the zeroth moment. The first temporal moment again relates to the average solute residence time, i.e., the travel time required to reach the monitoring point. The second temporal moment represents the spreading of the concentration wave along the path from the source to the monitoring point, expressed as a variance.

The properties of the temporal moments have been reviewed by Valocchi (1985) for a one-dimensional homogeneous medium, and by Goltz and Roberts (1986) for a three-dimensional homogeneous medium with unidirectional flow. The properties of the spatial moments have been summarized by Goltz and Roberts (1987). To recapitulate the important similarities and differences between the spatial and temporal moments: 1) the zeroth spatial moment captures only the dissolved solute mass, unless adjusted explicitly for partitioning, whereas the zeroth temporal moment quantifies the entire mass passing the sampling location; 2) the first spatial moment is affected by the extent of partitioning, including departure from sorption equilibrium caused by rate limitations, whereas the first temporal moment is affected by partitioning but is not influenced by rate limitation; and 3) both the spatial and temporal second moments can be affected by any of several factors that contribute to wave spreading, including hydrodynamic dispersion, kinetic rate limitation, and physical diffusion limitation.

As noted earlier, mass transfer limitations can lead to sharp breakthrough fronts and tailing elution, which will be reflected in greater variance (second moment). In the context of tailing elution, the second point above (theoretical constancy of temporal first moment) assumes full mass recovery. In practice, moment calculations of experimental data may not meet this constraint, especially under conditions of severe nonequilibrium. For such systems, a small fraction of the solute mass elutes at very low concentrations (possibly below detection limits) and at long times (possibly after monitoring has ceased). Temporal moment calculations based on experimental data will tend to be underestimates to the extent that they exclude the contribution of this small solute mass. Because of the increased weighting of very long-term results on higher moment calculations, errors in mass recovery are greatly magnified in computations of the first moment and are even more so for the second moment.

Focusing further on nonequilibrium behavior, it is apparent that deviations from equilibrium can be revealed by a temporal decrease in the unadjusted zeroth moment, by a temporal increase in the first spatial moment, and/or by spreading (characterized by the second spatial or temporal moment) to a degree in excess of that anticipated to result from hydrodynamic dispersion alone. However, identification of non-equilibrium behavior is not conclusive on the basis of on any single element of the moment analysis. A temporal decrease in the unadjusted zeroth spatial moment may indicate solute transformation. A temporal increase in the first spatial moment may arise from a coincidental increase of the distribution coefficient along the traversed path, or from the effect of nonlinear equilibrium in conjunction with the reduced solute concentrations resulting from spreading of the solute plume. Even analysis of the second moment does not distinguish among the various phenomena that can contribute to wave spreading—dispersion, kinetic limitation, and mass transfer limitation—because either form of rate limitation can be expressed in terms of an equivalent amount of dispersion (Passioura, 1971; Valocchi, 1985). Similarly, mass transfer phenomena at different hydrogeologic scales but with similar time scales (e.g., diffusion into low-permeability strata or slower diffusion into smaller intraaggregate regions) can have similar dispersive effects (Valocchi, 1989). Thus, except for certain simplified scenarios of transport in well-defined uniform media (e.g., Roberts et al., 1987), the moment analysis alone will not suffice to identify and quantify nonequilibrium behavior. Investigators must employ evidence of various kinds, including independent quantification of the pertinent rate parameters, to convincingly implicate deviation from local equilibrium as a significant factor in transport. It is a still greater challenge to provide estimation of relevant physical/chemical parameters in the detail and to the extent necessary to provide accurate prediction. In the following section, we discuss a field case study which illustrates this point. In the case presented, we believe that both reaction and nonequilibrium sorbing solute transport have been successfully inferred from a combination of field and laboratory analysis, but that prediction of many important aspects of the field behavior would not have been possible based on laboratory data alone.

Case Study and Field-Scale Simulation

A large-scale, long-term field experiment (Mackay et al., 1986) on solute transport under natural gradient conditions that was conducted at Borden, Ontario affords the opportunity for employing the moment analysis approach to characterize solute behavior and to quantify transport parameters. Known amounts of seven solutes were introduced as a block-like pulse, and subsequently followed by means of a three-dimensional monitoring array over a two-year period. The spatial moments of nonreactive tracers (bromide, chloride) were reported by Freyberg (1986), while the moments for the sorbing and reacting organic solutes (bromoform; carbon tetrachloride; tetrachloroethene; 1,2-dichlorobenzene; and hexachloroethane) were presented by Roberts et al. (1986). Aquifer solids from the site have also been well-characterized (Ball et al., 1990a).

Nonsorbing Tracer Behavior

As shown by Freyberg (1986), the zeroth spatial moment for the anionic tracers remained invariant over time, confirming that mass in solution was conserved as anticipated. The first moment analysis showed that the average groundwater velocity was constant in direction and magnitude, with a linear velocity of 0.091 m/day; this result agreed within 15% of the value predicted on the basis of the average hydraulic conductivity and the hydraulic gradient (Mackay et al., 1986). Freyberg (1986) interpreted the second moments of the tracer distribution to estimate the components of the dispersion tensor by means of a two-dimensional analysis of depth-averaged data, reporting average values of 0.36 m and 0.023 m for the longitudinal and transverse horizontal dispersivities, respectively. Sudicky (1986), employing detailed measurements of the spatial variability of the hydraulic conductivity, showed that the observed dispersion was commensurate with the stochastic theory of dispersion. It appeared, however, that the plume encountered a zone of low conductivity at the end of the second year, causing an abrupt increase in dispersion.

Rajaram and Gelhar (1991) employed a fully three-dimensional analysis of the same tracer data, which enabled them to quantify the vertical dispersivity. Rajaram and Gelhar reported values of 0.5 m, 0.05 m, and 0.002 m for the longitudinal, horizontal transverse, and vertical transverse dispersivities, respectively. Rajaram and Gelhar attributed their slightly larger values for the longitudinal and horizontal transverse dispersivities, compared to Freyberg (1986), to their procedure for accounting for the missing mass resulting from truncated sampling. The value of the vertical dispersivity was within the range anticipated for the effect of molecular diffusion. These findings also are in accordance with the expectation that longitudinal dispersivity typically exceeds transverse dispersivity by one to two orders of magnitude (Freeze and Cherry, 1979).

Organic Solutes

Mass Conservation

The zeroth spatial moments of the organic solutes (Roberts et al., 1986, Table 8 and Figure 9) agreed with the mass introduced at the outset of the experiment. Thereafter, the masses of some of the organic solutes began to decline, immediately in one case (hexachloroethane), but not until after a substantial lag time with two other compounds (bromoform and 1,2-dichlorobenzene). The masses of the remaining two organic solutes showed either no significant change (carbon tetrachloride) or a slight, but statistically significant, increase (tetrachloroethene). In all cases, the elementary zeroth moment calculation showed that mass disappeared from solution in accordance with the sorptive partitioning of the solutes; this is a feature of experiments in which the solutes are introduced as an aqueous pulse. Only after the dissolved mass had been adjusted for sorption, by the retardation factor inferred from the first moment, was it meaningful to compare the observed mass with the introduced mass, and thus to make inferences regarding possible transformations (Roberts et al., 1986).

The disappearance of bromoform and dichlorobenzene can readily be explained in terms of biotransformation, but both reactions are sensitive to redox conditions and these are uncertain in the field—while there is evidence that oxygen may be present in the mobile pore water, anoxic zones may occur in niches less influenced by advective transport. The transformation of dichlorobenzene under aerobic conditions has been shown by Bouwer and McCarty (1985) and Kuhn et al. (1985). The transformation of bromoform under anoxic conditions also has been demonstrated (Bouwer and McCarty, 1983). Carbon tetrachloride and tetrachloroethene (PCE) are known to transform under anoxic conditions (Bouwer and McCarty, 1983; Vogel et al., 1987), but degradation of these compounds at Borden was not supported by the field evidence. The abiotic reduction of hexachloroethane (HCA) to PCE had been reported prior to the field experiment, but it was not known whether this transformation would occur under subsurface conditions. Subsequently, Criddle et al. (1986) showed that HCA is biotransformed to PCE under aerobic conditions and in the presence of Borden solids, and Curtis (1991) demonstrated that the same transformation can occur abiotically at similar rates.

Modeling of the field-observed decline in the HCA zeroth moment was complicated by the potential interaction of biotic and abiotic effects, as well as by uncertainty with regard to the role of pH (Curtis, 1991). Analysis that adjusted pseudo-first-order rate constants for temperature but ignored the potential role of pH (8.3 in the laboratory versus 7.8 in the field) showed excellent agreement between abiotic degradation rates in the presence of Borden solids and those observed in the field, while attempts to adjust for pH suggest a potential three-fold rate enhancement in the field (Curtis, 1991).

With regard to the HCA reaction product (PCE), Figure 3.1 illustrates the zeroth moment computed assuming transport with reaction and rate-limited sorption as represented by Equation 3.16, and with HCA stoichiometrically converted to PCE (solid line). This prediction is compared with the case where no reaction was assumed to occur. The estimated moments were calculated on numerical model simulations using boundary conditions appropriate for a 1-D approximation of the Borden experiment and with transport and rate parameters as summarized elsewhere (Curtis, 1991). Both simulations are based on a sorption mass transfer rate estimated from fitting the field-scale first moment data, as described subsequently. After approximately 80 days, the two simulations bracket the observed field data. It is possible that the observed masses in solution in the field never approached the amount predicted from stoichiometric considerations because some of the mass in the field escaped below detection limits.

First Moments

The solute mobilities, as characterized by the first moments of the spatial distributions, behaved as anticipated in some important respects. The trajectories of the organic solute plume centers of mass followed the same path as the anionic tracers, but moved at a lower velocity (Roberts et al., 1986). The retardation factors, quantified as the inverse of the rates of movement of the plume centers of

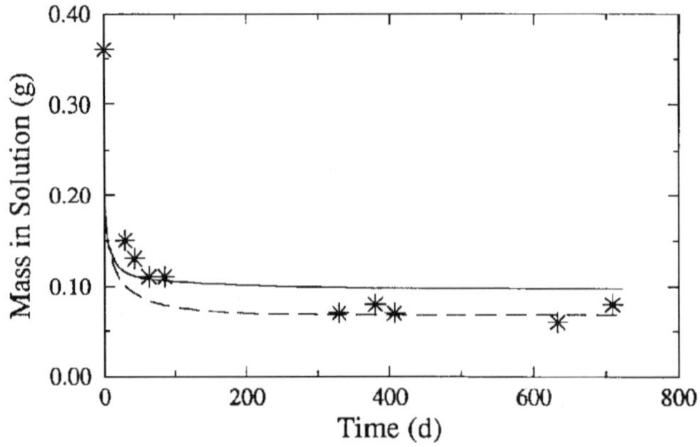

Figure 3.1. Comparison of field-observed tetrachloroethene (PCE) zeroth moments (Roberts et al., 1986) with calculated values for the case with (solid line) and without (broken line) inclusion of reductive dehalogenation of hexachloroethane (HCA). Calculations assume an initial aqueous mass equal to the measured mass injected (0.36 g).

mass relative to the anionic tracer, were greater for more hydrophobic solutes than for more soluble ones, i.e., the rank order of solute retardation factors was in accordance with the rank order of the solutes' octanol-water partition coefficients. Further, the short-term field-scale retardation factors agreed semi-quantitatively with the strength of sorption, characterized by the laboratory-determined 3-day distribution coefficients (Curtis et al., 1986b). However, sorption was found to be considerably stronger than anticipated on the basis of partitioning into the soil organic matter (Curtis et al., 1986b; Ball and Roberts, 1991a).

The estimated retardation factors inferred from the spatial data slightly exceeded the retardation factors estimated from the temporal data (Roberts et al., 1986). Nonetheless, the ranges of the estimates obtained from the spatial and temporal perspectives overlapped, or nearly so, and both sets of estimates increased with increasing solute hydrophobicity, in agreement with the partitioning concept.

Contrary to expectations for equilibrium sorption in a homogeneous aquifer, the analysis of the spatial moments suggested that the retardation of all of the organic solutes increased with elapsed time (Roberts et al., 1986). The retardation factors observed during the early part of the field experiment agreed closely with values computed from short-term (several days) laboratory sorption measurements, including both batch experiments and column experiments (Curtis et al., 1986; Ptacek and Gillham, 1992). However, the pattern of long-term increase in the retardation factors, which was observed for all of the organic solutes, suggested slow, rate-limited sorption that reached equilibrium only after contact times on the order of hundreds of days. Ball (1989) conducted such experiments with tetrachloroethylene, using an experimental protocol designed to minimize artifacts in long-term equilibrations (Ball and Roberts, 1991a) and showed that the ultimate sorption capacity of the Borden solids was indeed approximately twice as great as the value measured with one-day equilibration (Ball and Roberts, 1991b). Using the long-term estimate of the PCE sorption capacity, Ball et al. (Ball, 1989; Ball et al., 1990b; Ball and Roberts, 1991c) demonstrated that the measured long-term equilibrium data permitted a substantially better match to the first moment behavior than was obtained by Goltz and Roberts (1988) using sorption capacities based on short-term equilibrations. However, the rate parameters measured in the batch experiments predicted more rapid attainment of equilibrium than was observed in the field. Rate parameters, as adjusted to fit the field-observed first moments (Ball, 1989; Ball et al., 1990b; Quinodoz and Valocchi, 1993) are roughly 15 to 50 times lower than the batch estimate for bulk Borden sand (Ball, 1989; Ball and Roberts, 1991b), although the batch rates have been shown to very accurately predict short column breakthrough experiments (Young and Ball, 1994).

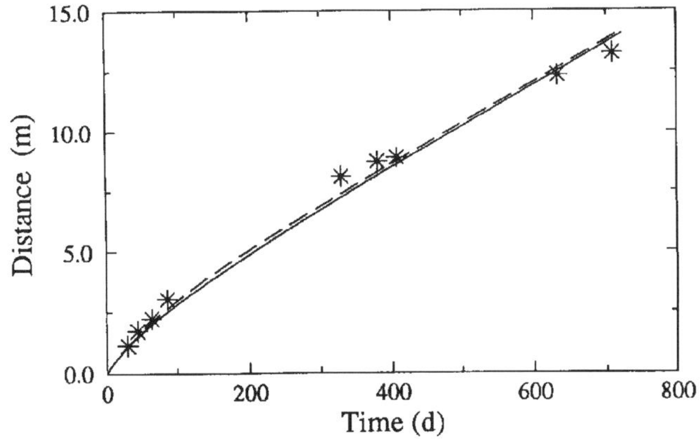

Figure 3.2. Comparison of field-observed PCE first moments (Roberts et al., 1986) with calculated values for the case with (solid line) and without (broken line) inclusion of reductive dehalogenation of HCA.

Figure 3.2 illustrates first moments of PCE transport at the Borden field site, including estimated moments computed from the same numerical simulations used for Figure 3.1 (Curtis, 1991). The simulations use the field-adjusted mass transfer rate of Ball (1989), which is 50 times less than the batch rate. Modeling with the faster (batch-measured) diffusion rates was also conducted by Curtis (1991) but, as with the non-reactive case considered by Ball et al. (Ball, 1989; Ball et al., 1991c; Quinodez and Valocchi, 1993) simulations failed to capture the field-scale first moment behavior. For the modeling shown in Figure 3.2 (field-adjusted transfer rate), the solid line represents the case where the transformation of HCA to PCE is considered, and the dashed line represents the case of no reaction, essentially as presented previously by Ball et al. The first moments in these two cases are nearly identical because the reaction of HCA is relatively rapid. The half-life of HCA of approximately 40 days inferred from the field (Roberts et al., 1986) and laboratory (Curtis, 1991) indicates that approximately half of the HCA was converted to PCE during the time that the original PCE was transported on the order of a meter, which is less than the initial pulse-width. In general, however, transformation reactions could have significant impacts on the first moments if the reactions are slower and if the molar amount of the reactant is large relative to the initial amount of reaction product.

Second Moments and Temporal Breakthrough

The amount of spatial plume spread can be summarized in terms of the standard deviation of the concentration measurements in the longitudinal direction. Quantitative interpretations of second moment data for the Borden organic solute plumes have been attempted (Goltz, 1986; Quinodoz and Valocchi, 1993). However, because of the extreme variability of the individual concentration measurements, mechanistic inferences from second moment data are highly uncertain. Qualitatively, the longitudinal plume spreading behaves according to the expectations of the advection-dispersion model (Roberts and Mackay, 1986). The compounds most strongly retarded by sorption (i.e., HCA, dichlorobenzene) tend to spread to a lesser extent than do those whose movement is retarded less strongly (carbon tetrachloride, bromoform) or not at all (bromide).

Goltz and Roberts (1986) analyzed several of the time series records for solute concentration breakthrough at individual monitoring points, with the objective of predicting the breakthrough dynamics using independently determined parameters. The computations demonstrated that a model accounting for the rate limitation, whether by means of the physical diffusion model or a first-order approximation, matched the breakthrough data appreciably better than did an equilibrium model. As with the first-moment modeling, temporal data fitting suggested mass transfer rates slower than those observed in the batch work. Brusseau (1992) also simulated selected Borden temporal data for PCE using Ball and Roberts' (1991a) long-term equilibrium K_d, and also observed that nonequilibrium aspects of behavior were dominated by a field-scale mass transfer rate. In this case, the field-scale transfer coefficient was obtained by fitting field-scale tracer data and, as with the fitted coefficient of Goltz and Roberts (1986), was slow in comparison with the batch-measured intraparticle diffusion rate. It is now also recognized that some of the PCE behavior, in particular excessive tailing, can be explained as an effect of the formation of PCE as a by-product of HCA decomposition (Curtis, 1991).

Conclusions

The moment analysis approach can be used to characterize solute behavior with regard to conservation of mass, sorptive partitioning, transport velocity, and plume spreading. In the case study summarized here, the application of the moment analysis enabled the investigators to discriminate among solutes that were transformed and those that were conserved, to quantify the strength of sorptive partitioning, to demonstrate the inverse relation between partitioning and mobility, and to illustrate the effects of nonequilibrium behavior. Interpretations such as these are helpful in testing hypotheses regarding mechanisms governing solute behavior, and in evaluating parameters to be used in simulating or predicting solute transport and fate.

SUMMARY

Clearly, the need for improved basic scientific understanding of physical/chemical processes for subsurface restoration is enormous. Understanding potential processes of concern requires knowledge in diverse aspects of chemistry, physics and the basic earth sciences, while proper understanding of heterogeneity and uncertainty also necessitates familiarity with the statistical sciences. Linking fundamental mechanistic understanding from these fields with processes of transport, fate, and treatment requires the expertise of applied scientists and engineers, while solution of resulting mathematical formulations often requires computer expertise.

In this chapter, we have addressed the title subject from the vantage point of the applied scientist and engineer, and have therefore attempted to put various processes into relative perspective through the application of conceptual modeling frameworks, in terms of both transport modeling equations and moment analyses of field data. Our emphasis has been on the physical and chemical processes affecting synthetic organic chemicals, with additional emphasis on those processes especially relevant to organic contaminant transport in groundwater: sorption, reaction, and related mass transfer.

Consistent with our ultimate goal of better understanding field-scale phenomena, we have highlighted the need to identify and characterize the unique processes which will be active at a given site. The Borden, Ontario, transport study has illustrated how laboratory experiment and field measurement can be used to improve our ability to understand and simulate the fate of contaminants in the subsurface. However, it is equally important to recognize that many scenarios of actual subsurface restoration are vastly more complex than the Borden case study, and that initial conditions of contamination are always less well characterized.

In this context, we cannot expect complete field-scale understanding or accurate predictive modeling in most field cases. Rather, it is our belief that increased awareness and understanding of potentially important physical/chemical mechanisms will help field-scale restoration efforts by guiding resource allocation toward those characterization efforts that are most important and toward those remediation approaches with the highest likelihood of success.

ACKNOWLEDGMENTS

The preparation of this chapter was supported by the Conference Sponsors, the National Science Foundation (Presidential Young Investigator Award Grant BCS-92-96241), and the Industrial-Hazardous Waste Division of Montgomery-Watson.

REFERENCES

Abriola, L.M. and G.F. Pinder. 1985. A Multiphase Approach to the Modeling of Porous Media Contamination by Organic Compounds. *Water Res. Res.*, **21**:11–18.

Alexander, M. and K.M. Scow. 1989. Kinetics of Biodegradation in Soils. In B.L. Sawhney and K. Brown, eds., *Reactions and Movements of Organic Chemicals in Soils; SSSA Publication No. 22*. Soil Sci. Soc. Am., Inc. and Am. Soc. of Agronomy, Inc., Madison WI, pp. 243–270.

Allen III, M.B. 1985. Numerical Modelling of Multiphase Flow in Porous Media. *Adv. Water Res.*, **8**:162–187.

Anderson, M.P. 1979. Using Models to Simulate the Movement of Contaminants Through Groundwater Flow Systems. *CRC Crit. Rev. Environ. Cont.*, **9**:97–156.

Aris, R. 1958. On the Dispersion of Linear Kinematic Waves. *Proc. R. Soc. London, Ser. A*, **245**:268–277.

Backhus, D.A. 1990. Colloids in Groundwater: Laboratory and Field Studies of their Influence on Hydrophobic Organic Contaminants. Ph.D. Dissertation. Massachusetts Institute of Technology, Cambridge, MA, 198 p.

Backhus, D.A. and P.M. Gschwend. 1990. Fluorescent Polycyclic Aromatic Hydrocarbons as Probes for Studying the Impact of Colloids on Pollutant Transport in Groundwater. *Environ. Sci. Technol.*, **24**:1214–1223.

Bahr, J.M. and J. Rubin. 1987. Direct Comparison of Kinetic and Local Equilibrium Formulations for Solute Transport Affected by Surface Reactions. *Water Res. Res.*, **23**:438–452.

Ball, W.P. 1989. Equilibrium Sorption and Diffusion Rate Studies with Halogenated Organic Chemicals and Sandy Aquifer Material. Ph.D. Dissertation. Stanford University, Stanford, CA, 356 p.

Ball, W.P. and P.V. Roberts. 1991a. Long-Term Sorption of Halogenated Organic Chemicals—Part 1. Equilibrium. *Environ. Sci. Technol.*, **25**:1223–1237.

Ball, W.P. and P.V. Roberts. 1991b. Long-Term Sorption of Halogenated Organic Chemicals—Part 2. Intraparticle Diffusion. *Environ. Sci. Technol.*, **25**:1237–1249.

Ball, W.P. and P.V. Roberts. 1991c. Diffusive Rate Limitations in the Sorption of Nonpolar Organic Chemicals by Natural Solids. In R.A. Baker, ed., *Organic Substances and Sediments in Water: Volume 2, Processes and Analytical*. Lewis Publishers, Boca Raton, FL, Chapter 13, pp. 273–310.

Ball, W.P., Ch. Buehler, T.C. Harmon, D.M. Mackay, and P.V. Roberts. 1990a. Characterization of a Sandy Aquifer Material at the Grain Scale. *J. Contam. Hydrol.*, **5**:253–295.

Ball, W.P., M.N. Goltz, and P.V. Roberts. 1990b. Sorption Rate Studies with Halogenated Organic Chemicals and Sandy Aquifer Material—Implications for Solute Transport and Groundwater Remediation. In O'Melia, C.R., ed., *Proceedings of the 1990 Specialty Conference on Environmental Engineering*. ASCE, NY, pp. 307–313.

Ball, W.P., M.N. Goltz, and P.V. Roberts. 1991. Comment on "Modeling the Transport of Solutes Influenced by Multiprocess Nonequilibrium" by M.L. Brusseau, R.E. Jessup, and P.S.C. Rao. *Water Res. Res.*, **27**:653–656.

Barbash, J.E. and M. Reinhard. 1989. Abiotic Dehalogenation of 1,2-Dichloroethane and 1,2-Dibromoethane in Aqueous Solution Containing Hydrogen Sulfide. *Environ. Sci. Technol.*, **23**:1349–1358.

Baveye, P. and A. Valocchi. 1989. An Evaluation of Mathematical Models of the Transport of Biologically Reacting Solutes in Saturated Solids and Aquifers. *Water Res. Res.*, **25**:1412–1421.

Bear, J. 1979. *Hydraulics of Groundwater*. McGraw Hill, NY, 569 p.

Bouwer, E.J. 1992. Bioremediation of Organic Contaminants in the Subsurface. *Environmental Microbiology*. Wiley-Liss, Inc., NY, pp. 287–318.

Bouwer, E.J. and P.L. McCarty. 1983. Transformations of 1- and 2-Carbon Halogenated Aliphatic Organic Compounds Under Methanogenic Conditions. *Appl. Environ. Microbiol.*, **45**:1286–1294.

Bouwer, E.J. and P.L. McCarty. 1985. Utilization Rates of Trace Halogenated Organic Compounds in Acetate-Grown Biofilms. *Biotech. Bioeng.*, **27**:1564–1571.

Bouwer, E.J. and J.P. Wright. 1988. Transformations of Trace Halogenated Aliphatics in Anoxic Biofilm Columns. *J. Contam. Hydrol.*, **2**:155–169.

Brusseau, M.L. 1991a. Application of a Multiprocess Nonequilibrium Sorption Model to Solute Transport in a Stratified Porous Medium. *Water Res. Res.*, **27**:589–595.

Brusseau, M.L. 1991b. Transport of Organic Chemicals by Gas Advection in Structured or Heterogeneous Porous Media: Development of a Model and Application to Column Experiments. *Water Res. Res.*, **27**:3189–3199.

Brusseau, M.L. 1992. Transport of Rate-Limited Sorbing Solutes in Heterogeneous Porous Media: Application of a One-Dimensional Multifactor Nonideality Model to Field Data. *Water Res. Res.*, **28**:2485–2497.

Brusseau, M.L. and P.S.C. Rao. 1989. Sorption Nonideality During Organic Contaminant Transport in Porous Media. *CRC Crit. Rev. Environ. Control*, **19**:33–99.

Brusseau, M.L., R.E. Jessup, and P.S.C. Rao. 1989. Modeling the Transport of Solutes Influenced by Multi-Process Nonequilibrium. *Water Res. Res.*, **25**:1971–1988.

Brusseau, M.L., R.E. Jessup, and P.S.C. Rao. 1990. Sorption Kinetics of Organic Chemicals: Evaluation of Gas-Purge and Miscible-Displacement Techniques. *Environ. Sci. Technol.*, **24**:727–735.

Brusseau, M.L., R.E. Jessup, and P.S.C. Rao. 1991a. Nonequilibrium Sorption of Organic Chemicals: Elucidation of Rate-Limiting Processes. *Environ. Sci. Technol.*, **25**:134–142.

Brusseau, M.L., A.L. Wood, and P.S.C. Rao. 1991b. Influence of Organic Cosolvents on the Sorption Kinetics of Hydrophobic Organic Chemicals. *Environ. Sci. Technol.*, **25**:903–910.

Brusseau, M.L., T. Larsen, and T.H. Christensen. 1991c. Rate-Limited Sorption and Nonequilibrium Transport of Organic Chemicals in Low Carbon Aquifer Materials. *Water Res. Res.*, **27**:1137–1145.

Brusseau, M.L., R.E. Jessup, and P.S.C. Rao. 1992. Modeling Solute Transport Influenced by Multiprocess Nonequilibrium and Transformation Reactions. *Water Res. Res.*, **28**:175–182.

Burris, D.R. and W.G. MacIntyre. 1986. A Thermodynamic Study of Solutions of Liquid Hydrocarbon Mixtures in Water. *Geochim. Cosmochim. Acta*, **25**:1545–1549.

Burris, D.R. and W.G. MacIntyre. 1987. Water Solubility Behavior of Hydrocarbon Mixtures—Implications for Petroleum Distillation. In J.H. Vandermeulen and S.E. Hrudey, eds., *Oil in Freshwater: Chemistry, Biology, Countermeasure Technology*. Pergamon Press, NY, pp. 85–94.

Celia, M.A., J.S. Kindred, and I. Herrera. 1989. Contaminant Transport and Biodegradation. I. A Numerical Model for Reactive Transport in Porous Media. *Water Res. Res.*, **25**:1141–1148.

Chen, Y.-M., L.M. Abriola, P.J.J. Alvarez, P.J. Anid, and T.M. Vogel. 1992. Modeling Transport and Biodegradation of Benzene and Toluene in Sandy Aquifer Material: Comparisons with Experimental Measurements. *Water Res. Res.*, **28**:1833–1847.

Cherry, J.A., S. Feenstra, and D.M. Mackay. 1992. Developing Rational Goals for In Situ Remedial Technologies. In C.H. Ward, ed., *Subsurface Restoration Conference Proceedings*. Rice University, Houston, TX, pp. 14–16.

Chin, Y-P and P.M. Gschwend. 1992. Partitioning of Polycyclic Aromatic Hydrocarbons to Marine Porewater Organic Colloids. *Environ. Sci. Technol.*, **26**:1621–1626.

Chiou, C.T. 1989. Theoretical Considerations of the Partition Uptake of Nonionic Organic Compounds by Soil Organic Matter. In B.L. Sawhney and K. Brown, eds., *Reactions and Movements of Organic Chemicals in Soils; SSSA Publication No. 22*. Soil Sci. Soc. Am., Inc. and Am. Soc. of Agronomy, Inc., Madison, WI, pp. 1–30.

Chiou, C.T., T. Shoup, and P.E. Porter. 1985. Mechanistic Roles of Soil Humus and Minerals in the Sorption of Nonionic Organic Compounds from Aqueous and Organic Solutions. *Org. Geochem.*, **8**:9–14.

Chiou, C.T., R.L. Malcolm, T.I. Brinton, and D.E. Kile. 1986. Water Solubility Enhancement of Some Organic Pollutants and Pesticides by Dissolved Humic and Fulvic Acids. *Environ. Sci. Technol.*, **20**:502–508.

Cho, H.J., P.R. Jaffe, and J.A. Smith. 1993. Simulating the Volatilization of Solvents in Unsaturated Soils During Laboratory and Field Infiltration Experiments, *Water Res. Res.*, **29**:3329–3342.

Chrysikopoulos, C.V., P.K. Kitanidas, and P.V. Roberts. 1990. Analysis of One-Dimensional Transport Through Porous Media with Spatially Variable Retardation Factor. *Water Res. Res.*, **26**:437–445.

Chrysikopoulos, C.V., P.K. Kitanidas, and P.V. Roberts. 1992. Macrodispersion of Sorbing Solutes in Heterogeneous Porous Formations with Spatially Periodic Retardation Factor and Velocity Field. *Water Res. Res.*, **28**:1517–1529.

Cline, P.V., J.J. Delfino, and P.S.C. Rao. 1991. Partitioning of Aromatic Constituents into Water from Gasoline and Other Complex Solvent Mixtures. *Environ. Sci. Technol.*, **25**:914–921.

Coats, K.H. and B.D. Smith. 1964. Dead-End Pore Volume and Dispersion in Porous Media. *Soc. Petrol. Eng. J.*, **4**:73–84.

Committee on Ground Water Modeling Assessment; Water Science and Technology Board; Commission on Physical Sciences, Mathematics, and Resources, National Research Council, 1990. *Ground Water Models: Scientific and Regulatory Applications*. National Academy Press, Washington, DC, 303 p.

Corapcioglu, M.Y. and S. Panday. 1991. Compositional Multiphase Flow Models. In Corapcioglu, M.Y., ed., *Advances in Porous Media, Vol. 1*. Elsevier, Amsterdam, Chapter 1.

Criddle, C.S., P.L. McCarty, M.C. Elliott, and J.F. Barker. 1986. Reduction of Hexachloroethane to Tetrachloroethylene in Groundwater. *J. Contam. Hydrol.*, **1**:133–142.

Criddle, C.S., L.M. Alvarez, and P.L. McCarty. 1991. Microbiological Processes in Porous Media. In J. Bear and M.Y. Corapcioglu, eds., *Transport Processes in Porous Media*. Kluwer Academic Publ. Co., Dordrecht, The Netherlands, pp. 641–691.

Crittenden, J.C., N.J. Hutzler, D.G. Geyer, J.L. Oravitz, and G. Friedman. 1986. Transport of Organic Compounds with Saturated Groundwater Flow: Model Development and Parameter Sensitivity. *Water Res. Res.*, **22**:271–284.

Curtis, G.P. 1991. Reductive Dehalogenation of Hexachloroethane and Carbon Tetrachloride by Aquifer Sand and Humic Acid. Ph.D. Dissertation. Stanford University, Stanford, CA, 228 p.

Curtis, G.P., M. Reinhard, and P.V. Roberts. 1986a. Sorption of Hydrophobic Organic Compounds by Sediments. In J.A. Davis and K.F. Hayes, eds., *Geochemical Processes at Mineral Surfaces, ACS Symposium Series 323*. American Chemical Society, Washington, DC, pp. 191–216.

Curtis, G.P., P.V. Roberts, and M. Reinhard. 1986b. A Natural Gradient Experiment on Solute Transport in a Sand Aquifer. 4. Sorption of Organic Solutes and its Influence on Mobility. *Water Res. Res.*, **22**:2059–2067.

Dagan, G. 1982. Stochastic Modeling of Groundwater Flow by Unconditional and Conditional Probabilities, 2, The Solute Transport. *Water Res. Res.*, **18**:835–848.

Dagan, G. 1986. Statistical Theory of Groundwater Flow and Transport: Pore to Laboratory, Laboratory to Formulation, and Formulation to Regional Scale. *Water Res. Res.*, **22**:120S–134S.

Dagan, G. 1990. Transport in Heterogeneous Porous Formations: Spatial Moments, Ergodicity and Effective Dispersion. *Water Res. Res.*, **26**:1281–1290.

Davis, J.A. and D.B. Kent. 1990. Surface Complexation Modeling in Aqueous Geochemistry. In M.F. Hochella and A.F. White, eds., *Mineral-Water Interface Geochemistry; Reviews in Mineralogy, Vol. 23*. Mineralogical Society of America, Washington, DC.

Dunnivant, F.M., R.P. Schwarzenbach, and D.L. Macalady. 1992. Reduction of Substituted Nitrobenzenes in Aqueous Solutions Containing Natural Organic Matter. *Environ. Sci. Technol.*, **26**:2133–2141.

Essaid, H.I., W.N. Herkelrath, and K.M. Hess. 1993. Simulation of Fluid Distributions Observed at a Crude Oil Spill Site Incorporating Hysteresis Oil Entrapment, Spatial Variability of Hydraulic Properties. *Water Res. Res.*, **29**:1753–1770.

EPA. 1989. Seminar Publication: Transport and Fate of Contaminants in the Subsurface. EPA/625/4-89/019. Center for Environmental Research Information; Office of Research and Development and R.S. Kerr ERL, U.S. EPA, Cincinnati, OH and Ada, OK, 147 p.

EPA. 1991. Seminar Publication: Site Characterization for Subsurface Remediation. EPA/625/4-91/026. Center for Environmental Research Information; Office of Research and Development, U.S. EPA, Cincinnati, OH, 259 p.

Etzler, F.M. and D.M. Fagundus. 1987. The Extent of Vicinal Water: Implications from Density of Water in Silica Pores *J. Colloid Interface Sci.*, **115**:513–519.

Freeze, R.A. and J.A. Cherry. 1979. *Groundwater*. Prentice-Hall, Inc., Englewood Cliffs, NJ, 604 p.

Freyberg, D.L. 1986. A Natural Gradient Experiment on Solute Transport in a Sand Aquifer. 2. Spatial Moments and the Advection and Dispersion of Non-Reactive Tracers. *Water Res. Res.*, **22**:2031–2046.

Friedly, J.C. 1991. Extent of Reaction for Open Systems with Multiple Heterogeneous Reactions. *AIChE J.*, **37**:687–693.

Friedly, J.C. and J. Rubin. 1992. Solute Transport with Multiple Equilibrium-Controlled/Kinetically Controlled Reactions *Water Res. Res.*, **28**:1935–1953.

Garabedian, S.P., L.W. Gelhar, and M.A. Celia. 1988. Large-Scale Dispersive Transport in Aquifers: Field Experiments and Reactive Transport Theory. Report 315. Ralph M. Parsons Lab. for Water Resources and Hydrodynamics., Massachusetts Institute of Technology, Cambridge.

Garbarini, D.R. and D.R. Lion. 1986. Influence of the Nature of Soil Organics on Sorption of Toluene and Trichloroethylene. *Environ. Sci. Technol.*, **20**:1263–1269.

Gauthier, T.D., W.R. Seltz, and C.L. Grant. 1987. Effects of Structural and Compositional Variations of Dissolved Humic Materials on Pyrene K_{oc} Values. *Environ. Sci. Technol.*, **21**:243–248.

Geller, J. T. and J. R. Hunt. 1993. Mass Transfer from Nonaqueous Phase Organic Liquids in Water-Saturated Porous Media. *Water Res. Res.*, **29**:833–845.

Gelhar, L.W. and C.L. Axness. 1983. Three Dimensional Stochastic Analysis of Macrodispersion in Aquifers. *Water Res. Res.*, **19**:161–180.

Gelhar, L.W. 1986. Stochastic Subsurface Hydrology—From Theory to Applications. *Water Res. Res.*, **22**:135S–145S.

Gelhar, L.W., C. Welty, and K.R. Rehfeldt. 1992. A Critical Review of Data on Field-Scale Dispersion in Aquifers. *Water Res. Res.*, **28**:1955–1975.

Gierke, J.S. , N.J. Hutzler, and J.C. Crittenden. 1990. Modeling the Movement of Volatile Organic Chemicals in Columns of Unsaturated Soil. *Water Res. Res.*, **26**:1529–1547.

Goltz, M.N. 1986. Three-Dimensional Analytical Modeling of Diffusion-Limited Solute Transport. Ph.D. Dissertation. Stanford University, Stanford, CA, 172 p.

Goltz, M.N. and P.V. Roberts. 1986. Interpreting Organic Solute Transport Data from a Field Experiment Using Physical Nonequilibrium Models. *J. Contam. Hydrol.*, **1**:77–93.

Goltz, M.N. and P.V. Roberts. 1987. Using the Method of Moments to Analyze Three-Dimensional Diffusion-Limited Solute Transport from Temporal and Spatial Perspectives. *Water Res. Res.*, **23**:1575–1585.

Goltz, M.N. and P.V. Roberts. 1988. Simulations of Physical Solute Transport Models: Application to a Large-Scale Field Experiment. *J. Contam. Hydrol.*, **3**:37–63.

Grathwohl, P. 1990. Influence of Organic Matter from Soils and Sediments from Various Origins on the Sorption of some Chlorinated Aliphatic Hydrocarbons: Implications on K_{oc} Correlations. *Environ. Sci. Technol.*, **24**:1687–1693.

Grathwohl, P., J. Farrell, and M. Reinhard. 1990. Desorption Kinetics of Volatile Organic Contaminants from Aquifer Materials. In R. Arendt, M. Hinsenveld, and W.J. van den Brink, eds., *Contaminated Soil '90*. Kluwer Academic Publishers, The Netherlands, pp. 343–350.

Grove, D.B. and K.G. Stollenwerk. 1987. Chemical Reactions Simulated by Ground-Water Quality Models. *Water Res. Bull.*, **23**:601–615.

Haag, W.R. and T. Mill. 1988a. Effect of Subsurface Sediment on Hydrolysis of Haloalkanes and Epoxides. *Environ. Sci. Technol.*, **22**:658–663.

Haag, W.R. and T. Mill. 1988b. Some Reactions of Naturally Occurring Nucleophiles with Haloalkanes in Water. *Environ. Toxic. Chem.*, **7**:917–924.

Harmon, T.C., W.P. Ball, and P.V. Roberts. 1989. Nonequilibrium Transport of Organic Contaminants in Groundwater. In B.L. Sawhney and K. Brown, eds., *Reactions and Movements of Organic Chemicals in Soils*. Soil Sci. Soc. Am., Inc. and Am. Soc. of Agronomy, Inc., Madison, WI, pp. 405–438.

Harmon, T.C., L. Semprini, and P.V. Roberts. 1991. Simulating Groundwater Solute Transport Using Independently Determined Sorption Parameters. *J. Environ. Eng.*, **118**:666–689.

Hess, K.M., W.N. Herkelrath, and H.I. Essaid. 1992. Determination of Subsurface Fluid Contents at a Crude-Oil Spill Site. *J. Contam. Hydrol.*, **10**:75–96.

Hunt, J.R., N. Sitar, and K.S. Udell. 1988. Nonaqueous Phase Liquid Transport and Cleanup. 1. Analysis of Mechanisms. *Water Res. Res.*, **24**:1247–1258.

Hutzler, N.J., J.C. Crittenden, and J.S. Gierke. 1986. Transport of Organic Compounds with Saturated Groundwater Flow: Experimental Results. *Water Res. Res.*, **22**:285–295.

Jafvert, C.T. and N.L. Wolfe. 1987. Degradation of Selected Halogenated Ethanes in Anoxic Sediment-Water Systems. *Environ. Toxic. Chem.*, **6**:827–837.

Johnson, R.L. and J.F. Pankow. 1992. Dissolution of Dense Chlorinated Solvents into Ground Water: 2. Source Functions for Pools of Solvent. *Environ. Sci. Technol.*, **26**:896–901.

Jury, W.A., D. Russo, G. Streile, and H. El Abd. 1990. Evaluation of Volatilization by Organic Chemicals Residing Below the Soil Surface. *Water Res. Res.*, **26**:13–20.

Jury, W.A., W.F. Spencer, and W.J. Farmer. 1983. Behavior Assessment Model for Trace Organics in Soil, I, Model Description. *J. Environ. Qual.*, **12**:558–564.

Karickhoff, S.W. 1984. Organic Pollutant Sorption in Aquatic Systems. *J. Hydraul. Eng., ASCE*, **110**:707–735.

Kriegman-King, M.R. and M. Reinhard. 1992. Transformation of Carbon Tetrachloride in the Presence of Sulfide, Biotite, and Vermiculite. *Environ. Sci. Technol.*, **26**:2198–2206.

Kuhn, E.L., P.C. Colberg, J.L. Schnoor, O. Wanner, A. Zehnder, and R.P. Schwarzenbach. 1985. Microbial Transformations of Substituted Benzenes During Infiltration of River Water to Groundwater: Laboratory Column Studies. *Environ. Sci. Technol.*, **19**:961–968.

Kung, K.-H.S. and M.B. McBride. 1991. Bonding of Chlorophenols on Iron and Aluminum Oxides. *Environ. Sci. Technol.*, **25**:702–709.

Larsen, T., T.H. Christensen, and M.L. Brusseau. 1992a. Predicting Nonequilibrium Transport of Naphthalene through Aquifer Materials Using Batch Determined Sorption Parameters. *Chemosphere*, **24**:141–153.

Larsen, T., P. Kjeldsen, and T.H. Christensen. 1992b. Sorption of Hydrophobic Hydrocarbons on Three Aquifer Materials in a Flow Through System. *Chemosphere*, **24**:439–451.

Lee, L.S., M. Hagwall, J.J. Delfino, and P.S.C. Rao. 1992. Partitioning of Polycyclic Aromatic Hydrocarbons from Diesel Fuel into Water. *Environ. Sci. Technol.*, **26**:2104–2210.

Lee, L.S., P.S.C. Rao, M.L. Brusseau, and R.A. Ogwaday. 1988. Nonequilibrium Sorption of Organic Contaminants During Flow through Columns of Aquifer Materials. *Environ. Toxicol. Chem.*, **7**:779–793.

Mabey, W. and T. Mill. 1978. Crit. Rev. of Hydrolysis of Organic Compounds in Water Under Environmental Conditions. *J. Phys. Chem. Ref. Data*, **7**:383–415.

Macalady, D.L. and N.L. Wolfe. 1985. Effects of Sediment Sorption on Abiotic Hydrolysis. I. Organophosphate Esters. *J. Agric. Food Chem.*, **33**:167–173.

Macalady, D.L., P.G. Tratnyek, and T.J. Grundl. 1986. Abiotic Reduction Reactions of Anthropogenic Organic Chemicals in Anaerobic Systems: A Critical Review. *J. Contam. Hydrol.*, **1**:1–28.

MacIntyre, W.G., T.B. Stauffer, and C.P. Antworth. 1991. A Comparison of Sorption Coefficients Determined by Batch, Column, and Box Methods on a Low Organic Carbon Aquifer Material. *Ground Water*, **29**:908–913.

Mackay, D.M. and J.A. Cherry. 1989. Groundwater Contamination: Pump-and-Treat Remediation. *Environ. Sci. Technol.*, **23**:630–636.

Mackay, D.M., D.L. Freyberg, P.V. Roberts, and J.A. Cherry. 1986. A Natural Gradient Experiment on Solute Transport in a Sand Aquifer. 1. Approach and Overview of Plume Movement. *Water Res. Res.*, **22**:2017–2029.

Mangold, D.C. and C. Tsang. 1991. A Summary of Subsurface Hydrological and Hydrochemical Models. *Rev. Geophysics*, **29**:51–79.

Massmann, J. and D.F. Farrier. 1992. Effects of Atmospheric Pressures on Gas Transport in the Vadose Zone. *Water Res. Res.*, **28**:777–791.

McCarthy, J.F. and J.M. Zachara. 1989. Subsurface Transport of Contaminants. *Environ. Sci. Technol.*, **23**:496–502.

McCarthy, K.A. and R.L. Johnson. 1993. Transport of Volatile Organic Compounds Across the Capillary Fringe, *Water Res. Res.*, **29**:1675–1683.

McCarthy, J.F. and C. DeGueldre. 1993. Sampling and Characterization of Colloids and Particles in Groundwater for Studying Their Role in Contaminant Transport, in J. Buffle and H.P. van Leeuwen (eds.), *Environmental Particles, Vol. 2*, Lewis Publishers, Boca Raton, FL, pp. 247–316..

Means, J.C., S.G. Wood, J.J. Hassett, and W.L. Banwart. 1982. Sorption of Amino- and Carboxy-Substituted Polynuclear Aromatic Hydrocarbons by Sediments and Soils. *Environ. Sci. Technol.*, **16**:93–97.

Mihelcic, J.R. and R.G. Luthy. 1991. Sorption and Microbial Degradation of Naphthalene in Soil-Water Suspensions under Denitrification Conditions. *Environ. Sci. Technol.*, **25**:169–177.

Mill, T. and W. Mabey. 1988. Hydrolysis of Organic Chemicals. In O. Hutzinger, ed., *The Handbook of Environmental Chemistry: Reactions and Processes, Vol. 2, Part D.* Springer-Verlag, Berlin, pp. 71–111.

Miller, M.E. and M. Alexander. 1991. Kinetics of Bacterial Degradation of Benzyl-amine in a Montmorillonite Suspension. *Environ. Sci. Technol.*, **25**:240–245.

Miller, C.T. and A.J. Rabideau. 1993. Development of Split-Operator, Petroc-Galerkin Methods to Simulate Transport and Diffusion Problems. *Water Res. Res.*, **29**:2227–2240.

Miller, C.T. and W.J. Weber, Jr. 1986. Sorption of Hydrophobic Organic Pollutants in Saturated Soil Systems. *J. Contam. Hydrol.*, **1**:243–261.

Miller, C.T. and W.J. Weber, Jr. 1988. Modeling the Sorption of Hydrophobic Contaminants by Aquifer Materials—II. Column Reactor Systems. *Water Res.*, **22**:465–474.

Miller, C.T., M.M. Poirier-McNeill, and A.S. Mayer. 1990. Dissolution of Trapped Nonaqueous Phase Liquids: Mass Transfer Characteristics. *Water Res. Res.*, **26**:2783–2796.

Molz, F.J., O. Guven, and J.G. Melville. 1983. An Examination of Scale-Dependent Dispersion Coefficients. *Ground Water*, **21**:715–725.

Molz, F.J., M.A. Widdowson, and L.D. Benefield. 1986. Simulation of Microbial Growth Dynamics Coupled to Nutrient and Oxygen Transport in Porous Media. *Water Res. Res.*, **22**:1207–1216.

Neuman, S.P., C.L. Winter, and C.P. Newman. 1987. Stochastic Theory of Field-Scale Dispersion in Anisotropic Porous Media. *Water Res. Res.*, **23**:453–466.

Nielsen, D.R., M.Th. van Genuchten, and J.W. Biggar. 1986. Water Flow and Solute Transport Processes in the Unsaturated Zone. *Water Res. Res.*, **22**:89S–108S.

Nkedi-Kizza, P., J.W. Biggar, M.Th. van Genuchten, P.J. Wierenga, H.M. Selim, J.M. Davidson, and D.R. Nielsen. 1983. Modeling Tritium and Chloride 36 Transport Through an Aggregated Oxisol. *Water Res. Res.*, **19**:691–700.

Nkedi-Kizza, P., J.W. Biggar, H.M. Selim, M.Th. van Genuchten, P.J. Wierenga, J.M. Davidson, and D.R. Nielsen. 1984. On the Equivalence of Two Conceptual Models for Describing Ion Exchange During Transport Through an Aggregated Oxisol. *Water Res. Res.*, **20**:1123–1130.

Ogram, A.V., R.E. Jessup, L.T. Ou, and P.S.C. Rao. 1985. Effects of Sorption on Biological Degradation Rates of (2,4-Dichlorophenoxy)acetic Acid in Soils. *Appl. Environ. Microbiol.*, **49**:582–587.

Pankow, J.F. and S.W. McKenzie, 1991. Parameterizing the Equilibrium Distribution of Chemicals between the Dissolved, Solid Particulate Matter and Colloidal Matter Compartments in Aqueous Systems. *Environ. Sci. Tech.*, **25**:2046–2053.

Parker, J.C. 1989. Multiphase Flow and Transport in Porous Media. *Rev. Geophysics*, **27**:311–328.

Parker, J.C., A.K. Katyal, J.J. Kaluarachchi, R.J. Lenhard, T.J. Johnson, and J.L. Zhu. 1991. Modeling Multiphase Organic Chemical Transport in Soils and Ground Water. EPA/600/S2-91/042. R.S. Kerr Environmental Research Laboratory, Ada, OK.

Passioura, J.B. 1971. Hydrodynamic Dispersion in Aggregated Media: 1. Theory. *Soil Sci.*, **111**:339–344.

Pedit, J.A. and C.T. Miller. 1992. Use of a Reactive Surface-Diffusion Model to Describe Apparent Sorption-Desorption Hysteresis and Abiotic Degradation of Lindane in a Subsurface Material. *Environ. Sci. Technol.*, **26**:1417–1427.

Pennell, K.D., R.D. Rhue, P.S.C. Rao, and C.T. Johnston. 1992. Vapor-Phase Sorption of p-Xylene and Water on Soils and Clay Minerals. *Environ. Sci. Technol.*, **26**:756–763.

Peterson, M.S., L.W. Lion, and C.A. Shoemaker. 1988. Influence of Vapor-Phase Sorption and Diffusion on the Fate of Trichloroethylene in an Unsaturated Aquifer System. *Environ. Sci. Technol.*, **22**:571–578.

Pickens, J.F. and G.E. Grisak. 1981. Scale-Dependent Dispersion in a Stratified Granular Aquifer. *Water Res. Res.*, **17**:1191–1211.

Pignatello, J.J. 1989. Sorption Dynamics of Organic Compounds in Soils and Sediments. In B.L. Sawhney and K. Brown, eds., *Reactions and Movements of Organic Chemicals in Soils; SSSA Publication No. 22.* Soil Sci. Soc. Am., Inc. and Am. Soc. of Agronomy, Inc., Madison, WI, pp. 15–80.

Pignatello, J.J. 1990a. Slowly Reversible Sorption of Aliphatic Halocarbons in Soils. I. Formation of Residual Fractions *Environ. Toxic. Chem.*, **9**:1107–1115.

Pignatello, J.J. 1990b. Slowly Reversible Sorption of Aliphatic Halocarbons in Soils. II. Mechanistic Aspects. *Environ. Toxic. Chem.*, **9**:1117–1126.

Pinder, G.F. and L.M. Abriola. 1986. On the Simulation of Nonaqueous Phase Organic Compounds in the Subsurface. *Water Res. Res.*, **22**:109S–119S.

Piwoni, M.D. and P. Banerjee. 1989. Sorption of Volatile Organic Solvents from Aqueous Solution onto Subsurface Solids. *J. Contam. Hydrol.*, **4**:163–179.

Poulsen, M.M. and B.H. Kueper. 1992. A Field Experiment to Study the Behavior of Tetrachloroethylene in Unsaturated Porous Media. *Environ. Sci. Technol.*, **26**:889–895.

Powell, R.M., B.E. Bledsoe, G.P. Curtis, and R.L. Johnson. 1989. Interlaboratory Methods Comparison for the Total Organic Carbon Analysis of Aquifer Materials. *Environ. Sci. Technol.*, **23**:1246–1249.

Powers, S.E., C.O. Loureiro, L.M. Abriola, and W.J. Weber, Jr. 1991. Theoretical Study of the Significance of Nonequilibrium Dissolution of Nonaqueous Phase Liquids in Subsurface Systems. *Water Res. Res.*, **27**:463–477.

Powers, S.E., L.M. Abriola, and W.J. Weber, Jr. 1992. An Experimental Investigation of Nonaqueous Phase Liquid Dissolution in Saturated Subsurface Systems: Steady State Mass Transfer Rates. *Water Res. Res.*, **28**:2691–2705.

Prausnitz, J.M. 1980. *Computer Calculations for Multicomponent Vapor-Liquid and Liquid-Liquid Equilibria.* Prentice-Hall, Inc., Englewood Cliffs, NJ.

Ptacek, C.J. and R.W. Gillham. 1992. Laboratory and Field Measurements of Non-Equilibrium Transport in the Borden Aquifer, Ontario, Canada. *J. Contam. Hydrol.*, **10**:119–158.

Quinodoz, H.A.M. and A.J. Valocchi. 1993. Stochastic Analysis of the Transport of Kinetically Sorbing Solutes in Aquifers with Randomly Heterogeneous Hydraulic Conductivity. *Water Res. Res.*, **29**:3227–3240.

Rajaram, H. and L.W. Gelhar. 1991. Three-Dimensional Spatial Moments Analysis of the Borden Tracer Test. *Water Res. Res.*, **27**:1239–1251.

Rao, P.S.C., J.M. Davidson, R.E. Jessup, and H.M. Selim. 1979. Evaluation of Conceptual Models for Describing Nonequilibrium Adsorption-Desorption of Pesticides During Steady-Flow in Soils. *Soil Sci. Soc. Am. J.*, **43**:22–28.

Rao, P.S.C., R.E. Jessup, D.E. Rolston, J.M. Davidson, and D.P. Kilcrease. 1980. Experimental and Mathematical Description of Nonadsorbed Solute Transfer by Diffusion in Spherical Aggregates. *Soil Sci. Soc. Am. J.*, **44**:684–688.

Rasmuson, A. and I. Neretnieks. 1980. Exact Solution of a Model for Diffusion in Particles and Longitudinal Dispersion in Packed Beds, *AIChEJ.*, **26**:686–690.

Rasmuson, A., T. Gimmi, and H. Fluhler. 1990. Modeling Reactive Gas Uptake, Transport and Transformation in Aggregated Soils. *Soil Sci. Soc. Am. J.*, **54**:1206–1213.

Rijnaarts, H.H.M., A. Bachmann, J.C. Jumelet, and A.J.B. Zehnder. 1990. Effect of Desorption and Intraparticle Mass Transfer on the Aerobic Biomineralization of alpha-Hexachlorocyclohexane in a Contaminated Calcareous Soil. *Environ. Sci. Tech.*, **24**:1349–1354.

Ripple, C.D., R.V. James, and J. Rubin. 1974. Packing-Induced Radial Particle-Size Segregation: Influence on Hydrodynamic Dispersion and Water Transfer Measurements. *Soil Sci. Soc. Am. J.*, **38**:219–222.

Roberts, P.V. and D.M. Mackay. 1986. A Natural Gradient Experiment on Solute Transport in a Sand Aquifer. Stanford University Technical Report No. 292. Department of Civil Engineering, Stanford University, Stanford, CA, 153 p.

Roberts, P.V., M.N. Goltz and D.M. Mackay. 1986. A Natural Gradient Experiment on Solute Transport in a Sand Aquifer. 3. Retardation Estimates and Mass Balances for Organic Solutes. *Water Res. Res.*, **22**:2047–2058.

Roberts, P.V., M.N. Goltz, R.S. Summers, J.S. Crittenden, and P. Nkedi-Kizza. 1987. The Influence of Mass Transfer on Solute Transport in Column Experiments with an Aggregated Soil. *J. Contam. Hydrol.*, **1**:375–393.

Roy, W.R., I.G. Krapac, S.F.J. Chou, and R.A. Griffin. 1991. Technical Resource Document: Batch-Type Procedures for Estimating Soil Adsorption of Chemicals. EPA/530-SW-87-006-F; NTIS: PB 92-188515. U.S. EPA, Risk Reduction Engineering Laboratory, Cincinnati, OH, 100 p.
Rubin, J. 1983. Transport of Reacting Solutes in Porous Media: Relation Between Mathematical Nature of Problem Formulation and Chemical Nature of Reactions. *Water Res. Res.*, **19**:1231–1252.
Rutherford, D.W., C.T. Chiou, and D.E. Kile. 1992. Influence of Soil Organic Matter Composition on the Partition of Organic Compounds. *Environ. Sci. Technol.*, **26**:336–340.
Schindler, P.W. 1990. Co-adsorption of Metal Ions and Organic Ligands. In M.F. Hochella and A.F. White, eds., *Mineral-Water Interface Geochemistry; Reviews in Mineralogy, Vol. 23*. Mineralogical Society of America, Washington DC.
Schwartz, F.W. 1977. Macroscopic Dispersion in Porous Media: The Controlling Factors. *Water Res. Res.*, **13**:743–752.
Schwarzenbach, R.P. and J. Westall. 1981. Transport of Nonpolar Organic Compounds from Surface Water to Groundwater. Laboratory Sorption Studies. *Environ. Sci. Technol.*, **15**:1360–1367.
Schwarzenbach, R.P., W. Giger, C. Schaffner, and O. Wanner. 1985. Groundwater Contamination by Volatile Halogenated Alkanes: Abiotic Formation of Volatile Sulfur Compounds under Anaerobic Conditions. *Environ. Sci. Technol.*, **19**:322–327.
Schwarzenbach, R.P., R. Stierli, K. Lanz, and J. Zeyer. 1990. Quinone and Iron Porphyrin Mediated Reduction of Nitroaromatic Compounds in Aqueous Solution. *Environ. Sci. Technol.*, **24**:1566–1575.
Scow, K.M. and M. Alexander. 1992. Effect of Diffusion on the Kinetics of Biodegradation: Experimental Results with Synthetic Aggregates. *Soil Sci. Soc. Am. Proc.*, **56**:128–134.
Shoemaker, C.A., T.B. Culver, L.W. Lion, and M.G. Peterson. 1990. Analytical Models of the Impact of Two-Phase Sorption on Subsurface Transport of Volatile Chemicals. *Water Res. Res.*, **26**:745–758.
Sleep, B.E. and J.F. Sykes. 1989. Modeling the Transport of Volatile Organics in Variably Saturated Media. *Water Res. Res.*, **25**:81–93.
Smith, L. and F.W. Schwartz. 1980. Mass Transport. 1. A Stochastic Analysis of Macroscopic Dispersion. *Water Res. Res.*, **16**:303–313.
Stauffer, T.B. and W.G. MacIntyre. 1986. Sorption of Low-Polarity Organic Compounds on Oxide Minerals and Aquifer Material. *Environ. Toxicol. Chem.*, **5**:949–955.
Steen, W.C., D.F. Paris, and G.L. Baughman. 1980. Effect of Sediment Sorption on Microbial Degradation of Toxic Substances. In R.A. Baker, ed., *Contaminants and Sediments, Vol. 1 Fate and Transport, Case Studies, Modeling, Toxicity*. Lewis Publishers, Boca Raton, FL, pp. 477–482.
Stone, A.T. 1989a. Enhanced Rates of Monophenyl Terephthalate Hydrolysis in Aluminum Oxide Suspensions. *J. Coll. Int. Sci.*, **127**:429–441.
Stone, A.T. 1989b. The Effect of Dismal Swamp Dissolved Organic Matter on the Adsorption and Surface-Enhanced Hydrolysis of Monophenyl Terephthalate Hydrolysis in Aluminum Oxide Suspensions. *J. Coll. Int. Sci.*, **132**:81–87.
Stone, A.T. 1991. Oxidation and Hydrolysis of Ionizable Organic Pollutants at Hydrous Metal Oxide Surfaces. *Rates of Soil Chemical Processes, Special Publ. No. 27*. Soil Sci. Soc. Am., Madison, WI, pp. 231–254.
Stone, A.T. and J.J. Morgan. 1984. Reduction and Dissolution of Manganese (III) and Manganese (IV) Oxides by Organics. 2. Survey of the Reactivity of Organics. *Environ. Sci. Technol.*, **18**:617–624.
Stone, A.T., A. Torrents, J. Smolen, D. Vasudevan, and J. Hadley. 1993. Adsorption of Organic Compounds Possessing Ligand Donor Groups at the Oxide/Water Interface. *Environ. Sci. Technol.*, **27**:895–909.
Stumm, W., R. Kummert, and L. Sigg. 1980. A Ligand Exchange Model for the Adsorption of Inorganic and Organic Ligands at the Hydrous Oxide Interface. *Croatica Chemica Acta*, **53**:291–312.
Sudicky, E.A. 1986. A Natural Gradient Experiment on Solute Transport in a Sand Aquifer. 5. Spatial Variability of Hydraulic Conductivity and Its Role in the Dispersion Process. *Water Res. Res.*, **22**:2069–2082.
Sudicky, E.A. and P.S. Huyakorn. 1991. Contaminant Migration in Imperfectly Known Heterogeneous Groundwater Systems. *Rev. Geophysics*, Supplement: 240–253.
Thorstenson, D.C. and D.W. Pollock. 1989a. Gas Transport in Unsaturated Porous Media: The Adequacy of Fick's Law. *Rev. Geophys.*, **27**:61–78.
Thorstenson, D.C. and D.W. Pollock. 1989b. Gas Transport in Unsaturated Porous Media: Multicomponent Systems and the Adequacy of Fick's Law. *Water Res. Res.*, **25**:477–507.
Tompson, A.F.B. and L.W. Gelhar. 1990. Numerical Simulation of Solute Transport in Three-Dimensional Randomly Heterogeneous Porous Media. *Water Res. Res.*, **26**:2541–2562.
Torrents, A. and A.T. Stone. 1991. Hydrolysis of Phenyl Picolinate at the Mineral/Water Interface. *Environ. Sci. Technol.*, **25**:143–149.
Tratnyek, P.G. and D.L. Macalady. 1989. Abiotic Reduction of Nitroaromatic Pesticides in Anaerobic Laboratory Systems. *J. Agric. Food Chem.*, **37**:248–254.

Turek, F. and R. Lange. 1981. Mass Transfer in Trickle-Bed Reactors at Low Reynolds Number. *Chem. Eng. Sci.*, **36**:569–579.

Valocchi, A.J. 1985. Validity of the Local Equilibrium Assumption for Modeling Sorbing Solute Transport Through Homogeneous Soils. *Water Res. Res.*, **21**:808–820.

Valocchi, A.J. 1989. Spatial Moment Analysis of the Transport of Kinetically Adsorbing Solutes Through Stratified Aquifers. *Water Res. Res.,* **25**:273–279.

van Genuchten, M.Th. and W.A. Jury. 1987. Progress in Unsaturated Flow and Transport Modeling. *Rev. Geophysics*, **25**:135–140.

van Genuchten, M.Th. and R.J. Wagenet. 1989. Two-Site/Two-Region Models for Pesticide Transport and Degradation: Theoretical Development and Analytical Solutions. *Soil Sci. Soc. Am. J.*, **53**:1303–1309.

van Genuchten, M.Th. and P.J. Wierenga. 1976. Mass Transfer Studies in Sorbing Porous Media: I. Analytical Solutions. *Soil Sci. Soc. Am. J.*, **40**:473–480.

van Genuchten, M.Th. and P.J. Wierenga. 1977. Mass Transfer Studies in Sorbing Porous Media: II. Experimental Evaluation with Tritium. *Soil Sci. Soc. Am. J.*, **41**:272–278.

Vogel, T.M., C.S. Criddle, and P.L. McCarty. 1987. Transformations of Halogenated Aliphatic Compounds. *Environ. Sci. Technol.*, **21**:722–736.

Weber, W.J., Jr., P.M. McGinley, and L.E. Katz. 1991. Sorption Phenomena in Subsurface Systems: Concepts, Models and Effects on Contaminant Fate and Transport. *Wat. Res.*, **25**:499–528.

Weber, W.J., Jr., P.M. McGinley, and L.E. Katz. 1992. A Distributed Reactivity Model for Sorption by Soils and Sediments. 1. Conceptual Basis and Equilibrium Assessments. *Environ. Sci. Technol.*, **26**:1955–1962.

Wolfe, N.L. and D.L. Macalady. 1992. New Perspectives in Aquatic Redox Chemistry: Abiotic Transformations of Pollutants in Groundwater and Sediments. *J. Contam. Hydrol.*, **9**:17–34.

Wood, A.L., D. Bouchard, M.L. Brusseau, and P.S.C. Rao. 1990. Cosolvent Effects on Sorption and Mobility of Organic Contaminants in Soils. *Chemosphere*, **21**:575–587.

Wu, S.-C. and P.M. Gschwend. 1986. Sorption Kinetics of Hydrophobic Organic Compounds to Natural Sediments and Soils. *Environ. Sci. Technol.*, **20**:717–725.

Yeh, G.T. and V.S. Tripathi. 1989. A Critical Evaluation of Recent Developments in Hydrogeochemical Transport Models of Reactive Multichemical Components. *Water Res. Res.*, **25**:93–108.

Young, D.F. and W.P. Ball. 1994. Intraparticle Diffusion During Tetrachloroethene Transport through Aquifer Material: A Priori Simulation of Column Behavior Using Batch Data. *Environ. Progress,* **13**:9–25.

Zachara, J.M., C.C. Ainsworth, C.E. Cowan, and R.L. Schmidt. 1990. Sorption of Aminonaphthalene and Quinoline on Amorphous Silica. *Environ. Sci. Technol.,* **24**:118–126.

Zaidel, J. and D. Russo. 1993. Analytical Models of Steady State Organic Species Transport in the Vadose Zone with Kinetically Controlled Volatilization and Dissolution. *Water Res. Res.*, **29**:3343–3356.

Chapter 4

Hydrology and Transport Processes

P.B. Bedient

INTRODUCTION

When a pollutant is released at or near the ground surface, it tends to migrate vertically downward through the unsaturated zone (vadose zone) to the water table. It can then move laterally in the direction of the hydraulic gradient in the saturated zone. As it is transported in the subsurface, its fate is controlled by a number of physical, chemical and biological processes. These include advection, dispersion, capillarity, biodegradation, retardation, and volatilization. Hydrologic, chemical, and biological fate and transport mechanisms must be evaluated at a waste site before any remedial action can be evaluated (Figure 4.1).

Freeze and Cherry (1979) in their classic text offer clear descriptions of many transport mechanisms and derive many of the governing transport equations. Domenico and Schwartz present a number of analytical models in their recent text, *Physical and Chemical Hydrogeology* (1990), and Anderson and Woessner (1992) present a modern treatment of numerical methods applied to flow modeling. This chapter describes solute transport concepts beginning with one- and two-dimensional solutions and including advection, dispersion, adsorption, and biodegradation for hydrologists and engineers working at hazardous waste sites. The concepts presented serve as a synthesis of existing theories, highlighted by examples and case studies for understanding and predicting migration and transport of solutes in groundwater from hazardous waste sources.

SITE CHARACTERIZATION

The majority of information regarding the geologic formations and related surfaces underlying a waste site will be obtained through sediment samples collected during drilling of soil borings and monitoring wells. It is worthwhile to describe all strata underlying the site to at least the maximum depth of known or potential contamination and generate a reliable and complete description of the subsurface geology. Continuous core samples can be collected using a variety of drilling methods.

One of the key elements affecting any remedial design is an accurate characterization of the groundwater flow system. This includes defining the physical parameters of the contaminated region, such as hydraulic conductivity, storage coefficient, and aquifer thickness. Wells can be used for pumping and slug tests to determine hydraulic conductivity and storativity. Recharge rates and pumping rates are required along with any physical and hydraulic boundaries which affect the rate and direction of flow. Groundwater movement can be analyzed through the measurement of water levels in wells and piezometers. It is helpful to categorize wells according to the elevation and geologic formation of the screened interval so that the horizontal and vertical gradients of hydraulic potential can be analyzed separately. If there are enough measuring points, a contour map of the potentiometric surface of each aquifer can be prepared. This map can then be evaluated to determine possible areas of groundwater recharge and discharge and to identify the direction of groundwater movement (Figure 4.2).

The rate of groundwater flow is determined by the porosity and the variation of hydraulic conductivity of the media through which it travels and by hydraulic gradients, which are influenced by recharge and discharge. Because contaminants are advected with moving groundwater, it is important to characterize both the rate and direction of groundwater flow from areas of recharge (commonly via rainfall, surface pits or ponds, or irrigation) to areas of discharge (surface water or wells). Groundwater flows are greatly affected by subsurface heterogeneities, (both vertical and horizontal), subsurface fractures, and other features which alter the average hydraulic conductivity of the aquifer. A typical plume of contamination in two dimensions is shown in Figure 4.3.

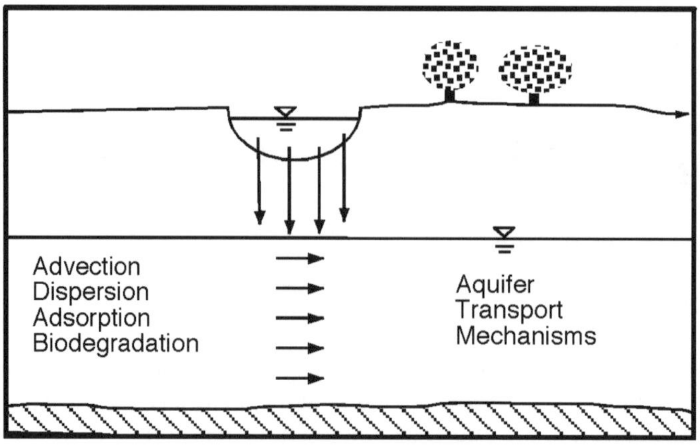

Figure 4.1. Transport mechanisms in the subsurface.

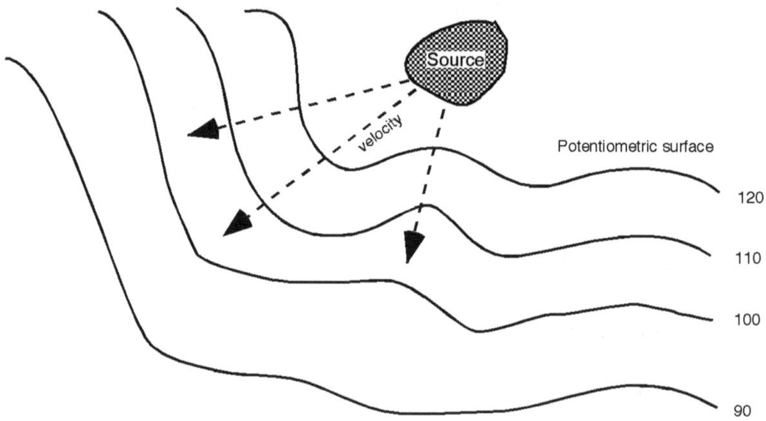

Figure 4.2. Potentiometric map of a site.

Quantitative characterization of the subsurface chemistry includes sampling both the vadose and saturated zones to determine the concentration distributions in the water and soil. A network of properly constructed monitoring wells needs to be installed to collect depth-discrete groundwater samples. Wells should be located in areas that will supply information on ambient (background) groundwater chemistry and on contamination in the plume. In the case of layered systems, it may be necessary to sample from wells properly completed in more than one vertical unit.

Soil and groundwater samples should be analyzed for the inorganic and organic parameters of concern from the waste stream. Of the various contaminants found in groundwater, the widely used industrial solvents and aromatic hydrocarbons from petroleum products are most common. Chemical properties of the plume are necessary to characterize the transport mechanisms of the chemicals and to evaluate the feasibility of a remedial system. A number of properties—including aqueous solubility, density, octanol-water partition coefficient, Henry's Law constant, and biodegradation rates and kinetics—influence the mobility of dissolved chemicals in groundwater and should be considered in detail for plume migration and cleanup.

SOURCE CHARACTERIZATION AND PLUME EVALUATION

A history of the contamination events should be prepared, to the extent possible, to define the types of waste and quantify their loadings to the system. It has become apparent in recent years that

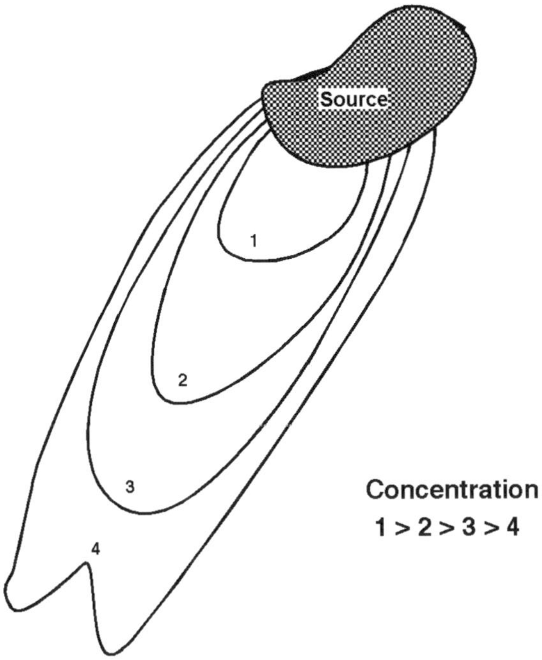

Figure 4.3. Typical contaminant plume (plan view).

the source definition is one of the most important factors affecting remediation success. It consists of defining: (1) the total mass or volume of chemicals released, (2) the total surface area affected by infiltration of chemicals, and (3) the time sequence of the release. Often, the release occurred so long ago that information is difficult to obtain, since many sites have been abandoned or records were not kept. In these cases, one is forced to use hydrogeologic and plume contaminant data to compute what the release must have been, based on a mass balance approach. In cases involving leaking tanks or pits, one can often approximate the volume of waste based on known pit or tank volumes and estimated infiltration rates. However, without detailed records, poor contaminant source definition is still the greatest source of error in designing a groundwater remediation scheme.

Information about the contaminant mix and spatial distribution of the plume is generally needed to select and analyze remedial alternatives during screening and detailed analysis phases. Physical and chemical properties of contaminants such as density, solubility, and partitioning should be assessed because they influence plume movement. It should be recognized that some contaminants may not be detected using routine analytical services, though they are present at levels that would be above cleanup levels, therefore; special analytical services may have to be used.

NONAQUEOUS PHASE LIQUIDS (NAPLS)

While soluble contaminant transport was the topic of major groundwater research in the decade of the eighties, a significant new problem was discovered at many sites near the end of the decade. Nonaqueous phase liquids (NAPLs) are undissolved or immiscible hydrocarbons in the subsurface that exhibit very different transport properties than simpler dissolved contaminant plumes. NAPLs constitute a separate, visible, oily phase in the subsurface whose migration through porous media is complex, governed by gravity, buoyancy, and capillary forces (Wilson et al., 1990). A portion of the NAPL can become trapped as small immobile globules, which cannot easily be removed by pumping. The decade of the nineties has seen enormous attention in the laboratory and at field scale in better understanding the properties and transport issues associated with NAPLs and residual contaminated sites.

NAPL Behavior

When released to the subsurface, NAPL migration is affected by a number of factors, including: 1) the volume of NAPL released, 2) the area of infiltration, 3) the duration of release, 4) NAPL fluid properties, 5) media properties, and 6) subsurface flow conditions (Mercer and Cohen, 1990). These fluids are usually characterized according to their specific gravity relative to water, where light NAPLs (LNAPLs) are lighter than water and dense NAPLs (DNAPLs) are heavier than water. LNAPLs, which mostly consist of fuel hydrocarbons associated with petroleum spills and leaks, move downward through the vadose zone, and can float on and depress the water table. A separate phase lens can form and spread laterally in the capillary fringe just above the saturated zone. As groundwater flows past the LNAPL zone and as the water table rises and falls with recharge, a dissolved contaminant plume can result. These hydrocarbon problems have created long-term LNAPL sources which can contribute to dissolved groundwater plumes for many years (see Figure 4.4).

DNAPLs are related to a wide variety of industrial activities, including facilities where degreasing, metal stripping, chemical manufacturing, pesticide production, coal gasification plants, creosote operations, or other activities involving chlorinated solvents were performed. Chlorinated solvents such as trichloroethylene (TCE), and dichloroethylene (1,1-DCE), and perchloroethylene (PCE) have created some of the largest plumes in existence, such as in Silicon Valley, California and in Arizona where hydrogeologic conditions allow for significant vertical migration hundreds of feet into the saturated zone (Pankow and Cherry, 1996).

The DNAPL problem is much more complex than LNAPL issues due to the rate of vertical migration which is often observed, and the fact that water supply wells often pump from great depths and can have an impact on this type of transport. The DNAPLs can penetrate to depth via preferential pathways or viscous fingering through small seams or fractures (Kueper and Frind, 1988). Once an aquifer is contaminated with DNAPL, it is almost impossible to remediate to drinking water standards (Pankow and Cherry, 1996). In addition, aquifers contaminated with significant DNAPLs are highly resistant to biodegradation.

Free phase or mobile NAPL flows as a continuous body under hydrostatic and gravity forces into the pores of the soil/aquifer matrix. Because a DNAPL is heavier than water, the pressure in a DNAPL pool can cause it to enter even very small pores and fractures in the subsurface as long as the NAPL supply is active. When the supply of NAPL is exhausted, capillary forces will cause small blobs (or ganglia) of NAPL to snap off or disconnect from the continuous NAPL body. These blobs can become trapped in individual pores or small groups of pores as *residual* NAPL, which is immobile under normal hydraulic gradients in the subsurface (Wilson and Conrad, 1984). The actual residual saturation value at a particular location is determined by the mix of chemicals of the NAPL, where the NAPLs are relative to the water table and the overall structure and hydrogeologic characteristics of the soil/aquifer matrix.

NAPLs have a negative impact on the remediation of contaminated aquifers, as it is very difficult or impossible to remove all of the residual NAPL blobs once they are trapped in individual pores. Many of the older pump and treat systems used for aquifer restoration in the U.S. were found to have failed due largely to the presence of free phase or residual NAPL contamination (NRC, 1994). Although many NAPL removal technologies are currently being tested, to date there have been few field demonstrations where sufficient NAPL has been successfully removed from the subsurface to restore an aquifer to drinking water quality (U.S. EPA, 1992a). The NAPL that remains trapped in the soil/aquifer matrix acts as a continuing source of dissolved contaminants to groundwater for many years, greatly complicating the restoration of many aquifer systems (Bedient et al., 1994).

Multiphase Flow Concepts

Many excellent references cover the details of multiphase flow in the subsurface, and only a brief review will be presented here. Several parameters of interest should first be defined. Saturation (S_i) is the fraction of total pore space occupied by fluid "i" in a representative volume of a porous media affected by NAPLs. Thus, saturations of all fluids in the medium add up to 1.0. Saturation values for free phase continuous masses of NAPLs are generally greater than 50% and can approach 100% in

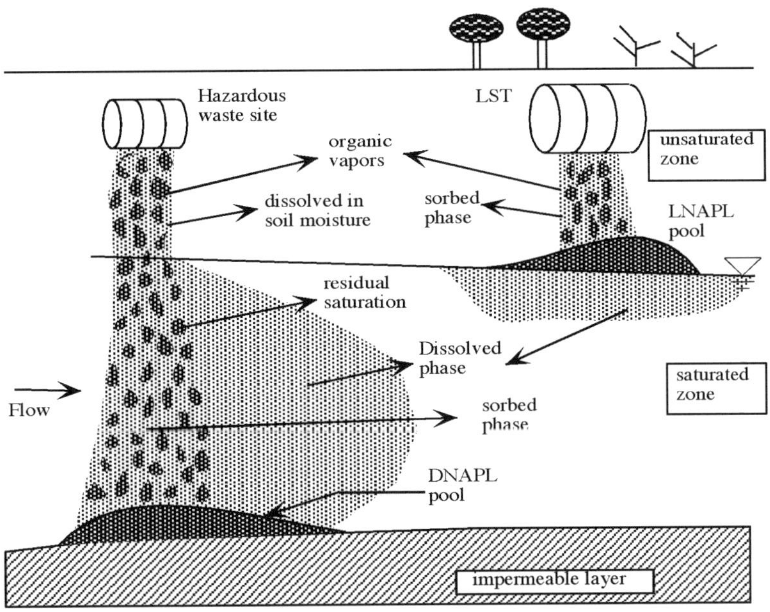

Figure 4.4. Contaminant migration pathways in the subsurface (modified from Wilson et al., 1990).

some systems. The saturation where a continuous NAPL becomes discontinuous and is immobilized by capillary forces is known as the residual saturation (S_{ir}). In unsaturated soils, residual saturation typically ranges from 5% to 20%. In the saturated zone, residual saturation values are typically higher because NAPL serves as a "non-wetting" fluid, with a layer of water, the "wetting" fluid between it and the soil matrix, and residual saturation values range from 15% to 50% (Mercer and Cohen, 1990; Schwille, 1988). Residual saturation appears to be relatively insensitive to the type of chemicals that comprise a NAPL, but is very sensitive to soil properties and heterogeneities (U.S. EPA, 1990).

At many sites the total contaminant mass in the NAPL phase is often many times larger than the total mass of even a large dissolved contaminant plume. Because many sites involve hundreds, thousands, or even hundreds of thousands of gallons of NAPLs released to the subsurface, the relative importance of the NAPL phase should not be overlooked. While the dissolved phase is important because it can migrate off-site relatively quickly, large quantities of NAPLs in the subsurface can hinder groundwater remediation efforts or cause them to fail completely.

Presence or Absence of NAPLs is Difficult to Confirm

Knowing whether a site is contaminated by NAPL is important in order to properly develop site characterization studies and to design reliable groundwater remediation systems. At some sites the presence of NAPLs can be confirmed by visually observing if free phase hydrocarbons have accumulated in wells or are present in soil cores. At many sites, however, the presence of NAPLs is difficult to confirm because residual NAPL will not flow into a monitoring well (it is trapped in the pores of the soil) and is even difficult to see by visual inspection of soil cores. Although several indirect methods are available to estimate the potential for NAPL occurrence, such as measuring the concentration of hydrocarbons on soil, it is still difficult to confirm the presence or absence of NAPLs at many sites.

Difficulty in Removing Trapped Residual NAPLs

Capillary forces make it very difficult or impossible to remove all the NAPLs that have been released to the subsurface. For example, oil-recovery operations at oil fields that employ pumping alone typically remove less than one-third of the oil in a petroleum reservoir. Enhanced oil recovery techniques, such as waterflooding or the application of surfactants, can bring 50%–80% of the original in-place oil (NAPL) to the surface under optimum conditions. These recovery rates are acceptable to the oil industry, where economics determine the ultimate amount of oil recovered from a reservoir. At hazardous waste sites, however, removal of much more than 99% of NAPL is probably required in order to restore a contaminated aquifer to drinking water standards (i.e., low part-per-billion range for most organics). At most sites, this level of recovery is impractical when using extraction and injection of water alone (i.e., pump-and-treat) (Wilson and Conrad, 1986), and is very difficult or impossible to achieve even when using such developing remediation technologies as *in situ* biodegradation. The overall level of aquifer cleanup that can be achieved is very site specific, and is dependent on the type and amount of NAPL present, the soil type, and the amount of resources that are devoted to the project.

Emerging technologies, described in later chapters in this book, are being tested to evaluate their efficacy at a number of field sites. These include soil flushing with chemical surfactants, cosolvents, and steam addition in an effort to mobilize trapped residual NAPL. Current efforts are underway at Hill Air Force Base in Utah to test these technologies in side-by-side cells which have been constructed in a contaminated source area at the base. Preliminary results to date have been encouraging (Rao et al., 1996).

TRANSPORT MECHANISMS

The main transport processes of concern in groundwater include: advection, diffusion, dispersion, adsorption, biodegradation, and chemical reaction. Advection is the movement of contaminants along with flowing groundwater at the seepage velocity in porous media. Diffusion is a molecular mass transport process in which solutes move from areas of higher concentration to areas of lower concentration. Dispersion is a mixing process caused by velocity variations in the porous media. Dispersion causes sharp fronts to spread out and results in the dilution of the solute at the advancing edge of the contaminant front. Adsorption is the partitioning of organic contaminants from the soluble phase onto the soil matrix. Biodegradation represents the transformation of certain organics to CO_2 and water in the presence of microbes in the subsurface. Recent attention also includes a consideration of nonaqueous phase liquids (NAPLs), which can be major sources of contamination in the subsurface. These are defined as separate phase liquids (free products), such as gasoline, which can float just above the water table, or chlorinated solvents (such as TCE), which are denser than water and can sink to the bottom of an aquifer. NAPLs can provide a source of contamination for many years.

Advection, diffusion, dispersion, adsorption, and biodegradation have been analyzed in some detail under both laboratory and field conditions. Soil columns have been set up in the laboratory and loaded with numerous organic compounds like those found at hazardous waste sites. Dispersion coefficients, rates of adsorption (retardation) and biodegradation have been evaluated with some success in these columns. Figure 4.5 shows the typical response (breakthrough curve) of a soil column loaded with a tracer or organic at one end and monitored at the outlet as a function of time.

Field efforts have not been as successful as laboratory studies in measuring dispersion, adsorption, and biodegradation mechanisms. These processes are the focus of current research efforts at many hazardous waste research sites. One of the most successful field sites for tracer studies has been the Borden landfill site in Canada, which resulted in a series of classic papers on dispersion and adsorption processes measured during a two-year field experiment (Mackay et al., 1986 and Roberts et al., 1986). Borden and Bedient (1986), Rifai et al. (1988), and Berry-Spark and Barker (1987) reported some success on measuring and modeling the biodegradation of contaminant plumes associated with a creosote site and BTEX in groundwater.

Advection Processes

Advection is the movement of a contaminant with the flowing groundwater according to the seepage velocity, v_x, defined as Darcy's velocity divided by porosity, n.

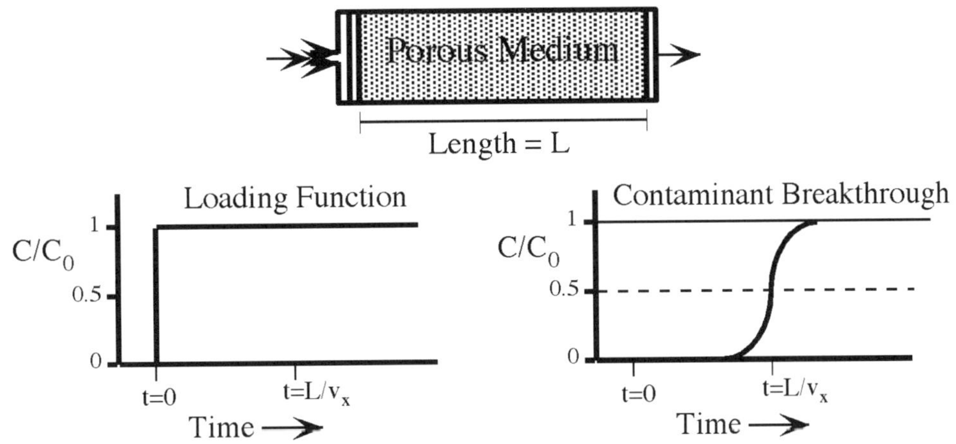

Figure 4.5. Soil column description and breakthrough curve for a one-dimensional continuous source.

It is important to realize that seepage velocity equals the average linear velocity of a contaminant in porous media and is the correct velocity for use in governing solute transport equations. The one-dimensional mass flux (F_x) due to advection equals the product of water flow and concentration (C) of solute, or $F_x = v_x nC$.

There are certain cases in the field where an advective model provides a useful estimate of contaminant transport. Some models include the concept of arrival time by integration along known streamlines (Nelson, 1977). Streamline models are used to solve for arrival times of particles which move along the streamlines at specified velocities, usually in a 2-D flow net. Others have set up an induced flow field through injection or pumping and evaluated breakthrough curves by numerical integration along flow lines. Dispersion is not directly considered in these models but results from the variation of velocity and arrival times in the flow field (Charbeneau, 1981, 1982). In cases where pumping of groundwater is dominating the flow field, one may be able to neglect dispersion processes without loss of accuracy. Advection can be treated numerically by particle tracking models, which represent the velocity of individual particles moving according to the local velocity field in the *x* or *y* direction (Konikow and Bredehoeft, 1978).

Diffusion and Dispersion Processes

Diffusion is a molecular-scale process which causes spreading due to concentration gradients and random motion. Diffusion causes a solute in water to move from an area of higher concentration to an area of lower concentration. Diffusive transport can occur in the absence of velocity. Mass transport in the subsurface due to diffusion in one dimension can be described by Fick's first law of diffusion:

$$F_x = -D_m(dC/dx) \tag{4.1}$$

where

F_x = mass flux ($M/L^2/T$),
D_m = diffusion coefficient (L^2/T), and
dC/dx = concentration gradient (M/L^4).

Dispersion is caused by heterogeneities in the medium that create variations in flow velocities and flow paths. These variations can occur due to friction within a single pore channel, velocity differences from one channel to another, or variable path length. Laboratory column studies indicate dispersion is a function of average linear velocity and a factor called dispersivity, α. Dispersivity in a soil column is on the order of centimeters, while values in field studies may be on the order of one to thousands of meters. Dispersion also follows the form of Fick's first law of diffusion, replacing the molecular diffusion coefficient, D_m, with a dispersion coefficient, $Dd = \alpha\, v_x$.

Diffusion is usually only a factor when very low dominate such as in a tight soil or clay liner, or in the case of mass transport involving very long time periods. Typical dispersion coefficients in groundwater are several orders of magnitude larger and tend to dominate the spreading process when higher velocities are present.

A soil column similar to the one used by Darcy can be used to introduce the concept of advective-dispersive transport. The column is loaded continuously with tracer at relative concentration of $C/C_0 = 1$ across the entire cross section. Figure 4.5 shows the resulting concentration versus time response measured at x = L, called the breakthrough curve. This response occurs in a one-dimensional (1-D) soil column for solute transport in the presence of advection and dispersion. The step function loading is begun at t = 0, and the breakthrough curve is smeared out due to dispersion processes which create a zone of mixing between the displacing fluid and the fluid being displaced. The advective front moves at the average linear velocity (or seepage velocity) and $C/C_0 = 0.5$ point, as shown in Figure 4.5. Velocity variations within the soil column cause some of the mass to leave the column in advance of the advective front and some to lag behind, producing a smooth breakthrough curve. The size of the mixing zone increases as the advective front moves farther from the source. At some distance behind the advective front the source concentration C_0 is encountered, and it remains at steady state in this region. If there were no dispersion, the shape of the breakthrough curve would be identical to the step input function.

Mass transport due to dispersion can also occur normal to the direction of flow. This transverse dispersion, D_T, results from diverging flowpaths in the porous media that cause mass to spread laterally from the main direction of flow. In most cases involving a two-dimensional plume of contamination, D_T is much less than D_L, and the shape of the plume tends to be elongated in the direction of flow (Figure 4.3).

Longitudinal dispersion causes spreading and decreases concentrations near the frontal portions of the plume. The main difference with the 1-D soil column results described above is that 2-D plumes have transverse spreading, which reduces concentrations everywhere behind the advective front.

ONE-DIMENSIONAL MODELS

The governing mass transport equation (a second-order partial differential equation) is difficult to solve in field cases of practical interest due to boundary irregularities and variations in aquifer characteristics, so numerical methods must generally be employed. There are, however, a limited number of relatively simple, 1-D problems for which analytical solutions exist. Several of these cases are presented below in order to gain insights into the effect of advection, dispersion, and adsorption on the overall patterns produced.

The governing assumptions include the following: (1) the tracer is ideal, with constant density and viscosity; (2) the fluid is incompressible; (3) the medium is homogeneous and isotropic; and (4) only saturated flow is considered. For the case of a nonreactive tracer in 1-D flow in the +x direction, the governing equation is:

$$D_x \frac{\partial^2 C}{\partial x^2} - v_x \frac{\partial C}{\partial x} = \frac{\partial C}{\partial t} \tag{4.2}$$

where

D_x = coefficient of hydrodynamic dispersion (L^2/T)
v_x = average seepage velocity (L/T)
C = contaminant concentration which varies in x and t (M/L^3)

Several different solutions can be derived for Equation 4.2 depending on initial and boundary conditions and whether the tracer input is a slug input or a continuous release. Initial conditions in a soil column are usually set to zero or to some constant background concentration. Boundary conditions must be specified at the two ends of the 1-D column. The concentration at $x = 0$ is set to C_0 for $t \geq 0$, and at $x = \infty$, the concentration is zero for all time.

Continuous Source in 1-D

For a semi-infinite column with background concentration of zero and input tracer concentration C_0, Bear (1961) solves the problem using the Laplace transform:

$$\frac{C(x,t)}{C_0} = \frac{1}{2}\left[\operatorname{erfc}\left(\frac{x - (v_x)t}{2\sqrt{D_x t}}\right) + \exp\left(\frac{v_x x}{D_x}\right)\operatorname{erfc}\left(\frac{x + v_x t}{2\sqrt{D_x t}}\right)\right] \tag{4.3}$$

where *erfc* is the complementary error function:

$$= 1 - \left(2/\sqrt{\pi}\right)\int_x^{\infty} e^{-u^2}\,du$$

The complementary error function is a tabulated function which varies from zero to 2.

The center of mass ($C/C_0 = 0.5$) of the breakthrough curve travels with the average linear velocity v_x and corresponds to the point where $x = v_x t$. The second term on the right side of Equation 4.3 can generally be neglected for most practical problems.

Instantaneous Source in 1-D

The corresponding solution to Equation 4.2 can be derived for the injection of a tracer pulse (instantaneous input) at $x = 0$ with background concentration $C_1 = 0$ in the column. As the slug of tracer moves downstream with seepage velocity v_x in the +x direction, its concentration profile obeys:

$$C(x,t) = \frac{M}{\left(4\pi D_x t\right)^{1/2}} \exp\left[-\frac{\left(x - (v_x)t\right)^2}{4 D_x t}\right] \tag{4.4}$$

where M is the injected mass per unit cross-sectional area (M/L^2). Figure 4.6 shows the resulting Gaussian distribution of concentration for the instantaneous pulse source in one dimension.

The differences between instantaneous (spike) source and continuous source transport problems are obvious in 1-D. The continuous source produces a response or breakthrough curve which starts at a low concentration and eventually levels off to the initial input concentration C_0 as a function of time. The spike source produces a normal or Gaussian distribution which continues to decrease in maximum concentration due to spreading out with time in the direction of flow. For the spike source, the amount of mass under each curve is identical if the tracer is conservative.

ADSORPTION EFFECTS

While there exist many reactions which can alter contaminant concentrations in groundwater, adsorption onto the soil matrix appears to be one of the dominant mechanisms. The concept of the

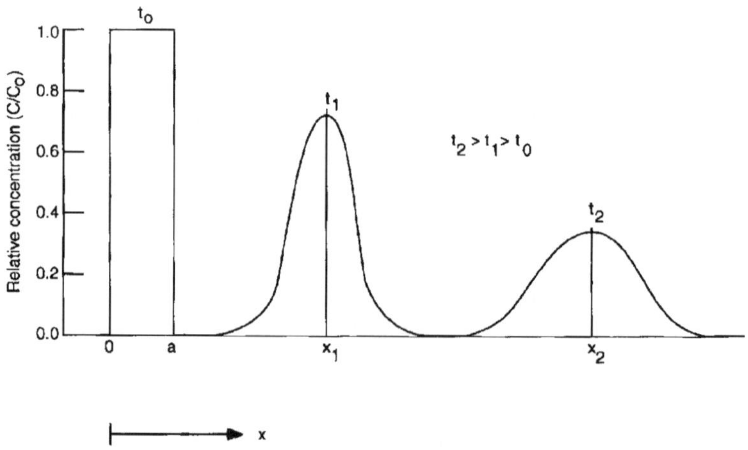

Figure 4.6. Instantaneous source in a soil column.

sorption isotherm is used to relate the amount of contaminant adsorbed to the solids S (mg/kg) to the concentration in solution C. One of the most commonly used isotherms is the Freundlich isotherm:

$$S = K_d C^b \tag{4.5}$$

where K_d is the partitioning coefficient (L^3/kg) and b = constant.

The use of the partitioning coefficient assumes that sorption of the solute to the soil is very fast relative to the rate of groundwater flow.

In the case of the linear isotherm (b = 1), the solution (Bear, 1972) for a 1-D column injected with a sorbing contaminant becomes:

$$C(x,t) = \frac{C_0}{2}\left[\operatorname{erfc}\left(\frac{Rx-(v_x)t}{2\sqrt{RD_x t}}\right) + \exp\left(\frac{v_x x}{D_x}\right)\operatorname{erfc}\left(\frac{Rx+(v_x)t}{2\sqrt{RD_x t}}\right)\right] \tag{4.6}$$

where $R = 1 + (\rho_b/m)K_d$, and ρ_b is the bulk density (kg/L^3).

Equation 4.4 or 4.6 can be used in laboratory studies to determine dispersion coefficients for nonreactive and adsorbing species. Retarded fronts can be derived from conservative fronts by adjusting $v_x \Rightarrow v_x/R$ and $\Rightarrow D_x/R$ in 1-D. Typical values of R for organics often encountered in field sites range from 2 to 10 (Roberts et al., 1986).

Transport and 1-D with First-Order Decay

One example of mass transport which includes a simple kinetic reaction is the first-order decay of a solute. This could be caused by radioactive decay, biodegradation, or hydrolosis, and is generally represented in the transport equation by adding the term $-\lambda C$, where λ is the first-order decay rate in units of t^{-1}. For example, Equation 4.4 or 4.6 would be multiplied by exp $(-\lambda t)$ in the resulting solution of an instantaneous source.

TWO-DIMENSIONAL MODELING

Two-dimensional (2-D) modeling is generally the most useful tool for analyzing and predicting solute transport in a field situation. This type of model is considerably more versatile than the 1-D models previously discussed. Horizontal 2-D modeling can be used where: (1) no vertical velocity components exist; or (2) observation wells or monitoring points are fully screened and provide averaged data in 2-D for a relatively uniform thick aquifer. Probably the most common reason for using a horizontal 2-D analysis is that the monitoring wells are typically distributed about the horizontal plane. Most wells are fully penetrating, and therefore any samples collected represent an approximate vertical average in that portion of the aquifer.

Analytical Models

Analytical models are developed by solving the transport equation for certain simplified boundary and initial conditions. Numerous analytical solutions are available in the literature (Bear, 1979; Hunt, 1978; Wilson and Miller, 1978; Cleary and Ungs, 1978; Shen 1976; Galya, 1987; Javandel et al., 1984) for pulse and continuous contaminant sources with boundary conditions ranging from no flow to constant heads. Processes that may be included in these models are advection, dispersion, adsorption, and first-order decay (biological or radioactive). Analytical solutions generally require constant parameters, simple geometries, and simple boundary conditions, but nevertheless provide useful insights to many groundwater contaminant problems at field sites.

Galya (1987) developed the horizontal plane source (HPS) model for a planar source geometry located in the x-y plane. The model assumes 1-D horizontal velocity, but allows for 3-D dispersion in the x, y, and z directions. The governing equations are solved for a continuous release, although a finite source can be handled through linear superposition in time. The model's geometry is useful in analyzing ponds or pits which leach contaminants from the surface down to the groundwater table.

Galya's model was derived through the use of Green's functions to simulate three-dimensional contaminant transport from a horizontal plane source. This analytical model incorporates retardation and decay, and can simulate varying source emission rates. Uses of the model include simulations of contaminant transport from landfills, waste lagoons, land treatment facilities, and areas of pesticide application. Comparison between HPS and point source solutions indicates that the HPS model will provide more accurate results than the point source solution, particularly near the source. Representative model applications indicate the model's sensitivity to variations in retardation, decay, and temporal period of source emission.

The semi-analytical approach has been used by Javandel et al. (1984) to show that concentration versus time data for a single well can be mapped to other observation points to estimate spatial distribution in concentration. The computer program RESSQ calculates 2-D contaminant transport by advection and adsorption in a steady-state flow field. Recharge wells and ponds act as sources, and pumping wells act as sinks. RESSQ calculates the streamline pattern in the aquifer, the location of contaminant fronts around sources at various times, and the variation of contaminant concentration with time at sinks. RESSQ was developed at the Lawrence Berkeley Laboratory based on a solution procedure by Gringarten and Sauty (1975); Javandel et al. (1984) present a detailed discussion, listing, and user's guide for the semi-analytical computer program RESSQ.

BIODEGRADATION

Biodegradation involves the conversion of subsurface hydrocarbons to CO_2 and H_2O by stimulating the native microbial population. The biodegradation potential of organic contaminants has been recognized and utilized in the wastewater treatment process for years (Lee et al., 1988). The ultimate goal of the biodegradation process is to convert organic wastes into biomass and CO_2, CH_4, and inorganic salts. Two essential criteria must be in place before biodegradation or bioremediation can occur: first, the subsurface geology must have a relatively large hydraulic conductivity (K) to allow the transport of oxygen and nutrients through the aquifer; second, microorganisms must be present in sufficient numbers and types to degrade the contaminants of interest. Formations with K values greater than 10^{-4} cm/sec are most amenable to *in situ* bioremediation.

ANAEROBIC BIODEGRADATION

More recently, anaerobic bioremediation processes have been investigated for use at sites contaminated with chlorinated solvents. Unlike aerobic processes that are typically limited by electron acceptor (O_2) availability, anaerobic dechlorination systems are typically limited by the availability of electron donors. The focus of anaerobic bioremediation systems today centers on the delivery of either organotrophic or lithotrophic substrates to the subsurface capable of providing reducing equivalents for anaerobic dechlorinator. At some sites, electron donors are found as co-contaminants with chlorinated solvents, and the addition of electron donors to these plumes may not be required. It is not uncommon to observe dechlorination products (e.g., vinyl chloride or DCE isomers from TCE) in this class of contaminant plumes. Anaerobic bioremediation processes have not yet seen widespread application; however field-scale systems are being operated at several sites across the U.S. An excellent treatment of bioremediation is available in Norris et al. (1994).

Bioremediation Processes

Enhanced aerobic bioremediation is essentially an engineered delivery of nutrients and oxygen, in the form of air, pure oxygen, or hydrogen peroxide, to the contaminated zone of an aquifer. The main advantage of biological treatment is that it offers partial or complete breakdown of the contaminant instead of simply transferring the contaminant from one phase in the environment to another. Lee et al. (1988) and Thomas et al. (1987) have discussed in detail the advantages and disadvantages of bioremediation.

Bioremediation can often be used to treat contaminants that are sorbed to soil or trapped in pore spaces. In addition to treatment of the saturated zone, organics held in the unsaturated and capillary zone can sometimes be treated when an infiltration gallery or soil flushing is used.

The time required to treat subsurface pollution using bioremediation can often be less than in simple pump-and-treat procedures. The basic steps involved in an *in situ* bioremediation program (Lee et al., 1988) are: (1) site investigation, (2) free product recovery, (3) microbial degradation study, (4) system design, (5) operation, and (6) maintenance. As in pump-and-treat systems, it is important to define the hydrogeology and the extent of contamination at the site prior to the initiation of any *in situ* effort (Figure 4.7).

After defining the hydrogeology, recovery of any free product at the site should be completed. For fuels less dense than water, the nonaqueous phase liquid (NAPL) can be removed using physical recovery techniques such as a single-pump system that produces water and hydrocarbon or a two-pump, two-well system that steepens the hydraulic gradient and recovers the accumulating hydrocarbon (Domenico and Schwartz, 1990).

Prior to the initiation of a bioremediation activity, it is important to conduct a feasibility study for the biodegradation of the contaminants present at the site. First, contaminant-degrading microorganisms must be present; and second, the response of these native microorganisms to the proposed treatment method must be evaluated. In addition, the feasibility study is conducted to determine the nutrient requirements of the microorganisms. These laboratory studies provide a reliable basis for performance at the field level only if they are performed under conditions that simulate the field.

The rate of aerobic biodegradation is usually limited by the amount of oxygen that can be transported to the organisms in the zone of contamination. A number of methods are available to supply oxygen to groundwater, including the addition of air, pure oxygen, or hydrogen peroxide, with increasing concentrations, respectively. Sparging the groundwater with air and pure oxygen can supply only 8 to 40 mg/L of oxygen depending on the temperature of the injection fluid (Lee et al., 1988). Hydrogen peroxide can provide more than 100 mg/L of equivalent oxygen.

NUMERICAL MODELING

In many instances, analytical models are not sufficient to describe the system being modeled. In these cases, one must turn to a numerical model. Numerical models approximate the differential equations on a user-defined numerical grid, and allow the user to enter temporally and/or spatially varying values for model parameters. Major parameters include hydraulic conductivity, aquifer thickness, boundary conditions, well pumpage rates, background concentrations, etc. Because of the usefulness of numerical models, significant effort has gone into developing and documenting models,

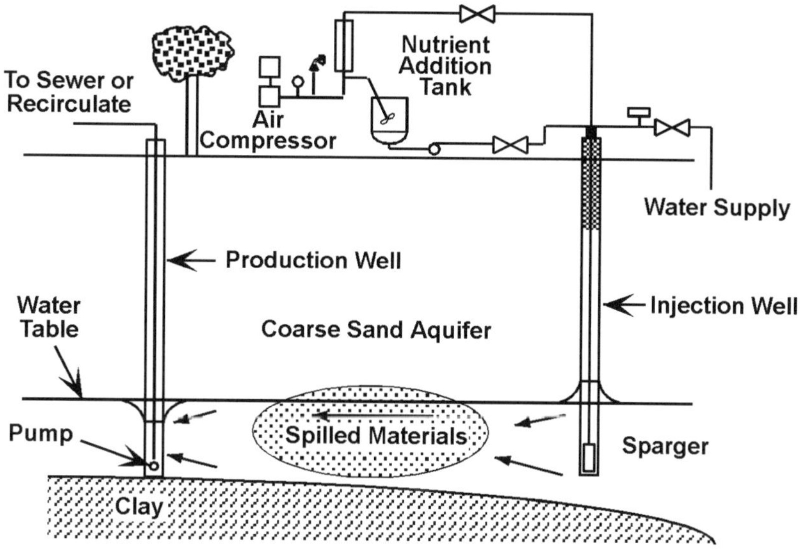

Figure 4.7. Oxygen injection sytsem for *in situ* bioremediation.

and, in recent years, to developing pre- and postprocessors for models. Numerical methods and modeling techniques are covered in detail in Anderson and Woessner (1992) and in Zheng and Bennett (1995).

Some of the more popular models include the USGS Method of Characteristics (MOC) code (Konikow and Bredehoeft, 1978), the USGS MODFLOW code (which simulates 3 D flow conditions) (MacDonald and Harbaugh, 1988), the USGS SUTRA code (Voss, 1984), the SWIFT code (Reeves et al., 1986; Ward, 1991), and the MT3D code (Zheng, 1990, 1992). Each of these codes has its own set of specific input and output requirements. In addition, models like MODPATH (Pollock, 1989) use the flow fields generated by MODFLOW to determine transport pathways from a source to a receptor.

The first numerical groundwater models were designed in the 1970s. Since that time, numerous advances have been made in the field of numerical modeling methods. In light of this, newer models have been designed to take advantage of the newer methods. One such model is FLOTRAN, written in the early 1990s. FLOTRAN solves Richard's equation for flow combined with a system of advection-diffusion-reaction equations describing contaminant transport. It assumes a three-dimensional, logically rectangular domain, but allows for fairly general geometrics by using smooth mappings between the logically rectangular domain and a rectangular computational domain. FLOTRAN's numerical methods are robust and fast (Dawson, 1991; Dawson, 1993).

The creation of a site model is a multi-step process. The user usually creates a conceptual model of the site, which includes defining all the aquifer parameters, sources, sinks, etc., then defines a grid upon which the model will be executed, and assigns the aquifer properties to the appropriate grid locations. In general, this method requires defining the grid as one of the first steps of the model preparation, since most of the other input values depend on grid location. In order to facilitate the creation of model input files and interpretation of model results, some programs allow the user to graphically enter input data. In addition, several programs allow outputs to be graphed and visualized on the screen. More recent preprocessors allow the user to create and make changes to a conceptual model of a site and create a model input files with a much lower level of effort than has been previously required. Some of the more popular pre- and/or postprocessors are GMS, developed by the ECGL at Brigham Young University, and Visual MODFLOW, developed by Waterloo Hydrogeologic, Inc. All of the above graphical and numerical improvements have promoted the application of complex models for sites that require it.

REFERENCES

Anderson, M.P. and W.W. Woessner. 1992. *Applied Groundwater Modeling.* Academic Press, Inc., San Diego, CA.

Annable, M.D., J.W. Jawitz, and P.S.C. Rao. 1996. "Field-Scale Test of Interfacial and Partitioning Tracers for Characterizing NAPL-Water Contact Area and Residual Saturation in Contaminated Aquifers: Preliminary Analysis of Results," *Proceedings of the Second Annual Symposium on Energy and the Environment,* Rice University, Houston, TX, 81–86.

Bear, J. 1961. Some experiments on dispersion. *J. Geophysical Res.*, **66**(8):2455–67.

Bear, J. 1972. *Dynamics of Fluids in Porous Media.* New York:Amer. Elsevier Pub. Co., 764 p.

Bear, J. 1979. *Hydraulics of Ground Water.* McGraw-Hill, New York, NY.

Bedient, P.B., H.S. Rifai, and C.J. Newell. 1994. *Ground Water Contamination: Transport and Remediation.* Prentice-Hall Publishing Co., Englewood Cliffs, NJ. 540 pp.

Berry-Spark, K. and J.F. Barker. 1987. Nitrate remediation of gasoline contaminated ground waters: results of a controlled field experiment, *in* Proc., Petroleum Hydrocarbons and Organic Chemicals in Ground Water, NWWA/API, pp. 3–10.

Borden, R.C. and P.B. Bedient. 1986. Transport of dissolved hydrocarbons influenced by reaeration and oxygen limited biodegradation: 1. Theoretical development. *Water Res. Res.*, **22**:1973–1982.

Borden, R.C. and P.B. Bedient. 1987. *In situ* measurement of adsorption and biotransformation at a hazardous waste site. *Water Res. Bull.*, **23**:629–36.

Charbeneau, R.J. 1981. Groundwater contaminant transport with adsorption and ion exchange chemistry: method of characteristics for the case without dispersion. *Water Res. Res.*, **17**(3):705–713.

Charbeneau, R.J. 1982. Calculation of pollutant removal during groundwater restoration with adsorption and ion exchange. *Water Res. Res.*, **18**(4):1117–25.

Cleary, R.W. and M.J. Ungs. 1978. Groundwater pollution and hydrology, mathematical models and computer programs, *Rep. 78-WR-15*, Water Res. Prog., Princeton University, Princeton, NJ.

Dawson, C.N. 1991. "Godunov-mixed methods for advective flow problems in one space dimension." *SIAM Journal of Numerical Analysis* **28**:1282–1309.

Dawson, C.N. 1993. "Godunov-mixed methods for advection-diffusion equations in multidimensions." *SIAM Journal of Numerical Analysis* **30**:1315–1332.

Domenico, P.A. and F.W. Schwartz. 1990. *Physical and Chemical Hydrogeology.* John Wiley, New York, NY.

Freeze, R.A. and J.A. Cherry. 1979. *Groundwater.* Prentice-Hall, Englewood Cliffs, NJ.

Freyberg, D.L. 1986. A natural gradient experiment on solute transport in a sand aquifer, (2) spatial moments and the advection and dispersion of nonreactive tracers. *Water Res. Res.*, **22**(13):2031–46.

Galya, D.P. 1987. A horizontal plane source model for ground-water transport, *Ground Water*, **25**:6.

Gringarten, A.C. and J.P. Sauty. 1975. A theoretical study of heat extraction from aquifers with uniform regional flow, *J. Geophysical Res.*, **80**(35):4956–62.

Hunt, B. 1978. Dispersive sources in uniform ground-water flow, *J. Hydrol. Div.* ASCE **104**:75–85.

Javandel, I., C. Doughty, and C.F. Tsang. 1984. *Groundwater Transport: Handbook of Mathematical Models.* Amer. Geophys. Union, Water Res. Monograph 10, Washington DC, 228 p.

Konikow, L.F. and J.D. Bredehoeft, 1978. "Computer Model of Two-Dimensional Solute Transport and Dispersion in Ground Water, Automated Data Processing and Computations," Tech. of Water Res. Investigations of the USGS, Book 7, Chapter C2, Washington, DC.

Kueper, B.H. and E.O. Frind. 1988. "An Overview of Immiscible Fingering in Porous Media," *J. Cont. Hydrology* **2**:95–110.

Lee, M.D., J.M. Thomas, R.C. Borden, P.B. Bedient, J.T. Wilson, and C.H. Ward. 1988. Biorestoration of aquifers contaminated with organic compounds. *CRC Critical Reviews in Environmental Control*, **18**(1):29–89.

Mackay, D.M., D.L. Freyberg, P.V. Roberts, and J.A. Cherry. 1986. A natural gradient experiment on solute transport in a sand aquifer, (1) Approach and overview of plume movement. *Water Res. Res.*, **22**(13):2017–29.

McDonald, M.G. and A.W. Harbaugh. 1988. *A Modular Three-Dimensional Finite-Difference Ground-Water Flow Model,* U.S. Geological Survey Techniques of Water Resources Investigations, Book 6, 586 pp.

Mercer, J.W. and R.M. Cohen. 1990. "A Review of Immiscible Fluids in the Subsurface: Properties, Models, Characterization and Remediation," *J. Cont. Hydrology,* **6**:107–163.

National Research Council (NRC). 1994. *Alternatives for Ground Water Cleanup,* National Academy Press, Washington, DC.

Nelson, R.W. 1977. Evaluating the environmental consequences of groundwater contamination. #1. An overview of contaminant arrival distributions as general evaluation requirements. *Water Res. Res.*, **14**:409–15.

Norris, R.D. [et al.] (Robert S. Kerr Environmental Research Laboratory, Ada, OK). 1994. *Handbook of Bioremediation,* Lewis Publishers, Boca Raton, FL. 257 pp.

Pankow, J.F. and J.A. Cherry. 1996. *Dense Chlorinated Solvents and other DNAPLs in Groundwater,* Waterloo Press, Portland, OR, 522 pp.

Pollock, D.W. 1989. *Documentation of Computer Programs to Compute and Display Pathlines Using Results from the U.S. Geological Survey Modular Three-Dimensional Finite-Difference Ground-Water Model,* U.S. Geological Survey Open-File Report, 89-381.

Reeves, M., D.S. Ward, N.D. Johns, and R.M. Cranwell. 1986. *Theory and Implementation for SWIFT II, The Sandia Waste-Isolation Flow and Transport Model for Fractured Media, Release 4.84,* NUREG/CR-3328 and SAND83-1159, Sandia National Laboratories, Albuquerque, NM.

Rifai, H.S., P.B. Bedient, J.T. Wilson, K.M. Miller, and J.M. Armstrong. 1988. "Biodegradation modeling at a jet fuel spill site," ASCE *J. Environ. Eng. Div.*, **114**:1007–19.

Roberts, P.V., M.N. Goltz, and D.M. Mackay. 1986. A natural gradient experiment on solute transport in a sand aquifer, (4), sorption of organic solutes and its influence on mobility. *Water Res. Res.*, **22**(13):2047–58.

Schwille, F. 1988. *Dense Chlorinated Solvents in Porous and Fractured Media: Model Experiments* (English translation). Lewis Publishers, Boca Raton, FL.

Shen, H.T. 1976. Transient Dispersion in Uniform Porous Media Flow." *J. Hydraul. Div.*, ASCE, **102:**707–716.

Sudicky, E.A. 1986. A natural gradient experiment on solute transport in a sand aquifer: spatial variability of hydraulic conductivity and its role in the dispersion process. *Water Res. Res.*, **22**(13):2069–82.

Thomas, J.M., M.D. Lee, P.B. Bedient, R.C. Borden, L.W. Canter, and C.H. Ward. 1987. "Leaking Underground Storage Tanks: Remediation with Emphasis on *In Situ* Biorestoration, NCGWR, Robert S. Kerr Environmental Research Laboratory," U.S. EPA, Ada, OK, EPA/600/2-87/008.

U.S. EPA. 1990. *Laboratory Investigation of Residual Liquid Organics from Spills, Leaks, and the Disposal of Hazardous Wastes in Groundwater,* EPA/600/6-90/004, U.S. Environmental Protection Agency, Washington, DC.

U.S. EPA. 1992a. *Estimating Potential for Occurrence of DNAPL at Superfund Sites,* EPA Quick Reference Fact Sheet, EPA Publication 9355.4-07FS, U.S. Environmental Protection Agency, Washington, DC.

Voss, C.I. 1984. SUTRA: *A Finite Element Simulation Model for Saturated-Unsaturated, Fluid-Density-Dependent Ground Water Flow with Energy Transport or Chemical Reactive Single-Species Solute Transport,* U.S. Geological Survey Water Resources Investigations Report 84-4369.

Ward, D.S. 1991. *Data Input Guide for SWIFT/386, Version 2.50.* GeoTrans, Inc. Sterling, VA.

Wilson, J.L. and P.J. Miller. 1978. Two-dimensional plume in uniform groundwater flow. *J. Hydraul. Div.* ASCE, **104**:503–514.

Wilson, J.L. and S.H. Conrad. 1984. "Is Physical Displacement of Residual Hydrocarbons a Realistic Possibility in Aquifer Restoration?" *Proc. Petroleum Hydrocarbon and Organic Chemicals in Ground Water,* National Water Well Association, Houston, TX.

Zheng, C. 1990. *MT3D: A Modular Three-Dimensional Transport Model for Simulation of Advection, Dispersion and Chemical Reactions of Contaminants in Groundwater Systems,* Report to the U.S. Environmental Protection Agency, Ada, OK, 170 pp.

Zheng, C. 1992. *MT3D version 1.8 Documentation and User's Guide,* S.S. Papadopoulos & Associates, Inc., Bethesda, MD.

Zheng, C. and G.D. Bennett. 1995. *Applied Contaminant Transport Modeling: Theory and Practice.* Van Nostrand Reinhold Publishing Co., New York, NY, 440 pp.

Chapter 5

Developing Rational Goals for *In Situ* Remedial Technologies

J.A. Cherry, S. Feenstra, and D.M. Mackay

INTRODUCTION

Remediation or remedial measures conducted at sites of subsurface contamination are engineered activities designed to protect present or future human health and maintain or improve environmental quality. Subsurface contamination may occur as soil contamination (i.e., contamination in the vadose zone) or groundwater contamination (i.e., contamination below the water table) or both. Subsurface contamination may present a variety of types of risks, including risks via groundwater usage, surface water usage, soil contact, vapor inhalation, or food consumption. Any of these risks may provide cause for subsurface remediation. However, the nature of the subsurface problem and the strategy for remediation depend greatly on which of these risks are significant and what degree of risk reduction or environmental protection is required.

This chapter provides a conceptual development of subsurface remediation goals that are rational and consistent with our present understanding of subsurface contamination caused by industrial organic liquids, particularly organic chemicals that are immiscible and heavier than water. These liquids are commonly known as DNAPLs (dense nonaqueous phase liquids), and commercial, industrial, and waste disposal sites where DNAPL is the major cause of groundwater contamination are referred to as DNAPL sites. In the 1980s, it became evident in North America and Europe that portions of many aquifers had become contaminated by a variety of chemicals of industrial, municipal, or agricultural origin, particularly nitrate, chloride and sulfate salts, metals, and industrial organic chemicals. Most sites that have received, or will soon receive, major remediation efforts are afflicted with DNAPLs, the most common of which are chlorinated solvents. Other important DNAPLs are: creosote, coal tar, PCB oils, some types of mixed organic industrial wastes, and some pesticides in pure-product form. Chlorinated solvents (such as trichloroethylene, tetrachloroethylene, trichloroethene, and methylene chloride) have been used by nearly all manufacturing industries in North America and continue to be used by many. Also, chlorinated solvents have been used for many decades in most dry cleaning facilities. They are much more common than was thought to be the case a few years ago. Most Superfund sites, most military facilities, and many RCRA sites are in this category. Many of the U.S. Department of Energy's sites have DNAPL in the subsurface, along with other types of contaminants. Realization that DNAPLs are common at industrial and military sites came gradually over a decade, evolving from the interpretations of site conditions by a few groundwater scientists in the early 1980s to broad recognition by the groundwater profession by the early 1990s. Although the occurrence of DNAPLs at many sites across the continent is now recognized, there is currently no consensus on how to deal with this immense groundwater problem. DNAPL sites pose much greater technical problems and financial burdens than other types of sites. Meeting regulatory cleanup criteria is elusive at most DNAPL sites. This chapter is intended to provide a structural scientific framework for conduct of debate on this issue.

Figure 5.1 shows schematic diagrams for three different conceptual models for groundwater contamination caused by industrial chemicals. In the simplest model (Figure 5.1a) plumes of contaminated groundwater are caused by leaching of contaminants from the soil or waste situated above the water table. Permanent restoration of such a site can be relatively simple, though not necessarily inexpensive: remove the contaminated soil or waste by excavation, or remove the contaminants from the soil or waste by *in situ* methods, and then extract the plume by groundwater pump-and-treat. It is now clear that this model fits only a very small proportion of those industrial sites where organic

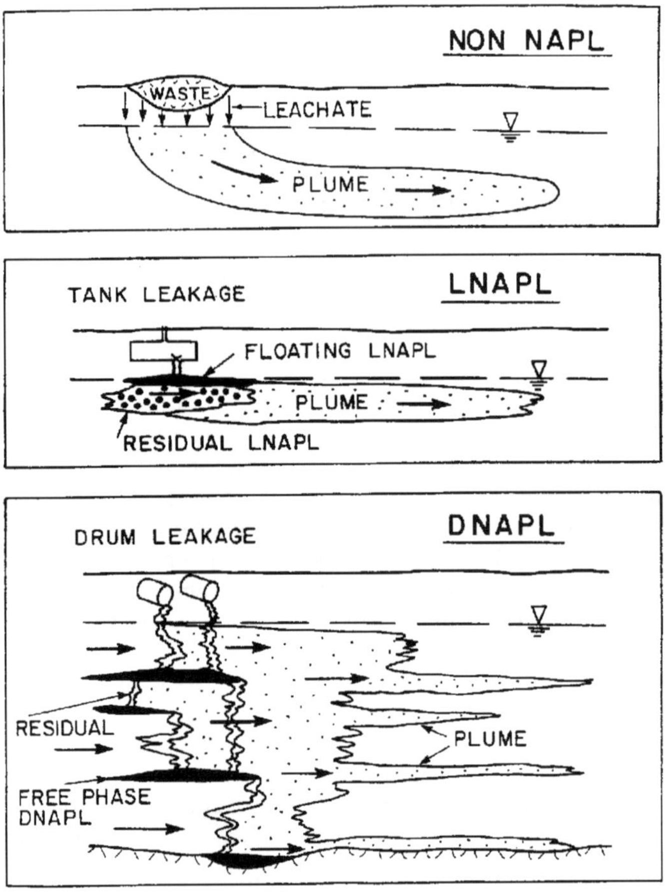

Figure 5.1. Three different conceptual models for groundwater contamination: (a) non-NAPL case; (b) LNAPL case; and (c) DNAPL case.

contaminants exist in groundwater. In the second model (Figure 5.1b), petroleum products such as gasoline, fuel oil, or kerosene (i.e., LNAPLs—light nonaqueous phase liquids, or immiscible liquids less dense than water), exist above the water table or in the zone of water table fluctuation. This subsurface mass of immiscible liquid emits dissolved contaminants to groundwater and soil water, causing a plume to form. Contaminant plumes from LNAPLs are usually not large, mainly because natural microbiological degradation of LNAPL-derived constituents in the plumes usually restricts growth of the plumes. Also, the fact that LNAPLs do not penetrate much below the water table tends to limit LNAPL-derived plumes to relatively shallow depths and therefore to restrict spread of plumes because shallow groundwater usually encounters local hydrologic boundaries. This is not to say that some plumes from LNAPL sources do not cause major problems, but rather that the problems are generally much more limited in geographic extent than other types of plumes, particularly plumes at DNAPL sites. To permanently clean a LNAPL site, the mass of immiscible phase liquid in the subsurface, above and below the water table, must be removed by excavation, or by extraction processes and *in situ* destruction, and the plume must be removed by pump-and-treat or *in situ* degradation.

The most difficult site to remediate is represented in Figure 5.1c. The primary cause of contaminant at DNAPL sites is slow dissolution of the mass of immiscible organic liquid situated below the water table. In Germany, Frederich Schwille recognized the importance of this type of groundwater contamination in the 1970s. Schwille (1981, 1984) conducted intensive laboratory research on chlorinated solvents in groundwater. Schwille's work was not known in North America until 1984. Villaume (1983) published the first North American paper describing DNAPL (creosote) in groundwater, but it was not until the mid- to late 1980s that the conceptual model displayed in Figure 5.1c and its extension into a DNAPL paradigm became known amongst groundwater scientists and engineers in North America (Cherry, 1984; Mackay et al., 1985; Mackay and Cherry, 1989; Feenstra, 1984). Specific consideration of DNAPL sites was not included in the late 1980s edition of the U.S. EPA guidance document for remedial investigations of Superfund sites (U.S. EPA, 1988). It was not until the early 1990s that the general implications of DNAPL for site monitoring and remediation were recognized by the U.S. Environmental Protection Agency (U.S. EPA, 1992). Such official recognition has not yet been made by regulatory agencies in Canada, Mexico, or Europe. The fact that it took about a decade for the new paradigm to move from creation by a small segment of the scientific community to widespread recognition is not surprising. Paradigm shifts in science and engineering often take a decade or more. Unfortunately, the large and costly effort of site investigation and subsurface remediation undertaken in the United States in the 1980s and continuing in the 1990s is based primarily on concepts for site conditions represented by non-DNAPL causes of contamination (Figures 5.1a and 5.1b). DNAPL sites typically have groundwater contamination that is deeper and much more widespread than is common at other sites. The main hazardous organic contaminants in groundwater at DNAPL sites tend to be more persistent than those at LNAPL sites.

Some industrial sites have very complex conditions of subsurface contamination better represented by a combination of the three conceptual models shown in Figure 5.1 rather than any one of the models. At these sites the DNAPL part of the problem is usually deeper in the subsurface and most difficult to remediate permanently.

REMEDIATION CONCEPTS AND TERMINOLOGY

Groundwater remediation technologies can be grouped in two general categories: proven or experimental. To be classified as a *proven technology*, a technology must have been applied previously at a sufficient number of sites so that a base of general knowledge has been developed to allow detailed design and reliable estimation of the cost and time required to achieve a specific remedial goal. *Experimental technologies* are those for which there is an expectation of effectiveness based on conceptual models, computer models, laboratory studies, or field trials, but for which performance cannot yet be predicted reliably.

Another useful categorization of remediation technologies distinguishes between those that are active or passive. *Active technologies* are those that require a considerable amount of ongoing engineering activity (operations, monitoring, and maintenance) to maintain adequate performance of the technology. Conventional groundwater pump-and-treat is an example of an active technology. In contrast, *passive technologies* are those that are put in place and subsequently require little or no ongoing energy inputs or maintenance other than performance monitoring. Low-permeability caps and covers to minimize infiltration and plume generation are examples of passive technologies. Active technologies have the disadvantage of transferring significant operating costs to future generations, whereas passive technologies minimize such transfer.

Subsurface chemical masses that emit persistent groundwater plumes are referred to as *sources* or *source zones*. Commonly, the most important sources at industrial sites are masses of heavier-than-water immiscible industrial liquids (DNAPLs) located below the water table (Figure 5.2). However, old drums or tanks containing chemicals or near-surface soils or sludges may also be active sources of groundwater contamination. At many sites, even when the drums, tanks, sludges, and soils are removed, the DNAPL that remains in the subsurface represents a significant continuing source of groundwater contamination. Until this subsurface source is removed, the plume will continue to receive contaminants leached from the DNAPL. Considering the relatively low solubilities of DNAPL chemicals and low drinking water standards for many DNAPL chemicals, plumes may be generated in this fashion for very long periods of time, even when the mass of DNAPL in the subsurface source zones is relatively small. Removal of the subsurface source zones to a degree that prevents continued formation of the plume is by far the most difficult task in aquifer restoration. The DNAPL mass that exists below the water table typically represents a much greater groundwater problem than any buried drums or buried sludge or soil that may exist at the site. The concept of the source of groundwater

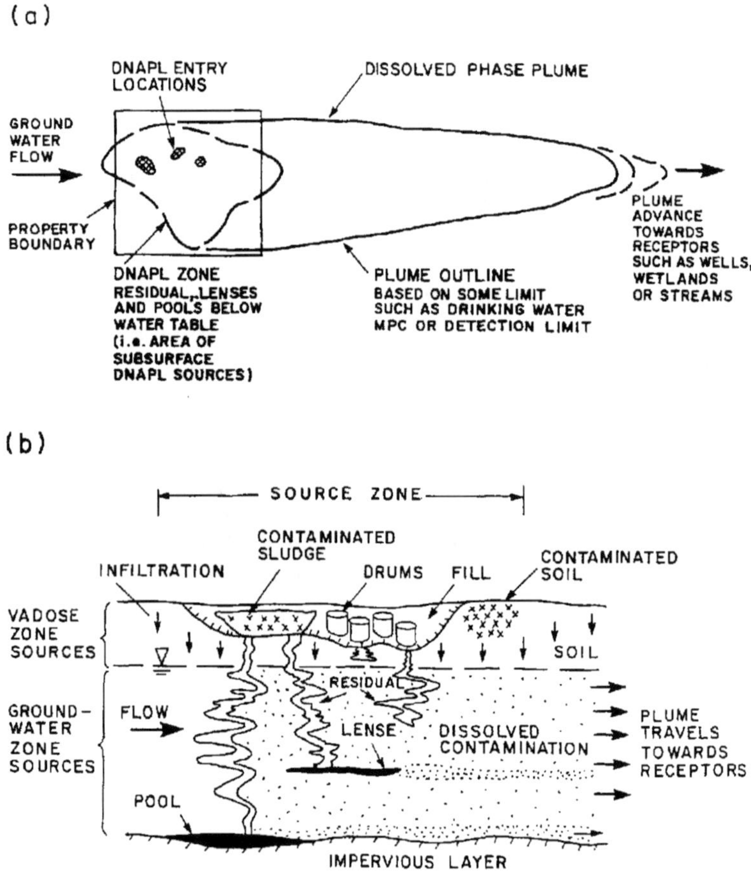

Figure 5.2. Anatomy of a DNAPL site: (a) plan view of source zone and plume; and (b) vertical section of source zone. Note contaminant sources (sludge, drums, contaminated soil, residual) above the water table and immiscible-phase DNAPL (residual, lenses, pool) source below the water table.

contamination at a DNAPL site represents a dichotomy: for groundwater scientists and engineers focused on the technical matters of groundwater remediation, the most significant *source* is generally the mass of immiscible liquid residing below the water table, whereas for the regulatory or the legal community, the "significant source" on the property is commonly the buried drums or sludge that still remains. The buried drums and sludge are typically all above the water table. The regulatory or legal viewpoint is what governs most remedial actions at industrial sites. Thus, at many DNAPL sites, particularly Superfund and RCRA sites, large remedial effort is commonly directed at the drum or sludge *source* and little effort at the below water table DNAPL source. Thus, at many DNAPL sites, the dichotomy in the concept of *source* results in "source remediation" for which benefits in risk reduction and improvement in groundwater quality may be minimal even after large expenditures have been made.

In the groundwater context, risk can be reduced or eliminated by controlling the migration of the aqueous phase and vapor phase plumes to achieve *plume containment*. A fully *contained plume* leads to no further expansion of contamination in the groundwater zone. Thus, around a contained plume, no additional portion of the groundwater resource is becoming contaminated. At many, if not most, DNAPL sites full containment is not achievable without engineered controls such as groundwater pump-and-treat (Figure 5.3a). In order to eliminate or reduce risk to an acceptable level, remedial measures for plume containment must continue for as long as the subsurface source zone persists. This typically requires long-term application of technologies such as groundwater pump-and-treat. In

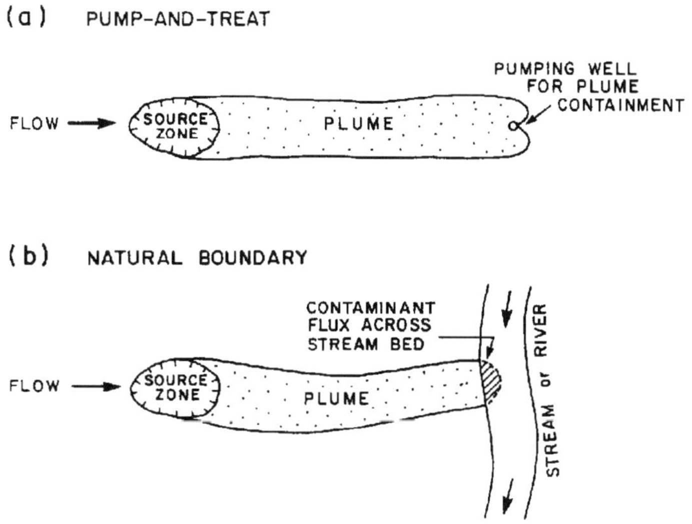

Figure 5.3. Containment of plumes at DNAPL site: (a) pump-and-treat for containment of plume; and (b) natural hydrologic boundary (river bed).

some cases, the condition of *full containment* is achieved by natural site circumstances. An example would be a plume that emanates from the DNAPL source and discharges into a nearby stream (Figure 5.3b). Although the plume may cause surface water contamination, the portion of the contaminated aquifer does not expand, and therefore in the context of the groundwater environment, the plume is contained naturally.

Aquifer restoration is a task much different from *plume containment.* Aquifer restoration requires source zone containment or source removal as well as plume removal to a degree sufficient to provide for the original beneficial use of the aquifer. This beneficial use is commonly drinking water supply. A *fully restored aquifer* is one that requires no further source zone containment or plume containment (Figure 5.4). Of the thousands of contaminated aquifers in North America, some have ongoing successful measures for plume containment, but very few have been fully restored to drinking water use because of difficulties with source-zone restoration. We know of no significant DNAPL site where full aquifer restoration has been achieved and substantiated with detailed monitoring data. Claims of full aquifer restoration have generally not been confirmed by adequate testing, or the cleanup standard has been lenient, not requiring restoration to drinking water use.

At the present time, the only remedial technologies that should be regarded as proven technologies for most organic industrial chemicals in most hydrogeologic settings are: (1) groundwater pump-and-treat for containment of groundwater plumes and source zones (but not source removal); (2) vertical cutoff walls coupled with hydraulic control inside the enclosure for source zone containment, and in some cases, coupled with caps for source containment; and (3) soil vapor extraction for containment of vapor plumes in the vadose zone, but not for complete removal of all significant contaminant from the vadose zone.

Ongoing elimination of risk in all areas outside of the source area and plume and prevention of further degradation of environmental quality can generally be achieved by plume containment without the necessity of significant reduction in contaminant mass in the source zone. It is possible to perform cost-benefit analyses for pump-and-treat for plume containment if the risks associated with the plume are known. In contrast, the lack of proven *in situ* technologies for source removal, and hence for full aquifer restoration, renders cost-benefit analysis for these technologies impossible because the ultimate cost of remediation cannot be defined. The difficult and unresolved issue in groundwater remediation therefore pertains to the degree of effort and the level of financial resources that should be directed towards the presently elusive goal of *full aquifer restoration.* This goal is elusive because of the extreme uncertainty in the performance of remediation technologies to achieve

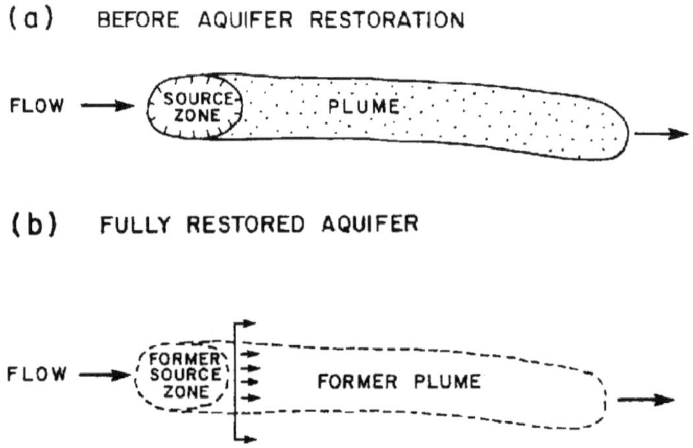

Figure 5.4. Conceptual representation of aquifer restoration. Source and plume removal sufficient to meet cleanup standard: (a) before aquifer restoration; and (b) conditions when fully restored; flux from source insufficient to produce plume.

common regulatory cleanup criteria. This uncertainty derives from the limitations inherent in the technologies and from the complications for technology performance imparted by subsurface geologic heterogeneity. A technology that performs well in a simple geologic site situation can be entirely inadequate in a complex geologic setting. There are many more complex sites than simple sites.

A *partially restored aquifer* is an aquifer in which the segment of the plume between the plume front and the downgradient periphery of the source zone (zone between A and B, Figure 5.5) has been removed, while the source zone is contained by engineered means such as cutoff walls or pump-and-treat to prevent dissolved contaminants emitted from the source from causing regrowth of the plume. Thus, partial restoration of the aquifer depends on long-term operation of the source zone containment system. Partial aquifer restoration makes available for use the portion of the aquifer in the area beyond the source zone control system. Partial aquifer restoration can be accomplished by one of two approaches: (1) plume removal by pump-and-treat (Figure 5.5); or (2) plume renovation by natural gradient groundwater flushing (Figure 5.6). Plume removal by pump-and-treat enables the plume to be removed more rapidly (Figure 5.6a), whereas plume renovation relies on natural groundwater flow to flush the plume from the aquifer. Once the source zone containment system is functioning, the plume exists only temporarily if there is an engineered or natural discharge point. After the plume separates from the source zone, it is flushed along the natural groundwater flowpaths toward points of groundwater discharge at streams, rivers, wetlands, or water supply wells. Under these constraints, natural gradient plume flushing results in a flux of contamination to the river, lake, or water supply well for a finite period of time.

The cumulative contaminant flux is the total initial contaminant mass in the plume (aqueous and sorbed phases) emitted from the groundwater flow system at the points of discharge over the length of time of the emission. This emission at points of discharge is typically small. Once separated from the source zone, plumes normally have a small total aqueous and sorbed mass. Mackay and Cherry (1989) provide examples of estimates of total aqueous phase contaminant mass in several large plumes in sand or gravel aquifers. In some cases, part or all of the plume mass may disappear due to biodegradation or abiotic degradation prior to arrival at groundwater discharge points. In many cases, the point-of-discharge emission causes insignificant environmental impact, and/or insignificant risk to actual or potential receptors. Plumes from many LNAPLs and DNAPLs, particularly chlorinated solvents, only rarely cause detectable or significant impact on rivers or lakes. Therefore, isolation of the source zone followed by natural flushing for plume removal may be a reasonable option in many cases when decision making can be governed solely by issues of science and economics. In some cases, the emission involves significant risk to receptors or causes a significant loss in use of water

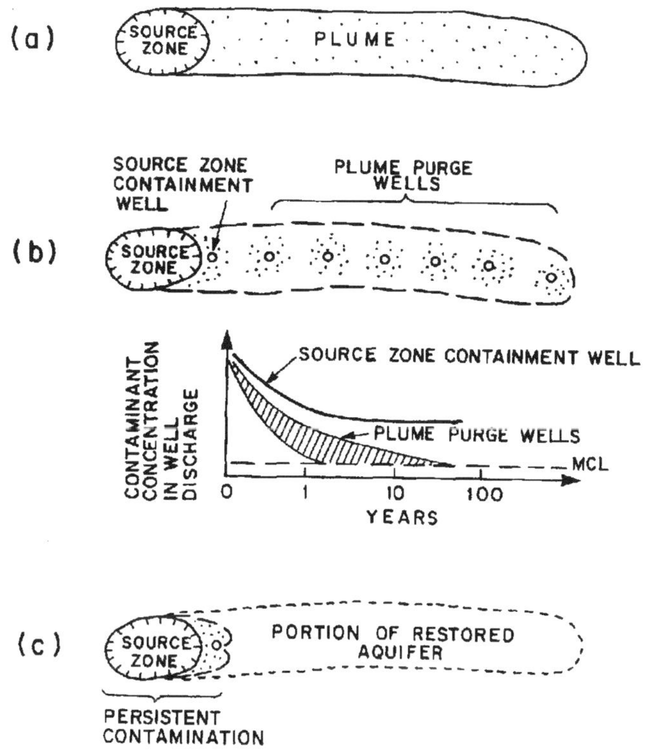

Figure 5.5. Conceptual representation of partial aquifer restoration using pump-and-treat: (a) initial condition; (b) performance of plume purge wells and contaminant well; and (c) partially restored aquifer with source zone containment to prevent regrowth of plume.

resources—surface water, groundwater, or both. In such cases, these risks can be prevented by *plume containment* or *plume removal.*

Full-scale plume removal is the higher-cost remedial option when compared to *natural-gradient plume renovation* or plume containment. In full-scale plume removal, expenses accrue from capital costs and operating and maintenance costs for the network of pumping wells and water treatment and for performance monitoring. For natural gradient plume renovation, the main expense is monitoring to ensure that the aquifer renovation rate and impact at discharge points such as rivers or lakes are within acceptable limits. The required source zone containment can usually be accomplished using only a small number of wells and a relatively small water-treatment system or cutoff wall enclosure around the source zone with little or no pumping.

At most sites where industrial organic contaminants are a groundwater problem, the majority of the expense for groundwater pump-and-treat is due to capital and operating costs for the portion of the system needed for full-scale plume removal rather than the portion needed for source containment. Therefore, selection of the full-scale plume removal option rather than the natural-gradient plume renovation option coupled with source containment is a selection of a higher-cost, lower-risk plan over one with a much lower cost and higher temporary risk. For the selection to be done in a cost-benefit context, it is necessary that remedial designs specific to each of these options be compared. Some pump-and-treat systems put into operation during the 1980s were compromises that accomplished neither full containment of the source zone nor removal of the plume.

A third option, intermediate between the two described above, is *natural flushing with capture.* In this option, the source zone is contained (Figure 5.6a) and the front of the plume is captured. The plume, thus detached from the source zone, is allowed to flush under only slightly accelerated above natural gradient conditions to the downgradient capture system. This is a lower-cost option than the full-plume removal option because of less pumping and lower water treatment costs. This option

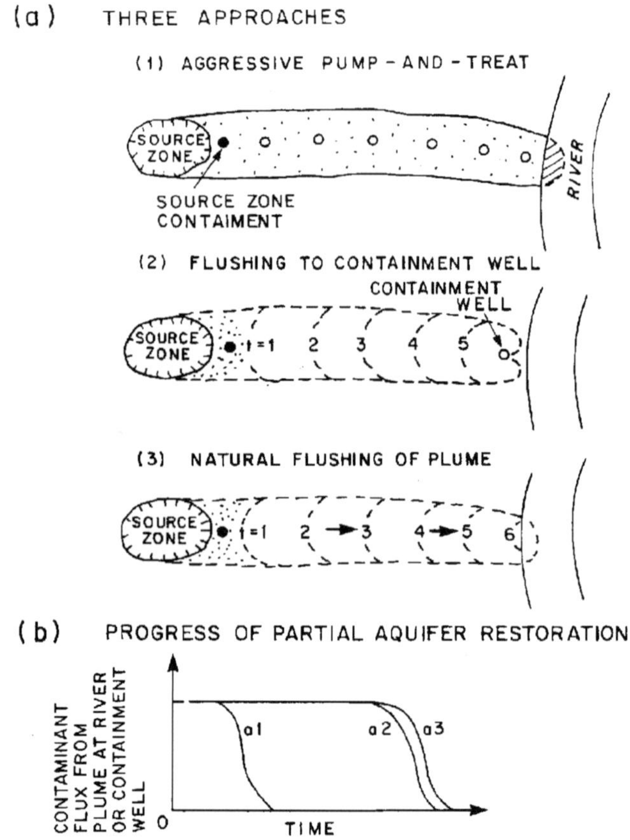

Figure 5.6. Three approaches for achieving partial aquifer restoration: (a) aggressive pump-and-treat; (b) flushing to containment well; and (c) natural flushing of plume to hydrologic discharge zone (contaminant flux to river).

accomplishes the same goal as the full-plume removal option, but it takes longer for the plume to disappear.

TASKS IN PLUME CONTAINMENT AND AQUIFER RESTORATION

Remediation of a DNAPL or LNAPL site can be considered on the basis of four fundamental tasks: plume-front containment, source-zone containment, plume removal, and source removal. The first three of these tasks can be undertaken to accomplish partial aquifer restoration but all four must be applied for an attempt at full aquifer restoration (Figure 5.2). Conventional pump-and-treat is a proven technology for accomplishing any of the first three of these tasks. Pump-and-treat generally accomplishes plume front or source zone containment with minimal difficulty if applied properly. Pump-and-treat is a viable technology for plume removal, but at some sites subsurface complexity causes large uncertainty in the time and cost to achieve the specified degree of plume removal. Thus, long-term cost estimates for this task are generally inaccurate or unreliable. Groundwater pump-and-treat is generally inadequate to accomplish the fourth task, source removal. Although conventional pump-and-treat can perform effectively for three of the tasks, it is not inexpensive and involves substantial ongoing costs because of energy use, maintenance, and monitoring. Therefore, there is incentive to develop less expensive and generally more passive means of accomplishing these tasks. Experimental technologies aimed at this goal are described briefly below.

Plume Front Capture

Most dissolved plumes are too large to be enclosed practically by cutoff wall barriers. The goal of plume-front containment is to capture the entire mass flux at the front of the plume during the entire duration of the flux. Contaminant fluxes at the plume front are commonly small, requiring a low annual rate of contaminant mass removal. Three potential engineering options for capture or containment of the plume-front are: (1) conventional pump-and-treat; (2) *in situ* treatment curtains; (3) *in situ* funnel-and-gates with *in situ* or *ex situ* treatment (Figure 5.7). The options for plume-front containment are nearly the same as those for source-zone containment, with the exception of cutoff wall barriers that can be used effectively at the source zone in many cases but generally not at the plume front.

An *in situ treatment curtain* is a reactive permeable material placed as a wall vertically across the plume (Figure 5.7b). As the plume passes through the curtain, physical, chemical, or microbial processes remove the contaminants from the plume. An effective treatment curtain avoids the need for removing water from the aquifer to achieve plume cutoff. The *in situ* treatment curtain is currently an unproven or experimental technology. Various versions of the technology for many contaminant species are being subjected to research and development. Burris and Cherry (1992) describe the status of this research. Gillham and Burris (this volume) provide a more recent review. Recently, prototypes of the technology have been installed at a few field sites.

A funnel-and-gate system, also known as a wall-and-gate system, uses segments of impervious wall to direct groundwater flow to engineered gaps in the wall, referred to as gates, through which the plume is channelled (Figure 5.8d). As the water flows through the gate, the contaminants are removed by comparable types of *in situ* treatment processes used in the *in situ* treatment curtains (Figure 5.8c). Plume containment may be achieved by a single funnel-and-gate placed across the entire plume, or if the plume is relatively wide, by a series of funnels-and-gates across the plume. Intensive research on *in situ* treatment curtains and funnel-and-gate systems began only recently. Starr and Cherry (1993) describe the hydraulic aspects of funnel-and-gate systems. The first prototype funnel-and-gate system was installed at a field site in Canada in late 1992. The funnel-and-gate configuration for *in situ* treatment has advantages over the curtain—ease of installation and possibility for repair or rejuvenation.

There are three general categories of *in situ* treatment zones, regardless of whether they are positioned as curtains or in gates: (1) physical systems (e.g., air sparging); (2) chemical (abiotic) systems; and (3) microbial systems (microbial filters). There are various types of air sparge systems for containing or cutting off plumes, three of which involve pumping air into or through a line of wells positioned in a row across the plume. In the other air sparge system, air is sparged upward through the plume as it passes through gates in a funnel-and-gate system (Pankow et al., 1993). In this type of funnel-and-gate, the gate is an open water column with no porous medium packed in it. The injected air strips volatile organic contaminants from the open water column in the gate. In each of the air sparge systems, air is used to strip volatile contaminants from the groundwater *in situ*. The contaminants are swept to the vadose zone, where they are degraded, or to ground surface, where they are released to the atmosphere, or collected and treated. In direct contrast to air sparge systems, which generally bring the contaminants to land surface, the chemical or microbial treatment zones are intended to destroy the contaminants in the treatment zone as the plume passes through it.

Source Zone Containment

Source zone containment may be accomplished by: (1) conventional pump-and-treat; (2) cutoff wall enclosure with or without interior pump-and-treat; (3) *in situ* treatment curtain or funnel-and-gate system immediately downgradient of the source zone; or (4) cutoff wall enclosure with *in situ* treatment gates as outlets from the enclosure (Figure 5.8). The first two approaches (Figures 5.8a and 5.8b) indicated above are proven technologies for many sites, whereas the others have components that are experimental. When conventional pump-and-treat is used for source zone containment, the pumping system need only capture the contaminant mass flux emanating from the source. It is not

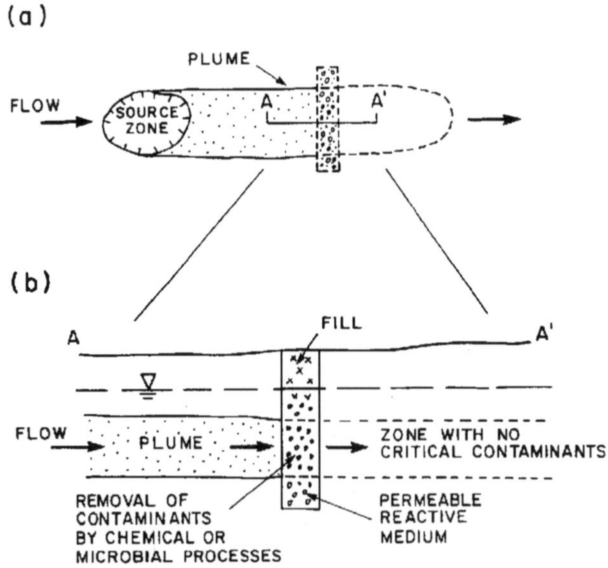

Figure 5.7. An *in situ* treatment curtain for plume control: (a) plan view; (b) vertical cross-section, both showing plume entering the treatment curtain and treated plume water exiting downgradient side.

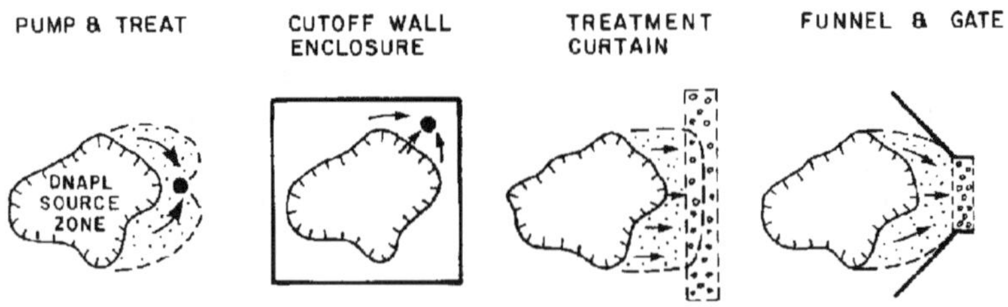

Figure 5.8. Options for source zone containment: (a) pump-and-treat; (b) cutoff wall with interior pumping; (c) *in situ* treatment curtain; and (d) funnel-and-gate. Solid dots are extraction wells, thick lines are impermeable walls, and areas with small circles are permeable media.

necessary to remove mass at a rate that depletes the source mass at a significantly higher rate than what is needed for source zone control. For source zone containment, as is also the case for plume front containment, the required mass removal rate needed may be relatively small. In some circumstances, hydrogeologic complexities and contaminant mass distributions are such that the only viable approach for source zone containment is pump-and-treat (Figure 5.9b). In other situations, cutoff wall enclosures are a proven technology. The choice can depend on total cost, cash flow factors, and future remedial intentions. Pump-and-treat involves lower capital cost (design and construction of the water treatment facility) and higher annual operating and maintenance costs than cutoff wall enclosures. Cutoff wall enclosures will have a higher capital cost but may provide the opportunity for reduced operating costs resulting from lower groundwater pumping rates. The construction of a cutoff wall enclosure may also facilitate further application of *in situ* remedial technologies, such as surfactant flushing.

Most DNAPLs are much more dense than water and, at many DNAPL sites, the bottom of the immiscible source mass is much deeper than non-DNAPL chemical substances would have penetrated. At some sites, the bottom of the immiscible source mass is hundreds of meters deep—in some cases deep into bedrock. Conventional cutoff walls generally cannot be installed to such depths. At many sites, the bottom of the DNAPL has not yet been determined. In some cases, shallower cutoff wall enclosures can be used to isolate the upper part of the source mass with minimal pump-and-treat, leaving the deeper part to be contained by pump-and-treat (Figure 5.9c).

Cutoff wall enclosures can be installed in fractured bedrock at most sites and to greatest depth by injection grouting. This technique involves injection of sealant (i.e., grout) into closely spaced boreholes to achieve sealing of fractures. The technique has been used for many decades in the geotechnical industry to reduce groundwater seepage at dam sites and other large-project construction sites. There is no doubt that considerable reduction in bulk hydraulic conductivity is achievable in nearly all types of fractured rock, but often at very high cost. The degree of hydraulic-conductivity reduction is generally much more difficult to predict and verify than is the case for conventional cutoff wall enclosures in overburden or at shallow depth in soft bedrock. As is the case for cutoff wall enclosures in overburden, decisions regarding selection of cutoff wall enclosures by borehole grouting should be founded on issues of cost and benefits.

There is no specific hydraulic conductivity that must be achieved in the cutoff wall enclosure in order to be effective. Lower bulk hydraulic conductivity of the wall provides lower leakage through the wall. Consequently a lower pumping rate inside the enclosure is required to maintain a hydraulic gradient directed inward across the wall. The inward gradient prevents outward lateral advection of dissolved contaminants. Lower pumping from the interior results in lower long-term cost of water treatment. The long-term cost of water treatment can be reduced further by capping or diverting surface runoff in the area inside the enclosure to minimize groundwater recharge. In some cases, much of the water entering the enclosure is upward seepage from below. Upward seepage at the bottom of the enclosure is generally desirable to prevent leakage of dissolved contaminants downward into deeper strata. However, in some cases, natural downward hydraulic gradients may be too large to allow gradient reversal to be imposed by pumping.

There are several technologies available for construction of low-permeability cutoff wall enclosures (described by Mitchell, this volume). These technologies are proven technologies in the sense that they can achieve specific, relatively low permeability values (less than 10^{-6} or 10^{-7} cm/s). There is very little published information pertaining to how long these barriers will maintain low permeability. There is a need for research in the longevity of low-permeability barriers. Many decades or centuries are generally expected. In the event that the permeability of the wall increases, the consequence would be an increase in the rate of pumping from the interior of the enclosure. If the increase in permeability of the wall is large, pump-and-treat inside the enclosure evolves towards the conventional pump-and-treat option without a barrier. The alternative would be to replace or repair the barrier. Repair of most vertical types of barriers is generally a much easier task than the initial construction of the barrier. At sites where no NAPL occurs, or where LNAPL but no DNAPL occurs, the depth to which the cutoff wall must be installed is generally not great. The hydrogeologic nature of the bottom of the subsurface source zone at these sites is less important than at DNAPL sites. For cutoff wall enclosures to be fully effective at DNAPL sites, it is generally necessary for the bottom of the wall to extend somewhat below the deepest zone of DNAPL. If the input of DNAPL to the groundwater zone has ceased prior to construction of the wall, it is expected that the DNAPL has already gone as deep as it will go under the existing conditions. It is important that the wall be installed and the hydraulics of the enclosure be operated in a manner such that DNAPL remobilization is avoided. Pumping within the enclosure that dewaters the DNAPL zones may cause further migration of the DNAPL downward (Figure 5.9c).

For source zone containment, an alternative to combining cutoff wall enclosures with interior pumping is to place *in situ* treatment gates in the cutoff wall and thereby avoid interior pumping. The water that enters the enclosure by infiltration of rain or snowmelt and by seepage into the enclosure has contaminants removed when it flows through the gates at the downgradient side of the enclosure. If the enclosure has a sufficiently low permeability, and if surface runoff from the source zone area is diverted, the flux through the gates will be small because groundwater recharge would be small. For most treatment processes within the gates, treatment efficiencies will be greater due to water flux rates through the gate. This fully passive approach to source zone containment is less suitable for sites where there is downward contaminant migration out of the bottom of the enclosure.

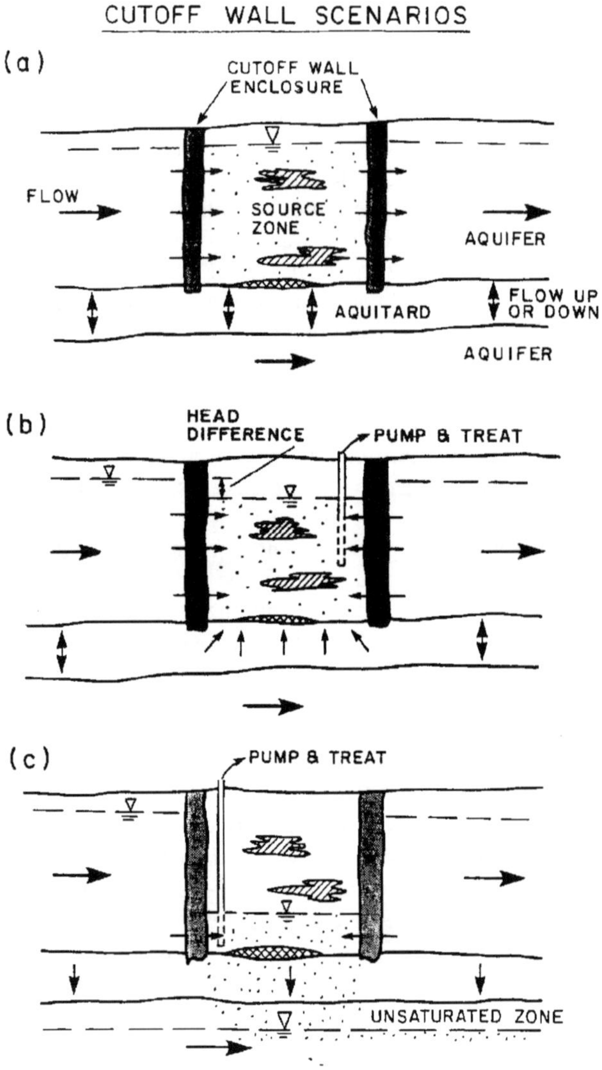

Figure 5.9. Cutoff wall enclosures around DNAPL source zones: (a) natural gradient case; (b) inward flow everywhere induced by pump-and-treat; (c) downward flow from bottom cannot be reversed by pump-and-treat.

Of the many experimental technologies being developed for groundwater remediation, those that will accomplish plume containment and source zone containment most passively and at lower cost than conventional pump-and-treat offer the best prospects for advancing in the next few years to the status of proven technologies. The technical challenge associated with restoration of source zones at DNAPL sites is much greater than that of source zone containment or plume capture. The technical limitations imposed by geologic complexity are severe for source zone restoration, but commonly much less severe for source zone containment or plume containment. Several experimental technologies now in the research stage, such as surfactant flush, alcohol flush, and chemical oxidant flush, are directed at removal of or *in situ* destruction of contaminant mass from source zones. A secondary advantage provided by cutoff wall enclosures is enhancement of the potential for more effective and environmentally safe application of the above-mentioned *in situ* flush technologies. The enclosure can effectively prevent lateral escape of chemical reagents or aqueous contaminants mobilized during immiscible phase removal. Also, the enclosure minimizes the mass of chemical reagent that must be added for effective enhancement of the mass removal or mass destruction processes. The source zone situated inside the enclosure can be manipulated in a matter analogous to a very large batch reactor rather than a groundwater flow system.

Plume Removal

Plumes can be removed using arrays of pumping wells positioned according to the well-established principles of aquifer hydraulics and solute transport. The initial mass of aqueous phase contamination in the plume can be removed by pumping but requires removal of much more groundwater than that equivalent to the volume of the plume. Additional plume volumes must be pumped to remove the mass of adsorbed phase contamination and the mass of dissolved and adsorbed phase stored in lower-permeability zones in the plume and in aquitards above or below the aquifer. Mathematical models can be used to determine optimal well positioning and pumping rates to remove the plume in the shortest possible time. However, predictions of the time required to achieve regulatory cleanup criteria are commonly unreliable due to processes which generally cannot be accounted for rigorously by the models (Figure 5.10), such as desorption effects, delayed yield from the capillary fringe, and diffusion out of low-permeability zones.

Possibilities exist for enhancing plume removal by combining pump-and-treat with aquifer heating, chemical flushing, or microbial processes. However, we can expect that plume removal is a task for which these new technologies may not provide much improvement in effectiveness or efficiency over conventional pump-and-treat. These enhancement processes for plume removal are intended to stimulate removal or destruction of the initially adsorbed portion of the plume mass. Efficiency of release of the adsorbed phase is expected to be inhibited at many sites by geologic complexity and slow diffusion from low-permeability zones.

Source Zone Restoration

Nearly all of the contaminant mass in the source zone is normally immiscible phase organic liquids relative to the other forms of contaminant mass: the dissolved and sorbed mass. If the entire plume mass, excluding the source mass, is removed, and if control of the source zone is not maintained thereafter, the plume will reestablish itself due to continued dissolution of immiscible phase into the groundwater. Conventional pump-and-treat is generally ineffective for removing much of the immiscible phase mass because the mass removal rate is insufficient due to dissolution and solubility constraints. Hydraulic systems such as pumping wells, and drainage trenches, with or without water injection, can in some situations remove considerable immiscible phase mass; however, most of the immiscible phase mass remains in the aquifer after these systems have reached their effectiveness limit. Thus, to accomplish source mass reduction beyond what can be accomplished by hydraulics alone, enhancement of the removal rate is required by means of experimental technologies such as chemically enhanced pump-and-treat, or other approaches such as steam flush or *in situ* chemical destruction of the NAPL.

Chemically enhanced pump-and-treat for removal of the immiscible phase involves flushing the DNAPL zone in the aquifer with water containing chemical additives or reagents (Figure 5.11). Research currently focuses on two categories of chemical additives: surfactants and alcohol. Surfactants added to water enhance the effective solubility of the DNAPL so that the contaminant mass removal rate by the surfactant-enhanced pump-and-treat is much greater than that of conventional pump-and-treat (Wunderlich et al., 1992). For cost-effectiveness, it is generally necessary to remove the contaminants from the mixture of water, surfactants, and contaminants in above-ground facilities so that the surfactant solution can be used for repeated aquifer flushes. The use of surfactants for removing DNAPL has recently progressed from laboratory studies to small-scale field prototype tests of flush systems.

For alcohol flushing to be effective for removal of immiscible phase DNAPL, flushing must be done with alcohol mixed with only a minor fraction of water. The cost of the alcohol is an important

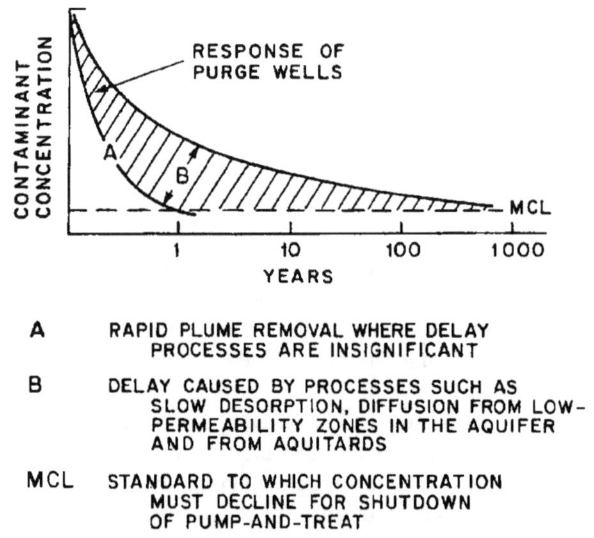

Figure 5.10. Influence of delay processes on time to completion of partial aquifer restoration by pump-and-treat.

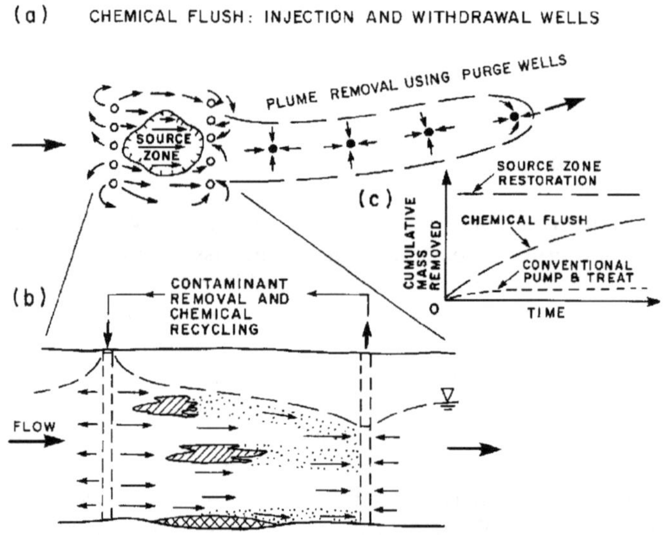

Figure 5.11. Restoration of DNAPL source zone using chemically enhanced flushing with injection with withdrawal wells: (a) plan, injection and withdrawal wells for chemical flush across source zone and purge wells to remove plume; and (b) section; and (c) comparison of mass removal with and without chemical enhancement.

factor, and therefore the efficiency of each flush and potential for reuse of the alcohol are critical factors. Alcohol flush for DNAPL removal is currently being assessed in laboratory studies and has not yet reached the field prototype stage.

Another technology for flushing DNAPL from aquifers relies on the use of steam. Steam injected into the source zone drives the contaminant in the immiscible phase as well as in the vapor phase to withdrawal wells. Udell and co-workers have advanced the understanding of the physics of steam flush, using theoretical and laboratory studies, as well as small-scale field trials. The effectiveness of steam flush for removing NAPL from below the water table is currently being assessed by field prototype trials.

The flush technologies described above have in common the process of moving the DNAPL mass from the aquifer via wells to above-ground facilities for treatment and disposal. An alternative approach is to flush the source zone with chemicals that destroy the DNAPL *in situ*. Schnarr and Farquhar (1992) conducted laboratory and prototype field trials at the Borden site in Canada of *in situ* chemical destruction of immiscible phase chlorinated solvents. Additional field trials are in progress. The destruction is accomplished using a strong chemical oxidant: potassium permanganate.

There are several factors that can severely limit the effectiveness of mass removal from source zones using surfactants, alcohol, steam, or chemical oxidants. First, the injectant must be very efficient for removal or destruction of the immiscible phase chemicals *when brought in contact with these chemicals*. Second, if pump-and-treat is used as part of the flush system, there must be effective facilities above ground for treating, recycling, or disposing of the injectant and reaction products. Third, the injectant chemicals and/or hazardous reaction products from the *in situ* processes must be prevented from escaping from the remediation zone into other parts of the aquifer. Finally, the injectant must be brought sufficiently in contact with the chemicals to enable the removal or destruction processes to be effective.

These latter two problems generally pose exceptional problems. At some sites, geologic complexity and the nature and amount of DNAPL make the task of complete source zone restoration impossible because the injected solution cannot be made to adequately contact the immiscible phase contaminant mass. For chemical flush or steam flush technologies to be effective, the spatial distribution of the immiscible phase DNAPL and the spatial distribution of hydraulic conductivity in the groundwater zone must be favorable for contact between the flush fluid and the immiscible phase. The contaminant mass removal rates or *in situ* destruction rates must be sufficient to produce source zone restoration in a practical time period, a period of months or years rather than decades. The spatial distributions of the immiscible phase and hydraulic conductivity derive their character primarily from geologic heterogeneity or complexity. Current capability to characterize DNAPL source zones adequately to allow prediction of the performance of technologies for *in situ* source removal or destruction is minimal. The problem of contact between the flush fluid and the immiscible phase mass is exacerbated by the fact that the DNAPL presence lowers considerably the hydraulic conductivity of the porous medium with respect to the aqueous solution in the portion of the aquifer where free-product DNAPL resides. A proposed advantage of the steam flush method is that, if pressures are released after initial heating, the NAPL caught in lower-permeability zones may be made to volatilize and advect into the more permeable strata (K. Udell, pers. comm.).

All technologies for *in situ* source removal or destruction of DNAPL have the potential to make the site conditions worse rather than better. A particularly adverse possibility is remobilization of DNAPLs, causing some of the DNAPL mass to sink deeper into the aquifer or to sink through an aquitard into an underlying aquifer not previously contaminated. Unfortunately, the state of knowledge of technologies for source mass removal is poor pertaining to their potential to cause DNAPL remobilization. Therefore, decisions to use *in situ* mass removal or destruction technologies in source zones should not be taken lightly. Poorly controlled experimentation at real sites can be hazardous to groundwater resources.

In the context of environmental safety, the ideal form of *in situ* source zone remediation is one that destroys the immiscible phase to an adequate degree, produces insignificant hazardous reaction products *in situ*, and causes no adverse mobilization of the immiscible phase. For chlorinated solvent DNAPLs, *in situ* destruction by chemical oxidation is the technology that currently comes closest to meeting these performance criteria. However, knowledge of the degree of effectiveness of this technology is currently limited to a few chlorinated solvents in relatively pure form. No trials at actual industrial sites have yet been undertaken.

In some types of porous fractured rock, such as sandstone and fractured clayey or silty overburden, all or much of the mass of immiscible phase organic liquid in the source zone may have transferred from the immiscible phase in the fractures to the aqueous and sorbed phase in the porous matrix over the years or decades since the immiscible mass entered the subsurface. Rather than a favorable change in state, this transfer has a major negative effect. The contaminants enter the matrix by molecular diffusion. The mass flux and time scales of the diffusion process back out of the porous matrix are such that no technology for source mass removal (other than excavation), whether it be conventional pump-and-treat or any of the chemically enhanced, heat-enhanced, or steam-enhanced experimental technologies when fully developed, can be expected to accomplish effective source mass removal except over extended periods of time (Parker et al., 1993).

DISCUSSION

Mass Removal

At DNAPL and LNAPL sites, nearly all of the contaminant mass occurs in the source zone as immiscible phase organic liquid (DNAPL and LNAPL). Therefore, when full aquifer restoration is the goal, essentially all of the immiscible phase mass must be removed. Conventional pump-and-treat accomplishes source mass removal too slowly, and therefore it is not a practical technology for source mass removal.

Some degree of mass removal is a consequence of most remedial measures. Containment of source zones or plumes does not require a specific degree of mass removal. Containment requires only that necessary changes in the groundwater flow pattern be imposed. However, removal of a source zone or a plume clearly requires a specific degree of mass removal. At DNAPL sites, it will seldom be possible to predict or even determine when a sufficient percentage of the total mass has been removed to achieve full source zone restoration. The cumulative mass removed by a remedial technology such as groundwater pump-and-treat or soil vapor extraction can generally be measured reliably. However, at most sites, the total initial mass in the plumes or the source zones is not known. Mass estimates for source zones, particularly the part of the source zone below the water table, commonly have uncertainties of one or more orders of magnitude. Consequently, it is not possible to reliably compare the mass removed to the mass remaining (Figure 5.12a), and therefore the percentage of mass removed cannot be determined.

For the most common type of DNAPL sites (chlorinated solvent sites), restoration of source zones requires reduction of bulk soil concentrations in the source zone from hundreds to thousands of parts per million to a part per million or lower. Restoration of the source zone to such low levels may often require ultimate mass removal effectiveness greater than 99.9%. This is a truly exceptional challenge for all *in situ* mass removal technologies. The best that the petroleum industry is able to do when applying advanced technologies for oil recovery is 40–50% of oil removal from the geologic stratum. The difficulty of this challenge generally has gone unrecognized. It may be viewed that any degree of mass removal is a beneficial step towards eventual achievement of source zone restoration. However, from a practical perspective, the benefits cannot be quantified unless the mass removed can be quantitatively related to the degree of restoration achieved, or unless the mass removed allows more cost-effective implementation or elimination of other remedial measures that might otherwise be required. Without such information, which is typically the case, the cost-benefit comparison for the mass removal action cannot be made.

There are currently great expectations in government and industry for technologies such as surfactant and steam flushing for restoration of DNAPL source zones. These technologies appear capable of removing DNAPL mass much more rapidly than conventional pump-and-treat, as illustrated in Figure 5.12b. The DNAPL mass removal graphs (Figures 5.12a and 5.12b) illustrate that source restoration is characterized by diminishing returns. As the restoration proceeds, the DNAPL mass remaining in the subsurface decreases; unfortunately, so does the efficiency of the extraction technology. Considering that the remaining DNAPL mass must be reduced to a small percentage of the original mass, the ultimate practicability of all enhanced removal technologies is in doubt and will remain in doubt for some time because the necessary research will be slow, difficult, and very expensive.

To select and compare rational options for remedial measures we must determine the appropriate conceptual model for the subsurface contamination given the type of chemicals, the nature of chemical release to the subsurface, and the hydrogeological setting. In simplest terms this would be the

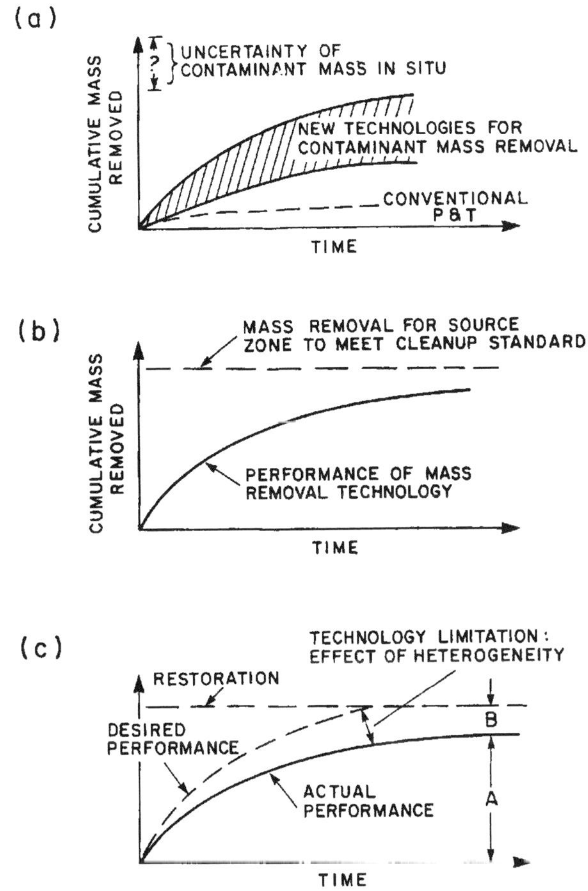

Figure 5.12. Schematic representation of cumulative mass removal for restoration of DNAPL source zones: (a) total mass in source zone is uncertain; (b) technology performance graph contaminant mass in source zone assumed known; and (c) discrepancy between desired performance and actual performance caused by heterogeneity, etc.

determination of whether the site has or likely has immiscible phase organic liquids (LNAPL or DNAPL) in subsurface source zones and the approximate spatial extent and character of such zones. Also, we must acquire sufficient site-specific information on the subsurface conditions and distribution of contamination to define or estimate the lateral and vertical extent of the migrating aqueous phase plumes. The two tasks indicated above need to be accomplished only to the level of detail appropriate for consideration of the performance of these technologies that appear, based on general knowledge of site use and conditions, to have potential applicability. It should be clearly recognized which technologies are considered proven technologies and which are experimental technologies in the site-specific context. A proven technology for one type of site may be unproven or experimental for other types of sites.

Experience with groundwater pump-and-treat for common industrial organic contaminants has shown that restoration of plumes (but not source zones) in sandy aquifers to achieve drinking water standards is possible in a practical time frame at some sites, but only after the source zones are adequately identified and contained. Full restoration of significant subsurface source zones has not yet been demonstrated. No technology or group of technologies for *in situ* removal or *in situ* destruction of chemical mass has yet advanced to the stage where it is possible to predict when, or even if, sufficient restoration of DNAPL source zones can be achieved.

Issues of Time and Benefits

Until source zone restoration is achieved, it will be necessary to maintain a source zone control system such as pump-and-treat or cutoff wall enclosures to prevent plume growth. With no special source mass removal activity, the pump-and-treat system at a typical DNAPL site may need to function for centuries or longer before the source mass will be dissipated sufficiently to meet an aquifer-restoration condition. When source mass removal is pursued by active engineering means such as chemical flush or steam flush or *in situ* destruction (i.e., enhanced source mass removal), the number of years that might be required to achieve restoration is reduced. Thus, if enhanced source mass removal achieves only partial source mass removal, some benefit is achieved in terms of the duration of pump-and-treat. However, this benefit is hypothetical only because its magnitude cannot be predicted, and it accrues at some undefinable future time. This hypothetical relationship of source mass removal to time of achievement of full aquifer restoration is displayed schematically in Figure 5.13. For example, with no enhanced source mass removal, the length of time that pump-and-treat would be necessary to achieve full aquifer restoration might be 1,000 years. If enhanced source mass removal removes half of the source mass and is then discontinued, the length of time necessary for pump-and-treat would be some fraction of 1,000 years, perhaps 500 years. A direct proportion assumption on time is optimistic; a more plausible expectation is a much more adverse time relationship. The key issue pertaining to partial source mass removal centers on the value of the ultimate benefits achieved in the distant future by such removal relative to the cost of the removal. If the benefit does not accrue for decades, then it has very little present worth, and the source removal may not be justified on solely economic grounds.

A major tenet of modern environmental ethics is the desirability of choosing courses of action that will lead most rapidly and/or assuredly to a sustainable environmental and economic condition. In a groundwater context, a sustainable environmental condition may be considered to be a condition where groundwater resources are not significantly diminished in quality due to continuing growth of contaminant plumes. Across North America, the ever-increasing number of pump-and-treat systems for source zone control represents large-scale action for risk control and groundwater protection that cannot be considered sustainable in the long term because pump-and-treat is an excessively active (and expensive) technology. Pump-and-treat requires continuing commitment by future generations of their societal resources to maintain the pump-and-treat facilities and for performance monitoring and energy use. Given the ephemeral nature of governmental policies and the transient existence of some of the corporations that carry some of the cost of the "in perpetuity" pump-and-treat burden, it is likely unreasonable to assume that the pump-and-treat facilities put into operation in this century will continue to be maintained properly or even maintained at all through the next century. Similarly, it may be unreasonable to expect that sufficient societal resources will be available to evaluate ever-increasing reams of monitoring data to determine if corrections and adjustments in the pump-and-treat systems are required.

Full aquifer restoration is generally not achievable at DNAPL sites and, therefore, unavoidably, our society must transfer responsibilities and risks associated with these sites to the future. Given this transferred burden, we should ask what more positive legacy we may be able to offer to future generations. Our best answer is that technologies for *in situ* aquifer remediation will be improved considerably in the next few decades but probably not in the next few years. Therefore, one of the goals of current aquifer remediation should be to pass to future generations site conditions that are better suited for application of new technologies than the present condition. Conventional pump-and-treat systems will likely contribute little toward this goal, particularly if their performance and effects are monitored as casually as is typically the case at present.

Conceptually, an array of remediation options are available for application to DNAPL sites (Figure 5.14), ranging from the most "active" option, which is aggressive, large-scale pump-and-treat, to the most passive, which has cutoff wall enclosures and *in situ* treatment systems without pump-and-treat. *In situ* treatment curtains and funnels-and-gates are new technologies still in the prototype field trial stage. However, we expect that these technologies will advance rapidly in the next few years. The *in situ* contaminant mass flux that these technologies must accommodate in most plumes is not

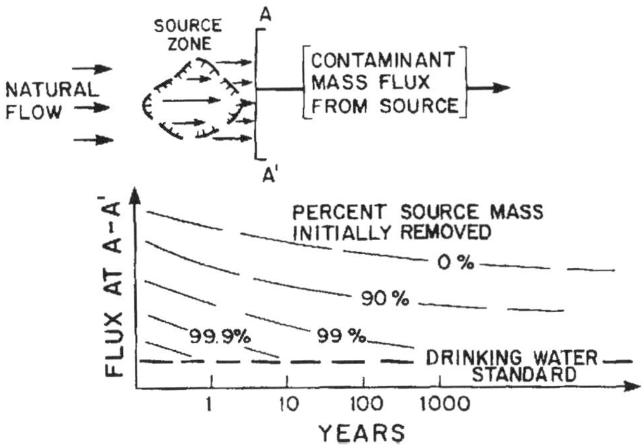

Figure 5.13. Relationship of percent contaminant mass removed from source zone during period of source zone restoration to time necessary to achieve full restoration of source zone by conventional pump-and-treat.

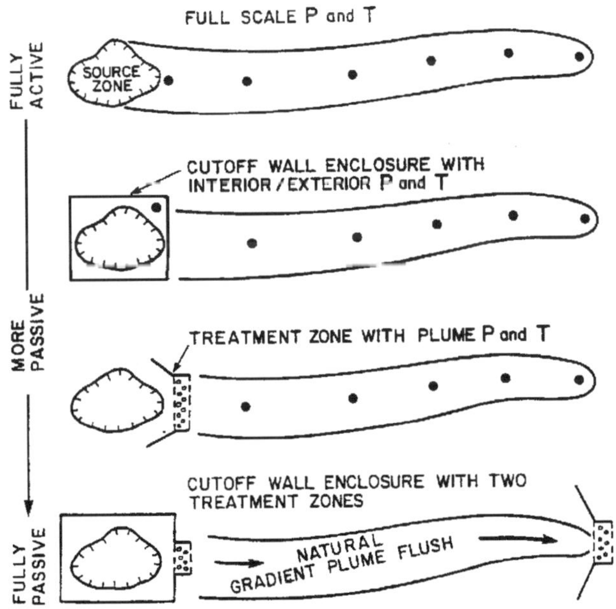

Figure 5.14. Variety of options for full aquifer restoration, ranging from fully active technology at top to fully passive at bottom.

large and therefore the technical challenge is not particularly difficult. However, even when these new technologies are fully developed, hydrogeologic conditions at many sites, particularly fractured rock sites, are such that they will not be suitable, leaving pump-and-treat as the only technically and economically viable option for groundwater remediation.

The Cap Dilemma

A common dilemma in decision making for remediation of waste disposal sites is whether or not to put a cap on the ground surface over the source zone. The construction of highly engineered low-permeability surface covers (i.e., caps) is commonly undertaken at waste disposal sites. In many cases, much of the expense of the cap derives from a specification that the cap prevent infiltration of rainwater and snowmelt through the waste. Prevention of infiltration through the waste is intended to prevent groundwater contamination. This intention makes sense if the site condition fits the conceptualization shown in Figure 5.1a where the entire source mass is situated above the water table. However, if the site is a DNAPL site (Figure 5.1c), allocation of large financial resources to construct a cap to achieve low infiltration to the water table is typically unwarranted on scientific grounds. At sites where appreciable DNAPL exists below the water table, it is generally not possible to achieve any significant reduction in plume growth or plume control by means of a cap. At DNAPL sites, the contribution of water to the plume from infiltration through the waste to the water table is typically very small in comparison to lateral groundwater flow through the DNAPL zone (Figure 5.15).

At sites where chlorinated solvents exist above the water table, even if only in small amounts, it is reasonable to expect that plumes of solvent vapor will occupy the vadose zone beneath the site and will move by diffusion away from the site. Solvent vapor in the vadose zone causes groundwater contamination by transfer of some vapor mass to the groundwater zone. Mendoza and McAlary (1989) and Hughes et al. (1993) describe solvent vapor migration in the vadose zone and its impact on groundwater. Placement of a low-permeability cap over the land area where immiscible phase solvents occur in the vadose zone does not necessarily diminish impacts of the solvents on groundwater. On the contrary, in many cases it is likely that the cap increases the tendency of vapor to contaminate groundwater by causing greater lateral spread of the vapor beneath and beyond the cap (Figure 5.16). This spread can be prevented by vacuum extraction of vapor; however, this imposes a need for long-term active engineering because full cleanup of vadose zone DNAPL by vacuum extraction typically takes decades. Another alternative is to connect the cap to vertical low-permeability walls (i.e., enclosures) that extend below the water table, with active or passive vapor venting.

In circumstances where source zone containment is implemented, the principal objective of a cap may be to prevent direct exposure to the contaminated soil and to resist erosion and weathering. These objectives may result in a cap design that is considerably different, and perhaps less costly, than one intended to reduce infiltration.

The main issue addressed here is that a remedial action such as a cap, which is a beneficial action at sites that have no solvent DNAPL, can be a beneficial option at solvent DNAPL sites unless the cap design is adapted to the special demands of the DNAPL problem. We expect that a detailed post mortem of the many caps that have been constructed at Superfund and RCRA sites would show, because of DNAPL issues, that little or no benefit was achieved by the groundwater environment, contrary to the expectation at the time that the cap was built. In fact, many caps have probably been detrimental to groundwater remediation.

In Situ Bioremediation

Of the various types of experimental technologies for *in situ* remediation at industrial sites, the one that has had the most research is *in situ* bioremediation. Unfortunately, the prognosis for most types of *in situ* bioremediation at DNAPL sites is not good.

Nearly all DNAPLs in the immiscible phase are toxic to microbes. Also, the groundwater in and near DNAPL source zones commonly has sufficiently high aqueous concentrations leached from the DNAPL to be toxic to microbes. Therefore, *in situ* microbial technologies are particularly ill-suited for removal of DNAPL mass from source zone. There is some hope for *in situ* microbial technologies to assist in the restoration of source zones at some stage after chemical or steam flush technologies or after *in situ* chemical destruction technologies have removed nearly all of the immiscible phase mass from the source zone. This assumes, however, that the primary technologies do not create a biological or geochemical environment that is inhospitable to *in situ* bioremediation.

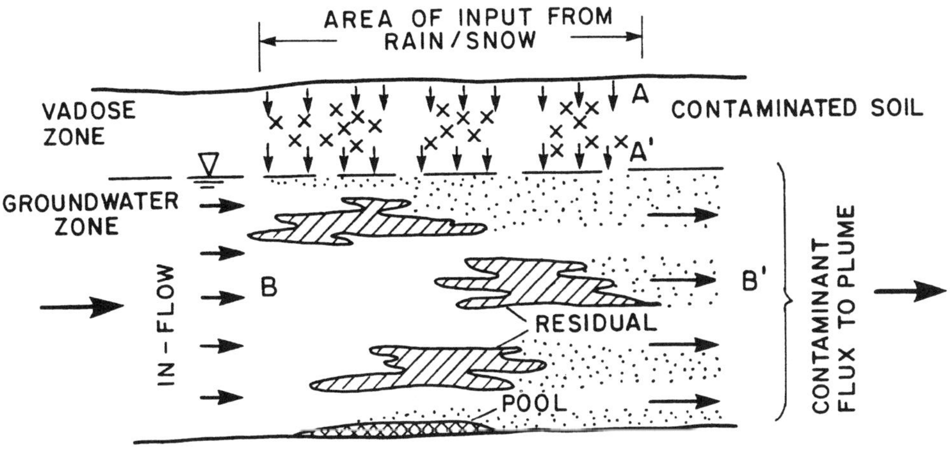

Figure 5.15. Contaminant contribution to groundwater plume from: A-A′, zone of leaching of contaminants from vadose zone to groundwater zone; and B-B′, zone of volatile contaminant from dissolution of DNAPL residual and pool causing main contaminant flux to plume. A′ flux is typically small relative to B′ flux.

The potential for *in situ* microbial methods to be useful is greatest in the plume at some distance from the DNAPL source zone where the aqueous phase contaminant concentrations are lower than levels toxic to microbes. However, the most common approaches suggested for *in situ* bioremediation of plumes involve injection of nutrients and electron acceptors (i.e., oxygen) into the plume to stimulate activity of natural microbes for destruction of contaminants that desorb from the porous medium. This approach suffers from various limitations related to delivery of the injectants to all zones where contamination is present. These limitations typically derive from geologic complexity or heterogeneity. An alternative approach is to stimulate microbial activity for *in situ* contaminant destruction in a permeable microbial curtain or in funnels-and-gates placed across the plume. In this *in situ* use of bacteria, the contaminants are brought to the microbial zone by natural groundwater flow, and therefore much of the injectant delivery problem is avoided. However, research on this type of *in situ* bioremediation is in its infancy. The main challenge is to develop relatively passive life support systems for the *in situ* microbial processes.

CONCLUSION

To date, when remedial measures are implemented to achieve source containment and/or plume containment at DNAPL sites, the measures nearly always include pump-and-treat. In most cases, these pump-and-treat systems will require operation "in perpetuity" because of our inability to achieve adequate removal or restoration of DNAPL source zones. This approach does not meet reasonable criteria for long-term environmental sustainability because pump-and-treat is an excessively active technology. A more passive approach for source zone containment would be the use of low-permeability cutoff wall enclosures, coupled with minimal pump-and-treat or, in some cases, no pump-and-treat. This approach involves higher up-front capital expenditures for construction but lower average annual long-term operations and maintenance costs. Since low-permeability cutoff wall enclosures as well as pump-and-treat systems are proven technologies for source zone containment, the choice is between one or the other of these two proven technologies. Issues of long-term cost and environmental sustainability should govern the technology choice. Cutoff wall enclosures may facilitate future use of promising *in situ* technologies for source mass removal that are more difficult or risky to use without such enclosures. There are many sites where significant mass of DNAPL exists too deep for use of enclosure or in fractured bedrock. Cutoff wall technologies are much less well developed for fractured rock than overburden, particularly hard rock. In hard rock such as granite, cutoff walls in the conventional sense cannot be constructed. Other types of walls made by injecting sealants into

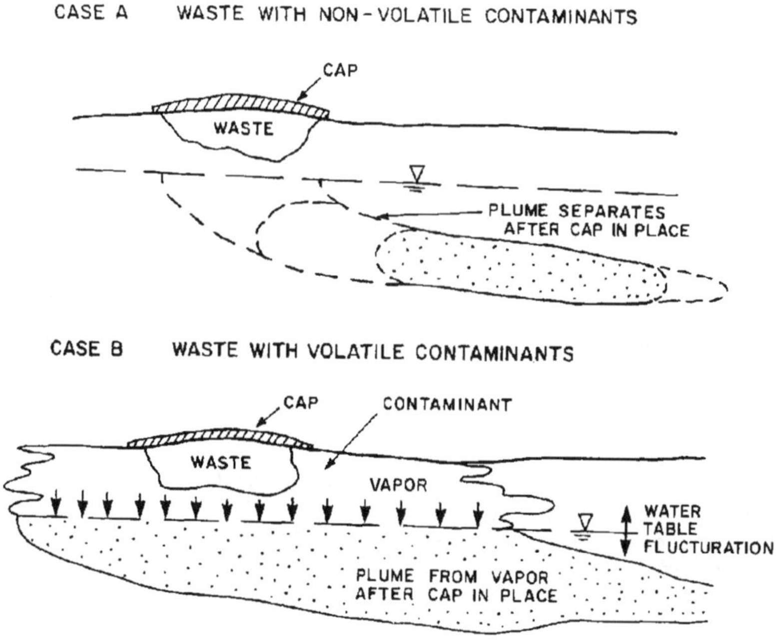

Figure 5.16. Comparison of the effect of caps on two source zones: (a) source materials exist only above the water table and emit no contaminant vapor; (b) volatile source materials exist above the water table, causing contaminant vapor to contaminate groundwater.

fractures via boreholes can be constructed, but reliable prediction of the performance prior to construction is generally not possible. In this situation, pump-and-treat is usually the only proven technology available for source zone or plume containment.

Apparent failures in the arena of subsurface restoration abound because of the lack of clear recognition of the difference between: (1) plume containment, (2) plume removal, and (3) full aquifer restoration. Many attempts at aquifer restoration have been made or are in progress. Remediation aimed directly at the goal of full aquifer restoration in this decade or even in the next decade or two should be recognized clearly as experimentation rather than actual remediation. In the United States, the Superfund process is intended to be primarily an aquifer restoration program, but it cannot perform as such because of the nature of the subsurface contamination at DNAPL sites and the inherent limitations of existing remedial technologies relative to what must be accomplished to achieve full aquifer restoration. In a groundwater context, most remedial efforts of the Superfund program may be regarded as numerous and costly experiments; unfortunately, in scientific terms, they are relatively unmonitored experiments with poorly defined hypotheses. In retrospect, the progress of these remedial efforts is best described as "trial and error." Opportunities to acquire information that would ultimately provide better remedial choices and designs and commensurate cost savings are being lost on a grand scale.

To derive benefits from careful site monitoring and experimentation, the experimentation must be designed specifically to track and assess in detail the progress of the restoration or plume containment attempts and to develop general scientific and engineering knowledge on the remediation technologies. This is rarely done. Rather, in the United States and Canada monitoring of remedial activities is generally implemented simply to satisfy basic regulatory requirements. Even the data acquired from such basic monitoring are rarely studied to derive guidance that will improve future efforts.

The acquisition of carefully obtained scientific and engineering knowledge from large-scale site experimentation will allow better design and implementation of remedial actions at other sites. However, the benefits may accrue mainly to future activities at other sites and to future generations of cleanup. Eventually, this class of large-scale experimentation is likely to pave the way to an era of

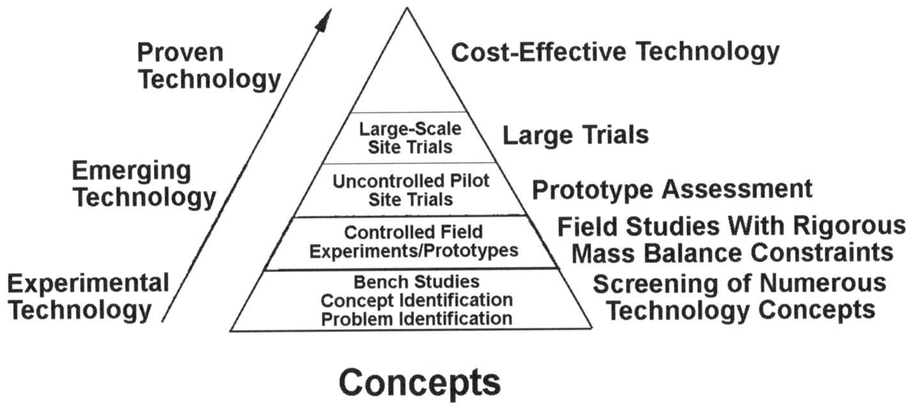

Figure 5.17. Stages in the development of groundwater remediation technology from concept to proven technology.

practical groundwater restoration where the stated goals are, in fact, achievable in a practical time scale.

Large-scale experimentation with technologies for subsurface source removal or aquifer restoration should normally be preceded by appropriate bench-scale studies, prototype technology trials, and then pilot scale technology assessments (Figure 5.17) in a variety of hydrogeological settings. In the rush towards aquifer restoration that is driven by public opinion and legislation, careful completion and critical review of each of these states in technology development is rarely accomplished, and hence large-scale experimentation is hampered by a weak base of fundamental knowledge even when large-scale field experimentation is accompanied by detailed monitoring. Rigorous methods or protocols for technology trials and pilot-scale trials are not well developed because of underemphasis of these essential steps in technology development.

It is truly unfortunate that, for most sites, today's technologies are inadequate for the task of source mass removal, and hence inadequate for full aquifer restoration. The current use of pump-and-treat or alternative technologies having the same intention, which can perform well for plume or source zone containment but not for source removal, must therefore continue until such time as more cost-effective mass removal and containment technologies are developed, or until source zone restoration through mass removal can be accomplished. The development of cost-effective source removal technologies will be slow, arduous, and expensive. Thus, we are transferring into the next century the task of operating and maintaining numerous active containment systems in order to prevent further growth or spread of contaminants from subsurface source zones. Although the transfer of this responsibility is presently unavoidable, today we have an opportunity to enhance possibilities for future remedial efforts by a systematic advance of scientific and engineering knowledge of the processes and performance of remediation technologies.

ACKNOWLEDGMENT

This chapter is a contribution from the University Solvents-In-Groundwater Research Program, Phase II, supported by funds from the Natural Sciences and Engineering Research Council (NSERC), the Ontario University Research Incentive Fund (URIF), and from the following corporations: Boeing, Ciba-Geigy, General Electric, Eastman Kodak, and Laidlaw Environmental Services.

REFERENCES

Burris, D.R. and J.A. Cherry. 1992. Emerging plume management technologies: *In situ* treatment zones. Paper presented at Subsurface Restoration Conference, 3rd International Conference on Ground Water Quality Research, Dallas, TX.

Cherry, J.A. 1984. Contaminant migration in groundwater: Processes and problems. *In* Proceedings of the 2nd National Water Conference—The Fate of Toxics in Surface and Groundwater.

Feenstra, S. 1984. Groundwater contamination from dense non-aqueous phase liquid (DNAPL) chemicals. Paper presented at the Geological Association of Canada Annual Meeting, London, Ontario.

Feenstra, S. and J.A. Cherry. 1988. Subsurface contamination by dense non-aqueous phase liquid (DNAPL) chemicals. Presented: International Groundwater Symposium, International Association of Hydrogeologists, Halifax, Nova Scotia.

Gillham, R.W. and D.R. Burris. 1992. *In situ* treatment walls—Chemical dehalogenation, denitrification, and bioaugmentation. Presented at Subsurface Restoration Conference, Dallas, TX.

Hughes, B.M., R.W. Gillham, and C.A. Mendoza. 1993. Transport of trichloroethylene vapours in the unsaturated zone: A field experiment. IAH Conference on Subsurface Contamination by Immiscible Fluids. Calgary, Alberta.

Mackay, D.M., P.V. Roberts, and J.A. Cherry. 1985. Transport of organic contaminants in groundwater. *Environ. Sci. Technol.,* **19**(5):384–392.

Mackay, D.M. and J.A. Cherry. 1989. Limitations of pump-and-treat remediation. *Environ. Sci. Technol.,* **23**(6):630–637.

Mendoza, C.A. and T.A. McAlary. 1989. Modelling of groundwater contamination caused by organic solvent vapours. *Ground Water,* **28**(2):199–206.

Pankow, J.F., R.L. Johnson, and J.A. Cherry. 1993. Air sparging in gate walls in cutoff walls and trenches for control of plumes of volatile organic compounds (VOCs). *Ground Water,* **31**(4).

Parker, B.L., R.W. Gillham, and J.A. Cherry. 1993. Diffusive disappearance of immiscible phase organic liquids in fractured geologic media. *Ground Water* (submitted).

Schnarr, M.J. and G.J. Farquhar. 1992. An *in situ* oxidation technique to destroy residual DNAPL from soil. Paper presented at Subsurface Restoration Conference, 3rd International Conference on Ground Water Quality Research, Dallas, TX.

Schwille, F. 1984. Volatile chlorinated hydrocarbons in porous and fractured media. Spec. Contrib. Germany Water Studies Annual Report No. 46, Federal Institute for Water Studies, Koblenz, Germany.

Schwille, F. 1981. Groundwater pollution in porous media by fluids immiscible in water. *The Science of the Total Environment,* **21**:173–185.

Starr, R.C. and J.A. Cherry. 1993. *In situ* remediation of contaminated groundwater: The Funnel-and-gate system. *Ground Water* (submitted).

U.S. EPA. 1992. Dense nonaqueous phase liquids: A workshop summary. Dallas, TX.

U.S. EPA. 1988. Guidance on remedial actions for contaminated ground water at Superfund sites. EPA/540/-88-003.

Villaume, J.F. 1983. Recovery of coal gasification wastes: An innovative approach. Proceedings, 3rd National Symposium on Aquifer Restoration and Ground Water Monitoring, National Water Well Association, Worthington, OH.

Wunderlich, R.W., J.C. Fountain, and R.E. Jackson. 1992. *In situ* remediation of aquifers contaminated with dense non-aqueous phase liquids by chemically enhanced solubilization. *J. Soil Contam.* **1**(4):361–378.

Chapter 6

Microbial Processes in the Subsurface

J.M. Thomas, J.T. Wilson, and C.H. Ward

ABSTRACT

Microorganisms have been detected at depths of 500 to 600 m below the surface. Many subsurface microorganisms are metabolically diverse, indicated by their ability to degrade many naturally occurring and synthetic organic compounds. Although many organic pollutants are known to biodegrade faster under aerobic than anaerobic conditions, anaerobic processes may be significant considering the slow movement of groundwater. Microbial degradation of pollutants using indigenous electron donors, electron acceptors and nutrients is defined variously as natural bioremediation or natural bioattenuation. When rates of natural bioremediation are too slow, limiting nutrients and electron donors or acceptors may be added to enhance biodegradation.

INTRODUCTION

The biosphere below the rhizosphere is inhabited predominantly by bacteria, although protozoa, fungi, and algae have been observed (Ghiorse and Wilson, 1988; Thomas and Ward, 1992). Microorganisms have been detected unequivocally as deep as 500 to 600 m and are most likely present at greater depths. Contrary to popular thought that microbial numbers decline with depth, total counts range from about 10^6 to 10^7 cells/g dry subsurface material (grams dry weight, or GDW). Although total counts do not vary appreciably with depth, counts are lower than those observed in surface soils, which range from 10^8 to 10^{10} cells/GDW. Variations in total cell counts are observed between strata. Sandy transmissive sediments contain the highest numbers and clayey nontransmissive sediments contain the lowest numbers. Viable counts are often one to two orders of magnitude lower but may be as high as the direct count. As with direct counts, higher viable counts are associated with sandy transmissive sediments rather than in those high in clay. Most likely, the low number of microorganisms associated with clays is related to the exclusion of cells from the small pore throats, and the low pH and water potential associated with these materials. Even though microbial populations are generally lower in the subsurface compared to the surface environment, the subsurface microflora constitutes a substantial biofilter at depths of 500 to 600 m, and possibly even deeper.

Several studies have indicated that microbial distribution, diversity and activity are related to the characteristics of the subsurface strata. In general, gram positive organisms are associated with clayey sediments whereas gram negative organisms are associated with sandy transmissive sediments (Kobel-Boelke et al., 1988; Bone and Balkwill, 1988; Sinclair and Ghiorse, 1989; Balkwill, 1989). Sinclair and Ghiorse (1989) speculate that the thick, highly anionic walls of gram positive cells are a fitness trait which allows survival in clayey materials where low pH, osmotic stress and metals can create a hostile environment. Similarly, the lipopolysaccharide cell envelope of gram negative bacteria is thought to be a fitness trait for survival in dilute aqueous environments. Diversity does not decline with depth; however, it does vary between geological formations (Balkwill, 1989). Preliminary investigations of samples from some deep subsurface materials (200 to 300 m) suggest that these aquifers are aerobic and that the microflora is predominantly oxidative rather than fermentative (Balkwill, 1989; Fredrickson et al., 1989; Madsen and Bollag, 1989).

The subsurface microflora is metabolically diverse, indicated by degradation of many naturally occurring and industrial-related organic compounds (Hicks and Fredrickson, 1989; Fredrickson et al., 1991). In addition, most microorganisms in uncontaminated subsurface materials grow best under low nutrient conditions (Balkwill, 1989), representing the oligotrophic nature of the pristine subsurface; but some grow under high nutrient conditions as well. These findings confirm the notion of oligotrophy in the subsurface since oligotrophic organisms are known to have low substrate specificity and high substrate affinity (Poindexter, 1981).

EFFECT OF CONTAMINATION ON SUBSURFACE MICROBIAL ECOLOGY

Several studies have been conducted to determine the effect of contamination on the microbial ecology and biodegradation potential in the subsurface (Table 6.1). The results of these studies suggest that the microbial ecology and activity in uncontaminated and contaminated samples from the same site are very different. In most samples, a stimulatory effect is observed provided that the contaminant concentration is not toxic. Direct counts, which indicate the number of active and inactive microorganisms, may not be substantially different in uncontaminated and contaminated samples; however, viable counts are usually much higher in contaminated material. When microorganisms are stimulated to grow as a result of a contamination event, the number of protozoa in these materials also increases. Smith et al. (1986) reported that gram positive organisms are associated with uncontaminated material and gram negative organisms are associated with contaminated material. In general, however, gram positive organisms are associated with materials high in clay and gram negative organisms are associated with sandy materials.

BIODEGRADATION OF ORGANIC COMPOUNDS

Subsurface microorganisms are able to degrade a variety of organic compounds that are classified as groundwater pollutants (Lee et al., 1987; Thomas and Ward, 1992). Compounds that have been shown to biodegrade in the subsurface belong to several classes of chemicals and include hydrocarbons, phenols, polar solvents, and the chlorinated aliphatic solvents. Most of the laboratory studies focused on biodegradation of these compounds have used samples of sediments collected from the shallow subsurface—3 to 6 m below the surface. However, some studies have investigated the potential for biodegradation in sediments collected from deeper regions (Hicks and Fredrickson, 1989; Fredrickson et al., 1990; Fredrickson et al., 1991). *In situ* biodegradation of groundwater pollutants using indigenous nutrients and electron acceptors is called natural bioremediation. As a result of natural bioremediation, many contamination events may go unnoticed (Thomas et al., 1992). However, when the rate of natural bioremediation is not fast enough to prevent migration of the contaminants and human health and the environment are at risk, remedial action must be taken. One of those options is enhanced bioremediation, which involves increasing the rate of natural bioremediation by adding and increasing the flux of essential nutrients and electron donors or acceptors.

NATURAL BIOREMEDIATION

There are three patterns of natural bioremediation of organic contaminants in the subsurface (U.S. EPA, 1993) (Table 6.2). The first pattern occurs during aerobic respiration or denitrification, where microorganisms rapidly adapt to contaminant degradation and reach high densities. Zero-order (Barker et al., 1987) and first-order (Rifai et al., 1988; Boggs et al., 1993) kinetics for this pattern have been reported. Barker et al. (1987) suggested that a variety of kinetics may be operational in an aquifer. The variety of kinetics results from the heterogeneous distribution of dissolved oxygen and an oxygen demand that exceeds the oxygen supply.

The natural bioremediation that occurred at an abandoned wood-preserving site at Conroe, Texas was observed to follow pattern 1 (Wilson et al., 1985). Several organic compounds found in creosote were tested for their biodegradability in samples of uncontaminated and contaminated material from the site under aerobic conditions; biodegradation occurred in the contaminated samples only, indicating that adaptation had occurred after exposure to the contaminants. In the field, however, a plume of contamination persisted in anaerobic groundwater. The authors concluded that the rate of biodegradation in the field was controlled by the rate at which oxygen mixed with the plume. These findings

Table 6.1. Microbial Ecology of Uncontaminated and Contaminated Samples from the Same Site (Thomas and Ward, 1992, copyright permission granted).

Spill	Variable	Uncontaminated	Contaminated	Reference
Creosote[a]	Predominant cell type	Gm+	Gm–	Smith et al., 1986
	Biomass[b] (nmol/GDW[c])	0.6(0.5)	1.0(0.8)–1.9(0.5)[d]	Smith et al., 1986
	Direct counts (10^6/GDW)	1.2(1.2)–16(8)	8(2)	Webster et al., 1985
		1.4(0.6)–2.3(0.9)	1.3(0.5)–2.4(0.8)	Thomas et al., 1989
			0.9(0.4)–2.1(0.8)	
	Viable counts[e] (10^6/GDW)	<10^2–0.1(0.02)	<10^2–2.9(0.2)	Thomas et al., 1989
	ATP (ng/GDW)	0.017	0.009	Webster et al., 1985
Coal tar	Direct counts (10^6/GDW)	1–100	1–700	Madsen et al., 1991
	Viable counts (10^2/GDW)	1–100	1–10,000	Madsen et al., 1991
	Protozoa (no./GDW)	<50	<50–19,000	Madsen et al., 1991
	Fungi (no./GDW)	10	10	Madsen et al., 1991
	Actinomycetes	10–1000	10–10,000	Madsen et al., 1991
Unleaded gasoline	Viable counts[e] (10^6/GDW)	1.6(0.1)–1.8(0.5)	0.3(0.03)–66(9.2)	Thomas et al., 1990
Jet fuel	Direct counts (10^6/mL)	0.001–0.02(0.008)	0.7(0.1)–13	Ehrlich et al., 1985
	Viable counts[e] (10^3/mL)	<1	62–1.400	Ehrlich et al., 1985
Oil products	Glucose uptake V_{max} (mg/L per h)	0	0.2	Fuksa, 1981

[a] All measurements of microbial numbers and activity in samples from a creosote spill are from the same site.
[b] Biomass measured as total extractable phospholipid.
[c] GDW: gram dry weight of subsurface material.
[d] Range of measurements can represent different depths in unsaturated and saturated zones and for the contaminated samples, different concentrations of contaminants.
[e] Range of counts on different types of media.

Table 6.2. Patterns of Natural Biodegradation in the Subsurface (adapted from U.S. EPA, 1993).

Pattern	Microbial Process	Parameter Limiting Biodegradation	Kinetics	Characteristics
1	Aerobic respiration; Denitrification	Rate of nutrient supply (usually the terminal electron acceptor)	Zero- or First-order	Microbial populations adapt rapidly and reach high densities
2	Methanogenesis; Sulfate-reducing; Iron-reducing	Reaction rate of micro-organisms and substrate concentration of electron acceptor is not limiting	First-order at concentrations below 1 mg/L (K_s) Zero-order at concentrations above K_s	Steady-state biomass. Adaptation is slow, i.e., months to years
3	Halogenated organic compound is the electron acceptor	Concentration of alternate electron acceptors and electron acceptor demand	First-order	Not well understood. Effective electron donors could be co-occurring contaminants, volatile fatty acids, dissolved organic carbon

were confirmed by field experiments conducted at the site (Borden et al., 1989). Contaminants that were injected into the plume in oxygenated water were rapidly degraded whereas those injected into oxygen-poor water were minimally degraded.

The second pattern occurs under iron-reducing, sulfate-reducing or methanogenic conditions where the rate of biodegradation is limited by the reaction rate of the microorganisms (U.S. EPA, 1993) (Table 6.2). Many aromatic contaminants can be degraded in the absence of oxygen (Grbic-Galic, 1990), which results in the accumulation of fatty acids, production of methane, solubilization of iron, and reduction of nitrate and/or sulfate. In contrast to pattern 1 in which adaptation is rapid and high densities are obtained, months to years are required for adaptation to occur and biomass to reach steady-state. The biodegradation rate is dependent on substrate concentration. When plotted, the rate appears as a hyperbolic function that is best described by Monod or Michaelis-Menton kinetics. The electron acceptor supply is not a limiting factor in biodegradation rates under methanogenesis and iron-reducing conditions since carbonate and iron minerals usually are ubiquitous. Supplies of sulfate are less widespread. Rates of biodegradation under sulfate-reducing conditions would be insignificant once sulfate is depleted.

Under sulfate-reducing conditions, Acton and Barker (1992) found that toluene and *m*-xylene, but not benzene, *o*-xylene, ethylbenzene or 1,2,4-trimethylbenzene, were degraded in a plume from an active landfill after the compounds were injected into the plume with 12 µg/L sulfate. At a different field site, Thierrin et al. (1992) reported that toluene, *p*-xylene and naphthalene but not benzene were degraded in a gasoline plume under sulfate-reducing conditions. Although both iron III and mixed valence iron minerals may be important electron sinks in anaerobic plumes, field evidence for the role of iron in contaminant biodegradation is not available. However, laboratory studies indicate that iron III can be used as the terminal electron acceptor in biodegradation of toluene, phenol and p-cresol (Lovley et al., 1989).

Many field studies have been conducted on the methanogenic transformations of the alkylbenzenes (Wilson et al., 1991; Cozzarelli et al., 1990; Wilson et al., 1986). Godsy et al. (1992a,b) published microcosm and field data on the methanogenic fate of phenolic compounds in a plume derived from creosote. In the microcosm study, the rate of biodegradation of several phenolic compounds was described using Monod kinetics (Godsy et al., 1992b). The K_s value for phenol was low (~1 mg/L), indicating that the microorganisms were adapted to oligotrophic conditions. In addition, the amount of biomass yield from phenol biodegradation (Y) was below that predicted by the Monod equation, suggesting some environmental factor, such as predators (U.S. EPA, 1993), limits biomass production. Anaerobic protozoa have been detected in subsurface materials contaminated with JP-4 jet fuel (Thomas et al, 1995;1997). These observations of low K_s and low biomass yields (Y) also may explain the behavior of methanogenic plumes of alkylbenzenes.

The third pattern occurs in the absence of oxygen when the organic compound is used as the electron acceptor rather than the carbon source or electron donor (U.S. EPA, 1993). This pattern can be observed for halogenated organic compounds (Table 6.2). The rate of biodegradation of the compound is dependent on the demand for electron donors and acceptors and the presence of conventional electron acceptors (Gossett and Zinder, 1996). The reductive dechlorination of tetrachloroethylene is an environmentally significant example of pattern 3 because the carcinogenic intermediate, vinyl chloride, is formed during this process (Vogel and McCarty, 1985; McCarty, 1996). A variety of organic compounds have been suitable electron donors in reductive dechlorination in laboratory studies: glucose (Friedman and Gossett 1989; Sharma and McCarty, 1996), methanol (DiStefano et al., 1992; Friedman and Gossett, 1989), formate (Friedman and Gossett, 1989; Sharma and McCarty, 1996), lactate (Bagley and Gossett, 1990; de Bruin et al., 1992; Gibson and Sewell, 1992; Sharma and McCarty, 1996), propionate, crotonate, butyrate, ethanol (Gibson and Sewell, 1992), acetate (Vogel and McCarty, 1985; Friedman and Gossett, 1989; Sharma and McCarty, 1996), benzoate (Scholz-Muramatsu et al., 1990), hydrogen (Friedman and Gossett, 1989; DiStefano et al., 1992), and toluene (Sewell and Gibson, 1991). DiStefano et al. (1992) and Maymo-Gatell et al. (1995) suggested that it is the hydrogen produced from the metabolism of substrates known to support reductive dechlorination that is used directly in the reductive dechlorination process.

ENVIRONMENTAL FACTORS AND MICROBIAL PROCESSES

The same conditions that affect microbial activity in aquatic and terrestrial environments apply in the subsurface (Table 6.3). These include the biogeochemistry of the site, presence of adapted populations, availability of adequate amounts of substrates and essential nutrients for maintenance and growth, tolerable amounts of toxicants and inhibitors, and moderate values for environmental parameters such as pH, temperature, salinity, and osmotic potential (Alexander, 1975).

Biogeochemistry

Biogeochemistry, which is the effect that living organisms have on the geochemistry of their environment, affects contaminant transport and fate. Nutrient cycling is a biogeochemical process, and contaminant biodegradation can be thought of as an extension of this process. An environmental condition which results from biological activities is the oxidation-reduction (redox) potential, measured as Eh. Microbial processes occur at different redox potentials using different electron acceptors (Table 6.4), which in turn determines the ultimate biological fate of contaminants that may be present. For instance, the fully chlorinated aliphatic solvents, like tetrachloroethene, are degraded by reductive dehalogenation under anaerobic conditions (de Bruin et al., 1992) but persist under aerobic conditions (Oldenhuis et al., 1989). Although aliphatic and aromatic hydrocarbons were once thought to be biodegraded only in the presence of molecular oxygen, more recent research indicates that some of these compounds can be biodegraded anaerobically (Hutchins, 1991; Aeckersberg, 1991).

The electron acceptors are more or less used in sequence of their energy yielding potential, which is the order of the microbial processes listed in Table 6.4. Therefore, contaminant degradation will be faster under aerobic than anaerobic conditions for most compounds (Table 6.5). The exception to this generality, of course, is indicated by the persistence of the heavily chlorinated compounds in aerobic environments. Even though more energy is derived from aerobic metabolism of contaminants, a significant amount of contaminant removal in the subsurface may be the result of anaerobic processes (Table 6.6). Once oxygen is depleted in a contaminated zone, anaerobic processes will be important because of the slow movement of groundwater and resultant slow mixing of oxygen-poor and oxygenated water at the margins of that zone (Suflita, 1989). The availability of the alternate electron acceptors will depend on the native fertility of the subsurface and their transport in groundwater.

The size of a contaminant plume in the subsurface is controlled by the various redox conditions dictated by the plume. The source and center of the plume are anaerobic, while the edges mix with oxygen from aerobic zones. For many contaminants, biotic processes confine the plume to a much narrower and shorter area than would be expected by the abiotic processes of advection, dispersion, and sorption. For a plume resulting from a landfill which received municipal and industrial wastes, redox status was inversely related to contaminant concentration (Lyngkilde et al., 1991). A methanogenic zone was located near the source, adjacent to a discontinuous limited zone of sulfate reduction, followed by discontinuous zones of manganese reduction, an extensive nitrate-reducing zone, and an aerobic zone.

Adaptation

Adaptation is defined as the amount of time between exposure of an organism to a substrate and the onset of detectable degradation of that substrate. Adaptation may be the time required for 1) the number of substrate-degrading organisms to increase, 2) enzyme induction, 3) genetic changes that confer degradative capabilities, and 4) depletion of a preferred substrate, or may be due to 5) predation of the substrate-degrading organisms, among others (Wiggins et al., 1987). The fate of a contaminant may differ depending on the length of its exposure to the contaminant-degrading organisms. More of the contaminant may be incorporated into biomass than is mineralized by microorganisms in uncontaminated subsurface environments (Swindoll et al., 1988). However, more of the contaminant may be mineralized than is incorporated into biomass by microorganisms in subsurface materials that have been contaminated for a long period of time (Aelion et al., 1989). The presence of adapted

Table 6.3. Critical Environmental Factors for Microbial Activity (Sims et al., 1989).

Environmental Factor	Optimum Levels
Available soil water	25% to 85% of water-holding capacity
Oxygen	Aerobic metabolism: greater than 0.2 mg/L dissolved oxygen, minimum air-filled pore space of 10%; anaerobic metabolism: O_2 concentrations less than 1%
Redox potential	Aerobes and facultative anaerobes: greater than 50mV; anaerobes: less than 50mV
pH	5.5 to 8.5
Nutrients	Sufficient nitrogen, phosphorus, and other nutrients so not limiting to microbial growth (suggested C:N:P ratio of 120:10:1)
Temperature	4° to 45°C

Table 6.4. Microbial Processes that Occur at Different Oxidation-Reduction Potentials (adapted from Zehnder and Stumm, 1988).

Microbial Process	Electron Acceptor	Products	Eh (mV)
Aerobic respiration	Oxygen	H_2O	+810
Denitrification	Nitrate, nitrite, nitrous oxide	N_2	+750
Manganese reduction	Mn^{4+}	Mn^{2+}	+550
Iron reduction	Fe^{3+}	Fe^{2+}	−100
Sulfate reduction	SO_4^{2-}	H_2S	−220
Methanogenesis	CO_2	CH_4	−250

Table 6.5. Degradation of *p*-Cresol under Various Redox Conditions (from Suflita, 1989).

Redox Conditions	Biodegradability	Lag Time	Relative Rate	Reference
Aerobic	+	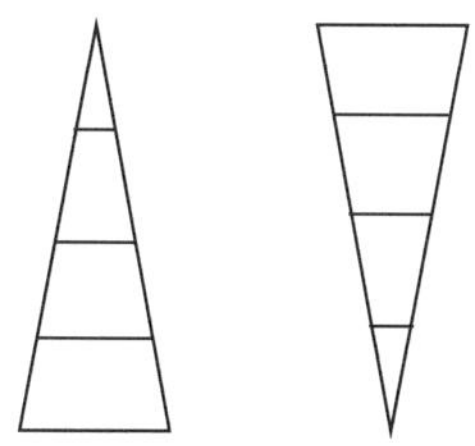		Hopper (1976, 1978)
Denitrifying	+			Bossert & Young (1986)
Sulfate-reducing	+			Bak & Widdel (1992) Smolenski & Suflita (1987)
Methanogenic	+			Smolenski & Suflita (1987) Godsy et al. (1983) Senior & Balba (1984)

microorganisms can be determined by assessing biodegradation potential in samples of contaminated and uncontaminated samples from the same site (Table 6.7). In most cases, prior exposure to the contaminant is required before its biodegradation can occur. In the one case when naphthalene was not degraded in contaminated material, the authors suggested that toxic levels of creosote were present (Thomas et al., 1989). In the cases in which toluene was degraded in uncontaminated as well as contaminated material, adapted organisms were already present in the uncontaminated material or the time required for adaptation was short.

Table 6.6. Compounds Biodegraded under Different Oxidation-Reduction Potentials.

Redox Conditions	Compounds	Reference
Denitrifying	Toluene, *m*-Xylene	Zeyer et al., 1986
	Ethylbenzene, *m*-, *o*-Xylenes, 1,2,4-Trimethylbenzene	Hutchins, 1991
Ferric iron–reducing	Toluene, Phenol, *p*-Cresol	Lovley et al., 1989
Sulfate-reducing	Toluene, *p*-Xylene	Haag et al., 1991
	Toluene	Beller et al., 1992
	Toluene, Xylene isomers	Edwards et al., 1992
	Cresol isomers, Tetrachloroethylene	Suflita et al., 1986
	C_{12}–C_{20} Alkanes	Aeckersberg et al., 1991
Methanogenic	Benzene, Ethylbenzene, *o*-Xylene, Toluene	Wilson et al., 1986
	Toluene, Benzene	Grbic-Galic & Vogel, 1987
	Chlorinated phenols, Phenol	Suflita & Miller, 1985
	Phenoxyacetic acids, Benzoate, Chlorinated benzoates, Phenol, Chlorinated phenols	Gibson & Suflita, 1986
	Cresol isomers	Smolenski & Suflita, 1987

Other Environmental Factors That Affect Biodegradation

Substrate concentration affects biodegradation. Threshold concentrations below which contaminant biodegradation does not occur may exist in nature (Bouwer, 1992). In contrast, contaminants may be present at concentrations that are toxic (Phelps et al., 1988; Thomas et al., 1989). Nutrient availability depends on the native fertility of the subsurface materials, recycle of nutrients from biomass, nitrogen fixation, and transport processes in groundwater which are slow. Several laboratory studies have indicated that the addition of inorganic nutrients to samples of subsurface materials either enhanced or had no effect on contaminant biodegradation (Swindoll et al., 1988; Thomas et al., 1989; Thomas et al., 1990).

The physical characteristics of the subsurface material also may affect biodegradation potential. Several investigators have reported that biodegradation potential is greater in sandy subsurface materials than in those with a high clay content (Phelps et al., 1988; Hicks and Fredrickson, 1989). The biodegradation of phenol and acetate in samples of subsurface material was positively correlated with pH and viable cell counts while negatively correlated with clay content (Hicks and Fredrickson, 1989).

The correlation of high biodegradation potential in sandy materials is not surprising since higher viable cell numbers usually are found in sandy materials than in those high in clay. However, this correlation may not exist for all compounds. Viable cell count was not positively correlated with biodegradation potential for the more recalcitrant compounds aniline, quinoline, pyridine, and t-butyl alcohol (Hickman and Novak, 1989; Fredrickson et al., 1990).

Table 6.7. Contaminant Biodegradation in Samples of Contaminated and Uncontaminated Subsurface Material from the Same Site (adapted from Thomas and Ward, 1992).

	Biodegradation[a]		
Contaminant	Uncontaminated	Contaminated	Reference
Naphthalene	no	yes	Wilson et al., 1985
	no	yes, no[b]	Thomas et al., 1987
Benzene	no	yes	Wilson et al., 1986
	no	yes	Thomas et al., 1990
Toluene	no	yes	Wilson et al., 1986
	yes	yes	Thomas et al., 1990
	yes	yes[c]	Aelion et al., 1989

[a] Biodegradation is defined as disappearance of parent compound or mineralization.
[b] Includes samples from several zones and locations.
[c] More was biodegraded in the contaminated than uncontaminated sample.

ENHANCED BIOREMEDIATION

When the rate of natural bioremediation is not fast enough to prevent migration of contaminants, and health and the environment are at risk, remedial action must be taken. One remedial option is enhanced bioremediation, which involves increasing the rate of natural bioremediation by adding or increasing the flux of essential nutrients and electron donors and acceptors. Several reviews on the design and operation of enhanced bioremediation have been published (Lee et al., 1988; Thomas and Ward, 1989; Sims et al., 1992). Because most organic compounds biodegrade faster under aerobic conditions, most treatment schemes have involved injecting limiting nutrients and a source of oxygen into the zone of contamination. Researchers have suggested that the rate of biodegradation of many groundwater contaminants is limited by the rate of oxygen transfer to the contaminant-degrading microorganisms (Wilson et al., 1985). Assuming the microorganisms have adapted to degrade the contaminants, biodegradation of many compounds can be considered instantaneous in the presence of oxygen.

SUMMARY

Microorganisms have been shown to exist at depths of 500 to 600 m below the surface and probably exist at even greater depths. In general, higher numbers are associated with sandy transmissive sediments rather than with those high in clay. Subsurface microorganisms have been found to degrade a variety of naturally occurring and synthetic organic compounds. Although many groundwater pollutants are thought to biodegrade faster and to a greater extent aerobically than anaerobically, anaerobic processes are known to contribute significantly to contaminant fate in the subsurface. Although the concentration of microorganisms is lower in the subsurface than in surface soils, subsurface microorganisms appear to be a significant biofilter for removing organic contaminants.

REFERENCES

Acton, D.W. and J.F. Barker. 1992. In situ biodegradation potential of aromatic hydrocarbons in anaerobic groundwater. *J. Contam. Hydrol.* **9**:325–352.

Aeckersberg, F., F. Bak, and F. Widdel. 1991. Anaerobic oxidation of saturated hydrocarbons to CO_2 by a new type of sulfate-reducing bacterium. *Arch. Microbiol.* **156**:5–14.

Aelion, C.M., D.C. Dobbins, and F.K. Pfaender. 1989. Adaptation of aquifer microbial communities to the biodegradation of xenobiotic compounds: influence of substrate concentration and pre-exposure. *Environ. Toxicol. Chem.* **8**:75–86.

Alexander, M. 1975. Environmental and microbiological problems arising from recalcitrant molecules. *Microb. Ecol.* **2**:17–27.

Bagley, D.M. and J.M. Gossett. 1990. Tetrachloroethene transformation to trichloroethene and cis-1,2-dichloroethene by sulfate-reducing enrichment cultures. *Appl. Environ. Microbiol.* **56**:2511–2516.

Bak, F. and F. Widdel. 1992. In situ biodegradation of phenol and phenol derivatives by *Desulfobacterium phenolicum* sp. nov. *Arch. Microbiol.* **146**:177–180.

Balkwill, D.L. 1989. Numbers, diversity, and morphological characteristics of aerobic, chemotrophic bacteria, in deep subsurface sediments from a site in South Carolina. *Geomicrobiol. J.* **7**:33–52.

Beller, H.R., D. Grbic-Galic, and M. Reinhard. 1992. Microbial degradation of toluene under sulfate-reducing conditions and the influence of iron on the process. *Appl. Environ. Microbiol.* **8**:786–793.

Boggs, J.N., L.M. Beard, V.R. Waldrop, T.B. Stauffer, W.G. MacIntyre, and C.P. Antworth. 1993. Transport of Tritium and Four Organic Compounds During a Natural-Gradient Experiment (MADE-2). Electric Power Research Institute, EPRI TR-101998.

Bone, T.L. and D.L. Balkwill. 1988. Morphological and cultural comparison of microorganisms in surface soil and subsurface sediments at a pristine study site in Oklahoma. *Microb. Ecol.* **16**:49–64.

Borden, R.C., M.D. Lee, J.M. Thomas, P.B. Bedient, and C.H. Ward. 1989. In situ measurement and numerical simulation of oxygen limited biotransformation. *Ground Water Mon. Rev.* **9**:83–91.

Bossert, I.D. and L.Y. Young. 1986. Anaerobic oxidation of *p*-cresol by a denitrifying bacterium. *Appl. Environ. Microbiol.* **52**:1117–1122.

Bouwer, E.J. 1992. Bioremediation of organic contaminants in the subsurface. In: R. Mitchell, (Ed.), *Environmental Microbiology*, John Wiley & Sons, Inc., p. 287–318.

Cozzarelli, I.M., R.P. Eganhouse, and M.J. Baedecker. 1990. Transformation of monoaromatic hydrocarbons to organic acids in anoxic groundwater environment. *Environ. Geol. Water Sci.* **16**:135–141.

de Bruin, W.P., M.J.J. Kotterman, M.A. Posthumus, G. Schraa, and A.J.B. Zehnder. 1992. Complete biological reductive transformation of tetrachloroethene to ethane. *Appl. Environ. Microbiol.* **58**:1996–2000.

DiStefano, T.D., J.M. Gossett, and S.H. Zinder. 1992. Hydrogen as an electron donor for dechlorination of tetrachloroethene by an anaerobic mixed culture. *Appl. Environ. Microbiol.* **58**:3622–3629.

Edwards, E.A., L.E. Wills, M. Reinhard, and D. Grbic-Galic. 1992. Anaerobic degradation of toluene and xylene by aquifer microorganisms under sulfate-reducing conditions. *Appl. Environ. Microbiol.* **58**:794–800.

Ehrlich, G.G., R.A. Schroeder, and P. Martin. 1985. Microbial Populations in a Jet-Fuel-Contaminated Shallow Aquifer at Tustin, California. U.S. Geological Survey 85-335, U.S. Geological Survey, Sacramento, CA, 14 pp.

Fredrickson, J.K., D.L. Balkwill, J.M. Zachara, S.-M.W. Li, F.J. Brockman, and M.A. Simmons. 1991. Physiological diversity and distributions of heterotrophic bacteria in **57**:402–411.

Fredrickson, J.K., F.J. Brockman, R.J. Hicks, and B.A. Denovan. 1990. Biodegradation of nitrogen-containing aromatic compounds in deep subsurface sediments. In: C.B. Fliermans and T.C. Hazen (Eds.), *Proc. First Int. Symp. on Microbiology of the Deep Subsurface*, January 15–19, Orlando, FL, WSRC Information Services, Aiken, SC, p. 6-27–6-44.

Fredrickson, J.K., T.R. Garland, R.J. Hicks, J.M. Thomas, S.W. Li, and K.M. McFadden. 1989. Lithotrophic and heterotrophic bacteria in deep subsurface sediments and their relation to sediment properties. *Geomicrobiol. J.* **7**:53–66.

Friedman, D.L. and J.M. Gossett. 1989. Biological reductive dechlorination of tetrachloroethylene and trichloroethylene to ethylene under methanogenic conditions. *Appl. Environ. Microbiol.* **55**:2144–2151.

Fuksa, J. 1981. Microorganisms in oil polluted groundwater environment. *Verh. Int. Verein. Limnol.* **21**:1376–1380.

Ghiorse, W.C. and J.T. Wilson. 1988. Microbial ecology of the subsurface. *Adv. Appl. Microbiol.* **33**:107–172.

Gibson, S.A. and G.W. Sewell. 1992. Stimulation of reductive dechlorination of tetrachloroethene in anaerobic aquifer microcosms by addition of short-chain organic acids and alcohols. *Appl. Environ. Microbiol.* **58**:1392–1393.

Gibson, S.A. and J.M. Suflita. 1986. Extrapolation of biodegradation results to groundwater aquifers: reductive dehalogenation of aromatic compounds. *Appl. Environ. Microbiol.* **52**:681–688.

Godsy, E.M., D.F. Goerlitz, and G.G. Ehrlich. 1983. Methanogenesis of phenolic compounds by a bacterial consortium from a contaminated aquifer in St. Louis Park, Minnesota. *Bull. Environ. Contam. Toxicol.* **30**:261–268.

Godsy, E.M., D.F. Goerlitz, and D. Grbic-Galic. 1992a. Methanogenic biodegradation of creosote contamination in natural and simulated ground-water ecosystems. *Ground Water* **30**:232–242.

Godsy, E.M., D.F. Goerlitz, and D. Grbic-Galic. 1992b. Methanogenic degradation kinetics of phenolic compounds in aquifer-derived microcosms. *Biodegradation* **2**:211–221.

Gossett, J.M. and S.H. Zinder. 1996. Microbiological aspects relevant to natural attenuation of chlorinated ethenes. In: Proceedings, Symposium on Natural Attenuation of Chlorinated Organics in Ground Water, Dallas, TX, EPA/540/R-96/509, pp. 10–13.

Grbic-Galic, D. 1990. Anaerobic microbial transformation of nonoxygenated aromatic and alicyclic compounds in soil, subsurface and freshwater sediments. In: J.M. Bollag and G. Stotzky (Eds.), *Soil Biochemistry*, New York, Marcel Dekker, Inc. p. 117.

Grbic-Galic, D. and T.M. Vogel. 1987. Transformation of toluene and benzene by mixed methanogenic cultures. *Appl. Environ. Microbiol.* **53**:254–260.

Haag, F., M. Reinhard, and P.L. McCarty. 1991. Degradation of toluene and *p*-xylene in anaerobic microcosms: evidence for sulfate as a terminal electron acceptor. *Environ. Toxicol. Chem.* **10**:1379–1389.

Hickman, G.T. and J.T. Novak. 1989. Relationship between subsurface biodegradation rates and microbial density. *Environ. Sci. Technol.* **23**:525–532.

Hicks, R.J. and J.K. Fredrickson. 1989. Aerobic metabolic potential of microbial populations indigenous to deep subsurface environments. *J. Geomicrobiol.* **7**:67–77.

Hopper, D.J. 1976. The hydroxylation of *p*-cresol and its conversion to *p*-hydroxybenzaldehyde in *Pseudomonas putida*. *Biochem. Biophys. Res. Commun.* **69**:462–468.

Hopper, D.J. 1978. Incorporation of [^{18}O] water in the formation of *p*-hydroxybenzyl alcohol by the *p*-cresol methylhydroxylase from *Pseudomonas putida*. *J. Biochem.* **175**:345–347.

Hutchins, S.R. 1991. Biodegradation of monoaromatic hydrocarbons by aquifer microorganisms using oxygen, nitrate, or nitrous oxide as the terminal electron acceptor. *Appl. Environ. Microbiol.* **57**:2403–2407.

Kobel-Boelke, J., E.-M. Anders, and A. Nehrkorn. 1988. Microbial communities in the saturated groundwater environment II: diversity of bacterial communities in a Pleistocene sand aquifer and their in vitro activities. *Microb. Ecol.* **16**:31–48.

Lee, M.D., J.M. Thomas, R.C. Borden, P.B. Bedient, J.T. Wilson, and C.H. Ward. 1988. Biorestoration of aquifers contaminated with organic compounds. *CRC Crit. Rev. Environ. Cont.* **18**:29–89.

Lovley, D.R., M.J. Baedecker, D.J. Lonergan, I.M. Cozzarelli, E.J.P. Phillips, and D.I. Siegel. 1989. Oxidation of aromatic contaminants coupled to microbial iron reduction. *Nature* **339**:297–300.

Lyngkilde, J., T.H. Christensen, B. Skov, and A. Foverskov. 1991. Redox zones downgradient of a landfill and implications for biodegradation of organic compounds. In: R.E. Hinchee and R.F. Olfenbuttel (Eds.), *In Situ Bioreclamation, Applications and Investigations for Hydrocarbon and Contaminated Site Remediation*, Butterworth-Heinemann, Boston, pp. 363–376.

Madsen, E.L. and J.-M. Bollag. 1989. Aerobic and anaerobic microbial activity in deep subsurface sediments from the Savannah River Plant. *Geomicrobiol. J.* **7**:93–101.

Madsen, E.L., J.L. Sinclair, and W.C. Ghiorse. 1991. In situ biodegradation: microbiological patterns in a contaminated aquifer. *Science* **252**:830–833.

Maymo-Gatell, X., V. Tandoi, J.M. Gossett, and S.H. Zinder. 1995. Characterization of an H_2-utilizing enrichment culture that reductively dechlorinates tetrachloroethene to vinyl chloride and ethene in the absence of methanogenesis and acetogenesis. *Appl. Environ. Microbiol.* **61**:3928–3933.

McCarty, P.L. 1996. Biotic and abiotic transformations of chlorinated solvents in ground water. In: Proceedings, Symposium on Natural Attenuation of Chlorinated Organics in Ground Water, Dallas, TX, EPA/540/R-96/509, pp. 5–9.

Oldenhuis, R., R.L.J.M. Vink, D.B. Janssen, and B. Witholt. 1989. Degradation of chlorinated aliphatic hydrocarbons by *Methylosinus trichosporium* OB3b expressing soluble methane monooxygenase. *Appl. Environ. Microbiol.* **55**:2819–2826.

Phelps, T.J., D. Ringelberg, D. Hedrick, J. Davis, C.B. Fliermans, and D.C. White. 1988. Microbial biomass and activities associated with surface environments contaminated with chlorinated hydrocarbons. *Geomicrobiol. J.* **6**:157–170.

Rifai, H.S., P.B. Bedient, J.T. Wilson, K.M. Miller, and J.M. Armstrong. 1988. Biodegradation modeling at an aviation fuel spill site. *ASCE J. Environ. Eng. Div.* **114**:1007–1029.

Scholz-Muramatsu, H.R., U. Szewzyk, and S. Gaiser. 1990. Tetrachloroethylene as electron acceptor for the anaerobic degradation of benzoate. *FEMS Microbiol. Lett.* **66**:81–86.

Senior, E. and M.T.M. Balba. 1984. The use of single stage and multi-stage fermenters to study the metabolism of xenobiotic and naturally occurring molecules by interacting microbial associations. In: J.M. Grainger and J.M. Lynch (Eds.), *Microbiological Methods for Environmental Biotechnology*. Society for Applied Bacteriology, Orlando, FL: Academic Press, Inc., p. 275–293.

Sewell, G.W. and S.A. Gibson. 1991. Stimulation of the reductive dechlorination of tetrachloroethene in anaerobic aquifer microcosms by the addition of toluene. *Environ. Sci. Technol.* **25**:982–984.

Sims, J.L., R.C. Sims, and J.E. Matthews. 1989. Bioremediation of contaminated surface soils. U.S. Environmental Protection Agency Document EPA-600/9-89/073, Ada, OK.

Sims, J.L., J.M. Suflita, and H.H. Russell. 1992. In-situ bioremediation of contaminated ground water. EPA/540/S-92/003, Environmental Protection Agency, R.S. Kerr Environmental Research Laboratory, Ada, OK.

Sinclair, J.L. and W.C. Ghiorse. 1989. Distribution of aerobic bacteria, protozoa, algae, and fungi in deep subsurface sediments, *Geomicrobiol. J.* **7**:15–31.

Sharma, P.K. and P.L. McCarty. 1966. Isolation and characterization of a facultatively aerobic bacterium that reductively dehalogenates tetrachloroethene to *cis*-1,2-dichloroethene. *Appl. Environ. Microbiol.* **62**:761–765.

Smith, G.A., J.S. Nickels, B.D. Kerger, J.D. Davis, S.P. Collins, J.T. Wilson, J.F. McNabb, and D.C. White. 1986. Quantitative characterization of microbial biomass and community structure in subsurface material: a procaryotic consortium responsive to organic contamination. *Can. J. Microbiol.* **32**:104–111.

Smolenski, W.J. and J.M. Suflita. 1987. Biodegradation of cresol isomers in anoxic aquifers. *Appl. Environ. Microbiol.* **53**:710–716.

Suflita, J.M. 1989. Microbial ecology and pollutant biodegradation in subsurface ecosystems. In: Transport and Fate of Contaminants in the Subsurface, EPA/625/4-89/019, Environmental Protection Agency, Center for Environmental Research, Cincinnati, OH, R.S. Kerr Environmental Research Laboratory, Ada, OK, p. 67–84.

Swindoll, C.M., C.M. Aelion, D.C. Dobbins, O. Jiang, S.C. Long, and F.K. Pfaender. 1988. Aerobic biodegradation of natural and xenobiotic organic compounds by subsurface microbial communities. *Environ. Toxicol. Chem.* **7**:291–299.

Swindoll, C.M., C.M. Aelion, and P.K. Pfaender. 1988. Influence of inorganic and organic nutrients on aerobic biodegradation and on the adaptation response of subsurface microbial communities. *Appl. Environ. Microbiol.* **54**:212–217.

Thierrin, J., G.B. Davis, C. Barber, and T.R. Power. 1992. Use of deuterated organic compounds as ground-water tracers for determination of natural degradation rates within a contaminated zone. In: *6th Intl. Symp. on Water Tracing,* Karlsruhe, Germany, Sept. 21–26, 1992.

Thomas, J.M., V.R. Gordy, C.L. Bruce, S.R. Hutchins, J.L. Sinclair, and C.H. Ward. 1995. Microbial activity in subsurface samples before and during nitrate-enhanced bioremediation. In: *Microbial Processes for Bioremediation.* R.E. Hinchee, C.M. Vogel, and F.J. Blockman (Eds.), Battelle Press, Columbus, OH, pp. 271–280.

Thomas, J.M., C.L. Bruce, V.R. Gordy, K.L. Duston, S.R. Hutchins, J.L. Sinclair, and C.H. Ward. 1997. Assessment of the microbial potential for nitrate-enhanced bioremediation of a JP-4 fuel contaminated aquifer. *J. Indus. Microbiol.* (in press).

Thomas, J.M., V.R. Gordy, S. Fiorenza, and C.H. Ward. 1990. Biodegradation of BTEX in subsurface materials contaminated with gasoline: Granger, Indiana. *Water Sci. Technol.* **22**:53–62.

Thomas, J.M., M.D. Lee, M.J. Scott, and C.H. Ward. 1989. Microbial ecology of the subsurface at an abandoned creosote waste site. *J. Indus. Microbiol.* 4:109–120.

Thomas, J.M., R.L. Raymond, J.T. Wilson, R.C. Loehr, and C.H. Ward. 1992. Bioremediation, In: J. Lederberg (Ed.), *Encyclopedia of Microbiology,* Academic Press, Inc. San Diego, CA, p. 369–385.

Thomas, J.M. and C.H. Ward. 1989. In situ biorestoration of organic contaminants in the subsurface. *Environ. Sci. Technol.* 23:760–766.

Thomas, J.M. and C.H. Ward. 1992. Subsurface microbial ecology and bioremediation. *J. Haz. Mat.* **32**:179–194.

U.S. Environmental Protection Agency (EPA). 1993. Seminars: Bioremediation of hazardous waste sites: Practical Approaches to Implementation. EPA 600/K-93/002.

Vogel, T.M. and P.L. McCarty. 1985. Biotransformation of tetrachloroethylene to trichloroethylene, dichloroethylene, vinyl chloride, and carbon dioxide under methanogenic conditions. *Appl. Environ. Microbiol.* **49**:1080–1083.

Webster, J.J., G.L. Hampton, J.T. Wilson, W.C. Ghiorse, and F.R. Leach. 1985. Determination of microbial cell numbers in subsurface samples. *Ground Water* **23**:17–25.

Wiggins, B.A., S.H. Jones, and M. Alexander. 1987. Explanations for the acclimation period preceding the mineralization of organic chemicals in aquatic environments. *Appl. Environ. Microbiol.* **53**:791–796.

Wilson, B.H., G.B. Smith, and J.F. Rees. 1986. Biotransformations of selected alkylbenzenes and halogenated aliphatic hydrocarbons in methanogenic aquifer material: a microcosm study. *Environ. Sci. Technol.* **20**:997–1002.

Wilson, B.H., J.T. Wilson, D.H. Kampbell, B.E. Bledsoe, and J.M. Armstrong. 1991. Biotransformation of monoaromatic and chlorinated hydrocarbons at an aviation gasoline spill site. *Geomicrobiol. J.* **8**:225–240.

Wilson, J.T., J.F. McNabb, J.W. Cochran, T.H. Wang, M.B. Tomson, and P.B. Bedient. 1985. Influence of microbial adaptation on the fate of organic pollutants in ground water. *Environ. Toxicol. Chem.* **4**:721–726.

Wilson, J.T., G.D. Miller, W.C. Ghiorse, and F.R. Leach. 1986. Relationship between the ATP content of subsurface material and the rate of biodegradation of alkylbenzenes and chlorobenzene. *J. Contam. Hydrol.* **1**:163–170.

Zehnder, A.J.B. and W. Stumm. 1988. Geochemistry and biogeochemistry of anaerobic habitats. In: A.J.B. Stumm (Ed.), *Biology of Anaerobic Microorganisms*, John Wiley & Sons, p. 1–38.

Zeyer, J., E.P. Kuhn, and R.P. Schwarzenbach. 1986. Rapid microbial mineralization of toluene and 1,3-dimethylbenzene in the absence of molecular oxygen. *Appl. Environ. Microbiol.* **52**:944–947.

Section III

Site Characterization

Chapter 7

A Site Characterization Strategy Based on Data Worth

R.A. Freeze

ABSTRACT

In the remediation of sites with contaminated groundwater, an assessment framework based on hydrogeological decision analysis espouses a sequential, iterative approach that alternates between data acquisition and economic analysis of alternative remedial designs. The economic analysis is based on a risk-cost-benefit objective function. Hydrogeological data have worth only if the risk reduction they provide exceeds the cost of obtaining them. It is possible to calculate the worth of a proposed measurement program before the measurements are taken. Three examples are provided: two generic, and one at an actual site. They illustrate the assessment of the worth of drillholes for the purpose of determining plume extent, and for the purpose of investigating aquitard continuity.

INTRODUCTION

The concepts and ideas presented in this chapter are based on a philosophical framework that is more fully developed in a series of four papers on hydrogeological decision analysis that appeared in *Ground Water* (Freeze et al., 1990; Massmann et al., 1991; Sperling et al., 1992; and Freeze et al., 1992) and in one paper on the worth of data that appeared in *Water Resources Research* (James and Freeze, 1993). This chapter provides an introduction to the framework and some examples of its application. The original papers provide the theoretical background, and an extended list of references that better reflects previous contributions on the topic.

The framework is based on the premise that hydrogeological data have worth if they can aid in the making of decisions between alternative remedial designs. The assessment of alternative designs is carried out under a risk-based design philosophy which requires the setting of performance criteria against which the alternative designs can be compared. It is recognized that each design alternative has associated with it a probability that it will not meet its performance objectives. This probability of failure, when coupled with the consequences of such a failure, constitutes the risk associated with each alternative. Risk reflects uncertainty in site characteristics, and uncertainty can be reduced by the collection of data. Data worth is assessed by comparing the cost of site characterization against the expected value of the risk reduction the data can provide.

The components of the design framework espoused in this chapter include a decision model, a hydrogeological simulation model, a geological uncertainty model, a parameter uncertainty model, an engineering reliability model, and a data acquisition system in the form of a field investigation program (Figure 7.1). The engineering reliability model will not be discussed in this chapter, nor will there be any detailed discussion of sophisticated field-investigation methodologies. Emphasis will be on data acquisition from the emplacement of boreholes. Site characterization strategy will revolve around the selection of the number, spacing, and locations of such boreholes.

REMEDIAL DESIGN FRAMEWORK

Consider the remedial strategy for a contaminated aquifer that involves migration control by means of a pump-and-treat system. The decision variables that differentiate alternatives will revolve around the number, location, depth, and pumping rates of extraction and injection wells. Failure will occur if the system does not create a capture zone that fully encompasses the contaminant plume.

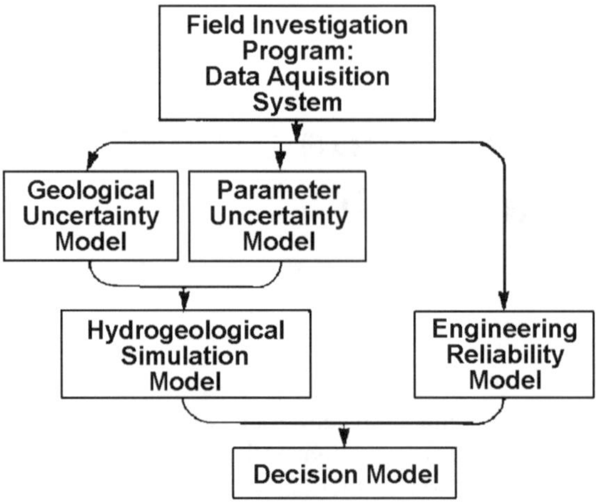

Figure 7.1. Components of a hydrogeological design framework.

The decision model allows for the comparison of alternative remedial designs. It is an economic analysis based on a risk-cost-benefit objective function. The objective function takes the form:

$$\phi_j = \sum_{t=0}^{T} \frac{1}{(1+i)^t} \left[B_j(t) - C_j(t) - R_j(t)\right] \tag{7.1}$$

where

ϕ_j = objective function for alternative j [$];
$B_j(t)$ = benefits of alternative j in year t [$];
$C_j(t)$ = costs of alternative j in year t [$];
$R_j(t)$ = risks of alternative j in year t [$];
T = time horizon [years]; and
i = discount rate [decimal fraction].

The risks R(t), in Equation 7.1 are defined as the expected costs associated with the probability of failure:

$$R(t) = P_f(t)\ C_f(t)\ \gamma(C_f) \tag{7.2}$$

where

$P_f(t)$ = probability of failure in year t [decimal fraction];
$C_f(t)$ = costs associated with failure in year t [$]; and
$\gamma(C_f)$ = normalized utility function [decimal fraction, $\gamma \geq 1$].

In Equation 7.1, the C(t) term represents the costs to the decision maker and the B(t) term represents the benefits. For remedial action at contaminated sites, benefits are zero; that is, there is no direct income to the decision maker (assumed here to be the owner of the site) from the remedial construction. The decision model therefore becomes one of risk-cost minimization. If the engineering time-horizon is short (so that the time value of money is not important), and if the owner is risk-neutral (so that $\gamma=1$ in Equation 7.2), then the objective function for any alternative simplifies to:

$$\phi = -\Sigma C - \Sigma P_f C_f \tag{7.3}$$

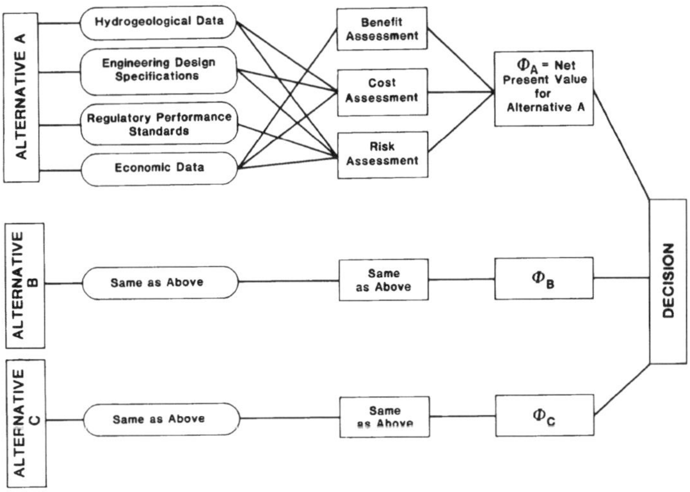

Figure 7.2. Application of a decision model to a set of design alternatives.

The probability of failure, P_f, is defined in terms of the ability or inability of a given remedial design to prevent plume migration toward a regulatory compliance point. The actual costs, C, are those associated with design, construction and operation of the remedial facility. There are also probabilistic costs, C_f, associated with potential failure; these include regulatory penalties, litigation costs, additional remedial costs, and loss of community goodwill. The best alternative is the one that maximizes the dollar value of the objective function, ϕ (Figure 7.2).

UNCERTAINTY REDUCTION AND DATA WORTH

Capture zone prediction is usually carried out with a hydrogeological simulation model. For simplicity, the analysis in this chapter is limited to conditions of advective transport of dissolved contaminants, in a steady saturated flow field, in an unconsolidated sand/clay aquifer/aquitard system. The input parameters to the simulation model for such conditions can be grouped into two classes: hydrogeologic parameter values, and boundary information. In the first class are the heterogeneous spatial distributions of hydraulic conductivity and porosity. In the second class are the location of impermeable boundaries, constant head boundaries, source area boundaries, plume boundaries, and boundaries between geological units. The latter define the conceptual stratigraphic model of the site hydrogeology.

Both these classes of input data are uncertain. Figure 7.3 provides schematic emphasis of these uncertainties. The degree of uncertainty is to a large degree dependent on the amount of available site characterization data. Additional data reduces uncertainty, reduces the probability of failure of the remedial alternatives, and reduces risk. The net worth of proposed additional data is the dollar value of the expected risk reduction minus the cost of obtaining the data. If the cost exceeds the anticipated risk reduction, the proposed data has no worth and should not be collected. The data worth methodology thus provides a stopping rule for site characterization activities.

The parameter uncertainty model in the design framework (Figure 7.1) allows for the updating of uncertainty in the hydrogeologic parameters when additional data become available from the site characterization program. The geological uncertainty model allows for the updating of uncertainty in geological boundaries. It is equally applicable to source boundaries and plume boundaries. The uncertainty modules use descriptors of uncertainty in the form of geostatistical models. Such models exist for either type of uncertainty. Much of the current literature addresses the parameter uncertainty question, especially with respect to hydraulic conductivity. The geostatistical models most commonly invoked utilize some form of kriging. The examples presented in this chapter involve the second type

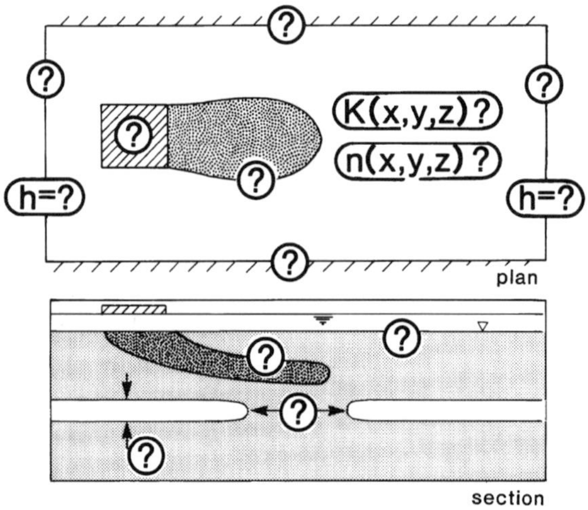

Figure 7.3. Uncertain input parameters to a hydrogeological simulation model.

of uncertainty: that associated with boundaries. At many sites, boundary uncertainties associated with plume extent or aquitard continuity (Figure 7.3) may be more important than parameter uncertainties associated with hydraulic conductivity. Boundary information is discrete: it comes in binary yes/no form. At any given borehole the plume (or aquitard) is either present or it is not. Geostatistical models based on search theory or indicator kriging are utilized in the uncertainty module.

The framework is applied to each design alternative. In each case, it requires application in a stochastic-simulation mode. The uncertainty models are used to generate a large number of equally likely realizations of the hydrogeological environment. The simulation model is then invoked in Monte Carlo mode, and the output is interpreted statistically, leading to an estimate of the probability of failure for the design alternative under analysis.

The data worth assessment is based on Bayesian updating concepts utilizing prior and posterior simulations. Prior uncertainty is reflected in a set of initial unconditional simulations, which are not conditioned on data points. The posterior uncertainty, reduced by the presence of data, is reflected in subsequent conditional simulations. Data worth is assessed in a pre-posterior analysis. It is possible to assess the worth of a proposed measurement program before the measurements are actually taken. This is because the uncertainty reduction is controlled by the number and location of the proposed measurements, not by the measured values themselves.

Figure 7.4 provides a summary of the decision framework presented in this chapter. It involves an iterative process between analysis and field measurement. A subjective prior interpretation of the hydrogeological environment is first analyzed with an unconditional stochastic simulation. The level of uncertainty reduction that can be achieved by field measurements is assessed with posterior conditional stochastic simulation. After each phase of field measurements, the available engineering alternatives are compared with an economic decision model. Before each new phase of field measurements, a data worth assessment is carried out to determine whether further measurements are economically justified. When the point is reached where they are not, the best alternative is selected as the design decision.

Emphasis in this chapter is on the data worth assessment. Three examples are presented. They include assessment of the worth of drillholes for the purpose of determining plume extent, and for the purpose of investigating aquitard continuity. The latter issue is addressed with both a generic example and at an actual site.

EXAMPLE 1: PLUME EXTENT

A five-spot pattern of monitoring wells in a shallow unconfined sand aquifer displays a contaminated well in the center surrounded by four clean wells (Figure 7.5). A downstream extraction well is

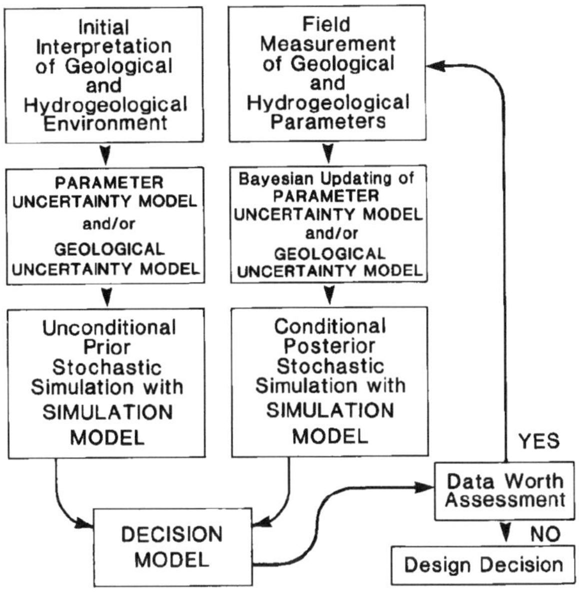

Figure 7.4. Summary of the decision framework.

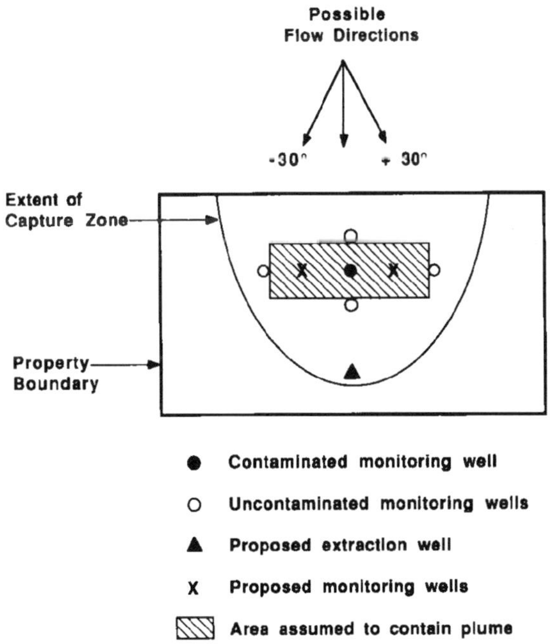

Figure 7.5. Plan view of a contaminated surficial aquifer and a proposed single-well migration-control system.

under design. It is hoped to create a capture zone that will encompass the contaminant plume, conservatively assumed to be rectangular in shape and extending to boundaries just shy of the clean monitoring wells. The decision variable is the pumping rate at the extraction well. Three rates are under consideration: 20 m^3/hr, 30 m^3/hr and 40 m^3/hr. The costs associated with these rates are $68,700, $84,400 and $97,700, respectively.

The direction and velocity of the groundwater flow at the site are uncertain, and an advective transport model has been utilized to calculate the conditional probabilities of failure for each pumping rate as a function of the direction and magnitude of flow.

A posterior application of the economic decision model shows that risk-cost minimization is achieved with the 40 m^3/hr pumping rate. This decision is robust over a range of possible failure costs, C_f, from $250,000 to $500,000.

The question arises: would it be worth drilling two additional monitoring wells inside the current five-spot? Posterior analysis shows that if these wells are clean, the reduced size of the proven plume will allow a reduction in pumping rate to 30 m^3/hr, with savings of $21,800. If the wells are dirty, no savings are possible. The data worth depends on the prior estimate of the probability of the new wells being dirty. If that probability is 0.5 then the worth of the new wells is $10,900. If they will cost more than that, they should not be installed. A more detailed presentation of the analysis can be found in Massmann et al. (1991).

EXAMPLE 2: AQUITARD CONTINUITY

Consider a three-layer stratigraphic hydrogeological environment in which a lower confined water supply aquifer is separated from an upper unconfined aquifer by a thin aquitard of uncertain continuity (Figure 7.6). A landfill is proposed on the land surface and two alternatives are under consideration: a lined landfill and an unlined landfill. The former costs $500K and the latter $300K. It is assumed (unrealistically, and for pedagogic purposes only) that P_f=0 for the lined case. Failure occurs for the unlined case only if there is a window through the aquitard and contaminants are drawn into the water supply wells in the confined aquifer. The cost associated with failure, C_f, is $1000K. The prior analysis indicates that the best alternative is the lined landfill.

Assuming the prior probability of encountering a window in the aquitard is 0.5, search theory is then invoked to calculate the pre-posterior worth of a grid of exploratory drillholes at 50-m spacings designed to identify the presence or absence of the aquitard at the drilled locations. Tables exist (Freeze et al., 1992) that provide the probability of not intercepting a target of diameter, a, with a grid of drillholes at spacing, d. Figure 7.7a indicates the geometry; Figure 7.7b graphs the probability of various ratios of d/a. This probability is of direct use in calculating the pre-posterior probability of encountering a window from the prior probability, using Baye's theorem. For example, a borehole network with 50-m spacing that does not encounter a window through the aquitard increases the probability that a 50-m-diameter window does not exist from 50% to 83% (Freeze et al., 1992).

These probabilities can then be used in prior and pre-posterior decision trees to calculate data worth. For the above example, the calculated data worth is $18K. If the proposed drillholes cost $1K each, and the area of interest can be covered by a grid of 18 holes or less at 50-m spacings, then the proposed drilling program should be carried out. If not, the proposed drilling program is not worthwhile. A more detailed presentation of the analysis can be found in Freeze et al. (1992).

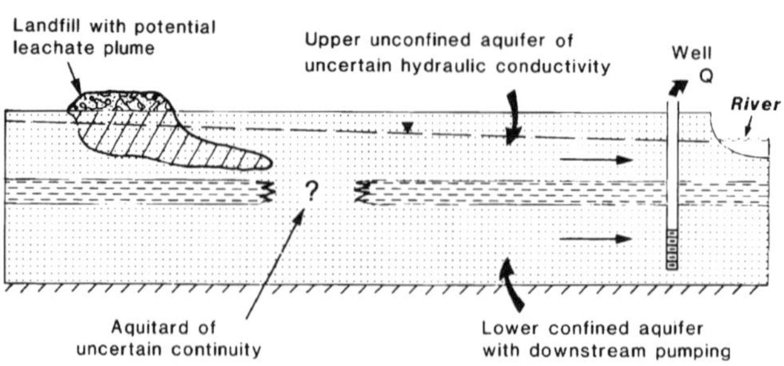

Figure 7.6. Section view of a three-layer stratigraphic section containing an aquitard of uncertain continuity.

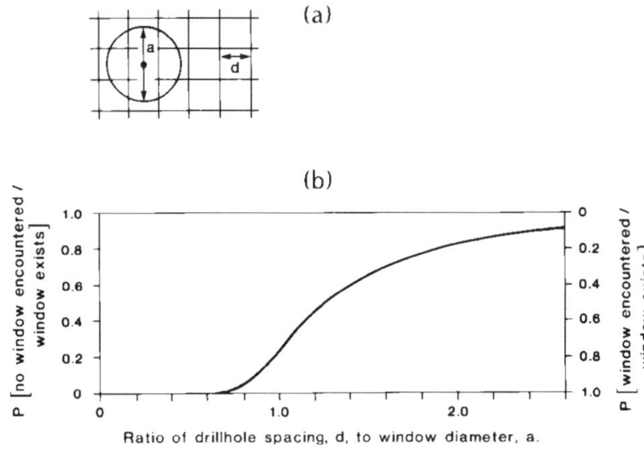

Figure 7.7. Search theory: (a) target of diameter, a, and drillhole grid of spacing, d; (b) probability of not encountering a window, given that one exists, for various ratios of d/a.

EXAMPLE 3: AQUITARD CONTINUITY, SAVANNAH RIVER SITE

The H-Area Seepage Basins are located in the General Separations Area of the Savannah River Site, South Carolina (Figure 7.8). The hydrogeological decision analysis framework was applied in a data worth context at this site (James and Freeze, 1993). The application was in the form of a post-audit of an earlier decision; it was not an on-line application.

The H-Area Seepage Basins are unlined earth basins that were used for the disposal of effluent water containing dissolved tritium and other radioactive and non-radioactive contaminant species. Groundwater seepage from the basins flows laterally to the south toward Four Mile Creek and to the east toward an unnamed creek (Figure 7.8) in an unconfined surficial aquifer. This aquifer is separated from the underlying Congaree Formation, which has regional water supply significance, by a thin (0–6 m) aquitard known as the Green Clay. As in the previous example, the major concern is the potential for contamination of the underlying aquifer through windows in the aquitard.

At the time the capping decision was under consideration, there were some data available from boreholes in the vicinity of the H-Area Seepage Basins (Figure 7.9). There was one borehole to the northeast considered to be a "hard" datum, because it included both a lithologic log and geophysical logs with a clear signature of the Green Clay. There were also several boreholes to the west, but these data were considered "soft" because they included only lithologic logs, and these were of questionable accuracy. All these hard and soft data points indicated the presence of Green Clay. However, there were no data immediately downgradient of the seepage basins, and it is for this area that the data worth assessment was carried out.

Two alternative closure options were considered for the H-Area Seepage Basins in this analysis. The first option was a no-action alternative; the second utilized a low-permeability clay cap. An indicator-kriging approach was utilized to assess the uncertainty reduction likely to be achieved by proposed drillholes with respect to aquitard continuity. The sampling programs that were assessed consisted of a single borehole at different locations, and various patterns of multiple boreholes. The results of the latter strategies are not reported here.

Figure 7.9 is a contoured plot of the net worth (risk reduction minus cost) of a single borehole in the vicinity of the H-Area Seepage Basins. The cost of a single borehole taken anywhere in the study area is assumed to be $20,000. This cost includes installing a piezometer down to the Congaree and taking a geophysical log of the borehole. The best location for a borehole is in the vicinity of point B where the net worth reaches a maximum of over $40,000. Note the position of the zero contour; it shows the conditioning effect of existing hard and soft data. It marks the zone where a single hard borehole is cost-effective. There is a zero contour around the one hard datum point and another bounding the region already covered by soft data points.

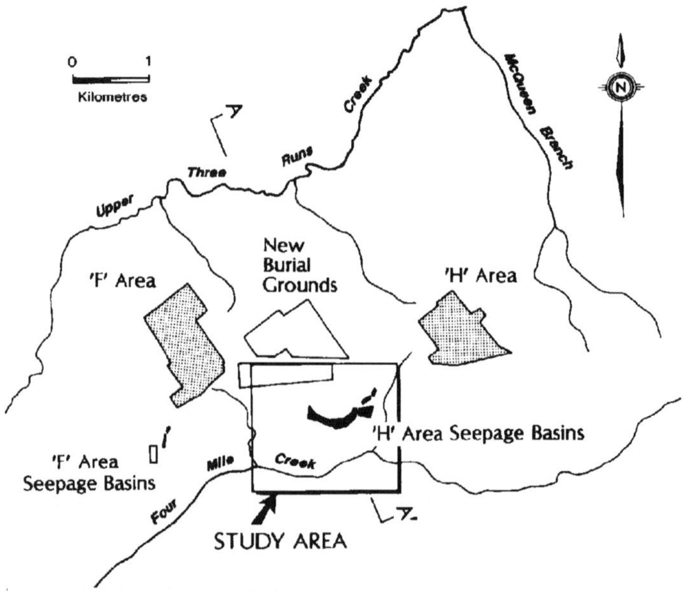

Figure 7.8. Location of the H-Area Seepage Basins in the General Separations Area of the Savannah River Site, South Carolina.

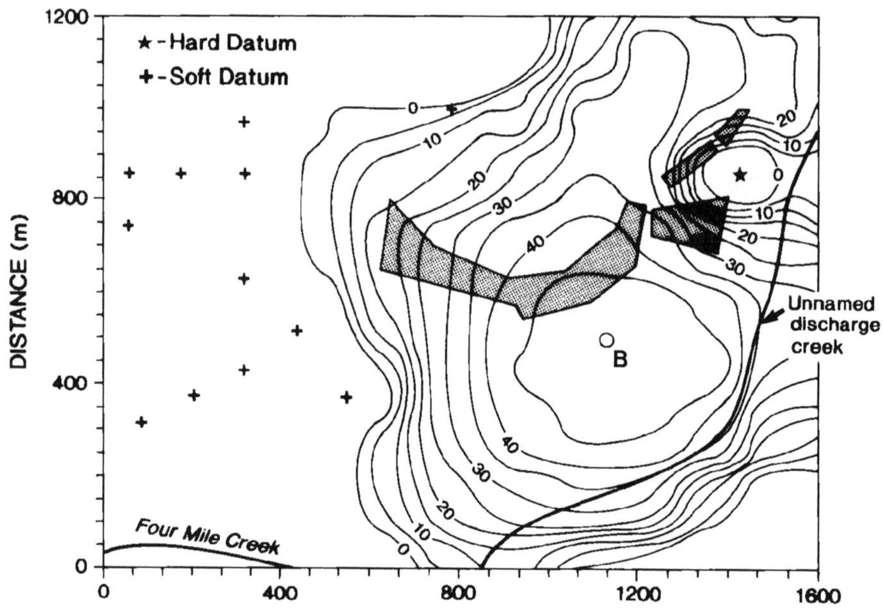

Figure 7.9. Contour map of the net worth of a single borehole in the vicinity of the H-Area Seepage Basins.

Sensitivity analysis showed that the net worth of a single borehole in the vicinity of point B on Figure 7.9 is very sensitive to the value of the cost associated with failure, C_f. For a value of $C_f = \$45$ million, the borehole is worth \$40,000; for $C_f = \$80$ million, it is worth \$200,000. In a generic sensitivity analysis (James and Freeze, 1993), it was found that data worth in an aquitard continuity context is generally more sensitive to economic parameters, such as the cost of failure and the discount rate, than it is to hydrogeological or geostatistical parameters.

The calculations that led to Figure 7.9 are very computer intensive. The Monte Carlo advective flow model had to be run for each alternative, for many locations (say 100 Monte Carlo runs for each of 100 locations, for each of 2 alternatives). The simulations can be run on a 386-level PC, but they take many hours of computer time.

The complete analysis can be found in James and Freeze (1993).

CONCLUSIONS

Site characterization activities should be closely integrated with remedial design activities. Hydrogeological data have worth if they can aid in the making of decisions between alternative remedial designs. The net worth of additional data is the value of the risk reduction they provide minus the cost of obtaining them.

A risk-based data worth methodology provides an economic stopping rule for site characterization. The methodology is complex and computer intensive. It involves the integration of a numerical contaminant transport model, a geostatistical module, a stochastic solution technique, and an economic analysis. Current applications are probably limited to relatively straightforward engineering projects in relatively simple hydrogeological environments. However, the framework is very general, and continued developments are to be expected in the next few years. They will lead to routine application of this or similar frameworks in due course.

ACKNOWLEDGMENTS

The decision analysis concepts outlined in this chapter have been developed together with a group of former students and colleagues in the graduate hydrogeology program, Geological Engineering Program, University of British Columbia. Specific thanks go to Leslie Smith, Joel Massmann, Tony Sperling, and Bruce James. The work was supported over the years by research grants from the Natural Sciences and Engineering Research Council of Canada.

REFERENCES

Freeze, R.A., J. Massmann, L. Smith, T. Sperling, and B. James. 1990. Hydrogeological decision analysis: 1. A framework, *Ground Water,* **28**:738–766.

Freeze, R.A., B. James, J. Massmann, T. Sperling, and L. Smith. 1992. Hydrogeological decision analysis: 4. The concept of data worth and its use in the development of site investigation strategies, *Ground Water,* **30**:574–588.

James, B.R. and R.A. Freeze. 1993. The worth of data in predicting aquitard continuity in hydrogeological design. *Water Resources Research,* **29**:2049–2065.

Massmann, J., R.A. Freeze, L. Smith, T. Sperling, and B. James 1991. Hydrogeological decision analysis: 2. Applications to groundwater contamination, *Ground Water,* **29**:536–548.

Sperling, T., R.A. Freeze, J. Massmann, L. Smith, and B. James 1992. Hydrogeological decision analysis: 3. Application to design of a ground-water control system at an open pit mine, *Ground Water,* **30**:376–389.

Chapter 8

Conceptual Models of Consolidated Sedimentary Aquifers

D.N. Lerner

INTRODUCTION

Aquifer Characteristics

The physical and chemical character of an aquifer will clearly influence the movement of pollutants through it. This influence will apply equally whether the pollutant is entering the aquifer and expanding its zone of influence, or whether attempts are being made to remove the pollutant. The important groups of characteristics in these contexts are porosity, permeability and mineralogy. All three groups are modified from their initial forms by the geological processes that the aquifer has undergone since deposition; these processes may be hydrochemical in origin or may arise from consolidation. This chapter reviews the net effect of these processes, with particular emphasis on physical characteristics affecting flow and diffusion.

Porosity is the fundamental characteristic of aquifers for water and pollutant movement, forming the basis of permeability and controlling solute velocities. Only interconnected porosity is of interest on a time scale of pollution transport, and represents the volume of water which is active in transport processes. An important second parameter is *kinematic porosity* (de Marsily, 1986), which represents the ratio between the velocity of the center of mass of an unretarded slug of tracer and the Darcian velocity of water. It is a measure of both the water that is free to move and the tortuosity of pore channels. Some authors use the term *effective porosity* (e.g., Domenico and Schwartz, 1990) for the same concept, but this usage is perhaps unfortunate as the term has other meanings (e.g., Bear, 1979, uses it for specific yield). As will be discussed below, the simple concept of kinematic porosity is not always applicable to fractured aquifers.

For pollution transport the variability of permeability is as important as its mean value. The variability occurs on all scales, from centimetric cross bedding structures to kilometric geological units. The ultimate effect of permeability variations is dispersion, the spreading out of a contaminant plume.

Aquifer mineralogy affects pollutant reactions, including sorption. As most organic pollutants are hydrophobic, sorption is normally dominated by partitioning from groundwater to the sedimentary organic matter in the aquifer. Thus organic carbon content is often the most important mineralogical property with organic pollutants. However, sorption of polar organics will be influenced by charged surfaces, such as clays. Biodegradation of pollutants is often controlled by pH and Eh potentials, which may in turn be controlled by the aquifer mineralogy through, for example, carbonate.

Processes of Consolidation

Consolidated sedimentary aquifers, such as sandstones, are, of course, formed from unconsolidated sediments. They retain many of their original features, such as bedding structures and bulk mineralogy. However, their hydrogeological characteristics may have been significantly altered by the geological processes which have operated since deposition. The important processes are:

- compaction, the squeezing of the deposit by overburden pressure, with consequent loss of porosity; there are often pressure-induced changes in mineralogy;
- cementation by minerals precipitated from pore fluids;
- diagenesis—that is, changes in mineralogy; and
- tectonic stressing, inducing fracturing.

The processes are interlinked with tectonic movement, causing burial, which in turn induces diagenesis and cementation through temperature and pressure changes (Schmidt and MacDonald, 1979). It has been convincingly argued that many deeply buried sandstones lose all their primary porosity during consolidation, and their observed porosity today is secondary, that is, created by fracturing or by redissolving cements (Hayes, 1979).

The net effects of consolidation cannot usefully be generalized. For example, permeability variations in the original sediments may remain, or may be dominated by secondary pathways both intergranular and fracturing. Cementation and diagenetic effects are also very varied and not to be generalized. Some possible important effects include:

- changes to carbonate content by dissolution or precipitation, affecting the buffering capacity available to moderate pH-sensitive reactions;
- deposition of iron hydroxides which can be important for sorption, and can control redox reactions if present in sufficient proportion;
- diagenetic creation and alteration of clays, with consequent effects on sorption and permeability;
- maturation or oxidation of organic matter, reducing sorption capacity for organic pollutants, and removing a carbon source which could be important in microbial removal of nitrates, for example; and
- hardening of strata, increasing brittleness and fracturing tendency.

A common feature of consolidated aquifers is fracturing. If present, fractures are likely to dominate the movement of pollutants in an aquifer, and the following sections describe the significance of fractures.

ROLE OF FRACTURES

Fractures provide preferential pathways for water and pollutant migration in consolidated aquifers. The parallel to fractures in unconsolidated aquifers is high-permeability lenses; however, there are several differences. The permeability contrast between fractures and matrix is (often) high, leading to much higher velocities in the fractures. Fractures can be vertical, offering bypass routes across the bedding structure and low-permeability layers. The area-to-volume ratio of a fracture is normally higher than that of a lens, so that surface processes, such as transverse diffusion, are more rapid.

The *dual porosity* conceptual model has been developed to describe the effect of fractures on pollutant movement in an aquifer which also has matrix porosity. The model takes account of the contrast in permeability between fracture and matrix, and of the high scope for diffusion (Figure 8.1). In its extreme form, the model only has advective water movement in the fractures; mechanical dispersion also only occurs in this domain. However, solutes diffuse between matrix and fracture pore waters over the large interface. This retards the movement of an invading pollutant, both reducing the size of the peak and delaying the center of mass.

SIGNIFICANCE OF FRACTURES

Effects on Hydraulic Conductivity

The hydraulic conductivity (m/s) of a parallel-sided planar fracture, K_f, is:

$$K_f = \rho g b^2/12\mu \qquad (8.1)$$

where ρ and μ are the density and dynamic viscosity of water, respectively, and b is fracture aperture (de Marsily, 1986). The importance of fractures to bulk conductivity of a sandstone will depend on both the fracture characteristics and the conductivity of the matrix. For example, regular fractures of 0.1 mm aperture at 0.67 m spacings have individual conductivities of 6.6×10^{-3} m/s (at 12°C, a typical United Kingdom groundwater temperature). Such fractures dominate the conductivity of a

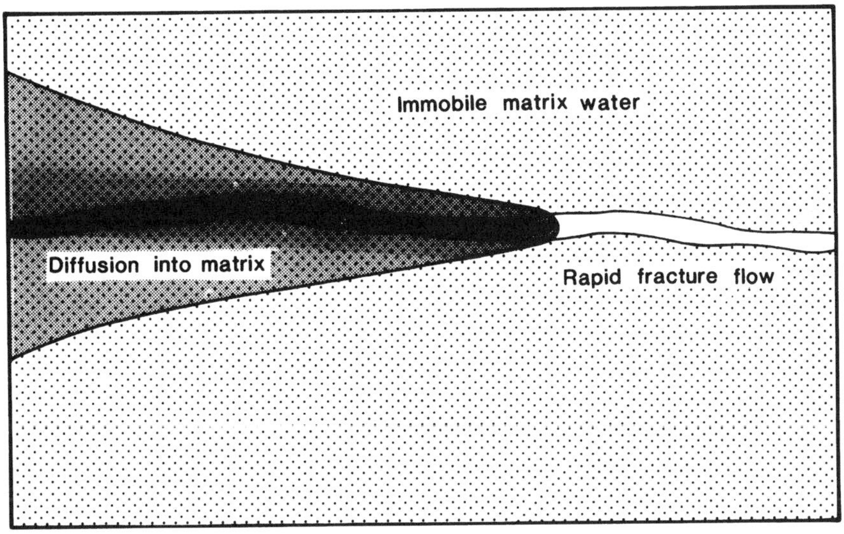

Figure 8.1. The dual porosity model of solute movement in fractured porous rocks.

Table 8.1. Effects of fractures on flow in sandstones.

Sandstone	Hydraulic conductivity (m/s) Matrix	Hydraulic conductivity (m/s) Bulk	Flow Ratio (R)	Flushing Ratio (F)	Diffusion (t_c - yrs)
A	10^{-7}	10.9×10^{-7}	10	10^4	18
B	10^{-5}	1.1×10^{-5}	0.1	100	18

Fracture aperture 0.1 mm, spacing 0.67 m. Temperature 12°C. Matrix porosity 0.15.
R is ratio of fracture to matrix flows (Equation 8.4).
F is relative flushing rate of fractures and matrix (Equation 8.5).
t_c is the characteristic diffusion time for an effective diffusion coefficient of 10^{-10} m^2/s (Equation 8.7).

sandstone with a low matrix conductivity (Sandstone A, Table 8.1) but are fairly insignificant in a sandstone with matrix conductivity two orders of magnitude higher (B).

Groundwater Flows

The flow of groundwater through parallel planar fractures in a unit cross-sectional area of aquifer, Q_f, is:

$$Q_f = K_f\, ib/B \tag{8.2}$$

where B is the fracture spacing and i is the hydraulic gradient.

Assuming that the same hydraulic gradient applies across the matrix, the flow in the matrix, Q_m, is:

$$Q_m = K_m i(B-b)/B \approx K_m i \tag{8.3}$$

The ratio of flow (R) contributed by the fractures and matrix, in natural or forced groundwater flow, is given by:

$$R = Q_f/Q_m = K_f\, b/K_m B = \rho g b^3/12\mu B K_m \tag{8.4}$$

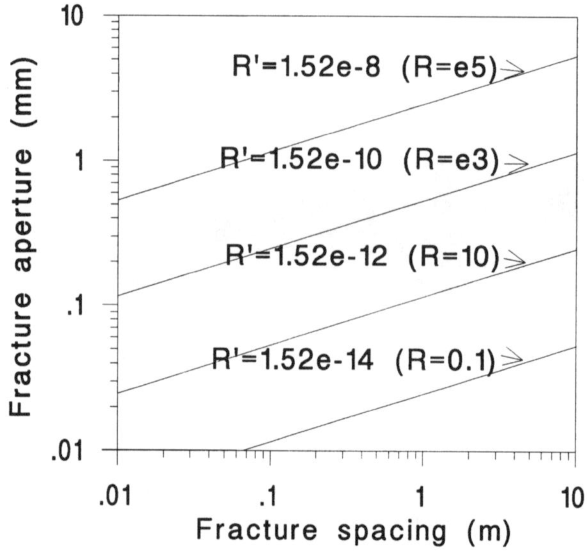

Figure 8.2. Ratio of fracture:matrix flow with parallel planar fractures. R′ ($R' = 12\mu K_m R/\rho g$) gives generalized values for all situations. R is fracture:matrix flow ratio, with example values given for matrix conductivity of 10^{-7} m/s.

This can be expressed in a generalized graphical form by a plot of $R' = 12\mu\, K_m R/\rho g = b^3/B$ on axes of b and B (Figure 8.2). For actual aquifers, values of R′ can be derived from R, and the example of an aquifer with matrix conductivity of 10^{-7} m/s is given in Figure 8.2. For the example sandstones in Table 8.1, the contribution of the matrix is 10% of the fractures for A, while the situation is reversed for B.

Flushing of Aquifer (Without Diffusion)

The contributions of flow do not reveal the relative rate at which the fractures and matrix are flushed, F. This is given by:

$$F = K_f\,\theta/K_m = \rho g b^2\theta/12\mu K_m \tag{8.5}$$

where θ is the matrix porosity.

F is independent of fracture spacing. Values of a generalized flushing rate $F' = 12\mu K_m F/\rho g\theta = b^2$ are shown in Figure 8.3 and can be used to estimate relative flushing rates for all aquifers given values of matrix conductivity and porosity. Example values of F are given for a reasonably porous but low-conductivity aquifer and show that fractures are flushed many more times than the matrix. The effect is less marked but still significant for a relatively high matrix conductivity (sandstone B in Table 8.1).

The simple analysis of flushing above ignores the effects of diffusion, permeability variations, and variable fracture apertures and spacing. Diffusion is discussed in the following section. Permeability variations in the matrix will extend the time required to flush the aquifer. In the extreme case, low-permeability zones will see so little advective flow that flushing will be primarily by diffusion towards adjacent higher-permeability zones, analogous to the dual porosity concept (Figure 8.1). Thus it can be argued that the dual porosity model is the limiting case of a more generalized model of solute transport in aquifers in which diffusion is a major process.

Variable fracture spacings have no effect on the results presented so far. Apertures are more significant, appearing as squared and cubed terms in Equations 8.5 and 8.4, respectively. Theoretical treatments of flow in fractures of variable aperture have been made (e.g., Wilson and Witherspoon, 1974). Few field data have been collected to quantify how apertures vary, both between and within fractures. This is an area requiring research, but imposing difficult field problems.

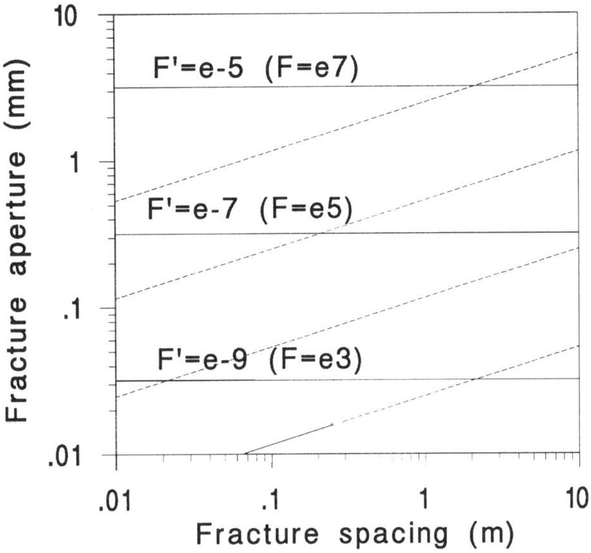

Figure 8.3. Ratio of fracture:matrix flushing rates, without diffusion. F′ ($F' = 12\mu K_m F/\rho g\theta$) gives generalized values. F is flushing ratio, with example values given for matrix conductivity of 10^{-7} m/s and porosity of 0.15. The dashed lines show the values of R′ given in Figure 8.2.

Diffusion

Figure 8.3 shows that flushing (or invasion) of the matrix by advection is insignificant for fractures over 30 µm. Field studies have shown that fracture apertures are typically 40 µm or over (J.A. Cherry, University of Waterloo, Ontario, personal communication), and hence, diffusion will dominate solute movement in most fractured materials.

The spread of diffusion processes can be judged by a characteristic time, based on the geometry of matrix blocks and the effective diffusion coefficient for the pollutant of interest (Barker, 1985). The mean square distance, a^2, travelled by diffusing molecules is:

$$a^2 = 2D^*t \tag{8.6}$$

where D^* is the effective diffusion coefficient and t is time. If a is taken to be a characteristic length of the matrix blocks, then the characteristic time for diffusion can be defined as:

$$t_c = a^2/2D^* \tag{8.7}$$

A common characteristic length is the ratio of volume to surface area for a block (Barker, 1985). For an infinite slab, a = B/2, and for a cube a = B/6. Barker (1985) gives expressions for characteristic times of other geometries, including a general form for irregular shapes.

The characteristic diffusive time is a measure of the time for significant diffusion to occur, rather than for it to be complete. This can be seen more clearly by solving the diffusion equation, assuming for simplicity that the matrix is at a uniform concentration initially and that concentrations drop suddenly to zero in the surrounding fractures. For slabs, the mean concentration falls to 23% of the initial value in time, $t_c = B^2/8D^*$ (Figure 8.4). For cubes, the mean concentration falls to 38% in the characteristic time, $t_c = B^2/72D^*$ (Figure 8.4). Alternative definitions of characteristic times could be used if lower values of mean concentration, or comparisons between different geometries, were required.

Characteristic times can be linked to the flow and flushing ratios discussed earlier. They are shown in Figure 8.5 in terms of $T' = t_c D^*$ for a slab geometry, as well as actual times for an effective diffusion

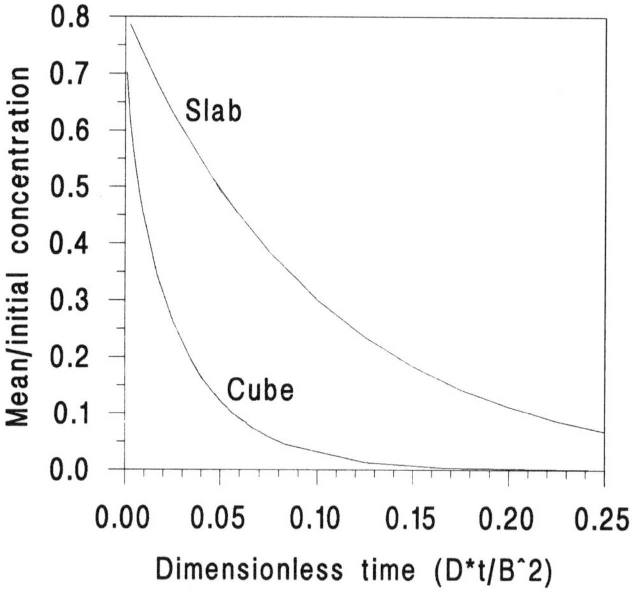

Figure 8.4. Decay of mean matrix concentration with time for two block geometries. Assumes fracture concentration fall to zero at t = 0 (Luikov, 1968).

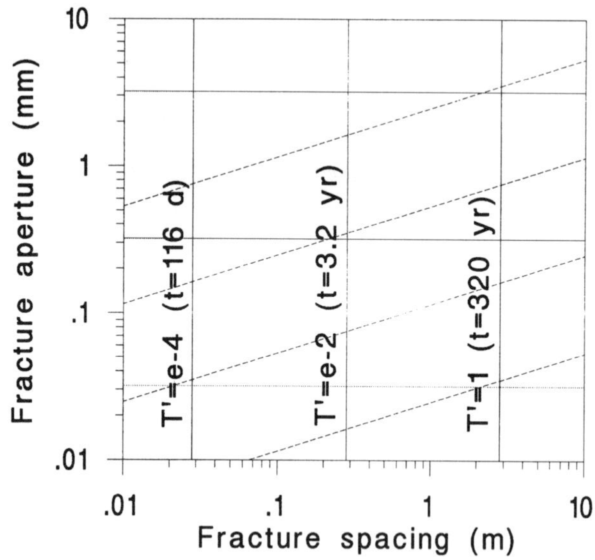

Figure 8.5. Characteristic diffusive time for slab geometry. T′ (T′ = t_cD^*) gives generalized values. t is characteristic time for an effective diffusion coefficient of 10^{-10} m^2/s. Dashed lines show R′ (Figure 8.2), dotted lines show F′ (Figure 8.3).

coefficient of 10^{-10} m^2/s. (This value has been measured for trichloroethylene in the Coventry sandstone discussed below; Chen and Lerner, 1992.) For the example fracture spacing of 0.67 m, the characteristic time for a slab geometry is 18 yr; for a cube geometry this falls to 2 yr. The fracture system will have been well flushed in this time, leading to large volumes of diluted water being pumped if a pump-and-treat style of pollution removal is attempted.

Kinematic Porosity

The concept of kinematic porosity is less useful for fractured porous aquifers than for intergranular systems, because the proportion of the porosity playing a part in solute transport varies. If groundwater flow is slow relative to the characteristic diffusion time, then all the interconnected porosity is active. Vogel and Giesel (1989) demonstrate this for parallel planar fractures and show that the kinematic porosity is ($b/B + \theta$). However, the kinematic porosity decreases as groundwater flow velocity increases, or as fracture spacings increase. In the limit, for example in a forced gradient tracer test, kinematic porosity approaches the fracture porosity. In pump-and-treat groundwater remediation, velocity increases towards the wells, and kinematic porosity is not a useful concept.

EXAMPLE SANDSTONE ENVIRONMENTS IN THE UNITED KINGDOM

In the UK, over 95% of homes receive a public water supply, and over 35% of public supplies are taken from groundwater. The role of groundwater varies considerably across the country, with 100% groundwater in much of the south and east, but only a few percent in Wales and Scotland.

The majority of aquifers in the UK are consolidated sedimentary formations. They range from limestones, including chalk, to a variety of sandstones (IGS, 1977). Unconsolidated alluvial and glacial deposits are present in much of the country, but rarely form extensive permeable deposits. Two contrasting sandstones have been selected to illustrate some of the features discussed above.

Permo-Triassic Sandstones

The Permo-Triassic sandstones are the second most important aquifer in the UK, and are also important aquifers in other parts of Europe. Their UK extent is shown in Figure 8.6. These red sandstones are exceptionally up to 2600 m thick, but more usually are in the range of 100–300 m thick. Three depositional environments are recognized in central England: aeolian, alluvial fan, and shallow lake. All are arid or semiarid deposits.

An exposure of the aeolian deposits is shown in Figure 8.7. The dune bedding is clearly visible. Vertical joint faces can be seen in the right-hand half of the picture, and occur at intervals of 1–10 m. In contrast, Figure 8.8 shows an exposure of a later fluvial deposit. No fractures are visible in this face. Occasional mudstones can be seen, although the whole sequence is dominated by sandstones as is usually the case in the formation.

Common cements are calcite and hematite, and hematite coatings to grains are very widespread. Calcite content is often insufficient to buffer landfill leachates (Mather, 1989). Organic carbon contents are low, typically 10^{-4} (Williamson, 1994). Diagenesis has often given rise to swelling clay minerals, but intergranular porosities and conductivities remain high (typically 24% and 10^{-5} m/s; Lovelock, 1977).

Overall, although there are many variations, the aquifer is an occasionally fractured medium; both fractures and matrix have significant flows. Mudstone lenses are rare but significant head gradients develop across them, indicating substantial areal extent.

Permo-Carboniferous Sandstones and Mudstones

Only some of the Carboniferous strata in the UK form aquifers. One such area is around Coventry (Figure 8.6). This area has been the subject of extensive recent researches on chlorinated solvent pollution (Burston et al., 1993), which has necessitated parallel studies of the hydrogeology (Lerner et al., 1993). The aquifer system is a complex arrangement of sandstone and mudstone lenses deposited in desert alluvial fan and river environments. The proportion of mudstones is much higher than in the Permo-Triassic rocks discussed above. Aquifer bodies cannot be correlated over large distances, although there is horizontal hydraulic interconnection.

An exposure of the Coventry sandstone formation is shown in Figure 8.9. The larger interbedded sandstones and mudstones are obvious, although laminae of a few millimeters thickness can be ob-

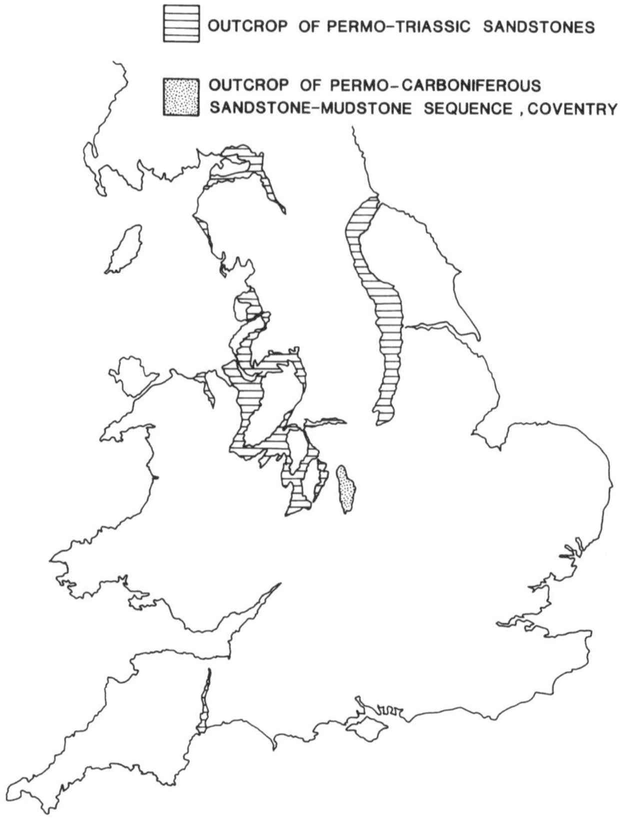

Figure 8.6. Distribution of example aquifers in England and Wales.

Figure 8.7. Exposure of Permo-Triassic dune sandstone near Bridgnorth.

served at close quarters; lenses greater than 10 m thick are rare. The sandstones are well cemented and extensive fracturing occurs. Porosities remain reasonably high at 15%, with fractures contributing a small fraction (0.4%). The mudstones do not appear to have open fractures at exposure, and can

Figure 8.8. Exposure of fluvially deposited Permo-Triassic sandstone (Marlbrook, Bromsgrove).

Figure 8.9. Exposure of Coventry sandstone (Carboniferous; Websters brickpit, Coventry).

sustain very high vertical hydraulic gradients. Carbonate content is high, and widespread hematite coatings give the red appearance. Organic carbon content is low. Overall the system behaves as a multi-layered dual-porosity aquifer.

DISCUSSION AND CONCLUSIONS

Consolidated aquifers differ from their unconsolidated ancestors, both geochemically and hydraulically. The geochemical changes will be important in pollution movement if reactions involve the mineral surfaces. Such reactions include sorption and biodegradation of organic species.

The emphasis of this contribution has been on the physical changes resulting from consolidation, in particular the role of fractures in flow and transport. The contrast between high hydraulic conductivity fractures and low conductivity matrix leads to a major role for diffusion in pollution movement. As diffusion is slow and cannot be influenced by man as easily as, say, groundwater flow, the remediation of pollution in such environments is inevitably slow.

Some simple non-dimensional parameters have been presented which illustrate the way fractures dominate groundwater flows. Even when the matrix conductivity is high enough to transport a reasonable flow, the fracture system will be flushed many times more often than the matrix. With characteristic diffusion times running into years, large volumes of water must be pumped if remediation is to be achieved by a pump-and-treat approach.

Two example sandstone aquifers in the UK have been briefly illustrated. Both have very low organic carbon contents and little potential to sorb organic pollutants. Mudstones are important in both, although they are much rarer in the Triassic sandstones. Fractures occur in both. The Carboniferous Coventry sandstones have low matrix conductivity; diffusion-controlled pump-and-treat remediation may be feasible because fractures occur frequently enough to keep characteristic diffusion times reasonable. By contrast, Triassic sandstones are infrequently fractured, but have higher matrix conductivities; remediation by advective flushing of matrix and fractures may be possible. (In this case variations of matrix conductivity will become important, and may lead to a diffusion-controlled process within matrix blocks.)

REFERENCES

Barker, J.H. 1985. Modelling the effects of matrix diffusion on transport in densely fissured media—*Memoires,* 18th IAH Congress, Cambridge UK, pp. 250–269.

Bear, J. 1979. *Hydraulics of Ground Water.* McGraw-Hill, New York.

Burston, M.W., M.M. Nazari, P.K. Bishop, and D.N. Lerner. 1993. Pollution of ground water in the Coventry region by chlorinated hydrocarbon solvents. *J. Hydrol.,* **149**:137–161.

Chen, T. and D.N. Lerner. 1992. Diffusion coefficient of trichloroethylene in Coventry sandstone. Internal Report, School of Earth Sciences, University of Birmingham, Birmingham, UK.

Cherry, J.A. Personal Communication. University of Waterloo, Ontario.

de Marsily, G. 1986. *Quantitative Hydrogeology.* Academic Press, San Diego.

Domenico, P.A. and F.W. Schwartz. 1990. *Physical and Chemical Hydrogeology.* John Wiley & Sons, New York.

Hayes, J.B. 1979. Sandstone diagenesis—the hole truth. *In* P.A. Scholle and P.R. Schlugger, Eds., *Aspects of Diagenesis,* Spec. Publ. 26, Society of Economic Paleontologists and Mineralogists, Tulsa, OK, pp. 127–139.

IGS (Institute of Geological Sciences). 1977. Hydrogeological maps of England and Wales. British Geological Survey, Keyworth, UK.

Lerner, D.N., M.W. Burston, and P.K. Bishop. 1993. Hydrogeology of the Coventry region (UK): an urbanised, multi-layer, dual-porosity aquifer system. *J. Hydrol.,* **149**:111–135.

Lovelock, P.E.R. 1977. Aquifer properties of the Permo-Triassic sandstones in the United Kingdom. Bulletin of the Geological Survey of Great Britain, 56 p.

Luikov, A.V. 1968. *Analytical Heat Diffusion Theory.* Academic Press, New York.

Mather, J.D. 1989. The attenuation of the organic component of landfill leachate in the unsaturated zone: a review. *Quar. J. of Eng. Geol.,* **22**:241–246.

Schmidt, V. and D.A. MacDonald. 1979. The role of secondary porosity in the course of diagenesis. *In* P.A. Scholle and P.R. Schlugger, Eds., *Aspects of Diagenesis,* Spec. Publ. 26, Society of Economic Paleontologists and Mineralogists, Tulsa, OK, pp. 175–207.

Vogel, P. and W. Giesel. 1989. Propagation of dissolved substances in rock: Theoretical considerations of the relationship between systems of parallel fractures and homogeneous aquifers. *In:* H.E. Kobus and W. Kinzelbach, Eds. *Contaminant Transport in Ground Water,* Balkema, Rotterdam, pp. 275–280.

Williamson, D.J. 1994. Controls on the transport of organic pollutants through UK aquifers. PhD thesis, School of Earth Sciences, University of Birmingham, UK.

Wilson, C.R. and P.A. Witherspoon. 1974. Steady state flow in rigid networks of fractures. *Water Resources Research,* **10**:328–335.

Chapter 9

Use of Site Characterization Data to Select Applicable Remediation Technologies

J.W. Mercer, R.M. Parker, and C.P. Spalding

ABSTRACT

Successful site remediation requires successful site characterization. Site characterization includes characterization of the contaminant sources, including the waste itself (e.g., residual separate phase liquid), and characterization of the media impacted by the waste. The media of interest for this discussion are soil (vadose zone) and groundwater. Both characterization and remediation depend on understanding the processes that control waste or chemical transport. Therefore, emphasis should be placed on determining the distribution of contaminants (including the chemical phase), contaminant properties, and the processes controlling their fate and transport. Selected remediation, where possible, should be consistent with the controlling transport process(es).

INTRODUCTION

Site characterization, a process following the scientific method, is performed in phases (Figure 9.1). In the initial phase, a hypothesis about the site or system behavior is made. Based on the hypothesis, a data collection program is designed in the second phase. Data are collected and an analysis or assessment follows. An iterative step of refining the hypothesis is performed using the results of the analysis, and additional data may be collected. As the knowledge of the site increases and becomes more complex, the working hypothesis may take the form of either a numerical or analytical model. Data collection continues until the hypothesis is proven sufficiently. Although possible remediations should be considered early in the site characterization process, the proof of the hypothesis forms the basis for final selection of applicable remediation technologies. During implementation of the remediation, the subsurface system often is stressed. This provides an opportunity (especially during pilot tests) to monitor and not only learn about the effectiveness of the remediation, but to learn more about the subsurface. Therefore, remediation should be considered part of the site characterization process, yielding data that may allow improvements to be made in the conduct of the remediation effort.

The first step in selecting a remediation is to define remedial objectives. These can range through the following contaminant management options: (1) plume or source containment, (2) aquifer restoration to specified cleanup levels, (3) point-of-use remediation, (4) well-head protection, and (5) source removal (includes soil). In setting these objectives, it may be necessary to distinguish between a dissolved plume and a contaminant source, such as residual separate phase liquid, because achievable goals may be different in the plume versus the source. The remediation selected depends on fluid flow data (site-specific information) and chemical data (chemical-specific information). In this chapter, *in situ* treatment technologies are emphasized. In the following discussion, various remediations are discussed based on these two types of data. These data control the processes that resulted in the distribution of contamination. A logical progression during the planning stages of a site investigation is to determine the contaminants and sources (based on site history), and then to estimate the contaminant properties from reference material. From these properties, the controlling fate and transport processes can be estimated, and a data collection program designed to obtain site-specific data to allow characterization of the controlling processes. During this planning stage, remediations can be considered that take advantage of these processes.

Data requirements are presented in this chapter as they relate to transport processes, and to appropriate soil remediation and groundwater remediation, consistent with the controlling processes. Al-

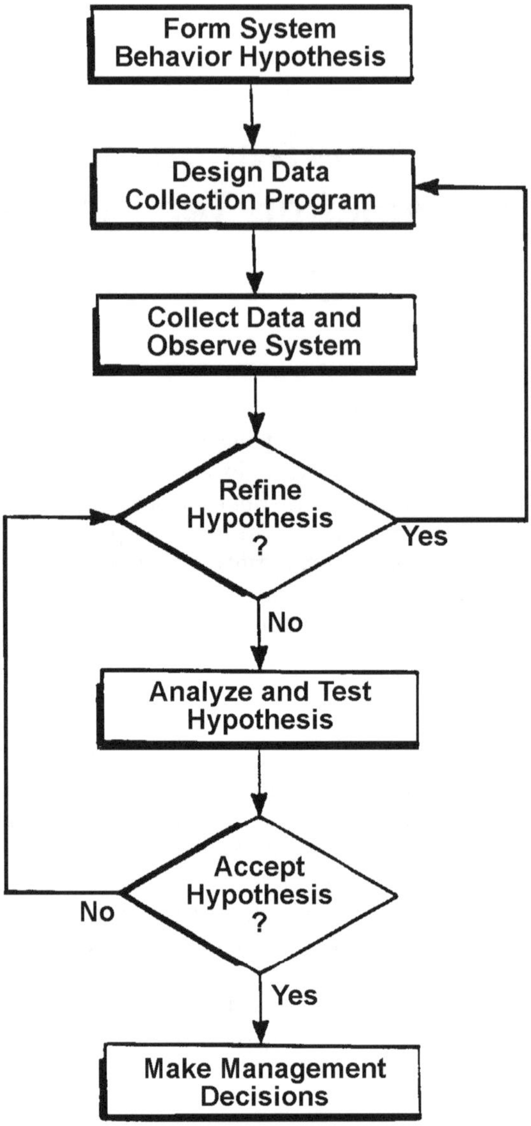

Figure 9.1. Site characterization phases (from Bouwer et al., 1988).

though discussed separately, site conditions may require that various remediations be combined to form a treatment train to address all aspects of the contamination problem (U.S. EPA, 1991). This is especially true where contaminant source areas can be delineated and have a contaminant distribution and/or phase that is different from a related contaminant plume.

REMEDIATION BASED ON CHEMICAL DATA

Remediations will be selected based, in part, on whether the chemical tends to partition into the soil air, tends to partition onto the soil matrix, or tends to stay dissolved in water, and/or tends to biodegrade. Therefore, the chemical properties related to each of these processes are discussed. In each subsection, data requirements are discussed, followed by soil remediation and treatment, and groundwater remediation and treatment. Only select remediations are discussed in this chapter; there is no attempt to be comprehensive.

Soil-Water Partitioning Characteristics

Soil-water partitioning characteristics include a variety of chemical properties that indicate a chemical's tendency to dissolve in water or sorb to the soil matrix. Removal of the chemical from the water results in retardation of the movement of the chemical front. Unfortunately, desorption (or leaching) during remediation can delay or prevent cleanup to drinking water goals. Also, contaminants can sorb to soil particles and/or colloids, which can be mobile, leading to facilitated transport. There are several mechanisms by which retardation occurs. These include precipitation and/or co-precipitation, ion exchange, and sorption, which includes phase partitioning.

Data Requirements

Numerous geochemical processes can affect groundwater chemistry and contaminant migration. Knowledge of these processes along with site- and chemical-specific information are critical in determining both the effectiveness and the efficiency of the chosen site remediation. Geochemical processes that affect inorganic groundwater chemistry include: ion speciation/complexation, acid/base reactions, precipitation/dissolution reactions, oxidation-reduction reactions, and adsorption of solutes.

Dissolved inorganic constituents in natural groundwaters can be subdivided into major, minor, and trace constituents (Table 9.1). Although groundwater contamination by trace constituents such as lead are of concern, measurement of the major and minor dissolved constituents are often overlooked. These constituents often play a major role in trace constituent dissolution and migration. An example of the control of groundwater geochemistry by the major dissolved constituents is the carbonate system.

The carbonate system is the most important acid/base system in natural groundwater. This system can neutralize strong acids or bases added to groundwater and tends to buffer the pH within a particular range. The pH, in turn, affects the dissolution of major, minor, and trace constituents in groundwater. To analyze the state of the carbonate system, determination of the pH, alkalinity, bicarbonate, carbonic acid, and carbonate concentrations and the presence of reactive minerals such as calcium is important. The partial pressure of CO_2 gas also is important in carbonate equilibrium.

The partial pressure of CO_2 gas in the atmosphere is $10^{-3.5}$ atm (Stumm, 1992). Under certain circumstances, groundwater will have partial pressures greater than the atmosphere. If these groundwaters are brought to the surface, CO_2 is off-gassed and this causes a rise in solution pH. For this reason, field measurement of pH in a groundwater sample is preferred over measurement of pH in the lab. An airstripping unit applied to groundwaters where partial pressures of CO_2 gas are greater in groundwater than in the atmosphere will become clogged by the precipitation of calcite with the rise of pH. By adding CO_2 gas to the airstripper, the pH of water can be maintained and calcite will not precipitate. Airstrippers may also be clogged due to the oxidation of iron and manganese to the Fe(+3) and Mn(+4) states that form colloidal precipitates. Recognizing and preventing clogging of the airstripper points out the importance of measuring those parameters which affect site-specific geochemical processes.

Important controlling variables for reduction-oxidation reactions include: the oxygen content of recharge water, the distribution and reactivity of organic matter and other potential reductants, the distribution of potential redox buffers of MnO_2, Fe_2O_3, $Fe(OH)_3$, and the circulation rate of groundwater (Longmire et al., 1992).

In reduction-oxidation reactions, the valence of the ion is an important determinator of the chemical state and thus degree of dissolution at a given pH. Groundwater analyses often only consider total concentration of dissolved constituents. Iron can exist in a dissolved state at a (+3) and (+2) valence, each having different chemical characteristics. Total iron concentrations will not allow determination of the chemical fate of iron in the groundwater system. Toxicity also varies according to valence. The toxicity of Cr(+6) is known to be greater than Cr(+3) (ICAIR, 1990). Eh-pH diagrams are useful in determining the stable form of the oxidized/reduced constituent. Direct laboratory analysis of the quantity of each form and the valence of the constituent are desirable.

Table 9.1. Classification of Dissolved Inorganic Constituents in Groundwater (Davis and DeWiest, 1966).

Major Constituents (greater than 5 mg/L) (balance useful for QA/QC of analysis)			
Bicarbonate	Chloride	Sodium	
Calcium	Magnesium	Sulfate	
Carbonic Acid	Silicon		
Minor Constituents (0.01–10.0 mg/L)			
Boron	Iron	Potassium	
Carbonate	Nitrate	Strontium	
Fluoride			
Trace Constituents (less than 0.1 mg/L)			
Aluminum	Copper	Phosphate	Tin
Arsenic	Lead	Platinum	Uranium
Barium	Manganese	Radium	Zinc
Bromide	Molybdenum	Selenium	
Chromium	Nickel	Thorium	

Most dissolved solutes in groundwater include calcium, magnesium, sodium, potassium, chloride, sulfate, silica, and bicarbonate ($6.4 \leq pH \leq 10.3$). Contaminated groundwaters contain other species (e.g., trace metals) greater than 0.1 mg/L.

The use of unfiltered samples to determine chemical balances can produce misleading results. In cases where only high-turbidity groundwater samples can be collected, samples preserved with acid can include dissolved aquifer material. Samples under these conditions would not be representative of the actual amount of material present in ambient site groundwaters. Puls and Powell (1992) suggest that low-rate purging and sampling (0.2–3.8 L/min) be utilized to minimize entrapment of aquifer material in samples. In some cases, e.g., karst, colloidal transport can be important, and both filtered and unfiltered samples may prove meaningful.

Aquifer matrix minerals present in minor and trace quantities often play a major role in determining groundwater chemistry. Table 9.2 presents the major reactive minerals potentially present in aquifer material. Determination of the presence or absence of these minerals is important in defining potential groundwater-soil interactions.

Adsorption is the partitioning of dissolved constituents onto soil matrix. The soil characteristics that can influence adsorption include the type and amount of adsorbent and surface area of adsorbent. Groundwater characteristics that can affect adsorption include the amount of constituent of interest, pH, Eh, complexing ligands, competing ions, temperature, and total dissolved solids and gases (EPRI, 1984).

Geochemical modeling of groundwaters has been developed as an attempt to handle the complex nature of soil-water chemical interaction for inorganic chemicals. Geochemical modeling utilizes existing thermodynamic data to determine the state of both dissolved and reactive soil mineralogy. Model input requirements are presented in Table 9.3 (water chemistry) and Table 9.4 (soil chemistry). Prior to data collection for assessment of a remediation, geochemical modeling can help determine the data required for site characterization. If sufficient data have been collected geochemical modeling can be useful in predicting the impact of future conditions on water and soil chemistry (Longmire et al., 1992).

The remainder of this discussion focuses on the sorption of organics. Sorption results from the affinity of a solute for a solid phase but may also occur from the lack of affinity of the solute for a solvent phase. In aqueous systems, this latter phenomenon is called hydrophobic adsorption. The hydrophobic character of these compounds also is responsible for the low solubility of these compounds in water.

The distribution coefficient, K_d, is a valid representation of the partitioning between the solution phase and the solid phase if the partitioning is fast (compared to the flow velocity) and reversible, and

Table 9.2. Reactive Minerals Characteristic of Aquifer Matrix Materials (Longmire et al., 1992).

Carbonates	**Oxides/Hydroxides**
Calcite ($CaCO_3$)	$Fe(OH)_3$
Dolomite ($CaMg(CO_3)_2$)	Goethite (FeOH)
Siderite ($FeCO_3$)	$Al(OH)_3$
Rhodochrosite ($MnCO_3$)	$Mn0_2$
Magnesian Calcite ($Ca_{1-x}Mg_xCO_3$)	
Sulfates	**Sulfides (reducing conditions)**
Gypsum ($CaSO_4 \times H_2O$)	Pyrite (FeS_2)
$KFe(SO_4)_2$	FeS
$Al(OH)SO_4$	As, Se, Mo
Silica (SiO_2)	**Silicates (majority of minerals)**
Cristobalite	Clays
Chalcedony	
Amorphous SiO_2	

Table 9.3. Data Requirements for Geochemical Modeling—Water Chemistry (Longmire et al., 1992).

Physical Parameters	
Temperature	Specific conductivity
pH	Total dissolved solids
Eh	Total suspended solids
Alkalinity	Color
Major Cations and Anions	
Ca, Mg, Na, K	
HCO_3^-, CO_3^-, Cl^-, SO_4^{2-}, NO_3^-	
Other Metals and Nonmetals	
Fe, Mn, Al, As, Se, Mo, NH_4, Ba, B, Ag, CN, H_2S, F, PO_4	
Stable Isotopes	
D/H, $^{18}O/^{16}O$, $^{13}C/^{12}C$	
Gases	
CH_4, CO_2, O_2	
Organics	
Volatiles, semivolatiles, humic and fulvic acids, herbicides, pesticides	

the isotherm is linear. Many organic compounds are reported to follow a linear adsorption isotherm. The distribution coefficient, K_d, is related to organic carbon content fraction according to the following:

$$K_d = (K_{oc})\,(f_{oc}) \tag{9.1}$$

where

K_{oc} is the organic carbon partition coefficient based on organic carbon content; and
f_{oc} is the fraction of total organic carbon content in terms of grams of organic carbon per gram of soil.

For many surface soils, a typical value for f_{oc} is 1.7%. The organic carbon partition coefficient is a function of the aqueous solubility (S) of a compound or the octanol-water partition coefficient (K_{ow}) of the compound, whereas f_{oc} is a property of the medium.

Table 9.4. Data Requirements for Geochemical Modeling—Soil/Rock Mineralogy and Chemistry (Longmire et al., 1992).

- Elemental analysis
- Organic compounds
- Mineralogy of major minerals and clays
- pH
- Total clay fraction
- Cation exchange capacity
- Exchangeable cations
- Carbonate mineral content
- Acid neutralizing capacity
- HCl-soluble Fe and Mn
- Organic matter content
- Pyrite content
- Gypsum content
- Petrographic analysis
  - Mineral identification
  - Mineral distribution
  - Proportion of matrix material vs. interstitial grains

The distribution coefficient is related to the retardation factor, R_f, for the saturated zone by the following:

$$R_f = 1 + (\rho_b/n)\, K_d \tag{9.2}$$

where n is porosity and ρ_b is the bulk density of the porous media. Bulk mass density is related to particle mass density, ρ_s, by the following:

$$\rho_b = \rho_s\,(1-n) \tag{9.3}$$

where $\rho_s = 2.65$ g/cm^3 for most mineral soils. The retardation factor for the vapor phase (vadose zone), R_a, is defined as the following:

$$R_a = 1 + n_w/(n_a K'_H) + \rho_b K_d/(n_a K'_H) \tag{9.4}$$

where

n_w is the water-filled porosity (moisture content);
n_a is the air-filled porosity; and
K'_H is Henry's Law constant (dimensionless form).

As may be seen, R_a increases with increasing water content and increasing distribution coefficient, and decreases with increasing air-filled porosity and increasing Henry's Law constant. More mobile chemicals have low retardation factors.

For example, consider trichloroethene (TCE), which has a K_{oc} = 126 mL/g. Using Equation 9.1 and a f_{oc} value of 1.7%, the distribution coefficient is 2.14 mL/g. If porosity is 0.3, then a typical ρ_b is calculated to be 1.85 g/cm^3, and the retardation in groundwater (using Equation 9.2) is 14.20. For unretarded chemical movement, R=1. Thus, for this situation, TCE moves at a rate of only about 7% that of the groundwater. Using the same data, but assuming the porous media is only 50% water filled, a retardation factor for the vapor phase can be calculated for TCE. If K'_H is 0.379, then using Equation 9.4, the R_a for TCE is 73.26.

Table 9.5 shows the chemical information required to determine sorption/desorption or leaching potential including: (1) water solubility, (2) octanol-water partition coefficient, (3) organic carbon

Table 9.5. Chemical and Site Information Required to Evaluate Environmental Processes.

Chemical Information	Site Information
Sorption or Leaching Potential	
Water solubility	Hydraulic conductivity distribution
Octanol-water partition coefficient	Fraction of organic carbon
Organic carbon partition coefficient	Eh-pH environment
	Moisture content
	Recharge rates
Volatilization Potential	
Vapor pressure	Hydraulic conductivity distribution (air)
Henry's Law constant	Moisture content
	Soil temperature
Degradation Potential	
Half-life	Dissolved oxygen distribution
Degradation products	Soil gas carbon dioxide
Concentration	Soil temperature
	Moisture content
	pH
	Carbon content
Chemical Reactivity	
Hydrolysis half-life	Media matrix mineralogy
Chemical kinetics	Eh-pH environment
	Major/minor ion chemistry
	Fraction of organic carbon

partition coefficient, and (4) Henry's Law constant (for the vadose zone). Site-specific data required to determine the sorption or leaching potential includes: (1) fraction of organic carbon, as well as information on (2) chemical distribution, (3) hydraulic conductivity distribution, (4) Eh-pH environment, (5) moisture content (for vadose zone), (6) recharge and/or flow rates, (7) porosity, and (8) bulk density. Of these data requirements, perhaps the most important are S, K_{oc}, K_{ow}, and f_{oc}, which are discussed below.

Water solubility governs the extent to which a separate phase or nonaqueous phase liquid (NAPL) contaminant will partition into the aqueous phase. Solubility indicates maximum possible dissolved concentration at a specific temperature. High solubility indicates low sorption tendencies. More soluble contaminants are expected to migrate further in the subsurface than less soluble compounds. The greater the water solubility of a compound, the greater the tendency for that compound to migrate with the aqueous advective flow system. Contaminants with higher water solubilities are more amenable to removal from the saturated zone by pump-and-treat technology. These same compounds are more likely to migrate through the vadose zone to groundwater. Table 9.6 shows the effect water solubility has on various remediation selection parameters.

The octanol-water partition coefficient is a measure of the extent to which a contaminant partitions between octanol and water. It is the ratio of the concentration of the compound in octanol to the concentration of the compound in water. The K_{ow} provides an indication of the extent to which a compound will sorb to a soil or an aquifer solid, particularly organic material. The greater the K_{ow} value of a compound, the greater its tendency to be sorbed in the subsurface. Table 9.7 shows how K_{ow} affects remediation selection parameters. The organic carbon partition coefficient is the ratio of the amount of chemical sorbed per unit weight of organic carbon in the soil to the concentration of the chemical in solution at equilibrium. The K_{oc} is similar to the K_{ow}. A measure of the organic carbon in the soil is f_{oc}; a higher f_{oc} value results in greater sorption.

Table 9.6. Effect of Water Solubility (mg/L) on Remediation Selection Parameters (Modified from Ney, 1990).

Remediation Selection Parameter	Water Solubility Low <10	Medium 10–1,000	High >1,000
Mobility	Minor	Either way	Yes
Adsorption	Yes	Either way	Minor
Biodegradation	Maybe	Either way	Yes
Persistence	Yes	Either way	Minor

Table 9.7. Effect of Octanol-Water Partition Coefficient on Remediation Selection Parameters (Modified from Ney, 1990).

Remediation Selection Parameter	Octanol-Water Partition Coefficient Low K_{ow} < 500	Medium K_{ow} 500–1,000	High K_{ow} > 1,000
Mobility	Yes	Either way	Minor
Adsorption	Minor	Either way	Yes
Biodegradation	Yes	Either way	Slowly
Persistence	Minor	Either way	Yes

Soil Remediation and Treatment

This section emphasizes *in situ* vadose zone remediations that may be utilized on chemicals and site situations where sorption is high. Possible remediations in this situation include: (1) solidification/stabilization, (2) containment, and (3) soil flushing. Basically, the first two remediations involve contaminant immobilization and the third one involves contaminant mobilization.

Solidification and stabilization represent immobilization techniques designed to accomplish one or more of the following (Sims, 1990): (1) produce a solid from a liquid or semi-solid waste, (2) reduce contaminant solubility, and/or (3) decrease the exposed surface area across which transfer may occur. The stabilization process reduces the hazardous potential of waste material by converting the contaminants into their least soluble, mobile, or toxic form. Data required for solidification/stabilization include reagent viscosity, permeability of soils, porosity of waste materials and soil, distribution of waste in surrounding material (rocks, soils, etc.), and rate of reaction. Table 9.8 lists other data requirements.

Containment technologies are used to isolate contaminated areas in the subsurface from the surrounding uncontaminated environment. Containment usually involves installation of a low impermeable barrier around the affected area. The barrier may take the form of a slurry wall (e.g., soil-bentonite wall or cement-bentonite wall), a grout curtain, or sheet piling cutoffs. The selection of the barrier material should take into account the compatibility of the material with the contaminant(s) and native chemistry. Containment also may include installation of a cap over the contaminated area to impede infiltration of water into that area.

Soil flushing (Table 9.8) technology involves the use of extractant solvents to remove organic and/or inorganic contaminants from soils in the subsurface. Extractant solvents may include water, water-surfactant mixtures, acids, bases, chelating agents, oxidizing agents, and reducing agents. The extractants used should be limited to those that exhibit low toxicity. Proper control measures should be exercised to prevent migration of extractant-contaminant mixtures from the area of concern. Applications are limited to soils with adequate hydraulic conductivity ($K>10^{-5}$ cm/s) and a reasonable degree of homogeneity (U.S. EPA, 1990a). The effectiveness of soil flushing relative to other vadose zone remedial technologies is not clear. Depending on the solvent used, the potential exists for mobilization of naturally occurring trace elements above drinking water standards or undesired NAPL

Table 9.8. Characterization Parameters for *In Situ* Vadose Zone Treatment Techniques (Modified from U.S. EPA, 1989).

Treatment Technique	Parameter	Purposes and Comments
Vapor Extraction		
Vacuum extraction Steam-enhanced	Vapor pressure of contaminants and moisture content	To estimate ease of volatilization
Hot air–enhanced	Permeability, porosity, particle size distribution	To determine if the soil matrix will allow adequate fluid movement
	Depth of contamination and water table	To determine relative distances; technology applicable in vadose zone
Solidification/Stabilization (undisturbed)		
Pozzolanic Polymerization Precipitation	Presence of subsurface barriers (e.g., drums, debris, geologic formations)	To assess the feasibility of adequately delivering and mixing the S/S agents
	Depth to first confining layer	To determine required depth of treatment
	Chemistry	To determine feasibility
Soil Flushing		
Steam/hot water Surfactant Solvent	Presence of subsurface barriers (e.g., drums, debris, geologic formation) and permeability	To assess the feasibility of adequately delivering the flushing solution
	Porosity and moisture content (for vadose zone)	To calculate pore volume to determine rate of treatment
	Octanol-water partition coefficient and organic carbon content of soil; CEC	To assess removal efficiency; to evaluate potential for contaminant flushing
	Alkalinity of soil	To estimate the likelihood of precipitation
	Major cations/anions	To estimate the likelihood of precipitation and plugging
Microbial Degradation		
Aerobic Anaerobic	Permeability	To determine ability to deliver nutrients and/or oxygen to matrix and to allow movement of microbes
	Contaminant concentrations and toxicity	To determine viability of microbial population
	Contaminant properties	To assess feasibility of degradation

mobilization. Due to the potential environmental impact of *in situ* soil flushing, the technology should be used in situations where other remediation technologies of lower potential environmental impact are not appropriate (U.S. EPA, 1990a).

Groundwater Remediation and Treatment

In this section, saturated zone remediations for high-sorption situations are emphasized. Possible remediations in this situation include: (1) hydraulic containment using pump-and-treat technology and (2) chemical extraction. Depending on other chemical properties, other remediations discussed in subsequent sections may also be applicable. Table 9.9 provides a summary of these remediations.

Standard pump-and-treat technology refers to the extraction of contaminated groundwater from the subsurface and subsequent treatment of the extracted water at the surface. The extraction process

Table 9.9. Characterization Parameters for Groundwater Treatment Technologies.

Treatment Technique	Parameter	Purposes and Comments
Containment		
Physical barriers	Depths to water table and first confining layer	To assess feasibility of physical barriers
Hydraulic barriers	Distribution of contaminants	To locate extraction points or extent of containment
Immobilization techniques	Permeability; flow system definition; recharge	To design hydraulic containment system
	Contaminant properties	To design immobilization or surface treatment
Removal		
Extraction wells	Distribution of contaminants	To locate extraction points
Drains	Permeability and flow system definition	To design extraction system
Combination (e.g., with injection)	Organic carbon partition coefficient and fraction of organic carbon	To assess removal efficiency and cleanup times
	NAPL presence and characteristics	To assess feasibility of removal
	Major cations/anions	To estimate likelihood of precipitation and plugging (for injection and treatment)
In Situ		
Biodegradation	Permeability	To determine ability to deliver nutrients, oxygen or EOR fluids to matrix and to allow movement of microbes or contaminants
Enhanced Oil Recovery	Contaminant concentration and toxicity	To determine viability of microbial population
	Contaminant properties	To assess feasibility of degradation
	NAPL presence and characteristics	To assess feasibility of removal and EOR efficiency

is accomplished using wells, the locations and depths of which require prior delineation of the contaminant plume and knowledge of aquifer properties and the groundwater flow system. Standard pump-and-treat is best suited for media that are homogeneous and have a hydraulic conductivity greater than about 10^{-5} cm/s (Mercer et al., 1990). Injection wells may be used to enhance contaminant recovery by increasing the hydraulic gradient toward extraction wells.

Standard pump-and-treat technology is best suited to recover dissolved mobile chemicals that have log K_{oc} or log K_{ow} values less than 3.0 and 3.5, respectively (U.S. EPA, 1990a). For chemicals that have values higher than these, and thus tend to sorb, pump-and-treat technology may be used for hydraulic containment. Hydraulic containment involves retardation of movement of a contaminant plume by using pumping wells to control groundwater hydraulic gradients. Control of hydraulic gradients can also be considered in conjunction with physical containment options (e.g., bentonite slurry walls) to limit the in-flow of water that would need to be treated.

A modification of standard pump-and-treat technology is pulsed pumping, which involves regular or periodic cessation of pumping to optimize groundwater cleanup. Pulsed pumping may be more efficient in cases where desorption and/or dissolution of contaminants in the subsurface is relatively slow. Another application is in fractured media where diffusion out of the porous blocks between fracture zones is slow. Pulsed pumping may be appropriate for relatively homogeneous hydrogeologic settings containing contaminants with log K_{oc} values between 2.0 and 4.0 (or log K_{ow} values between 2.5 and 4.5) or heterogeneous formations containing contaminants with log K_{oc} and log K_{ow} values less than 3.0 and 3.5, respectively (U.S. EPA, 1990a).

Reinjection can be used in combination with standard pump-and-treat technology or with pulsed pumping. It generally refers to injection of treated groundwater back into the subsurface. Reinjection may be accomplished through the use of injection wells or infiltration galleries. In special cases, the injectate may be amended with surfactants or extractant solvents (chemical extraction) to enhance desorption or solubilization of contaminants. Chemical extraction is most applicable in cases where contaminants are not easily mobilized or removed with water alone, i.e., strongly sorbed to aquifer solids (U.S. EPA, 1990a). Caution should be exercised during chemical extraction to prevent potential adverse impact due to the introduction of the extraction chemicals into the subsurface.

Air-Water Partitioning Characteristics

Air-water partitioning characteristics include a variety of chemical properties that indicate a chemical's tendency to volatilize into the soil air or remain in the liquid phase (dissolved in water or as a separate phase liquid). The extent to which the released chemical products partition into the vapor phase is dependent upon the characteristics of the chemical, the nature of the subsurface environment, and the elapsed time since the release occurred (Pedersen and Curtis, 1991). Only the chemical characteristics are discussed in this section.

Data Requirements

Volatilization refers to the transfer of matter from liquid and soil to the gaseous phase. Thus, chemicals in the soil gas may indicate the presence of separate phase or dissolved chemicals. Chemical/physical properties affecting volatilization from the separate phase (NAPL) chemical are vapor pressure, solubility in water, and mole fraction of the separate phase chemical; for the dissolved compound, Henry's Law constant. These are the primary factors that control volatilization (i.e., mass transfer to the soil gas); there are other properties that control vapor transport and hence affect the volatilization rate. Other factors affecting vapor transport and volatilization rate include: concentration in the soil; soil moisture content; soil air movement; sorptive and diffusion characteristics of the soil; temperature; and bulk properties of the soil such as organic carbon content, porosity, density and clay content (Lyman et al., 1982).

Raoult's law has been used to quantify the ideal reference state for the equilibrium between NAPL and air (Corapcioglu and Baehr, 1987). Raoult's law relates the ideal vapor pressure and the relative concentration of a chemical in solution to its vapor pressure over the solution:

$$P_A = X_A P_A^O \tag{9.5}$$

where

P_A is the contribution to the total vapor pressure of A in the solution;
X_A is the mole fraction; and
P_A^O is the vapor pressure of the pure compound.

Thus, a NAPL generates chemical vapors according to Raoult's Law, and the soil gas concentration near the NAPL can be calculated. That is, the volume concentration in the gas phase is:

$$C_g = \text{mol/vol} = n/V$$

where n is the number of moles of gas and V is volume. From the ideal gas equation,

$$PV = nRT$$

or

$$C_g = n/V = P/RT$$

where T is the temperature (°K), P is the vapor pressure and R is the ideal gas constant (8.20575×10^{-5} atm-m^3/mol-K). But P is total pressure or the sum of the partial pressures of each component. Thus, from Raoult's Law,

$$C_{gA} = X_A \, P_A^O / RT \tag{9.6}$$

which gives the concentration of compound A in the soil gas adjacent to the NAPL.

As an example, consider the solvent trichloroethene (TCE), which has a vapor pressure (at 20°C) of 0.079 atm and a molecular weight of 131.5 g/mol. For pure TCE, the mole fraction is 1. Thus, using Equation 9.6, the soil gas concentration is

$$\begin{aligned} C_{gA} &= (1)(0.079 \text{ atm}) / [(8.20575 \times 10^{-5} \text{ atm-m}^3 / \text{mol°-K})(293.15\text{° K})] \\ &= 3.28 \text{ mol} / \text{m}^3 \, (131.5 \text{ g} / \text{mol}) \\ &= 432.63 \text{ mg} / \text{L} \end{aligned}$$

This is the theoretical soil gas concentration of TCE near a separate phase source at 20°C; this concentration will decrease with distance from the source due to vapor transport processes. The concentration also will vary with temperature changes; vapor pressures and thus concentrations will increase with increasing temperatures.

Volatilization can also occur from a dissolved chemical. In this case, Henry's Law relates the concentration of a dissolved chemical in water to the partial pressure of the chemical in the gas:

$$P = K_H C_W \tag{9.7}$$

where

P is the partial pressure of the chemical in the gas phase (atm);
C_W is the concentration of the chemical in water (mol/m^3); and
K_H is Henry's Law constant (atm-m^3/mol).

A second method of defining Henry's Law constant is:

$$K'_H = C_a / C_W \tag{9.8}$$

where

C_a = molar concentration in air, mole/m^3;
C_W = molar concentration in water, mole/m^3; and
K'_H = alternative form of Henry's Law constant, dimensionless.

Equations 9.8 and 9.7 are related using the ideal gas law as follows:

$$K'_H = K_H/(RT) = 41.6 \, K_H \text{ at } 20\text{°C}. \tag{9.9}$$

The Henry's Law constant for TCE is 9.10×10^{-3} atm-m^3/mol at 20°C. Using Equation 9.9 this converts to a dimensionless Henry's Law constant of 0.379. Substituting this into Equation 9.8 indicates that a gas concentration of only about 2 μg/L would be in equilibrium with groundwater containing 5 μg/L TCE. The water containing TCE might be a small layer on top of the water table. Thus, monitor well results that sample over a larger vertical distance would yield lower concentrations.

In summary, important chemical properties that indicate a chemical's tendency to volatilize are vapor pressure and Henry's Law constant. The vapor pressure of a compound provides an indication of the extent to which the compound will volatilize. The tendency of a compound to volatilize will rise proportionately with its vapor pressure. Henry's Law constant provides an indication of the extent to which a compound will volatilize from an aqueous solution. Henry's Law constant is directly proportional to the vapor pressure of the compound and inversely proportional to the water solubility of the compound. The greater the Henry's Law constant of a compound, the greater will be the tendency of the compound to volatilize from an aqueous solution.

Soil Remediation and Treatment

This section emphasizes vadose zone remediations that may be utilized on chemicals and site situations where volatilization is high. Possible remediations in this situation include: (1) soil vapor extraction (SVE) and (2) *in situ* steam extraction.

SVE, considered a proven technology, involves the enhanced removal of chemicals in the subsurface through application of a vacuum. The applied vacuum enhances volatilization of compounds from soil and pore water. The technology is particularly applicable to relatively volatile organic compounds (Henry's Law constant > 10^{-3} atm-m^3/mol) residing in the vadose zone (U.S. EPA, 1990a). The technology also is applicable for removal of volatile light nonaqueous phase liquids (LNAPLs) floating on the water table or entrained in the capillary fringe, where the chemicals of concern have high vapor pressures. The process involves installation of vacuum extraction wells at strategic locations and depths. The spacing of extraction wells depends on soil properties such as permeabilities and porosity (Table 9.8). The technology is applicable to most soil types although removal efficiency will generally decrease with decreasing soil permeability and increasing subsurface heterogeneity. According to Reidy et al. (1990), the conditions where SVE are more successful are (1) soil air conductivity greater than 10^{-4} cm/s; (2) soil temperature greater than 20°C; (3) moisture content less than 10% by volume; (4) soil surface area less than 0.1 m^2/g; (5) carbon content less than 1% by weight; (6) vapor pressure greater than 100 mm Hg; and (7) water solubility less than 100 mg/L.

During SVE, contaminant removal is expected to be enhanced by decreasing soil moisture. However, when soil is very dry, volatile chemicals may sorb directly onto mineral surfaces, where the magnitude of sorption is increased and consequently volatilization is decreased. Henry's Law constant is not appropriate under these conditions, because partitioning is between air and soil phases. When moisture is added to soil, the effect is reversible (Sims, 1990). For contamination by hydrocarbons with multiple components, volatilization will proceed such that lower-molecular-weight chemicals will preferentially volatilize compared to higher-molecular-weight compounds. Through this process of weathering (selective volatilization) of the waste-soil mixture, SVE extraction efficiency is observed to decrease. As the lower-molecular-weight compounds are removed, the boiling point of the mixture increases, requiring either heat input or higher vacuum to maintain extraction efficiency.

Heterogeneous flow conditions also affect the efficiency of contaminant removal, with higher flow zones (preferential flow zones) cleaning up faster than low flow zones (less permeable zones). Due to release of volatile chemicals from the soil matrix when extraction wells are temporarily turned off, concentrations increase in soil air (referred to as "rebound effect"), with an equilibrium concentration determined by Henry's Law constant. When blowers are turned on, an increase in the concentration of extracted vapor from the soil will be observed. Diffusive release from less permeable zones will cause the slow continual release of chemicals into the soil gas (Sims, 1990).

SVE also can serve a dual purpose by enhancing removal of subsurface organic contaminants through stimulation of aerobic biodegradative processes. This is accomplished by adding oxygen for use by indigenous subsurface microorganisms. SVE also may be enhanced with hot air or *in situ* steam extraction. Steam extraction may enhance the recovery of organic chemicals, including NAPLs, from the vadose zone.

In situ steam extraction, an emerging technology, facilitates the removal of moderately volatile (vapor pressure between approximately 0.001 to 1.0 mm Hg) residual organics, including NAPLs, from the vadose zone. Steam extraction technology utilizes injection of pressurized steam to the contaminant horizon to thermally enhance the evaporative rate (by increasing the compound's vapor or partial pressure) of the contaminant and its subsequent removal. Injection of steam also can be expected to enhance removal of residual NAPLs in the vadose zone by decreasing their viscosities. Condensing steam adds water to the system that must be considered. Steam extraction is generally used in conjunction with SVE.

Groundwater Remediation and Treatment

In this section, saturated zone remediations for volatile chemicals are emphasized. In addition to pump-and-treat technologies previously discussed, air sparging may be used.

Air sparging generally involves use of injection wells to inject gas (typically air) into the saturated zone below areas of contamination. Ideally, dissolved, separate phase and sorbed contaminants will partition into the injected air, effectively creating an *in situ* air-stripping system. This can take place within a single-well system (Herrling et al., 1990) or the stripped contaminants can be transported in the gas phase to the vadose zone and collected by SVE wells (Marley et al., 1992). The injection wells can be vertical or inclined, ranging to horizontal. Although air sparging is a physical/chemical treatment process, the addition of air has the potential to promote biodegradation.

Design parameters include contaminant type and site hydrogeology and geochemistry. The compounds most amenable to air sparging are ones that are volatile and meet the condition for SVE. As with most other *in situ* methods, air sparging works best for homogeneous conditions and where sufficient permeability allows ease of flow. Heterogeneities allow for (preferential flow) bypassing zones of contamination. Precipitation of dissolved minerals, affecting permeability, is a concern due to changes in oxidation reduction (redox) potential. The air flow through the saturated porous media is governed, in part, by the physics of multiphase flow, which requires additional data discussed in the section on fluid flow data.

Degradation Characteristics

This section considers chemicals and their properties that relate to a chemical's tendency to degrade in subsurface environments. Chemicals released to the environment are susceptible to several degradation pathways. Emphasis in this discussion is on biodegradation.

Data Requirements

Biodegradation refers to biologically mediated processes that chemically alter a solute or substrate. Primary biodegradation refers to any biologically induced structural transformation in the parent compound that changes its molecular integrity. Microorganisms are responsible for biodegradation. Reactions take place both in the presence (aerobic) and in the absence (anaerobic) of oxygen.

In all biochemical reactions, the number of electrons must be conserved. When one reaction product has carbon atoms in a higher oxidized state through the loss of electrons, then another product must be present in a reduced state by having more electrons/carbon atoms. The reduced product (electron acceptor) often is not part of the original substrate. In aerobic reactions, oxygen is the electron acceptor and is reduced to water. In anaerobic systems, NO_3^- can be the electron acceptor and is reduced to NO_2^-, or N_2O, or N_2. If SO_4^{2-} is present, it also can accept electrons and be reduced to H_2S. Even CO_2 is utilized as an electron acceptor by methane bacteria to form CH_4.

Although each reaction causes the primary degradation of a compound, different reactions affect the toxicity of a compound in different ways. Mineralization refers to the complete degradation of an organic compound to inorganic products. In many reactions, however, organic products remain. Detoxication reactions produce innocuous metabolites from a toxic substance. In activation reactions, microbes convert an innocuous compound into a toxic compound. The defusing of potentially hazardous compounds occurs when biota produce an innocuous compound before the parent compound's harmful form is generated. Finally, a toxic compound may be transformed chemically but retain its toxicity.

To determine the potential for biodegradation, first determine if the contaminants of concern degrade using indigenous microbes. An indication of this may be given by looking at dissolved oxygen and soil gas carbon dioxide distributions (Suchomel et al., 1990). If dissolved oxygen in the contaminant plume is lower than in surrounding groundwater, and soil gas carbon dioxide above the plume is higher than in surrounding soil, then degradation is likely occurring. Biodegradability tests may also be performed and many common contaminants have their tendency to degrade discussed in reference texts (e.g., Montgomery and Welkom, 1990, and Montgomery, 1991). If the contaminant does degrade, the next required piece of information is to determine the transformation or degradation products, which are also listed in reference texts. Another chemical parameter is the chemical's degradation rate, or half-life.

If the decay rate is logarithmic (straight line on log C vs. t plot), then the rate of degradation is proportional to the amount of substance present and is given by the following:

$$N = N_o e^{-\lambda t}$$

where N is the number of atoms remaining after time t, and N_o is the original number present at time t=0. The coefficient λ is the decay constant and has units of t^{-1}.

Information is usually given on chemical half-lives, instead of on λ. The relationship between half-life and decay constant is straightforward. Since the half-life, $t_{1/2}$, is simply the time at which the number of atoms remaining is just one-half the original number, the above equation may be evaluated as follows:

$$N/N_o = 1/2 = e^{-\lambda t_{1/2}}$$

or

$$\begin{aligned} t_{1/2} &= \ln 2 / \lambda \\ \lambda &= 0.693 / t_{1/2} \end{aligned} \tag{9.10}$$

Therefore, to determine the decay constant from half-life data, the half-life value can be divided into 0.693. The smaller the half-life value, the faster the chemical will degrade.

Hydrolysis, an abiotic transformation, is another first-order degradation process. Certain chemicals undergo OH^--induced hydrolysis by dehydrohalogenation. At a fixed pH, the overall rate process is pseudo first-order and the half-life of the chemical is independent of its concentration, as indicated by Equation 9.10. Hydrolytic rate constants (half-lives) for some chemicals are provided in the literature.

Soil Remediation and Treatment

Four approaches are generally used for *in situ* biological treatment, including (Sims, 1990): (1) enhancement of biochemical mechanisms for detoxifying or degrading chemicals, (2) augmentation with exogenous acclimated or specialized microorganisms, (3) application of cell-free enzymes, and (4) vegetative update. Enhancement of biochemical mechanisms may involve: (1) control of soil factors such as moisture, pH, nutrients, and temperature to optimize microbial activity, (2) addition of organic amendments to stimulate cooxidation or cometabolism, (3) control of soil oxygen by moisture control to accomplish aerobic or anaerobic biodegradation, and (4) addition of colloidal gas aprons (microscopic bubbles of gas) to increase the concentration of terminal electron acceptor (oxygen) in the soil and thereby enhance aerobic biodegradation (U.S. EPA, 1990b).

Microbial ecologists have identified ranges of critical environmental conditions that affect the activity of soil microorganisms (Table 9.10). These conditions can be changed to enhance potential biodegradation. Water is necessary for microbial life, and the soil water matrix potential is the energy required by microorganisms to extract water from the soil pores to overcome capillary and adsorptive forces. Soil water also serves as the transport medium through which many nutrients and organic constituents diffuse to the microbial cell, and through which metabolic waste products are removed (U.S. EPA, 1991). Soil water affects soil aeration status, nature and amount of soluble materials, soil-water osmotic pressure, and the pH of the soil solution (Paul and Clark, 1989). Heterogeneous sub-

Table 9.10. Critical Environmental Factors for Microbial Activity (Modified from U.S. EPA, 1991).

Environmental Factor	Optimum Levels
Flow conditions	Homogeneous condition; hydraulic conductivity greater than 10^{-3} cm/s.
Available soil water	25–85% of water holding capacity; –0.01 MPa.
Oxygen	Aerobic metabolism: Greater than 0.2 mg/L dissolved oxygen, minimum air-filled pore space of 10%.
	Anaerobic metabolism: O_2 concentrations less than 1% in soil gas.
Redox potential	Aerobes and facultative anaerobes; greater than 50 millivolts.
	Anaerobes: less than 50 millivolts.
pH	5.5–8.5
Nutrients	Sufficient nitrogen, phosphorus, and other nutrients so not limiting to microbial growth (suggested C:N:P ratio of 120:10:1).
Temperature	15–45°C (Mesophiles).
Microbial parameters	Adequate cell count and metabolic activities.

surface flow conditions may limit water (and nutrient) access to low hydraulic conductivity zones that are contaminated, thus limiting biodegradation.

Respiration of microbes, plant roots, and other organisms removes oxygen from soil gas and enriches it with carbon dioxide. Gases diffuse into the soil from the air above it, and gases in the soil diffuse into the air. However, oxygen concentration in a soil may be much less than in air, while carbon dioxide concentrations may be many times that of air (U.S. EPA, 1991). Even so, a large fraction of the microbial population within the soil depends on oxygen as the terminal electron acceptor in metabolism. When soil pores become filled with water, the diffusion of gases through the soil is restricted. Oxygen may be consumed faster than it can be replaced by diffusion from the atmosphere, and the soil may become anaerobic. Clay content of soil and the presence of organic matter also may affect oxygen content in soil. Clayey soils tend to retain a higher moisture content, which restricts oxygen diffusion, while organic matter may increase microbial activity and deplete available oxygen.

One way to provide oxygen to the vadose zone is through SVE wells. SVE has the potential to enhance the biodegradation of volatile and semivolatile chemicals in the soil by reinjection of air. Vacuum extraction provides air to the vadose zone, and thus carries oxygen that can be used by soil microorganisms to biodegrade contaminants. This is referred to as bioventing. Hinchee (1989) and Hinchee and Downey (1990) successfully applied SVE for the enhancement of biodegradation of petroleum hydrocarbons in JP-4 jet fuel at Hill Air Force Base, Ogden, Utah by increasing subsurface oxygen concentrations. Soil moisture was found to be a sensitive variable affecting biodegradation, with increased soil moisture related to increased biodegradation. Monitoring temperature and carbon dioxide and oxygen concentrations, as well as estimating the mass of chemical biodegraded, is recommended for evaluation of potential enhanced biodegradation.

Groundwater Remediation and Treatment

Wilson et al. (1986) present an overview of bioreclamation as a groundwater remediation technique. In the saturated zone, oxygen needed for aerobic biodegradation is available in a dissolved state at relatively low concentrations. Many of the bioreclamation systems are derived from a patented process at Suntech, Marcus Hook, Pennsylvania, U.S.A. (Raymond, 1974). The first step in this process is to recover mobile NAPL (free product), if present. Next, laboratory studies are conducted to determine if the native microbial population can degrade the contaminants. Finally, a combination of oxygen (from air or supplied by chemicals) and nutrients (nitrogen, phosphorus and potassium-NPK) is injected into the contaminated zone to enhance natural biodegradation (biostimulation). Thus, bioremediation involves enhancing biodegradation of contaminants in the subsurface through the artificial stimulation of indigenous soil and groundwater microbial populations. Controlling groundwater flow in the saturated zone is critical for transporting oxygen and nutrients to the contaminated zone and enhancing the degradation process. Therefore, the limiting process in bioreclamation is convective transport of oxygen and nutrients, not rate limitations due to degradation. Hence, site characterization providing a detailed definition of hydraulic conductivity is required for a successful bioreclamation design.

In situ bioremediation can be applied as a specialized form of pump-and-treat technology. The successful execution of an *in situ* bioremediation depends on: (1) amenability of the organic compounds to biodegrade, (2) a hydraulic conductivity greater than 10^{-3} cm/s (Reidy et al., 1990), (3) simple site stratigraphy, (4) ability of the delivered oxygen and nutrients to reach the contaminated area, (5) groundwater temperature greater than 10°C and a pH between 6–8 (Reidy et al., 1990), and (6) microbial parameters (e.g., total cell count, denitrifiers, hydrocarbon degraders). Determination of iron and manganese concentrations is important because high concentrations of these metals can cause precipitation under aerobic conditions, caused by the infiltration of oxygen during the bioremediation process. Other heavy metals can be important, especially at contaminated sites, because at toxic levels, they can inhibit the activity of microorganisms. Inorganic nutrients like nitrogen and phosphorus may be limiting when the carbon/nitrogen/phosphorus (C:N:P) ratio is unfavorable. Determination of ammonium, nitrate, and nitrite gives insight in the stage of the conditions in the subsoil (Staps, 1990). Lack of sufficient microorganisms may be overcome by bioaugmentation—introduction of exogenous bacteria—especially acclimated to specific contaminants.

REMEDIATION BASED ON FLUID FLOW DATA

Fluid flow data includes the permeable media characteristics (e.g., porosity, fractures, stratigraphy) and the fluid characteristics (e.g., air, water, nonaqueous phase liquid—NAPL). Therefore, media and fluid characteristics are discussed. Because many remediations have been discussed previously, data requirements and only those remediations that have not been previously discussed are covered in this section.

Permeable Media Characteristics

A conceptual model of the nature and scope of a groundwater contamination problem is needed before site characterization and an appropriate remedial action can be determined. This model is used to begin the first phase of a site investigation and is modified as new data and information develop.

Data Requirements

One of the key elements affecting remedial design is the characterization of the groundwater flow system (Mercer et al., 1990). This includes: the hydraulic properties of the contaminated region (e.g., hydraulic conductivity, porosity, and storage coefficient distributions); system stresses (e.g., recharge, discharge, pumping rates, and hydraulic gradient); and system geometry (e.g., physical and hydraulic boundaries, stratigraphy, lithology, structural geology, and heterogeneities). For long-term pumping, the storage coefficient is less significant than the hydraulic conductivity. By understanding where groundwater recharges and discharges (mass balance), the laws governing flow (e.g., Darcy's Law), and the geological framework through which this flow occurs, it is possible to determine these characteristics. It is important to portray the flow system accurately so the impact of installing a remedial system can be properly analyzed. Table 9.11 lists the information typically used for remedial design requiring hydrogeological and engineering expertise.

Because migrating miscible (dissolved) contaminants travel with moving groundwater, it is important to characterize groundwater flow. Groundwater flows from areas of recharge (commonly via

Table 9.11. Site/Hydrogeological Data Considered in Remediation.

- Load bearing capacity
- Subsidence—past, potential
- Buried objects—size, what, location
- Vertical extent of contamination—rate of movement
- Horizontal extent of contamination—rate of movement
- Earthquake zone
- 50-100-200 year flood plains
- Maximum-normal-minimum water table elevations
- Lithology
- Lithology elevations (profile view—gradients)
- Aquifers—number, depth, flow rate/direction, thickness
- Integrity of confining layers
- Potential receptors—time to reach
- Air/water extraction rates
- Preferential pathways
- Recharge sources and rates
- Porosity (pore volume quantity)
- Feasibility of physical containment (100–125 feet maximum depth)
- Media saturation (actual, residual)
- Potential to seep through surface
- Potential of surface gas releases
- Heat capacity of media (specific heat)
- Cone of influence—extraction wells
- Mounding—reinjection wells
- Site development—paving, buildings
- Site activity—operational, dormant
- Climatological—wind data, temperatures, precipitation
- Surrounding area use
- Population density
- Cleanup goals
- Time constraints

rainfall, surface water bodies, or irrigation) to areas of discharge (surface water or wells). Along the way, subsurface heterogeneities (such as fractures) influence its direction. The rate of groundwater flow is controlled by the porosity and hydraulic conductivity of the media through which it travels and by hydraulic gradients, which are influenced by recharge and discharge (see e.g., Freeze and Cherry, 1979).

Pumping wells influence the flow system. If contamination is detected in a water supply, there has been a tendency to close the well. This alters the flow system and causes the contaminant's plume to migrate elsewhere. Depending on the site, it may be advantageous to install well-head treatment and keep the well on-line to prevent further plume migration. Conversely, it may be advantageous to close the well if it is believed further pumping might exacerbate spreading of the plume. This interim remedial action may be consistent with and can become part of a final remedy.

It is important to conduct a site characterization quickly; however, groundwater flow systems vary with time. Seasonal variations in water levels, which often are several feet, can adversely impact remediation. For example, at one site, an intercept drain was constructed to collect contaminated groundwater but was designed based on only one survey of water levels. Subsequent monitoring revealed that the water levels represented a seasonal high. Thus, for most of the year, the groundwater intercept drain was above the water table and did not collect the contaminated groundwater. Groundwater transport is generally a slow process with plumes moving a few tens of feet per year or less. Therefore, adequate time should be taken to characterize the groundwater flow system.

A particularly difficult problem to characterize and remediate is heterogeneity in aquifer properties, especially permeability. One of the most heterogeneous systems is fractured rock. A complete description of a contaminated fractured rock system would include data on (U.S. EPA, 1991): (1) the dimensions of the system; (2) individual fracture length, aperture width, location, and orientation; (3) the hydraulic head throughout the system; (4) the porosity and permeability of the rock matrix; (5) the sources of water and contaminants; (6) the nature and concentrations of the contaminants throughout the system; and (7) the chemical interactions between the contaminants and rock matrix. Presently, collection of such detailed information is neither technically possible nor economically feasible at the scale of most contaminated sites.

Soil Remediation and Treatment

Typical remediations for soils include (Sims, 1990): biological treatment technologies, physical/chemical treatment technologies, immobilization technologies, thermal treatment technologies, and *in situ* treatment technologies. Just limiting the discussion to *in situ* treatment technologies for soil, the options include vapor extraction, solidification/stabilization, soil flushing, and biodegradation, which have been discussed previously.

Fluid flow characterization data required to evaluate these soil remediations include a variety of media and fluid properties, in addition to the contaminant distribution. Permeability (including heterogeneities), porosity, and particle size distribution are needed to determine if the soil matrix will allow adequate fluid movement. This includes movement of vapors for vapor extraction, fluids containing nutrients and/or oxygen for biodegradation, and flushing solution for soil flushing. The depth to water, moisture content, and recharge rate need to be determined to assess the applicability of certain technologies as well as help estimate fluid movement. To assess the effects of sorption on removal efficiency, the organic carbon content of the soil (for organic contaminants) and the cation exchange capacity (for inorganics) need to be determined. To estimate the likelihood of chemical precipitation, soil alkalinity, major cation/anion chemistry, and reactive soil mineralogy should be determined. Finally, for solidification/stabilization, the depth to the first confining layer may need to be estimated through soil borings and/or well completions.

Groundwater Remediation and Treatment

There are three basic approaches to solve a groundwater contamination problem: containment to prevent the problem from spreading; removal and treatment of the extracted water; and *in situ* treatment, whereby contaminants are transformed underground to innocuous substances (Table 9.9). For containment, the options include physical barriers, hydraulic barriers, and immobilization techniques. To assess the feasibility of physical barriers, the depths to the water table and the first confining layer need to be determined. To design a hydraulic containment system, the permeability and flow system need to be defined.

For removal and treatment, only the removal portion is considered, and includes extraction wells, drains or a combination including injection wells or infiltration galleries. To design an efficient extraction system, the permeability and flow system must be defined. To assess removal efficiency and cleanup times (extended due to possible sorption), the fraction of organic carbon needs to be determined. To assess the feasibility of removal, the presence of NAPL and its characteristics need to be estimated. For injection wells and surface treatment, the likelihood of chemical precipitation and plugging need to be estimated using major cation/anion chemistry.

For heterogeneous conditions such as fractured media, advected water will sweep through the more permeable zones (fractures), removing contamination from those zones. Movement of contaminants out of the less permeable zones is a slower process than advective transport in the more permeable zones. The contaminants either are slowly exchanged by diffusion with the flow water present in the larger pores or move at relatively slower velocities in the smaller pores. A rule of thumb is that the longer the site has been contaminated and the more lenticular (layered) the geologic material, the longer will be the tailing effect. The water and contaminants residing in the more permeable zones are those first mobilized during pumping. Thus, pump-and-treat technologies work in heterogeneous media, but cleanup times will be longer and more difficult to estimate than for similar systems in more homogeneous media.

Fluid Characteristics

The phase of the chemical needs to be determined; that is, whether the chemical is dissolved in water, contained in soil vapor, sorbed, or in a nonaqueous phase (NAPL). The phase(s) of the contaminant may vary depending on the proximity to a source area.

Data Requirements

Contaminant information includes: (1) source characterization, (2) concentration distribution of contamination and naturally occurring chemicals, and (3) data associated with the processes that affect plume development. Source characterization consists of the following: (1) the chemical volume released, (2) the area infiltrated, and (3) the time of and duration of release. Often, the release occurred so long ago that information is difficult to obtain. If the release (e.g., a leaking tank) is ongoing, the source should be located and the release eliminated.

The presence of a separate nonaqueous phase greatly complicates the contaminant characterization (Mercer et al., 1990). Movement of a contaminant as a separate, immiscible phase is not well understood in either the saturated or vadose zones. A nonaqueous phase moves in response to pressure gradients and gravity. Its movement and, hence, recovery, is influenced by interfacial tension and by the processes of volatilization and dissolution.

The additional data requirements for NAPLs include: (1) fluid specific gravity (density), (2) fluid viscosity, (3) residual saturation, (4) relative permeability-saturation-capillary pressure relationships, (5) NAPL thickness and distribution, (6) NAPL composition, and (7) solubility in groundwater. Following a spill or release, light NAPLs tend to spread over the water table. Dense nonaqueous phase liquids (DNAPLs) tend to move downward through the water table until reaching a low-permeability barrier, such as a confining bed. Examples of DNAPLs include 1,1,1-trichloroethane, carbon tetrachloride, pentachlorophenols, dichlorobenzene, tetrachloroethene, and creosote; examples of LNAPLs include gasoline, heating oil, kerosene, jet fuel, and aviation gas. Following a spill, a product of low viscosity will penetrate more rapidly into the soil than a product with high viscosity.

NAPL properties (data requirements) are listed in Table 9.12 and selected properties are discussed below. The density of a NAPL indicates whether the NAPL is heavier or lighter than water. The density of water is approximately 1.0 g/cm^3. Dynamic viscosity provides an indication of the ease with which a NAPL will flow. The mobility of the NAPL is inversely proportional to its dynamic viscosity. The dynamic viscosity of water is approximately 1.0 centipoise. The kinematic viscosity, the dynamic viscosity divided by density, provides an indication of the ease with which the NAPL will percolate through the subsurface. The lower the kinematic viscosity, the greater will be the tendency to migrate. The kinematic viscosity of water is approximately 1.0 centistoke.

Another useful property is the melting point. The melting point of a compound indicates the physical state of a pure compound at field temperatures. Compounds with melting points lower than 30°C may be present as mobile NAPL. Compounds with melting points above 30°C, for example, would be expected to be immobile in pure form, but would represent a source for dissolved concentration. Thus, such compounds would be of primary concern when in the dissolved phase, either in water or other solvent.

Residual saturation, also known as irreducible saturation, is the saturation below which fluid drainage will not occur. The residual saturation depends mainly on two factors: (1) the distribution of soil pore sizes, and (2) the type of immiscible fluid involved. Residual saturations are difficult to estimate accurately and are subject to considerable error.

The residual saturation of hydrocarbons has important consequences on soil cleanup, petroleum product recovery, and groundwater contamination. As oil moves through a soil, it leaves oil trapped at residual saturation. The amount of oil retained in the soil is normally between 15 and 40 liters per cubic meter (Fussell et al., 1981). According to API (1980), this trapped oil can last for many years as

Table 9.12. Properties of NAPL and Media that Influence Contaminant Migration (Mercer and Waddell, 1993).

Property	Definition	Range of Values
Saturation	The volume fraction of the total void volume occupied by that fluid at a point.	Zero to one.
Residual saturation	Saturation at which NAPL becomes discontinuous and is immobilized by capillary forces.	Approximately 0.1 to 0.5.
Interfacial tension	The free surface energy at the interface formed between two immiscible or nearly immiscible liquids. Surface tension is the interfacial tension between a liquid and its own vapor.	Values of interfacial and surface tensions for NAPL-forming chemicals generally range between 0.015–0.5 N/m.
Wettability	Describes the preferential spreading of one fluid over solid surfaces in a two-fluid system; it depends on interfacial tension.	Determined through contact angle studies. Commonly, in the vadose zone, NAPL is the wetting fluid; in the saturated zone, water is the wetting fluid.
Capillary pressure	The difference between the nonwetting fluid pressure and the wetting fluid pressure. It is a measure of the tendency of a porous medium to attract the wetting fluid and repel the nonwetting fluid.	Depends on the interfacial tension, contact angle, and pore size.
Relative permeability	The ratio of the effective permeability of a fluid at a fixed saturation to the intrinsic permeability.	0.0 to 1.0, depending on the fluid saturation.
Solubility	The maximum concentration of the chemical that will dissolve under a set of chemical and physical conditions.	Varies widely depending on chemical and aquifer conditions.
Volatilization	The transfer of matter from liquid and soil to the gaseous phase.	Depends on organic partitioning between water and air, and NAPL and air.
Density	Mass per unit volume of a substance. Specific gravity is the ratio of a substance's density to that of some standard substance, usually water.	Specific gravities of petroleum products may be as low as 0.7, whereas chlorinated aliphatic compounds can be as high as 1.2 to 1.5.
Viscosity	The internal friction within a fluid that causes it to resist flow.	Varies depending on fluid and temperature.

the oil slowly degrades. While residual saturation has the effect of depleting a plume of oil, thus reducing the contamination impact of pure product migrating within the saturated zone, it has the detrimental effect of providing a long-term source of miscible contaminants. For NAPLs subject to water-table fluctuations, residual saturations can occur below the water table. This has detrimental consequences for a remediation system.

When more than one fluid exists in a porous medium, the flowing fluids compete for pore space. The net result is that the mobility is reduced for each fluid. The reduction can be quantified by multiplying the intrinsic permeability by a dimensionless ratio, known as relative permeability, k_r. Relative permeability is the ratio of the effective permeability of a fluid at a fixed saturation to the intrinsic permeability. Relative permeability varies from zero to one and can be represented as a single-valued function of phase saturation, S. At residual saturation, S_r, the respective relative permeability becomes zero; that is, flow ceases to occur and product recovery stops.

Remediation and Treatment

For *in situ* treatment of soils, SVE, bioventing and other possible remediations have been discussed. For *in situ* treatment of groundwater, the options considered are pumping mobile NAPL, enhanced oil recovery (EOR) technologies, and biodegradation (discussed previously). Permeability and associated heterogeneities need to be estimated to determine the ability to deliver nutrients, oxygen, or EOR fluids to the matrix and to allow movement of microbes or contaminants. The presence of NAPL and its physical and chemical characteristics (e.g., density, viscosity) need to be determined to assess the feasibility of removal and EOR efficiency.

Experience in the recovery of mobile NAPL currently is limited and pertains almost exclusively to the recovery of lenses of LNAPLs, most notably petroleum products floating on the surface of the water table. Blake and Lewis (1983) provide a summary of the special considerations and procedures involved in the recovery of LNAPLs. Other references describing recovery techniques include de Pastrovich et al. (1979), API (1980, 1989), and Fussell et al. (1981). Examples of DNAPL recovery are provided by Villaume et al. (1983a,b) for coal tar from gravel, and by Ferry et al. (1986) for DNAPL from fractured rock.

Once mobile NAPL is collected, immobile NAPL remains at residual saturation. Residual saturation decreases with decreasing capillary pressure; therefore, enhanced oil recovery (EOR) methods have been suggested for residual NAPL removal because they reduce the interfacial tension and/or NAPL viscosity by flooding with one of the following (Mercer and Cohen, 1990): (1) hot water or steam, (2) carbon dioxide, (3) surfactant, (4) alcohol, (5) alkali, or (6) polymers. For further information on these topics, see Shah (1981). If EOR is used to mobilize NAPL at residual saturation, NAPL flow must be controlled carefully. Otherwise, previously clean portions of the subsurface may become contaminated during remediation.

EOR techniques are largely untested for environmental problems. There are technical problems, especially when dealing with shallow water table aquifers where injection pressure is limited by aquifer thickness. Laboratory studies and small-scale field experiments are likely to yield overoptimistic expectations (Mackay and Cherry, 1989). In the field, unexpected spatial variability in porous medium parameters can detrimentally affect the outcome of an EOR application. In addition to the technical problems, high costs and regulatory restrictions are associated with injection. Finally, primary recovery typically removes ~30–40% of the NAPL. Secondary and tertiary recovery, such as EOR, if successful, may remove only an additional 30–50%. Thus, as much as 10–40% of the NAPL can remain in the subsurface after successful application of the EOR methods (Mercer and Cohen, 1990).

TREATMENT TRAINS

It may be necessary to implement a combination of treatment technologies (treatment train) to effectively remediate or control subsurface contamination. An example of such a combination is pump-and-treat with *in situ* bioremediation or chemical extraction (U.S. EPA, 1990a). Another example is a different treatment for the dilute portion of a plume than that used for the higher-concentration portion of the plume. A treatment train may be appropriate at sites where contaminants are sorbed and pump-and-treat alone is expected to have limited success. *In situ* bioremediation or chemical extraction could facilitate removal of the sorbed contaminants, thereby enhancing the overall remediation effort. In general, *in situ* bioremediation or chemical extraction is most effective after initial recovery efforts using pump-and-treat alone have been completed (U.S. EPA, 1990a).

Another useful treatment combination involves pump-and-treat and containment, such as walls to reduce the influx of water requiring treatment. This combination may be of interest in cases where DNAPL is distributed in a spatially discontinuous manner. DNAPL recovery in such a case will be very difficult, and the only option may be to control and/or contain the contamination. Pump-and-treat could initially be used to reduce the size of the aqueous phase contaminant plume generated by the DNAPL. Physical containment could then be used to isolate the DNAPL source area.

An additional treatment combination, which may be of interest, is aquifer dewatering using pump-and-treat followed by soil vapor extraction. This combination of technologies may be used in cases where an aquifer is contaminated with volatile organics and dewatering portions of the aquifer is feasible. Pumping could be used to dewater a portion of the aquifer so that SVE could be applied to enhance volatilization and biodegradation of the volatile organic contaminants in the dewatered zone.

Combinations involving more than two treatment technologies also should be considered in efforts to optimize cleanup of subsurface contamination. These combinations of remediation can be applied in parallel or in series. If applied in series, a determination is required as to when to advance to the next treatment. This decision should incorporate several factors, including economics. Focusing only on data, however, the decision can be made by plotting concentration or mass removal rate versus time. Such plots often demonstrate that concentration and/or mass removal rate approaches an asymptote. When this occurs, additional efforts using the current remediation produce little additional cleanup. At this point, the next step in a possible treatment train should be considered, i.e., revision of operating parameters of the remedial method in current use, or substitution of an entirely different unit process/operation.

CONCLUSIONS

Both site characterization and remediation should be performed in phases, where data collected is used to improve site understanding and conceptualization. Early in the site characterization process, the contaminants of concern need to be identified. Contaminant properties should be determined and used to estimate controlling fate and transport processes. A data collection strategy should be designed to obtain site-specific data to allow characterization of the controlling processes, and remediations should be considered during site characterization that take advantage of these processes.

The fate and transport processes of the contaminants along with the hydrogeology determine the ease of cleanup at hazardous waste sites (Table 9.13). In this table, the complexity of contaminant chemistry increases from left to right and the complexity of hydrogeology increases from top to bottom. In this scheme, a dissolved, mobile chemical that degrades and/or volatilizes and is contained within a single subsurface layer that is homogeneous (and isotropic) is the easiest to clean up, as indicated by the ranking of 1 in the table. In general, conditions that are given the ranking of 1 can be

Table 9.13. Ease of Cleanup at Hazardous Waste Sites as a Function of Contaminant Chemistry and Hydrogeology.

	Contaminant Chemistry					
Hydrogeology	Mobile, Dissolved (degrades/ volatilizes)	Mobile, Dissolved	Sorbed, Dissolved (degrades/ volatilizes)	Sorbed, Dissolved	Separate-Phase LNAPL	Separate-Phase DNAPL
Homogeneous, single layer	1[a]	1	2	2	3	3
Homogeneous, multiple layers	1	1	2	2	3	3
Heterogeneous, single layer	2	2	3	3	3	4
Heterogeneous, multiple layers	2	2	3	3	3	4
Fractured	3	3	3	3	4	4

[a] Relative ease of cleanup where 1 is easiest and 4 is most difficult.

cleaned up within reasonable time frames using standard pump-and-treat systems, SVE in the vadose zone for chemicals that volatilize, or biodegradation for chemicals that degrade.

As the ability to clean up becomes more difficult, the ranking is increased, with the rank of 4 given to the most difficult conditions. The capability of pump-and-treat systems, or any other groundwater remediation, to achieve low, risk-based concentrations within reasonable time frames is diminished with increasing rank in Table 9.13. The physical-chemical subsurface system must allow the contaminants to flow to the extraction wells. Consequently, the subsurface must have sufficient hydraulic conductivity to allow fluid to flow readily and uniformly, and the chemicals must be transportable by the fluid, thus making the use of remediation highly site specific. For many conditions ranked 3, and all conditions ranked 4, combinations of remediations, or treatment trains, may be required, with containment being the most feasible option presently available for certain conditions (e.g., DNAPL in heterogeneous media).

REFERENCES

American Petroleum Institute. 1980. *Underground Spill Cleanup Manual.* American Petroleum Institute, Washington, DC, Publication No. 1628.

American Petroleum Institute. 1989. *A Guide to the Assessment and Remediation of Underground Petroleum Releases.* American Petroleum Institute, Washington, DC, Publication No. 1628, 2nd edition.

Blake, S.B. and R.W. Lewis. 1983. Underground oil recovery. *Ground Water Monitoring Review.* **3**:40–46.

Bouwer, E., J. Mercer, M. Kavanaugh, and F. DiGiano. 1988. Coping with ground water contamination. *J. Wat. Poll. Cont. Fed.* Aug., Washington, DC, pp. 1415–1427.

Corapcioglu, M.Y. and A.L. Baehr. 1987. A compositional multiphase model for groundwater contamination by petroleum products, 1. Theoretical considerations. *Water Resources Research.* **23**:191–200.

Davis, S.N. and R.J.M. DeWiest. 1966. *Hydrogeology.* John Wiley & Sons, Inc., NY.

de Pastrovich, T.L, Y. Baradat, R. Barthel, A. Chiarelli, and D.R. Fussell. 1979. *Protection of Groundwater from Oil Pollution. CONCAWE* (Conservation of Clean Air and Water—Europe), The Hague.

Electric Power Research Institute (EPRI). 1984. *Solute Migration in Ground Water at Utility Waste Disposal Sites.*

Ferry, J.P., P.J. Dougherty, J.B. Moser, and R.M. Schuller. 1986. Occurrence and recovery of a DNAPL in a low-yielding bedrock aquifer. *Proceedings of Petroleum Hydrocarbons and Organic Chemicals in Ground Water: Prevention, Detection and Restoration,* National Water Well Association, Houston, TX, November 12–14, pp. 722–733.

Freeze, R.A. and J.A. Cherry. 1979. *Groundwater.* Prentice-Hall, Englewood Cliffs, NJ.

Fussell, D.R., H. Godjen, P. Hayward, R.H. Lilie, A. Marco, and C. Panisi. 1981. *Revised Inland Oil Spill Clean-up Manual. CONCAWE* (Conservation of Clean Air and Water—Europe), The Hague, Rep. No. 7/81.

Herrling, B., W. Buermann, and J. Stamm. 1990. In-situ remediation of volatile contaminants in groundwater by a new system of "Under-pressure Vaporizer-Wells." *Proceedings of the Conference on Subsurface Contamination by Immiscible Fluids,* K.U. Weyer (ed.), Calgary, April 18–20, A.A. Balkema Publisher, Rotterdam.

Hinchee, R. 1989. Enhanced biodegradation through soil venting. *Proceedings of the Workshop on Soil Vacuum Extraction,* U.S. Environmental Protection Agency, Robert S. Kerr Environmental Research Laboratory, Ada, OK, April 27–28.

Hinchee, R. and D. Downey. 1990. In situ enhanced biodegradation of petroleum distillates in the vadose zone. *Proceedings of International Symposium on Hazardous Waste Treatment: Treatment of Contaminated Soils,* Air and Waste Management Association, U.S. Environmental Protection Agency, Risk Reduction Engineering Laboratory, Cincinnati, OH, February 5–8.

ICAIR, Life Systems, Inc. 1990. The drinking water criteria document on chromium. TR-1242-64A, NTIS, PB91-142844, p. I-5.

Longmire, P., W. Deutsch, and R. Meyerhein. 1992. Course notes introduction to groundwater geochemistry short course, AGSWE, Denver, CO (February 25–27).

Lyman, W.J., W.F. Reehl, and D.H. Rosenblatt. 1982. *Handbook of Chemical Property Estimation Methods—Environmental Behavior of Organic Compounds.* McGraw-Hill, NY.

Mackay, D.M. and J.A. Cherry. 1989. Groundwater contamination: Pump-and-treat remediation. *Environmental Science and Technology.* **23**:620–636.

Marley, M.C., D.J. Hazebrouck, and M.T. Walsh. 1992. The application of in situ air sparging as an innovative soils and ground water remediation technology. *Ground Water Monitoring Review.* **12**:137–145.

Mercer, J.W. and R.M. Cohen. 1990. A review of immiscible fluids in the subsurface: Properties, models, characterization and remediation. *J. Contam. Hydrol.* **6**:107–163.

Mercer, J.W., D.C. Skipp, and D. Giffin. 1990. Basics of pump-and-treat ground-water remediation technology. U.S. EPA Report EPA/600/8-90/003.

Mercer, J.W. and R.K. Waddell. 1993. Contaminant transport in groundwater (Chapter 16). In D.R. Maidment (ed.), *Handbook of Hydrology*, McGraw-Hill, NY.

Montgomery, J.H. and L.M. Welkom. 1990. *Groundwater Chemicals Desk Reference.* Lewis Publishers, Inc., Boca Raton, FL.

Montgomery, J.H. 1991. *Groundwater Chemicals Desk Reference*, Volume 2. Lewis Publishers, Inc., Boca Raton, FL.

Ney, R.E., Jr. 1990. *Where Did That Chemical Go?* Van Nostrand Reinhold, NY.

Paul, E.A. and F.E. Clark. 1989. *Soil Microbiology and Biochemistry.* Academic Press, Inc., San Diego, CA.

Pederson, T.A. and J.T. Curtis. 1991. *Soil Vapor Extraction Technology Reference Handbook.* Office of Research and Development, EPA/540/2-91/003, Washington, DC.

Puls, R.W. and R.M. Powell. 1992. Acquisition of representative ground water samples for metals. *Ground Water Monitoring Review.* **12**:167–176.

Raymond, R.L. 1974. Reclamation of hydrocarbon contaminated groundwaters. U.S. Patent Office, Washington, DC, 3,846,290, ptd November 5, 1974.

Reidy, P.J., W.J. Lyman, and D.C. Noonan. 1990. Assessing UST corrective action technologies: Early screening of clean-up technologies for the saturated zone. Risk Reduction Engineering Laboratory, EPA/600/2-90/027, Cincinnati, OH.

Shah, D.O. (Ed.). 1981. *Surface Phenomena in Enhanced Oil Recovery.* Plenum Press, NY.

Sims, R.C. 1990. Soil remediation techniques at uncontrolled hazardous waste sites. *Journal of the Air & Waste Management Association Reprint Series RS-15.* Pittsburgh, PA. pp. 3–30.

Staps, S.J.J.M. 1990. International evaluation of in-situ biorestoration of contaminated soil and groundwater. Office of Emergency and Remedial Response, EPA/540/2-90/012, Washington, DC.

Stumm, W. 1992. *Chemistry of the Solid-Water Interface.* Wiley-Interscience, New York, p. 58.

Suchomel, K.H., D.K. Kreamer, and A. Long. 1990. Production and transport of carbon dioxide in a contaminated vadose zone: A stable and radioactive carbon isotope study. *Environ. Sci. Technol.* **24**:1824–1831.

U.S. Environmental Protection Agency. 1989. Guide for conducting treatability studies under CERCLA. Office of Solid and Emergency Response and Office of Research and Development, Washington, DC, Contract No. 68-03-3413, November.

U.S. Environmental Protection Agency. 1990a. Subsurface contamination reference guide. Office of Emergency and Remedial Response, EPA/540/2-90/011, Washington, DC.

U.S. Environmental Protection Agency. 1990b. *Handbook on In Situ Treatment of Hazardous Waste-Contaminated Soils.* Risk Reduction Engineering Laboratory, EPA/540/2-90-002, Cincinnati, OH.

U.S. Environmental Protection Agency. 1991. Site characterization for subsurface remediation. U.S. EPA Report EPA/625/4-91/026, ORD, Washington, DC.

Villaume, J.F., P.C. Lowe, and G.P. Lennon. 1983a. Coal tar recovery from a gravel aquifer: Stroudsburg, Pennsylvania. *Proceedings of the Conference on the Disposal of Solid, Liquid and Hazardous Wastes*, Bethlehem, PA, American Society of Civil Engineers—Lehigh University, pp. 12-1 to 12-18.

Villaume, J.F., P.C. Lowe, and D.F. Unites. 1983b. Recovery of coal gasification wastes: An innovative approach. *Proceedings of the Third National Symposium on Aquifer Restoration and Ground Water Monitoring.* National Water Well Association, Worthington, OH, pp. 434–445.

Wilson, J.R., L.E. Leach, M. Henson, and J.N. Jones. 1986. In situ biorestoration as a groundwater remediation technique. *Ground Water Monitoring Review.* **6**:56–64.

Chapter 10

Hydraulic Control for Contaminant Containment

G. de Marsily

INTRODUCTION

Hydraulics, in a broad sense, are considered here as one of the means by which contaminants that have leaked into the ground (or are threatening to do so) can be kept under control. Engineered structures acting on groundwater flow can be used to immobilize and contain, e.g.:

(i) As a remedial technique in the early stages, easy to install and able to stop the progression of a contaminant plume toward the environment.
(ii) As a temporary measure for protecting the environment while larger-scale remedial action is in progress, such as: excavation of the polluted soils, building of impervious barriers, etc.
(iii) As a more or less permanent method for containing the contaminants, in general in conjunction with other containment methods (impervious barriers, etc.).
(iv) Finally, as a means of allowing hydraulics to be instrumental in remedial action inside a polluted zone by letting fluids bring in chemicals or bacteria or nutrients to enhance the extraction/transformation/degradation of pollutants *in situ*.

In this chapter we briefly review the different engineering techniques that can be used in hydraulic containment and describe a case study where many of these techniques are being used, but where the emphasis is clearly on the *in situ* remediation of a polluted waste disposal site.

ENGINEERING CONTAINMENT TECHNIQUES: HYDRAULICS

Flow of water in the ground is influenced by a number of forces or gradients:

- gravity;
- pressure gradients;
- concentration gradients (osmotic effects);
- electrical fields;
- thermal fields;
- magnetic fields, etc.

Engineers can act on most of these forces in order to control the direction and velocity of the groundwater flow (de Marsily, 1986). For example:

(i) **The gravity field can hardly be modified *in situ*.** However, in the laboratory, this can easily be done in centrifuges. Although this is not feasible for *in situ* remediation, it can be useful for measuring rock properties when designing remedial schemes. For example, the method has been used with significant success to measure hydraulic conductivity on low-permeability cores (Conca and Apted, 1992). Since high pressure gradients have an additional mechanical effect on the sample and produce biased estimates of the hydraulic conductivity, this conductivity can be accurately determined on small samples of low conductivity by increasing the gravity, and the velocity, in centrifuges.
(ii) **Modifying the pressure field** is, of course, the most common way of controlling groundwater flow. For contaminant containment, the following techniques are used: wells and drains.

(iii) **Draining curtains:** trenches are dug in the soil, in the same way as when concrete or slurry walls are constructed, but the trenches are filled with gravel to make them pervious. In the gravel material, wells or horizontal drains are installed so that the level of water in the trench can be controlled by pumping or gravity drainage. Such draining curtains are often used in conjunction with an impervious wall, on one or both sides of the impermeable wall. The draining curtain collects the clean or polluted water, which would otherwise flow across the area now protected by the impervious wall.

(iv) **Vertical drains:** many small-diameter boreholes are drilled or pushed into the formation. The fluid can then be withdrawn either by an individual submerged pump in each hole or, if the water table is shallow (less than 10 m), by connecting all the boreholes to an aspiration pump which lowers the pressure, e.g., down to 1/10th of the atmospheric pressure, and "sucks" the water out.

(v) **Horizontal or slanted drains:** boreholes are drilled into the formation, sometimes from the bottom of a large-diameter shaft reaching the substratum of the aquifer. The drains are pushed or drilled into the formation over very long horizontal distances (up to 100 m or more). Thus, a very large volume of rock can be drained from a single spot, the water being collected by gravity at the bottom of the shaft. Such drains can also be built from tunnels bored or blasted beneath the zone to be drained (as demonstrated in the case study).

In most cases such extraction equipment is used to drain the contaminated zone; by locally lowering the pressure, the groundwater flow is made to converge on the drains, carrying with it the contaminants, and most importantly, preventing them from escaping from the zone in question. Thus, hydraulic containment is not, strictly speaking, a containment; it means forcing the flow toward a prescribed zone, preventing its escape and extracting the contaminated water in order to treat it before releasing it into the environment.

When nonaqueous phase liquids (NAPL) are present, hydraulic containment can be a little more subtle. Pumping or draining the water is generally not enough. The separated phase liquids (NAPL) must be controlled individually. If the NAPL are lighter than water, they tend to float on top of the water table. A well will then create a cone, in the water table, and gravity causes the NAPL flow toward the center of the cone, leaving behind a residual NAPL saturation in the pores. The NAPL must therefore be collected (skimmed) from the surface of the well at the center of the cone. This may require a second pump whose aspiration is maintained just above the surface of the water. Alternatively, if the flow of NAPL is small, a mobile pump can be used to periodically empty the NAPL accumulated on top of the water. A new and efficient method uses an oleophilic metallic or plastic turning belt, which continuously moves up and down in the well and carries with it a thin layer of NAPL that is swept away by a blade raking the surface.

For NAPL with a density higher than that of water, the situation is reversed: the separated phase may lie as lenses at the bottom of the aquifer, under the water or simply present at residual saturation over the whole thickness of the aquifer. Pumping out the water will also create a convergence of the DNAPL lenses toward the well, with a cone pointing upward, reproducing the cone in the water table with an amplifying factor given by the relative density contrast $\Delta\rho/\rho$ of DNAPL and water (this is known as "upconing" in the oil industry and in wells above a seawater-freshwater interface in coastal aquifers). If these DNAPL are to be extracted (again leaving residual saturation of DNAPL in the pores), a second pump must be installed in the well in order to withdraw the separated phase from the bottom of the aquifer.

In NAPL control, it is, however, common to use hydraulics only for the water, leaving the NAPL *in situ* and extracting only the small fraction of NAPL that can become dissolved in the water, since, in most cases, when water is contained hydraulically, so are the NAPLs. In the case of DNAPL, however, this is more uncertain, as the slope of the aquifer substratum may have a greater effect (by gravity) on the flow direction of the DNAPL than the pressure gradient imposed by hydraulics. It is indeed easy to show that for a thin DNAPL layer resting on a slanted aquifer substratum, the flow direction is given by the sine of the difference:

$$dh/dx - (\Delta\rho/\rho) \sin \alpha$$

where dh/dx is the hydraulic gradient parallel to the slope, $\Delta\rho$ is the difference in density between DNAPL and water, ρ is the density of water and α is the slope of the substratum. For instance, for a DNAPL of density 1,600 kg/m^3, a substratum slope of 1° can make the DNAPL move downslope even if the hydraulic gradient pulls in the opposite direction with a magnitude of up to approximately 1%, which is a very high value!

Hydraulic barriers can also be created by injecting clean fluids into the aquifer to prevent contaminated water from flowing, e.g., toward a well. This is not exactly confinement, since it will, on the contrary, increase the spreading of the contaminant in the environment away from the injection wells, boreholes or curtains. The method has been applied, e.g., to coastal aquifers where fresh water has been used to push the seawater back toward the sea, thus permitting an intense exploitation of the freshwater behind the injection barrier. However, part of the extracted water is lost in the injection barrier. To my knowledge, it has not so far been used extensively on contaminated sites. It could however be utilized in conjunction with "draining hydraulics," with injection barriers "channelling" the flow of contaminated water toward a "draining" system.

When several wells or drains are pumping or injecting water from or into an aquifer, it is possible to optimize the location, flow rate and sign of injection/withdrawal rates in an aquifer. This is done with the help of a groundwater flow model, assuming that the response functions depend linearly on the flow rates and setting the optimization problem as a linear programming question. The unknowns are the flow rates; the objective function is the total cost or the total extraction/injection; the constraints are the magnitude and direction of the velocity or the hydraulic head (Maddock, 1972; Marsily et al., 1978; Illangasekare and Morel-Seytoux; 1982; Hubert, 1984). Thus, it is possible to prescribe the velocity field and to force a contaminant to flow from a point A to a point B, even in heterogeneous aquifers.

Finally, a technique for confining waste by hydraulics is presented in this volume by Gillham and Burris (Chapter 21): it involves channelling the flow of contaminated water through a draining curtain filled with an absorbing material acting as a reactive filter. Instead of the flow being confined, the polluted water is reclaimed *in situ* by its passing through the curtain and supposedly emerging clean. The function of the hydraulic engineering is to ensure that all of the contaminated flow passes through the filter and that the containment is obtained by the action of the filter material. This can be used as a passive control system, provided that the filter is sufficiently effective. If necessary, the filter material can be periodically excavated and rejuvenated to maintain its efficiency.

There are also "unconventional" engineering techniques, such as the use of electric fields (Lancelot et al., 1990). The work by Casagrande (1952) established that the movement of ionic species in solution inside an electric field also transports water, particularly in porous media, because the hydration layers of anions and cations are different: globally, there is a net flow of water toward the cathode. In low-permeability media, this effect has been used, in very special cases, to increase the drainage rate of clayey soils. It has not yet been used for hydraulic containment of contaminated sites, as far as I know. It should be noted, however, that the application of electric fields to a contaminated site, where ions are in solution, could also speed up the decontamination process by moving the ionic species toward the electrodes (Dewière et al., 1987). Alternatively, it has been suggested that electric fields could be used to transport ionic species *in situ*, e.g., to provoke reactions with contaminants and transform them into immobile or less toxic species.

The use of thermal or magnetic fields as a means of containing or displacing fluids or contaminants has not been considered feasible so far, although such mechanisms can play a role in nature (Jamet et al., 1992).

Similarly, osmotic pressure is potentially a natural flow and transport mechanism. It has been suggested that it might, in certain circumstances, provide an explanation of abnormal pressures in deep layers in sedimentary basins (Marine, 1985). Clay layers may play the role of osmotic membranes and let water from adjacent freshwater aquifers converge toward a central brackish one. However, no engineering equivalent of this mechanism has yet been used to contain contaminants. It may have to be considered when clayey slurry walls are used to confine concentrated solutions since a net flow of water may osmotically seep into the confined zone from the adjacent protected zones.

To sum up, contaminant containment by hydraulic methods is very common: a variety of draining/ injecting techniques are available and well developed industrially. They are often used in conjunction with impervious barriers or waste immobilization (NRC, 1996). In the following, we present a case study where many of these techniques are employed.

A CASE STUDY

The site of Montchanin, Saôme-et-Loire in central France, which is described below, is a toxic waste disposal site that has been in operation for about 10 years. The waste accepted here was mostly industrial chemical waste. It was disposed of in a "honeycomb" mode, in a former clay quarry, without any lining of the bottom and sides, since they were assumed to be perfectly impervious. No liquid waste was accepted, but wastes were sometimes solidified onsite before disposal through mixing with clays extracted from the area. Each cell of the honeycomb was filled with layers of waste and covered with clay. When filled, the cell was covered with compacted clays to prevent infiltration of rain water. A final capping with more impervious materials (liners, dense clays, etc.) was planned for the end of the site operating phase. About 400,000 T of waste had been emplaced when the site was suddenly closed in the late eighties.

The decision to close the site (when only about half of its disposal capacity had been used up) was taken by the Administration, prompted by the abnormal and dangerous behavior of the system, chiefly in two respects:

(i) Production of very large amounts of toxic gases, which found their way to the surface in a populated area and made the air nauseous over several kilometers. More than 300 different chemicals were found in samples collected on the site, many of them mercaptans, volatile solvents, and methane. Respiratory problems were detected by the local medical authorities and potential threats to public health could not be ruled out.
(ii) Dissolved chemicals were also found in nearby lakes and streams. Later monitoring of groundwater in the vicinity of the site also revealed a very significant pollution by the toxic waste.

It was later established that both phenomena are linked to the poor quality of the site. Although it was originally considered very tight (as a former clay quarry), a subsequent thorough analysis of the geology showed that the clay layers (which had an almost vertical dip because of local tectonics) were actually both fractured and interbedded with thin sandstone layers (Figure 10.1). Due to an inadequate site survey, this had remained unnoticed when the site was selected. While the site was being filled, water seeped in from the sides as well as from above, since the initial compacted clay cover was also too pervious. As a consequence, the water content of the layered clay-waste mix became too high, resulting in intense anoxic biodegradation, which produced the methane and the toxic gases that escaped through the cover and leakage of the contaminated liquids into the environment. The question then arose: How could this site be restored?

RESTORATION

The measures that are described in the following have now been implemented. For brevity, only the general ideas leading up to the restoration plans will be given. Certain site-specific details are omitted since they have no universal significance, whereas the general solution that was adopted can have very significant implications for other sites.

Several propositions were examined and rejected by an advisory panel assembled for this study:

- **Excavation and incineration of the waste**: this was rejected on the grounds that the danger associated with excavation was too great. Very large amounts of toxic gases might be released, and the risk of a fire starting in the waste could not be ruled out. It is well known that when a fire has started in a waste dump (e.g., in a coal mine dump), it is almost impossible to put it out.
- ***In situ* reclamation**: the waste would be excavated and mixed *in situ* with lime, and then remain in place. The risks would be the same as those outlined above (escape of gases and fire) and furthermore, the panel considered that it could not be demonstrated that this mixture would be stable in the long term nor guaranteed against leaching.

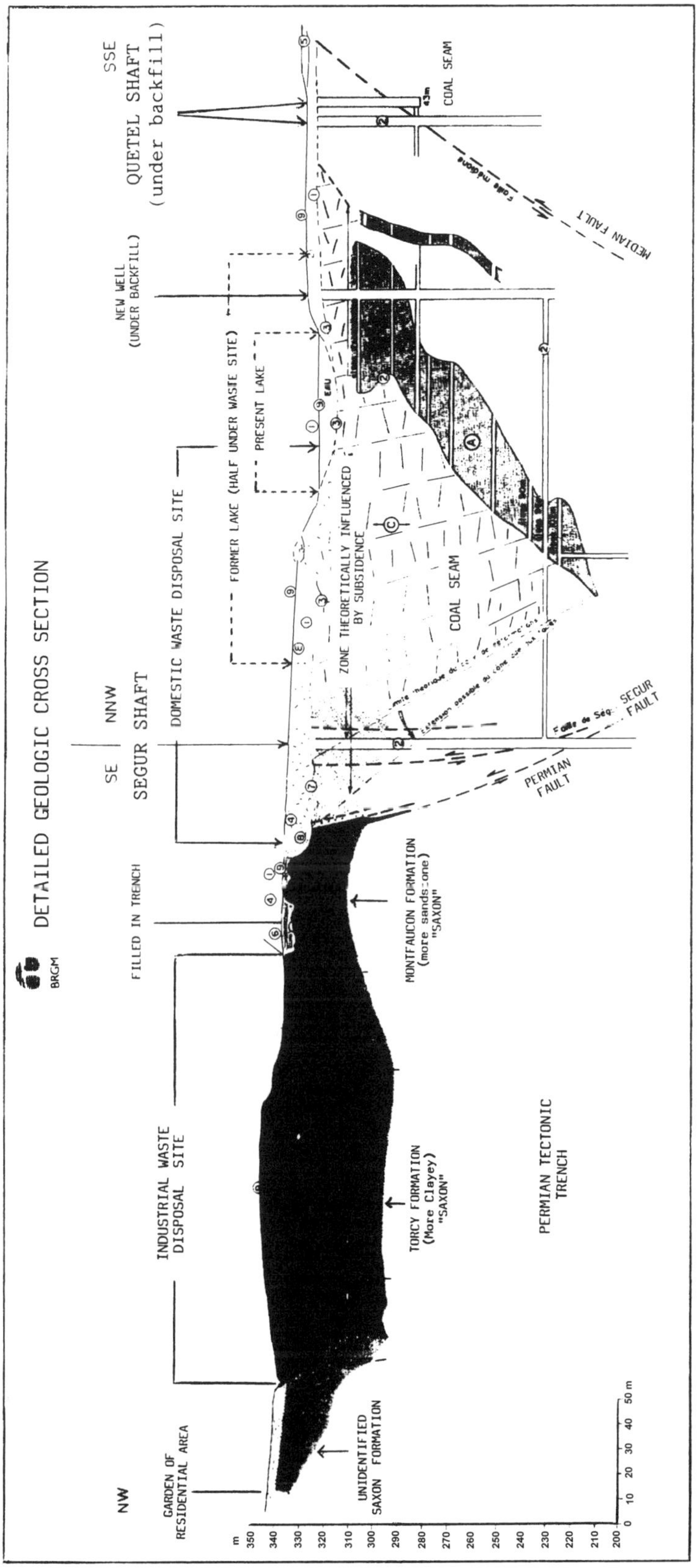

Figure 10.1. Detailed geologic cross-section of the Montchanin Waste Disposal Site.

- **Confinement**: most of the proposals received by the panel suggested that the waste be covered by a liner surmounting a gas drainage network and that the toxic gases be extracted (by depression) and incinerated. The site would also be surrounded by slurry trenches to stop all in- and outflow of contaminated liquids. Although feasible, this solution was not considered satisfactory because it would not prevent the bottom of the site from leaking, and the toxicity of the waste inside the "sarcophagus" would remain unchanged. Eventually, after some tens of years, the top liner or the slurry walls would deteriorate, allowing the system to begin decomposing or to resume leaking.

THE SOLUTION

To overcome these drawbacks, two different concepts were introduced, which go hand in hand: (i) *in situ* bioremediation and (ii) bottom drainage. Simplified representations of "the solution" are given in Figures 10.2, 10.3, and 10.4.

Bioremediation

The very fact that gases are being produced naturally indicates that reactions, mostly of biodegradation, are occurring inside the waste site. Rather than trying to stop these reactions, it is preferable to try to enhance them and collect their by-products (gases and liquids). Gases can easily be collected under the top liner, but liquids have to be drained from below (see next section). How do we enhance the reactions? First, it is necessary to keep the water content of the waste under control. Once the top liner and lateral slurry walls are built, the water content will probably decrease. In order to keep it at the correct level, a system of pipes was installed below the liner in a different set of trenches from that collecting the gases. Fluids can be injected under pressure into these pipes, honeycomb cell per honeycomb cell, in order to adjust the water content. Nutrients, new types of bacteria, etc., can also be injected. Although bioremediation may still be in its infancy, it is believed that allowing the system to be controllable by fluids will make it possible to engineer the *in situ* degradation of a large fraction of the waste in the long term (Thomas and Ward, 1996). If the waste is systematically, and for a long period of time, "leached" by a controlled flux of fluids, the waste products that cannot be extracted by drainage from below or transformed onsite by bioreactions will become stable in the environment of the site, and therefore no longer present a threat to the environment. This *in situ* reclamation may take tens of years, but once the collecting system (liquids and gases) is installed, it could become a low-cost operation. The incineration at high temperature (1200°C for 2 seconds) can be discontinuous, if the gas production rate is small. The drained fluids could be transported by truck and treated offsite if their volume becomes small enough to allow it. The important point is not to aim at confinement but instead at controlled extraction of whatever is mobile or can be made mobile. The waste dump becomes an active reactor defusing the danger, not a "sarcophagus" which eventually may become a time bomb.

Bottom Drainage

The above proposal cannot work without an effective way of draining and controlling the liquids in the system. It was found that one of the most effective ways of achieving this would be to build a lined tunnel beneath the dump site, in the underlying formation. Since this formation is polluted, the tunnel is to be built with a remote-controlled tunnelling machine, while fluids and gases are collected and monitored during construction. Alternatively, two parallel tunnels, one on each side of the dump, but not directly below it, could be considered. More conventional, and perhaps cheaper, tunnelling methods could then be used with appropriate ventilation and control of the incoming fluids. Once the tunnel is built and equipped with an impervious concrete liner, a forest of drains can be drilled from it, above and below: above, into the waste itself, and at its base, to drain fluids and possibly to give access from below for injecting other fluids, nutrients, bacteria, etc., in areas which cannot be controlled from above. Measurements (temperature, moisture content, nature of the gas production, etc.)

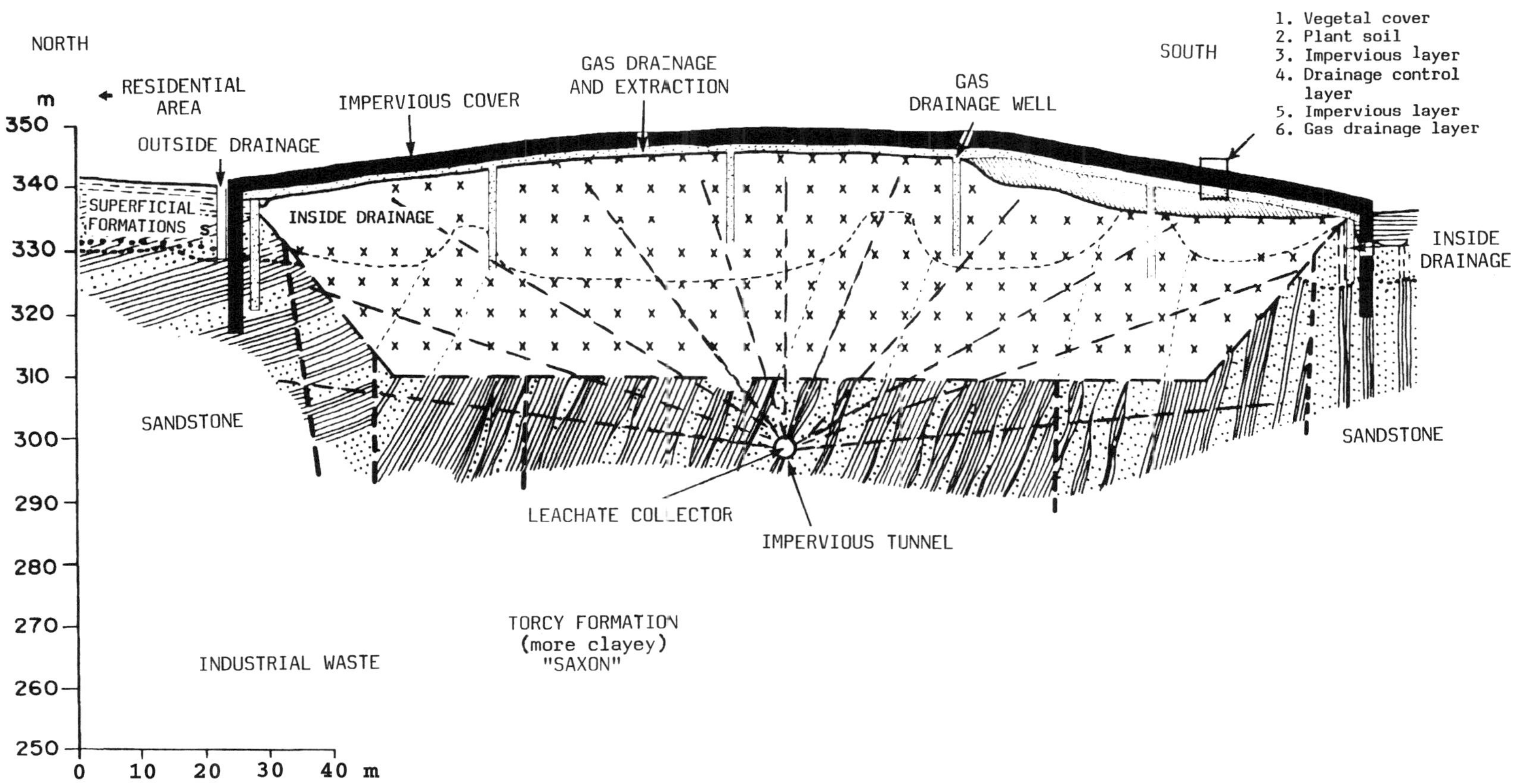

Figure 10.2. Cross-section of the Montchanin Waste Disposal Site with proposed drainage system by boreholes drilled from a tunnel (or, alternatively, from vertical shafts).

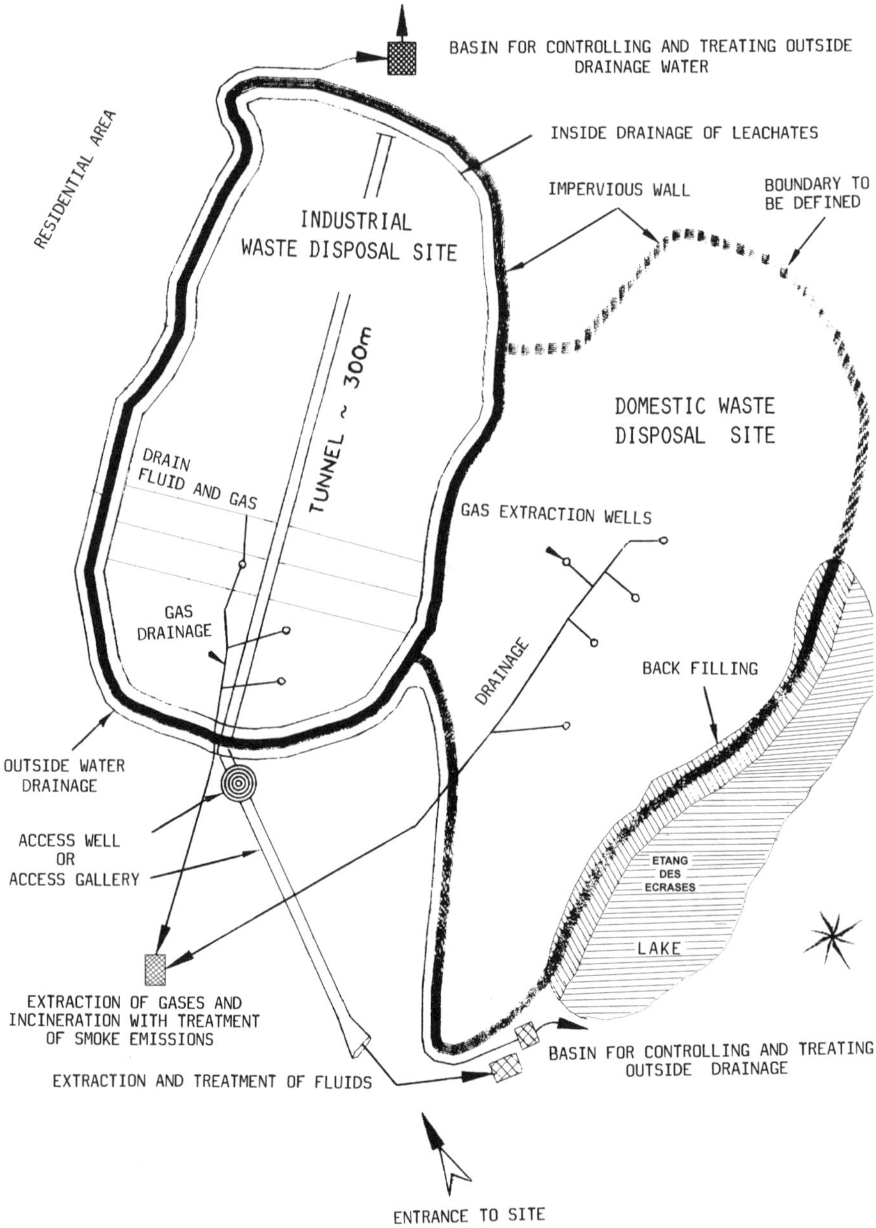

Figure 10.3. The remediation of the Montchanin Waste Disposal Site: slurry walls and drainage system.

can also be made and used when decisions are made on how to operate the "*in situ* reactor." Drains would also be installed below the tunnel, to collect and dispose of incoming fluids seeping from the pervious bottom or from beneath the slurry walls (two draining curtains are installed outside and inside the slurry walls, to collect this water). Gases can also be extracted from some of the drains. Great flexibility is provided by making the drains accessible from the tunnel for maintenance and measurements. Drains can be pressure-washed and, if necessary, rebored if clogged. The amount of drainage can be monitored in each zone of the site. Water from some of the drains (e.g., those draining the fluids from below) may eventually become uncontaminated and fit to be released into the environment. The liquids requiring special treatment can be collected separately, without being mixed with unpolluted water, thus reducing the treatment costs.

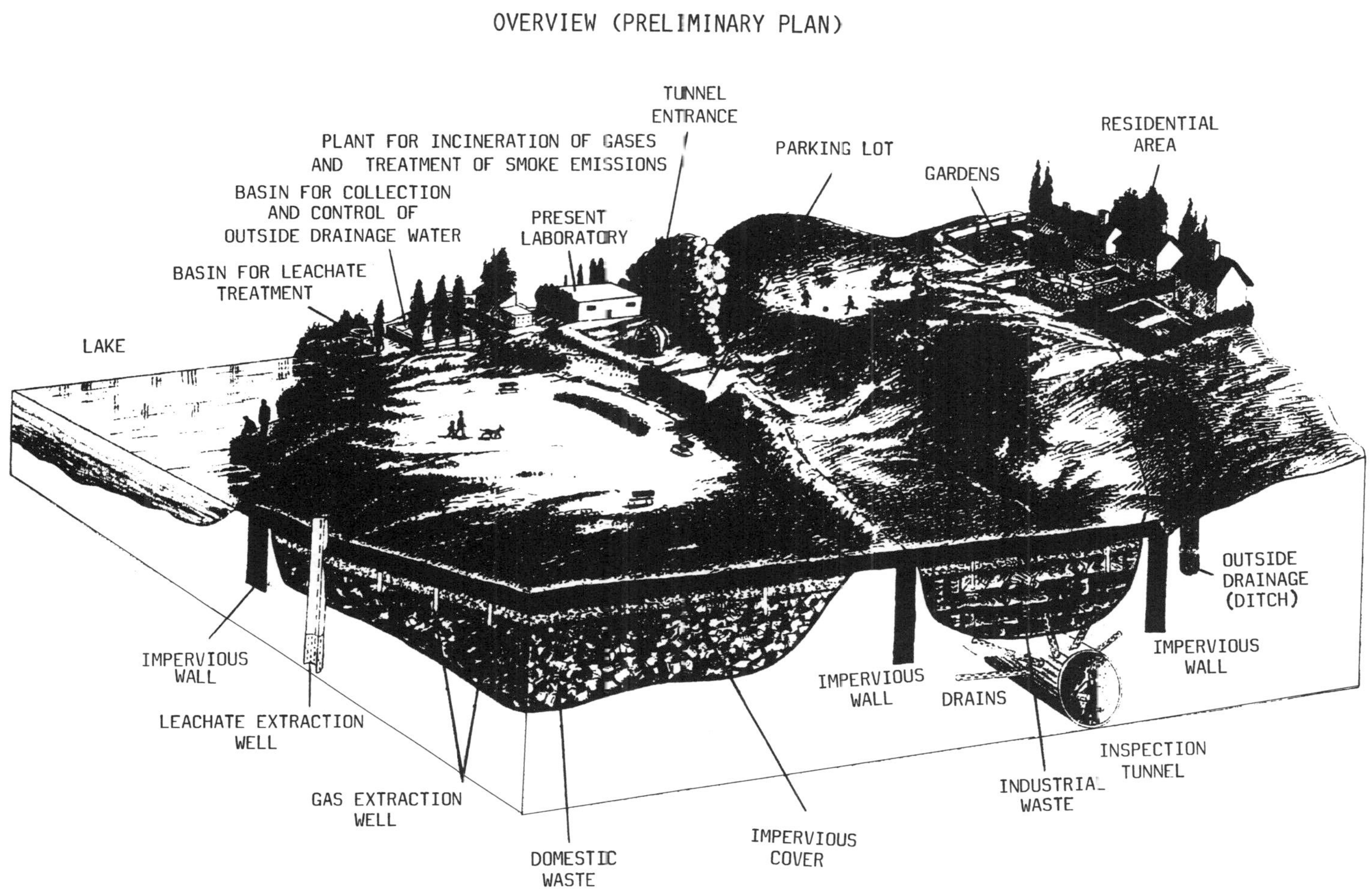

Figure 10.4. Artist's view of the Montchanin Waste Disposal Site after remediation.

The alternative to tunnels is vertical shafts with similar drains drilled horizontally from the bottom of the shaft. For economic reasons, this solution was preferred in the end.

CONCLUSION

The prescribed topic of this chapter was "contaminant immobilization and containment hydraulics." We have made a brief survey of the techniques that can be used for this purpose. However, looking a little further, we realize that fighting contamination by immobilization and containment may, in the end, not be the best policy. The case study, outlined above, clearly shows that hydraulics, if properly engineered, can be more effective than confinement as an overall solution to problems of site restoration. Hydraulics can be used in conjunction with impervious barriers: (i) to prevent leaks and contamination of the environment, (ii) considering the contaminated site as a reactor where biological or chemical remediation can be engineered, they can be instrumental in placing the necessary ingredients where they are needed, and collecting the resulting degradation or solubilization products. Once the site has been "leached" in this way for a sufficiently long time, whatever remains *in situ* is either inert or rendered immobile by the prevailing conditions (pH, Eh, etc.). It no longer presents a threat to the environment, and the site can then really be said to have been "restored."

At the Montchanin site, where these ideas have been implemented, a long-term monitoring program is now planned to observe the *in situ* evolution of the wastes and the fluxes of gases and leachates.

ACKNOWLEDGMENTS

The selection of the recommended remediation plan was made on the basis of proposals received from seventeen companies or consulting firms worldwide; the selection committee, which I chaired, included G. Grognet, C. Guillemin, J.M. Junger, A. Navarro, M. Soyez, D. Sreel, C. Tillier, and A. Zmirou. Figures presented in this chapter were drafted by Bureau de Recherches Géologiques et Minières, Orléans, France, under the guidance of H. Bonin.

REFERENCES

Casagrande, L. 1952. Electro-osmosis stabilization of soils. *Boston Soc. Civ. Eng. Contrib. Soil Mech. 1941–1953*, Boston, MA.

Conca, J.L. and M. Apted. 1992. Report to Power Reactor and Nuclear Fuel Development of Japan. INTERA, Inc., Austin, TX.

de Marsily, G., E. Ledoux, A. Levassor, D. Poitrinal, and A. Salem. 1978. Modelling of large multilayered aquifer systems: theory and application. *J. Hydrol.,* **36**:1–33.

de Marsily, G. 1986. Quantitative hydrogeology: groundwater hydrology for engineers. Academic Press, New York, 440 p.

Dewière, L., E. Ledoux, and G. de Marsily. 1987. Estimation and detection of leaks beneath landfills. *Proc. Symp.* "Land Disposal of Hazardous Waste," Cambridge, UK, September 1987, published by Hartwood, 1988.

Gillham, R.W. and D.R. Burris. 1997. Recent developments in permeable in situ treatment walls for remediation of contaminated groundwater. This volume, Chapter 21.

Hubert, P. 1984. Exploitation d'un système de deux puits en situation de compétition et en situation de coordination. *J. Hydrol.,* **70**:353–367.

Illangasekare, T.H. and H. Morel-Seytoux. 1982. Stream-aquifer influence coefficients as tools for simulation and management. *Water Res. Res.,* **18(1)**:168–176.

Jamet, Ph., D. Fargue, P. Costeseque, G. de Marsily, and A. Cernes. 1992. Thermogravitational effect in porous media. A modelling approach. *Transport in Porous Media,* **9**:223–240.

Lancelot, F., H. Londiche, and G. de Marsily. 1990. Experimental results on the influence of electric fields on the migration of oil, ionic species and water in porous media. *J. Petrol. Sci. Eng.,* **4**:67–74.

Maddock, T., III. 1972. Algebraic technological function for a simulation model. *Water Res. Res.,* **8**(1):129–134.

Marine, W. 1985. Discussion, Int. Symp. on the Hydrology of Rocks of Low Permeability, Tucson AZ, Memoirs, Vol. XX, Int. Assoc. of Hydrogeologists.

National Research Council (NRC). 1996. Environmental Management Technology Development Program at the Department of Energy. 1995 Review National Academy Press, Washington, DC. 131 p.
Thomas, J.M. and C.H. Ward. 1996. Microbial processes in the subsurface. This volume, Chapter 6.

Section IV

Technologies for Contaminant Immobilization and Containment

Chapter 11

Barrier Design and Installation: Walls and Covers

J.K. Mitchell and W.A.N. van Court

ABSTRACT

In-ground walls and surface covers are used to prevent flow of water and/or contaminants into or out of specific regions of the ground or to divert contaminated flows to extraction points. Covers are used for waste source control, whereas barrier walls can be used to contain wastes at their source and to contain or divert plumes. Cutoff walls may be constructed using soil, cement or other chemicals, geosynthetics and sheet piles. Covers are usually composite layered systems containing compacted soil, geomembranes, filters and drainage layers. Types, applications and characteristics of different barrier types and materials are reviewed. Present understandings of the compatibility between barrier and waste materials is summarized, and consideration is given to flows to be expected through properly constructed barriers.

INTRODUCTION

Physical barriers for the isolation and containment of waste are generally configured in the manner shown in Figure 11.1. The functions of these barriers are to prevent flow of water and/or contaminants into or out of the enclosed region (Mitchell, 1986; Starr and Cherry, 1990). Barriers can be important elements for permanent waste containment and for a site restoration. As illustrated by Figure 11.1, there are two categories of barriers: vertical and horizontal. Vertical barriers, usually in the form of walls, are used to block horizontal flows. Horizontal barriers, i.e., covers, liners and bottom seals, prevent vertical flows. Covers are used to prevent water flow into waste repositories and landfills and for gas containment and air pollution control. Bottom seals are sometimes required beneath the contaminated ground to prevent the escape of leachate or inflow of groundwater. Liner systems for new waste containment facilities are outside of the scope of this chapter. Comprehensive recent descriptions of waste containment barrier systems are contained in Rumer and Ryan (1995) and Rumer and Mitchell (1996).

The potential for interactions between the barrier materials and the chemicals that they are built to contain requires careful evaluation. Special test programs may be necessary to determine the effects of the contaminants on the barrier materials and the effects of prolonged exposures.

The selection of a barrier and its design depends on the nature of the contamination and on the specific site conditions. Barriers for source control may be different than barriers for plume control. Containment at the source is used to prevent plume development. For plume control, the barriers are typically used to contain pollutants within property boundaries and to minimize the volumes of groundwater that need treatment by diverting clean flows around a region or contaminated plumes to extraction wells or drains. Near a source, the contaminants are usually more concentrated, and this may affect the selection of the barrier material.

Owing to the variability of the subsoil and the fact that barrier walls and bottom seals are constructed below ground, direct examination and evaluation of the effectiveness of the completed work is impossible. Thus, quality control must be done indirectly using measurements of such quantities as grout takes; electrical or hydraulic conductivity; and density, water content, and strength tests on samples. Long-term quality assurance requires continual monitoring of the groundwater for predetermined parameters at strategically placed locations.

No barrier type or barrier material is totally impervious to chemical transport through it; however, properly designed and constructed barriers can be very effective in limiting chemical transport by

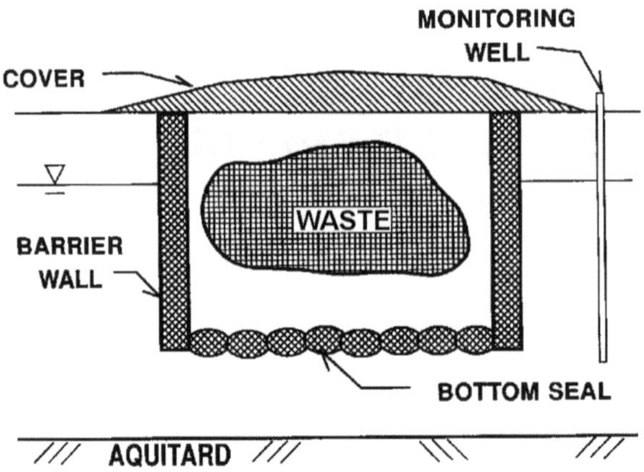

Figure 11.1. General configuration of physical barriers for waste containment.

advection and diffusion. Comparisons of the relative effectiveness of different barrier types should be on the basis of their capability to prevent the advective and diffusive flows of the contaminants present and their long-term stability.

VERTICAL BARRIERS

Compacted Clay Walls

Compacted clay walls are formed by excavation of a trench that then is backfilled with a clay soil compacted in thin layers to a high density (Mitchell, 1986; Mitchell and Brandon, 1985), as shown in Figure 11.2. Such walls are suitable where there is a shallow depth to the aquitard layer, a low flow rate of water into the open trench, and the ground can be excavated without collapse of the side walls. In addition, an adequate supply of suitable clay must be readily available. An advantage of these barriers is that the construction can be monitored and controlled as the clay is compacted into the trench; however, compacted clay barriers are difficult to construct to depths of more than a few meters.

Slurry Trench Cutoff Walls

Slurry trench cutoff walls can be installed below the water table, as shown in Figure 11.3, to depths over 30 meters (D'Appolonia, 1980; Millet and Perez, 1981; Ryan, 1987). They are typically made by excavating a vertical trench that is 0.5 to 2.5 meters wide. The open trench is stabilized against collapse during the excavation by filling it with a slurry of bentonite, attapulgite clay or biodegradable mud. The depth of the slurry in the trench is maintained at a level above the existing water table so a net outward hydrostatic pressure acts against the trench walls. For wall stability the slurry should have a minimum density of 1025 kg/m^3 (Hayward-Baker, 1991). The suspended material in the slurry clogs the pores in the soil and minimizes flow of the slurry from the trench into the ground. The permanent seepage barrier is then constructed by backfilling the trench with a low-permeability backfill which displaces the slurry. The slurry will be readily displaced if its maximum density is at least 240 kg/m^3 less than the backfill material and the viscosity, as determined by API 13B-1 (API, 1990), is a minimum of 40 second-Marsh (D'Appolonia, 1980). Backfill materials include soil-bentonite mixtures, soil-cement-bentonite, cement-bentonite, concrete and geomembranes. Soil-bentonite (SB) cutoff walls are made by backfilling the slurry trench with a soil-bentonite mixture (Geo-Con, 1990), as shown in Figure 11.4. Typically the excavated material is mixed with up to 5% bentonite to reduce the hydraulic conductivity to less than 1×10^{-9} m/s. The mix is then dozed into the trench. Attapulgite

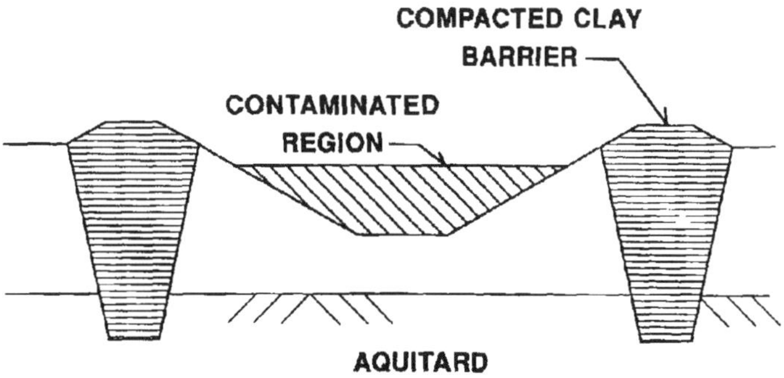

Figure 11.2. Compacted clay barrier.

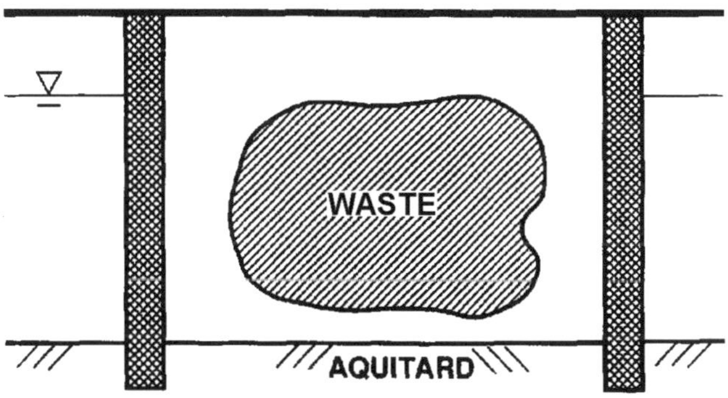

Figure 11.3. Slurry trench cutoff wall.

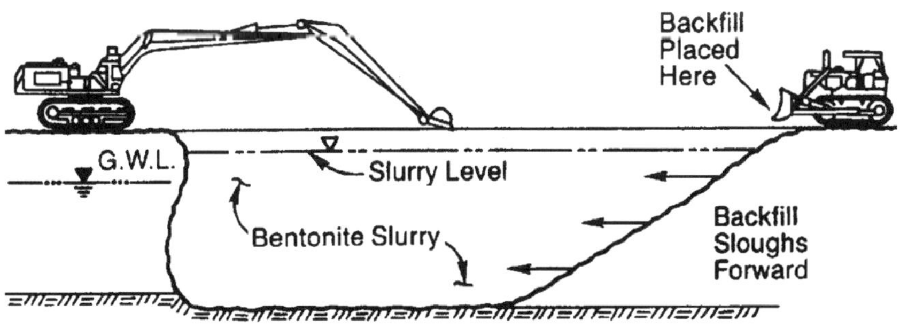

Figure 11.4. Construction of soil-bentonite cutoff wall. (*Source:* Geo-Con Inc., 1990)

clay can be substituted for the bentonite when better chemical stability is necessary (Ryan, 1987; Tobin and Wild, 1986).

Cement-bentonite (CB) cutoff walls are formed by mixing cement into the bentonite slurry that stabilizes the trench (Evans, 1986; Ryan, 1987). The slurry sets in place and forms a low-permeability barrier. This technique is useful at sites where there is not adequate space to mix the soil and bentonite for an SB wall. Another advantage of this technique is that no backfilling of the trench is required so construction defects are minimized (Evans, 1986). On the other hand, the CB walls usually are more expensive than the SB walls and CB walls have higher hydraulic conductivities, on the order of 1×10^{-8} m/s. Also, the excavated material from CB wall construction requires disposal, a problem which may be exacerbated if this material is contaminated.

Another form of slurry trench wall is the concrete cutoff wall. In this method the trench is backfilled with tremied concrete (Bell and Mitchell, 1986; Starr and Cherry, 1990; Evans, 1986; Evans et al., 1987), i.e., concrete that is pumped through hoses to fill the trench from the bottom up. The concrete can be rigid, which uses a conventional mix, or plastic concrete, which is a concrete-bentonite mixture. The plastic concrete forms a more flexible and less permeable barrier (Evans, 1986). Precast concrete panels can also be used, but the integrity of the joints between the panels must be ensured (Bell and Mitchell, 1986; Starr and Cherry, 1990). Concrete barriers are very expensive, but their use can be justified if they are also to be used as structural walls for future construction (Evans, 1986).

Geomembrane cutoff walls have seen increasing application (Druback and Arlotta, 1985; Mitchell, 1986). In one approach a geomembrane envelope, configured as shown in Figure 11.5, is placed in the slurry trench and filled with sand and water. The pressure of the sand and water in the envelope forces the geomembrane to the sides and bottom of the trench. Additional security can be attained by setting monitoring wells within the sand in the envelope. If contaminants are observed in the well, then the spread of contamination can be prevented by pumping from the well. The main disadvantage with this type of wall is that it is difficult to construct, especially around large or irregularly shaped sites. An additional concern is that there may be leakage under the wall because thick geomembranes may not mold well to the bottom of the trench to form an adequate seal. More usually a geomembrane is simply inserted as a single panel in a trench filled with bentonite slurry.

Grout Curtains

Grout curtains are formed by injecting particulate or chemical grouts from a borehole to fill the void spaces and fissures in the ground (Bell and Mitchell, 1986; Mitchell, 1986; Starr and Cherry, 1990; USEPA, 1985). The holes are spaced so that the treated zones overlap to form a continuous barrier, as shown in Figure 11.6. Grout curtains can be formed in both soil and fractured rock. Often two or three rows of grout holes are necessary to achieve the required low permeability (USEPA, 1985). In addition to the compatibility between the grout and the contaminants present, the choice of grout depends on the size of the soil particles or fracture widths (Bell and Mitchell, 1986), as shown

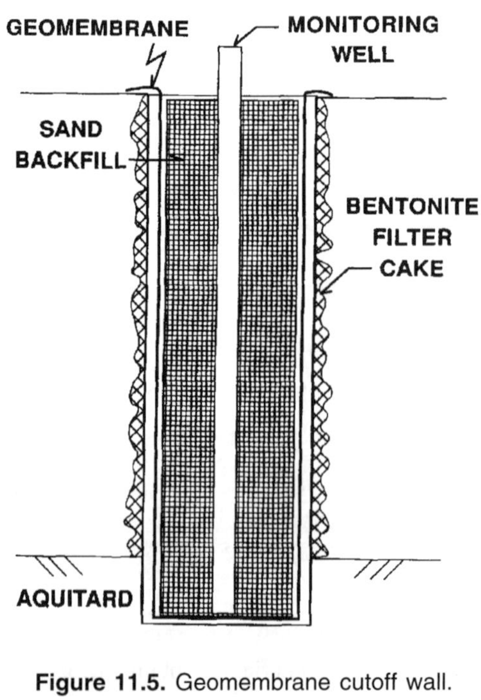

Figure 11.5. Geomembrane cutoff wall.

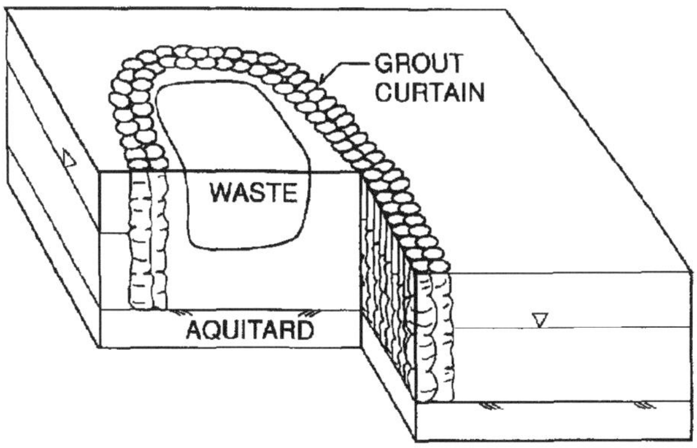

Figure 11.6. General configuration of a grout curtain for waste containment.

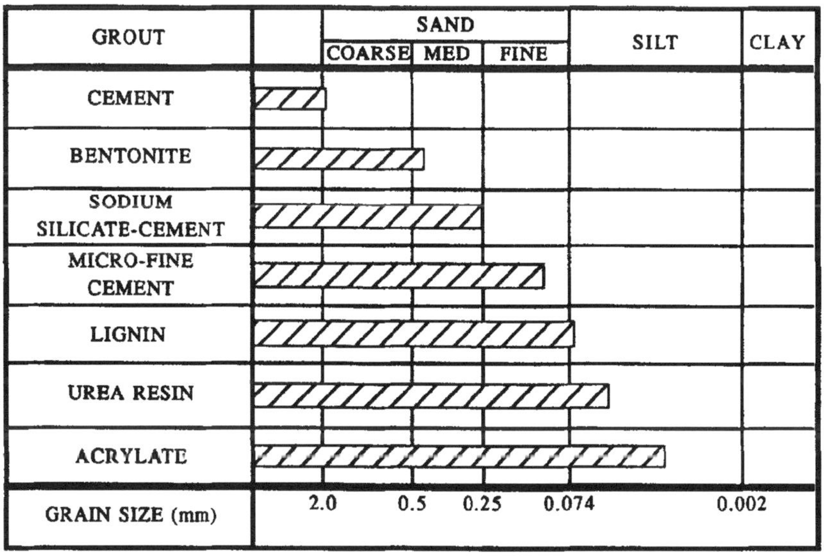

Figure 11.7. Effective grain size ranges for different grouts.

in Figure 11.7. Particulate grouts are used for sands and gravels or to fill wide, open fractures; chemical grouts are usually used in fine sands and silts. Figure 11.7 also indicates how the development of microfine cements has extended the range of the particulate grouts into finer soil sizes. Different grout types are sometimes used in combination. For example, larger voids can be first filled with one type of grout, i.e., cement, and then a second application with a chemical grout can be used to fill the remaining smaller voids. A concern with these barriers is that it is difficult to ensure adequate overlap of the grout columns and to evaluate the overall integrity of the curtain.

Mixed-in-Place Seepage Barriers

Mixed-in-place seepage barriers are constructed using deep soil mixing or jet grouting techniques (Geo-Con, 1990; Taki et al., 1990; USEPA, 1990). Cement and bentonite are the most commonly used additives to reduce the permeability. The resulting treated columns are overlapped to make a continuous wall about one meter thick. An advantage of mix-in-place techniques is that disturbance of adjacent areas is minimized.

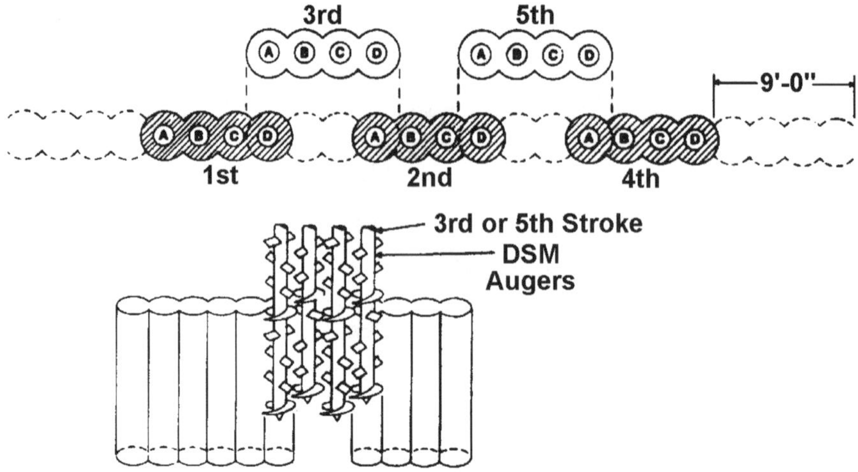

Figure 11.8. Construction details of soil mixed wall. *Source:* Geo-Con, Inc., 1989.

Deep soil mixing (DSM) barriers, also known as soil mixed walls (SMW), use specialized augers that inject and mix the additives into the ground. The augers are often mounted in groups so several columns can be treated at one time, as shown in Figure 11.8 (Geo-Con, 1989). Figure 11.8 also illustrates how the columns or groups of columns are made in a "leapfrog" manner to ensure the continuity of the barrier. This technique can be used in a variety of soils, from sands and clays to soft or weathered rock (USEPA, 1990). Recent research has reported favorable results for the use of admixtures, such as latex and asphalt emulsions, to increase the flexibility and decrease the permeability of soil mixed walls (McFarlane and Holtz, 1992), but the use of admixtures with the soil additives is not part of current practice.

Jet grouting uses a very high-pressure fluid jet to erode the soil for removal or for in situ mixing of the soil with additives (Mitchell, 1981; Mitchell and van Court, 1992). The effective range of the fluid jet can be increased by enclosing it in a concentric cone of air, which allows this technique to be effective below the water table. Figure 11.9 illustrates the steps for constructing a treated column of soil by jet grouting. If the jetting rod is not rotated while being withdrawn, then a panel of soil can be formed instead of a column. The resulting barrier is a continuous row of overlapping columns or panels. As with DSM, several jetting rods can be mounted together and used to treat several columns with each pass. The effective radius of the jet and, therefore, the effectiveness of jet grouting, depends on the type of soil at the site, as well as on the pressure of the jet. It may be difficult to determine the size and shape of the treated columns and ensure adequate overlap (May et al., 1986).

Vibrating Beam Walls

The construction of vibrating beam walls is performed by driving an "H" beam into the ground with a vibratory hammer and filling the void with grout as the beam is withdrawn (Jepsen and Place, 1985; Starr and Cherry, 1990). The driving and filling process is repeated in an overlapping pattern to form a continuous barrier for containing the waste or contamination. This process produces a very thin, 0.025 to 0.10 meter, barrier (Jepsen and Place, 1985). In addition, slight deflections of the beams from vertical during construction will result in a discontinuous barrier. Because of these disadvantages, vibrating beam walls are not recommended for long-term containment of pollutants.

Sheet Piling

Sheet pilings are often used at construction sites to control groundwater flows into the excavation (Bell and Mitchell, 1986; Starr and Cherry, 1990). The pilings are typically steel sheets that are

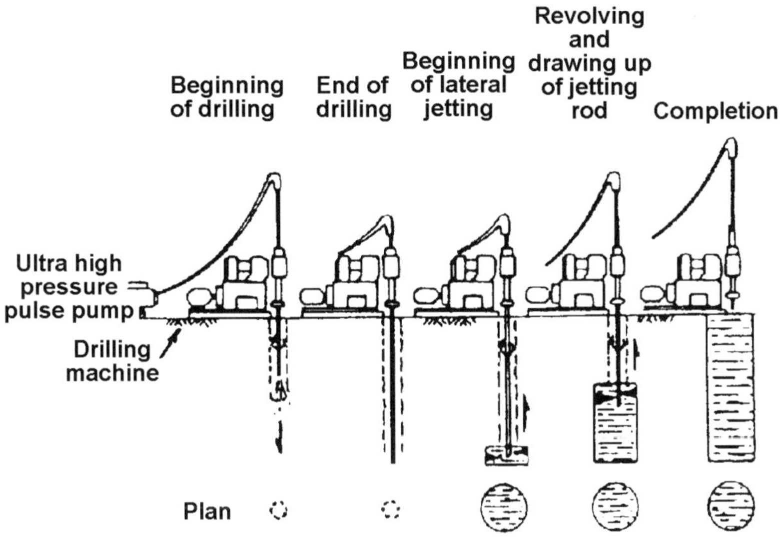

Figure 11.9. Steps for using jet grouting to treat a soil column.

connected by interlocks. The advantages of sheet piles include their ready availability and rapid placement; however, this is offset by the vibrations caused by driving and leakage through the interlocks (Starr and Cherry, 1990). Sheet pile barriers usually have high permeability until the ground they support stresses the piles and wedges them together at the interlocks (Bell and Mitchell, 1986). Also, in rubble or soil with boulders the sheet piles can come out of the interlocks or tear, so it is difficult to ensure continuity. To address the problem of high flows through the interlocks, two new piles types have recently become available for use in waste containment applications, the Waterloo Sealable Joint Piling (WSJP) and geomembrane sheets with sealing interlocking joints.

The WSJP comes in two styles; the first involves a method for attaching a sealable joint to conventional hot or cold rolled sheet piling, and the other is a cold rolled piling with an integral joint (Starr et al., 1991). The method for attaching a sealable joint is to modify conventional sheet piles, as shown in Figure 11.10, by welding an angle iron over the joint. The angle iron provides a cavity over the joint that can be filled with grout to reduce the joint permeability. The piling with the built-in joint, shown in the center of Figure 11.10, has the grout cavity as part of its design. In each case the bottom of the cavity is designed to prevent soil from entering the cavity as the piles are driven. After the piles are placed the cavity is washed out with water and then filled with grout. Starr et al. (1991) report that the hydraulic conductivities of these barriers are less than 1×10^{-9} m/s.

The other alternative is to use geomembranes with interlocking joints (Gundle, 1991). The details of the joint are shown at the bottom of Figure 11.10. The cavity in the joint is filled during placement with a material that swells on contact with water to ensure an adequate seal. By using an insertion plate to support the sheets while driving, these sheets can be placed with a vibratory hammer in the same manner as conventional sheet piling. At some sites the sheets have been placed to depths of 12 m by this method. When the insertion plate is used, the problems associated with vibrations during driving continue to be a disadvantage with this barrier, as well as with the WSJP barriers. The interlocking geomembranes can also be inserted up to 30 m deep in slurry trench cutoff walls. This not only enhances the effectiveness of the slurry wall, but also there are no vibrations during installation.

Ground Freezing

Temporary barriers to fluid flow can be made by freezing the ground around the contaminated site (ASCE, 1991; Bell and Mitchell, 1986; Iskander and Jenkins, 1985). To use this technique in unsat-

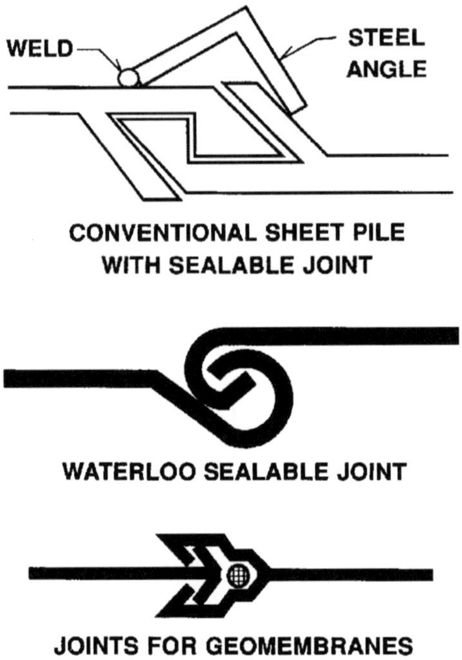

Figure 11.10. Details of sealable sheet piling joints.

urated soils, water may need to be added during the ground freezing. A frozen soil barrier is made by inserting refrigeration pipes into the ground to form a circle or ellipse around the site. In some cases the pipes can be angled inward to get under the site for complete containment (ASCE, 1991). A refrigeration brine at –20°C to –40°C is used. Freezing is not usually effective if the groundwater flow velocity is greater than about 1.5 meters per day, because so much heat is entering the system that efficient ground freezing is not possible (Bell and Mitchell, 1986).

Once the ice barrier is formed, the refrigeration units are still required to maintain it; however, the barrier will remain frozen for a long time should power be accidentally lost. Another disadvantage is that after the ground thaws the permeability may be higher than before due to the volume expansion of water that was caused by freezing (Iskander and Jenkins, 1985).

Hydraulic Barriers

Hydraulic modifications can be useful in manipulating contaminant plumes (Mackay and Cherry, 1989). Hydraulic barriers can be used in combination with other barrier types or by themselves (Mitchell, 1986; Mitchell and van Court, 1992). After a site has been enclosed with a vertical barrier, pumping from within the enclosed region will maintain an inward gradient to prevent the spread of pollutants, and it will keep the pollutants away from the barrier, thus minimizing chemical-barrier interaction problems.

Hydraulic barriers can also be used by themselves to prevent the spread of contamination by creating a pressure trough and ridge system (Mitchell, 1986). The inward gradient at the extraction wells creates the pressure trough. The water is treated and then put back into the ground at the injection well to create a mound or ridge. There are two main concerns with hydraulic barriers. First, they are not passive barrier systems. The water must be continuously pumped to maintain the proper gradients to keep the system working effectively. Second, the water that is extracted from the ground will typically need to be treated to reduce the contamination to specified levels.

WALL PLACEMENT

Determination of the type of vertical barrier is only part of the problem. After deciding upon the wall type, the placement of the wall needs to be determined (Mitchell, 1986; Starr and Cherry, 1990; USEPA, 1985). Many factors enter this decision, including whether the barrier is to control a source or a plume. For a source of contamination, a circumferential barrier can be built to completely surround the site and minimize the volumes of soil and groundwater subsequently affected. To ensure complete containment in this situation, the barrier should be keyed into an underlying aquitard. For plume control, the barriers are typically installed up- or downgradient of the site to minimize the volumes of groundwater that need treatment by diverting clean flows around a region or contaminated plumes to extraction wells or drains. It may not be necessary to construct these barriers down to an aquitard layer beneath the site. Proper modeling of the groundwater flow pattern at a site and the influences of barrier and extraction well locations is necessary for evaluation of the effectiveness of any proper design.

Circumferential Barriers

A circumferential barrier is built to completely surround a contaminated area to prevent the horizontal spread of pollution. This configuration also prevents the horizontal flow of clean groundwater into the site, thus limiting the amount of contaminated water to be treated. At sites where pump-and-treat technology is employed a circumferential barrier may not be recommended, because some flow of groundwater through the soil is needed to wash out the pollutants. Alternatively, at pump-and-treat sites where circumferential barriers are used, injection of treated groundwater can be employed to enhance the flushing of the contaminated soil.

Upgradient Barriers

In some situations it may only be necessary to divert the clean groundwater from flowing into the site rather than to completely isolate the site. Upgradient barrier placement can reduce the pumping volumes required for pollution containment in areas with high groundwater flow rates.

Downgradient Barriers

Instead of diverting clean flows around a site the contaminated groundwater could be diverted to an extraction point by a downgradient barrier. Such barriers can enhance the effectiveness of pump-and-treat systems by allowing the groundwater to flow through the contaminated region and wash out the pollutants which are then collected and treated. This type of barrier will also minimize any inflow of clean water from downgradient which would dilute the contaminants and increase the volumes to be treated.

Keyed-In Barriers

A keyed-in barrier extends into an aquitard layer that is continuous beneath the site. By keying the wall into the aquitard, the wastes can be completely contained, provided an adequate seal can be developed between the wall material and the aquitard material. Unfortunately, it is not always possible to construct a keyed-in wall, because the depth to a continuous aquitard layer may be so great that the construction is not feasible, or there may be insufficient geotechnical data to confidently establish the continuity of the aquitard. On the other hand, it is not necessary to key in the wall if a positive gradient can be maintained into the enclosed area.

Hanging Walls

A hanging wall is not keyed into an underlying aquitard layer, as shown in Figure 11.11 (Starr and Cherry, 1990). A hanging wall can contain the pollution by maintaining an inward groundwater gradient using drains or wells. In addition, there are situations when it may not be necessary to build a keyed-in wall to achieve adequate containment. The migration of lighter-than-water nonaqueous phase liquids and gaseous phase contaminants in the vadose zone can be minimized by a barrier that extends beneath the groundwater table, though the contaminant can continue to be transported beneath the wall by solubilizing in the groundwater (Starr and Cherry, 1990). This type of barrier should be constructed so that it will remain submerged during the lowest level of fluctuation of the water table.

HORIZONTAL BARRIERS

Horizontal barriers are used to block vertical flows, and include caps or covers and bottom seals. Caps will minimize the infiltration of precipitation into the site and prevent the migration of gases from the site. Bottom seals may be required beneath the contaminated ground to prevent the downward escape of leachate or the inflow of groundwater. However, the construction of bottom barriers is likely to be both difficult and expensive.

Covers

Covers, also called caps, are placed over contaminated sites or waste landfills to prevent the infiltration of water, which can lead to the generation of additional leachates and gases, and to prevent the vertical migration of gases from the site (Koerner and Daniel, 1992; Mitchell, 1986; USEPA, 1979; USEPA, 1986). These barriers are typically composite systems, similar to that shown in Figure 11.12, which include low-permeability layers to prevent vertical flows and drainage layers to direct the water from the site. The waste and daily cover material are typically capped with a compacted clay layer that is 0.61 m thick and has a hydraulic conductivity less than 1×10^{-9} m/s. Some sites may require a gas collection layer between the waste and the clay layer. On top of the compacted clay is a geomembrane and then the drainage layer. The hydraulic conductivity of the drainage layer should be greater than 1×10^{-4} m/s to allow the water to flow easily away from the cap. A geofabric filter separates the drainage layer from the topsoil which covers the cap. The topsoil layer should be thick enough to prevent frost penetration to the drainage layer so free drainage will be maintained year round. The cap is then vegetated to minimize erosion and improve aesthetics. The cap should be designed to have a final slope, after any settlements of the site, that is at least 2% to promote drainage from the site and prevent ponding (Jessberger and Stone, 1991; Koerner and Daniel, 1992).

Bottom Seals

Bottom seals can be used to develop a flow barrier beneath a site that, when combined with barrier walls, will completely isolate the contaminated region. The base of a site can be sealed using either penetration grouting or jet grouting techniques (Mitchell and van Court, 1992; Tausch, 1992). Care must be taken to ensure that the contamination does not spread to lower layers during the seal installation.

In coarse-grained soils, e.g., sands and gravels, penetration grouting can be used to build a layer of overlapping grout bulbs beneath the site (May et al., 1986; Tausch, 1992). The grout is injected at low pressures so it will displace the water in the void spaces and maintain the soil structure. The grout can be placed by sleeve pipe ("tube-à-manchette") or vibration techniques (Tausch, 1992). For the "tube-à-manchette" method, the grout tubes are set in drill holes. Packers or other control measures are used to inject the grout within the desired depth range.

The technique of vibration grouting has been employed successfully at construction sites to prevent groundwater from entering the excavations (Tausch, 1992). In the case histories presented by Tausch (1992) vibration grouting resulted in lower permeabilities than the "tube-à-manchette" method.

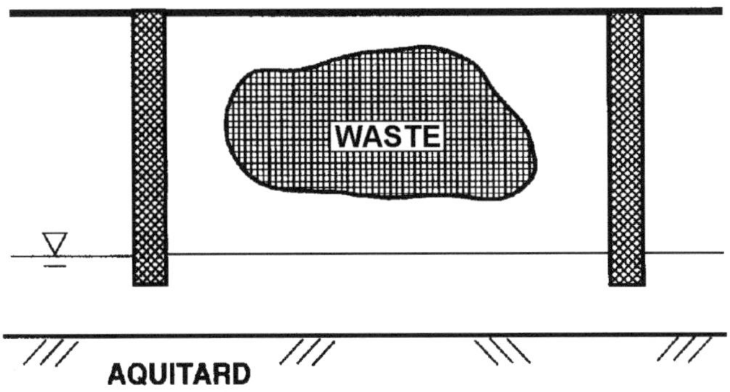

Figure 11.11. Hanging wall, not keyed into underlying aquitard.

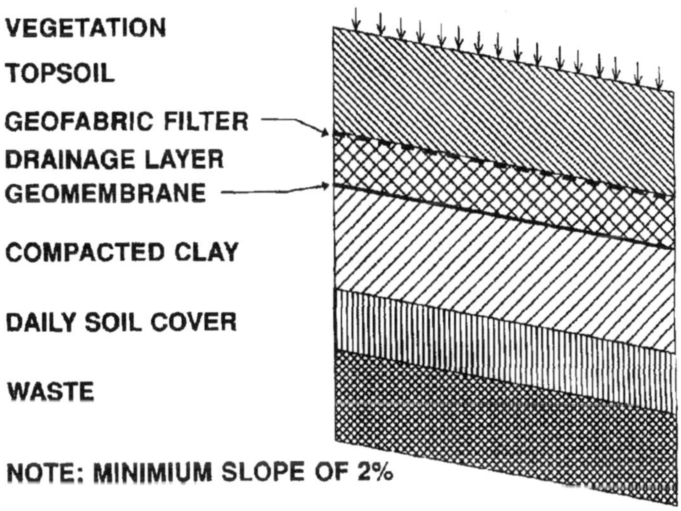

Figure 11.12. Layers in a typical composite cover system.

Some details of vibration grouting are illustrated in Figure 11.13. First the grout valves are welded to steel plates. Then a vibratory hammer and a drive pipe pile are used to drive the plates with the grout valves to the required depth. As the pipe pile is withdrawn, the grout valve remains in place. During withdrawal the soil sloughs into the opening and is compacted around the valve and the polyethylene tube that connects the valve with the surface. The ground can then be grouted to seal the bottom of the site. Tausch (1992) noted that bottom seals with hydraulic conductivities of 1×10^{-7} to 1×10^{-8} m/s were achieved in sands and gravels using this method. Jet grouting, as described earlier, is another technique for bottom sealing. The jets are used to treat short columns or disks that will overlap each other to form a seal beneath the contaminated region (May et al., 1986; Mitchell and van Court, 1992; Tausch, 1992). The effectiveness of this technique will depend on the soils encountered, though a sensor can be used to determine the diameter of the treated disk (May et al., 1986).

COMPATIBILITY ISSUES

The compatibility of the proposed barrier materials for a site and the specific contaminants must be determined before the barrier construction begins, as the hydraulic conductivity or other properties of the barrier materials may be adversely affected by the contaminants present at the site. The effects of contaminants on clay barriers have been reported by Alther et al. 1984, Anderson et al.

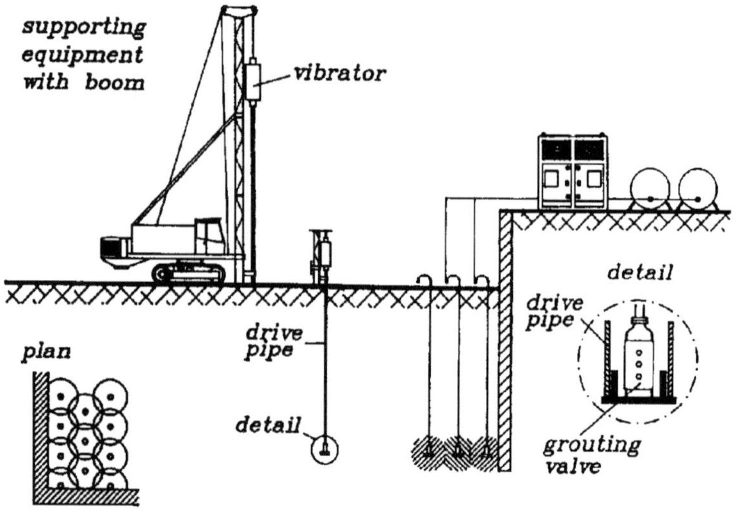

Figure 11.13. Details of vibration grouting to form a bottom seal. *Source:* Tausch, 1992.

(1985), Bowders and Daniel (1987), Daniel (1987), Evans et al. (1985), Madsen and Mitchell (1989), Peterson and Gee (1985) and others. Studies regarding the compatibility of grouts with contaminants were performed by Bodocsi et al. (1987), Jefferis (1992), Siluwa and Krizek (1992), and Spooner et al. (1982, 1984). The suitability of geomembranes in various chemical environments was reported by Koerner (1990a, 1990b), Koerner et al. (1991, 1992) and White and Alen (1991). When possible compatibility tests should be performed for a length of time sufficient to pass an adequate volume of permeant to achieve a chemical equilibrium between the material and the contaminant (Daniel, 1987; Jefferis, 1992). The general compatibilities described in this section should be used only as a guide to determining the suitability of a particular barrier material. Site specific compatibility tests on the proposed barrier materials should be performed under representative conditions before the final barrier design is determined.

The effects of the contaminant in the field may be different than the effects in laboratory tests due to scale effects and differences between sample and field construction. However, the constructed barrier cannot be tested with the actual contaminant plume. Under California regulations the comparison of the hydraulic conductivities determined for clean and contaminated water in the laboratory is used to correct the results of field tests run with clean water (CCR Title 26).

Clay Materials

Many factors can affect the influence of a contaminant on the clay materials used to construct a barrier. The type of hydraulic conductivity tests and how they are performed can produce differing results (Daniel et al., 1985; Madsen and Mitchell, 1989). Each type of hydraulic conductivity test is representative of a specific situation and should be used appropriately, as described by Daniel et al. (1985).

The influence of contaminants on clay materials also depends on the type of clay and its water content (Evans et al., 1985; Madsen and Mitchell, 1989). Attapulgite clays can be specified where a high level of chemical stability of clay slurries is required (Ryan, 1987; Tobin and Wild, 1986) as their properties are less affected by chemical charges than those of smectite clays. At higher water contents, such as in soil-bentonite backfills, the clay particles are more free to move than when the material is compacted at lower water content and porosity.

The state of stress of the clay in the barrier also influences the compatibility of the clay with the contaminant (Boynton and Daniel, 1985). A clay that would tend to crack in the presence of a given contaminant may not do so if there are adequate confining stresses to keep the cracks from forming and opening. This ability of the cracks to remain closed or to close after forming is referred to as the self-healing ability of the clay. The effects that inorganic compounds have on clays depend on their effect on the diffuse double layer and the interparticle forces (Madsen and Mitchell, 1989). The thicker the diffuse double layer, then the more dispersed and less permeable the clay is likely to be at a given density. A thinner diffuse double layer will cause a more flocculated and more permeable structure. Table 11.1 summarizes the effect of changes in the permeating fluid composition on the clay structure and hydraulic conductivity.

Changes in the pH of the permeating fluid can also affect the hydraulic conductivity of a clay. In acidic solutions the alumina at the particle edges ionizes positively and can attach to the negatively charged surfaces of the clay particles to form a flocculated structure (Madsen and Mitchell, 1989). Low pH solutions can also attack the octahedral sheets in the clay minerals, leading to dissolution and increased permeability. Other components in the soil, e.g., carbonates or iron oxides, may also be dissolved by high acidity. At high pH the alumina ionizes negatively, and the interparticle force balances may cause the clay to form a more dispersed structure. However, the overall effects of a basic solution on the clay are hard to predict, because the increases in the electrolyte concentration and high cation valences can cause formation of a flocculated structure. Strong base solutions can also attack the tetrahedral sheets in the clay minerals and dissolve the clay.

The effects of organic compounds on clay barriers depend on the water solubility of the compound, its dielectric constant and polarity, and whether a pure phase is present or not (Alther et al., 1984; Anderson et al., 1985; Bowders and Daniel, 1987; Evans et al., 1985; Madsen and Mitchell, 1989). Hydrophobic organic compounds cannot easily displace the water in the soil, but can cause changes in the larger pores and fissuring of the clay (Daniel et al., 1985; Madsen and Mitchell, 1989). Water-soluble compounds can more readily displace the water in the clay and affect the entire structure.

The thickness of the diffuse double layer is affected by the dielectric constant of the permeating fluid. Fluids with high dielectric constants usually cause a more dispersed and less permeable structure (Evans et al., 1985; Madsen and Mitchell, 1989). The dielectric constant of many organic compounds is much less than that of water, so the result is a tendency toward a more flocculated structure with a higher hydraulic conductivity. For water-soluble organics the change in the dielectric constant does not have a noticeable effect on the structure until the concentration of the organic compound is greater than 75 to 80% (Bowders and Daniel, 1987; Madsen and Mitchell, 1989). Hydrophobic organics do not have a significant effect on the hydraulic conductivity at their solubility limits. However, pure phase compounds can result in large detrimental changes in the hydraulic conductivity.

Table 11.1. Effect of Changes in Pore Fluid Parameters in Clay. *Source:* Madsen and Mitchell, 1989.

Pore Fluid Parameter	Change in Parameter	Change in Double Layer Thickness	Fabric Change Tendency	Usual Effect on Hydraulic Conductivity
Dielectric	increase	increase	dispersed	decrease
constant	decrease	decrease	flocculated	increase
Electrolyte	increase	decrease	flocculated	increase
concentration	decrease	increase	dispersed	decrease
Cation valence	increase	decrease	flocculated	increase
	decrease	increase	dispersed	decrease
Cation size	increase	increase	dispersed	decrease
	decrease	decrease	flocculated	increase
pH	increase	increase	dispersed	decrease
	decrease	decrease	flocculated	increase
Anion	increase	increase	dispersed	decrease
adsorption	decrease	decrease	flocculated	increase

The effect of organic acids and bases in the pure phase is similar to the effect of inorganic acids and bases (Madsen and Mitchell, 1989). The effect of dilute organic acids and bases on the hydraulic conductivity of clays appears to be negligible.

Grouts, Including Cement

Contaminants can affect the set time of grouts and the long-term properties of the barrier, so the testing program should assess both effects of the contaminants (Bodocsi et al., 1987; Spooner et al., 1982 and 1984). The influence on the set time can be to retard it to a point where the grout will not set, or the set time can be shortened to such a point that a flash set results, and the grout will not effectively penetrate the ground. The effects of many contaminants on grouts are well known; for example, sulfates cause the formation of expansive minerals in type I portland cement which leads to cracking and strength loss. Other contaminant and grout combinations have not been investigated as thoroughly. The effects of some contaminants on the set times and durability of different types of grouts are summarized in Table 11.2.

Geomembranes

The synthetic materials used to make geomembranes and other geosynthetics can be degraded by chemical compounds and by ultraviolet light (Koerner, 1990a,b; Koerner et al., 1991 and 1992; White and Alen, 1991). In most barriers the geomembranes are buried beneath the ground surface, so the effects of relatively short exposures to UV radiation during construction typically do not cause any problems. Table 11.3 summarizes the general geomembrane/contaminant compatibility between the various materials used to make geomembranes and some common chemicals. As the use of geomembranes for the construction of barriers began only relatively recently, the long-term effects of various chemical compounds and mixtures of compounds on these materials is not well known (Koerner, 1990a; Koerner et al., 1991 and 1992). Due to the rapid development of new geomembrane materials, the geomembranes in current use may have significantly different properties than those used previously. New testing procedures are under development to better determine the contaminant compatibility and long-term behavior of geosynthetics (Koerner, 1990a; White and Alen, 1991).

Table 11.2. Grout Compatibility. ***Source:*** **Spooner et al., 1982.**

	Acid	Base	Heavy Metal	Non-Polar Solvent	Polar Solvent	Inorganic Salt
Bitumen	?A	?A	??	?D	?D	?D
Portland Cement, Type I	1D	1D	1A	2C	2C	2C
Portland Cement, Types II and V	1D	1A[a]	??	2?	2?	2A
Bentonite	?C	?C	?D	?D	?D	2D
Clay-Cement	?C	?D	??	??	??	?D[b]
Silicate	3A	2C	??	??	??	3?
Acrylamide	2C	3D	3?	?A	??	3D
Phenolic	?A[c]	?D	??	?A	?A	?D
Urethane	2C	?D	??	?A	?A	?D
Urea Resin	1A	2C	??	2?	2?	?A
Epoxy	?A[c]	?A	??	?D	?D	?A
Polyester	?D	?D	??	?D	?D	??

1 No significant effect on set time
2 Increase in set time
3 Decrease in set time

A No significant effect on durability
B Increase in durability
C Rapid decrease in durability
D Long-term decrease in durability

[a] KOH & NaOH are 1D
[b] Sulfates are ?C
[c] Non-oxidizing

Table 11.3. General Geomembrane/Contaminant Compatibility. ***Source:*** Koerner, 1990.

	Geomembrane Type																	
	Butyl Rubber		Chlorinated Polyethylene (CPE)		Chloro-sulfonated Polyethylene (CSPE)		Elasticized Polyolefin		Epichlorohydrin Rubber		Ethylene Propylene Diene Monomer (EPDM)		Polychloroprene (Neoprene)		Polyethylene		Polyvinyl Chloride (PVC)	
Chemical	100°F	158°F	100°F	158°F	100°F	158°F	100°F	158°F	100°F	158°F	100°F	158°F	100°F	158°F	100°F	158°F	100°F	158°F
General																		
Aliphatic hydrocarbons			x	x			x		x	x			x	x	x	x		
Aromatic hydrocarbons							x		x	x			x	x	x	x		
Chlorinated solvents	x	x					x		x	x	x		x		x	x		
Oxygenated solvents	x	x					x		x		x	x	x	x	x	x		
Crude petroleum solvents			x	x			x		x	x			x	x	x	x		
Alcohols	x	x	x	x			x		x	x	x	x	x	x	x	x	x	x
Acids																		
Organic	x	x	x	x	x		x		x		x	x	x	x	x	x	x	x
Inorganic	x	x	x	x	x		x		x		x	x	x	x	x	x	x	x
Bases																		
Organic	x	x	x	x	x		x		x	x	x	x	x	x	x	x	x	x
Inorganic	x	x	x	x	x		x		x	x	x	x	x	x	x	x	x	x
Heavy metals	x	x	x	x	x		x	x	x	x	x	x	x	x	x	x	x	x
Salts	x	x	x	x	x		x	x	x	x	x	x	x	x	x	x	x	x

x = generally good.

BARRIER EFFECTIVENESS EVALUATION

Owing to the variability of the subsoil and the fact that barrier walls and bottom seals are constructed below ground, direct examination of the completed work is impossible. In addition, it is very difficult to detect small changes in the hydraulic conductivity that may mean the difference between an acceptable barrier and one that is not acceptable. Prediction of the long-term performance of a barrier is usually based on the results of laboratory tests and analyses. During construction, quality control programs are instituted to ensure that the design criteria are being met. Long-term quality assurance requires continual monitoring of the chemical constituents in strategically placed monitoring wells. For some barriers, such as slurry trench cutoff walls or compacted clay barriers, it may be possible to site monitoring wells within the wall and determine the hydraulic conductivity through the wall in the vicinity of the well (Teeter and Clemence, 1986). Wells built in the wall can also monitor contaminant flow through the wall. Unfortunately, however, it is difficult to install such wells in such a way as to ensure the desired alignment, and the probability of intercepting high-permeability windows is low.

The quality control program used for a given project will depend on the type of barrier being built. Measurements of quantities, such as grout takes or added bentonite, are often performed (Joseph et al. 1992; Tausch, 1992). In some cases the properties of the barrier materials, e.g., water content, slump or density, are related to barrier performance and monitored during construction (Daniel, 1987; Ryan, 1987). Also, laboratory analyses can be performed on samples collected during construction or core samples taken after completion of the barrier.

Researchers are continuing to investigate a variety of field methods for evaluating the effectiveness of *in situ* barriers. One method, called geotomography, uses high-frequency electromagnetic waves propagated between boreholes to form a "picture," or geotomogram, of the ground. The geotomograms from before the grouting are compared to those taken afterwards to determine the location and extent of the grout injected into the ground (Laine, et al., 1980; Ramirez and Daily, 1990). Another technique being investigated is to use the change in the electrical resistivity of the ground to determine the volume of grout that occupies the void spaces in sandy soils (Komine, 1992). Research in acoustic emissions monitoring has had promising results for determining the extent of grout penetration or locating seepage through barriers. Acoustic emissions are generated as a result of the deformation of the ground during the injection of grout (Koerner, et al., 1981, 1984; Majer, 1989). The location where these waves are generated is correlated with the extent of the grout penetration. The same type of emissions also result from seepage through the ground and can be used to locate leaks. More research is needed to develop all of these methods to a point where they are precise enough to accurately evaluate hydraulic barriers and achieve general acceptance.

The best available measure of the effectiveness of a barrier is by long-term monitoring at the contaminated site. If specific chemical constituents are detected in the groundwater, then additional containment measures should be employed. Continuous monitoring allows the observational procedure to be applied to the design and construction of barrier systems. The observational procedure (Terzaghi and Peck, 1967) recognizes that variations in the soil properties and construction of the barrier cannot be perfectly known. Thus, development of the project is based on most likely conditions, but allows for changes as information about the site is collected during and after the construction. Using this procedure, the barrier can be built in stages and the effectiveness of each stage can be determined. Subsequent containment measures would only be employed if the monitoring indicates the necessity for them. This can result in reduced construction costs and more effective barrier systems.

FLOWS THROUGH BARRIERS

No barrier type or barrier material is totally impervious to chemical transport through it (Mitchell, 1986; Starr and Cherry, 1990). Nevertheless, the barriers that are now available can be built to meet the hydraulic conductivity requirements of the governing regulatory authorities and be resistant to chemical flow by advection (flow of dissolved and suspended materials with the groundwater) and

Table 11.4. Hydraulic Conductivities and Permittivities of Barrier Materials.

Material	Hydraulic Conductivity (m/s)	Typical Thickness (m)	Permittivity ($m^3/s/m^2/m$)
Compacted clay	1×10^{-9}	0.9	1.1×10^{-9}
Soil-bentonite	1×10^{-9}	0.5–1.5	$0.7\text{-}2 \times 10^{-9}$
Cement-bentonite	1×10^{-8}	0.5–1.5	$0.7\text{-}2 \times 10^{-8}$
Soil mixed wall	1×10^{-9}	1.0–2.0	$0.5\text{-}1 \times 10^{-9}$
Grout barrier	1×10^{-8}	1.0–2.0	$0.5\text{-}1 \times 10^{-8}$
Jet grout barrier	1×10^{-8}	1.0–2.0	$0.5\text{-}1 \times 10^{-8}$
Geomembrane	1×10^{-14}	0.001–0.002	$0.5\text{-}1 \times 10^{-11}$
Geosynthetic clay layer	1×10^{-12}–1×10^{-10}	0.012	8×10^{-11}–8×10^{-9}
Steel sheet piling*	1×10^{-7}	0.01–0.015	$0.7\text{-}1 \times 10^{-5}$
Drainage layer*	1×10^{-4}	0.3–0.6	$1.5\text{-}3 \times 10^{-4}$

* For comparison.

diffusion (chemical flow under a chemical concentration gradient). Typical values for the hydraulic conductivity of different barrier materials are given in Table 11.4. This table also shows the typical thickness for a barrier made with the given materials. The ratio of the hydraulic conductivity to the thickness of the barrier gives the *permittivity* which, according to Darcy's Law, is the flow rate for a unit difference in head across the barrier. The permittivity provides a useful basis for comparison of the relative effectiveness of different barrier types in preventing the advective flow of contaminants (Starr and Cherry, 1990).

In materials with low hydraulic conductivities, the chemical diffusion of the contaminants can be very important relative to advection, because the flow rate of water through the material is so low. Chemical diffusion coefficients are typically in the range of 2×10^{-10} to 2×10^{-9} m^2/s (Mitchell, 1991) for soils, and on the order of 3×10^{-14} m^2/s (Daniel and Schackelford, 1988) for geomembranes.

The relative importance of advective and diffusive flows for contaminant transport through a typical compacted clay barrier was demonstrated by Schackelford (1988). Figure 11.14 shows a plot of the transit time versus the hydraulic conductivity for compacted clay that is 0.91 m thick with a porosity of 0.5 and acted on by a hydraulic gradient of 1.33. The transit time is defined as the time required for the solute concentration on the downgradient side of the barrier to be 50% of the concentration on the upgradient side of the barrier. Figure 11.14 also shows the plots for the transit time required if diffusion were the only transport process, and the transit time for advection and diffusion combined. A molecular diffusion coefficient of 6×10^{-10} m/s was assumed in this example. This example demonstrates that when the hydraulic conductivity is less than 2×10^{-9} m/s diffusion starts to have a noticeable effect on the contaminant transport. If the diffusion rate is ignored in low hydraulic conductivity materials, less than 2×10^{-9} m/s in this case, then the transit time through the barrier would be greatly overestimated, since the transit time for diffusion alone is approximately 49 years. For most barriers the transit time would be a function of both advection and diffusion, which is shown as the advection-dispersion line in Figure 11.14. These times can be significantly less than the times due to either pure advection or pure diffusion.

Another consideration in estimating the flow rate through a barrier is the presence of holes or imperfections (Starr and Cherry, 1990). Figure 11.15 shows that relatively small imperfections, compared to the size of the barrier, greatly affect the hydraulic conductivity of the barrier. It was assumed that the imperfection had a hydraulic conductivity of 10^{-4} m/s, the same as a typical clean sand. If the imperfections account for one millionth; i.e., –6 on the X axis, of the total barrier area (height × length) then the bulk or effective hydraulic conductivity is approximately the same for all walls with a hydraulic conductivity less than 10^{-10} m/s. Starr and Cherry (1990) note that higher-conductivity barriers that can be built with few imperfections and adequate inspections are preferable to barriers that have a lower nominal hydraulic conductivity but are more subject to imperfections during installation.

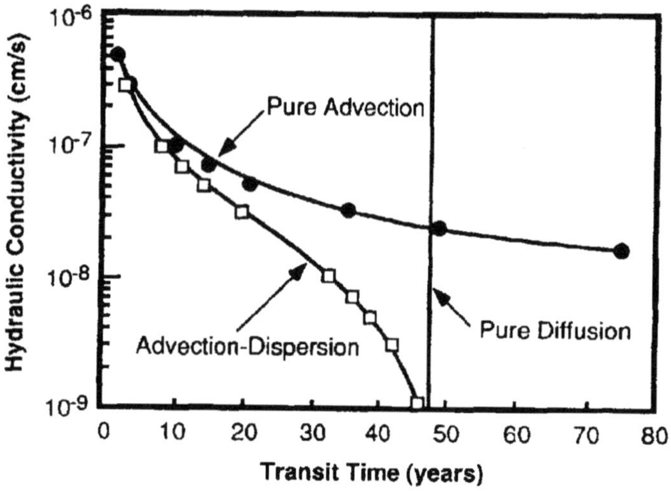

Figure 11.14. Transit time through a barrier due to advection, diffusion and the combination of advection and diffusion (advection-dispersion). *Source:* Schackelford, 1988.

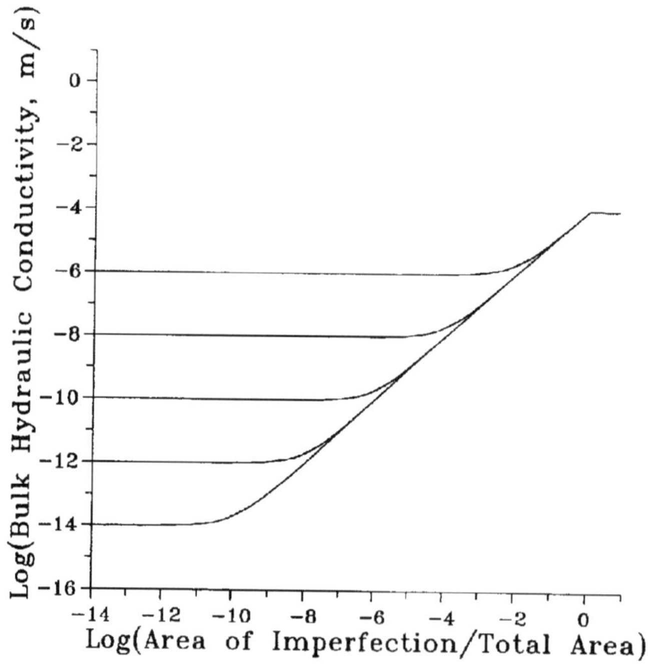

Figure 11.15. The effects of high-permeability imperfections on the bulk hydraulic conductivity of a barrier. The hydraulic conductivity of the imperfections is 0.0001 m/s. *Source:* Starr and Cherry, 1990.

REFERENCES

Alther, G., J.C. Evans, H.Y. Fang, and K. Witmer. 1984. Influence of inorganic permeants upon the permeability of bentonite, A.I. Johnson, R.K. Frobel, N.J. Cavalli, and C.B. Pettersson (Eds.), *Hydraulic Barriers in Soil and Rock*, ASTM STP 874, Philadelphia, PA, pp. 64–74.

Anderson, D.C., W. Crawley, and J.D. Zabcik. 1985. Effects of various liquids on clay soil: bentonite slurry mixtures, A.I. Johnson, R.K. Frobel, N.J. Cavalli, and C.B. Pettersson (Eds.), *Hydraulic Barriers in Soil and Rock*, ASTM STP 874, Philadelphia, PA, pp. 93–101.

API. 1990. Recommended practice standard procedure for field testing water-based drilling fluids, API Recommended Practice 13B-1, 1st ed., API Production Department, American Petroleum Institute, Washington, DC.

ASCE News. 1991. Company wants to put hazwaste "on ice," *Civil Engineering,* **59**(5).

Bell, F.G. and J.K. Mitchell. 1986. Control of groundwater by exclusion, J.C. Cripps, F.G. Bell, and M.G. Culshaw (Eds.), *Groundwater in Engineering Geology*, Geological Society Engineering Geology Special Publication No. 3, pp. 429–443.

Bodocsi, A., M.T. Bowers, and R.A. Sherer. 1987. Reactivity of various grouts to hazardous wastes and leachates, EPA Contract Number 68-03-3210, Cincinnati, OH.

Bowders, J.J. and D.E. Daniel. 1987. Hydraulic conductivity of compacted clay to dilute organic chemicals, *ASCE Journal of Geotechnical Engineering Division,* **113**(12):1432–1448.

Boynton, S.S. and D.E. Daniel. 1985. Hydraulic conductivity tests on compacted clay, *ASCE Journal of Geotechnical Engineering Division* 111, No. 4:465–478 California Code of Regulations Title 26, Section 23-2541.

Daniel, D.E. 1987. Earthen liners for land disposal facilities, *Geotechnical Practice for Waste Disposal '87,* ASCE Geotechnical Special Publication No. 13, pp. 21–39.

Daniel, D.E., D.C. Anderson, and S.S. Boynton. 1985. Fixed-wall versus flexible-wall permeameters, A.I. Johnson, R.K. Frobel, N.J. Cavalli, and C.B. Pettersson (Eds.), *Hydraulic Barriers in Soil and Rock*, ASTM STP 874, Philadelphia, PA, pp. 107–126.

Daniel, D.E. and C.D. Schackelford. 1988. Disposal barriers that release contaminants only by molecular diffusion, *Nuclear Chem. Waste Management,* **8**:299–305.

D'Appolonia, D.J. 1980. Soil-bentonite slurry trench cutoffs, *ASCE Journal of Geotechnical Engineering Division,* **106**(4):399–417.

Druback, G.W. and S.V. Arlotta, Jr. 1985. Subsurface pollution containment using a composite system vertical cutoff barrier, A.I. Johnson, R.K. Frobel, N.J. Cavalli, and C.B. Pettersson (Eds.), *Hydraulic Barriers in Soil and Rock*, ASTM STP 874, Philadelphia, PA, pp. 24–33.

Evans, J.C. 1986. Slurry trench cutoff walls for waste containment, Proc. International Symposium on Environmental Geotechnology, H.Y. Fang (Ed.), Envo Publishing Co., Inc., pp. 303–311.

Evans, J.C., I.J. Kugelman, and H.Y. Fang. 1985. Organic fluids effects on the strength, deformation and permeability of soil-bentonite walls, Toxic and Hazardous Wastes, Proc. 17th Mid-Atlantic Industrial Waste Conference, pp. 275–291.

Evans, J.C., E.D. Stahl, and E. Droof. 1987. Plastic concrete cutoff walls, *Geotechnical Practice for Waste Disposal '87*, ASCE Geotechnical Special Publication No. 13, pp. 462–472.

Geo-Con, Inc. 1989. Technical brief: deep soil mixing, T-DSM-01-89, Pittsburgh, PA.

Geo-Con, Inc. 1990. Technical brief: soil-bentonite slurry walls, T-SW-01-85, Pittsburgh, PA.

Gundle Lining Systems, Inc. 1991. Gundwall subsurface polyethylene vertical barrier, 20 CHP 0991, Houston, TX.

Hayward-Baker. 1991. Guide Specifications for Cement-Bentonite Slurry Trench Cut-Off Construction, Hayward-Baker, Odenton, MD.

Iskander, J.K. and T.F. Jenkins. 1985. Potential use of artificial ground freezing for contaminant immobilization, Proc. International Conf. on New Frontiers for Hazardous Waste Management, Pittsburgh, PA.

Jefferis, S.A. 1992. Contaminant-grout interaction, R.H. Borden, R.D. Holtz, and I. Juran (Eds.), *Grouting, Soil Improvement and Geosynthetics*, ASCE Geotechnical Special Publication No. 30, New Orleans, LA, pp. 1393–1402.

Jepsen, C.P. and M. Place. 1985. Evaluation of two methods for constructing vertical cutoff walls at waste containment sites, A.I. Johnson, R.K. Frobel, N.J. Cavalli, and C.B. Pettersson (Eds.), *Hydraulic Barriers in Soil and Rock*, ASTM STP 874, Philadelphia, PA, pp. 45–60.

Jessberger, H.L. and K.J.L. Stone. 1991. Subsidence effects on clay barriers, *Geotechnique* **41**(2):185–194.

Joseph, K.L., H. Rachid, and E.B. Perry. 1992. Methods to estimate composition of jet grout bodies, R.H. Borden, R.D. Holtz, and I. Juran (Eds.), *Grouting, Soil Improvement and Geosynthetics*, ASCE Geotechnical Special Publication No. 30, New Orleans, LA, pp. 194–205.

Koerner, R.M. 1990a. *Designing With Geosynthetics*, 2nd Edition, Prentice Hall, Englewood Cliffs, NJ.

Koerner, R.M. 1990b. Preservation of the environment via geosynthetic containment systems, *Geotechnical Fabrics Report,* **8**(5):18–22 and (6):31–47.

Koerner, R.M., W.M. McCabe, and L.F. Baldivieso. 1981. Acoustic emission monitoring of seepage, *ASCE Journal of Geotechnical Engineering Division,* **107**(4):521–526.

Koerner, R.M., J.D. Laird, and J.P. Welsh. 1984. Uses of acoustic emissions as a non-destructive testing method to monitor grouting, *Innovative Cement Grouting*, American Concrete Institute Publication SP-83, pp. 85–102.

Koerner, R.M., Y. Halse-Hsuan, and A.E. Lord. 1991. Long-term durability of geomembranes, *Civil Engineering,* **61**(4):56–58.

Koerner, R.M. and D.E. Daniel. 1992. Better cover-ups, *Civil Engineering,* **62**(5):55–57.

Koerner, R.M., Y. Hsuan, and A.E. Lord. 1992. Remaining technical barriers to obtain general acceptance of geosynthetics, R.H. Borden, R.D. Holtz, and I. Juran (Eds.) *Grouting, Soil Improvement and Geosynthetics*, ASCE Geotechnical Special Publication No. 30, New Orleans, LA, pp. 63–109.

Komine, H. 1992. Estimation of chemical grout void filling by electrical resistivity, R.H. Borden, R.D. Holtz, and I. Juran (Eds.), *Grouting, Soil Improvement and Geosynthetics*, ASCE Geotechnical Special Publication No. 30, New Orleans, LA, pp. 372–383.

Laine, E.F., R.J. Lytle, and J.T. Okada. 1980. Cross-borehole observation of soil grouting, *ASCE Journal of Geotechnical Engineering Division,* **106**(8):871–875.

Mackay, D.M. and J.A. Cherry. 1989. Groundwater contamination: pump-and-treat remediation, *Environmental Science and Technology,* **23**(6):630–636.

Madsen, F.T. and J.K. Mitchell. 1989. Chemical effects on clay hydraulic conductivity and their determination, Nr.135 Mitteilungen des Institutes fur Grundbau und Bodenmechanik Eidgenossische Technische Hochschule, Zurich.

Majer, E.L. 1989. The application of high frequency seismic monitoring methods for the mapping of grout injections, *Int. Journal of Rock Mechanics, Mineral Science and Geomechanic Abstracts,* **26**(3/4):249–256.

May, J.H., R.H. Larson, P.G. Malone, J.A. Boa, and D.L. Bean. 1986. Grouting techniques in bottom sealing of hazardous waste sites, USEPA Hazardous Waste Engineering Research Laboratory, Cincinnati, OH.

McFarlane, T.S. and R.D. Holtz. 1992. Selection and laboratory evaluation of modifying additives for soil-cement-bentonite, R.H. Borden, R.D. Holtz, and I. Juran (Eds.), *Grouting, Soil Improvement and Geosynthetics*, ASCE Geotechnical Special Publication No. 30, New Orleans, LA, pp. 1006–1018.

Millet, R.A. and J.Y. Perez. 1981. Current U.S.A. practice: slurry wall specifications, *ASCE Journal of Geotechnical Engineering Division,* **107**(8):1041–1056.

Mitchell, J.K. 1981. Soil improvement: state-of-the-art, Proc. 10th ICSMFE, Stockholm, Sweden, pp. 508–565.

Mitchell, J.K. 1986. Hazardous waste containment, J.C. Cripps, F.G. Bell, and M.G. Culshaw (Eds.), *Groundwater in Engineering Geology*, Geological Society Engineering Geology Special Publication No. 3, pp. 145–157.

Mitchell, J.K. 1991. Conduction phenomena: from theory to geotechnical practice, *Geotechnique,* **41**(3):299–340.

Mitchell, J.K. and T.L. Brandon. 1985. Containment barriers for land disposed solvent bearing hazardous wastes, Environmental Institute for Waste Management Studies, University of Alabama.

Mitchell, J.K. and W.A. van Court. 1992. The role of soil modification in environmental engineering applications, R.H. Borden, R.D. Holtz, and I. Juran (Eds.), *Grouting, Soil Improvement and Geosynthetics*, ASCE Geotechnical Special Publication No. 30, New Orleans, LA, pp. 110–143.

Peterson, S.R. and G.W. Gee. 1985. Interactions between acidic solutions and clay liners: permeability and neutralization, A.I. Johnson, R.K. Frobel, N.J. Cavalli, and C.B. Pettersson (Eds.), *Hydraulic Barriers in Soil and Rock,* ASTM STP 874, Philadelphia, PA, pp. 229–245.

Ramirez, A.L. and W.D. Daily. 1990. Geotomography of grout penetration in fractured granite, *Int. Journal of Rock Mechanics, Mineral Science and Geomechanic Abstracts,* **27**(1):57–63.

Rumer, R.R. and J.K. Mitchell. 1996. Assessment of Barrier Containment Technologies—A Comprehensive Treatment for Environmental Remediation Applications. NTIS Publication #PB96-180583, 437 p. (This publication is a product of the International Containment Technology Workshop, Baltimore, MD, Aug. 29–31, 1995.)

Rumer, R.R. and M.E. Ryan. 1995. Barrier Containment Technologies for Environmental Remediation Applications. John Wiley & Sons, Inc. NY, 170 p.

Ryan, C.R. 1987. Vertical barriers in soil for pollution containment, *Geotechnical Practice for Waste Disposal '87*, ASCE Geotechnical Special Publication No. 13, pp. 182–204.

Schackelford, C.D. 1988. Diffusion as a transport process in fine-grained barrier materials, *Geotechnical News,* **6**(3):24–27.

Siluwa, J.M. and R.J. Krizek. 1992. Permanence of grouted sands exposed to various water chemistries, R.H. Borden, R.D. Holtz, and I. Juran (Eds.) *Grouting, Soil Improvement and Geosynthetics*, ASCE Geotechnical Special Publication No. 30, New Orleans, LA, pp. 1403–1419.

Spooner, P.A., G.E. Hunt, V.E. Hodge, and P.M. Wagner. 1982. Collection of information on the compatibility of grouts with hazardous wastes, USEPA Municipal Environmental Research Laboratory, Cincinnati, OH.

Spooner, P.A., G.E. Hunt, V.E. Hodge, P.M. Wagner, and I.R. Melnyk. 1984. Compatibility of grouts with hazardous wastes, USEPA Municipal Environmental Research Laboratory, Cincinnati, OH.

Starr, R.C. and J.A. Cherry. 1990. In situ barriers for groundwater pollution control, *Prevention and Treatment of Soil and Groundwater Contamination in the Petroleum Refining and Distribution Industry*, Montreal, Quebec.

Starr, R.C., J.A. Cherry, and E.S. Vales. 1991. Sealable joint sheet pile cutoff walls for preventing and remediating groundwater contamination, Technology Transfer Conference, Ontario Ministry of the Environment, Toronto.

Taki, O., S. Takeshima, and D.S. Yang. 1990. Pollution containment using soil-cement mixed walls, Proc. Hazmacon '90, Anaheim, CA.

Tausch, N. 1992. Recent European developments in constructing grouted slabs, R.H. Borden, R.D. Holtz, and I. Juran, (Eds.), *Grouting, Soil Improvement and Geosynthetics*, ASCE Geotechnical Special Publication No. 30, New Orleans, LA, pp. 301–312.

Teeter, R.M. and S.A. Clemence. 1986. In-place permeability measurement of slurry trench cutoff walls, *Use of In Situ Tests in Geotechnical Engineering*, Geotechnical Special Pub. No. 6, Proc. In Situ '86, pp. 1049–1061.

Terzaghi, K. and R.B. Peck. 1967. *Soil Mechanics in Engineering Practice*, 2nd ed., John Wiley & Sons, New York City, NY.

Tobin, W.R. and P.R. Wild. 1986. Attapulgite: A clay liner solution, *Civil Engineering,* **56**(2):56–58.

U.S. Environmental Protection Agency. 1979. Design and construction of covers for solid waste landfills, EPA-600/2-79-165, Cincinnati, OH.

U.S. Environmental Protection Agency. 1985. Handbook of remedial action at waste disposal sites, EPA, Cincinnati, OH.

U.S. Environmental Protection Agency. 1986. Covers for uncontrolled hazardous waste sites, EPA-540/2-85-002.

U.S. Environmental Protection Agency. 1990. The Superfund innovative technology evaluation program: Technology Profiles, EPA/540/5-90/006, Cincinnati, OH.

White, D.F. and S.R. Alen. 1991. Evaluating the chemical resistance of geosynthetics, *Geotechnical Fabrics Report,* **9**(4):30–33.

Section V

Technologies for Contaminant Removal

Chapter 12

Removal of Nonaqueous Phase Liquids Using Surfactants

J.C. Fountain

INTRODUCTION

Organic pollutants such as the common chlorinated solvents trichloroethylene, perchloroethylene, carbon tetrachloride and methylene chloride are characterized by low aqueous solubilities, densities greater than water and low viscosities (Feenstra and Cherry, 1988). When released in the subsurface these liquids may move rapidly downward due to their high densities. Due to their low aqueous solubilities, only small volumes of these pollutants can dissolve in groundwater. Thus when any significant volume of organic contaminant reaches the water table, it will not totally dissolve but will remain as a separate phase, referred to as a dense nonaqueous phase liquid (DNAPL).

Pump-and-treat is ineffective at restoration of aquifers contaminated with DNAPL due to a combination of the low solubility of the DNAPL components and large capillary forces that immobilize the nonaqueous phase (Feenstra and Cherry, 1988; MacKay and Cherry, 1989; Mercer and Cohen, 1990). Pumping of contaminated water extracts only the small amount of contaminants present as a dissolved phase. The clean water moving into the contaminated zone is then immediately contaminated by the nonaqueous phase that remains in the aquifer. Since little of the DNAPL dissolves in each pore volume of water that is extracted, a very large volume of water would have to be pumped to dissolve the DNAPL. Typically pumping times of a hundred years or more are required.

Surfactants have the potential to increase the efficiency of pump-and-treat systems both by increasing the solubility of organic pollutants and by decreasing the interfacial tensions between DNAPLs and groundwater (Vigon and Rubin, 1989; Abdul et al., 1992; Liu et al., 1992; West and Harwell, 1992; Baran et al., 1994). These two approaches, although not necessarily mutually exclusive, involve significantly different mechanisms and will thus be discussed separately.

SURFACTANT-ENHANCED MOBILIZATION

The use of surfactants to lower interfacial tensions and hence increase the mobility of the nonaqueous phase is a direct extension of the use of surfactants in enhanced oil recovery (Lake, 1989; Pope and Wade, 1995). Under normal conditions, large interfacial tensions between DNAPL and water result in large capillary forces that immobilize the DNAPL, preventing it from being driven to recovery wells by the pumping of groundwater (Wilson and Conrad, 1984; Hunt et al., 1988; Pennell et al., 1996). Surfactants can lower the interfacial tension by four orders of magnitude, which would reduce capillary forces sufficiently so that the traction of pumped groundwater could theoretically move the DNAPL to an extraction well.

This approach has been attempted at a field test at a wood-treating facility in Wyoming (Sale et al., 1989). In this test, a significant volume of DNAPL was recovered, but high levels of contamination remained in the soil. Inhomogeneities in an aquifer cause variations in sweep efficiencies; thus total DNAPL removal cannot be expected by mobilization. This approach may, nonetheless, be an effective method of rapid removal of large volumes of DNAPL.

The principal limitation of this approach for DNAPL remediation is the possibility of inducing unwanted vertical mobility of the contaminants. Reduction of interfacial tensions, and hence reduction of capillary forces, promotes vertical migration of DNAPL as much as it promotes horizontal mobility.

DNAPLs tend to move downward through the groundwater in an aquifer because they are denser than water. Their flow is governed by capillary forces. To enter each water-filled pore the DNAPL must overcome a capillary displacement (entry) pressure, and when leaving each pore, capillary forces retain some of the DNAPL (Kueper and Frind, 1991; Kueper and McWhorter, 1991; Pankow and Cherry, 1996). Since the displacement pressure is higher in finer-grained layers, they act as barriers to downward migration. DNAPL will accumulate on the top of the layer until it accumulates to a sufficient thickness to generate the required displacement pressure.

The lowering of interfacial tensions increases the horizontal and vertical mobility by lowering capillary forces. The displacement pressure decreases directly with interfacial tension; thus a large reduction of interfacial tension may allow DNAPL to penetrate fine-grained layers that it had not previously been able to penetrate. This may allow the DNAPL to move downward through an aquitard, contaminating underlying zones that had previously been clean.

The risk of downward mobility is also enhanced simply by inducing horizontal movement of the DNAPL. Aquitards may have fractures, sand lenses, root holes or other pathways within them and inducing horizontal flow of the DNAPL increases the chance that such pathways may be encountered by the DNAPL. Thus reduction of interfacial tensions required to mobilize the DNAPL increases the ability of the DNAPL to penetrate the pathways and increases the chance such pathways will be encountered.

In view of the probability that such pathways occur in any aquitard, the risk of mobilization will generally be unacceptable. The risk is highest for dense, low-viscosity DNAPLs such as the halogenated solvents and lower for low-density, viscous DNAPLs such as creosote. The risk is highly site dependent and must be evaluated for each site.

SURFACTANT-ENHANCED SOLUBILIZATION

The use of surfactants to increase the solubility of contaminants can be viewed as a method of enhancing the efficiency of conventional pump-and-treat technology. Surfactant solution is injected into the contaminated zone, recovered and treated as with conventional pump-and-treat technology. The objective of adding surfactants is to increase the solubility of the contaminants and its effect is to increase the concentration of the contaminants in the extracted water.

The ability of surfactants to increase the solubility of hydrophobic compounds derives from their formation of aggregates, known as micelles (Rosen, 1989). Micelles are capable of corporating molecules of normally hydrophobic compounds, thus greatly increasing their solubility. Micelles form because surfactant molecules are characterized by a hydrophobic moiety and a hydrophilic moiety. In aqueous solution, the molecules arrange themselves in spherical aggregates with their polar, hydrophilic portions outward and their hydrophobic, nonpolar ends in the center (Figure 12.1). The nonpolar core of the micelles is capable of incorporating nonpolar molecules, thus increasing the solubility of virtually any hydrophobic contaminant. Large solubility increases have been demonstrated for chlorinated solvents, hydrocarbons, PCBs, PAHs and many other compounds (Saito and Shinoda, 1967; Attwood and Florence, 1983; Luthy et al., 1991; Fountain, 1992; Baran et al., 1994; Adeel and Luthy, 1995) and should occur for all hydrophobic organic compounds (Wayt and Wilson, 1989).

Since increasing the solubility of the components of a DNAPL is the objective of this use of surfactants, the degree of solubilization is the principal criterion for surfactant selection. Although virtually any contaminant can be solubilized by the appropriate surfactant, different surfactants are required for optimum solubilization of different contaminants.

Other criteria for surfactant selection include: toxicity, biodegradability, surfactant sorption, surfactant solubility and compatibility with the separation process. Toxicity is a concern since the surfactants will be added to groundwater. Many common surfactants have relatively low toxicities: several types are certified for use in food products and many others for use in food preparation. Surfactants are the active ingredients in dishwashing detergents and shampoos, and are thus occasionally ingested. Biodegradability is also of concern since it must be assumed that if a large volume of surfactants is injected into an aquifer, not all of it will be recovered. Surfactants are generally readily biodegradable since they are often disposed of in wastewater and hence have been formulated for

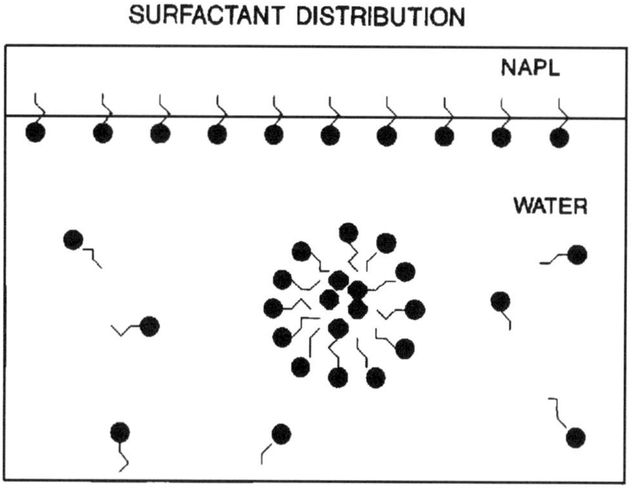

Figure 12.1. In nonaqueous phase liquid/ground water systems surfactant monomers will concentrate along the interface with their non-polar tails (bent lines) in the NAPL and the polar ends (filled circles) in the water. This produces a reduction of interfacial tension. When the concentration of surfactant molecules reaches the critical micellar concentration, the monomers form micelles. In the micelle, the surfactant molecules arrange themselves with their tails in the center, forming a nonpolar environment. The micelle is thus capable of incorporating molecules of NAPL into its center.

breakdown in conventional wastewater treatment plants. The biodegradability of surfactants has been reviewed by Swisher (1987). The typically rapid biodegradation of surfactants may prove to be a hindrance to surfactant flushing projects since significant surfactant loss due to degradation may occur in surface tanks and biofouling may occur in injection well systems.

Sorption of surfactant to aquifer solids can be significant (Scamehorn et al., 1982; Von Rybinski and Schwuger, 1987; Liu et al., 1992), which increases the cost of remediation due to loss of surfactant to sorption. The extent of sorption varies with surfactant type, aquifer composition and water chemistry, generally increasing with fraction organic content (Liu et al., 1992). It may be necessary to use a surfactant that has a lower solubilization ability than some other surfactants but also has lower sorption if the aquifer to be treated has a high organic carbon content. Generally, anionic surfactants, which have negatively charged functional groups, have lower sorption than nonionic surfactants (Rouse et al., 1992).

Typically, many pore volumes of surfactant solution will have to be circulated to remove all DNAPL from a contaminated zone; thus the surfactant must be recirculated for the process to be economical. This requires separation of the contaminant from the surfactant solution. Air stripping can be used for volatile contaminants (surfactants are nonvolatile). We have successfully stripped tetrachloroethylene and carbon tetrachloride from surfactant solutions in separate field tests (Fountain and Hodge, 1992; Fountain, 1993; Fountain et al., 1995). Other techniques that can be used include solvent extraction, sorption and ultrafiltration (Lipe et al., 1992; 1996).

Whichever technique is used, the surfactant must be compatible with the separation method. For example, air stripping requires low-foaming surfactants. Solvent extraction requires a surfactant that does not dissolve significantly in the organic solvent used for extraction. If sorption is used, which would be feasible only for pollutants that are strongly sorbed on organic material (such as PCBs), sorption of the surfactant must be minimal.

Interfacial tension is also an important parameter. Surfactants can reduce the interfacial tension between DNAPL and water by as little as a factor of three to as much as four orders of magnitude. As previously discussed, reduction of interfacial tension reduces capillary forces and hence increases the mobility of contaminants. Interfacial tension reduction should be minimized where site geology indicates there is a significant risk from mobilization, or it could be maximized where there is little risk from mobilization and there is a large DNAPL pool that could be extracted through mobilization. In general, surfactants that reduce the interfacial tension by the smallest amount are preferred.

THE BORDEN FIELD TEST

The ability of surfactants to effectively remove DNAPL from an aquifer was demonstrated at a field test recently conducted at the Canadian Forces Base Borden test site operated by the University of Waterloo Centre for Groundwater Research. The results of this test also illustrate some of the limitations of the technique.

A 3 m by 3 m test cell was constructed by driving sheet piling walls through the surficial unconsolidated sand aquifer into a clay aquitard that is found approximately 4 m below the surface. Two hundred and thirty one liters of tetrachloroethylene (also known as perchloroethylene or PCE) were introduced into the test cell in a controlled release designed to study the patterns of DNAPL migration (Kueper et al., 1993). Subsequently the cell was excavated to a depth of 1 m, a layer of bentonite was added to confine the remaining contaminated sand, and the cell was backfilled with clean sand. A surfactant flood was then utilized to remediate the remaining 3 m of contaminated sand.

Surfactant solution was injected into a row of five injection wells on one side of the cell and was extracted from a second row of five wells located on the other side (Figure 12.2). Injection and extraction wells were screened over the entire 3 m interval from the aquitard to the bentonite layer. After extraction, the solution was sent through an air stripper to remove the PCE. The solution was then brought up to the desired 2% (by weight) surfactant concentration in a mixing tank and transferred to a third tank for reinjection. The surfactant mixture used was a 1:1 mixture of a nonylphenol ethoxylate (Witconol NP100 or Alkaryl NP10) and a phosphated ester of a nonylphenol ethoxylate (Rexophos 25/97). Batch tests determined that the 2% mixture used was capable of solubilizing approximately 11,500 ppm of PCE at 10°C compared to an aqueous solubility of about 150 ppm at this temperature.

Prior to starting the surfactant flushing, DNAPL was removed by direct pumping from the five injection and five extraction wells. DNAPL was recovered by pumping through small-diameter tubing that was lowered to the bottom of each well. After DNAPL production dropped to nearly zero, water flooding was begun to determine how much DNAPL could be driven to the extraction wells by pumping groundwater. Less than 3 L of PCE was recovered (as free product) during the water flooding (Table 12.1), which illustrates the immobility of DNAPL under normal conditions. DNAPL recovery did not significantly increase after the addition of surfactants; less than 1 L of free product (this does not include dissolved PCE) was recovered during the entire surfactant flood (Table 12.1).

Surfactant injection was begun on November 10, 1990. The concentration of PCE in the effluent rose to over 4000 ppm when the surfactant broke through (Figure 12.3). The irregular spikes on the effluent concentration curve correspond to periods of restarting after shutdowns. The system was down approximately 50% of the time for the first four months either for equipment repair or for holiday leave.

Breakthrough of the surfactant occurred at the same volume as breakthrough of bromide during an earlier tracer experiment, indicating sorption was insignificant, as expected from batch tests. The aquifer consists of sand with virtually no clay nor organic matter, thus sorption was expected to be very low.

In the succeeding 6 months a total of 34,441 gal, equivalent of approximately 14 pore volumes, was circulated. A total of about 62 L of PCE was removed as a dissolved phase. Samples were taken regularly from six sampling points on each of the five multi-level monitoring wells near the extraction wells (Figure 12.1). Individual sampling points reached over 10,000 ppm, near the maximum solubilization capacity of the solution (11,500 ppm). The system was then shut down for an extended period (over five months) during which time the surfactant, through a combination of hydrolysis and biodegradation, decomposed. The system was then operated without surfactant; as expected, concentrations decreased rapidly.

Three cores were taken prior to surfactant flooding and analyzed to determine both hydraulic conductivity and DNAPL distribution prior to surfactant treatment. Additional cores were taken after

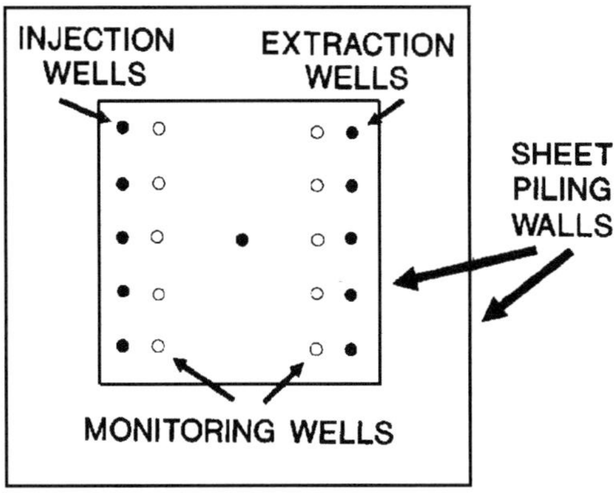

Figure 12.2. Design of the test cell at CFB Borden. The inner cell is 3 m square.

Table 12.1. Volume of Perchloroethylene Recovered as a Nonaqueous Phase

Operation	Vol. PCE Recovered, L
Direct Pumping	45.86
Water Flooding	2.35
Surfactant Flooding	0.87

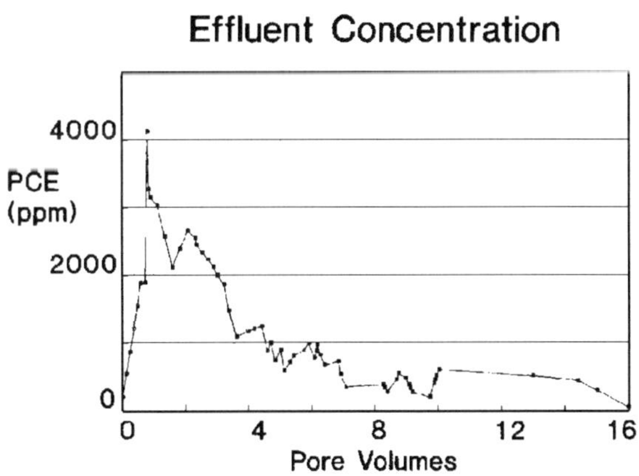

Figure 12.3. Effluent concentration from the Borden field experiment. The effluent concentration increased to over 20 times the aqueous solubility of PCE after surfactant injection began. After 14.4 pore volumes, the system was changed to water circulation.

7 and 14 pore volumes. The relative saturation of PCE in the cell, as revealed by analyses of samples from these cores, shows a rapid reduction in the volume of PCE by the surfactant flooding (Figure 12.4). Both the maximum saturation of PCE and the thickness of the contaminated zone decreased dramatically.

Three important observations from the core data are: 1) the surfactant solution rapidly reduced the amount of DNAPL present, 2) the zones of high DNAPL saturation remained at the same elevation, indicating little vertical mobilization and 3) the cleaning was not complete at 14 pore volumes (evidenced by the significant saturations still found on the aquitard).

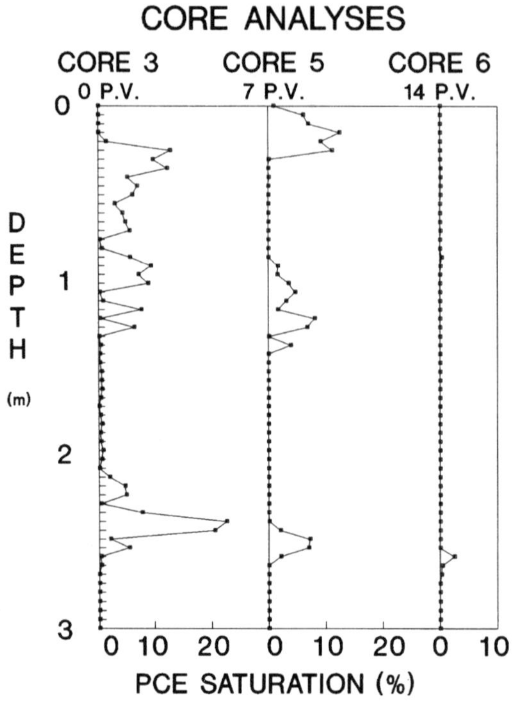

Figure 12.4. PCE saturation at the start of surfactant flooding (core 3), after seven pore volumes (core 5) and after 14 pore volumes were circulated (core 6). Depths are distances below the bentonite layer. The aquitard is located at about 2.6 m, just below the lowest DNAPL pool. Analyses done at the University of Waterloo Centre for Groundwater Research (Kueper et al., 1993).

The relation of PCE saturation to hydraulic conductivity is shown in Figure 12.5. PCE saturation, expressed as the percentage of pore space occupied by PCE, is plotted on the left and hydraulic conductivity is plotted on the right. The conductivity is very uniform, varying by only a factor of five among most samples. PCE accumulated on several low-conductivity layers just over 1 m below the top of the aquifer and on the clay aquitard (Figure 12.5). Since DNAPL must accumulate a certain displacement pressure to enter a saturated layer, and the required pressure is higher for fine grained layers, any decrease in grain size has the potential to cause horizontal spreading. Thus significant horizontal spreading occurred even within the relatively uniform sands (Kueper et al., 1993). Although the general pattern of saturation can be explained from the hydraulic conductivity, a detailed correlation would not be expected due to the fact that the scale of layering within the aquifer is on the order of millimeters; thus conductivities averaged over a length of centimeters may not reveal the effect of the fine layering.

In August the cell was excavated jointly by teams from the University at Buffalo and the University of Waterloo. The sand was removed by hand shovel in 30 cm lifts. Small (2 cc) core samples were taken on a 30 cm grid, 100 samples per lift, on every lift. Over 1000 samples were taken during excavation to determine the mass distribution of PCE within the cell. These analyses are only partially completed at the time of writing of this chapter; however, the data is sufficient at this point to clearly reveal the extent of remediation within the cell. Additional data is presented in Fountain et al., 1996.

PCE DISTRIBUTION AT THE CONCLUSION OF THE FIELD TEST

Excavation revealed that the majority of the cell was clean, but there were three areas with significant PCE remaining. Thus there were four distinct zones encountered during excavation:

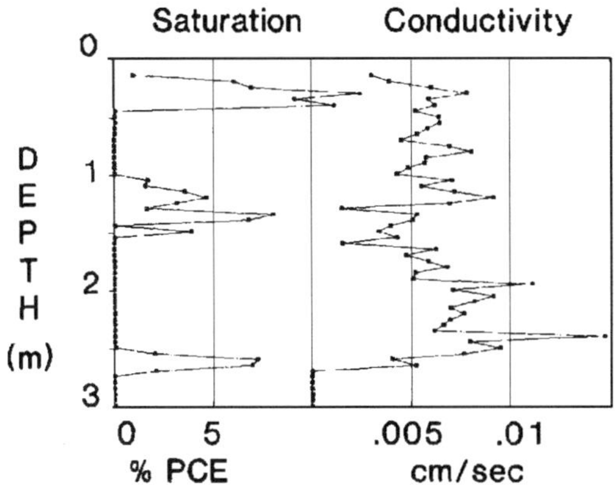

Figure 12.5. Comparison of the distribution of DNAPL with hydraulic conductivity. Saturation of PCE is expressed as a percentage of the total pore space. Depth is expressed as distance below the bentonite layer which marks the top of the contaminated sand.

1. The majority of the cell, from about 20 cm below the top of the cell to within a few cm of the aquitard, was clean. In nearly all of the 700 samples taken over this interval, the PCE concentration was below the limit of detection, which is about 5 ppm for the gas chromatograph method used. These data indicate no significant DNAPL remained in this interval.
2. Near the top of the cell, about 15 cm below the bentonite, a distinctly pink zone with an area of about 0.7 m^2 was encountered. (The PCE had been dyed red with Sudan IV dye). Calculated PCE saturation values ranged from 1% to 30% within this zone; a total of approximately 1.5 L of PCE remained in this layer.
   The persistence of PCE in this zone suggests the zone had not been flushed by a sufficient volume of surfactant solution. The zone appeared to have an average hydraulic conductivity, but flow may have been limited because the zone was above the top of the well screens for both the injection and extraction wells.
   It is significant that PCE saturations remained high in this zone from the start of the experiment to the end, indicating that this pool at least was not vertically mobilized by the surfactant.
3. A DNAPL pool on the aquitard. On the surface of aquitard, which was formed of a silt layer underlain by a clay layer, a thin pool (typically a few mm) of DNAPL was encountered over approximately one half of the cell. At the start of surfactant flooding, there was a pool with high saturations that was over 20 cm thick (Figure 12.4). Thus the pool was reduced in thickness from 20 cm to less than 1 cm in 14 pore volumes. We believe that the pool would have been removed completely in another few pore volumes. The continued rapid cleaning of this layer was indicated by the concentrations of PCE in multi-level samples from Level 6, which was located just at the aquitard. Samples from Level 6 of the three center monitoring wells averaged over 3800 ppm at the time surfactant circulation was stopped, indicating pool removal was continuing.
4. DNAPL was able to enter the aquitard in a number of locations, through sand lenses that reached the surface of the aquitard or via pathways provided by the numerous wells and probes that penetrated the aquitard. Where PCE was able to enter the silt, it was not removed by the surfactant since flow though the silt layer is negligible.

SUMMARY AND CONCLUSIONS

The results of the field test illustrate both the capabilities and the limitations of aquifer remediation using chemically enhanced solubilization. Four conclusions may be drawn.

1. The high concentrations of PCE in the effluent show that surfactants greatly increase the efficiency of pump-and-treat where DNAPLs are present.

2. The large cleaned zone, much of which had initial PCE saturations in excess of 10% of its pore volume, shows that surfactant flooding can effectively remove large volumes of DNAPL.
3. The persistence of DNAPL in the top zone illustrates that where flow is low, a larger number of pore volumes is required. Typically this may occur where contamination is in a zone of moderate hydraulic conductivity and a zone of higher conductivity is also present. Since flow velocity is higher in the higher-conductivity layers, significantly more solution must be pumped than would be required in a homogeneous system.
4. The sand lenses within the silt illustrate that inevitably there will be zones of little or no flow. Where aquifer heterogeneities produce such zones, cleaning will be very slow. It may be expected that some such heterogeneities will occur at all sites, and thus a tail of extended cleaning will occur. The volumetric importance of this tail will be site dependent.

In addition to the tailing caused by heterogeneities, diffusion will also cause tailing. During the time that the DNAPL is in contact with a fine-grained layer, it will diffuse into the water in the pores of the layer. If the DNAPL is removed, the PCE will diffuse back out of the trapped pore water, providing a slow flux of contaminant to the aquifer. The concentrations along the interface will be diluted by many orders of magnitude by the larger volume of water that does not contact the fine-grained layers. Thus diffusion may not raise the concentration of the average water within an aquifer to significant values. Since the dilution is dependent upon the distribution of fine-grained layers within the system, the relative effect of diffusion will be highly site dependent.

In summary, the use of surfactants has been shown to be an effective method of removing the majority of DNAPL from an aquifer. The volume of solution required to remove the DNAPL will be a function the volume of the source and the specific hydrogeology of the site. Heterogeneities can be expected to cause a tailing in the removal of DNAPL and diffusion will cause an additional tailing when, and if, the DNAPL has been removed.

ACKNOWLEDGMENTS

The Borden field test was supported by the New York State Center for Hazardous Waste Management and the Solvents in Ground-Water Program of the Waterloo Centre for Groundwater Research. Bob Starr, Bernie Kueper, Mette Poulsen and Dave Redman provided significant input to the study. Tom Middleton and Mike Beikirch operated the test and the entire surfactant group at the University at Buffalo assisted in excavation.

REFERENCES

Abdul, A.S., T.L. Gibson, C.C. Ang, J.C. Smith, and R.E. Sobczynski. 1992. *In situ* Surfactant Washing of Polychlorinated Biphenyls and Oils from a Contaminated Site. *Ground Water* **30**:219–231.

Adeel, Z. and R.G. Luthy. 1995. Sorption and Transport Kinetics of a Nonionic Surfactant Through an Aquifer Sediment. *Environ. Sci. Technol.,* **29**:1032–1042.

Attwood, D. and A.T. Florence. 1983. *Surfactant Systems.* Chapman & Hall, New York, New York, 794 p.

Baran, J.R.J., G.A. Pope, W.H. Wade, V. Weerasoorlya, and A. Yapa. 1994. Microemulsion Formation with Chlorinated Hydrocarbons of Differing Polarity. *Environ. Sci. Technol.,* **28**(7):1361–1365.

Feenstra, S. and J.A. Cherry. 1988. Subsurface Contamination by Dense Nonaqueous Phase Liquid (DNAPL) Chemicals. *Proceedings*, The International Ground Water Symposium, Halifax, Nova Scotia, Canada.

Fountain, J.C. 1992. Field Tests of Surfactant Flooding: Mobility Control of Dense Nonaqueous-Phase Liquids. *In* D.A. Sabatini and R.C. Knox, *Transport and Remediation of Subsurface Contaminants*, American Chemical Society, Washington, DC, pp. 182–191.

Fountain, J.C. and D.S. Hodge. 1992. Extraction of Organic Pollutants Using Enhanced Surfactant Flushing—Initial Field Test (Part 1): *New York State Center for Hazardous Waste Management, State University of New York at Buffalo, Project Summary*, 9 p.

Fountain, J.C. 1993. Extraction of Organic Pollutants Using Enhanced Surfactant Flooding: Pilot Scale Field Trial at Corpus Christi, Texas. *New York State Center for Hazardous Waste Management, State University of New York at Buffalo, Project Summary,* 9 p.

Fountain, J.C., R.S. Starr, T. Middleton, M. Beikirch, C. Taylor, and D.S. Hodge. 1996. A Controlled Field Test of Surfactant Enhanced Aquifer Remediation. *Ground Water,* 34:910–916.

Fountain, J.C. C. Waddell-Sheets, A. Lagowski, C. Taylor, D. Frazier, and M. Byrne. 1995. Enhanced Removal of Dense Nonaqueous-Phase Liquid Using Surfactants: Capabilities and Limitations from Field Trials. In Sabatini, D.A., R.C. Knox, and J.H. Harwell, Eds. *Surfactant-Enhanced Subsurface Remediation. American Chemical Society Surface Sciences, Symposium Series,* 594, 177.

Hunt, J.R., M. Sitar, and K.S. Udell. 1988. Nonaqueous Phase Liquid Transport and Cleanup 1. Analysis of Mechanisms. *Water Resources Research,* **24**:1247–1259.

Kan, A.T. and M.B. Tomson. 1990. Ground Water Transport of Hydrophobic Organic Compounds in the Presence of Dissolved Organic Matter. *Environ. Tox. Chem.*, **9**:253–263.

Kueper, B.H. and E.O. Frind. 1991. Two-Phase Flow in Heterogeneous Porous Media, 1. Model Development. *Water Resources Research*, **27**:1049–1057.

Kueper, B.H. and D.B. McWhorter. 1991. The Behavior of Dense Non-Aqueous Phase Liquids in Fractured Clay and Rock. *Ground Water,* **29**:716–728.

Kueper, B.H., D. Redman, R.C. Starr, S. Reitsma, and M. Mah. 1993. A Field Experiment to Study the Behavior of Tetrachloroethylene Below the Watertable: Spatial Distribution of Residual and Pooled DNAPL. *Ground Water,* **31**:756–766.

Lake, L.W. 1989. *Enhanced Oil Recovery*. Prentice Hall, Englewood Cliffs, NJ.

Lipe, K.M., D. Sabatini, and J. Harwell. 1992. Surfactant Recovery Using MEUF in Subsurface NAPL Remediation. *Proceedings*. Subsurface Restoration Conference, pp. 244–246.

Lipe, K.M., D.A. Sabatini, M.A. Hasegawa, and J.H. Harwell. 1996. Micellar Enhanced Ultrafiltration and Air Stripping for Surfactant-Contaminant Separation and Surfactant Reuse. *Ground Water Monitoring and Research,* Winter, 85–92.

Liu, Z., D.A. Edwards, and R.G. Luthy. 1992. Sorption of Non-Ionic Surfactants onto Soil. *Wat. Res.,* **26**:1337–1345.

Luthy, R.G., Z. Liu, and D.A. Edwards. 1991. Solubilization of Polycyclic Aromatic Hydrocarbons in Micellar Nonionic Surfactant Solutions. *Environ. Sci. Technol.*, **25**:127–133.

Mackay, D.M. and J.A. Cherry. 1989. Ground Water Contamination: Pump-and-Treat Remediation. *Environ. Sci. Technol.,* **23**:630–636.

Mercer, J.W. and R.M. Cohen, 1990. A Review of Immiscible Fluids in the Subsurface: Properties, Models, Characterization and Remediation, *Journal of Contaminant Hydrology,* **6**:107–163.

Pankow, J.F. and J.A. Cherry. 1996. Dense Chlorinated Solvents and Other DNAPLs in Groundwater. Waterloo Press, Portland, OR, 522 p.

Pennell, K.D., G.A. Pope, and L.M. Abriola. 1996. Influence of Viscous and Capillary Forces on the Mobilization of Residual Tetrachloroethylene During Surfactant Flushing. *Environ. Sci. Technol.*

Pope, G.A. and W.H. Wade. 1995. Lessons from Enhanced Oil Recovery Research for Surfactant-Enhanced Aquifer Remediation. In: Sabatini, D.A., R.C. Knox, and J.H. Harwell, Eds. Surfactant-Enhanced Subsurface Remediation. *American Chemical Society Surface Sciences, Symposium Series,* **594**, 142.

Rosen, M.J. 1989. *Surfactants and Interfacial Phenomena,* 2nd ed., John Wiley & Sons, New York, New York.

Rouse, J.D., D.A. Sabatini, and J.H. Harwell. 1992. Minimizing Surfactant Losses During Surfactant Enhanced NAPL Remediation. *Proceedings*. Subsurface Restoration Conference, pp. 260–262.

Saito, H. and K. Shinoda. 1967. The Solubilization of Hydrocarbons in Aqueous Solutions of Nonionic Surfactants. *J. Colloid. Interface Science*, **24**:10–15.

Sale, T., M. Pitts, and K.R Piontek. 1989. Chemically Enhanced *In situ* Soil Washing. *Proceedings.* Petroleum Hydrocarbons and Organic Chemicals in Ground Water, Worthington, OH, pp. 419–422.

Scamehorn, J.F., R.S. Schechter, and W.H. Wade. 1982. Adsorption of Surfactants on Mineral Oxide Surfaces from Aqueous Solutions. *Journal of Colloid. Interface Science*, **85**:463–501.

Swisher, R.D. 1987. *Surfactant Biodegradation*, 2nd ed., Marcel Dekker, Inc., New York, NY.

Vigon, B.W. and A.J. Rubin. 1989. Practical Considerations in the Surfactant-aided Mobilization of Contaminants in Aquifers. *J. Water Pollution Control Federation*, **61**:1233–1240.

von Rybinski, W. and M.J. Schwuger. 1987. Adsorption and Wetting. M.J. Schick (ed.), *Nonionic Surfactants: Physical Chemistry*, Vol. 23-Surfactant Science Series. Marcel Dekker, Inc., New York, NY.

Wayt, H.J. and D.J. Wilson. 1989. Soil Clean Up by In-Situ Surfactant Flushing. *Separation Science & Technology*, **24**:905–937.

West, C.C. and J.H. Harwell. 1992. Surfactants and Subsurface Remediation. *Environ. Sci. Technol.,* **26**:2324.

Wilson, J.L. and S.H. Conrad. 1984. Is Physical Displacement of Residual Hydrocarbons a Realistic Possibility in Aquifer Restoration? *Proceeding of NWWA/API,* Conference on Petroleum Hydrocarbons and Organic Chemicals in Ground Water, Worthington, OH, pp. 274–298.

Chapter 13

Considerations for the Design of *In Situ* Vapor Extraction Systems: Radius of Influence vs. Zone of Remediation

P.C. Johnson and R.A. Ettinger

ABSTRACT

In practice, a typical vapor extraction design is based on the site-specific "radius of influence" of a vapor extraction well, which is selected from analysis of pilot-test vapor phase pressure distribution data or experience. The number of wells installed at any site is then chosen to ensure that the zone of contamination is encompassed by the radii of influence of the system of wells. In this chapter, we examine common methods used to determine the radius of influence, and through examples illustrate implications of this design approach relative to remedial effectiveness. The most significant conclusions are that: a) systems designed by radius of influence–based approaches may never achieve desired remedial goals, and b) systems designed by radius of influence–based approaches may result in longer operation times and higher total costs than a system design that incorporates remedial goals and some level of fundamentally based predictive modeling. At best, the radius of influence–based approach ensures containment of contaminant vapors.

INTRODUCTION

In situ soil vapor extraction, or soil venting, is a process especially suited for the remediation of volatile compounds from permeable soils. Its use at leaking underground fuel tank locations has escalated, and many "successful" applications are reported. Based on the success and acceptance of this technology, a number of other developing *in situ* vapor extraction–based processes are rapidly being implemented, including bioventing, air sparging, and thermally enhanced vapor extraction. A common feature of each of these technologies is the extraction of vapors from the subsurface.

In practice, vapor extraction system designs are based on a parameter commonly known as the "radius of influence" R_I [m], which is assumed to be some measure of the volume of soil expected to be remediated by a single well. Once a value for this parameter is selected, well locations are chosen so that circles of radius R_I drawn around the well locations encompass the zone of contamination, as shown in Figure 13.1. This practice is very similar to that used in the design of groundwater capture systems; there the "radius of influence" corresponds to the distance at which some minimum groundwater drawdown is measured or predicted to occur.

How one determines R_I is often shrouded in mystery (many contractors regard this to be proprietary information); however, three approaches are routinely used. In the first, the vapor extraction system designer makes an "educated guess" based on previous experience. Reported values (Hutzler et al., 1989) generally fall within the range 10–60 m, and conventional thinking is that shallower wells, less permeable soils, and lower applied vacuums correspond to smaller R_I values. In the second and third approaches, R_I is determined from an analysis of pilot test data (Johnson et al., 1990). Ideally, the pilot test consists of withdrawing vapors from a well screened such that vapor flow through the contaminated zone is maximized; during the test, the extraction vapor flow rate and applied vacuum are measured at the extraction well-head, while the transient and steady-state vapor phase vacuum response is monitored at dedicated vadose zone monitoring points (screened across narrow

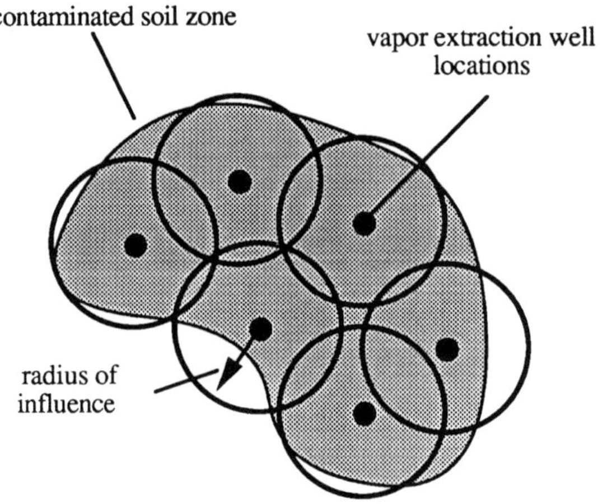

Figure 13.1. Illustration of common soil vapor extraction design process based on R_I.

intervals). In practice, however, the test often consists of withdrawing vapors from an existing groundwater monitoring well that has a screened area extending into the unsaturated zone, and subsurface vapor phase pressures are typically monitored in other existing monitoring wells (that may have very broad screened intervals).

Once the data are obtained from the pilot test, the value for R_I is extracted either from steady-state vapor phase pressure distribution data, or from measured extraction vapor flow rates and applied well-head vacuums. In the former, steady-state subsurface vapor phase pressure data are plotted as a function of distance from the extraction well on linear, semi-log, or log-log graph paper, and a best-fit line is drawn through the data points. The radius of influence is then chosen to be the distance from the extraction well where the gauge vacuum decreases below some arbitrary criterion; ≤1 in. H_2O, ≤0.1 in. H_2O, and 1% of the applied vacuum at the well-head are criteria that are commonly used. Figure 13.2 illustrates this procedure. Sometimes the data are correlated by the following analytical expression corresponding to the subsurface vapor phase pressure distribution for one-dimensional compressible radial flow to a vertical extraction well (Johnson et al., 1990):

$$P^2(r) - P_{Atm}^2 = \left(P_W^2 - P_{Atm}^2\right) + \left(P_W^2 - P_{Atm}^2\right)\frac{\ln\left(R_W / r\right)}{\ln\left(R_I / R_W\right)} \tag{13.1}$$

where P(r) [kg/m-s^2], P_{Atm} [kg/m-s^2], and P_W [kg/m-s^2] denote the absolute pressures measured a distance r [m] from the extraction well, in the atmosphere, and at the extraction well of bore-hole radius R_W [m], respectively. For $P_W/P_{Atm} \approx 1$, it is predicted that the gauge vacuum (= P_{Atm} – P(r)) plotted against a logarithmic distance axis will produce a line that intersects the r-axis at the quantity R_I defined by Equation 13.1.

Sometimes the vapor extraction flow rate Q [m^3/s], and applied well-head pressure P_w [kg/m-s^2] are used to determine R_I. In these cases the one-dimensional steady-state radial flow expression relating flowrate Q to a quantity defined as R_I is used:

$$Q = H\pi\left(\frac{k_v}{\mu}\right)P_W \frac{\left[1 - \left(P_{Atm} / P_W\right)^2\right]}{\ln\left(R_W / R_I\right)} \tag{13.2}$$

where H [m] is the thickness of the vapor flow interval, k_v [m^2] is the permeability to air flow, and μ [kg/m-s] is the viscosity of air.

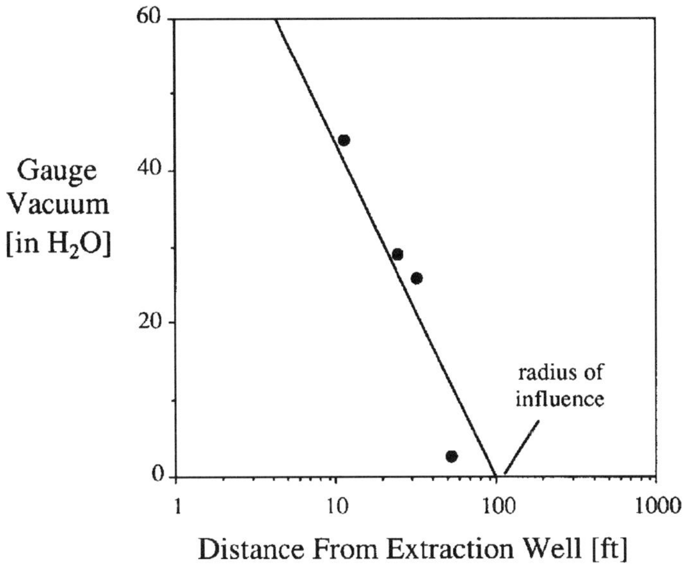

Figure 13.2. Typical method for determining R_I from pilot data.

It is worthwhile mentioning at this point that the quantity R_I appearing in Equations 13.1 and 13.2 is an artifact of the mathematical derivation of the equations, as this key point is often overlooked by practitioners who extend these expressions to uses for which they are not intended. In these equations the quantity R_I arises from a boundary condition that mathematically defines R_I to be the distance away from the extraction well where the subsurface vapor phase pressure equals the atmospheric pressure. As we have artificially constrained the flow field to be one-dimensional in the derivation, it is necessary to set this boundary condition at some finite distance or else no flow can occur (extending this boundary to infinity implies infinite flow resistance). As can be seen in the work of Baehr and Hult (1991) and Shan et al. (1992), it is not necessary to impose this boundary condition when considering two-dimensional flow fields, and there is no term equivalent to R_I appearing in their analytical flowfield solution. Thus, it is important to remember that the term R_I appearing in Equations 13.1 and 13.2 does not represent some inherent physical characteristic of an extraction well. Despite the artificially constrained flow field, we find that the resulting Equations 13.1 and 13.2 are useful screening-level predictors of flow rate and pressure distributions because they are both relatively insensitive to significant changes in R_I. For the same reason, however, the quantity R_I predicted by these equations is very sensitive to small changes in flow rate or pressure distributions, and seemingly non-physical results can be predicted (i.e., R_I>1000 m).

A screening-level alternative to designing systems with the R_I-based design approach was introduced by Johnson et al. (1990, 1991). Based solely on equilibrium considerations, there is a minimum volume of vapor (air) required to achieve any desired level of remediation, independent of contaminant distribution and vapor flow paths. Thus, the maximum degree of remediation achieved by vapor extraction is related to the cumulative volume of air contacted with contaminated soil. Johnson et al. (1990, 1991) built on this concept to develop an alternative screening method for selecting the number of extraction wells based on ideal model predictions and optimizing the air flow through the contaminated soil zone. In this approach, the minimum number of wells N_{wells} is based on the mass of contaminant present M_{cont} [kg], desired time for remediation T_R [s], estimated flow rate to a single well Q_{well} [m^3/s], and a parameter a [m^3-vapor/kg-contaminant] corresponding to the minimum volume of air per unit contaminant mass required to achieve the desired remediation under ideal conditions (for single-component systems and ideal flow conditions, 1/a is equivalent to the saturated vapor concentration for the compound of interest):

$$N_{wells} = \frac{\alpha M_{cont}}{W_{well} T_R} \tag{13.3}$$

In the event that the prediction gives a number of wells less than that required to induce vapor flow through the entire contaminated zone, then N_{wells} can be based on geometrical arguments related to R_I as depicted in Figure 13.1. This screening-level approach can also be combined with economic considerations to form a screening-level economically based design/feasibility approach (Johnson et al., 1993).

An interesting feature of the R_I-based design practice is that one never tries to quantify expected system performance and how this relates to remedial goals, such as the time to achieve some desired level of contaminant reduction. Inherent in R_I-based design practices, therefore, is the assumption that the extent and rate of remediation are somehow dependent only upon the magnitude of the subsurface vacuum measured at any location. In other words, it is assumed that a measurable subsurface vacuum some distance from a vapor extraction well is an indicator of significant remediation between that point and the vapor extraction well. Essentially, we treat the quantity R_I based on subsurface vapor phase pressure measurements as a "radius of remediation." Below, we explore the validity of this assumption through the use of two example problems.

ILLUSTRATIVE EXAMPLES

The following two idealized example problems have been chosen to illustrate specific limitations of conventional vapor extraction system design practices. In the first we investigate whether or not pressure distributions are valid indicators of remediation, while in the second we examine the predicted performance of a system designed based on the radius of influence approach.

Example 1: Layered Confined Soils

Suppose two homogeneous soil layers are confined between impermeable zones as shown in Figure 13.3. Each layer contains equal levels of a contaminant, and the contaminant is uniformly distributed throughout each layer. To effect cleanup, trenches are placed a distance L apart, and a vacuum P_{Trench} is applied to one trench while the other is left open.

For this scenario the steady-state governing flow equations for each layer (Baehr and Hult, 1991) reduce to Laplace's Equation for P^2 ($-P^2 = 0$). Solution of these equations, subject to the boundary

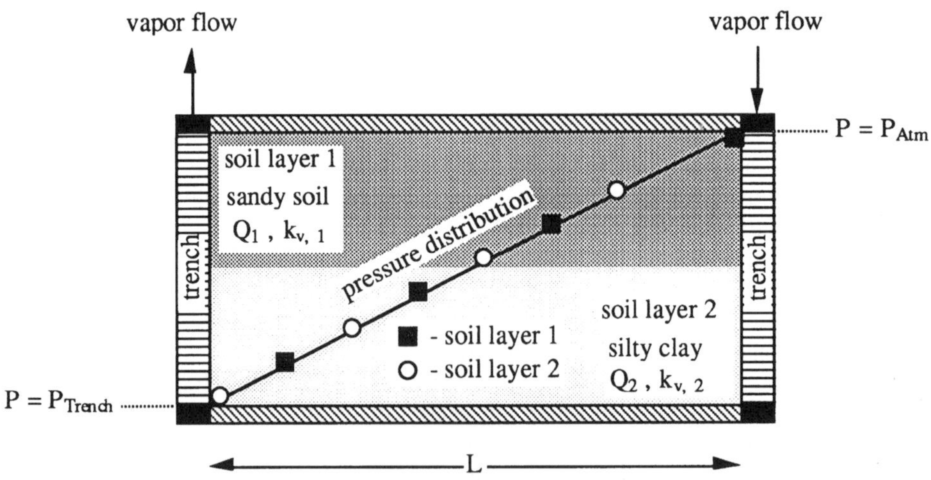

Figure 13.3. Steady-state pressure distribution for confined layers.

conditions depicted in Figure 13.3, yields identical steady-state vapor phase pressure distributions in each layer:

$$\frac{P_{Atm}^2(x) - P^2}{P_{Atm}^2 - P_{Trench}^2} = \left(1 - \frac{x}{L}\right) \tag{13.4}$$

When $(P_{Trench}/P_{Atm}) \approx 1$, then Equation 13.4 predicts the linear pressure distribution shown superimposed on Figure 13.3. The steady-state pressure distribution in each layer must be the same because boundary conditions for both layers are identical and the system is confined (note that the transient propagation of the pressure field may be different in each layer). As determined by Darcy's Law, however, air flow rates through each layer Q_i [m^3/s] are different and the ratio of vapor flow rates is equivalent to the ratio of permeabilities to vapor flow ($k_{v,i}$):

$$\frac{Q_1}{Q_2} = \frac{k_{v,1}}{k_{v,2}} \tag{13.5}$$

Now consider the case where layer 1 is a sandy soil and layer 2 a silty clay ($k_{v,1}/k_{v,2} \approx 10^4$). Equation 13.5 (and intuition) indicates that a larger volume of air passes through the sandy zone than the clayey zone in any given period of time. Consequently, if the degree of remediation is dependent upon the volume of air passed through each soil zone, then remediation of the more permeable sandy zone occurs at a much accelerated rate in comparison with the less permeable clayey zone.

The analysis above demonstrates that pressure distribution data alone are not sufficient to correctly assess the remedial effectiveness of any vapor extraction system. Clearly, the presence, or absence, of a measurable vacuum at some point in the soil is no measure of remediation; in fact it also is not a measure of significant vapor flow. The magnitude of vapor flow through any zone of soil is a function of the permeability of the soil to vapor flow and the pressure gradient, as dictated by Darcy's Law. Thus, we see that there is a fatal flaw with conventional design practice; any design approach based only on a pressure criterion is limited, at best, to ensuring containment of vapors.

Example 2: Vapor Extraction Trench

Figure 13.4 depicts a scenario where a surface spill has caused the uniform contamination from ground surface to a relatively shallow depth in a homogeneous formation (note that only one-half of the formation is shown here). A very long vapor extraction trench has been installed to conduct pilot tests so that the spacing between additional trenches can be determined. During the test a 100-in. H_2O gauge vacuum is applied to the trench, and the resulting pressure response is measured 5 ft below ground surface at a number of monitoring points. Results from this test are presented in Figure 13.5. Following the common practice of plotting the data and extrapolating the results to some low value (<1 in H_2O) of the gauge vacuum, we see that the radius of influence might be selected to be $R_I \approx 30$ ft. Based on this value, a conventional design approach would call for additional trenches spaced $2R_I \approx 60$ ft apart.

The "measured" pressure distribution shown in Figure 13.5 is that predicted by solution of the governing flow equations for a homogeneous, isotropic formation. In this case we treat the flow as two-dimensional and use a conformal mapping analysis to generate a semi-analytical solution for the streamlines, isobars, and vapor flow rate (Ettinger and Johnson, 1993). This particular example was selected for illustration purposes because flowlines and pressure distributions are easily computed; they are superimposed on Figure 13.4 and indicate the path air follows as it travels from ground surface to the extraction trench. A key feature to note here is that flowlines are plotted such that the volumetric rate of air flow between adjacent flowlines is constant and equal to 7.5% of the total flow Q (where 50% of the total flow comes from the half-plane shown in Figure 13.4).

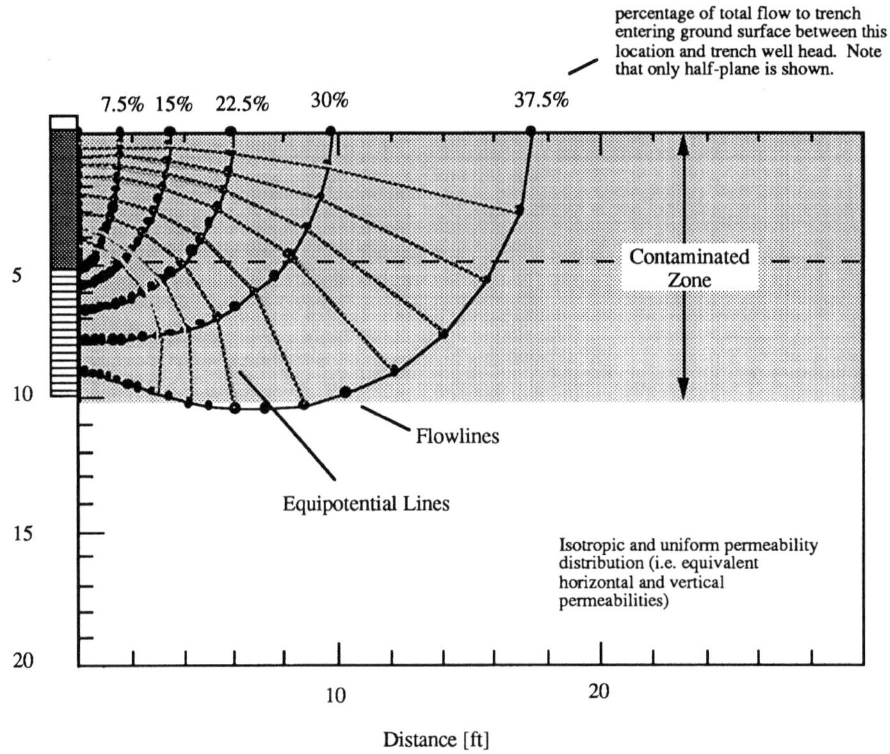

Figure 13.4. Vapor extraction trench scenario.

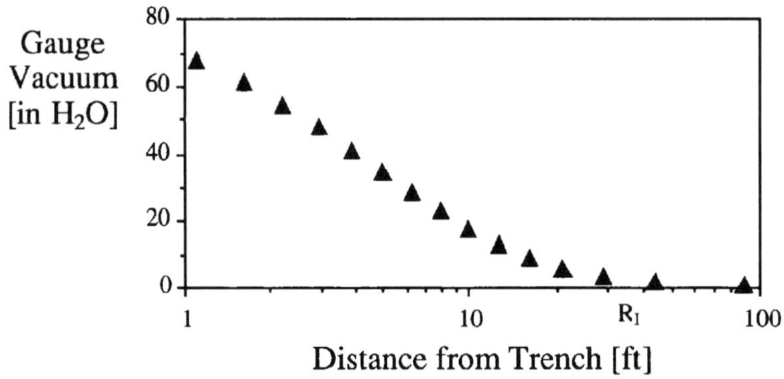

Figure 13.5. Subsurface pressure distribution away from extraction trench at a depth of 5 ft below ground surface.

As discussed above, equilibrium partitioning considerations (Johnson et al., 1990, 1991) dictate that the maximum degree of remediation achieved by a vapor extraction system is limited by the volume of air flowed through, or past, the contaminated soil per unit mass of contaminant present. Based on this concept, we define the parameter g as follows:

$$g = \frac{Q_t}{\alpha \, C_{soil} \, \rho_{soil} \, V_{soil}} \tag{13.6}$$

where V_{soil} [kg soil] denotes a soil volume of bulk density ρ_{soil} [kg-soil/m^3-soil] containing a contaminant at concentration C_{soil} [kg contaminant/kg soil] through which vapor flows at a rate Q

[m^3 vapor/s]. As discussed previously, α [m^3 vapor/kg contaminant] denotes the minimum volume of air per unit mass of contaminant required to achieve the required degree of contaminant reduction, and is determined through modeling calculations or laboratory-scale studies (Johnson et al., 1990, 1991). Thus, g is a dimensionless measure of the progress of remediation; under ideal conditions, g=1 at the time when the desired degree of remediation is achieved.

For those interested in extending this approach to a more complete two-dimensional representation of remediation, this can be accomplished by defining differential soil volumes about the flowlines and setting g to be the dimensionless distance along any given flowline (measured from ground surface) where the remedial goals have been met. This approach is described in Johnson et al. (1993)

With respect to the vapor extraction trench results presented in Figure 13.4, we can use the given streamlines to define the boundaries of different soil volumes. Qualitatively, we can already see that areas close to the trench will be remediated quicker as the volume of soil between streamlines decreases as we move towards the trench (while the contaminant concentration and vapor flow rate are the same for all volumes). As a quantitative measure, Figure 13.6 presents computed values of g as a function of distance from the trench for 90 d, 180 d, and 360 d after startup. For this example r_{soil}=1700 kg/m^3, C_{soil}=10,000 mg/kg, α=100 m^3 air/kg contaminant, and the total vapor flow rate to the trench Q=0.0031 m^3/s/m trench length ($\approx$2 ft^3/min/ft trench length). Obviously, some liberties have been taken to reduce the true two-dimensional progression of remediation to a one-dimensional representation seen in Figure 13.6; the details are as follows: a) the mass of contaminant (per unit width of trench) between adjacent flowlines was computed by multiplying the area between flowlines by the contaminant concentration and soil bulk density; b) the flow rate of air (per unit width of trench) between flowlines was determined to be 0.075Q for flowlines shown in Figure 13.4; c) for each area between flowlines, a value for g was computed according to Equation 13.6 and assigned to the distance where the outermost flowline for that area intersects the 5 ft depth line shown in Figure 13.4. Calculated in this manner the results do not truly reflect the two-dimensional nature of the remediation process (g is a function of depth as well as distance from the trench); however, for this example the g-values still provide a valuable qualitative measure of the remediation progress.

At this point it is useful to introduce the distance Z_R, which roughly defines the zone remediated to a desired level at any given time ($g \geq 1$). Thus, for this example case $Z_R \approx 9$ ft after one year, as indicated in Figure 13.6. If we now try to answer the question: "How close should additional trenches be placed to ensure remediation in one year?," we see that this distance should be at most (as the predictions are for ideal conditions) $2Z_R$=18 ft. Given this information we can now evaluate the effectiveness of the system that would have been installed under conventional design practices with a trench spacing $2R_I \approx 60$ ft. Based on Figure 13.6 it is obvious that remedial goals would not be achieved; in fact for the case described above, it would take $\approx$10 yr (under ideal conditions) for the trench to remediate a zone extending 30 ft from one side of the trench.

CONCLUSIONS

In summary, it should be clear that remedial goals (degree of remediation, time of operation) are not considered in common "radius of influence"–based approaches for *in situ* vapor extraction system design. The examples given above illustrate the fatal flaw inherent in current practice: *the presence (or absence) of a measurable vapor phase vacuum does not ensure significant vapor flow or remediation.* Systems designed by radius of influence–based approaches may result in longer operation times and higher total costs than a system design that incorporates remedial goals and some level of fundamentally based predictive modeling. *At best, the radius of influence–based approach ensures containment of contaminant vapors.* It should be noted that the results presented in Figures 13.4 through 13.6 are based on specific site parameters and operating conditions, and should not be extrapolated to other scenarios that are markedly different. The shape of the g-vs-distance curve can change for different scenarios; e.g., for one-dimensional radial flow to a well, remediation occurs initially at the distance R_I and proceeds towards the well because clean air is constrained to enter the system only at the distance R_I (g is a maximum at the distance R_I, then decreases in the direction of the extraction well).

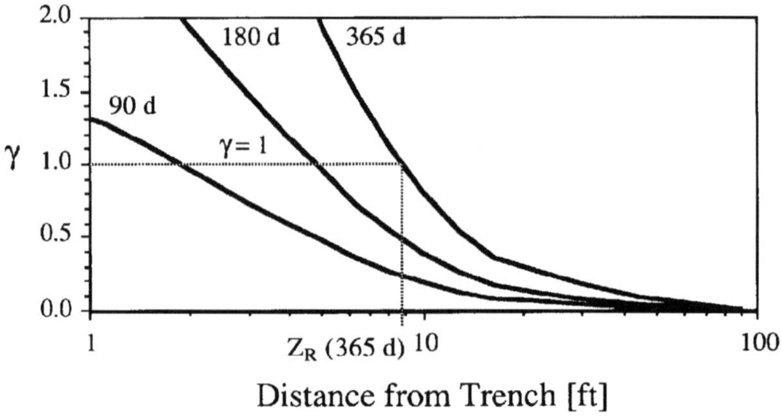

Figure 13.6. Predicted zone of remediation for vapor extraction trench example.

REFERENCES

Baehr, A.L. and M.F. Hult. 1991. Evaluation of unsaturated zone air permeability through pneumatic tests. *Water Resources Research.* **27**(10).

Ettinger, R.A. and P.C. Johnson. 1993. Conformal Mapping Analyses for Vapor Recovery Trenches. In Preparation.

Hutzler, N.J., B.E. Murphy, and J.S. Gierke. 1989. *State of Technology Review: Soil Vapor Extraction Systems.* EPA/600/2-89/024. U.S. Department of Commerce NTIS document PB89-195184.

Johnson, P.C., G.E. Hoag, R.H. Hinchee, R.A. Brown, and A.L. Baehr. 1993. *Vapor Extraction-Based Technologies.* In the WASTECH Monograph Series of Remedial Technologies.

Johnson, P.C., M.W. Kemblowski, and J.D. Colthart. 1990. Quantitative Analysis for the Cleanup of Hydrocarbon-Contaminated Soils by In Situ Soil Venting. *Ground Water.* **3**(28):413–429.

Johnson, P.C. and A.J. Stabenau. 1991. *HyperVentilate—A Software Guidance System Created for Vapor Extraction Applications.* Shell Oil Company (see also USEPA 510-CB-92-001 for software orders).

Johnson, P.C., C.C. Stanley, M.W. Kemblowski, D.L. Byers, and J.D. Colthart. 1990. A Practical Approach to the Design, Operation, and Monitoring of In Situ Soil-Venting Systems. *Ground Water Monit. Rev.* Spring 1990.

Johnson, P.C., C.C. Stanley, D.L. Byers, D.A. Benson, and M.A. Acton. 1991. Soil Venting at a California Site: Field Data Reconciled with Theory. *Hydrocarbon Contaminated Soils and Groundwater: Analysis, Fate, Environmental and Public Health Effects, Remediation, Vol. I.* P.T. Kostecki and E.J. Calabrese (eds.). Lewis Publishers, Boca Raton, FL.

Shan, C.S., R.W. Falta, and I. Javandel. 1992. Analytical Solutions for Steady Gas Flow to a Soil Vapor Extraction Well. *Water Resources Res.* **28**(4):1105–1120.

Chapter 14

Chemically Enhanced Removal of Metals from the Subsurface

C.D. Palmer and W. Fish

ABSTRACT

The injection of reactive agents into the subsurface to enhance the transfer of contaminants from the aquifer matrix to the mobile groundwater is a viable method for accelerating the removal of metal contaminants from the subsurface. The "reactive agents" may be 1) acids and bases, 2) oxidants and reductants, 3) competitive adsorbates, 4) complexing agents, or 5) solid phase substituents. However, rational design of such a chemical enhancement system must consider delivery of the reactive agent to the contaminated areas of the aquifer, the ability of the reactive agent to enhance removal of the contaminant from the aquifer matrix, the removal of the reactive agent from the subsurface, treatment of the extracted waters, and ultimate disposal of the sludges. The most suitable reactive agent depends on the key processes limiting remediation by pump-and-treat.

INTRODUCTION

Metals-contaminated sites are relatively common and are found in every region of the United States (Palmer et al., 1988). Pump-and-treat, the most commonly used technique for groundwater remediation (Travis and Doty, 1989), is often employed at these sites. A key concern in pump-and-treat is that the rate of mass removal via the extraction wells will not be sufficient to achieve remediation on a reasonable time scale. In most cases, large quantities of water must pass through the contaminated sections of the aquifer before it is remediated (Palmer and Fish, 1992).

The extended periods of time required to achieve remediation by conventional pump-and-treat has led many contaminant hydrologists to consider enhancing the pump-and-treat process by injecting into the subsurface "reactive agents" that accelerate the rate of removal of the target metals. The applicability of such "chemical enhancement" technology depends on the target contaminants and the specific processes that limit their transfer from the aquifer matrix to the groundwater. Not all metal contamination sites are amenable to chemical enhancements. In particular, when prolonged remediation is controlled by physical transport limitations rather than by chemical processes, chemical enhancement has little advantage over conventional pump-and-treat. Nonetheless, certain types of contamination are amenable to well-conceived applications of reactive agents. In this chapter, we will identify chemical processes that limit the removal of metal contaminants from the subsurface and identify general types of reactive agents that can be used to enhance the removal of metals from the subsurface.

LIMITATIONS OF THE REMOVAL OF METALS BY PUMP-AND-TREAT

Metal contaminants (Table 14.1) in the subsurface are dissolved in the water, associated with mobile colloids, or bound to the immobile aquifer matrix as adsorbed ions or precipitates (Figure 14.1). Dissolved metals and metals associated with mobile colloids can be readily removed by pump-and-treat. However, the matrix-associated contaminants cannot be removed directly by pump-and-treat; they first must transfer from the matrix to the mobile groundwater. The limitation posed by this mass transfer can be a major impediment to a successful pump-and-treat operation, even when the mobile phase can be assumed to equilibrate instantly with the solid phase. For example, if the concentration of metal in the extraction water is controlled primarily by the solubility of a solid phase, the

Table 14.1. Priority Inorganic Contaminants and Their Common Forms.

Element	Common Redox States	Common Forms
Cd	+2	Cd^{2+}, $CdCO_{3(s)}$
Cu	+1	$Cu_2O_{(s)}$
	+2	Cu^{2+}, $CuO_{(s)}$, $Cu_2(CO_3)(OH)_2$
Ni	+2	Ni^{2+}, $Ni(OH)_{2(s)}$
Zn	+2	Zn^{2+}, $ZnCO_{3(s)}$, $ZnO_{(s)}$
Pb	+2	Pb^{2+}, $PbSO_{4(s)}$, $PbCO_{3(s)}$, $PbS_{(s)}$
Ba	+2	Ba^{2+}, $BaSO_{4(s)}$
Hg	0	$Hg_{(l)}$
	+1	Hg_2Cl_2
	+2	Hg^{2+}, $HgCl_4^{2-}$, $HgO_{(s)}$, $HgCH_3^+$, $Hg(CH_3)_2^o$
Co	+2	Co^{2+}, $CoCO_{3(s)}$, $Co(OH)_{2(s)}$, CoS
Ag	+1	Ag^+, $AgCl_{(s)}$
As	+3	$H_2AsO_4^{3-}$, $HAsO_4^{4-}$, AsO_4^{5-}
	+5	$H_3AsO_3^{+2}$, $H_2AsO_3^{+1}$, $As_2S_{5(s)}$, $FeAsS_{4(s)}$
Cr	+3	Cr^{3+}, $Cr(OH)^{2+}$, $Cr(OH)_{3(am)}$
	+6	CrO_4^{2-}, $HCrO_4^-$, $Cr_2O_7^{2-}$, $BaCrO_{4(s)}$, $PbCrO_{4(s)}$
Se	−2	H_2Se^o, HSe^-
	0	Se^o
	+4	SeO_3^{2-}, $HSeO_3^-$, $H_2SeO_3^o$
	+6	SeO_4^{2-}, $HSeO_4^-$
Sb	+3	$Sb(OH)_2^+$, $Sb(OH)_3^o$, $Sb(OH)_4^-$, $Sb(OH)_{3(s)}$, $Sb_2O_{3(s)}$, $Sb_2S_{3(s)}$
	+5	$Sb(OH)_5^o$, $Sb(OH)_6^-$
Tl	+1	Tl^+, $Tl_2S_{(s)}$
	+3	Tl^{3+}, $Tl_2O_{3(s)}$
	+4	$Tl_2O_{4(s)}$

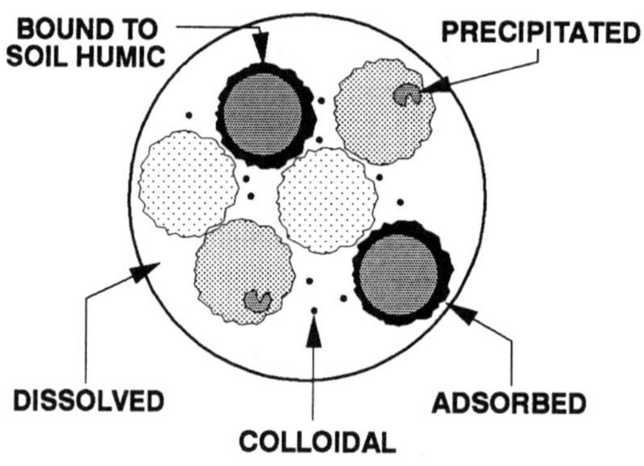

Figure 14.1. Metal contaminants in soils may be either mobile (dissolved or colloidal) or immobile (precipitated, adsorbed, or bound to soil matrix).

number of times the pores must be flushed with uncontaminated water to remove all of the contaminant from the soil matrix, the retardation factor, R, varies inversely with the equilibrium concentration in the extraction water:

$$R = 1 + \frac{C_{mtx}\rho_b / \theta}{C_{eq}} \qquad (14.1)$$

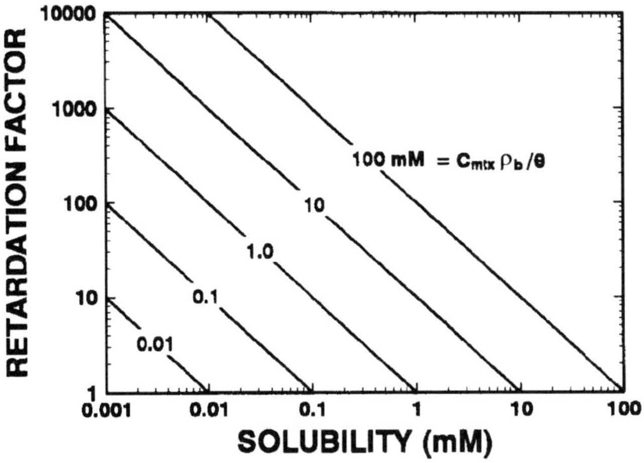

Figure 14.2. Number of pore volumes that must be flushed through a soil (retardation factor) to remove a mineral containing a metal contaminant as a function of solubility for different concentrations of contaminants in the soil under equilibrium conditions. C_{mtx} is the metal-contaminant concentration in mol/kg, ρ_b is the dry bulk density of the soil, and θ is the volumetric water content.

where

C_{mtx} is the concentration of the solid in mol/kg;
ρ_b is the dry bulk density of the soil (g cm^{-3});
θ is the volumetric water content (dimensionless); and
C_{eq} is the equilibrium concentration of the contaminant in the groundwater (M).

If the solid containing the contaminant is fairly insoluble, i.e., if C_{eq} is small relative to the concentration of the contaminant in the soil matrix, then thousands of pore volumes may have to be flushed through the soil (Figure 14.2).

Under the ideal conditions posed in this example, the equilibrated concentration in the extraction water was assumed to remain constant; such invariant concentrations would not be achieved when retardation is due to adsorption of the contaminant to the soil matrix. For example, the adsorption of chromate ion onto 1.5×10^{-3} mole of hydrous ferric oxide per liter of water at pH 6.5 and ionic strength of 0.1 M was simulated using the equilibrium constants reported by Dzombak and Morel (1990) and the geochemical model MINTEQA2 (Allison et al., 1991). The retardation factors, R, calculated by

$$R = [1-(\text{fraction adsorbed})]^{-1} \tag{14.2}$$

are close to unity when the aqueous concentrations are greater than 2×10^{-4} M and increase to 27 as the aqueous concentrations decline to the drinking water standard of 10^{-6} M (Figure 14.3). Thus, it is relatively easy to remove the higher concentrations of chromate from the subsurface, but the more successful the pump-and-treat, i.e., the lower the concentrations in the extraction water, the more difficult it is to obtain the next decrement in concentration.

CHEMICAL ENHANCEMENTS

Four key areas must be addressed in the design stages of a successful chemical enhancement method for pump-and-treat remediation (Figures 14.4a and b): 1) delivery of the reactive agent to contaminated portions of the aquifer; 2) the interaction between the contaminant and the reactive

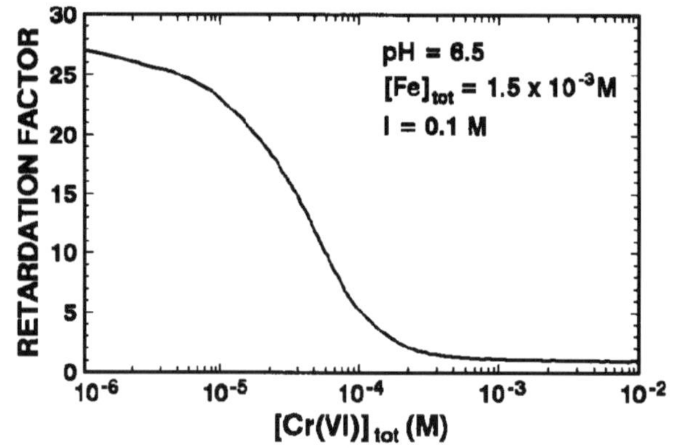

Figure 14.3. Retardation factors for Cr(VI) adsorption onto hydrous ferric oxide as a function of $[Cr(VI)]_{tot}$.

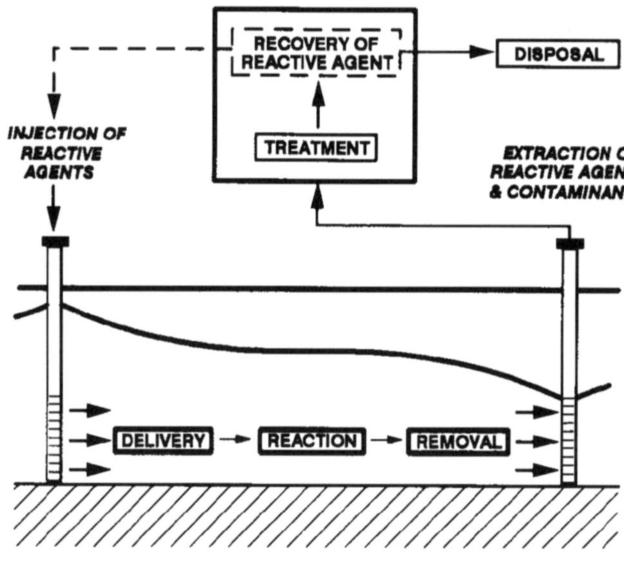

Figure 14.4a. Schematic representation of chemical enhancement of a pump-and-treat operation using injection wells. Key areas of concern are shown in boxes. In some cases the reactive agent will be recovered and reused.

agent; 3) removal of the contaminant, reactive agent, and reaction by-products from the subsurface; and 4) treatment of the extracted water, recovery of the reactive agent, and disposal of sludges (Palmer and Fish, 1992).

Delivery of the reactive agents to the contaminated portions of the aquifer can be achieved by injection via either vertical or horizontal wells (Figure 14.4a) or by gravity flow through infiltration galleries (Figure 14.4b). Regardless of the method utilized, several key factors must be considered. First, the water amended with the reactive agents must enter the aquifer without clogging the area near the injection well or the surface of the infiltration gallery with particles, chemical precipitates, or biomass. Additives may be required to prevent precipitation. Problems with particles in the injection water can be overcome by flocculation and filtration prior to injection. Microbiological activity can be inhibited by removal of nutrients or dissolved oxygen. If these problems are not controlled, periodic treatment such as surging or acid treatment of the injection wells or the mechanical raking of the infiltration bed may be necessary. Once in the aquifer, the solution of reactive agents must move in

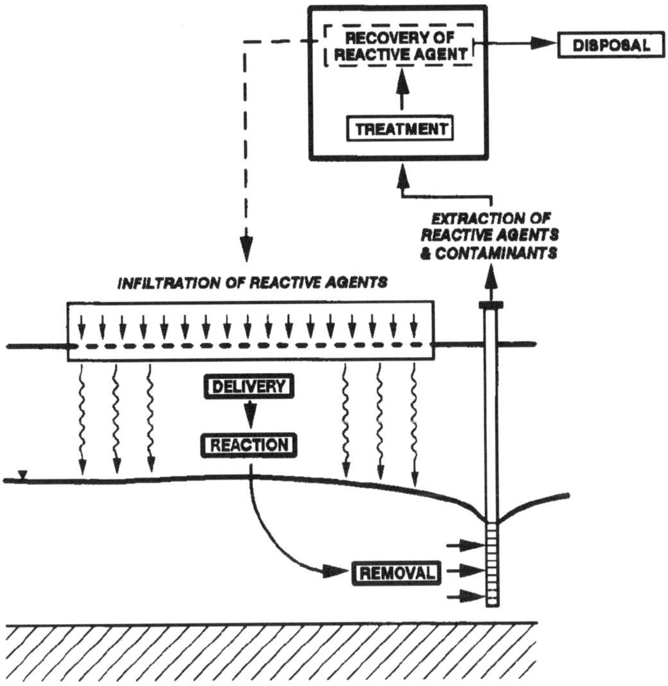

Figure 14.4b. Schematic representation of a chemical enhancement of a pump-and-treat operation using infiltration galleries.

some reasonable period of time to the contaminated portions of the aquifer. The rate of movement into those areas will be affected by the heterogeneity of the aquifer and to a certain extent it will be controlled by the rate, mode, and scheduling for the injection wells, infiltration events, and pumping wells (Palmer and Fish, 1992).

The basic concept of chemical enhancement to pump-and-treat is to increase the mobility of the contaminants in the subsurface so they may be more easily removed via extraction wells. Removal of the contaminants from the subsurface in a chemical enhancement scheme, therefore, requires decisions about the spatial density, placement, and pumping rates of these extraction wells. Many aspects of such an enhanced extraction system design will be similar to those utilized in conventional pump-and-treat systems. Use of a reactive agent in a pump-and-treat scheme introduces one or more new chemical constituents into the subsurface. To be effective, reactive agents generally must be added to an aquifer at non-trace concentrations. Even if the reactive agent is harmless to human health, state and federal regulations often will require that the concentration of the reactive agent be lowered to some permissible level. Thus, not only must the contaminants be removed from the subsurface, but in many cases the reactive agent must also be removed. Requirements for such removal may affect the choice of the reactive agent.

Even when a chemical enhancement plan is completely satisfactory in terms of subsurface deployment and removal of injected solutions, technical difficulties in the treatment and handling of the extracted wastes may render the chemical enhancement effort useless. A key requirement is that the reactive agent should not interfere with the treatment of the target contaminant. In addition, state and federal regulations may require the reactive agent to be removed from the extracted water before discharge. The removal of the reactive agent must be technically feasible and economically viable. Finally, the recovery of the reactive agent for reuse in the enhanced extraction system may be desirable, particularly if the reactive agent is expensive and is amenable to cost-effective recovery.

The design of a successful chemical enhancement scheme ultimately depends on choosing an appropriate reactive agent. Different reactive agents can be chosen depending upon the processes that control the transfer of the contaminant mass from the aquifer matrix to the groundwater. Thus, it is important to identify the key limiting processes during characterization of the site. We will focus on potential reactive agents that can aid in the removal of metal contaminants from the subsurface.

REACTIVE AGENTS

Reactive agents that enhance the removal of metal contaminants from the subsurface can be categorized as 1) acids and bases, 2) oxidants and reductants, 3) competitive adsorbates, 4) complexing agents, and 5) solid phase substituents. Choice of the optimal reactive agent requires knowledge of the key processes that limit removal of the metal contaminants by pump-and-treat remediation. These processes can be desorption from mineral/oxide surfaces, desorption from soil humic materials, or dissolution of solid phase precipitates containing the target metal contaminant. Some classes of reactive agents can assist in the removal of several metal contaminants while others are effective at removing only one type of contaminant.

Acids and Bases

Many metal contaminants (Pb, Zn, Cu, Cd, Ni, Co, As, Se, Cr) precipitate as metal oxides, carbonates, sulfides, and other sparingly soluble precipitates (e.g., $BaCrO_4$, $FeAsO_4$). The most likely form in the subsurface depends on the pH, and for redox active contaminants, the oxidation-reduction potential, or Eh. For example, Cd^{2+} is likely to be the prevalent form of cadmium in the soil when the pH is low (Figure 14.5), but in slightly acidic to alkaline waters most of the cadmium is associated with $CdCO_{3(s)}$ (otavite). The rate of removal of such precipitates varies with the solubility of the solid phase, its dissolution kinetics, and the concentration of that phase in the soil. The solubility often depends on pH. For example, for a soil with $C_{mtx}\rho_b/\theta$ of 5×10^{-6} M $CdCO_{3(s)}$ and 0.008 M of dissolved inorganic carbon (DIC) at an ionic strength of 0.1 M, and pH is greater than 6.7, more than 100 pore volumes would have to be flushed through the soil to remove $CdCO_{3(s)}$ (Figure 14.6). If the pH of the soil is decreased to below 5.5, the $CdCO_{3(s)}$ could be removed in one or two pore volumes. If the concentration of $CdCO_{3(s)}$ were increased tenfold, the pH would have to be lowered to below 5 to achieve removal in less than two pore volumes.

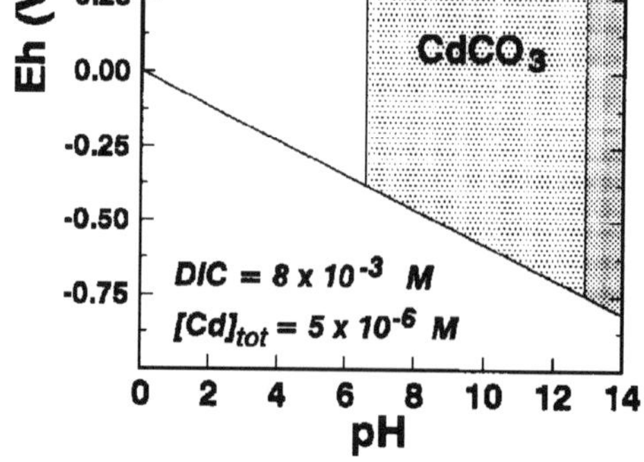

Figure 14.5. Eh-pH diagram for the Cd-C-H_2O system at 25°C.

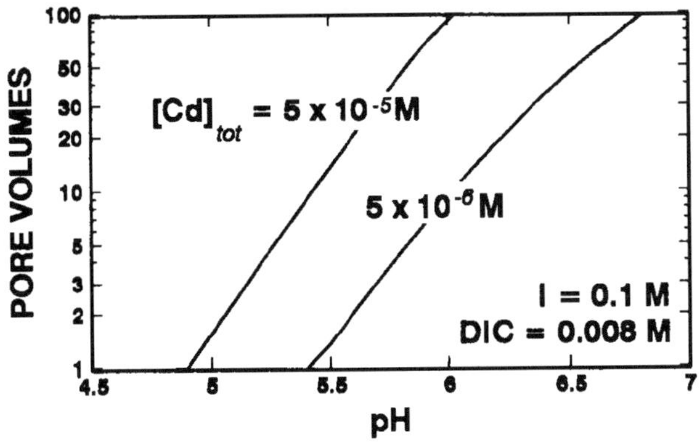

Figure 14.6. Number of pore volumes required to remove $CdCO_{3(s)}$ from a soil as a function of pH.

In addition to precipitating as mineral phases, metal ions adsorb to mineral and oxide surfaces. The fraction of the metal contaminant adsorbed to a surface depends on the concentration of the metal contaminant (adsorbate), the concentration and adsorption density of the adsorbent, the ionic strength, the concentration of other ions that compete for the same adsorption sites, and the pH of the groundwater. For metals such as Pb, Cu, and Ni, whose dominant aqueous forms are cations, the fraction of the metal adsorbed to the aquifer matrix increases with increasing pH (Figure 14.7). For oxyanions such as selenate and chromate, the fraction generally decreases with increasing pH (Figure 14.8). These properties of adsorbing metal ions are described by surface complexation models (Dzombak and Morel, 1990).

In principle, the removal of adsorbed metals from the subsurface can be enhanced by altering the pH via the addition of acids or bases. We illustrate the importance of pH to metal mobility by com puting the retardation factors, R, for 10^{-4} M Cr(VI) on different concentrations of hydrous ferric oxide at ionic strength of 0.1 M (Figure 14.9). For pH > 8, R is close to unity and the chromate behaves like a conservative tracer; that is, it can be removed in a single pore volume. R increases to a maximum near pH 5 and then decreases as the pH drops below 5. Raising the pH of a chromium-contaminated aquifer thus seems a likely desirable option for enhancing remediation.

Increasing pH can enhance the removal of adsorbed anionic contaminants from the subsurface; however, it can result in the precipitation of a solid phase that impedes the rate of remediation (e.g., $BaCrO_4$). Lowering the pH enhances the removal of cationic contaminants such as Zn, Cd, Cu, and Pb from many environments by dissolving carbonates and oxides and by diminishing the fraction of cations adsorbed to oxide and mineral surfaces. However, it is extremely difficult to adjust the pH of most subsoils because the buffer intensity of soil-water systems is considerable.

The abundant proton-exchange sites on clays and oxide minerals act as natural buffers that resist perturbation in solution pH. In addition, acid-promoted dissolution of metal oxides consumes protons and thereby adds to the buffer intensity of a porous matrix. Conversely, precipitation of metal oxides can consume most of the alkalinity added to a system. Finally, aquifers containing even minor amounts of carbonate minerals have a large buffer intensity until the carbonates are completely dissolved. Thus, for example, it is virtually impossible to acidify the water in a limestone aquifer. For these reasons, pH adjustment is rarely a practical option.

A further complication is that acid/base extraction of contaminants from the subsurface is not selective. Acidification will dissolve carbonates and metal oxides. Extreme pH can enhance the weathering of silicates and increase the concentration of H_4SiO_4, Al, Ca, Mg, Na, and K. In many cases, the contaminants are present as part of a solid solution series with some nontoxic mineral component (e.g., $Cd_{1-x}Ca_xCO_3$) rather than a pure mineral (e.g., $CdCO_3$). Because acid extraction removes the bulk mineral phase rather than the contaminant component of the mineral, nontoxic substances can constitute most of the sludge removed from treated extraction water.

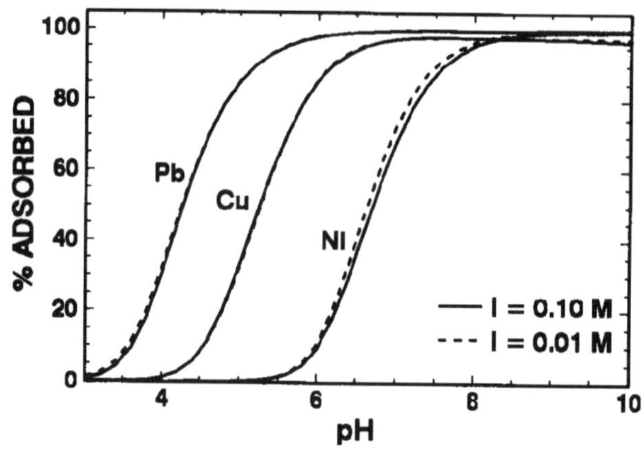

Figure 14.7. pH-adsorption edges for the cations Pb^{2+}, Cu^{2+}, and Ni^{2+} on hydrous ferric oxide ($C_{Fe}\rho_b/\theta = 2 \times 10^{-3}$ M, $[Me]_{tot} = 10^{-5}$ M).

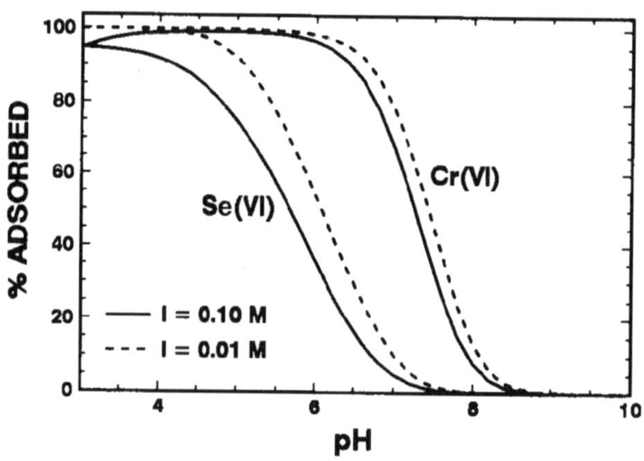

Figure 14.8. pH-adsorption edges for the oxyanions Cr(VI) and Se(VI) on hydrous ferric oxide ($C_{Fe}\rho_b/\theta = 2 \times 10^{-3}$ M, $[Me]_{tot} = 10^{-4}$ M).

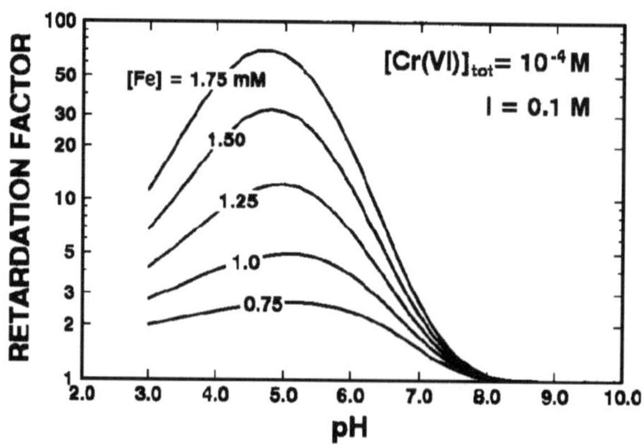

Figure 14.9. Retardation factors for adsorption of 10^{-4} M Cr(VI) onto different concentrations of hydrous ferric oxide as a function of pH.

Oxidants and Reductants

With the exception of Zn, Cd, and Ni, the oxidation states of inorganic priority pollutants can, in principle, be altered by using oxidants and reductants as reactive agents. Such changes in redox state can 1) decrease the toxicity of the contaminant, 2) change its mobility, or 3) increase its susceptibility to complexing agents. The stability of many solid phases depends not only on the redox state of the system but also on the pH. Also, the rate of many redox reactions is strongly dependent on pH. Finally, because redox reactions generally involve the production or consumption of alkalinity, the pH is likely to change during modification of redox state. Therefore, the use of oxidants and reductants may also require the use of acids and bases.

Oxidants and reductants are nonspecific in their reactivity. If a reductant is added to the subsurface, it can reduce ferric and manganese oxides in addition to the target contaminant. The reductive dissolution of these oxides from the aquifer matrix can significantly enhance metal removal by decreasing the concentration of the adsorbent. However, the consumption of reductants by these constituents will retard the rate of transport of the reducing agent to the contaminated portions of the aquifer. Similarly, minerals containing reduced elements such as Fe(II) can impede the transport of an oxidant added to the subsurface. In addition, removal of these nontoxic constituents will increase the volume of sludge that must be disposed following treatment of the extracted water. Another problem is defining the optimal concentration of oxidant or reductant. The redox buffering intensity of the subsurface system must be taken into account when determining the concentration and total amount of reactive agent required (Barcelona and Holm, 1991). Finally, it should be borne in mind that many redox reactions are slow relative to the desired time scales for the transport of the reactive agent.

Competitive Adsorbates

One strategy for removing adsorbed contaminants is to desorb them with a competing adsorbate. Strongly binding adsorbates can effectively compete with contaminants for adsorption sites when the reactive agent concentration is greater than the adsorption density of the soil. This is illustrated for the competitive adsorption of phosphate and chromate onto 3×10^{-3} M hydrous ferric oxide at pH 6 and ionic strength of 0.1 M (Figure 14.10). When $[P]_{tot}$ is 10^{-5} M, the retardation factor for Cr(VI) is 69; at greater $[P]_{tot}$, the retardation factor for Cr(VI) decreases and is close to unity when $[P]_{tot}$ is greater than 3×10^{-4} M. However, strongly binding adsorbates are highly retarded unless their concentrations are very high. For $[P]_{tot} < 10^{-4}$ M, $R > 1000$ and even when $[P]_{tot}$ is 10^{-3} M, R is 2. Thus, to deliver the reactive agent to the contaminated areas of the aquifer, either many pore volumes of injection water must be flushed through the soil or high P concentrations ($> 3 \times 10^{-3}$ M) must be used in the injection water. Recovery of the reactive agent will require even greater numbers of pore volumes. Therefore, despite their seeming advantages, strongly binding adsorbates are not practical reactive agents.

Although it may seem counterintuitive, weakly binding adsorbates may be more useful for competitive adsorption. Weakly binding adsorbates can compete with metal contaminants if their concentrations are an order of magnitude or more greater than the adsorption density of the soil. For example, the weakly binding SO_4^{2-} ion can compete with Cr(VI) for adsorption sites on hydrous ferric oxide (Figure 14.11). At 10^{-4} M $[SO_4^{2-}]$, Cr(VI) retardation factors are close to 50. As $[SO_4^{2-}]$ is increased to 0.02 M, the retardation factor for Cr(VI) decreases below 4. Because SO_4^{2-} is only weakly binding, only a small fraction of the total amount of the reactive agent is adsorbed onto the soil and R for SO_4^{2-} is close to unity. Therefore, it is possible to transport the reactive agent to the contaminated areas of the aquifers in very few pore volumes. Although they bind only weakly, these competitive adsorbates can dramatically increase the fraction of the metal ion in the mobile groundwater. If high adsorbate concentrations are used, the retardation factors of both the contaminant and the reactive agent will be low and the aquifer can be remediated after only a few pore volumes of the amended injection water have been flushed through the aquifer. Such an application of SO_4^{2-} may be feasible because of its high solubility, low cost, and negligible toxicity.

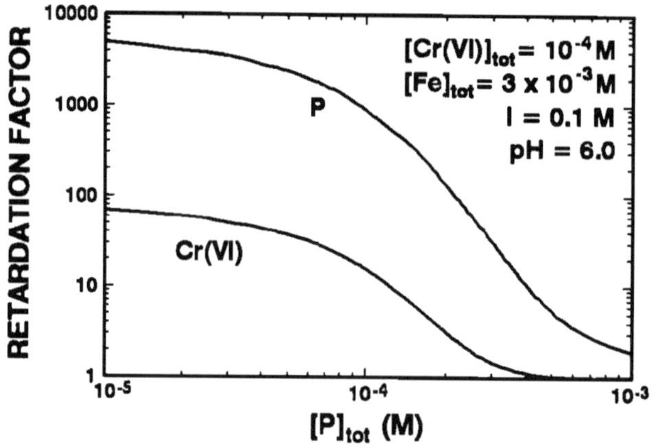

Figure 14.10. Retardation factors for Cr(VI) and phosphate as a function of $[P]_{tot}$ ($(Cr(VI))_{tot} = 10^{-4}$ M, $[Fe]_{tot} = C_{Fe}\rho_b/\theta = 3 \times 10^{-3}$ M, I = 0.1 M, pH = 6.0).

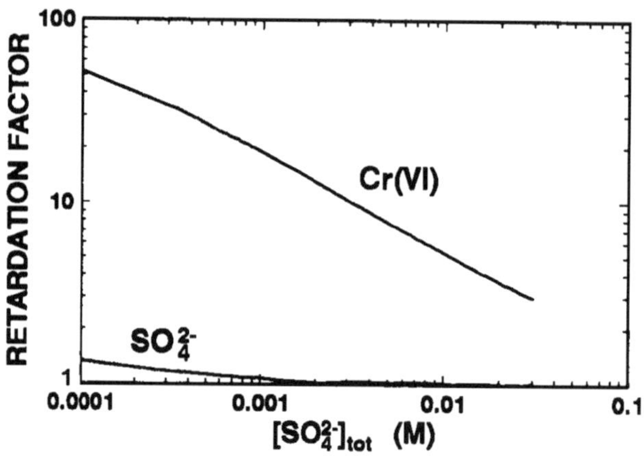

Figure 14.11. Retardation factors for Cr(VI) and SO_4^{2-} as a function of $[SO_4^{2-}]_{tot}$ ($(Cr(VI))_{tot} = 10^{-4}$, $[Fe]_{tot} = C_{Fe}\rho_b/\theta = 3 \times 10^{-3}$ M, I = 0.1 M, pH = 6.0).

Complexing Agents

Complexing agents can significantly increase the solubility of metal contaminants by complexing the metal. Certain metal-ligand complexes do not significantly adsorb to the aquifer matrix and therefore can be easily extracted from the surface. For example, the solubility of $Cr(OH)_{3(am)}$ is increased by three orders of magnitude as the concentration of phthalate is increased from 10^{-4} to 10^{-1} M (Figure 14.12). However, most environmentally safe complexing agents are relatively weak and nonspecific; they bind Fe(III), Al, and alkaline earth metals in addition to the target contaminants. Therefore, high concentrations of the reactive agent must be used to enhance removal of the target contaminants. The complexation of the metals in oxyhydroxides can be used as part of a strategy for removing contaminants from the subsurface. These oxide surfaces adsorb metal contaminants, thereby retarding their removal from the subsurface. Decreasing the concentration of these adsorbents will mobilize the target contaminants and expedite their removal. The choice of complexing agents depends on the target contaminant, pH, and the concentrations of interfering ions. Biodegradation of some organic ligands may decrease the efficiency of the extraction. However, biodegradation of organic ligands remaining in the aquifer following extraction of the metals may be desirable.

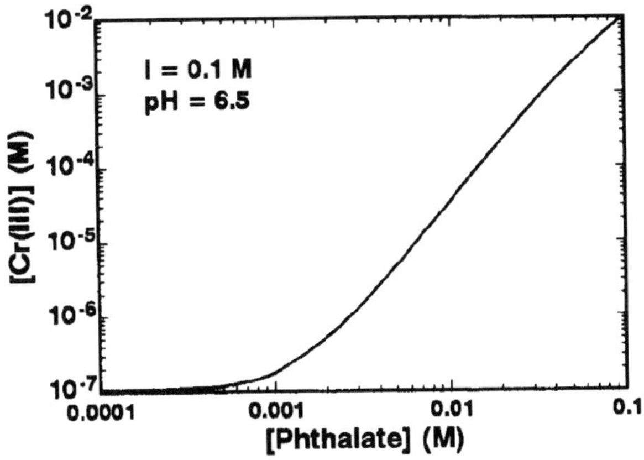

Figure 14.12. Increased solubility of $Cr(OH)_{3(am)}$ with increasing concentration of the complexing agent phthalate (I=0.1 M; pH=6.5).

Surfactants are currently being investigated as a method of enhancing the removal of metals from the subsurface. Laboratory experiments have shown that surfactant (sodium dodecyl sulfate, sulfosuccinated and sulfonated anionics) increases the concentration of Cr(VI) by factors of 2 to 3 compared to using water alone (Thirumalainivas et al., 1994). Factors of 10 to 12 increases in Cr(VI) concentrations are obtained when the micelles are used to deliver a complexing agent (diphenylcarbazide) to the contaminated soil. The mechanisms by which this enhancement occurs are not well understood at this time and research is continuing (Dr. Sabatini, University of Oklahoma, personal communication).

Solid phase Substituents

Metal contaminants in solid phases necessarily contain one or more counterions. In certain situations, the counterions can be precipitated as a harmless phase by the addition of a suitable reactive agent. The formation of the new phase draws the counterions from the contaminant phase, thereby solubilizing the contaminant so that it can be removed from the subsurface. We illustrate this process using a specific example of sulfate extraction of $BaCrO_4$ that we have been investigating (Ward and Palmer, 1994).

Rai et al. (1988) suggest that $BaCrO_4$ may affect the transport and fate of Cr(VI) in the subsurface. Geochemical studies at the United Chrome Products site (UCP) in Corvallis, Oregon found that most of the contaminated groundwaters are near equilibrium with $BaCrO_4$. Palmer and Wittbrodt (1990) suggest that 25 to 50 pore volumes may be required to leach this solid from the soil via dissolution into unamended groundwater.

If $BaCrO_4$ limits the remediation of Cr(VI) contaminated sites, the addition of a chemical constituent that enhances the solubility of the $BaCrO_4$ component should shorten the overall length of time for remediation. This can be done by finding a reactive agent that precipitates the counterion, Ba^{2+}, in a new solid phase. One such reactive agent is SO_4^{2-}. If sufficient SO_4^{2-} is added to the groundwater, we can expect the release of Cr(VI) to the groundwater by dissolution of the $BaCrO_4$ and precipitation of Ba^{2+} in $BaSO_4$:

$$BaCrO_4 + SO_4^{2-} \rightleftarrows BaSO_4 + CrO_4^{2-}$$

This reaction is a simplification of the actual process. $BaCrO_4$ and $BaSO_4$ form a complete solid solution series, and it is possible that solids with compositions intermediate between these two end-members may form.

To determine the concentration of SO_4^{2-} large enough to release most of the Cr(VI) into the groundwater, we computed chromate concentrations in a system in equilibrium with a fixed amount of $BaCrO_4$ with various concentrations of SO_4^{2-} at a constant pH of 6.5 and at a fixed ionic strength of 0.1 M. When the solutions were supersaturated with respect to $BaSO_4$ the composition was computed for equilibrium with both $BaSO_4$ and the finite amount of $BaCrO_4$. A retardation factor for the removal of the Cr(VI) was then obtained using Equation 14.1. For chromate, the retardation factor is equivalent to the number of pore volumes required to dissolve all of the $BaCrO_4$ under equilibrium conditions. The amount of $BaCrO_4$ in the soil was assumed to be 0.108 mM/100 g of soil. For a saturated soil with a 47% porosity and a particle density of 2.63 g/cm^3, this concentration of $BaCrO_4$ is equivalent to 3.2 mmol of $BaCrO_4$ per liter of groundwater. The results of the computations (Figure 14.13) show that in the absence of sulfate, the chromate retardation factor is 67. The retardation factor decreases monotonically with increasing $[SO_4^{2-}]$. At a sulfate concentration of about 0.003 M, the retardation factor is close to unity, indicating that nearly all of the $BaCrO_4$ could be brought into solution at this SO_4^{2-} concentration.

This retardation factor for Cr(VI) is for a point within the aquifer assuming the SO_4^{2-} has been successfully delivered to that location. During the reaction with the $BaCrO_4$, SO_4^{2-} is precipitated as $BaSO_4$; thus, the SO_4^{2-} will be retarded as it migrates through the aquifer. We calculated the retardation factor for SO_4^{2-} as the ratio of the total (injected) SO_4^{2-} concentration ($[SO_4^{2-}]_{tot}$) to the aqueous SO_4^{2-} after equilibration with $BaCrO_4$ and $BaSO_4$ (when the initial solution is supersaturated). A graph of this retardation factor plotted against the $[SO_4^{2-}]_{tot}$ (Figure 14.13) shows that when $[SO_4^{2-}]_{tot}$ is low, the solution is undersaturated with $BaSO_4$ and there is no loss of SO_4^{2-}; i.e., the retardation factor is equal to unity. As $[SO_4^{2-}]_{tot}$ increases, a larger fraction of the SO_4^{2-} is precipitated as $BaSO_4$ and the retardation factor increases until $[SO_4^{2-}]_{tot}$ exceeds the concentration of the $BaCrO_4$ in the soil. After this point, the fraction of $[SO_4^{2-}]_{tot}$ that is precipitated in $BaSO_4$ becomes smaller with increasing $[SO_4^{2-}]_{tot}$ and the retardation factor sharply decreases back to unity.

In a general sense, the retardation factors for Cr(VI) and SO_4^{2-} are related to the effectiveness of a given $[SO_4^{2-}]_{tot}$ to remove the $BaCrO_4$ from a point in the subsurface and how quickly we can deliver that concentration to that point, respectively. Although we have discussed these two ideas separately, they are intrinsically coupled. In practical field application, the retardation of both the Cr(VI) and SO_4^{2-} need to be considered. However, the best scenario, the removal of the $BaCrO_4$ in one or two flushes of the SO_4^{2-}-amended water, will require the retardation factor for SO_4^{2-} to be close to unity.

We note that Ba^{2+} is a priority pollutant and that the SO_4^{2-} extraction does not address its removal. However, the Ba^{2+} present will in many cases be a natural soil constituent and not from an anthropogenic source. Also, equilibria with $BaSO_4$ and natural cation exchangers will keep Ba^{2+} below the MCL. Even under the extreme conditions at the UCP site, Ba^{2+} in solution rarely exceeded 0.3 mg/L (Palmer and Wittbrodt, 1990).

SUMMARY

Removal of metal contaminants from the subsurface can be greatly impeded by slow transfer of the contaminant from the immobile soil matrix to the mobile groundwater. Under these conditions, well-conceived chemical enhancement methods can accelerate contaminant extraction by increasing this mass transfer.

A rational chemical enhancement scheme must deliver the reactive agent to the contaminated areas of the aquifer in a relatively short period of time, address the key chemical process limiting removal of the metal contaminants, allow for rapid removal of the reactive agent and reaction by-products, and consider the treatment and disposal of the extracted waters. Potential reactive agents include 1) acids and bases, 2) oxidants and reductants, 3) competitive adsorbates, 4) complexing agents, or 5) solid phase substituents.

By altering the pH of the subsurface it is possible to increase the solubility of minerals containing the target contaminant and to decrease the fraction of the target contaminant adsorbed to the soil matrix. However, the high buffer intensity of many aquifers may render this impractical. Changing the redox conditions within the soil can change the mobility of metal contaminants either by changing their redox state or by reductively dissolving oxide surfaces that impede metal removal. Weakly

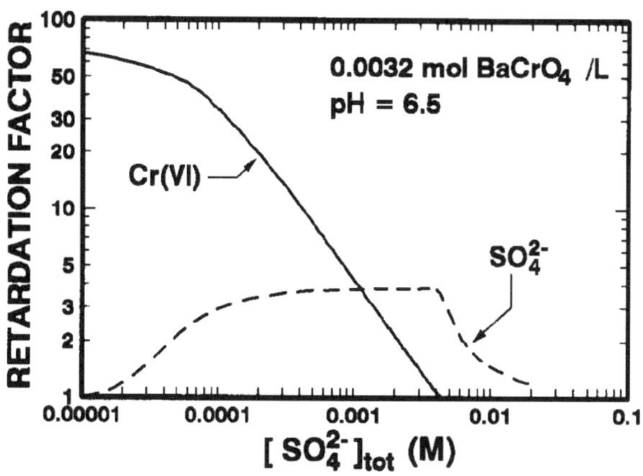

Figure 14.13. Retardation factors for Cr(VI) and SO_4^{2-} at different $[SO_4^{2-}]_{tot}$ for a model system containing 3.2 mM $BaCrO_4$ and pH 6.5.

binding adsorbates can successfully compete with metals for adsorption sites if sufficiently high concentrations of the reactive agent are used. Complexing agents can bind both target metal contaminants and adsorptive metal oxides, thereby greatly increasing the contaminant concentration in the mobile groundwater. Finally, some reactive agents can bind the counterions in minerals containing metal contaminants, thereby releasing the metal to the mobile groundwater.

A common aspect of many reactive agents is the lack of specificity. The reactive agent will often remove nontoxic constituents in addition to the target contaminants. This can be advantageous when the nontoxic substance is the primary impediment for the removal of the metal contaminant. However, the presence of these nontoxic constituents in the treatment train and their disposal need to be included in the design and estimated costs of the system.

Chemical enhancement methods have the potential of decreasing the time for contaminant removal by one or more orders of magnitude compared to conventional pump-and-treat. The optimal choice of reactive agent depends on the key process limiting the transfer of the metal contaminant to the groundwater. Thus, identification of these limiting processes must be part of site characterization prior to design of any chemical enhancement system. A well-conceived chemical enhancement design may provide an attractive alternative for groundwater remediation. The practical application of these methods needs to be explored and efficient and cost-effective methods must be developed.

REFERENCES

Allison, J.D., D.S. Brown, and K.J. Novo-Gradac. 1991. *MINTEQA2/PRODEFA2, A Geochemical Assessment Model for Environmental Systems: Version 3.0 User's Manual.* EPA/600/3-91/021.

Barcelona, M.J. and T.R. Holm. 1991. Oxidation-Reduction Capacities of Aquifer Solids. *Environ. Sci. Technol.*, **25**(9):1565–1572.

Dzombak, D.A. and F.M.M. Morel. 1990. *Surface Complexation Modeling, Hydrous Ferric Oxide.* Wiley-Interscience Publication, New York, 393 p.

Palmer, C.D. and W. Fish. 1992. Chemical Enhancements to Pump-and-Treat Remediation. Ground Water Issue Paper, U.S. Environmental Protection Agency, EPA/540/S-92/001.

Palmer, C.D., W. Fish, and J.F. Keely. 1988. Inorganic contaminants: recognizing the problem. Proceedings of the Second National Outdoor Conference on Aquifer Restoration, Ground Water Monitoring, and Geophysical Methods, May 23–26, Las Vegas, NV, 555–579.

Palmer, C.D. and P.R. Wittbrodt. 1990. Geochemical Characterization of the United Chrome Products Site, Final Report. IN: Stage 2 Deep Aquifer Drilling Technical Report, United Chrome Products Site, Corvallis, OR, September 28, 1990, CH2M Hill, Corvallis, OR.

Palmer, C.D. and P.R. Wittbrodt. 1991. Processes Affecting the Remediation of Chromium-Contaminated Sites. *Environmental Health Perspectives*, 92:25–40.

Rai, D., J.M. Zachara, L.E. Ainsworth, C.C. Amonette, J.E. Cowan, C.E. Szelmeczka, R.W. Resch, C.T. Schmidt, S.C. Smith, and D.C. Girvin. 1988. Chromium reactions in geologic materials. Electric Power Institute Research Institute Report EA-5741, Palo Alto, CA, 287 pp.

Thirumalainivas, B.B.J., D. Shiau, D.A. Sabatini, and J.H. Harwell. 1994. Surfactant enhanced extraction of Cr(VI) using U.S. Coast Guard Media, Elizabeth City, North Carolina. Groundwater Research Seminar, Robert S. Kerr Environmental Research Laboratory and the National Center for Ground Water Research, Oklahoma City, Oklahoma, June 1–3, 1994 (Abstract).

Travis, C.C. and C.B. Doty. 1989. Superfund: A Program without Priorities. *Environ. Sci. Technol.*, 23(11):1333–1334.

Ward, C.H. and C.D. Palmer. 1994. Sulfate Extraction of Chromate Contaminated Soil: Column Studies of Enhanced Barium Chromate Removal. National Ground Water Association, Outdoor Action Conference/Expo, Minneapolis MN, May 23–25, 1994.

Chapter 15

Remediation of Soils and Aquifers Contaminated with Hydrophobic Organic Chemicals: Theoretical Basis for the Use of Cosolvents

D.C.M. Augustijn, L.S. Lee, R.E. Jessup, P.S.C. Rao, M.D. Annable, and A.L. Wood

INTRODUCTION

Solubility and sorption play an important role in governing the behavior of organic chemicals in soils and aquifers. Solubility determines the maximum concentration in the solution phase, while sorption by the solid phase results in a retardation of solute transport. The effects on solubility and sorption of organic solutes upon addition of one or more organic cosolvents to an aqueous phase, referred to as *cosolvency*, have now been well established based on considerable experimental data and theoretical analysis (Rao et al., 1991). Organic solvents (e.g., alcohols) may be present at waste disposal/spill sites as components of complex wastes, or introduced at these sites by codisposal. Presence of organic cosolvents will enhance contaminant migration away from the source area by increasing the solubility and reducing sorption. Although the presence of cosolvents can exacerbate environmental problems, the principles of cosolvency can also be applied in designing remedial strategies at contaminated sites. For example, removal of contaminated soils from a waste disposal/spill site and extraction with solvent mixtures, commonly referred to as *solvent washing*, is one remediation technique that is receiving considerable attention (U.S. EPA, 1990). Another potential use of cosolvents is a technique referred to as *solvent flushing*, where mixed solvents are used for *in situ* extraction at contaminated sites (U.S. EPA, 1991).

In this chapter, we will briefly review the basic principles of the cosolvency theory, and discuss how key parameters for predicting sorption and transport of organic chemicals are affected by the presence of solvent mixtures. Use of cosolvents in the design of remediation techniques involving solvent washing and solvent flushing techniques will be illustrated with some model simulations.

COSOLVENCY THEORY

Equilibrium Dissolution from Single-Component Wastes

Dissolution from a single-component waste, either a solid or a liquid, is of considerable interest in many remediation scenarios. Cleanup of soils or groundwater contaminated with residual amounts of organic immiscible liquids (OILs),[1] such as chlorinated solvents, is a prominent example.

By adding an organic solvent to water, the polarity of the solvent mixture decreases, resulting in an increased solubility of nonpolar organic chemicals. The solubility of a nonpolar organic solute in a binary solvent mixture (S_m) increases in a log-linear manner with increasing volume fraction of cosolvent (f_c) (Yalkowsky and Roseman, 1981; Fu and Luthy, 1986a; Morris et al., 1988):

$$\log S_m = \log S_w + \beta \sigma f_c \tag{15.1}$$

[1] The term OIL is preferred here over the frequently used term nonaqueous phase liquid (NAPL) because mixed solvents, the focus of this article, can also be considered as a NAPL.

where

$$\sigma = \log(S_c/S_w) \tag{15.2}$$

β is an empirical coefficient that accounts for water-cosolvent interactions; S_c is solubility in a neat solvent; S_w is solubility in water; and σ is the cosolvency power. When it is assumed that the cosolvent effects are additive, Equation 15.1 can be generalized as follows for a mixture of water and several cosolvents:

$$\log S_m = \log S_w + \Sigma\, \beta_i\, \sigma_i\, f_{c,i} \tag{15.3}$$

where the subscript i=1,2,...N designates the values for the ith cosolvent, and N is the number of cosolvents in the mixture.

The log-linear model stated in Equations 15.1 and 15.3 is adequate for describing the solubility profiles of a number of nonpolar organic chemicals in a variety of solvent-water mixtures. However, when water-cosolvent interactions are significant, deviations are found from the log-linear solubility profiles predicted by Equation 15.3. Pinal et al. (1991) proposed modifications to the log-linear model to account for these deviations.

Equilibrium Dissolution from Complex Wastes

Dissolution from complex waste mixtures (e.g., gasoline, diesel, creosote, coal tar), which consist of multiple organic constituents, will also be enhanced by the presence of an organic cosolvent. The equilibrium concentration of component i (C_i) in a solution which is in contact with a complex mixture can be predicted using Raoult's law (Banerjee, 1984):

$$C_i = \chi_i\, S_i \tag{15.4}$$

where χ_i is the mole fraction of component i in the complex mixture and S_i is the liquid solubility of the ith component. Since S increases in a log-linear manner with increasing cosolvent content as shown in Equations 15.1 and 15.3, it follows that C_i obeys the same log-linear model. Thus, cosolvent addition to a multi-component waste mixture can be expected to result in an exponential increase in solution phase concentrations of organic constituents of the waste. This prediction is consistent with the data reported by Lane and Loehr (1992) for the dissolution of several aromatic hydrocarbons from tar-contaminated soils into binary mixed solvents. Peter and Luthy (1993, 1994) have presented an evaluation of a semi-empirical thermodynamic model using experimental data for enhanced dissolution of coal tar components into water-miscible cosolvent mixtures. Similar studies with other complex wastes are needed.

Equilibrium Sorption

Sorption of nonpolar organic compounds can generally be described by a linear sorption isotherm, characterized by an equilibrium partition coefficient (K) (Chiou et al., 1983). The change in sorption of organic chemicals from mixed solvents is directly related to cosolvent effects on the solubility of the organic solute. Since sorption is inversely related to the solubility, an increase in solubility, upon addition of cosolvents, leads to a proportional decrease in sorption (Rao et al., 1985; Nkedi-Kizza et al., 1985; Fu and Luthy, 1986b):

$$\log K_m = \log K_w - \Sigma\, \alpha_i\, \beta_i\, \sigma_i\, f_{c,i} \tag{15.5}$$

where α is an empirical coefficient that accounts for solvent-sorbent interactions (Rao et al., 1991), the subscripts m and w denote mixed solvent and water, respectively, and all other terms are as defined previously.

Sorption Nonequilibrium

In many field-scale applications, equilibrium approaches prove to be inadequate in estimating the dissolution or sorption processes during transport since nonequilibrium conditions usually prevail. Failure of pump-and-treat or bioremediation techniques has been attributed to nonequilibrium at grain-scale, local-scale, and aquifer-scale. Brusseau and Rao (1989a) discussed various factors that limit the rate of approach to sorption equilibrium, and reviewed various conceptual/mathematical models proposed to describe sorption nonequilibrium. In this section, we will briefly examine sorption and dissolution kinetics as affected by addition of cosolvents.

Sorption nonequilibrium has generally been described by a bicontinuum model (Brusseau and Rao, 1989a), where sorption is considered to take place in two steps. The first step is considered to occur instantaneously in domains that are readily accessible for the sorbate (S_1), while the second step is rate-limited, often attributed to a diffusion controlled process (S_2). Some researchers have proposed retarded intra-particle diffusion as a mechanism to describe rate-limited sorption (Wu and Gschwend, 1986; Ball and Roberts, 1991) while others have described sorption nonequilibrium by intra-organic matter diffusion (Brusseau et al., 1991a).

The bicontinuum model can be conceptualized as follows:

$$C \xleftrightarrow{FK} S_1 \xleftrightarrow{\lambda} S_2 \tag{15.6}$$

where F is the fraction of sorption occurring in the instantaneous regions; and λ is a first-order mass transfer rate coefficient that describes the rate-limited sorption. Sorption kinetics are more commonly expressed in terms of k_2, the reverse first-order mass transfer rate coefficient, defined as $\lambda/(1\text{-}F)$.

Several researchers found an increase in sorption rate coefficients with increasing cosolvent fractions (Nkedi-Kizza et al., 1989; Wood et al., 1990; Brusseau et al., 1991b). Brusseau et al. (1991b) attributed this to a swelling of organic matter in the presence of an organic cosolvent, thus enhancing sorbent permeation by the sorbate. In case of retarded intra-particle diffusion this phenomenon can be explained by reduced sorption of the solute in the micro-porous particles.

Dissolution Nonequilibrium

Dissolution processes are generally considered to be limited by diffusion of the component through a boundary layer from the interface into the bulk solution (Pfannkuch, 1984). The rate of dissolution from an OIL ($\partial m/\partial t$) is commonly described by a first-order rate law:

$$\partial m/\partial t = \theta\, k\, (C_{eq} - C) \tag{15.7}$$

where m is the mass in the OIL phase; t is time; θ is the volumetric water (or mixed solvent) content; k is the first-order dissolution rate coefficient; and C is the solution phase concentration. C_{eq} is the equilibrium concentration which equals the solubility for single-component OILs, or the equilibrium concentration based on Raoult's law (Equation 15.4) for multi-component OILs. The mass transfer coefficient for dissolution, k, often expressed as the dimensionless Sherwood number, appears to have positive correlation with pore water velocity, OIL saturation, and fluid properties such as dynamic viscosity and density (Powers et al., 1991).

When a cosolvent is added to a soil contaminated with an OIL, the solubility will increase according to Equation 15.3, enhancing the dissolution rate. In addition, a decrease in surface tension upon addition of an organic cosolvent may mobilize the residual OIL. In viscous, multi-component liquids like coal tar or crude oil, diffusive transport within the organic phase is most likely the mechanism controlling the dissolution kinetics. Since no information is available on the effect of cosolvents on the actual mass transfer rate coefficient for dissolution, and very little research has been done on the dissolution kinetics from complex mixtures, this process will not be considered in the following sections.

PARAMETER ESTIMATION

The following discussion focuses on methods for estimating several parameters needed in the use of the cosolvency model for soil cleanup. All equations are summarized in Table 15.1. It should be noted that most of the approaches presented below were not derived from first principles, but are based on empirical correlations using measured data.

Cosolvency Power

The cosolvency power (σ) is probably the most important parameter in the cosolvency theory. It is a function of the solute molecular surface area and the interfacial free energy of the cosolvent, and can be derived from thermodynamic principles (Rao et al., 1985). The solubility of a hydrophobic solute will increase with decreasing polarity of the solvent. For this reason, values can be estimated from polarity or hydrophobicity indices.

One hydrophobicity index for which a large database has been established over the past several years is the octanol-water partition coefficient (K_{ow}). Since the solubility in a cosolvent is proportional to the solubility in octanol, the cosolvency power can be estimated from (Morris et al., 1988):

$$\sigma = A \log K_{ow} + B \tag{15.8}$$

where A and B are empirical constants, unique to each cosolvent, and represent the slope and intercept of the σ vs. log K_{ow} plot. Ideally, A=1, and B=log(S_c/S_o) where S_c is the solute solubility in the neat cosolvent and S_o its solubility in octanol. In general, the solubility in an organic solvent and octanol are of similar magnitude; hence, B is relatively small. As a first approximation, the cosolvency power can therefore be estimated as follows:

$$\sigma \approx \log K_{ow} \tag{15.9}$$

In reality, B is not a true constant because different compounds may have nonproportional solubilities in the solvent and octanol, depending on their structure. In addition, nonideal behavior may affect the values of A and B. Morris et al. (1988) performed regression analysis to estimate A and B values using data for solubility of several solutes measured in mixtures of water and organic solvents that are completely miscible in water. A summary of their results is given in Table 15.2.

Table 15.1. Summary of Equations to Estimate Important Parameters Related to the Cosolvency Theory.

Parameter	Estimation	Reference
S_m	$\log S_m = \log S_w + \Sigma\, \beta_i\, \sigma_i\, f_{c,i}$	Yalkowsky and Roseman, 1981
K_m	$\log K_m = \log K_w - \Sigma\, \alpha_i\, \beta_i\, \sigma_i\, f_{c,i}$	Rao et al., 1985
$k_{2,m}$	$\log k_{2,m} = \log k_{2,w} + 0.668\, \Sigma\, \alpha_i\, \beta_i\, \sigma_i\, f_{c,i}$	Brusseau et al., 1991b
σ	$\sigma = A \log K_{ow} + B$	Morris et al., 1988[a]
S_{scl}	$\log S_{scl} = \log S_w + 2.95(T_m - T)/T$	Yalkowsky, 1979
K_w	$\log (K_w/f_{oc}) = \log K_{ow} - 0.21$	Karickhoff et al., 1979
R_m	$\log (R_m - 1) = \log (R_w - 1) - \Sigma\, \alpha_i\, \beta_i\, \sigma_i\, f_{c,i}$	Nkedi-Kizza et al., 1987
$k_{2,w}$	$\log k_{2,w} = 0.301 - 0.668 \log K_w$	Brusseau and Rao, 1989b

As an initial estimate F = 0.5, α = 1 and β = 1

[a] For values of A and B see Table 15.2.

Table 15.2. Summary of Regression Parameters for Terminal Slope (s) vs. Log K_{ow} in Each Binary Solvent System, Reported by Morris et al. (1988).

Cosolvent	r^2	n[a]	A	B
Acetone	0.93	14	1.00	0.48
Acetonitrile	0.90	8	1.03	0.35
Dimethylacetamide	0.95	7	0.86	0.89
Dimethylformamide	0.94	7	0.87	0.87
Dimethlysulfoxide	0.95	7	0.89	0.87
Ethylene glycol	0.75	7	0.36	1.04
Ethanol	0.94	11	0.85	0.81
Isopropanol	0.85	9	0.89	0.63
Methanol	0.84	16	0.68	1.07
Propylene glycol	0.96	8	0.62	0.77

[a] Number of data points.

Solubility

Values for the solubility can generally be found in chemical reference books. For solids dissolved in a complex liquid waste (e.g., naphthalene in coal tar), S in Equation 15.4 represents the solubility of a hypothetical super-cooled liquid (S_{scl}). The super-cooled liquid solubility is larger than the crystal solubility (S_{cr}) since the energy required to destroy the crystal structure before it can dissolve is eliminated. Direct determination of S_{scl} is not possible; however, several estimation methods are available. One of the simplest and most widely used methods is given by (Yalkowsky, 1979; Martin et al., 1979):

$$\log S_{scl} = \log S_{cr} + 2.95(T_m - T)/T \qquad (15.10)$$

where T is the ambient temperature and T_m the melting point. The constant 2.95 is the ratio of the entropy of fusion (assumed to be 13.5 cal/mole K for all compounds) and the gas constant (1.987 cal/mole K).

Sorption

Sorption of nonpolar organic solutes by soils from solvent mixtures can be estimated using Equation 15.5. Values for four parameters are required: aqueous equilibrium sorption coefficient (K_w); cosolvency power (σ); and the empirical coefficients α and β. Among the techniques available for estimating K_w for nonpolar organic chemicals, the most popular is perhaps the correlation with K_{ow} (Karickhoff et al., 1979):

$$\log K_{oc} = 1.00 \log K_{ow} - 0.21 \qquad (15.11)$$

where K_{oc} is K_w (mL/g) normalized to the soil organic carbon content (f_{oc}). Since this relationship was first published, similar relationships have been developed for specific groups of compounds (cf., Gerstl, 1989).

Water-cosolvent interactions generally lead to positive deviations from the log-linear model; thus, β values are expected to be larger than or equal to 1. Cosolvent-sorbent interactions may result in either positive deviations ($\alpha > 1$) or negative deviations ($\alpha < 1$). The product $\alpha\beta$ accounts for deviations arising from both cosolvent-water and cosolvent-sorbent interactions. Values for α and β vary with cosolvent, soil, and chemical. Since limited information is available for estimating α or β values for specific cases, as a first approximation, the product $\alpha\beta$ can be assumed equal to 1.

Retardation

For transport problems, equilibrium sorption is commonly expressed in terms of a retardation factor (R) which represents the residence time of a chemical in pore volumes:

$$R = 1 + \rho K/\theta \qquad (15.12)$$

where ρ is the dry bulk density. Combination of Equation 15.5 and 15.12 gives (Nkedi-Kizza et al., 1987):

$$\log(R_m - 1) = \log(R_w - 1) + \Sigma\, \alpha_i\, \beta_i\, \sigma_i\, f_{c,i} \qquad (15.13)$$

The validity of the log-linear model for describing the impact of cosolvents on the mobility of organic solutes is confirmed by column data (Nkedi-Kizza et al., 1987, 1989; Wood et al., 1990; Brusseau et al., 1991b; Wood, 1995).

Sorption Nonequilibrium

Brusseau and Rao (1989b) derived an empirical log-log linear inverse relationship between K (mL/g) and k_2 (hr^{-1}) for a wide range of nonpolar organic compounds:

$$\log k_2 = 0.301 - 0.668 \log K \qquad (15.14)$$

Equation 15.14 states that with increasing sorption the mass transfer constraints also increase (i.e., smaller k_2). Brusseau et al. (1991a) proposed that this decrease in k_2 was due to increasing diffusion constraints for larger molecules. When this relationship is substituted in Equation 15.5, a log-linear relationship follows between k_2 and f_c:

$$\log k_{2,m} = \log k_{2,w} + 0.668\, \Sigma\, \alpha_i\, \beta_i\, \sigma_i\, f_{c,i} \qquad (15.15)$$

Brusseau et al. (1991b) found this relationship to be useful up to a cosolvent volume fraction of 70%.

Another important parameter describing kinetics in the bicontinuum model is the fraction of instantaneous sorption domains (F). Brusseau et al. (1991b) found that F generally is unaffected up to 20% cosolvent and then decreases with increasing cosolvent fraction. General applicability of these findings is uncertain, and more research is needed to develop a clear understanding of the cosolvent effects on F. Since reported values of F for aqueous data usually vary between 0.3 and 0.7, the best first guess would be F=0.5 in the absence of specific data.

APPLICATIONS

The cosolvency theory has been tested in laboratory experiments to dramatically enhance the desorption and dissolution of hydrophobic compounds that have small aqueous solubility and large sorption coefficients (Rao et al., 1991; Luthy et al., 1992; Augustijn et al., 1994; Roy et al., 1995; Wood, 1995). Similarly, enhanced mobilization and displacement of OIL ganglia and banks has also been demonstrated (e.g., Farley et al., 1993; Grubb and Sitar, 1994). Information on commercial use of cosolvents for remediation of contaminated soils is very limited. It is evident from the discussion in the previous sections that addition of cosolvents can help in faster cleanup of contaminated soils for four reasons: (1) increased solubility due to reduced polarity of the solvent mixture; (2) decreased sorption or retardation due to enhanced solubility; (3) decreased nonequilibrium conditions due to faster sorption/dissolution kinetics; and (4) mobilization of trapped OIL due to reduced interfacial tension.

Cosolvents could be used to clean up contaminated soils excavated from a site or for *in situ* remediation. The first technique is often referred to as *solvent washing* or *solvent extraction*, while the latter process is referred to as *solvent flushing*. Pilot-scale and commercial-scale applications of solvent extraction technology have been documented (U.S. EPA, 1990) for treating organic contaminants (such as PCBs, VOCs, halogenated solvents, and petroleum components) from a wide range of matrices (such as soils, sediments, and sludges) and a spectrum of waste types (e.g., paint wastes, coal tar wastes, wood treating wastes, and pesticide wastes).

In this section, the use of the cosolvency theory for both types of remediation techniques will be illustrated using model simulations. For solvent washing, we assume here that the contaminated soil is placed in a reactor and extracted with a solvent mixture. This reactor could either be a continuously stirred tank reactor (CSTR) or a packed bed. Solvent flushing involves the *in situ* injection of a cosolvent at a site with high contamination levels, and recovery of the cosolvent once it has moved through the contaminated zone. Since solvent extraction from a packed-bed reactor and *in situ* solvent flushing are very similar from a modeling perspective, both cases will be treated the same as a one-dimensional steady flow through a finite column. In the case of solvent washing, we refer to the treatment of excavated contaminated soil in a CSTR by pumping through a mixed solvent at a steady flow rate.

Three hypothetical remediation scenarios are considered as case studies. Case 1 involves remediation of a soil contaminated with low levels of a highly sorptive single constituent (e.g., PCBs, DDT); here, simulations using an equilibrium sorption model and a bicontinuum sorption model are contrasted. Case 2 represents a contaminated soil containing an excess of an OIL (e.g., TCE, PCE); here, the equilibrium dissolution from a single-component waste is of interest. Finally, Case 3 was chosen to represent dissolution from a multi-component liquid waste (e.g., coal tar). These three hypothetical cases cover a range of remediation scenarios that require the use of increasingly more complex models to describe the processes of contaminant release into the environment. Also, note that while Case 1 is typical for remediation of "far-field" regions, Cases 2 and 3 represent "near-field" regions. A summary of the three case studies with each of the two remediation techniques is given in Table 15.3. Parameter values used to run the simulations for each case are given in Table 15.4. The values listed in the right column for case 1 and 2 are estimated based on the information in Table 15.1 and the chosen parameters listed in the column on the left. The mass-to-volume ratio for the CSTR is chosen equal to the mass-to-volume ratio in the column to allow comparison in removal efficiency between CSTRs and packed-bed reactors. In reality, however, the mass-to-volume ratio in a CSTR is much smaller. This does not change the efficiency or total volume required to clean up a given amount of contaminant, but it will affect the total time of operation. Although these values are strictly hypothetical, they were chosen in the same range as for compounds for which cosolvents most likely would be used. The simulations are performed for an arbitrary cosolvent for which we assume Equation 15.9 applies.

Case 1A

This case represents the removal of a highly sorptive compound (i.e., large log K_{ow} value, such as observed for PCBs) from soil or aquifer material in a CSTR. The equations that govern the CSTR reactions are given in Appendix A of this chapter. Figure 15.1 shows the simulation results under equilibrium and nonequilibrium conditions for extraction with water and a 50% cosolvent solution. The concentration is scaled to the total mass in the system, such that the area under the concentration profiles equals unity. For remediation purposes the removal of mass is usually also important; therefore, graphs showing the fraction of mass removed from the system (i.e., integrated area under the elution curves) are also included. When desorption is rate limited (dashed line), an initially more rapid decrease in concentrations was observed, with much longer tailing compared to the equilibrium profile, which is typical for nonequilibrium conditions. Adding cosolvents to the feed solution of the CSTR decreases the retardation factor and consequently increases the concentration in the effluent, which will result in a more efficient removal of the contaminant. With increasing cosolvent fraction the recovery of the contaminant becomes more efficient (i.e., fewer pore volumes are required to completely remove the compound from the reactor). The efficiency increases more rapidly under equilibrium conditions; however, additional modeling results indi-

Table 15.3. Summary of Remediation Scenarios.

Case No.	Hypothetical Scenario Considered	A: *In situ* or CSTR	B: packed bed
1	Desorption under equilibrium and nonequilibrium conditions	Case 1A Figure 15.1	Case 1B Figure 15.2
2	Dissolution of single-component OIL and desorption under equilibrium conditions	Case 2A Figure 15.3	Case 2B Figure 15.3
3	Multi-component dissolution from a complex mixture under equilibrium conditions	Case 3A Figure 15.4	Case 3B Figure 15.5

Table 15.4. Parameter Values for Each Remediation Scenario Used for Simulations.

Case 1

$\text{Log } K_{ow} = 6$	$K_w = 600$ mL/g
$f_{oc} = 0.001$	$k_2 = 0.03\ \text{hr}^{-1}$
$M/V = \rho/\theta = 4$ g/mL	$\sigma = 6$
$q/V = v/L = 125\ \text{hr}^{-1}$	$F = 0.5$
$P = 100$	$\alpha = \beta = 1$

Case 2

$\log K_{ow} = 2.5$	$K_w = 0.2$ mL/g
$S_w = 500$ mg/mL	$\sigma = 2.5$
$f_{oc} = 0.001$	$\alpha = \beta = 1$
$M_{OIL} = 60\ \text{mg/cm}^3$	
$M/V = \theta/\rho = 4$ g/mL	
$P = 50$	

Case 3

Compound	S_{scl}[a] (mg/L)	S_w[a] (mg/L)	MW[a] (g/mole)	Mass (mg/cm^3)	σ ($\log K_{ow}$)[b]
Naphthalene	114	32	128.2	0.28	3.35
1-Methylnaphthalene	27	27	142.2	0.10	3.87
2-Methylnaphthalene	34	26	142.2	0.20	4.00
Fluorene	16	1.9	166.2	0.04	4.18
Phenanthrene	5.6	1.0	178.2	0.08	4.52
Anthracene	5.7	0.07	178.2	0.02	4.50
Fluoranthene	1.3	0.27	202.0	0.02	5.20
Pyrene	2.9	0.16	202.0	0.03	5.00
Benz(a)anthracene	0.12	0.006	228.2	0.02	5.91
Chrysene	1.2	0.006	228.2	0.01	5.86

[a]From: Lee et al. (1992) [b]From: Sangster (1989)
$m_p = 3.2\ \text{mg/cm}^3$
$MW_p = 300$ g/mole
$V = \theta = 0.4\ \text{mL/cm}^3$
$P = 50$

cated that at large cosolvent fractions ($f_c > 0.75$) the equilibrium and nonequilibrium model both approach the elution curve for a nonsorbed solute (i.e., simple dilution effects). This is a result of the exponential decrease of K, with K approaching zero (i.e., no sorption) at high cosolvent fractions. The nonequilibrium model also approaches the equilibrium model at larger values of the Damkohler number (ω). This can be achieved when longer contact times between sorbent and solution are allowed by reducing the flow rate. When more compounds are present in the system,

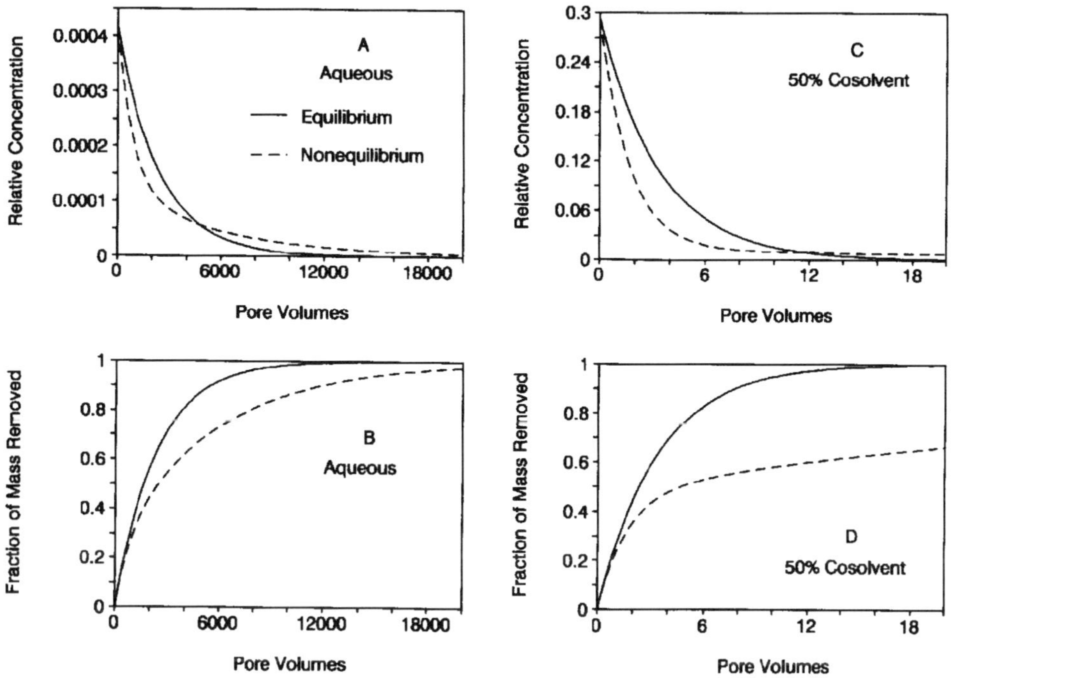

Figure 15.1. Simulated concentration and mass removal profiles from a CSTR for elution of a strongly sorbed compound (Case 1A) with an aqueous and 50% cosolvent solution.

the desorption curves for each compound can be derived in the same manner, assuming no competition for sorption among the nonpolar organic compounds.

Case 1B

Augustijn et al. (1994) developed a one-dimensional transport model, including sorption kinetics, that may serve as a first approximation for the elution profiles from a packed-bed reactor or *in situ* solvent flushing as affected by cosolvents. The equations on which the model is based are summarized in Appendix B of this chapter. Figure 15.2 shows the simulated equilibrium and nonequilibrium concentration and mass removal profiles using this model for elution with water and a 50% cosolvent mixture and for the same compound modeled in case 1A. It was assumed that the porous medium was initially in equilibrium with an aqueous solution. For extraction with water, the concentration is maintained at a constant level for a large number of pore volumes. This is a result of the large retardation factor in an aqueous system. Under nonequilibrium conditions, the concentration decreases earlier, but exhibits much longer tailing, similar to the elution profiles from the CSTR. When a solvent mixture is applied, the first effluent represents the concentration in the aqueous phase ($C^* = 1/R$). As the solvent front proceeds through the column, values for R (and ω under nonequilibrium conditions) change, increasing the concentration in the solution phase and therefore enhancing the elution of the contaminant. The concentrations in the column effluent are even higher than for the CSTR, because mass is accumulated at the mixed solvent front. When the equilibrium mass removal curves for the CSTR (Figures 15.1B and 15.1D) and column (Figures 15.2B and 15.2D) are compared, it can be seen that *in situ* remediation or remediation from a packed-bed reactor is more efficient than contaminant removal from a CSTR. This is a direct result from the mixing properties in a CSTR. However, the main advantage of a CSTR is the destruction of possible stagnant water zones. Such zones can cause physical nonequilibrium conditions, which may be the most important limitation regarding applications of cosolvents in the field.

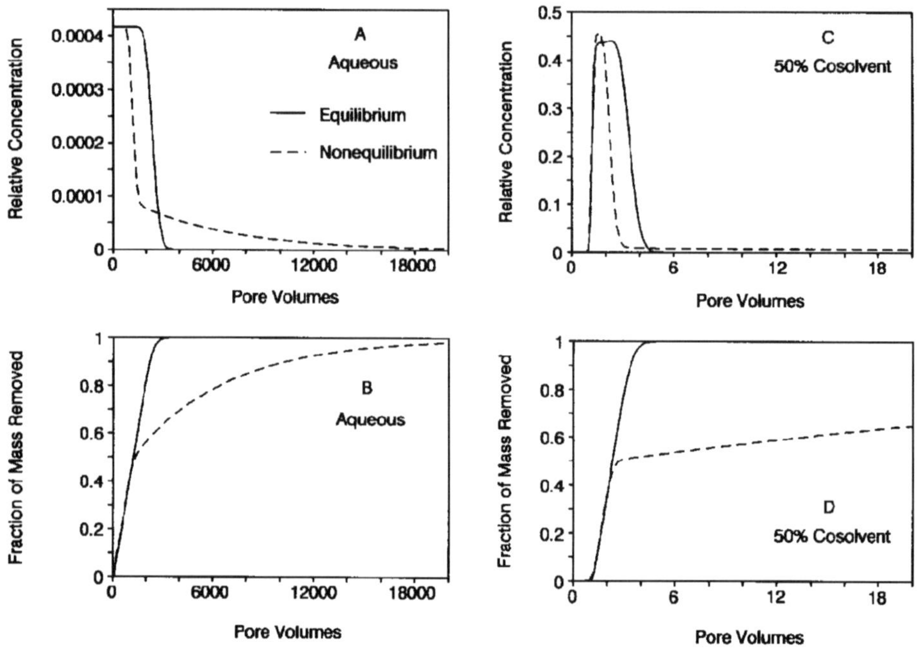

Figure 15.2. Simulated concentration and mass removal profiles from a column for elution of a strongly sorbed compound (Case 1B) with an aqueous and 50% cosolvent solution.

Cases 2A and 2B

Dissolution processes control the concentration in the solution when the contaminant(s) are present in a separate phase. During dissolution, solutes may be sorbed onto the soil. Organic immiscible liquids (OILs) may also interact with the solid matrix of the soil, which will affect the dissolution kinetics as well as the sorption of solutes. In the following discussion, however, we assume that interactions between the OIL and the soil do not impact dissolution or sorption processes. Further, we will assume that sorption of hydrophobic organic chemicals can be described by a linear isotherm over the entire concentration range, from zero to solubility.

For dissolution of a single-component OIL, the concentration in the solution will remain at the solubility limit until the separate phase is completely dissolved. Beyond this point, the contaminant is only present in the solution and sorbed phase and the concentration profiles behave the same as discussed in Case 1A and 1B. Figure 15.3 shows the enhanced effect of cosolvents on the removal of a residual hypothetical compound under equilibrium conditions for elution from a CSTR. The profiles for dissolution from a column are not much different from those presented in Figure 15.3. The parameter values of this hypothetical compound, given in Table 15.4, are in the same range as for TCE and PCE. The mass of OIL corresponds to a residual saturation of approximately 10% assuming a density for the OIL of 1.5 g/mL. The change in volumetric water content (V or θ) when the OIL dissolves away, is assumed to be small enough such that it can be treated as a constant. The Peclet number (P) is also assumed constant. The blockage of pores by the OIL and creation of immobile zones may cause more hydrodynamic dispersion; therefore, a smaller P value was chosen for Cases 2 and 3. Initially, the concentration equals the compound's solubility in the respective solvent or solvent mixture.

Once the residual phase is depleted, the contaminant concentration decreases very rapidly to zero, due to limited sorption, especially at higher cosolvent contents. For this same reason, the removal of contaminant mass may be approximated by a straight line with a slope equal to VS/M for a CSTR or θS/M for a column. Note that this only holds for equilibrium dissolution and desorption conditions and for low sorption. As evident from Figure 15.3, the plateau at which the concentration equals the

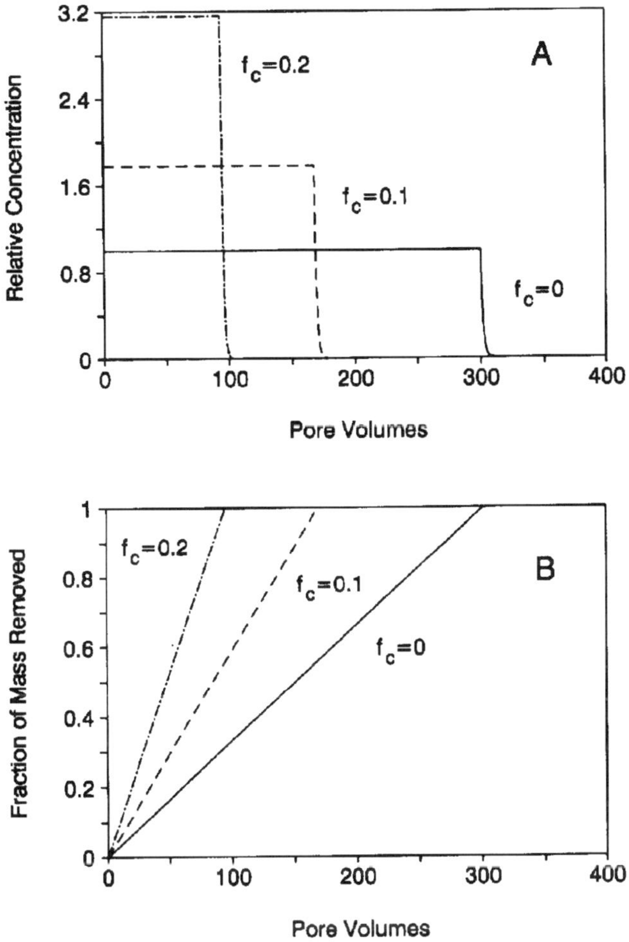

Figure 15.3. Simulated concentration and mass removal profiles from a CSTR for elution of an OIL phase (Case 2A) with three different solvent mixtures.

solubility becomes shorter with increasing f_c because the solubility limit increases exponentially with f_c. Depending on the solubility and total OIL mass in the system, it is possible that at high cosolvent fractions all mass will instantaneously dissolve into the solution phase. This would require only one pore volume to clean up the soil from a packed-bed reactor (assuming low dispersion-induced spreading) or by *in situ* solvent flushing, while approximately six pore volumes are needed to remove essentially all of the OIL mass from a CSTR.

Case 3A

Remediation of soils contaminated with multi-component liquid waste mixtures is receiving increasingly more attention. Besides the complex dissolution processes, many uncertainties may exist regarding the composition of the waste as well as the size and distribution of the source at a site. This problem is not specific for this case but does also apply to Cases 1 and 2. In the following discussion we assume a certain amount of coal tar of known composition (Table 15.4) that is homogeneously distributed in the soil. In addition, we assume equilibrium dissolution conditions described by Equation 15.4.

When some of the contaminant mass is removed by dissolution, the total mass of the complex mixture will decrease, while the average molecular weight will increase, assuming that only the larger,

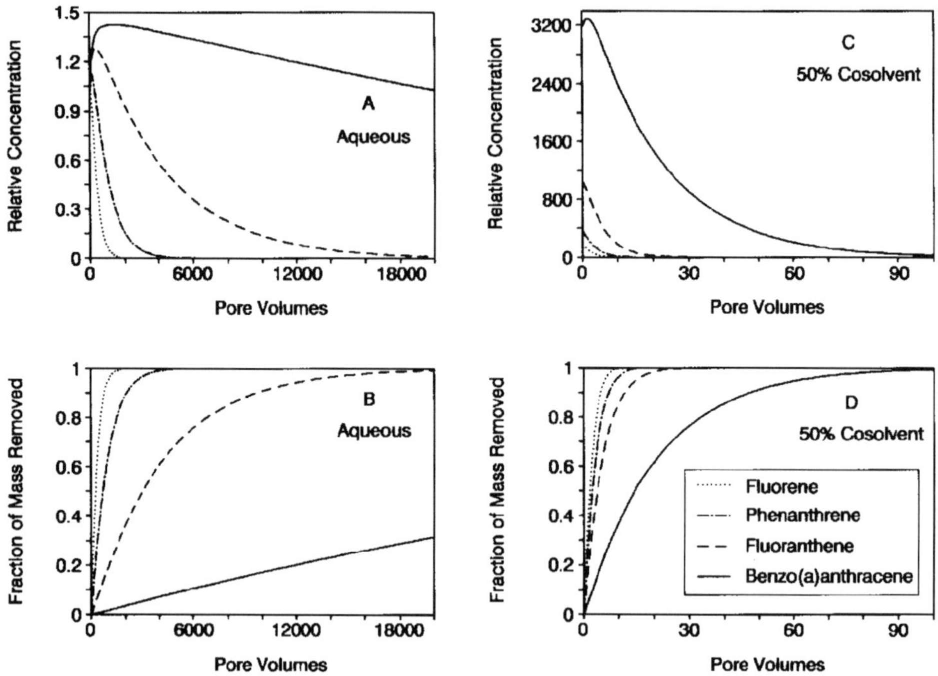

Figure 15.4. Simulated concentration and mass removal profiles from a CSTR for elution of PAHs from a soil contaminated with coal tar (Case 3A) with an aqueous and 50% cosolvent solution.

less soluble constituents remain. This process was simulated by a numerical procedure based on the equations for a CSTR (Appendix A of this chapter) and Equation 15.4. A simulation was performed for the dissolution of polyaromatic hydrocarbons (PAHs) from the hypothetical coal tar depicted in Table 15.4. The initial mass of coal tar corresponds roughly to a residual saturation of 1%. Figure 15.4 shows elution profiles of several PAHs from a CSTR for water and 50% cosolvent. The concentration of each compound was normalized to its initial aqueous concentration. The concentrations of smaller compounds, like fluorene, decrease nearly exponentially, while larger compounds like fluoranthene and benzo(a)anthracene show an increase in concentrations before concentrations eventually decrease. This results from the changing mass and molecular weight of the coal tar during dissolution. In 50% cosolvent, the PAH concentrations have increased considerably and the number of pore volumes required to completely remove the contaminants has been drastically reduced. Note that large cosolvency power (σ) values result in relatively larger increases in solute concentration, as shown in Figure 15.4 for benzo(a)anthracene.

An additional complication is that a large part of coal tar consists of large insoluble compounds, so-called "pitch," which was here assumed to be 80% of the mass of the hypothetical coal tar. During the dissolution process the consistency of coal tar changes from a viscous liquid to a more solid-like material. This is due to the loss of more-soluble compounds and suspected polymerization of the coal tar. The molecular weight of the pitch controls the contaminant concentration in solution; the larger the molecular weight of the pitch the more efficient is the recovery of the soluble compounds. For more-soluble organic liquids like gasoline, all compounds eventually dissolve, especially when cosolvents are used.

Case 3B

The equations that govern the multi-component dissolution from a column are given in Appendix C of this chapter. The simulated elution profiles with solute concentrations normalized to their initial value in water are shown in Figure 15.5. Abrupt changes can be noted in the contaminant concentra-

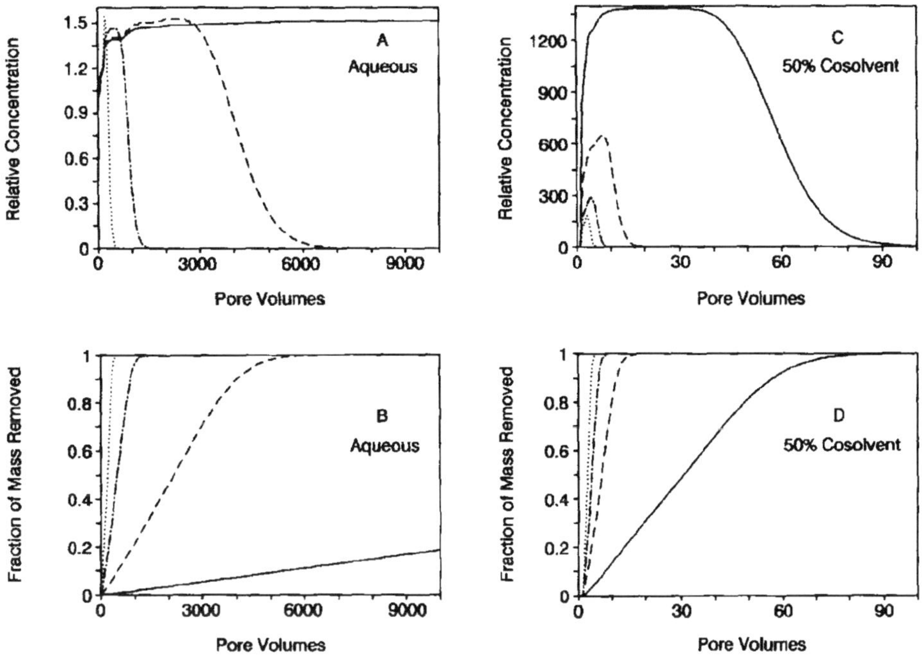

Figure 15.5. Simulated concentration and mass removal profiles from a column for elution of a soil contaminated with coal tar (Case 3B) with an aqueous and 50% cosolvent solution (for legend see Figure 15.4).

tions in the effluent, which are more prominent for elution with aqueous solutions (Figure 15.5A) than with 50% cosolvent (Figure 15.5C). Note that such changes are accentuated due to chromatographic separation of contaminants in a packed-bed reactor, in comparison to a well-mixed CSTR (Figure 15.5 vs. 15.4). These step changes are associated with increases in the mass fractions of all constituents remaining in coal tar as a more soluble compound is depleted. As in previous scenarios, addition of cosolvents enhances the contaminant recovery. Also, the recovery of contaminant in a packed-bed reactor or from *in situ* treatment is more efficient than from a CSTR.

FIELD-SCALE TESTING OF *IN SITU* COSOLVENT FLUSHING

The use of *in situ* cosolvent flushing for enhanced remediation of OIL-contaminated aquifers was evaluated at a field site at Hill AFB, Utah. The experiment was conducted in a hydraulically isolated test cell (4.3 meters × 3.5 meters) established in a surficial sand-gravel-cobble aquifer (see Figure 15.6), which was contaminated with a mixture of jet fuel and chlorinated solvents. At the site, a clay confining unit was present at about 7 meters below ground surface (b.g.s.), and the interlocking sheet piles with grout-sealed joints were driven about 3 meters into the clay unit to install the test cell. This type of containment is also referred to as the Waterloo Barrier system (Starr et al., 1992).

The source of contamination at this Superfund site existed as a residual saturation of lighter-than-water OIL present over a 1.5-meter smear zone located both above and below the water table. The multi-component OIL is a mixture of aviation fuel hydrocarbons (JP-4) and chlorinated solvents (chloroalkenes and chlorobenzenes) disposed in two chemical disposal pits at the site; JP-4 from a fire training area may also have contributed to contamination at this site. The residual OIL and the associated groundwater contaminant plume extend over an area of about 8 ha at the site. The principle target constituents of concern include: BTEX, TCE, TCA, PCE, PAHs, and n-alkanes. The water table inside the test cell was raised to 5 meters b.g.s. to saturate the entire OIL smear zone. All of the tracer tests and the *in situ* cosolvent flushing tests were conducted under saturated, steady flow conditions.

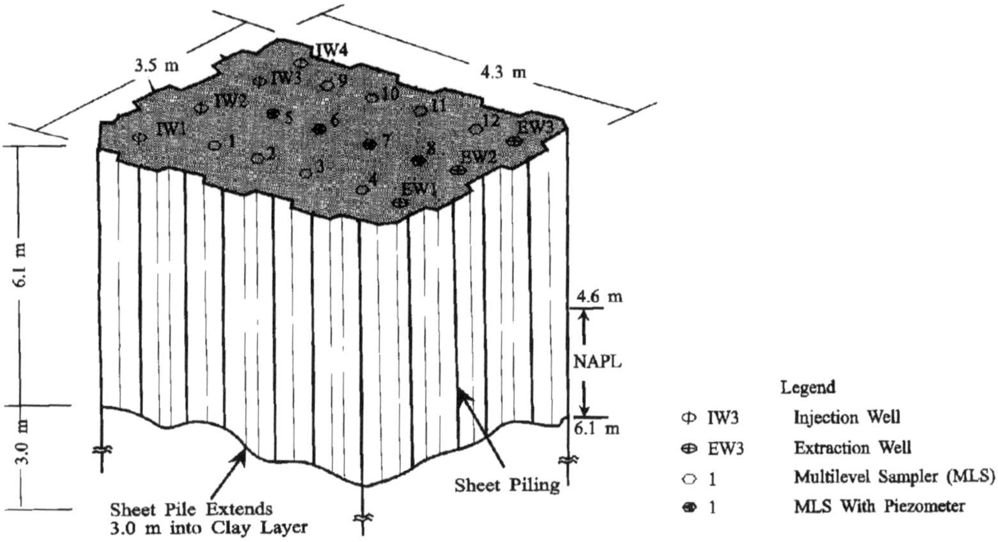

Figure 15.6. A schematic representation of the hydraulically isolated test cell in which field-scale test of *in situ* cosolvent flushing was conducted at the Hill AFB, Utah (modified from Annable et al., 1996a).

The following series of tests were conducted in the test cell during 1994–1995 period. First, an inter-well partitioning tracers (IWPT) test (Jin et al., 1995) was conducted to quantify the total amount and the spatial distribution of the residual OIL within the test cell. This test consisted of displacing a small pulse of a bromide (as a non-reactive tracer) and a mixture of methyl alcohols (as the partitioning tracers) through the cell, and monitoring the concentration breakthrough curves at 60 internal multi-level sampling points (12 locations, 5 or 6 depths at each location) and the three fully screened extraction wells. These data were used to quantify the average residual OIL saturation within the test cell (Annable et al., 1996a) as well as its spatial distribution within the test cell (Annable et al., 1996c).

The second test—the *in situ* cosolvent flushing test—consisted of pumping about 40,000 L (approximately ten pore volumes) of 70% ethanol, 12% n-pentanol, and 18% water mixture through the test cell over a period of 15 days. The cosolvent mixture was introduced at one end of the cell through three fully screened injection wells, and extracted through four fully screened extraction wells at the opposite end of the cell. Based on data from companion laboratory studies, the ethanol-pentanol-water mixture selected was expected to provide substantial solubility enhancement of the OIL constituents, but not lower than the oil-water interfacial tension enough to mobilize the residual OIL. During the experiment, samples were collected from 60 internal multi-level monitoring points and three fully screened extraction wells. These samples were analyzed for OIL constituents to evaluate the efficiency of NAPL solubilization. Alcohol concentrations were also measured in these samples to evaluate the efficiency of hydrodynamic displacement during the cosolvent flushing (Rao et al., 1996; Annable et al., 1996b; Sillan et al., 1997). The final test was another IWPT test to ascertain the changes in the OIL amount and distribution resulting from cosolvent flushing. Soil cores were also collected before and after the flushing tests (Annable et al., 1996b,c; Rao et al., 1996).

The effectiveness of the *in situ* cosolvent flushing technique was evaluated using the following methods: (1) OIL constituent concentrations measured by extraction and analysis of pre- and post-flushing soil core samples; (2) cumulative mass of several target contaminants removed, as determined from analysis of the extraction well fluids; (3) analysis of the data from the pre- and post-flushing interwell partitioning tracer tests; and (4) pre- and post-flushing analysis of groundwater samples for target contaminant concentrations. The results obtained by the various methods agree, and indicate that >90% mass of the several target contaminants and >85% of the total OIL mass present in the test

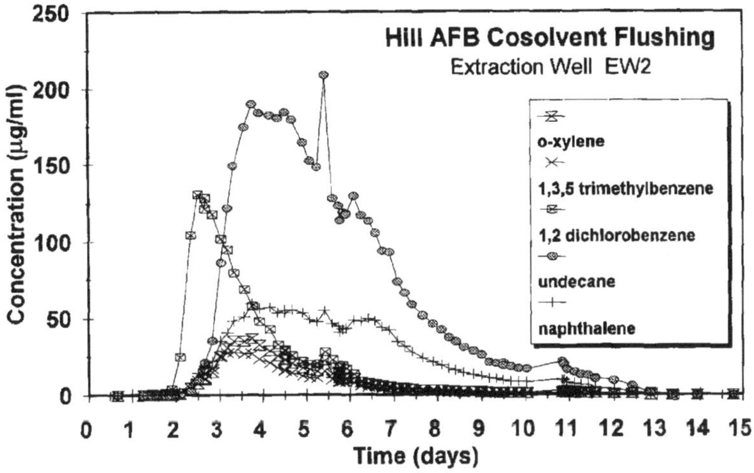

Figure 15.7. Breakthrough curves for selected analytes in samples collected at Extraction Well 2 during *in situ* cosolvent flushing in a test cell installed within an OIL-contaminated field site at Hill AFB, Utah (modified from Rao et al., 1996).

cell was removed as a result of the cosolvent flushing (Annable et al., 1996b,c; Rao et al., 1996). Representative data are presented in Figure 15.7 for enhanced solubilization of OIL constituents, as determined from GC-FID analysis of samples taken at one of the three extraction wells (EW2), during the cosolvent flood. Further analyses of the samples and the data are continuing. The analysis and interpretation of the results from the IWPT tests and pre- and post-flushing soil core analyses suggest that the partitioning tracer method is a very useful method for characterizing the subsurface OIL sources (Annable et al., 1996a,c). Within the test cell, the average residual OIL content was determined to be about 5% of the porosity, with the highest OIL content located at the bottom (clay-sand boundary) and the lowest OIL content located near the top (water table). Soil core and groundwater analyses for OIL constituents confirmed these conclusions (Annable et al., 1996a; Rao et al., 1996).

The Hill AFB study represented the first field-scale test of the IWPT technique for applications in environmental hydrology, and the first field-scale test of the *in situ* cosolvent flushing technique at a Superfund site. The primary goals were therefore to conduct this small-scale study as a "proof-of-concept" test to produce a reliable database for research and development purposes. The cosolvent pumping strategies were not optimized based on cost considerations, nor was an attempt made to recover and recycle the cosolvents or to separate the OIL in order to minimize waste designated for disposal. Both aspects will have to be considered in designing larger-scale demonstration tests. The results from the field tests done at Hill AFB are encouraging, and suggest that additional pilot-scale tests should be conducted at other sites with different geohydrology and contaminant sources—especially source regions that contain free phase and/or residual saturations of denser-than-water OILs—to further test the utility of the *in situ* cosolvent flushing technology for enhanced aquifer remediation. The results from the completed study in the test cell can be used for designing a full-scale demonstration at the Hill AFB site. The residual OIL contamination at the site offers an ideal site for a full-scale demonstration of *in situ* flushing technologies, especially if such tests were coordinated with plans for source area containment (probably with slurry wall) at this site. Since the residual OIL contamination is confined primarily to an elongated sand/gravel channel (about 100 meters by 200 meters), optimal well configurations can be designed to achieve cost-effective OIL recovery.

SUMMARY

In this chapter we presented a brief overview of the cosolvency theory and indicated how important parameters can be estimated by simple empirical relationships. Based on the properties of cosolvents it was suggested that cosolvents can be used as chemical additives to enhance the remediation of contaminated soils. The use of cosolvents for two remediation techniques and in three remediation scenarios was evaluated. In all cases the use of cosolvents reduced the remediation time drastically; the higher the cosolvent fraction and the higher the cosolvency power, the greater was the effect.

Although this study supports the potential application of cosolvents in remediation techniques, several problems still need careful consideration before implementation. This is particularly true for *in situ* flushing scenarios irrespective of the additive (i.e., cosolvents or surfactants). Such problems include: (1) delivery of the additive to the source; (2) reduced efficiency of the flushing treatment due to soil heterogeneities and non-uniform distribution of the contaminants; (3) transport of dissolved contaminants offsite; (4) recovery and waste management; and (5) costs. These problems are more manageable for *ex situ* applications (solvent washing) involving the use of cosolvents or surfactants, and the simulations presented here should give a reasonable estimation of the impact of cosolvents. Experimental data and small-scale field studies are needed to confirm presented simulations for both solvent washing and solvent flushing.

ACKNOWLEDGMENTS

Research support from the U.S. Environmental Protection Agency (CR-814512) and the Electric Power Research Institute (RP-2879-7) is gratefully acknowledged. Although the research described in this chapter has been funded by the U.S. EPA through a cooperative agreement, it has not been subjected to the agency's required peer and policy review, and therefore does not necessarily reflect the views of the agency and no official endorsement should be inferred. Approved for publication as Florida Agricultural Experiment Station Journal Series No. R-02834.

The field study of cosolvent flushing was funded by the Strategic Environmental Research and Development Program (SERDP), which is a collaborative effort involving the U.S. EPA, U.S. DOE, and U.S. DOD. Supplemental financial and technical assistance was provided by the Environmental Restoration and Management Division, Hill Air Force Base, Utah. Although SERDP and the U.S. Air Force funds were used to support the field study, this document has not been subjected to peer review within either agency, and the conclusions stated here do not necessarily reflect the official views of either agency, nor does this document constitute an official endorsement by either agency. The authors thank Mr. Kevin Bourne and Mr. Robert Elliott, both with the Environmental Management Directorate, Hill AFB, for providing access to the site, partial funding, invaluable technical support, and cooperation for the field study. Dr. Carl Enfield and his colleagues at the U.S. EPA National Risk Management Research Laboratory, Ada, Oklahoma played a major role in designing and conducting the field study.

APPENDIX A

The change in total mass in a CSTR is given by:

$$\partial M/\partial t = -qC \tag{A.1}$$

For the effluent concentration (C) of a single compound desorbing from a contaminated soil under equilibrium conditions, the following solution can be derived:

$$C = C_o \, \mathrm{Exp}(-p/R) \tag{A.2}$$

where p is dimensionless time expressed in pore volumes, defined as:

$$p = q\,t/V \tag{A.3}$$

and

$$R = 1 + M_s\,K/V \tag{A.4}$$

The initial equilibrium concentration in the mixed solvent (C_o) is equal to M/VR. Equation A.2 assumes that the contaminated soil in the CSTR is initially in equilibrium with the particular solvent or solvent mixture. No losses by volatilization or degradation are considered.

Using the bicontinuum model to represent sorption nonequilibrium, the following solution to Equation A.1 is obtained:

$$C=C_o Q_+Q_- ((1/\omega+1/Q_-) Exp(Q_-p)-(1/\omega+1/Q_+) Exp(Q_+p))/(Q_+-A_-) \quad (A.5)$$

where

$$Q_+=(-\omega R-1+(\omega^2R^2+1+2\omega(R-2-2F(R-1))^{0.5})/(2+2F(R-1)) \quad (A.6a)$$

$$Q_-=(-\omega R-1-(\omega^2R^2+1+2\omega(R-2-2F(R-1))^{0.5})/(2+2F(R-1)) \quad (A.6b)$$

and

$$\omega = k_2 V/q \quad (A.7)$$

The pore volume and retardation factor are as defined in Equations A.3 and A.4. The Damkohler number (ω) is the ratio of the advective residence time in the CSTR to the characteristic desorption time, and serves as an indicator for nonequilibrium conditions.

APPENDIX B

The one-dimensional solute transport equation including nonequilibrium sorption is given by (Augustijn et al., 1994):

$$\partial C^*/\partial p + \partial F(R-1)C^*/\partial p + \partial S^*/\partial p = (1/P)\,\partial^2 C^*/\partial X^2 - \partial C^*/\partial X \quad (B.1)$$

and

$$\partial S^*/\partial p = \omega\,((1-F)(R-1)(C^*-S^*) \quad (B.2)$$

where

$$C^* = \theta C/M \quad (B.3)$$

$$S^* = \rho S_2/M \quad (B.4)$$

$$p = v\,t/L \quad (B.5)$$

$$R = 1 + \rho K/\theta \quad (B.6)$$

$$X = x/L \quad (B.7)$$

$$P = v\,L/D \quad (B.8)$$

$$\omega = k_2\,L/v \quad (B.9)$$

APPENDIX C

In order to write the transport equation for a system containing a multi-component OIL it is convenient to rewrite Raoult's law (Equation 15.4) as follows:

$$C_i = m_i\,S_i/(MOL_{cm}\,MW_i) \quad (C.1)$$

where

$$MOL_{cm} = m_p/MW_p + S\ (m_i/MW_i) \qquad (C.2)$$

and MOL_{cm} is the total number of moles in the complex mixture; MW is molecular weight; subscript p stands for pitch; and all other parameters are as defined before.

Using Equation (C.1), one-dimensional, multi-component solute transport including dissolution from a complex mixture can be written as follows:

$$\partial C_i/\partial p + \partial(MOL_{cm}/S_i)C_i/q\partial p = (1/P)\ \partial^2 C_i/\partial X^2 - \partial C_i/\partial X \qquad (C.3)$$

When m_p=0, and only one component is considered, Equation C.3 describes the dissolution process of a single-component OIL without sorption.

GLOSSARY

The following symbols are used in this chapter:

A = empirical constant to estimate σ;
B = empirical constant to estimate σ;
C = concentration in solution (g/mL);
C_{eq} = equilibrium concentration (g/mL);
C_o = initial concentration (g/mL);
C* = normalized concentration;
D = hydrodynamic dispersion coefficient (cm^2/s);
f_c = volume fraction of cosolvent;
f_{cm} = mass fraction of complex mixture;
f_{oc} = fraction of organic carbon;
F = fraction of instantaneous sorption domains;
k = dissolution rate coefficient (hr^{-1});
k_2 = reverse first-order sorption rate coefficient (hr^{-1});
K = equilibrium sorption coefficient (mL/g);
K_{oc} = K normalized to f_{oc};
K_{ow} = octanol-water partition coefficient;
L = column length (cm);
m = mass of a component in complex mixture (g/cm^3);
m_p = mass of pitch (g/cm^3);
M = total contaminant mass in the system (g/cm^3);
M_{cm} = mass of complex mixture (g/cm^3);
M_s = mass of soil in CSTR (g/cm^3);
MOL_{cm} = total number of moles in complex mixture ($moles/cm^3$);
MW = molecular weight (g/mole);
MW_{cm} = average molecular weight of complex mixture (g/mole);
p = dimensionless time in pore volumes;
P = Peclet number;
q = Darcy flux (cm/hr);
Q = dummy variable in Appendix A;
R = retardation factor;
S = solubility (g/mL);
S* = normalized sorbed concentration;
S_{cr} = crystal solubility (g/mL);
S_{scl} = super-cooled liquid solubility (g/mL);
S_1 = sorbed concentration in the instantaneous sorption domain (g/g);
S_2 = sorbed concentration in the rate-limited domain (g/g);
t = time (hr);
T = ambient temperature (^oK);

T_m = melting point (°K);
v = pore water velocity (cm/hr);
V = relative volume of water/solvent mixture in CSTR (mL/cm^3);
x = distance in the column (cm);
X = relative distance;
α = empirical coefficient accounting for solvent-sorbent interactions;
β = empirical coefficient accounting for solvent-cosolvent interactions;
λ = first-order mass transfer rate coefficient (hr^{-1});
θ = volumetric liquid content of the soil (mL/cm^3);
ρ = dry bulk density (g/cm^3);
σ = cosolvency power;
χ = mole fraction in complex mixture; and
ω = Damkohler number.

Subscripts *w* and *m* refer to water and mixed solvent, respectively. Subscript *i* refers to the ith component.

REFERENCES

Annable, M.D., P.S.C. Rao, K.H. Hatfield, W.D. Graham, A.L. Wood, and C.G. Enfield. 1996a. Use of partitioning tracers for measuring residual NAPL: Results from field-scale test. *J. Env. Eng.*

Annable, M.D., P.S.C. Rao, R.K. Sillan, K.H. Hatfield, W.D. Graham, A.L. Wood, and C.G. Enfield. 1996b. Field-scale application of in situ cosolvent flushing: Evaluation of approaches. *Amer. Soc. Civil Eng.* Conference on NAPLs, November 10–14, 1996.

Annable, M.D., P.S.C. Rao, R.K. Sillan, A.L. Wood, and C.F. Enfield. 1996c. Partitioning tracers for evaluation of in-situ cosolvent flushing. Manuscript prepared for submission to *Environ. Sci. Technol.*

Augustijn, D.C.M., R.E. Jessup, P.S.C. Rao, and A.L. Wood. 1994. Remediation of contaminated soils by solvent flushing. *J. Environ. Eng.* **120**:42–57.

Ball, W.P. and P.V. Roberts. 1991. Long-term sorption of halogenated organic chemicals by aquifer material. II. intra-particle diffusion. *Environ. Sci. Technol.* **25**:1237–1249.

Banerjee, S. 1984. Solubility of organic mixtures in water. *Environ. Sci. Technol.* **18**:587–591.

Boyd, G.R. and K.J. Farley. 1992. NAPL removal from groundwater by alcohol flooding: Laboratory studies and applications. In: *Hydrocarbon Contaminated Soils and Groundwater, Vol. 2.* E.J. Calabrese and P.T. Kostecki (Eds.). Lewis Publishers, Inc., Boca Raton, FL.

Brusseau, M.L. and P.S.C. Rao. 1989a. Sorption nonideality during organic contaminant transport in porous media. *Crit. Rev. Environ. Control* **19**:33–99.

Brusseau, M.L. and P.S.C. Rao. 1989b. The influence of sorbate-organic matter interactions on sorption nonequilibrium. *Chemosphere* **18**:1691–1706.

Brusseau, M.L., R.E Jessup, and P.S.C. Rao. 1991a. Nonequilibrium sorption of organic chemicals: Elucidation of rate-limiting processes. *Environ. Sci. Technol.* **25**:134–142.

Brusseau, M.L., A.L. Wood, and P.S.C Rao. 1991b. Influence of organic cosolvents on the sorption kinetics of hydrophobic organic chemicals. *Environ. Sci. Technol.* **25**:903–910.

Chiou C.T., P.E. Porter, and D.W. Schmedding. 1983. Partition equilibria of nonionic organic compounds between soil organic matter and water. *Environ. Sci. Technol.* **17**:227–231.

Farley, K.J., R.W. Falta, D. Brandes, J.T. Millazzo, and S.E. Brame. 1993. Remediation of hydrocarbon contaminated groundwater by aquifer flooding. Technical Report, Hazardous Waste Management Research Fund, University of South Carolina, SC.

Fu, J. and R.G. Luthy. 1986a. Aromatic compound solubility in solvent/water mixtures. *J. Environ. Eng.* **112**:328–345.

Fu, J. and R.G. Luthy. 1986b. Effect of organic solvent on sorption of aromatic solutes onto soils. *J. Environ. Eng.* **112**:346–366.

Gerstl, Z. 1989. Predicting the mobility and availability of toxic organic chemicals. *In* Z. Gerstl, Y. Chen, U. Mingelgrin, and B. Yaron, eds., *Toxic Organic Chemicals in Porous Media*, Vol. 73-Ecological Studies, Springer-Verlag, New York, NY, pp. 151–162.

Grubb, D.G. and N. Sitar. 1994. Evaluation of technologies for in situ cleanup of DNAPL contaminated sites. EPA/600/R-94/120, August 1994, U.S. EPA, Ada, OK.

Jin, M., M. Delshad, V. Dwarakanath, D.C. McKinney, G.A. Pope, K. Seperhrnoori, C. Tilburg, and R.E. Jackson. 1995. Partitioning tracer test for detection, estimation, and remediation performance assessment of subsurface nonaqueous phase liquids. *Water Res. Res.,* **31**(5):1201–1211.

Karickhoff, S.W., D.S. Brown, and T.A. Scott. 1979. Sorption of hydrophobic pollutants on natural sediments. *Water Res.* **13**:241–248.

Lane, W.F. and R.C. Loehr. 1992. Estimating the equilibrium aqueous concentrations of polynuclear aromatic hydrocarbons in complex mixtures. *Environ. Sci. Technol.* **26**:983–990.

Lee, L.S., P.S.C. Rao, and I. Okuda. 1992. Equilibrium partitioning of polycyclic aromatic hydrocarbons from coal tar into water. *Environ. Sci. Technol.* **26**:2110–2115.

Luthy, R.G., D.A. Dzombak, C.A. Peters, M.A. Ali, and S.B. Roy. 1992. Solvent extraction for remediation of manufactured gas plant sites. EPRI TR-101845. Electric Power Research Institute, Palo Alto, CA.

Martin, E., S.H. Yalkowsky, and J.E. Wells. 1979. Fusion of disubstituted benzenes. *J. Pharm. Sci.* **68**:565–568.

Morris, K.R., R. Abramowitz, R. Pinal, P. Davis, and S.H. Yalkowsky. 1988. Solubility of aromatic pollutants in mixed solvents. *Chemosphere* **17**:285–298.

Nkedi-Kizza, P., P.S.C. Rao, and A.G. Hornsby. 1985. Influence of organic cosolvents on sorption of hydrophobic organic chemicals by soils. *Environ. Sci. Technol.* **19**:975–979.

Nkedi-Kizza, P., P.S.C. Rao, and A.G. Hornsby. 1987. Influence of organic cosolvents on leaching of hydrophobic organic chemicals through soils. *Environ. Sci. Technol.* **21**:1107–1111.

Nkedi-Kizza, P., M.L. Brusseau, P.S.C. Rao, and A.G Hornsby. 1989. Nonequilibrium sorption during displacement of hydrophobic organic chemicals and 45Ca through soil columns with aqueous and mixed solvents. *Environ. Sci. Technol.* **23**:814–820.

Peter, C.A. and R.G. Luthy. 1993. Coal tar dissolution in water-miscible solvents: Experimental evaluation. *Environ. Sci. Technol.* **27**:2831–2843.

Peter, C.A. and R.G. Luthy. 1994. Semi-empirical thermodynamic modeling for liquid-liquid phase equilibrium: Coal tar dissolution in water-miscible solvents. *Environ. Sci. Technol.* **28**(7):1331–1340.

Pfannkuch, H.O. 1984. Mass-exchange processes at the petroleum-water interface. *U.S. Geol. Surv. Water Res. Invest. Rep.*, 84-4188, pp. 23–48.

Pinal, R., L.S. Lee, and P.S.C. Rao. 1991. Prediction of the solubility of hydrophobic compounds in nonideal solvent mixtures. *Chemosphere* **22**:939–951.

Powers, S.E., C.O. Loureiro, L.M. Abriola, and W.J. Weber. 1991. Theoretical study of the significance of nonequilibrium dissolution of nonaqueous phase liquids in subsurface systems. *Water Res. Res.* **27**:463–477.

Rao, P.S.C., A.G. Hornsby, D.P. Kilcrease, and P. Nkedi-Kizza. 1985. Sorption and transport of hydrophobic organic chemicals in aqueous and mixed solvent systems: Model development and preliminary evaluation. *J. Environ. Qual.* **14**:376–383.

Rao, P.S.C., L.S. Lee, and A.L. Wood. 1991. Solubility, sorption, and transport of hydrophobic organic chemicals in complex mixtures. *Environ. Res. Brief*, EPA-600/M-91/009, U.S. EPA, Ada, OK., 1991.

Rao, P.S.C., M.D. Annable, R.K. Sillan, D.P. Dai, K.H. Hatfield, W.D. Graham, A.L. Wood, and C.G. Enfield. 1997. Field-scale evaluation of in situ cosolvent flushing for remediation of a shallow, unconfined aquifer contaminated with residual LNAPL. *Environ. Sci. Technol.* (in review).

Roy, S.B., D.A. Dzombak, and M.A. Ali. 1995. Assessment of in situ solvent extraction for remediation of coal tar sites: Column studies. *Water Environ. Res.*, **67**(1):4–15.

Sangster, J. 1989. Octanol-water partition coefficients of simple organic compounds. *J. Phys. Chem. Ref. Data* **18**:1111–1229.

Sillan, R.K., M.D. Annable, D.P. Dai, and P.S.C. Rao. 1997. Evaluation of in situ cosolvent flushing for enhanced remediation of a LNAPL-contaminated aquifer: Analysis of spatial patterns of dissolution in a field test cell during the cosolvent flood. *J. Contam. Hydrol.* (in preparation).

Starr, R.C., J.A. Cherry, and E.S. Vales. 1992. A new sheet piling with sealed joints for groundwater pollution control. 45th Canadian Geotechnical Conference, Toronto, Ontario, Canada.

U.S. EPA. 1990. Solvent extraction treatment. *Eng. Bulletin*, EPA/540/2-90/013, Office of Emergency and Remedial Response, U.S. EPA, Washington, DC.

U.S. EPA. 1991. *In situ* soil flushing. *Eng. Bulletin*, EPA/540/2-91/021, Office of Emergency and Remedial Response, U.S. EPA, Washington, DC.

Wood, A.L. 1995. Influence of cosolvents on the transport of hydrophobic organic chemicals in soils under isocratic and gradient conditions. Ph.D. Dissertation, University of Oklahoma, Norman, OK.

Wood, A.L., D.C. Bouchard, M.L. Brusseau, and P.S.C. Rao. 1990. Cosolvent effects on sorption and mobility of organic contaminants in soils. *Chemosphere* **21**:575–587.

Wu, S. and P.M. Gschwend. 1986. Sorption kinetics of hydrophobic organic compounds to natural sediments and soils. *Environ. Sci. Technol.* **20**:717–725.

Yalkowsky, S.H. 1979. Estimation of entropies of fusion of organic compounds. *I&EC Fundam.* **18**:108–111.

Yalkowsky, S.H. and T. Roseman. 1981. Solubilization of drugs by cosolvents. *In* S.H. Yalkowsky, ed., *Techniques of Solubilization of Drugs*, Marcel Dekker Inc., NY, pp. 91–134.

Chapter 16

Thermally Enhanced Removal of Liquid Hydrocarbon Contaminants from Soils and Groundwater

K.S. Udell

ABSTRACT

The acceleration of recovery rates of second phase liquid contaminants from the subsurface during gas or water pumping operations is realized by increasing the soil and groundwater temperature. Several methods of delivery of thermal energy to soils and groundwater are possible. Of these thermal methods, combined steam injection and vacuum extraction appears to be the most economical and versatile technique to recover volatile, semivolatile, and nonvolatile contaminants from the subsurface. One-dimensional experiments have shown effective removal of both volatile and semivolatile second liquid phase chemicals, as well as nonvolatile aqueous phase contaminants from sand packs. Two-dimensional experiments with homogeneous and layered sand packs have shown rapid recovery of volatile and semivolatile hydrocarbon liquids. The enhancement of mass transfer from lower-permeability regions during the depressurization mode of operation has been observed to be effective in the removal of residual contaminants. Field-scale studies of steam injection and vacuum extraction confirm the effectiveness of this technique and its applicability to remove contaminants found both above and below the water table.

INTRODUCTION

The *in situ* recovery of contaminants from soils is required in many locations where spills have occurred. For volatile contaminants in the unsaturated zone, soil venting has obvious application due to the low viscosity of the carrier fluid (air) and the low capital requirements for implementation. In addition, the high molecular diffusivity of most of the volatile chemicals in air allows successful operation in real geologic environments where mass transfer limitations are certain to occur. However, for low-volatility contaminants or for regions below the water table, air venting will not be effective. Beneath the water table, groundwater pumping has been used extensively. However, cleanup rates with pumping have been found to be slow, particularly in the later stages of cleanup when the dissolution rates become mass transfer limited. In such a situation, the recovery rates are proportional to the aqueous phase diffusivities of the contaminants, which are typically quite small.

The use of thermal techniques to enhance recovery rates of *in situ* "pump-and-treat" operations is of major interest due to the enhanced mass transfer rates that can be obtained. Higher recovery rates mean shorter treatment times and lower volumes of fluid that would be produced and treated. In most remediation projects, the contaminant will consist of many chemical components of varying thermodynamic properties contained in both the high- and low-permeability zones within the saturated and unsaturated zones. Inevitably, during gas venting operations, the rate of evaporation of the contaminant into the vented air will be mass transfer limited. As such, the evaporation rate of each component will be proportional to its gaseous phase concentration at equilibrium with the multi-component liquid. This concentration is proportional to the vapor pressure multiplied by the constituent molecular weight, adjusted for its molar fraction in the multi-component liquid. By Raoult's law, this adjustment factor is equal to the component's molar fraction for ideal mixtures. By increasing the temperature from ambient conditions to a higher temperature, significant increases in the vapor pressures and evaporation rates of the contaminants will be realized. Figure 16.1 shows the equilibrium concentra-

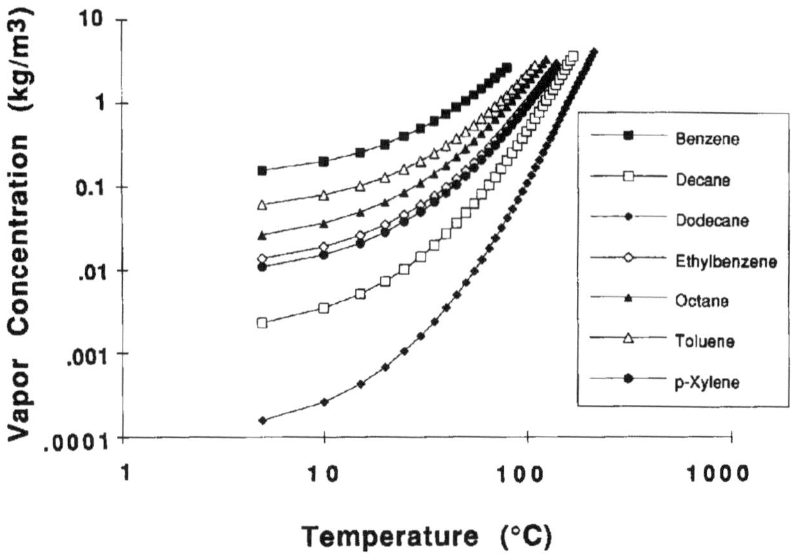

Figure 16.1. Vapor concentrations of various hydrocarbons versus temperature.

tion of several contaminants in air at atmospheric pressure as a function of temperature. The increase in evaporation rate that can be realized by increasing temperature is simply the ratio of the concentration at 100°C to that at 20°C for equal vapor flow rates and subsurface fluid flow distributions. For the compounds shown in Figure 16.1, these ratios are from 25 to 40. The concentration ratios are inversely proportional to the remediation time scale. For compounds with boiling points greater than that of the water at near atmospheric pressure, complete vaporization of a liquid phase contaminant can occur with reasonable energy cost. Semivolatile and volatile compounds are particularly amenable to thermally enhanced extraction due to the large change in vapor pressure with temperature. Furthermore, for longer-chained "nonvolatile" components such as naphthalene, evaporation can occur at significant rates only if the temperature is increased. However, for low-volatility petroleum hydrocarbons such as diesel or lubricating oils, vaporization rates will be limited by low vapor pressures even with thermal enhancement.

In situ thermal techniques include electrical heating, radio frequency heating, hot air injection, hot water injection, and steam injection. Electrical heating and radio frequency heating would typically be used in conjunction with either soil gas or groundwater pumping. The increase in vapor pressure and molecular diffusivities with temperature would be exploited to increase the contaminant recovery rates. Since the thermal energy is supplied to the subsurface without the injection of fluids when radio frequency or joule heating is applied, heating occurs preferentially in zones having the highest radio frequency absorptivity or electrical conductivity, respectively. Therefore, they are well suited to low-permeability soils. Hot gas injection would have application to the enhancement of recovery rates of semivolatile contaminants in the unsaturated zone, and possibly below the water table. However, the low specific heat of air does not allow high-energy input rates unless the gas is very hot. For example, for equivalent mass injection rates, air would need to be injected at over 2200°C to equal the energy input of steam at 100°C. Hot water injection may find application to the acceleration of the cleanup of contaminants below the water table due to the trends of decreasing adsorption, decreasing interfacial tension, decreasing density, and increasing solubilities with increasing temperature. But the inherently slow aqueous phase transport processes plaguing cleanup by groundwater pumping will still limit the rates.

Combined steam injection and fluid extraction is applicable to the *in situ* cleanup of the subsurface contaminants found above or below the water table. Steam effectively heats and displaces water from the medium- to high-permeability zones. Low-permeability zones can be effectively heated by thermal conduction if they are surrounded by steam-bearing regions. This range of applicability is the motivation for the interest this process has received during the last few years. As with other thermal techniques, significant gains in the rates of contaminant recovery can be realized by the heating of the soil to steam temperatures. But the injection of steam is unique in that it provides for the dilution and displacement of nonvolatile contaminants from regions below the water table, effective dewatering of contaminant-laden strata providing additional contaminant-steam contact, and enhanced mass transfer from the low-permeability zones during depressurization. The last mechanism, while thermodynamically subtle, will have dramatic influence on the recovery of contaminants from heterogeneous environments, including fractured media. The use of steam injection to exploit these mechanisms constitutes the use of the steam-enhanced extraction process (Udell et al., 1991) discussed herein.

The use of thermal techniques for the recovery of volatile and semivolatile liquids from porous media is not new. A large body of research on steam injection for enhanced oil recovery by viscosity reduction and distillation can be found in the petroleum literature (i.e., Mandl and Volek, 1969; Volek and Pryor, 1972; and Konopnicki et al., 1979). However, the use of thermal processes for the *in situ* recovery of contaminants has a very short history. Early exploratory field work on steam injection was carried out in the Netherlands in the early 1980s (Hilberts, 1985). Radio frequency heating has been tested (Dev, 1986) as has *in situ* vitrification by resistance heating (Fitzpatrick et al., 1986). More recently, combined steam injection and vacuum recovery for solvent recovery was pilot tested by Udell and Stewart (1989). A field-scale cleanup of a diesel spill followed (Van Sickle, 1992). The process was applied to the full-scale cleanup of a gasoline spill at Lawrence Livermore National Laboratory (LLNL) in 1993. Summary results of the LLNL project are presented in this chapter.

More definitive laboratory experiments on steam injection for volatile and semivolatile liquid contaminant recovery have been reported. Lord et al. (1987) reported kerosene removal efficiencies as functions of time of steam contact and soil type. Hunt et al. (1988) reported removal of pure trichloroethene, a mixture of benzene and toluene, and gasoline as separate liquid phases flowing ahead of the steam condensation front. Stewart and Udell (1988) developed three phase flow solutions to the problem of steam displacement of volatile and nonvolatile liquid contaminants and verified the theory with experimental data. Yuan (1991) and Yuan and Udell (1993) showed the conditions necessary to build a bank of separate phase liquid ahead of the steam condensation front based on criteria related to the vapor pressure of the contaminant components.

In addition to a presentation of the more relevant information found in the literature, new data obtained from laboratory and field research is reported herein. In the following sections, data and discussion are presented first in a one-dimensional context. Results of two-dimensional visualization experiments are then reported. Finally, the results of two field demonstrations will be presented and discussed. The data are used to illustrate the dominant mechanisms that are exploited in the use of combined steam injection and vacuum extraction for the cleanup of soil and groundwater contaminated by second liquid phase contaminants.

One-Dimensional Studies

A series of one-dimensional studies have been completed, including the steam displacement of TCE, a benzene-toluene mixture, and gasoline with essentially complete recovery and separate phase hydrocarbon recovery immediately ahead of the steam condensation front (Hunt et al., 1988). Stewart and Udell (1988) reported experimental data showing that low-volatility mineral oil is not displaced by steam, although evidence of steam distillation in the steamed region was found. This result was consistent with the data of Lord et al. (1987) showing recovery of kerosene in the effluent steam. Yuan (1991) and Yuan and Udell (1993) showed that for contaminant components with vapor pressure greater than that of decane at steam temperatures, a bank of separate phase liquid may develop ahead of the steam condensation front. Vaughan et al. (1993) examined salt displacement by steam injection using electrical conductivity measurements to gain spatial information on salt concentrations ahead of the steam condensation front. These experimental and theoretical studies provided a great deal of insight into the mechanisms governing the removal of volatile, semivolatile, and nonvolatile contaminants from porous media. The dominant mechanisms are detailed and illustrated in the following sections.

Displacement of Aqueous Phase and Nonaqueous Phase Contaminants

A simple one-dimensional conceptual model aids the understanding of the unique characteristics of steam injection. When saturated steam is injected into an initially cool porous medium, the steam condenses, giving up the latent heat of vaporization to heat the porous solid and interstitial fluids. With continued injection, three distinct zones will develop, as illustrated in Figure 16.2: a nearly isothermal steam zone, a variable temperature zone, and an isothermal zone at the ambient medium temperature. Under adiabatic conditions, the rate of expansion of the steam zone is directly proportional to the injected steam enthalpy flux if the length of the variable temperature zone remains constant with time (Hunt et al., 1988, and Stewart and Udell, 1988). The interface between the steam zone and the variable temperature zone is essentially the location of the steam condensation front if the temperature gradient in the steam zone is not significant. For injection pressures near atmospheric and soil thermal properties and temperatures characteristic of field conditions, approximately 350 pore volumes of steam will be required to expand the steam zone an additional unit volume. Thus, steam injection is characterized by large vapor flow rates and pressure gradients in addition to high temperatures.

The displacement of the original fluids by the propagation of the steam condensation front is due to the high steam pressure gradients. These high gradients are sufficient to reduce the water saturation to its residual value in distances of less than a meter (Menegus and Udell, 1985) and provide the force necessary to overcome the capillary forces trapping the hydrocarbon phase (Stewart and Udell, 1988). As compared to the injection of a non-condensing gas into a liquid-saturated porous medium, hydrodynamic fingering does not occur due to the requirement of over 300 pore volumes of steam to provide the thermal energy to propagate the steam zone into each additional volume of media. Thus, higher pressure gradients are observed in the displacing phase (steam) than in the displaced phases (water and hydrocarbon liquid). Saffman and Taylor (1958) used this condition to define stable displacement. Additional stability of the condensation front has also been attributed to the heat losses from incipient fingers (Stewart and Udell, 1989, Choke, 1975). These stabilizing mechanisms do not prevent larger-scale channeling in heterogeneous media however. Rather than the pore-scale fingering that would be observed if gas were injected into a water-saturated porous media, only meter-scale steam channels are expected during steam injection (Basel and Udell, 1991).

The combination of effective steam displacement of liquid water and the addition of steam condensate at the condensation front produces an unexpected bonus: removal of nearly all aqueous phase solute. Large reductions in aqueous phase solute concentrations due to steam displacement and dilution have been quantified in a series of one-dimensional experiments. The solute chosen was sodium chloride because of its low adsorption on quartz and nonvolatility. In an initial series of experiments, a dilute saline solution of 14.0 g/kg concentration was injected into a sand pack until the effluent fluid electrical resistivity matched that of the injected fluid. The 5.04-cm-diameter sand pack was then subjected to a steady steam flow rate of 2 gm/min. As the steam condensation front moved steadily through the sand pack, the effluent was collected in a graduated cylinder. To maintain one-dimensional conditions in the sand pack, a series of heater tapes were sequentially turned on and adjusted in accordance with a predetermined calibration of heat loss versus sand pack/room temperature difference. After steam was observed in the outlet tubing, noted as time of steam breakthrough, injection continued until the effluent mass flux equaled the steam injection rate. The water remaining in the sand pack was thus considered to be immobile and at the irreducible saturation, which was determined from a mass balance to be 12.9%. A measured volume of distilled water was then injected into the steam zone as the column was allowed to cool. At the end of this procedure, the sand pack was completely saturated with water. The system was allowed to rest for 48 hours, allowing the salt remaining in the sand pack to diffuse into the injected water. The water was then allowed to drain by gravity from the sand pack through the injection port. All effluent solutes were analyzed for salt concentrations by atomic adsorption spectroscopy.

Of the 10.1 g of salt initially in solution in the 0.724 L of initial pore water, 9.8 g were recovered in the 0.803 L of effluent water at the time of breakthrough. An additional 0.0133 g of salt were removed after breakthrough, accounting for the remaining mobile water which flowed from the sand pack from the steam zone. From the measured volumes and salt concentrations of the drained water

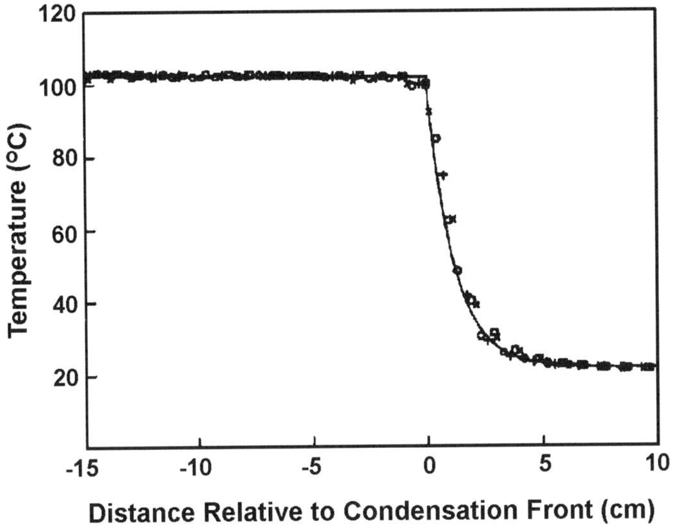

Figure 16.2. Temperature profile near the steam condensation front.

recovered after the water-flood and resting period, a total salt mass of 0.177 g was determined to have remained in the sand pack after steam injection. Therefore, while 87.1% of the saline water was displaced by the steam, a decrease in salt concentration of 98.5% was observed, as would be calculated on a dry sand mass basis. Thus, the residual water remaining in the steam zone was diluted on average by a factor of 7.3. Subsequent experiments reported by Vaughan et al. (1993) show similar behavior. Through electrical conductivity measurements, a dilution of the salt concentration of the immobile water remaining in the steam zone by a factor of 10 gave reasonable agreement with measured electrical conductivities of the partially saturated sand. While the economics of using steam displacement to recover nonvolatile, aqueous phase contaminants from the subsurface are questionable, this displacement mechanism may be exploited in cleanup operations targeting mixed waste sites.

Distillation in the Steam Zone

Depending on the volatility of a liquid phase contaminant, it will either vaporize directly behind the steam condensation front or at a location far behind the front in the steam zone. The conditions that delineate which situation would be found have been outlined in Yuan and Udell (1993). In general, compounds with large vapor pressures or low residual saturations would be expected to be vaporized at a rate that is high enough that the hydrocarbon distillation wave velocity would be greater than the steam condensation front velocity. As such, the hydrocarbon liquid would appear to be displaced by the steam condensation front since any hydrocarbon liquid left in the steam zone would vaporize, be convected to the steam condensation front, co-condense, and coalesce with any other hydrocarbon liquid in the vicinity. For hydrocarbon compounds with lower vapor pressures and/or high residual saturations, the distillation front would move though the soil with a velocity less than that of the steam condensation front. In that case, hydrocarbon liquids will remain in the steam zone and will be recovered through continuing steam distillation processes. Figure 16.3 from Yuan and Udell (1993) shows the various evaporation wave velocities of example liquid hydrocarbons as functions of the residual hydrocarbon saturations. Also shown in this figure as the horizontal dotted lines are two typical steam condensation front velocities. The first situation of an apparent complete removal of the contaminant near the condensation front corresponds to conditions above the horizontal lines in Figure 16.3. This condition would be typical of the removal of compounds such as toluene and those in gasoline which are relatively volatile. Steam injection into a liquid gasoline- or toluene-

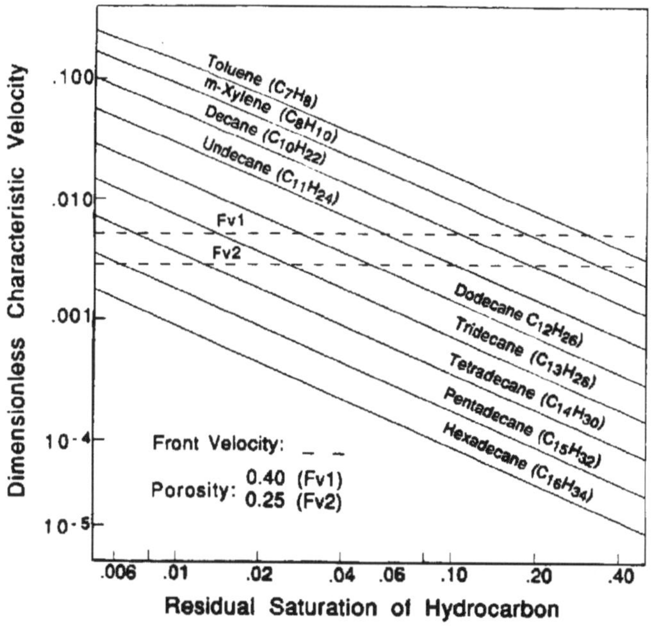

Figure 16.3. Velocities of evaporation waves of various hydrocarbons versus hydrocarbon content in the soil. Also shown are the condensation front velocities for two different porosities. If the velocity of the evaporation wave is greater than the steam condensation front velocity, then the hydrocarbon will appear to be completely displaced by the steam.

contaminated sand forms a liquid hydrocarbon bank ahead of the steam condensation front (Hunt et al., 1988). However, steam injection was not observed to develop a mobile liquid hydrocarbon bank during steam injection experiments with liquid diesel (Basel, 1991). Since diesel hydrocarbons are typically of lower vapor pressure than decane, one would not expect a growing bank of diesel to appear ahead of the steam front according to theory represented in Figure 16.3.

For nonaqueous phase contaminants with boiling points lower than that of steam at pressures near atmospheric, vaporization will occur in the rapidly decreasing temperature zone ahead of the steam condensation front (see Figure 16.2). The resultant contaminant vapor would then migrate in a direction determined by a combination of buoyancy and pressure forces, and the local temperature field. In one dimension, the low-boiling-point liquid will be mobilized ahead of the steam condensation front in spite of capillary forces which would otherwise trap the second phase liquid. As evidenced by the experimental observation of second phase contaminants flowing from the exit in non-emulsion form during the experiments reported by Hunt et al. (1988), it is expected that the contaminant vapor either condenses upon the second phase liquid or forms emulsions which coalesce to the second phase liquid bank. Without doubt, pure separate liquid phase contaminants with boiling points less than that of water will be completely removed from the steam zone except for the small amount which is adsorbed to solid surfaces or is dissolved in liquid water which will be present in the steam zone.

Desorption of Contaminants

When fluids containing a contaminant come into contact with a solid such as a porous matrix, there is a tendency for a fraction of the contaminant to adsorb onto the solid surface. While the mass of adsorbed contaminant is determined by the solid specific surface area, the fraction of organic carbon and the partitioning of the contaminant in the various fluid phases, it is also a strong function of temperature. Adsorption arises because of intermolecular forces between the contaminant and the solid surface. In soil systems the attractive forces between organic carbon and organic contaminants are particularly large. For an isothermal system, an increase in pressure (or fugacity) will result in an increase in the contaminant adsorption as predicted by a Langmuir adsorption isotherm. Yet, for this adsorption to occur, a considerable amount of heat must be released, which is called the latent heat of adsorption. Likewise, for desorption to occur, energy must be supplied to the system. The energy necessary for desorption to occur in contaminated soils can be provided by steam condensation. Also, an increase in the temperature of the contaminated soil increases the intermolecular forces required for a surface to hold a molecule. Thus, the adsorbed mass of contaminant will decrease for an increase in temperature for a fixed fluid composition. As an example, the use of low-pressure steam to regenerate activated carbon used in solvent recovery systems has been common in industry for decades (Scheflan and Jacobs, 1953).

In the context of the steam injection/vacuum extraction process, desorption of volatile compounds is expected to occur in the steam zone in a wave-like manner analogous to that of second phase contaminant evaporation. Also, the desorption of compounds from solid surfaces to the aqueous phase in the region ahead of the condensation front is expected due to the increase in temperature as the steam zone approaches. However, high contaminant concentrations near the steam and contaminant condensation front may lead to re-adsorption, depending on the adsorption characteristics of the contaminants and soil at steam temperatures.

Boiling of Interstitial Water

The thermodynamic system consisting of steam, water, semivolatile contaminant, and porous matrix found within the steam zone is complex and unusual. Because of capillary forces stemming from the radius of curvature of the water-vapor interface, the water is in a stable superheated state in equilibrium with a slightly superheated vapor (Udell, 1983). This situation produces an effective liquid-to-vapor phase-change mechanism where a decrease in the local vapor pressure will promote distributed pore-scale evaporation of the interstitial water at a rate determined primarily by the rate of depressurization and heat transfer from adjacent solids. From energy and mass balances on an adiabatic control volume containing steam, water and a porous matrix, it is evident that the removal of the enthalpy of a given mass of steam will require a corresponding decrease of the system internal energy as well as a decrease in the interstitial water mass. The decrease in the porous matrix internal energy provides the necessary heat for the vaporization of the liquid water. With continued steam withdrawal, the system temperature will continue to drop until the liquid water content is reduced to zero. The significance of this effect is that the vacuum withdrawal of steam from the steam zone can produce a vapor flux from microscopic and macroscopic regions that would not be otherwise contacted by flowing steam. This mechanism can be brought to bear in a field process by simply stopping steam injection to a contaminated region while maintaining a vacuum withdrawal of the vapor phase and continuing liquid pumping. Reduction of the water content to zero would require high steam pressures and minimal liquid water re-infiltration. One-dimensional data supporting this theory was presented previously (Udell and Stewart, 1992). Two-dimensional and field data showing the higher recovery rates by this mechanism is presented in the following sections.

Two-Dimensional Studies

Two-dimensional steam injection experiments have been performed with a 5-cm-thick sand pack, a meter in height and a meter in width, with visualization capabilities. The gravitational orientation of the experiments allowed visual studies of fluid distributions that might be expected under field conditions as well as the effects of gravity on the movement of the steam condensation front. These studies have shown that the conclusions drawn from the one-dimensional studies also pertain to the recovery of nonaqueous phase contaminants from much more complex situations. Specifically, the differences in the distillation wave velocities for various compounds predicted by theory (Yuan, 1991; Yuan and Udell, 1993) have been visually observed in the two-dimensional experiment. Also, the limitation of steam channel lengths due to heat losses to adjacent material has been confirmed (Basel, 1991; Basel and Udell, 1991). However, other important mechanisms influence the movement of the contaminants in a gravity field. Most important of these effects is the phenomenon termed

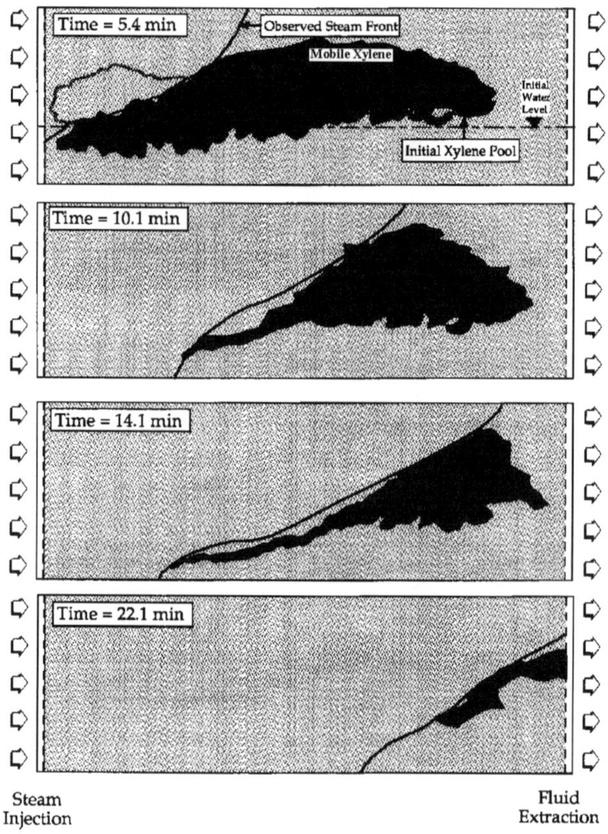

Figure 16.4. Displacement of a pool of xylene floating on the water table (from Basel, 1991). As the steam front reaches the edge of the pool of xylene (5.4 min), the xylene begins to vaporize and be displaced by the steam. After 10 minutes of injection, the pool of xylene begins to mound on the bank of water being displaced by the steam. After 18 minutes, the pool of xylene has been displaced to the exit port and is being recovered as free product. Thirty minutes was required to completely recover the pool.

steam override. If the applied horizontal steam pressure gradient is much less than the hydrostatic pressure gradient, then the steam has been observed to move to the top of the confined sand pack (Basel and Udell, 1989). If the ratio of the horizontal pressure gradient to the hydrostatic pressure gradient is about one, gravitation control of the steam zone is obtained. For this ratio equal to one, the angle of the steam condensation front interface in a homogeneous medium has been predicted and observed to be 45° from horizontal. For higher steam pressure gradients, the steam front becomes more vertical. Layered media have been found to give additional gravitational stability to the steam condensation front.

The removal of second phase liquid contaminants from the two-dimensional sand packs was also studied using this apparatus. Figure 16.4 shows the displacement of *o*-xylene in response to the moving steam front. The conditions of this experiment are scaled to represent a spill at 3.5 m depth, with 10 m spacing between wells. The displacement required about 350 pore volumes of steam to remove the xylene from the sand pack. In this case, the steam appeared to mobilize the xylene as a free product bank ahead of the steam condensation front. Different results were obtained using lower-volatility diesel fuel. A layered sand pack was used in this experiment to examine the effects of larger-scale heterogeneities on the movement of steam and the contaminant liquid. Figure 16.5 shows the initial location of the diesel pool, the movement of the diesel as a mobile separate phase and the region where residual diesel liquid was observed after steam breakthrough. As indicated in these

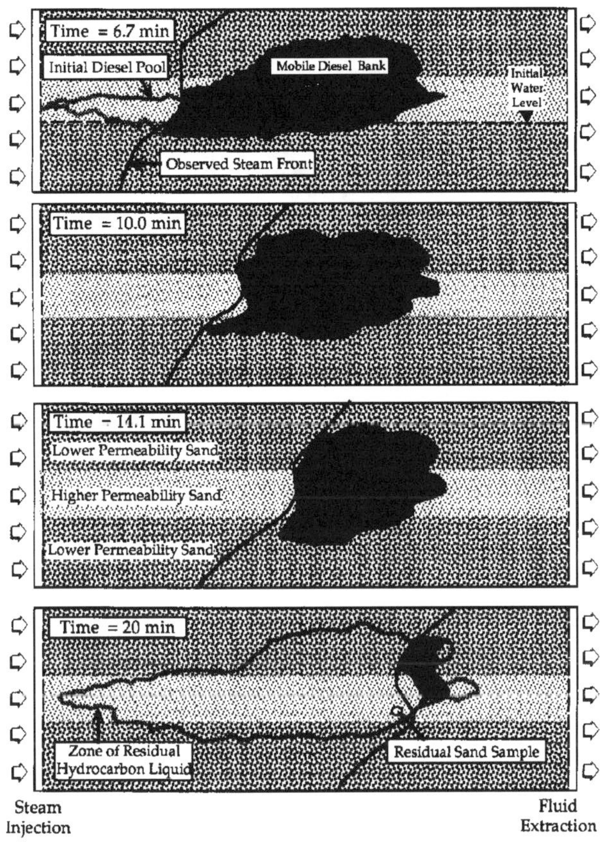

Figure 16.5. Displacement of a pool of diesel floating on the water table in a heterogeneous sand pack (from Basel, 1991). While there was some initial steam displacement of the free product diesel, the volume of the mobile pool decreased with time and residual hydrocarbon liquids were observed in all regions that had contained liquid diesel. Due to the low volatility of diesel components, the appearance of a co-condensed hydrocarbon bank at, and ahead of, the steam condensation front would not be expected. Diesel compounds were recovered predominantly in the vapor phase by steam distillation.

figures, there was not significant physical displacement of the diesel by the steam condensation front. While there was some initial steam displacement of the free product diesel, the volume of the mobile pool decreased with time and residual hydrocarbon liquids were observed in all regions that had contained liquid diesel. These results are in agreement with the theory of Yuan and Udell (1993). The higher volatility of the xylene would have allowed a velocity of the xylene vaporization wave to be greater than the steam condensation front velocity, providing the appearance of complete displacement. The evaporation wave velocities of components such as those found in diesel fuel are slower than the steam condensation front velocity, thus their removal by steam injection is then dominated by steam distillation in the steam zone. In such a situation, one would expect multiple evaporation wave velocities—one for each component (Yuan, 1991). This is borne out in the gas chromatographic analysis of effluent liquids and the residual diesel left in the sand taken from the location indicated in Figure 16.5. Figures 16.6a–16.6c show the chromatograms of two effluent samples and a residual sand sample. It is clear from examination of these chromatograms that the evaporation wave velocities of the more volatile compounds (shorter retention times) were fast enough to be removed in the time frame of the experiment. The less volatile (longer retention times) components remain until the evaporation wave of that compound passes each location.

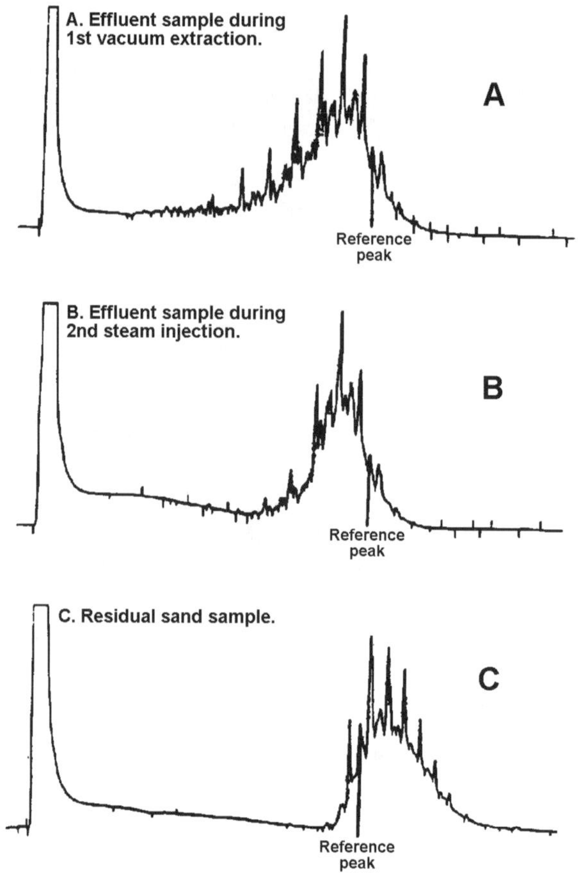

Figure 16.6. Gas chromatograms of diesel removed in the vapor phase (A and B) and of the residual remaining after steam contact (C). Evidence of preferential steam distillation of the more volatile diesel components is shown (from Basel, 1991).

Also of interest in these series of studies was the observation of the enhancement of the mass transfer of the contaminant from the low-permeability regions during the depressurization mode of operation. In this mode, the porous media becomes an *in situ* steam generator, producing a convective flow of steam from the low-permeability regions to the higher-permeability regions, thus enhancing the mass transfer rates. Figure 16.7 shows the effluent concentrations of a four-component pseudo-gasoline mixture of pentane, hexane, toluene, and decane during both steam injection and vacuum depressurization modes of operation. While high recovery rates were observed just before steam breakthrough, indicating displacement similar to that of pure xylene, the removal rates of the hydrocarbon mixture during the vacuum mode were of comparable magnitude.

Field Studies

Two field studies have been operated to examine the effectiveness of steam injection to recover contaminants from spill sites. The first site demonstration at Solvent Service, Inc. was a pilot study of small dimensions with multiple contaminants found in the vadose zone. The second field demonstration, Lawrence Livermore National Laboratories, was a gasoline spill with second phase liquid contaminants found both above and below the water table. The gasoline was found to extend from about 17 m to 45 m from the surface. The general process schematic for the process used at both sites is shown in Figure 16.8.

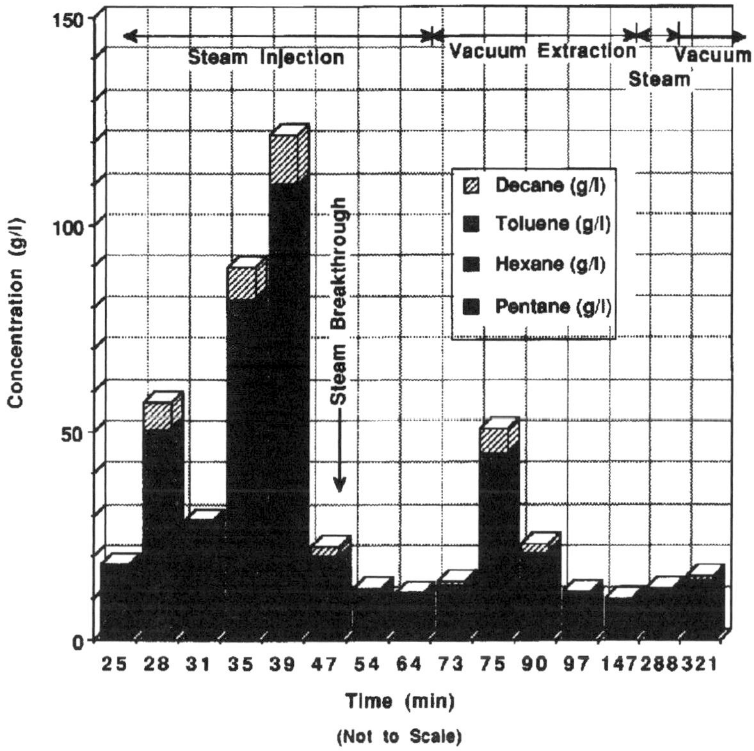

Figure 16.7. Effluent concentrations of a hydrocarbon mixture of pentane, hexane, toluene and decane from a two-dimensional, heterogeneous sand pack during steam injection and vacuum depressurization modes of operation. The highest concentrations were observed just before steam breakthrough in each of the high-permeability zones, and during the period of depressurization immediately after cessation of steam injection.

Solvent Service, Inc., San Jose, California

- The site chosen for the study had been used as a solvent recycling and acid treatment facility since 1973. Soil contamination at the site resulted from leaking underground storage tanks and surface spills associated with the handling of these industrial wastes. Twenty organic volatile compounds and six nonvolatile compounds have been identified at the site. Fourteen of the organic compounds have been detected in concentrations exceeding 1.5 ppm. The principal contaminants of the soil are xylenes, ethylbenzene, 1,2-dichlorobenzene, 1,1,1-trichloroethane and acetone which, along with trichloroethene and tetrachloroethene, occur with maximum concentrations greater than 500 ppm.

A clay barrier was located approximately 20 ft below the surface which inhibited the further vertical migration of the contaminants. Perched water was found in a 1-ft zone on top of the clay layer. The soil was unsaturated above the perched water. In characterizing the extent of soil contamination, the complications associated with mixtures were apparent. High concentrations of acetone were found throughout the site, which is not surprising given its high volatility and complete miscibility in water. Most of the other contaminants at this site were miscible or highly soluble in acetone. Thus, the acetone provided a mechanism for contaminants with low water solubilities and mobilities to migrate with the water.

The location of the pilot study was in the region of highest contaminant concentrations. Total contaminant concentrations were 10,000 ppm in some locations. The pattern of injection and recovery wells used for the pilot study was selected as a prototype for a design to remediate a spill site

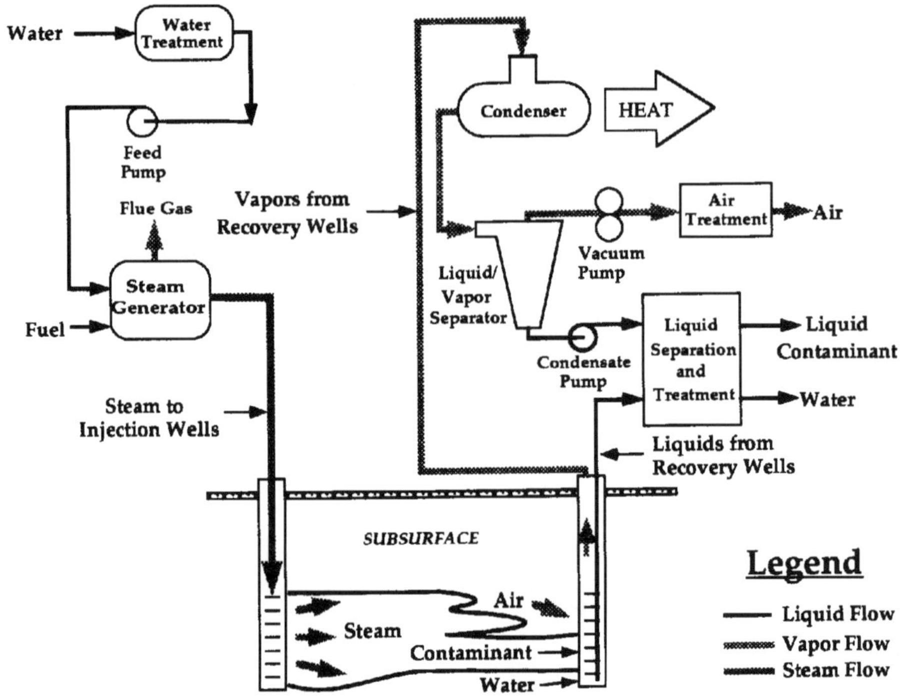

Figure 16.8. Process schematic of steam-enhanced extraction.

similar to the LLNL gas pad. The system consisted of a central recovery well surrounded by six injection wells evenly spaced on a 3.1 m (10 ft) diameter circle. The injection wells were drilled to 6 m (19 ft), but completed over only the bottom 3 ft. The recovery well was completed throughout and extended into the clay layer. Liquids were removed from the recovery well via a jack pump and vapors were drawn out through application of sub-atmospheric pressure provided by a vacuum pump placed on an extraction line downstream of a condenser. Also, 12 temperature observation wells were placed inside the pattern, 3 on the edge, and 2 outside the pattern.

After 140 hours of steam injection and vacuum extraction, nearly 400 kg of contaminant was recovered, accounting for over 90% of the original contaminant mass. Post steam injection soil sampling showed low levels (< ppm total) of contaminants in the high-permeability zones which had ample steam contact. However, high levels (near initial values) of some contaminants (particularly high water-solubility compounds) were found within the lower-permeability zones. Given the short process operation time, such results were expected. It should also be mentioned that the recovery rates at the end of the 140 hours remained high. Further details of this project can be found elsewhere (Udell and Stewart, 1989).

Injection cycling at the end of the pilot did produce contaminant vapors in higher concentrations in the gas stream than those measured during steady steam injection. The effectiveness of this phase of the steam injection/vacuum extraction process for recovering contaminants from the low-permeability regions could not be adequately tested at this site without recontaminating the steamed region by drawing in contaminants from outside the test pattern.

Lawrence Livermore National Laboratories, Livermore, California

In a more recent study, steam was injected both above and below the water table at Lawrence Livermore National Laboratory in an effort to clean up gasoline-contaminated soil within water-bearing zones. This project is referred to as Dynamic Underground Stripping. The site plan and hydrogeologic section are shown in Figures 16.9 and 16.10, respectively. The subsurface geology

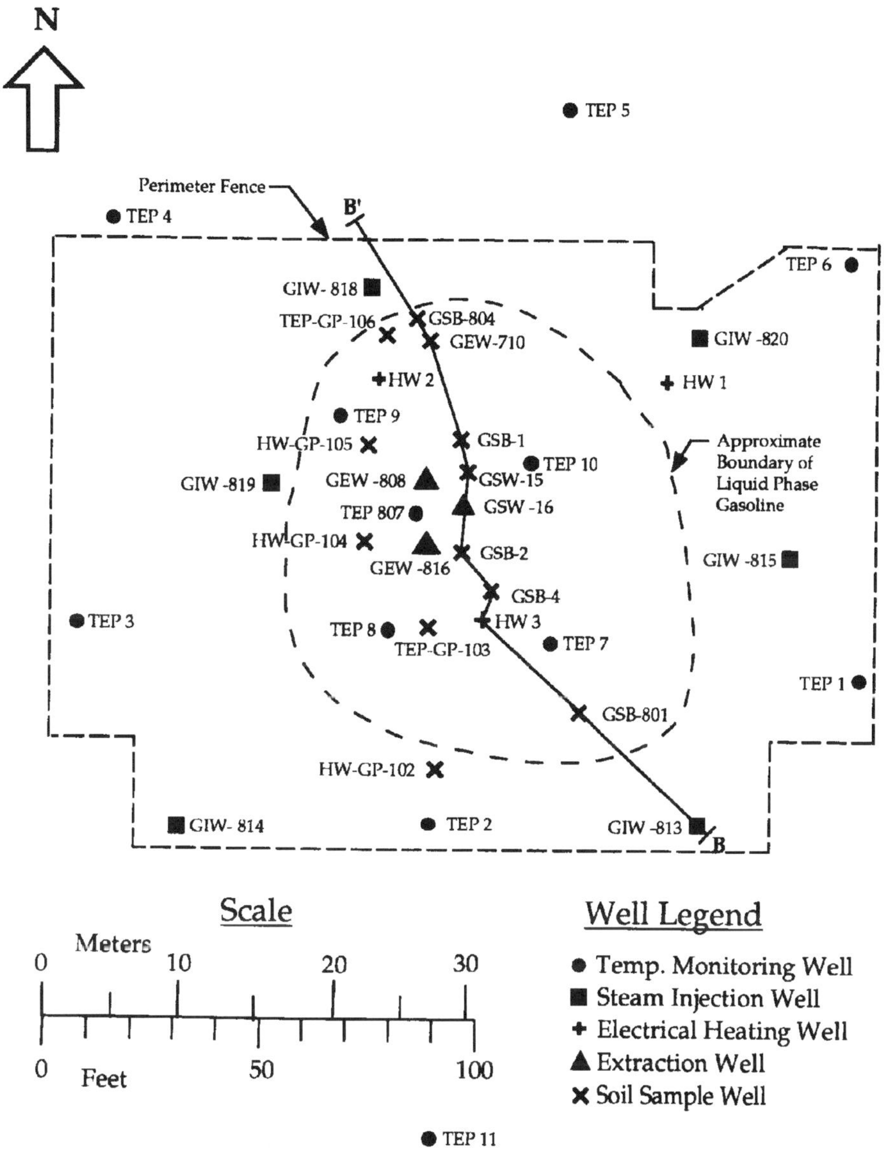

Figure 16.9. Site plan of the LLNL demonstration project.

was of alluvial origin, with layers of high permeability interspersed with clays and clayey silts. The horizontal correlation of the high-permeability layers was moderate except for a thick gravel layer found below the water table from about 35 m (140 ft) to 42 m (150 ft). This gravel layer was found in each well surveyed. The gasoline was detected as a second phase in an area about 50 m (150 ft) in radius as shown in Figure 16.9. The vertical distribution ranged from 17 m (50 ft) to about 45 m (137 ft) in depth as shown in Figure 16.10, with significant spreading due to major groundwater elevation fluctuations. Indeed, separate phase gasoline was found in the deep water-bearing zone nearly 8 m below the water table as shown. This situation rendered other technologies such as ambient temperature vacuum extraction to be impracticable, and groundwater extraction to be time-prohibitive. There were two major flow zones at this site: the deep gravel layer and an upper unsaturated zone consisting of intermingled sands, silts, clays and gravels. A contaminated clay layer of variable thickness separated the two permeable zones.

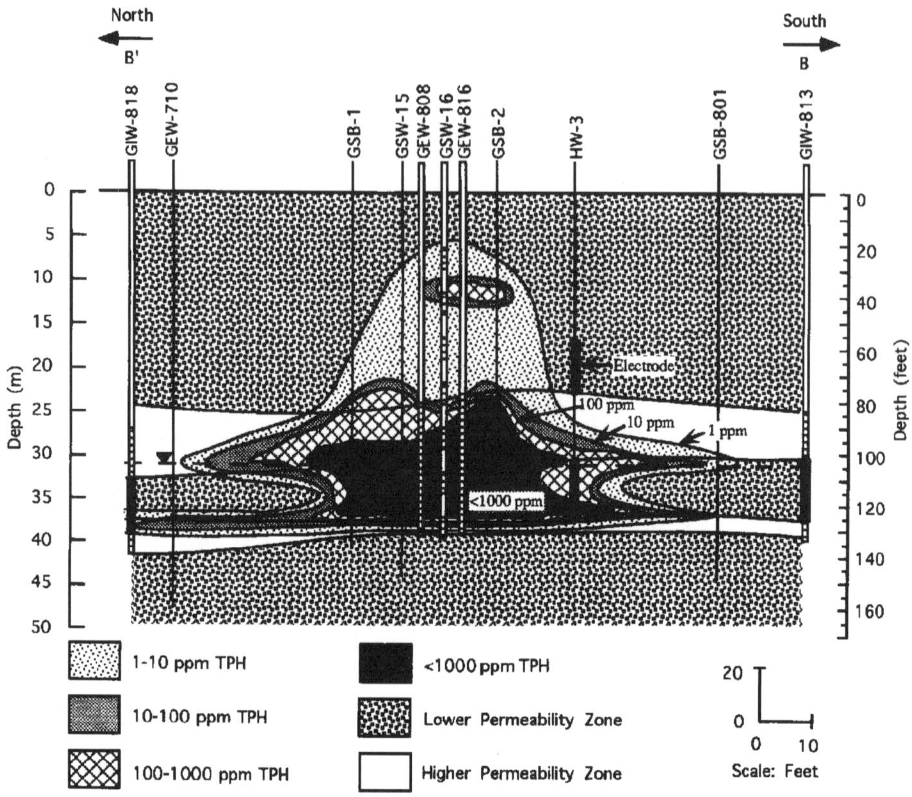

Figure 16.10. Simplified geologic cross-section showing approximate initial distribution of gasoline, the upper and lower steam zones, the injection and extraction well screened intervals, and the water table elevation.

Six injection wells were placed around the perimeter of the area contaminated by separate phase gasoline; three recovery wells were installed near the center of the plume as shown in the site plan of Figure 16.9. The injection wells were constructed to allow separate steam injection into the upper and/or lower permeable regions. The recovery wells were completed over the entire height of the contaminated zone. Also shown in Figure 16.9 are the placement of the 11 temperature observation wells.

During the first steam pass, steam was injected first into the lower gravel layer below the water table for two weeks. Steam broke through at the recovery wells about 10 days after injection. After the first two weeks, steam was injected into the upper zone, while continuing steam injection into the lower zone at a lower rate. This lower rate did not provide sufficient energy into the gravel layer below the water table to maintain the steam zone. Thus the steam flow rate into the recovery wells dropped to zero until steam broke through to the recovery wells in the upper zone on the 23rd day from the initiation of steam injection. From day 21 to day 30, steam was injected only into the upper zone. From day 30 to day 37, steam was injected only into the lower zone. Steam broke through again into the lower zone on day 32. Steam injection and fluid extraction ceased at the end of day 37 for a period of 3 months while higher-capacity treatment equipment was installed. Steam injection resumed for a second pass on 6/2/93 after 9 days of groundwater extraction and vacuum pumping. The steam injection schedule was periodic, with periods of steam pressurization and depressurization. Steam was injected a total of 21 days during this 46 days of gas and groundwater extraction. Vacuum extraction and groundwater pumping resumed in October and continued for a period of 70 days. Electrical heating occurred during November for a period of 10 days. No steam was injected during this final 70-day phase.

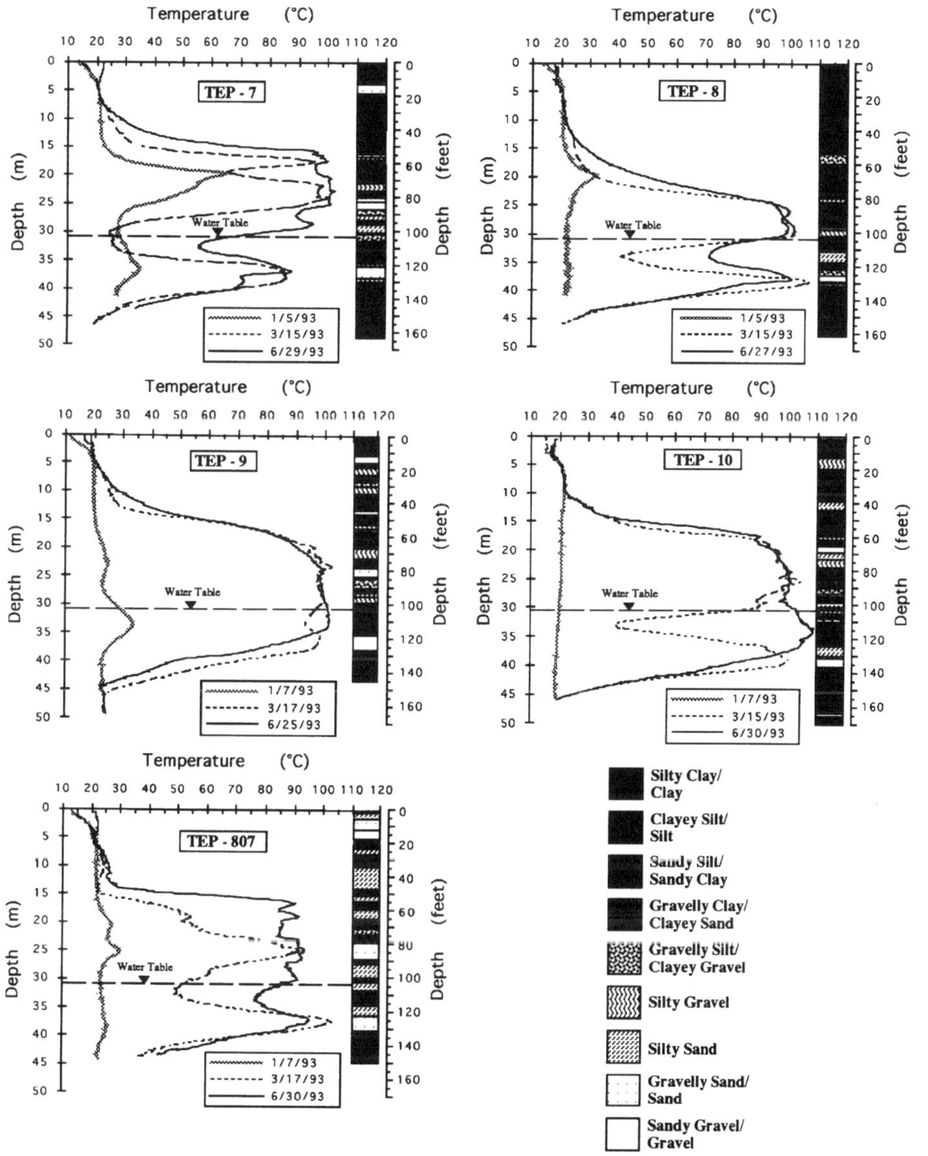

Figure 16.11. Temperature profiles in wells inside of steam injection ring. The dotted lines represent the temperature profiles near the end of the first pass while the solid lines represent the temperature profiles near the end of the second pass. The subsurface lithologies are superimposed to show correlations between the subsurface geology and the thermal field during steam injection.

Temperature logs in each of the 11 temperature monitoring wells were obtained daily during the first two phases of this demonstration. Sample temperature profiles inside the ring of injection wells at the end of the first and second pass of steam injection are shown in Figure 16.11. For reference, the lithology log is superimposed to show the correlation between the geologic features and the development of the temperature field. Electrical heating prior to steam injection raised temperatures by up to 50°C, and seemed to preferentially heat those zones that were not readily heated by steam (TEP-7). While the steam initially entered the highest-permeability zones during the first pass, conductive heating brought the smaller lower-permeability zones to steam temperatures by the end of the second pass. Uniform temperatures in the target zones were observed in TEP-9 and TEP-10 by the end of the

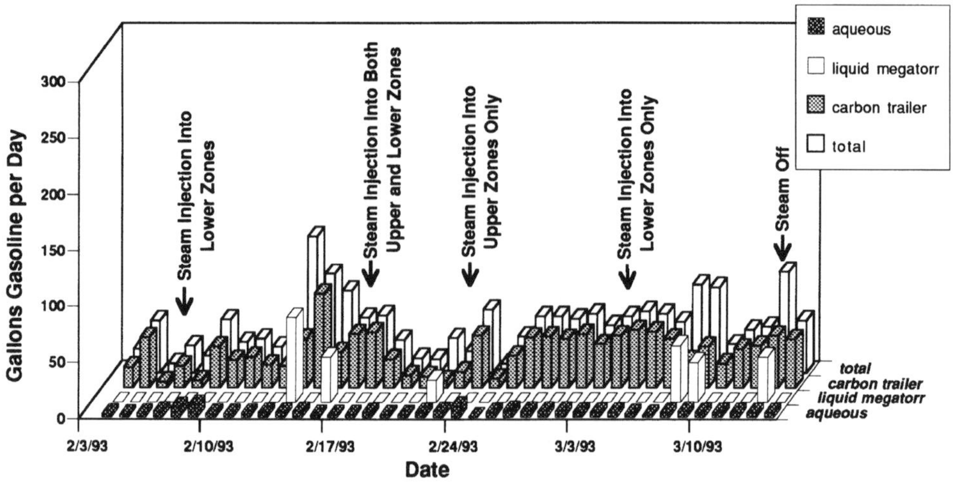

Figure 16.12. Recovery rates of gasoline from the extraction wells during the first pass.

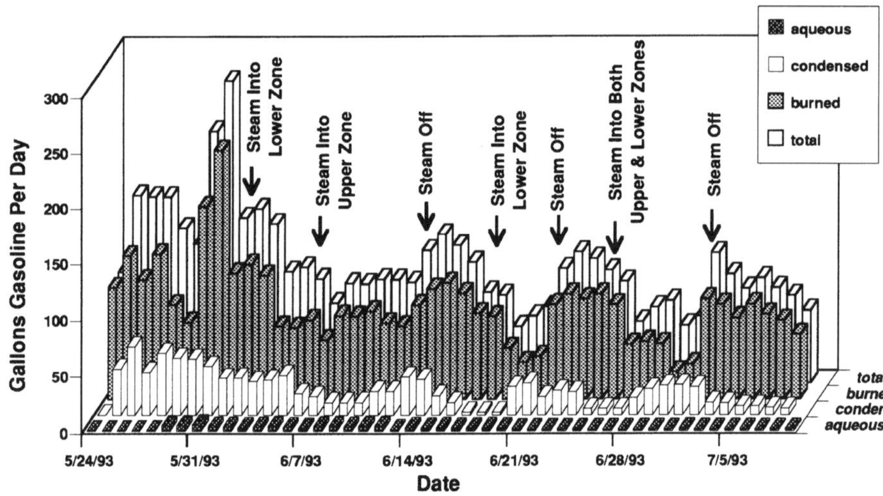

Figure 16.13. Recovery rates of gasoline during the second pass.

second pass. However, the lower-permeability zone separating the two steam zones near TEP-7 and TEP-8 had not yet reached steam temperature by the end of the second pass. An interesting detail seen in each of the plots is the effect of groundwater pumping on the temperature profiles in the lower steam zone. Since the energy injection rate into the lower zone was less in the second pass than in the first pass, and the pumping rates were somewhat higher, convective cooling of the lower steam zone occurred due to the infiltration of cold water from unheated zones.

The recovery rates of gasoline are shown as a function of time in Figures 16.12, 16.13 and 16.14. The process was operated continuously during three different intervals, with system upgrades and additional field work performed during the down-time. During the 1st pass (Figure 16.12), increases in recovery rates were observed in solution, in the gas phase, and as a separate phase liquid within the first few days of steam injection, with the highest rate occurring on the day of steam breakthrough into the lower interval of the extraction wells (2/15/93 and 3/3/93). Otherwise, gasoline recovery rates were limited by the use of the regenerated carbon gas treatment system with a capacity of less than 50 gallons of gasoline per day. A total of approximately 1700 gallons of gasoline was recovered during the first phase. However, direct measurements of the volumes of the liquid gasoline removed

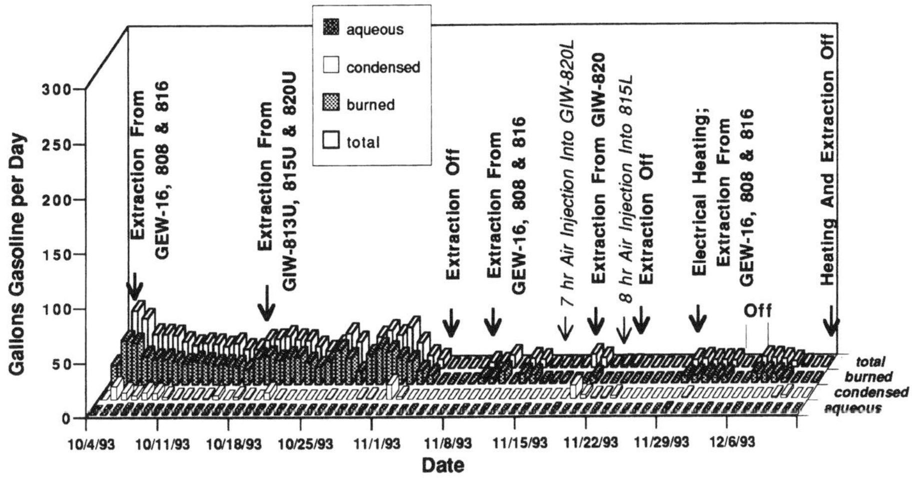

Figure 16.14. Recovery rates of gasoline during the final operation of pumping system.

from the condenser could not be made. Therefore those amounts are not shown in Figure 16.12, except as a portion of the liquid imprecisely measured in the liquid megatorr which included the pumped free product. While a greater amount of displaced free product was expected before steam breakthrough in either the upper or lower zones, the recoveries were restricted by the limitations on liquid pumping rates imposed by the liquid treatment system and the intrinsic difficulty of capturing a free product bank with the two liquid pumping wells. Higher gas and liquid handling capacities in the surface equipment could have allowed increased recovery rates. After five weeks of operation, the system was shut down to install improved liquid gasoline measuring systems and higher capacity fluid handling equipment.

The recovery rates observed during the second pass were much higher, as shown in Figure 16.13. The increased rates were due to higher-capacity gas and water treatment systems, as well as the increased subsurface temperatures and cyclic steam enhancement. The recovery rates were highest before steam injection began. Thereafter, periodic peaks in the recovery rates were measured corresponding to times when steam injection ceased and depressurization began. A total of 4900 gallons of gasoline was recovered during the second pass. The 2nd pass was terminated due to permitting restrictions on the boiler while recovery rates remained high and may have continued with additional cycles of steam injection. During the next 80 days, soil samples were collected, the steam boiler was taken from the site, and additional electrical heating wells were installed. The treatment system was shut down during this period.

The soil concentrations found after the second pass are shown in Figure 16.15. In general, gasoline was effectively removed from the hot steam-bearing zones and concentrations were significantly reduced in the low-permeability zone separating the upper and lower permeable units. An estimate of 750 gallons of gasoline remaining in the soil after the end of the second pass was made from the soil concentrations shown in Figure 16.15.

The recovery rates during the final phase are shown in Figure 16.14. While significant rates of recovery were measured during the beginning of pumping, the rates fell with time, and the magnitudes were much less than those observed during the second pass. Some improvements in the recovery rates were achieved by converting upper zone injection wells into vapor recovery wells. However, little could be done to increase recovery rates during the final few days of operation—including electrical heating. Regardless, an additional 1000 gallons of gasoline were recovered in this final phase before the recovery rates dropped to small values. The cumulative total of gasoline removed was 7600 gallons at the end of operation.

Subsequent gas and water sampling is showing decreasing concentrations of gasoline components with time, implying that there is no separate phase gasoline remaining at this site. Ethylbenzene,

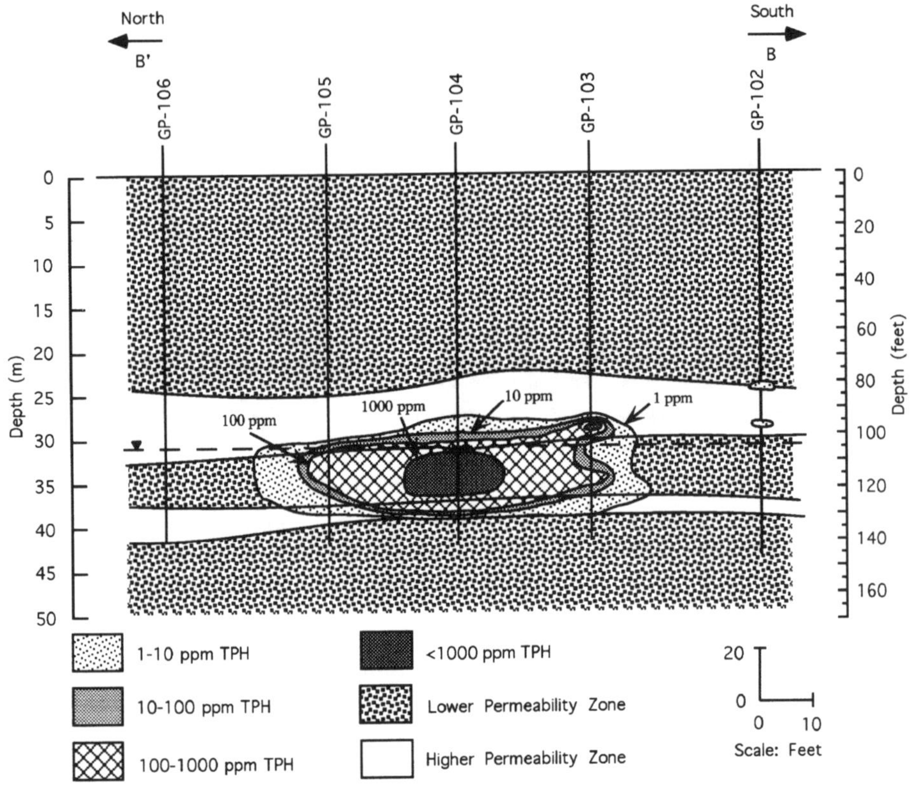

Figure 16.15. Simplified geologic cross-section showing the approximate distribution of gasoline after the second pass.

toluene, and xylene concentrations in the water have dropped to below MCL. Benzene concentrations continue to decrease. Hydrocarbon-degrading biological activity was found in the zones subjected to steam temperatures, indicating that the application of steam did not leave the site sterile. The culture makeup, however, has been dramatically altered. It is expected that no further treatment will be required. Details of all aspects of the Livermore field demonstration can be found in reports to follow (Newmark, 1994).

CONCLUSION

The results of several years of experimental, theoretical, and field research have shown the applicability of steam injection and fluid extraction to the *in situ* cleanup of second liquid phase contaminants found both above and below the water table. The primary mechanisms responsible for the effectiveness of steam injection and fluid pumping are: the acceleration of the vaporization of volatile and semivolatile compounds, the displacement of the pore fluids by the steam condensation front, desorption of contaminant adsorbed to the solids, and the enhanced vaporization of contaminants in the low-permeability zones during the cyclic steam injection operation. With proper design and operation, this method of soil and groundwater remediation is found in practice to be rapid, robust, and relatively predictable.

ACKNOWLEDGMENTS

Support for this work was provided to the University of California by the U.S. Environmental Protection Agency, Grant No. R-815247-01-0, and the National Institute of Environmental Health

Sciences, Grant No. 3 P42ES04705-0251. Support for the Lawrence Livermore National Laboratory field work was provided by the Department of Energy, Office of Technology Development, EM 50, under contract W-7405-Eng-48. The major contributions of the following LLNL scientists and engineers in the LLNL field demonstration are acknowledged: Roger Aines, Robin Newmark, William Siegel, William Daily, Abe Ramirez, Ed Folsom, Michael Buettner, and John Ziagos. The contributions of UCB students Ron Goldman, David Kayes, and Kent Kenneally in developing and operating the temperature logging system at LLNL are also acknowledged.

REFERENCES

Basel, M.D. 1991. "Two-Dimensional Propagation of Steam Through Partially Saturated Porous Media," Ph.D. Dissertation, University of California at Berkeley.

Basel, M.D. and K.S. Udell. 1989. "Two-Dimensional Study of Steam Injection into Porous Media," *Multiphase Transport in Porous Media*, ASME HTD-127.

Basel, M.D. and K.S. Udell. 1991. "Effect of Heterogeneities on the Shape of Condensation Fronts in Porous Media," *Heat Transfer in Geophysical Media*, ASME HTD-172, pp. 63-70.

Dev, H. 1986. "Radio Frequency Enhanced *In situ* Decontamination of Soils Contaminated with Halogenated Hydrocarbons," *Proc. of EPA Conf. on Land Disposal, Remedial Action, Incineration and Treatment of Hazardous Waste*, Cincinnati, Ohio, April 1986, published by EPA Haz. Waste Engin. Res. Lab., Cincinnati, Ohio.

Fitzpatrick, V.F., C.L. Timmerman, and J.L. Buelt. 1986. "*In situ* Vitrification—A Candidate Process for *In situ* Destruction of Hazardous Waste," *Proceedings, Seventh Superfund Conference*, Washington, DC, 1986, published by HMCRI.

Hilberts, B. 1985. "*In situ* Steam Stripping," Assink, J.W. and W.J. Van Den Brink, Eds., *Contaminated Soil*, Proc. of First Intern. TNO Conf. on Contaminated Soil, Utrecht, The Netherlands, November 11–15, 1985, pp. 680–687.

Hunt, J.R., N. Sitar, and K.S. Udell. 1988. Non-Aqueous Phase Liquid Transport and Cleanup, Part I. Analysis of Mechanisms, Part II. Experimental Studies. *Water Resources Research*, **24**:1247–1269.

Konopnicki, D.T., E.F. Traverse, A. Brown, and A.D. Deibert. 1979. Design and Evaluation of the Shiells Canyon Field Steam-Distillation Drive Pilot Project. *J. Pet. Tech.*, **31**:546–552.

Lord, A.E., Jr., R.M. Koerner, V.P. Murphy, and J.E. Brugger. 1987. "Vacuum-Assisted, *In situ* Steam Stripping to Remove Pollutants from Contaminated Soil," *Proc. of EPA Conf. on Land Disposal, Remedial Action, Incineration and Treatment of Hazardous Waste*, Cincinnati, Ohio, May 6–8, 1987, p. 511.

Mandl, G. and C.W. Volek. 1969. Heat and Mass Transport in Steam-Drive Processes. *Soc. Pet. Eng. J.*, **9**:59–79.

Menegus, D.K. and K.S. Udell. 1985. "A Study of Steam Injection into Water Saturated Capillary Porous Media," *Heat Transfer in Porous Media and Particulate Flows*, ASME HTD, 46.

Newmark, R.L. (Ed.). 1994. "Dynamic Underground Stripping Project: LLNL Gasoline Spill Demonstration Report," LLNL UCRL-ID-116964.

Saffman, P.G. and G.I. Taylor. 1958. The Penetration of a Fluid into a Porous Medium or Hele-Shaw Cell Containing a More Viscous Liquid. *Proc. Roy. Soc.* (London), Series A, **245**:312–329.

Scheflan, L. and M.B. Jacobs. 1953. *The Handbook of Solvents*, D. Van Nostrand Co., New York.

Stewart, L.D., Jr. and K.S. Udell. 1988. Mechanisms of Residual Oil Displacement by Steam Injection. *SPE Reservoir Engineering*, **3**:1233–1242.

Stewart, L.D., Jr. and K.S. Udell. 1989. "The Effects of Gravity and Multiphase Flow on the Stability of Condensation Fronts in Porous Media," *Multiphase Transport in Porous Media,* ASME HTD-127.

Udell, K.S. 1983. Heat Transfer in Porous Media Heated from Above with Evaporation, Condensation and Capillary Effects. *J. of Heat Transfer*, **105**:485–492.

Udell, K.S. and L.D. Stewart, Jr. 1989. "Field Study of *In situ* Steam Injection and Vacuum Extraction for Recovery of Volatile Organic Solvents," Sanitary Engineering and Environmental Health Research Laboratory, University of California, Berkeley, UCB-SEEHRL Report No. 89-2.

Udell, K.S. and L.D. Stewart, Jr. 1992. "Combined Steam Injection and Vacuum Extraction for Aquifer Cleanup," *Subsurface Contamination by Immiscible Fluids*, Weyer (ed.) A.A. Balkema, Rotterdam, Netherlands, pp. 327–336.

Udell, K.S., J.R. Hunt, N. Sitar, and L.D. Stewart, Jr. 1991. "Process for *In situ* Decontamination of Subsurface Soil and Ground Water," U.S. Patent #5,018,576.

Van Sickle, W.R. 1992. "The Steam Enhanced Recovery Process," Abstract Proceedings: Fourth Forum on Innovative Hazardous Waste Treatment Technologies: Domestic and International, EPA/540/R-92/081.

Vaughan, P.J., K.S. Udell, and M.J. Wilt. 1993. The Effects of Steam Injection on the Electrical Conductivity of an Unconsolidated Sand Saturated With a Salt Solution. *J. Geophysical Res.*, **98**:509–518.

Volek, C.W. and J.A. Pryor. 1972. Steam Distillation Drive—Brea Field, California. *J. Pet. Tech.*, 24:899–906.

Yuan, Z.-.G. 1991. "Steam Distillation of Multicomponent Hydrocarbon Liquids in Porous Media," Ph.D. Dissertation, University of California at Berkeley.

Yuan, Z.-G. and K.S. Udell. 1993. Steam Distillation of Single Component Hydrocarbon Liquids in Porous Media. *International Journal of Heat and Mass Transfer*, **36**:887–897.

Chapter 17

Removal of Aqueous Phase Dissolved Contamination: Non-Chemically Enhanced Pump-and-Treat

J.L. Wilson

ABSTRACT

Pump-and-treat systems have been used over the last decade in an attempt to remediate contaminated aquifers by flushing out aqueous phase dissolved contamination. They have failed to restore aquifers to drinking water standards. The failure can be attributed, at least in part, to the presence of aquifer heterogeneities and non-equilibrium chemistry, both associated with diffusion control of processes. In pumping wells, concentrations level off or show long tails, and when the pumping is stopped concentrations rise again. The tailing can be explained by non-uniform flow caused by aquifer heterogeneities and the geometry of pumping and injection facilities. Part of the tailing, and essentially all of the recovery of concentrations back toward their pre-remediation value, is caused by the slow bleeding of contaminant from aquifer heterogeneities under diffusion control, and from slow chemical desorption. At the field scale the important time constants are months, years and possibly decades long, making relevant mass transfer rates difficult to observe in the laboratory, or even in short-duration field-scale characterization studies.

INTRODUCTION

Pump and treat systems are the most popular aquifer remediation technology. When these systems were first employed for the remediation of contaminated aquifers there was little pretense that aquifers would be restored to pristine conditions. Consider one of the early remediation conferences, held just ten years ago (National Water Well Association, 1984). A series of papers were assembled under the heading of pump-and-treat. But read their titles. They contain these terms: "hydraulic containment" (Wilson, 1984), "groundwater extraction," and "purge well effectiveness" (Whiffen and Bahr, 1984).

Three related papers also talk about "containment" and "hydraulic barriers." The emphasis in these papers is first on control of groundwater contamination, by controlling groundwater flow, and only secondarily on aquifer remediation. This was also the effective emphasis in the early days of the federal Superfund Program which unrealistically sought short-term solutions to contamination. Since short-term remediation was out of the question, the pump-and-treat systems of the time effectively emphasized containment and/or well-head treatment (Doty and Travis, 1991). That ended with the amendments of 1986, which called for permanent remediation, meeting drinking water standards. "Pump-and-treat" became the remediation method of choice (Doty and Travis, 1991; Haley et al., 1991). Few people paid attention to its inherent limitations as a remediation method. Today most, if not all installed pump-and-treat systems are failing in this mission, although they generally continue to do a nice job controlling the further migration of contamination (EPA, 1989; Mackay and Cherry, 1989; Doty and Travis, 1991; Haley et al., 1991; National Research Council, 1994). "We were unable to locate an aquifer in the U.S. that had been confirmed to be successfully restored through pumping" (Doty and Travis, 1991).

Existing pump-and-treat systems share two important characteristics. It is observed at extraction wells that initially high dissolved contamination concentration levels fall off with time and eventually level out (Mackay and Cherry, 1989; Mercer, et al., 1990; Doty and Travis, 1991; Haley et al., 1991).

We'll refer to this asymptotic behavior of the system as "tailing." Unfortunately, these concentrations often level out at levels far above drinking water standards (Doty and Travis, 1991). Even if standards are met, when the pumps are turned off concentration levels rise again (Whiffin and Bahr, 1984; Mackay and Cherry, 1989; Mercer, et al., 1990; Doty and Travis, 1991; Haley et al., 1991). We'll refer to this as the "recovery" of contaminant concentration. It may take months or years, but the concentrations may even recover to levels approximating those before remediation.

There are a variety of contributions to the asymptotic tailing and recovery behavior of pump-and-treat systems that lead to their failure as a method of aquifer remediation (EPA, 1989; Mackay and Cherry, 1989; Mercer et al., 1990; Doty and Travis, 1991; Haley et al., 1991; National Research Council, 1994). One of the most prominent is the possible presence of a nonaqueous contaminant phase, trapped by capillary forces within the pore space (Mackay and Cherry, 1989; Wilson et al., 1990; Conrad et al., 1992; Environmental Protection Agency, 1991). Slowly dissolving, it can act as a source of dissolved organic pollutants for years, if not decades. In a pump-and-treat application this leads to dissolved contaminant concentrations with the very same asymptotic and recovery behavior observed in practice (Mackay and Cherry, 1989, Mercer et al., 1990). It's considered by some as the primary explanation for the failure of pump-and-treat: "In my view, most groundwater problems are DNAPL (dense nonaqueous phase liquid) problems, and most DNAPL problems are chlorinated solvents" (J. Cherry, personal communication, 1992). However, there appears to be evidence for failure, even in the absence of a nonaqueous phase (Whiffin and Bahr, 1984; Mercer, et al., 1990; Doty and Travis, 1991; Haley et al., 1991; Mackay and Cherry, 1989). In this chapter we investigate three other reasons for this failure.

Our focus is on the *in situ* hydraulics of the aqueous phase, and the removal of its dissolved contamination. We leave to other chapters in this volume the more complex problems associated with nonaqueous phase contamination, chemically enhanced *in situ* treatment, and surface treatment operations. Of course, the issues that affect the performance of pump-and-treat are also important to the characterization of the contamination and, once recognized, can be used to improve the design and operation of the groundwater control or remediation system. We also briefly address these topics. Although the discussion in this chapter is aimed at the saturated zone beneath the water table, it is also relevant to the vadose zone, since essentially the same set of issues occur with vapor extraction schemes.

FACTORS INFLUENCING FAILURE

Besides the possible presence of nonaqueous phase liquids there are three major factors that lead to failure of pump-and-treat systems and the observed features of contaminant concentration tailing and recovery. These are diffusion, heterogeneity and chemistry. Even the failures due to heterogeneity and chemistry are probably controlled by diffusion.

Aquifer Heterogeneity

Aquifer heterogeneities are the result of geological depositional, diagenetic and structural processes. These lead to heterogeneities of hydraulic conductivity, porosity and capillarity. In the saturated zone, hydraulic conductivity variations are the most important for dissolved contamination. Pump-and-treat extraction wells and injection facilities preferentially communicate along higher-conductivity paths, while contamination in low-conductivity zones is bypassed, leading to the characteristic tailing associated with pump-and-treat. The bypassed contamination then slowly "leaks" out, providing a source for long-term contamination, adding to the tailing behavior and leading to concentration recovery.

Let's start with a simple example that we have simulated using a groundwater flow and transport computer code (Sudicky, 1989). Figure 17.1 is a cross-sectional diagram of a portion of an idealized fluvial aquifer. Saturated with water, the sandy aquifer is bounded above and below by silt aquitards and contains several imbedded silt lenses of lower hydraulic conductivity. The hydraulic conductivity is 50 m/d in the sand and 0.005 m/d in the silt. This four-order-of-magnitude variation is typical

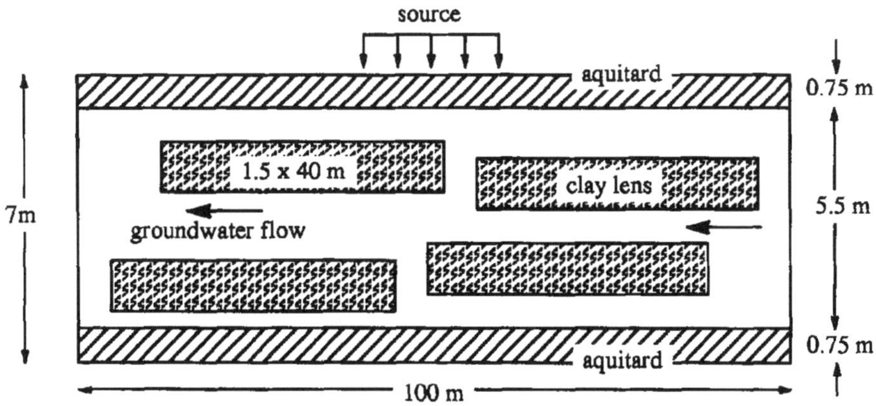

Figure 17.1. Cross-section of idealized aquifer used in the contaminant plume simulations for pump-and-treat.

for fluvial deposits. We've assumed that the following transport properties are the same in both: porosity = 0.356, longitudinal dispersivity = 0.5 m, transverse dispersivity = 0.01 m, and effective molecular diffusion coefficient = $2.0 \times 10^{-4} m^2/d$. The natural groundwater flow is from right to left, under a mean hydraulic gradient of 10^{-4}. In the simulations we used a horizontal grid spacing of 0.5 m and a vertical grid spacing of 0.25 m. Truncation error and mass balance were tested with grid spacings of 0.25 and 0.1 m, with no appreciable difference.

Let's assume that some time ago the aquifer was contaminated from above by an industrial wastewater disposal process that led to the injection of substantial amounts ($0.22 m^2/d$) of contaminated water into the aquifer, dramatically altering the groundwater flow field. Figure 17.2a shows the contamination plume after the first year of operation. If C_o is the concentration of contamination in the wastewater, then the contour lines in this figure describe $0.9C_o$, $0.8C_o$, ...$0.1C_o$, and are shaded from black ($>0.9C_o$) to white ($<0.1C_o$). Disposal continued for 10 years and then ceased, leaving the contamination pictured in Figure 17.2b. During this time enough wastewater was injected to fill up the pictured portion of the aquifer three times over. But not all of the aquifer system was contaminated the same. The lower-conductivity aquitards and lenses were essentially bypassed during the first year (see Figure 17.2a) and after ten years (see Figure 17.2b) their concentration still lagged far below that of the surrounding sand. Their hydraulic conductivity is four orders of magnitude lower than the sand, and except for the portion of upper aquitard through which the wastewater entered, there was essentially no groundwater flow through these units. Transport in the low-conductivity materials was controlled by diffusion, and even after 10 years not enough time had elapsed for diffusion to fill these materials with contamination.

Over the next five years not much happened, although there was some redistribution of the contamination. The modest natural groundwater gradient moved the plume to the left and, more importantly, diffusion continued. At the end of this period concentrations in the low-conductivity units approached those in the sands, as shown in Figure 17.2c. It is at this point that a pump-and-treat system was designed. Fully penetrating pumping and injection wells were placed at the left and right sides, respectively, of the pictured cross-section. Clean water was injected into the aquifer, displacing the plume into the pumping well on the left, under a mean hydraulic gradient of 0.01.

Simulated aquifer concentrations after the first 60 days of remediation are shown in Figure 17.2d. Within 150 days concentrations in the pumping well were no longer changing significantly and the remediation project was halted after 300 days. Concentrations along the left side of the pictured aquifer are graphed in Figure 17.3 (solid line) for the entire period of contamination, redistribution, remediation, and later recovery. During remediation this curve represents the pumping well concentrations, which are also highlighted in the upper right corner of the figure. The classical tailing of pumping well concentrations is evident even in this simple example. The main figure illustrates the original gradual contamination buildup, and its relatively sudden reduction by pump-and-treat. When

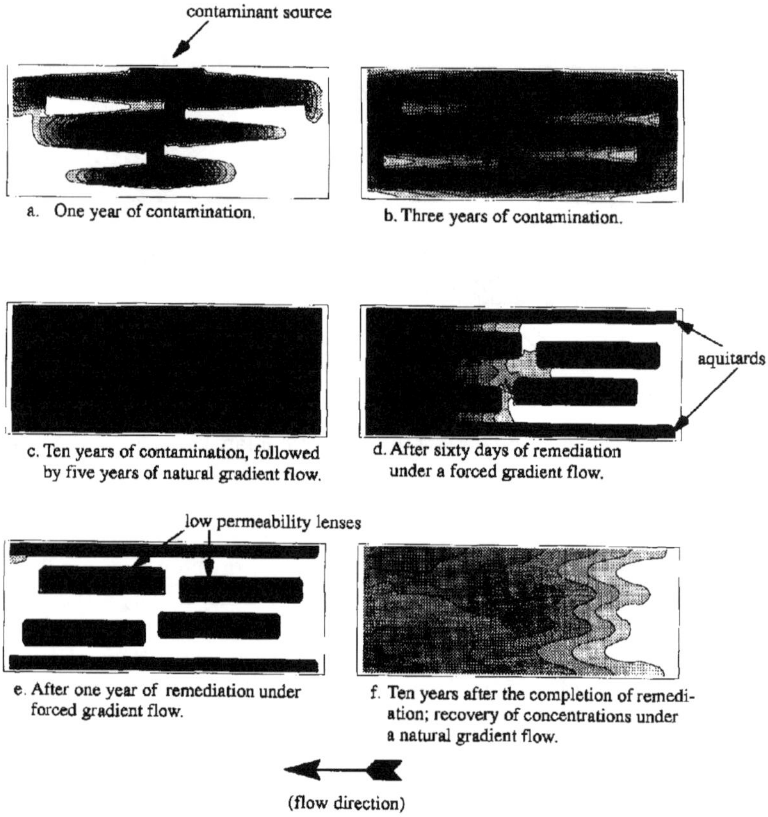

Figure 17.2. Contamination and remediation of a heterogeneous fluvial aquifer.

the remediation system was turned off, some contamination remained in the slowly responding silt aquitards and lenses, as shown in Figure 17.2e. Slowly it diffused out, recontaminating the natural groundwater flow, which is shown in Figure 17.2f ten years later. Concentrations in the aquifer recover, reaching 40% of their pre-remediation values at the downstream boundary. Even this is an optimistic scenario. Figure 17.2f assumes that the upgradient portion of the aquifer was somehow completely remediated, so only "clean" water flows into the study area during this recovery period.

Suppose monitoring wells were installed in the aquifer and could track the progress of these events from the beginning. Concentrations from two such wells are shown in Figure 17.4. Monitoring well 1 samples the sand facies located between two silt lenses. It responds relatively quickly to the original contamination. Monitoring well 2 samples from within the silt lens above. It responds much more slowly, controlled by diffusion from the sand through the silt to the sampling point located at the center of the lens. Even after the source is cut off, concentrations in this well continue to rise slightly. During remediation the sand well (#1) responds quickly, and indicates that the aquifer has been remediated. But the silt well (#2) is still rising, even at the end of remediation. When it does respond, it takes 5 years to drop by half, and it acts as a source of contamination in the sand (#1), where concentrations recover to 40% of their pre-remediation value. Of course we seldom have monitoring wells that sample from the beginning of a contamination event. We also seldom have monitoring wells established in the lower hydraulic conductivity units. All we normally would have are the concentrations from the sand well (#1) starting only a brief time before the remediation period. In this well we'd initially see an apparently successful pump-and-treat process, but if we continue to monitor, it's obvious that it's not completely successful.

What are the important issues in the tailing and recovery of concentrations in this example? One of the most important parameters is the volume of aquifer in which contamination can be "stored" while the pump-and-treat system operates, only to be released later. In the case of heterogeneities, this is the

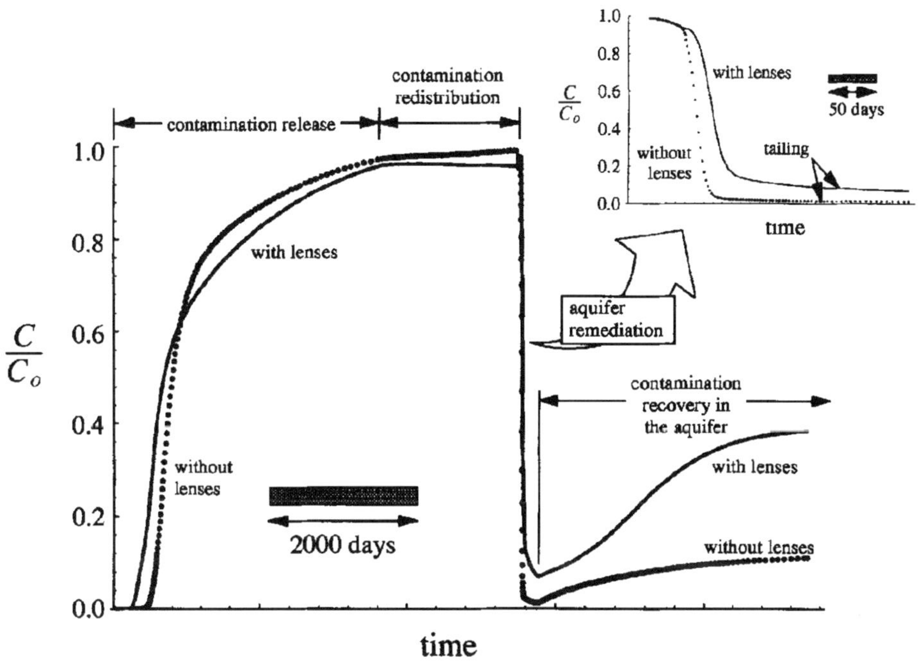

Figure 17.3. Aqueous phase concentrations at the left boundary of the domain pictured in Figure 17.1, for the contamination scenario shown in Figure 17.2. During the remediation this is the concentration in the pumping well. The pumping well concentrations are plotted with an exaggerated time scale in the upper right.

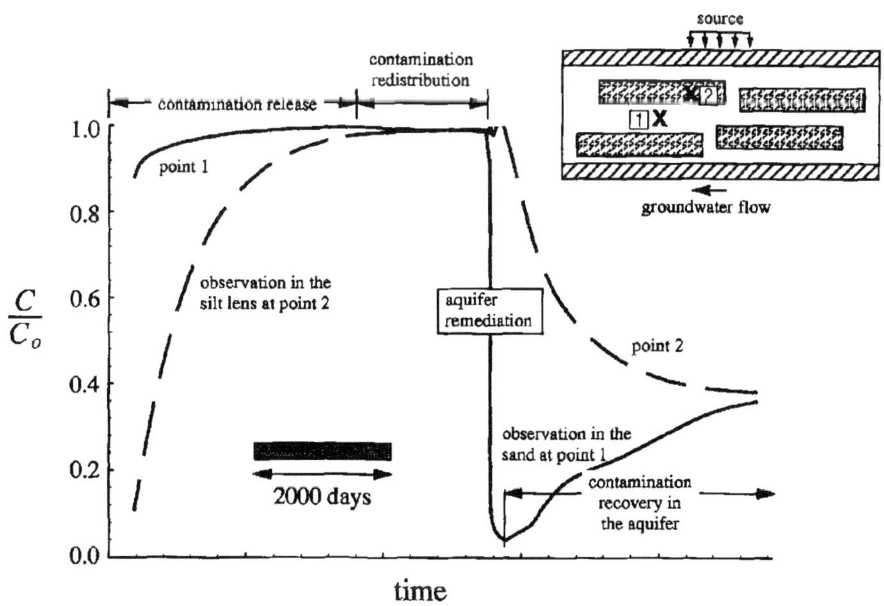

Figure 17.4. Aqueous phase concentration in monitoring wells located in sand (well #1) and silt (well #2) facies.

volume of aquifer in which diffusion controls the local transport. Twenty one percent of the system pore volume is in the aquitards, and 34% is in the silt lenses, both controlled mostly by diffusion. If the source concentration is $C_o = 1$ mg/L, then these silt units can store 138.8 g of solute in their pore space within the domain shown in Figures 17.1 or 17.2. If we could reduce this volume, say by removing the silt lenses and replacing them with sand, then we would expect less tailing and less recovery. This is shown in Figure 17.3 as the dotted line. If we went further and removed the aquitards, replacing them with impervious non-diffusive boundaries, then there would be no storage, a little tailing caused by dispersion, and no recovery.

The second important issue is non-uniform flow caused by the spatial pattern of heterogeneities. Let's return to the heterogeneous aquifer of Figures 17.1 and 17.2, but assume that diffusion did not exist. If there is no diffusion, there is no storage. The plume would enter the aquifer much as pictured in Figure 17.2a, but never move into the aquitards—except near the source area at top—or the lenses. During remediation, concentrations would drop and then tail out modestly, more so than shown by the dotted line in the upper right corner of Figure 17.3. Why the tailing? The low-conductivity silt lenses provide a barrier to flow, and because they are discontinuous, lead to non-uniform flow in the aquifer. In the sand some flowlines are longer than others, and there are low-velocity stagnation points, causing concentrations at the pumping well to tail out. This effect of the heterogeneities is the subject of intense current study, usually under the label "macro-dispersion." In our idealized example, with no diffusion, what happens when the remediation is shut down? Concentrations do not recover, but continue to fall, albeit more slowly. Thus the influence of heterogeneity on non-uniform flow cannot explain the failure of pump-and-treat. Diffusion-controlled processes must be involved.

The third important issue concerns how quickly the diffusively controlled heterogeneities can respond to change. It is these units where the storage takes place. For example, we found that concentrations in the monitoring well located in the silt lens took 5 years to change by a factor of two. This "half-life" corresponds to a classical e-fold time constant of about 3 years (3 years for a falling concentration to drop from its original value to 1/e times that value). We can use dimensional analysis to find this time constant directly, without running the simulation model. Assume that transport in the silt lenses is controlled by diffusion across in its narrowest dimension. Then the time constant is given by

$$\tau_k = \frac{l_z^2}{g_k D_o} \tag{17.1}$$

where l_z is the characteristic vertical dimension of the lens, g_k is a geometric factor, and D_o is the effective molecular diffusion coefficient. If we take the characteristic dimension as half the lens thickness, or 0.75 m, then the geometric factor is around 3 (e.g., Carslaw and Jaegger, 1959). The calculated time constant is roughly 3 years, essentially the same as the simulated value. The aquitards have the same characteristic length and the same time constant. The original contamination source lasted 10 years, which together with the additional 5 years for redistribution gave the aquifer 15 years or 5 time constants to thoroughly contaminate the silt lenses and the aquitards. But the remediation scheme took only 300 days to "clean up" the aquifer, much less than the time constant for the low-conductivity "storage zones." They were effectively untouched. But 9 to 10 years and three time constants later they had significantly recontaminated the aquifer.

The spatial pattern of heterogeneities leads to a wide range of time constants within an aquifer. Let's assume that diffusion control is the issue. Then the pattern of interest is that part of the aquifer in which the Peclet number is roughly one or less, where the Peclet number is defined as

$$Pe = \frac{va}{D_o} \tag{17.2}$$

In this expression a is the dispersivity and v is the seepage velocity. The choice of longitudinal or transverse dispersivity in Equation 17.2 depends on the direction of the seepage velocity and the concentration gradient. Let's assume that transverse dispersion is typically more the issue. Then in our example aquifer diffusion controls wherever the seepage velocity drops below $D_o/a = 2$ cm/d. If longitudinal mixing is the issue, then diffusion controls wherever the velocity drops below 0.4 mm/d. Obviously diffusion control is a more important control on transverse mixing, as in the silts, lenses and aquitards, than it is on longitudinal mixing.

There is another level of aquifer heterogeneity, and it occurs at the pore level. Suppose that the aquifer material has an internal porosity. Aggregated porous media consist of particles or grains that are themselves permeable. Dissolved contaminants can then diffuse from the primary porosity into the intragranular porosity within the grains (Brusseau et al., 1989), and back out again. This takes time, with the result that removing contaminants from the primary porosity does not clean up the aquifer. Contaminants stored inside the grains continue to diffuse out for some period of time. We can estimate a time constant for this process too, and it looks something like

$$\tau_m = \frac{1}{\alpha_m} = \frac{r^2}{g_m T D} \qquad (17.3)$$

where α is the mass transfer coefficient, g_m is a geometric factor, r is a characteristic grain radius, T is the intragranular tortuosity, and D_o is the molecular diffusion coefficient in bulk water. In the primary porosity of sand, a typical value of tortuosity might be about 2/3. Assuming a similar value for internal porosity, it is easy to show that for non-reactive contaminants intraparticle diffusion is not important (Brusseau et al., 1989). The time constant is simply too short, even though there could be significant storage available internally. Another way to look at it is to contrast the length scale l_z of the heterogeneities with the much smaller grain radius, and square the ratio. Heterogeneities dominate the physics. However, we can change things by adding some chemistry.

Chemistry

The aquifer solid matrix can chemically adsorb and later release contamination, leading to observations of concentrations in pumping wells that are similar in character to those described above (Whiffin and Bahr, 1984; Mackay and Cherry, 1989; Mercer, et al., 1990; Doty and Travis, 1991; Haley et al., 1991). Concentrations level off or show long tails, and when the pumping is stopped concentrations rise again (Whiffin and Bahr, 1984; Goltz and Oxley, 1991). These effects are usually attributed to non-equilibrium rate-limited sorption-desorption on the grain surface, where the sorption is due to presence of a clay, oxide or organic coating (Brusseau et al., 1989; Mackay and Cherry, 1989).

Let's return to our hypothetical aquifer to investigate this hypothesis. We added non-equilibrium linear sorption to the code and removed the silt lenses, although the aquitards remain. The contaminant in the wastewater has an equilibrium sorption coefficient $K_d = 0.198$ cm^3/gm in the aquifer sand, and the sand has bulk density of $\rho_b = 1.8$ g/m^3, yielding a retardation coefficient of R = 2 under equilibrium chemistry conditions. In the silt aquitards these parameters are $K_d = 0.668$ cm^3/g, $\rho_b = 1.6$ g/m^3, and R = 4. The mass transfer rates for linear non-equilibrium sorption are $\alpha = 9.1 \times 10^{-4}$/d in the sand and $\alpha = 5.0 \times 10^{-3}$/d in the silt. The effective sorption-desorption time constants for these rates are 3 years and 200 days, respectively. That is, we've chosen the sorption time constant in the aquifer equal to the estimated diffusion time constant for the silt lenses and aquitards in our previous example.

The plume history is shown in Figure 17.5. The left column is the aqueous phase concentration, where for convenience we've assumed an injection concentration of 1 mg/L. The right column shows the sorbed concentration, in tenths of a µg/g of soil. The maximum possible sorbed concentration would then be $S_{max} = K_d C_{max} = 0.2$ µg/g in the sand and 0.67 µg/g in the silt. A black contour interval in the right column of the figure represents 0.9 µg/g, or more, and white is less than 0.1 µg/g.

The first row of the figure shows conditions just one year after contamination first entered the aquifer system from the source. The aqueous phase plume moved deep into the aquifer and migrated slightly downgradient (Figure 17.5a). There was minor upgradient plume migration because there were no heterogeneities to redirect the wastewater recharge in that direction. Little contaminant was

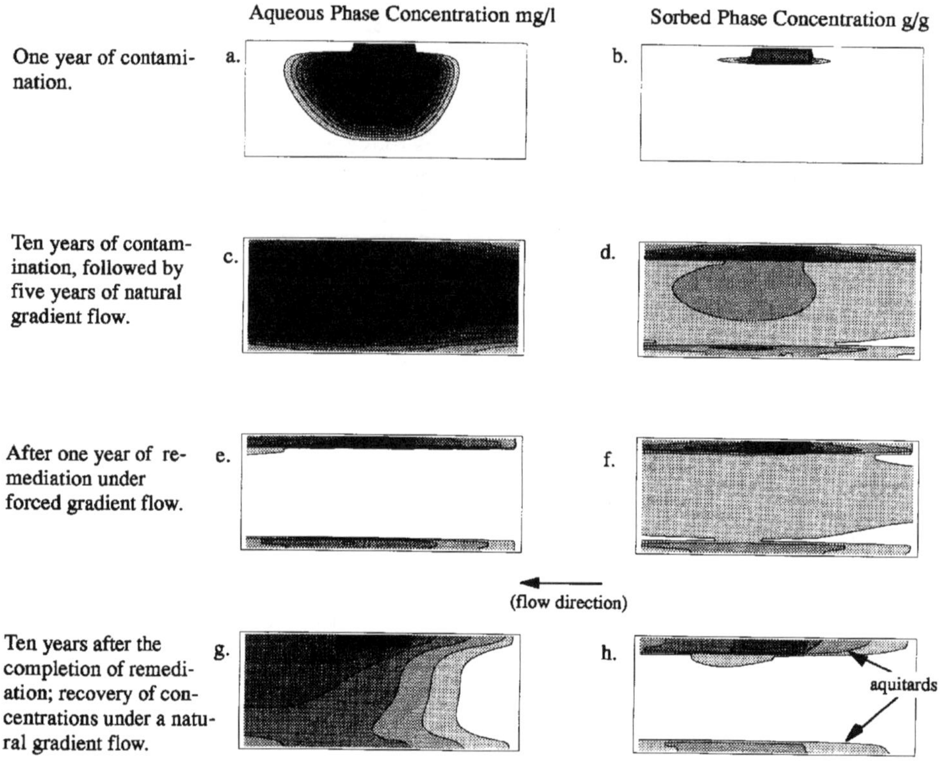

Figure 17.5. Contamination and remediation of a homogeneous aquifer experiencing non-equilibrium sorption with a time constant of 3 years in the aquifer sand.

sorbed onto aquifer material (Figure 17.5b). Its rate coefficient was too small, and its sorption time constant too large. In the upper aquitard, beneath the source, the two phases were closer to equilibrium because the mass transfer coefficient was higher here. The second row pictures the contaminant distribution just before the remediation system was initiated. Contamination had spread to the entire aquifer system (Figure 17.5c), although not so much as in the previous example where the heterogeneities led to upgradient migration. There was also less diffusion into the aquitards, because of the added presence of sorption. The sorbed concentration just before remediation is shown in Figure 17.5d. Concentrations in the sand were substantially above 0.1 µg/g, and in the middle of the aquifer they were close to the saturated value of 0.2 µg/g. The pattern mimics the pattern of aqueous phase concentration, indicating that conditions, after 15 years and 5 sorption time constants, were much closer to equilibrium.

The 200-day sorption time constant in the aquitard is much shorter than its diffusion-controlled time constant of 3 years. This suggests that sorption in the silt might be regarded as essentially instantaneous, and thus at equilibrium. At equilibrium the effective time constant for the aquitards is determined by both sorption and diffusion:

$$\tau_k = \frac{R l_z^2}{g_k D_o} \tag{17.4}$$

The already long 3-year diffusion-controlled time constant would extend to four times that—or 12 years—if sorption were in equilibrium. We can check this assumption by calculating the Damkohler II number. It contrasts the rate of sorption to the rate of diffusion in the aquitards:

$$\text{Damkohler II} = \frac{\alpha l_z^2}{D_o} \tag{17.5}$$

The resulting value of 14 >> 1 suggests that sorption in the aquitard will approach equilibrium. Thus we should expect a time constant closer to 12 years than 3 years, consistent with Figure 17.5c.

Three hundred days of remediation had no trouble sweeping out the aqueous phase contamination in the aquifer, but it had much less effect on the more slowly responding sorbed concentrations and aquitards, as shown in Figure 17.5e and 17.5f. The aqueous phase concentration history along the downgradient boundary is shown in Figure 17.6. The aqueous concentration rose slowly but steadily to an asymptotic value, controlled by dispersion, of about 80% of the source. During remediation it promptly dropped to a low value and tailed out. After the remediation wells were shut off, the contaminant desorbed from the sand grains and diffused out of the aquitards. Aqueous phase concentrations recovered almost 50% of their pre-remediation value.

In discussing heterogeneity and diffusion we mentioned that "storage" and "time constant" were two important issues. In the present case the major storage is in the sorption capacity of the solids. If the mass storage in the aqueous phase is given a value of one, the mass sorption capacity at equilibrium is *R*–1. In the sand aquifer the sorption capacity is the same as the pore space in the aqueous phase (195 g for an injection concentration of 1 mg/L), and in the aquitards it's three times higher (161 g), for a total of 356 g of storage. Compare to storage in the diffusion-controlled pore spaces: the aquitards, which have an additional 53.4 g of storage, and the heterogeneous lenses of the previous example could store an additional 85.4 g. Sorption provides abundant storage space, explaining the significant recovery of the aqueous concentrations after remediation. But it is not sufficient. The time constant remains an issue.

A three-year time constant for sorption is almost unobservable in the laboratory using conventional laboratory protocols. Contaminants and materials with such a long time constant would normally be identified as nonreactive. Laboratory tests more commonly identify sorption time constants in the range of 0 to 30 days. In our aquifer these are, for all practical purposes, instantaneous. We can check this by calculating the Damkohler I number for the aquifer. It contrasts the rate of sorption to the rate of advection:

$$\text{Damkohler I} = \frac{\alpha L}{v} = \frac{L}{\tau_s v} \tag{17.6}$$

where

L is the length of the aquifer, 100 m;
v is the average mean seepage velocity in the aquifer; and
τ_s is the aquifer sorption time constant.

In the situation illustrated in Figure 17.5 the aquifer sorption time constant is 3 years. The Damkohler I number has a value of 6.5 under natural gradient conditions, indicating mass transfer control approaching equilibrium (Brusseau, 1992). During remediation it drops to 0.065, indicating advection control, and thus no effective desorption. If the aquifer sorption time constant is more like, say, 10 days, then the Damkohler I number is over 700 under the natural gradient conditions, and at equilibrium, and over 7 during remediation, and approaching equilibrium.

Let's examine the contamination history under equilibrium chemistry conditions $\alpha = 0$, $\tau_s = \infty$ by focusing on the aqueous phase concentration history at the downstream boundary. It's illustrated in Figure 17.6 by the dashed line. Although there is some tailing, there is little recovery, and it's due to the desorption-limited diffusion out of the aquitards. That is, equilibrium chemistry cannot explain the failure of pump-and-treat. The sorption time constant and sorption capacity are both important.

How is it possible for sorption processes to react so slowly that their reaction rates and time constants become important on an aquifer scale, especially given the competing ever-present effects

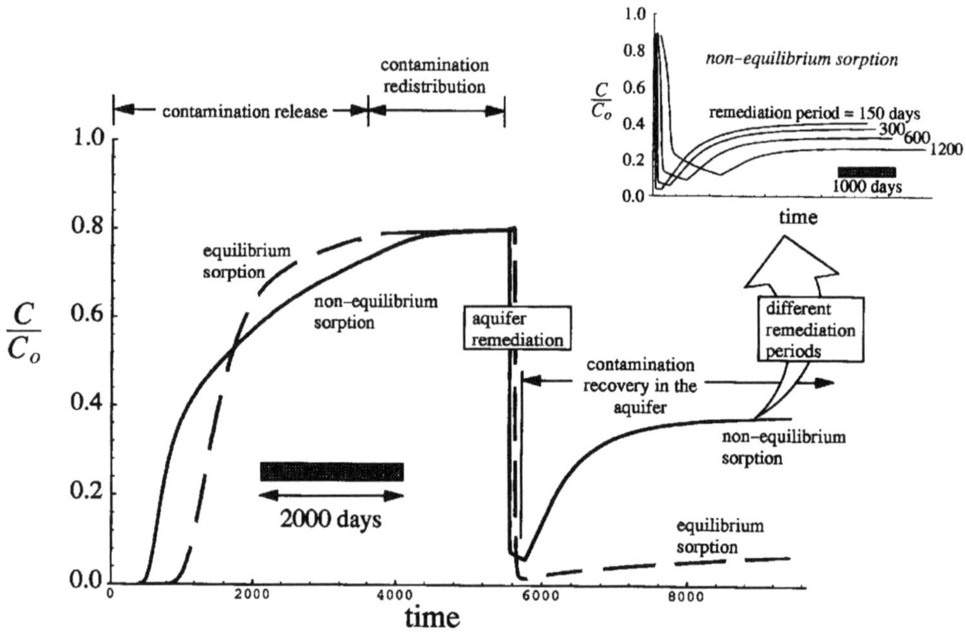

Figure 17.6. Aqueous phase concentrations at the left boundary of the domain pictured in Figure 17.1, for the non-equilibrium sorption contamination scenario shown in Figure 17.5. Concentrations for rate-limited (solid line) and equilibrium (dashed line) conditions are shown. In the upper right insert concentrations for several different remediation durations are plotted with an exaggerated time scale.

of heterogeneity? Diffusion must be in control again, but on the grain scale. Recall that some aquifer material may have internal porosity. Suppose that there are internal sorption sites at which equilibrium (or non-equilibrium) adsorption-desorption occurs (Brusseau, 1992). Then with a sufficiently large internal retardation factor R_m it is possible that the time scale of intraparticle diffusion and retardation can match the time scale of the effects of heterogeneity, producing an effect observable during pump-and-treat:

$$\tau_m = \frac{1}{\alpha_m} = \frac{R_m r^2}{g_m TD} \tag{17.7}$$

Another contribution to large time constants for intraparticle diffusion is the tortuosity factor. It can be much smaller than the typical value observed for primary porosity. Intragranular porosity is composed of microscopic pores, approaching in size the mean free path or even the molecular radius of some of the chemicals we are interested in, particularly some of the larger organic molecules (Brusseau, 1993). Consequently, intragranular tortuosity can be very small, and the sorption time constant in Equation 17.7 grows even larger. It is not actually necessary to have intragranular porosity to achieve high sorption coefficients. Patchy organic matter coating the grains can achieve the same effect, with low effective intrapatch diffusion coefficients (Brusseau et al., 1989). The rate coefficient for desorption may also be much slower than adsorption, especially in organic matter. Thus an adsorption process may have a short time constant, and effectively be in equilibrium, but its desorption time constant can be much longer. Standard laboratory tests may misinterpret this slow but important desorption rate as irreversible sorption, leading to overly optimistic assessments for success of a pump-and-treat system. Finally, sorption may be non-linear, further complicating the issue. All of these issues occur together and are usually called non-ideal sorption. Worse, there is no reason to assume that sorption properties or processes are constant in space, as we have assumed.

Hydraulic conductivity is not the only property that is heterogeneous. We know very little about the heterogeneity of sorption and other chemical properties.

Dueling with Time Constants

An aquifer contamination event evolves through stages. In our example the stages were original contamination, redistribution, remediation, and post-remediation with the very real possibility of concentration levels recovering. The success of remediation depends on the duration of each of these stages relative to the aquifer time constants. We've characterized the aquifer in terms of time constants for diffusion and sorption. Both are influenced by heterogeneities. Contamination present in the aquifer for only a short time, relative to the aquifer time constants, can be remediated. It doesn't have an opportunity to enter the slowly responding "storage." Contamination resident in the aquifer for many time constants finds that storage and cannot be remediated quickly, if at all. Assuming that the amount of contamination in slow storage is significant, a rule of thumb is that pump-and-treat will take three time constants to "clean up" an aquifer. This is of course complicated by the vast range of time constants caused by the combination of aquifer heterogeneity, diffusion and sorption.

The upper right insert in Figure 17.6 illustrates the issue for the non-equilibrium sorption example. Suppose we simply change the remediation time, but pump the same total amount of water. The upper right corner of the figure shows that the longer the remediation period, the less the concentration recovery, even though the same total amount of water is being produced and treated. Contaminant mass removed from the aquifer improves with the slower schemes. It's given by the integral of the well concentration over the remediation period. Slower is better. It is sometimes speculated that pulse pumping would work better, by quickly removing the mobile contaminant, and increasing concentration gradients from the bypassed or sorbed contaminant back into the main flow (Keeley, 1984). However, it doesn't necessarily work this way, at least for linear systems (Haggerty and Gorelick, 1994). The time constant can't be fooled. Pulse pumping's only real value is operational: pumps don't have to run continuously; produced water must be collected and treated only periodically; etc. As far as the aquifer is concerned, it is not the pulsing that matters but the total duration of the remediation period.

Another example of the duel with time constants is shown in Figure 17.7. Four different contamination source scenarios, with different durations, have different impact on the aquifer. The solid line is the non-equilibrium sorption case we examined above, with a 10-year-long chronic source, followed by five years of contamination redistribution. The other cases represent shorter-duration sources, but with the same total amount of mass injected. That is, they involve the same amount of wastewater with the same concentration, just injected for different lengths of time. Each source period is followed by five years of redistribution before remediation. The shorter source periods lead to higher concentrations faster, but more natural attenuation due to redistribution and offsite migration during the redistribution period, related to their higher concentration gradients. Remediation is more or less similar in each case, but contaminant recoveries are substantially smaller for the more acute contamination events. The material simply hasn't been in the aquifer long enough to experience the same level of adsorption.

Characterization

To remove contamination it is necessary to know where it is and to understand the aquifer well enough to accurately predict how it will behave during the remediation. Aquifer heterogeneity is important, but it is not the only issue. Much of the effort to characterize remediation sites has been wasted preparing data for litigation. Little of the data collection effort is optimized for the identification and estimation of parameters and processes of interest to the remediation designer. Even when such an effort is made, the result can be less than perfect.

Let's take a look at one such effort conducted at the Borden Site in Ontario, Canada, operated by the University of Waterloo. This may well be the most studied site in North America. We've been characterizing the capture zone of a pumping well in a shallow aquifer natural gradient flow field

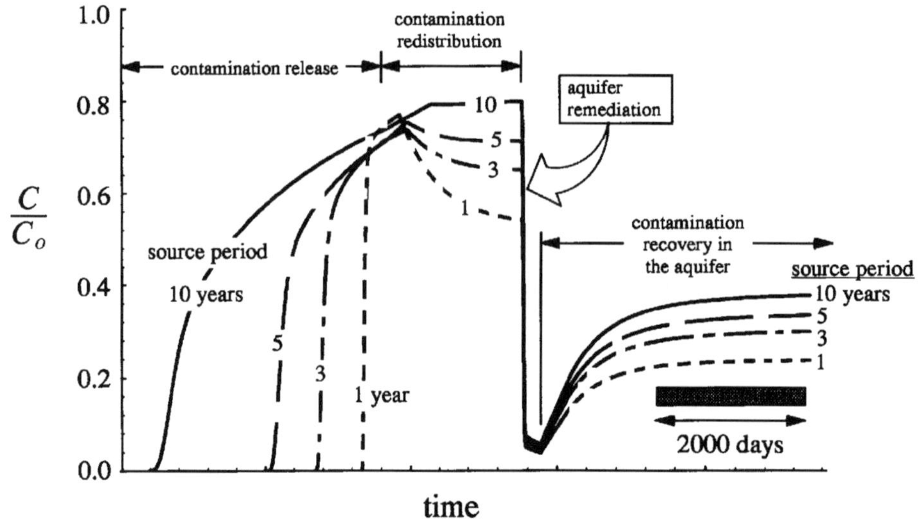

Figure 17.7. Aqueous phase concentrations at the left boundary of the domain pictured in Figure 17.1, for the non-equilibrium sorption contamination scenarios involving the same mass of contamination and volume of injected water, but with different source durations. The aquifer sorption time constant is 3 years. The solid line represents the chronic contamination condition pictured in Figures 17.5 and 17.6. The short dashed line represents an acute contamination event lasting only 1 year.

(Linderfelt and Wilson, 1994a,b). The well captures an 8-m-wide swath of the natural gradient flow which can then be treated. The idea is to design the capture zone to encompass the contaminated zone (Keeley, 1984; Javandel and Tsang, 1986). This is just about the simplest possible pump-and-treat scenario, especially since this aquifer is considered fairly homogeneous. The pumping well was installed in the immediate vicinity of previous natural and forced gradient plume studies, and in the midst of an intense multiple-core investigation of heterogeneity. A model of the capture zone was developed using the parameters estimated from the plethora of data available. Using this model, together with groundwater gradient information from a three-year-long natural gradient plume experiment by Stanford University and the University of Waterloo, as modified by some contemporary water level measurements, we predicted the capture zone shown in Figure 17.8. It extended upgradient, toward the southwest.

We designed a set of tracer injection experiments, utilizing eight different tracers at fifteen injection points, to confirm this hypothesis. Tracer arrivals were monitored in the pumping well and along a downgradient transect of multilevel samplers, located to observe any tracer bypassing the pumping well. The figure shows the normalized amount of tracer mass injected at each injection well that was recovered at the pumping well. These recoveries define the actual capture zone. The prediction was not accurate. The actual capture zone had a more southerly orientation.

As part of this field study we installed additional instrumentation to carefully monitor the flow field over the period of the test. We observed previously unrecognized temporal variations of groundwater flow directions that are 45 to 70 degrees off from the direction assumed in all of the previous analyses of natural gradient plumes at this site. This kind of variation can have a profound effect on a pump-and-treat system. Imagine for example, a contaminant plume originally located at the center of the predicted capture zone. Perhaps a third of it would have migrated north, past the pumping well, and not have been captured. Typical model studies of pump-and-treat (Keeley, 1984; Javandel and Tsang, 1986; Mercer et al., 1990) would have assumed one flow direction and ignored the possibility of failure due to this uncertainty.

The frightening thing about this prediction is that it was made at the best studied site in North America. Obviously the flow field had not been characterized well enough. Using hindsight, and the water levels observed at 21 piezometers during the experiment, it was possible to reconstruct the capture zone much more accurately. That reconstruction also revealed a subtle but significant vertical

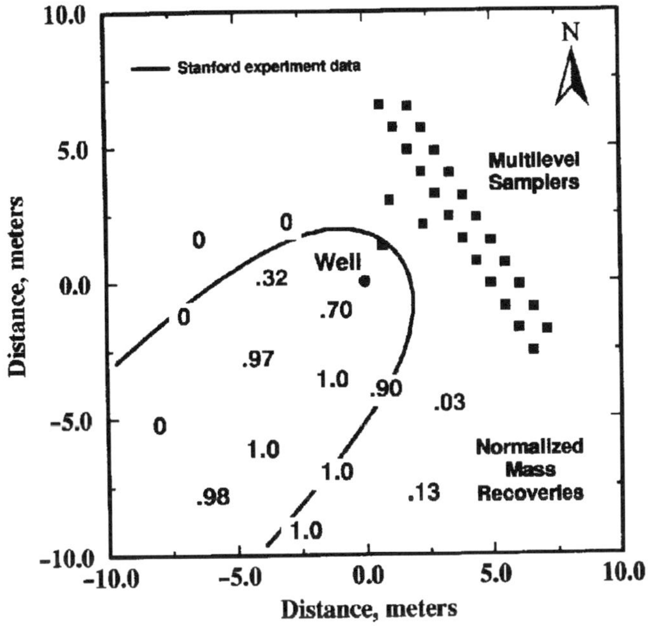

Figure 17.8. Predicted and measured capture zone in the Borden Sand Pit. The fully penetrating well pumped at 1.5 L/min for 4 months. The natural groundwater seepage velocity is roughly 10 cm/d. Approximately 60 g of tracer was injected over the upper 3 m at each injection site, in this 7- to 8-m deep sandy aquifer. Tracer recovery in the well was segmented into 7 different 1-m-long intervals, and measured daily.

stratification of hydraulic conductivity that had been suspected but not verified. For example, note the injection well just 4 m to the west of the pumping well, where 32% of the injected tracer was recovered by the well. Only the bottom portion of the injected tracer was recovered; the top portion bypassed the well and moved downgradient (Linderfelt and Wilson, 1994a). The boundary of a capture zone is very sensitive to subtle variations in heterogeneity.

Design and Operation

There are a myriad of design and operation problems (Haley et al., 1991), most of which have technological fixes. For example, the clogging of injection wells can be controlled through the proper treatment of water before injection. There are other problems more difficult to deal with. The heterogeneities and resulting preferred flow paths imply that a pump-and-treat system will dilute the contaminated water with ambient aquifer water, and/or already treated injected water (Wilson, 1984; Haley et al., 1991; Doty and Travis, 1991). Eventually the scheme is recontaminating far more water than it is cleaning up. Even in the absence of heterogeneities it is impossible to achieve the ideal linear displacement typical for cross-sectional models of homogeneous aquifers, like that in Figure 17.5. In a real aquifer flow converges toward a pumping well and away from injection facilities. The resulting non-uniform flow field leads to inefficiencies and tailing (Satkin and Bedient, 1988; Palmer and Fish, 1992).

Without resorting to chemical means to control preferred flow paths, the designer has only the geometry of the system to work with (Satkin and Bedient, 1988). The five-spot pattern is the petroleum industry's response to this issue (e.g., Schafer-Perini and Wilson, 1991). Several years ago I suggested another, the so-called "double-cell" method (Wilson, 1984). A small-scale inner recirculation pump-and-treat cell is designed to flush out the contamination for treatment, while an outer recirculation cell, created by a set of distant extraction and injection wells, circulates a much larger volume of untreated water. The presence of the outer cell contains the inner cell, reducing the volume of water pumped from the inner cell for treatment, maximizing concentrations, and improving efficiency. A more systematic approach is to characterize the aquifer and its contamination, and then to design a remediation scheme that moves with the contamination, pumping and injecting at appropriate times and locations. Optimization approaches are being applied to this problem (Ahlfeld and Sawyer, 1990; Gailey and Gorelick, 1993; Haggerty and Gorelick, 1994), but only for relatively simple conceptual models of aquifer heterogeneities and processes.

CONCLUSIONS

Pump-and-treat systems have failed to restore aquifers to water quality standards. Concentrations level off or show long tails, and when the pumping is stopped concentrations rise again. Aquifer heterogeneities under diffusion control are a principle reason for this failure. Non-equilibrium chemistry controlled by diffusion at the grain scale also contributes. Together they make it difficult to characterize the aquifer and its contamination, and to design an efficient and effective remediation system. The systems that result may be able to contain and control contamination, but unless we are willing to wait for decades of operation, they cannot restore aquifer water.

There are three important issues controlling the tailing and recovery of aqueous phase contaminant concentrations. The first issue is the aquifer system's capacity to store contamination while the aquifer is supposed to be flushed out. This storage can be in the pores of diffusion controlled by low-conductivity heterogeneities, or it can occur as sorption on the solid material. Significant amounts of storage can exist through both means which, in any event, will occur in combination. The second issue concerns the time constants for this storage. Quickly responding storage can be remediated. Slowly responding storage cannot be hurried. Diffusion and sorption time constants duel with the duration of the original contamination event, its redistribution through the aquifer, and the remediation itself. A successful remediation recognizes where the contamination is stored, and plans a duration long enough to access it. A rule of thumb is to identify the longest time constant controlling significant amounts of contamination, and plan on the remediation lasting three times longer. The third issue concerns the non-uniform flow field created by hydraulic conductivity heterogeneities and the geometry of the pump-and-treat system. Together they lead to preferred flow paths and significant bypassing of aquifer fluid and solids. As time goes on more and more water is produced in the pumping well that does not need treatment. Even if there is some flow through the bypassed zones, so that they are not diffusively controlled, concentrations in the pumping well(s) will still tail out, and the remediation system will become less efficient.

In the field these issues are difficult to characterize. A typical field-scale study will usually fail to identify any of the important processes, simply because they are almost unobservable on a short time scale unless you go looking for them. Of particular concern is the possible presence of non-equilibrium chemistry. It is not clear that it is recognized when it is most important. At the aquifer scale the important desorption time scale may be months, years, or even decades long. Most laboratory procedures would assume that the contaminant is irreversibly adsorbed. Thus it appears that this important issue is likely to go unrecognized, leading to overly optimistic assessments for the success of a pump-and-treat system.

ACKNOWLEDGMENTS

This work was supported by the Geophysical Research Center of the New Mexico Institute of Mining and Technology. Jun Lin assisted with the computer simulations presented in Figures 17.2 through 17.7. Bill Linderfelt conducted the field experiment and simulations of the Borden capture zone, in collaboration with the University of Waterloo with support from the Environmental Protection Agency.

REFERENCES

Ahlfeld, D.P. and C.S. Sawyer. 1990. Well location in capture zone design using simulation and optimization techniques. *Ground Water*, **28**(4)507–512.

Brusseau, M.L. 1992. Transport of rate-limited sorbing solutes in heterogeneous porous media: Application of a one dimensional multifactor nonideality model to field data. *Water Res. Res.*, **28**(9)2485–2497.

Brusseau, M.L. 1993. The influence of solute size, pore water velocity, and intraparticle porosity on solute dispersion and transport in soil. *Water Res. Res.*, **29**(3)1071–1080.

Brusseau, M.L., R.E. Jessup, and P.S.C. Rao. 1989. Modeling the transport of solutes influenced by multiprocess nonequilibrium. *Water Res. Res.*, **25**(9)1971–1989.

Carslaw, H.S. and J.C. Jaegger. 1959. Conduction of Heat in Solids. 2nd ed., Oxford Univ. Press., London.

Conrad, S.H., J.L. Wilson, W. Mason, and W. Peplinski. 1992. Visualization of residual organic liquid trapped in aquifers. *Water Res. Res.*, **28**(2)467–478.

Doty, C.B. and C.C. Travis. 1991. The effectiveness of groundwater pumping as a restoration technology. ONRL/TM-11866, Oak Ridge National Laboratory, Oak Ridge, TN.

Environmental Protection Agency. 1989. Evaluation of groundwater extraction remedies. EPA/540/0289/054, Office of Solid Waste and Emergency Response, Washington, DC.

Environmental Protection Agency. 1991. Dense nonaqueous phase liquids—A workshop summary. EPA/600/R-92/030, R.S. Kerr Lab., Environ. Protection Agency, Ada, OK.

Gaily, R.M. and S.M. Gorelick. 1993. Design of optimal, reliable plume capture schemes; Application to the Gloucester landfill groundwater contamination problem. *Ground Water*, **31**(1)107–114.

Goltz, M.N. and M.E. Oxley. 1991. Analytical modeling of aquifer decontamination by pumping when transport is affected by rate-limited sorption. *Water Res. Res.*, **27**(4)547–556.

Haggerty, R. and S.M. Gorelick. 1994. Design of multiple contaminant remediation: Sensitivity to rate-limited mass transfer. *Water Res. Res.*, **30**(2)435–446.

Haley, J.L., B. Hanson, C. Enfield, and J. Glass. 1991. Evaluating the effectiveness of ground water extraction systems. *Ground Water Monitoring Review*, Winter.

Javandel, I. and C.F. Tsang. 1986. Capture-zone type curves: A tool for aquifer cleanup. *Ground Water*, **24**(5)616–625.

Keeley, J.F. 1984. Optimizing pumping strategies for contamination studies and remedial actions. *Ground Water*, **4**(3)63–67.

Linderfelt, W.R. and J.L. Wilson. 1994a. Capture zone delineation: Models and experiments. Project completion report, in review, R.S. Kerr Lab., Environ. Protection Agency, Ada, OK.

Linderfelt, W.R. and J.L. Wilson. 1994b. Field study of capture zones in a shallow sand aquifer. *Transport and Reactive Processes in Aquifers*. T.H. Dracos and F. Stauffer, Eds. Balkema, Rotterdam, 289–294.

Mackay, D.M. and J.A. Cherry. 1989. Remediation of subsurface contamination: Limitations of pump and treat programs, *Environ. Sci. Technol.*, **23**(6)630–636.

Mercer, J.W., D.C. Skipp, and D. Giffin. 1990. Basics of pump-and-treat ground-water remediation technology. EPA/600/8-90/003, R.S. Kerr Lab., Environ. Protection Agency, Ada, OK.

National Research Council. 1994. Groundwater Remediation Alternatives, National Academy of Science Press, Washington, DC.

National Water Well Association. 1984. Proceedings of The Fourth National Symposium and Exposition on Aquifer Restoration and Ground Water Monitoring, NWWA, Columbus, OH.

Palmer, C.D. and W. Fish. 1992. Chemical enhancements to pump-and-treat remediation. EPA/540/S-92/001, R.S. Kerr Lab., U.S. EPA, Ada, OK.

Satkin, R.L. and P.B. Bedient. 1988. Effectiveness of various aquifer restoration schemes under variable hydrogeologic conditions. *Ground Water*, **26**(4)488–499.

Schafer-Perini, A.L. and J.L. Wilson. 1991. 3-dimensional stochastic flow and displacement in a 5-spot pattern. SPE 21243 *Proceedings*, 11th SPE Symposium on Reservoir Simulation, Soc. of Pet. Eng., Anaheim, CA, 405–423.

Sudicky, E.A. 1989. The Laplace transform Galerkin technique: A time-continuous finite element theory and application to mass transport in groundwater. *Water Res. Res.*, **25**(8)1833–1846.

Whiffen, R.B. and J.M. Bahr. 1984. Assessment of purge well effectiveness for aquifer decontamination. *Proceedings*, The Fourth National Symposium and Exposition on Aquifer Restoration and Ground Water Monitoring, National Water Well Assoc., Houston, TX, 75–81.

Wilson, J.L. 1984. Double-cell hydraulic containment of pollutant plumes. *Proceedings*, The 4th National Symposium & Exposition on Aquifer Restoration and Ground Water Monitoring, National Water Well Assoc., Houston, TX.

Wilson, J.L., S.H. Conrad, W.R. Mason, W. Peplinski, and E. Hagan. 1990. Laboratory investigation of residual liquid organics. EPA/600/6-90/004, R.S. Kerr Lab., Environ. Protection Agency, Ada, OK.

Section VI

Technologies for Contaminant Destruction

Chapter 18

Natural Biological Attenuation of Aromatic Hydrocarbons Under Anaerobic Conditions

J.F. Barker and J.T. Wilson

INTRODUCTION

Aromatic organics are common contaminants in groundwaters impacted by landfill leachate (Barker, 1987; Reinhard et al., 1984), coal tar creosote (Meuller et al., 1989) and petroleum product spills (Hult, 1987, for example). Of particular concern are the alkylbenzenes (benzene, toluene, ethylbenzene, and xylene isomers—as a group termed BTEX—and trimethylbenzenes), chlorinated benzenes, and 2-ring polynuclear aromatic hydrocarbons (PAHs) such as naphthalene due to their high aqueous solubilities (30 to 1780 mg/L) and relatively low drinking water standards (e.g., 5 µg/L for benzene).

In groundwater contaminant plumes some concentration reductions occur away from the source. This natural attenuation results from dispersion for all solutes, sorption for many solutes, and chemical or biological reaction for some solutes. For organic contaminants of interest here, dispersion and sorption often provide inadequate protection of groundwater resources. This chapter examines the field evidence that anaerobic biotransformation might provide additional protection.

Microbially mediated biotransformations use a variety of electron acceptors in the oxidation of aromatics. A subsurface environment can be conveniently defined by the electron acceptor being used. For aromatics, the preferred electron acceptor is O_2 and the environment is termed aerobic. In groundwaters O_2 entry is usually limited and so aromatic contaminants are often found in anaerobic environments. Long (1–3 km) plumes of BTEX-contaminated groundwater are generally anaerobic (Wilson et al., 1989; Chiang et al., 1989, for example) supporting the concept of aromatics persistence under anaerobic conditions. The sequence of preferred electron acceptors used under anaerobic conditions generally is:

NO_3^-: denitrification
Fe^{3+}: iron reduction
Mn^{4+}: manganese reduction
SO_4^{2-}: sulfate reduction
CO_2: methanogenesis (CH_4 is the product)

Anaerobic fermentation (organic matter is both electron acceptor and donor) probably occurs under all anaerobic conditions. Thus, we will speak of biotransformation occurring under methanogenic conditions, for example, implying that CO_2 reduction to CH_4 is a definitive redox reaction occurring, but recognizing that fermentations are also likely occurring.

This chapter examines the field evidence for anaerobic biotransformation of aromatics. An expectation can be drawn from controlled laboratory experiments, but it is necessary to seek verification that the potential is attained in contaminated groundwaters. This field evidence is organized in terms of the likely electron acceptor involved.

The approximate, apparent half-lives for biotransformation will be estimated from many of these field studies. In some cases, reported first-order or zero-order rates are converted into half-lives. In other cases, it was assumed that reactions were zero-order and the time for half the mass or concentration to be lost includes any lag period. A more detailed analysis seems unwarranted given the uncertainty in many field studies and the likely site-specific nature of biotransformation kinetics.

DENITRIFICATION

From what is known of their population ecology and growth strategies, the majority of denitrifiers are heterotrophic bacteria that are inherently aerobes (Tiedje, 1988). The capacity to denitrify does not come into play unless nitrate is available and oxygen is limiting. Denitrifiers are often facultative anaerobes and utilize oxygen as an electron acceptor when it is available.

The biodegradation of aromatic hydrocarbons has been observed in laboratory studies under nitrate-reducing conditions. Table 18.1 provides a summary of which compounds were evaluated and which were reported to degrade. For benzene, in particular, these studies generate contradictory expectations. Some limitations (e.g., substrate concentrations, nutrient additions, presence of other aromatics, potential for oxygen leakage into the system) in the experimental practices could be the cause of some of the different behavior.

Reinhard et al. (1991) established an *in situ* bioreactor at Seal Beach, California. The near-coast sand contained both dissolved and free product gasoline contamination. In the initial column experiments NO_3^- was consumed, but SO_4^{2-} was not, so the system was thought to be operating under denitrifying conditions. Influent BTEX concentrations ranged from 1 to 1600 μg/L with benzene usually above 500 μg/L and ethylbenzene usually <30 μg/L—too low for conclusions to be drawn about ethylbenzene's behavior. Although preliminary, the results suggest biotransformation of all BTEX compounds with the sequence of removal being toluene > *m*- and *p*-xylene > *o*-xylene > benzene.

Barbaro et al. (1992) injected gasoline-contacted groundwater into two adjacent monitored sections of the anaerobic landfill leachate plume at CFB Borden, Ontario. One injection was amended with 340 mg NO_3^-/L (1988 experiment) or 280 mg NO_3^-/L (1989) and the other injection was not amended with NO_3^-. Benzene, toluene, ethylbenzene, and *m*-, *p*-, and *o*-xylene concentrations were about 4.5, 2.4, 0.3, 0.8, 0.3, and 0.5 mg/L, respectively. Repeated weekly injections created an essentially continuous plume which was monitored at selected points over 100 to 320 days and at 1, 2, and 5 m from the injection wells. Groundwater moves horizontally at about 9 cm/day at this site. Denitrification was evident from the decline in nitrate levels along the flow system and was confirmed by the production of N_2O when acetylene was briefly included in the injected water.

Nitrate-amended groundwaters demonstrated extensive toluene loss after a short (40-day) lag, but toluene concentrations did not fall below 50 μg/L even after 5 m travel and 400 days. Ethylbenzene biotransformation followed a longer lag and was almost as extensive. Removal of *m*- and *p*-xylene was more extensive than that of *o*-xylene. Significantly, benzene was recalcitrant, even when it was the sole aromatic introduced into the system for 4 weeks toward the end of the field experiment.

Barbaro et al. (1992) and Barbaro (1990) evaluated the mass loss that occurred in flow tubes from the injection well to selected wells 1 m downgradient using a breakthrough curve analysis technique. This was done for two sequential injection experiments and the biotransformation half-lives were calculated for both. The initial experiment encountered an unacclimated microbial population, while the last could have encountered bacteria adapted to gasoline hydrocarbon degradation. The residence time for groundwaters reaching these wells were estimated from Br^- breakthrough plots and averaged 15 days. Given the calculated mass loss over this residence time (Barbaro et al., Table 18.4 and Barbaro, Table 18.4), apparent biotransformation half-lives were calculated and the mean values for each experiment shown in Table 18.3, experiments 1 and 4.

Hutchins and Wilson (1991) and Hutchins et al. (1991a) reported the results of profusing dissolved nitrate (10 mg N/L), phosphate and ammonium salts through JP-4-contaminated sands at a U.S. Coast Guard facility at Traverse City, Michigan. Groundwater was recirculated through a contaminated zone using an infiltration gallery. Unamended, but oxygen-bearing groundwaters were recirculated through the target zone near the water table for 2 weeks prior to adding the amendments. Nitrate removal commenced in the contaminated zone about 20 days after its introduction. Cored material showed increased numbers of denitrifiers after nitrate addition.

Unfortunately, benzene was hydraulically removed from groundwaters in the test zone prior to nitrate addition. Toluene present in the infiltrating groundwater was reduced only slightly for about 10 days, but then was reduced from about 200–80 μg/L (infiltration gallery) to <1 μg/L (1 m into contaminated zone) within 20 days of commencement of nitrate addition, at a time when nitrate

Table 18.1. Fate of Selected Aromatic Hydrocarbons Under Denitrifying Conditions—Laboratory Microcosm Studies.

Aromatic Compound	Biotransformation Result Reported: Biotransformed	Not Biotransformed
Benzene	2, 9	1, 6, 7, 8, 10
Toluene	1, 2, 5, 7, 8, 9, 10, 11	6
Ethylbenzene	1, 7	4
m-Xylene	1, 2, 5, 7, 10, 11	8
o-Xylene	1, 2, 5	6, 7, 8
p-Xylene	1, 2	7, 8
Trimethylbenzenes	1	
Chlorobenzene		4
Dichlorobenzenes		4, 6
Naphthalene	3	4, 6, 8
Indene		6
Acenaphthene	3	

Microcosm experiments: 1) Hutchins et al., 1991: 2) Gersberg et al., 1991; 3) Mihelic and Luthy, 1988; 4) Bouwer and McCarty, 1983; 5) Evans et al., 1991; 6) Lyngkilde et al., 1988; 7) Ball et al., 1991; 8) Flyvbjerg et al., 1993; 9) Major et al., 1988.
Dynamic column experiments: 10) Kuhn et al., 1988; 11: Zeyer et al., 1986.

removal began to be evident. A similar lag was observed for *m*- and *p*-xylene, but when biotransformation commenced after 50 days, *m*- and *p*-xylene concentrations declined from about 200–50 µg/L to only about 10 µg/L. Meanwhile, *o*-xylene concentrations did not decline over the same interval. Ethylbenzene showed similar behavior to *m*- and *p*-xylene. These interpretations may be complicated by the likely presence of JP-4 residuals in the study zone. These results were generally supported by laboratory microcosm experiments using contaminated material from this site (Hutchins et al., 1991b). Ranges of half-life of aromatics have been estimated from zero-order rate constants given in Table 18.1 of Hutchins and Wilson (1991). These values are only approximate, given the changing infiltration concentrations, but are included in Table 18.3, experiment 2.

Monitoring contaminated sites where aromatics occur under denitrifying conditions provides additional evidence. Monitoring from a river infiltration site (Schwarzenbach et al., 1983) found *m*-xylene removal was rapid, presumably under both aerobic and denitrifying conditions, but 1,4-dichlorobenzene removal was less complete, presumably due to persistence under denitrifying conditions. Barber et al. (1988) also noted the persistence of dichlorobenzenes through a zone of denitrification in sewage-contaminated groundwater.

The field evidence is that toluene, xylenes and ethylbenzene are degradable under denitrifying conditions in groundwaters. Chlorobenzenes appear somewhat less degradable. Benzene is problematic—only one field experiment found benzene to be slowly biotransformed.

IRON/MANGANESE REDUCTION

Lovely and Lonergan (1990) and Lovely et al. (1989) reported the anaerobic biotransformation of aromatics with Fe^{3+} serving as the electron acceptor. Field observations are reviewed below.

Lyngkilde and Christensen (1992) studied the fate of aromatics in various redox zones in the landfill leachate plume at Vejen, Denmark. Although the redox zones were not absolutely defined, much of the attenuation of aromatics appeared to have occurred between 19 and 39 m downgradient of the landfill in a zone of iron reduction. In this zone chloride concentration declined from 984 to 589 mg/L. This is attributed to dispersive dilution and would produce a similar decline of about 40% in organic solute concentrations. Sorptive losses are considered minimal in this organic-poor sand aquifer. Taking the average groundwater velocity as 175 m per year, the apparent half-lives of the target aromatics were calculated and are shown in Table 18.3. Note that essentially all aromatics were below detection by the 39 m well, and so the apparent half-lives could be even shorter.

There is too little experience with these conditions in the field to categorize the likely biotransformation potential, but it certainly warrants detailed research, particularly given the potentially large pool of electron acceptor in Fe(III) minerals.

UNSPECIFIED ELECTRON ACCEPTORS

Barbaro et al. (1992) also conducted a natural gradient experiment without added electron acceptors ("unamended control" experiment) in the anaerobic Borden landfill plume. At the test site the plume contains no oxygen, no nitrate, more than 10 mg/L Fe (mostly Fe^{2+}), about 200 mg/L SO_4^{2-} and no methane. The persistence of SO_4^{2-} is attributed to the low DOC (<10 mg/L). This system is poised at a redox level below denitrification but above sulfate reduction, with the most likely electron acceptor being Fe^{3+}.

In the unamended control experiment no BTEX mass loss was observed up to 1 m from the injection, but toluene declined from an injection concentration of about 2.4 mg/L to about 200 µg/L at 5 m from the injection well. Significant toluene loss was observed only after toluene had been detected at the 5 m fence for about 30 days. The reason for this lag time is unknown. Biotransformation removed 92% of the toluene over a travel time of about 45 days, between the 1 m and 5 m fence. This yields an apparent half-life of 24 days for toluene under these anaerobic conditions.

Acton and Barker (1992) established an *in situ* column (termed column C) in this plume amended with anaerobic groundwater, about 150 µg/L various aromatics and 300 mg/L SO_4^{2-}. In the 80-day experiment, SO_4^{2-} concentrations declined, as did the Br^- tracer, suggesting sulfate reduction was not significant. The dominant electron acceptor is not known. Figure 18.1 shows the behavior of several target aromatics. Sorption may be the major cause of immediate declines in concentrations of hydrophobic compounds such as trimethylbenzene and naphthalene. The rapid removal of some less-hydrophobic aromatics like toluene suggests aerobic biotransformation possibly resulting from oxygen entering the *in situ* column. However, reductive dechlorination of carbon tetrachloride was observed during this time, a reaction requiring anoxic conditions. Therefore, the post-day-2 removals shown in Figure 18.1 are attributed to anaerobic biotransformation. Note the sequential utilization pattern with toluene and ethylbenzene used and only then significant removal of *m*- and *p*-xylene, cumene, benzene, and chlorobenzene. 1,2,4-Trimethylbenzene and naphthalene showed initial concentration declines attributed to sorption, but persisted through the 80-day experiment. Apparent half-lives were calculated as the time when the day 2 concentration, reduced due to sorption, was halved and assumed a smoothed line through the data points in Figure 18.1. Apparent half-lives are listed under experiment 5 in Table 18.3.

SULFATE REDUCTION

Sulfate-reducing bacteria couple the reduction of sulfate (SO_4^{2-}) to H_2S or HS^- with the conversion of organic compounds to CO_2, H_2O, etc. This reaction is limited to very reducing anaerobic environments where most aqueous species have already been reduced (i.e., NO_3^- to N_2; Fe^{3+} to Fe^{2+}; etc.).

Laboratory evidence of aromatic compound degradation under sulfate-reducing conditions is rapidly accumulating. Haag et al. (1991) reported the biotransformation of toluene under apparently sulfate-reducing conditions in batch experiments using sediment from a site contaminated by gasoline. Other reported biotransformations include Beller et al. (1992)—toluene, Edwards et al. (1991, 1992)—toluene, *p*- and *o*-xylene, and Edwards and Grbic-Galic (1992)—benzene. Thierrin et al. (1992a) conducted large-scale column (200 cm long, 15 cm dia.) experiments with Bassendean sand and anaerobic groundwater in which toluene and sulfate degraded simultaneously. No benzene, ethylbenzene, *p*-xylene, or *o*-xylene degradation was observed.

These laboratory experiments suggest that most aromatic hydrocarbons could degrade under sulfate-reducing conditions. The field experience is examined next.

Acton and Barker (1992) reported *in situ* column experiments where sulfate reduction was stimulated by the addition of acetate or lactate plus yeast extract (columns D and F). Under these condi-

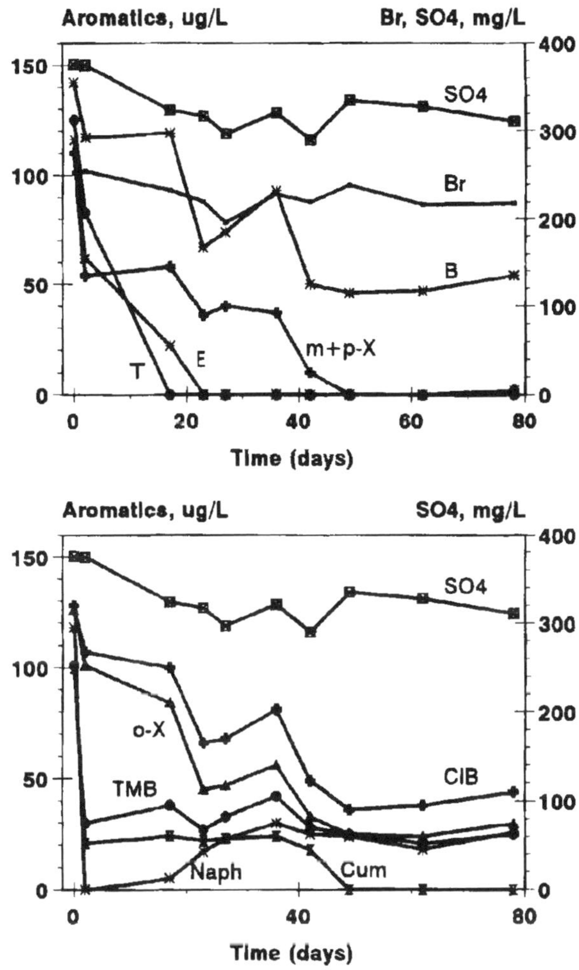

Figure 18.1. Changing concentrations of aromatics in an *in situ* column operating under an unspecified anaerobic condition (modified from Acton and Barker, 1992). Key: B = benzene, T = toluene, E = ethylbenzene, m+p–X = *m*- and *p*-xylene, o-X = *o*-xylene, ClB = chlorobenzene, TMB = 1,2,4-trimethylbenzene, Cum = cumene (1-methylethylbenzene), Naph = naphthalene.

tions, essentially no biotransformation of target aromatics was found. It appears that the readily utilized carbon substrate is used in preference to the aromatics. It also suggests that biotransformation of specific aromatics depends greatly on the other carbon substrates available—monoaromatics do not seem to be preferred substrates.

Thierrin et al. (1992a,b) described the bioattenuation of aromatics in a groundwater plume from a gasoline spill in Perth, West Australia. Benzene was detected in groundwater at least 400 m from the spill site, while toluene was not found beyond 200 m. Sulfate is thought to be the electron acceptor in the anaerobic plume. Biotransformation of toluene, *m*- and *p*-xylene, *o*-xylene, 1,3,5-trimethylbenzene, and naphthalene, but not of benzene, was observed. Their preliminary first-order biotransformation half-lives (Thierrin et al., 1992a) are reported in Table 18.3.

Thierrin et al. (1992a, b) also reported a unique tracer test conducted within this plume. About 400 L of groundwater from the plume was amended with a bromide tracer and fully deuterated aromatics and reinjected back into the plume about 80 m from the spill site. The deuterated aromatics could be distinguished from the aromatics present in the plume using mass spectrometry. Their calculated half-lives in the sulfate-reducing zone for first-order decay are included as is in Table 18.3, experiment 7.

IRON AND MANGANESE REDUCTION AND/OR METHANOGENESIS

Fe^{3+} is often available in aquifer minerals and so its reduction is expected to continue in anaerobic aquifers until no more Fe^{3+} is available from the solids. When methane is also being produced *in situ,* such an aquifer segment is probably best described as methanogenic.

Baedecker et al. (1988) and Cozzarelli et al. (1990) discuss the distribution of redox-sensitive species and monoaromatic hydrocarbons, respectively, in groundwater impacted by a crude oil pipeline rupture near Bemidji, Minnesota. A thin, shallow zone of anoxic groundwater was produced downgradient of a highly contaminated area by the consumption of electron acceptors by petroleum-degrading microorganisms. This zone (zone 5 of Baedecker et al.) appears to have active methanogenesis and active Fe and Mn reduction. Cozzarelli et al. considered 1,2,3,4-tetramethylbenzene as a recalcitrant, mobile tracer and concluded that biotransformation of other aromatics could be inferred from their greater attenuation compared to this tracer. Their Figure A-11 shows rapid, dramatic attenuation of toluene and *o*-xylene over about 12 m laterally. Benzene attenuation was almost as great, while ethylbenzene, n-propylbenzene and cumene appeared recalcitrant. All of these monoaromatics had been reduced to below detection limits by 100 m downgradient. Some oxygen could have been involved in the final removal, but the initial biotransformations would likely have required much more oxygen than would be available from the nearby (<0.5 m for some wells) unsaturated zone. The apparent attenuation of aromatics is converted into apparent biotransformation half-lives, by examining the relative attenuation from about 26 m to 35 m downgradient (Cozzarelli et al., Figure A-11) and considering it to have occurred over about 11 days (9 m travel distance at a velocity of 0.8 m/d). These half-lives are reported in Table 18.2, experiment 8.

Wilson et al. (1990) also report rapid biotransformation of monoaromatic hydrocarbons in a methanogenic/iron-reducing, shallow sand aquifer contaminated by aviation fuel, near Traverse City, Michigan. Rates of BTX biotransformation were estimated in two ways. In one, the changes in concentration along a presumed groundwater flow path were divided by the groundwater residence time between the two sampling points to obtain a biotransformation rate. In the second method, the changing concentrations over time were noted for a zone which had become essentially stagnant after a purging well was put into operation. The first-order half-lives are listed in Table 18.3, experiments 9 and 10, respectively. Anaerobic, methanogenic laboratory microcosms constructed with aquifer material produced similar biotransformation rates.

SULFATE REDUCTION AND/OR METHANOGENESIS

A number of laboratory and field studies were unable to clearly establish the dominant electron acceptor being utilized during aromatic hydrocarbon biotransformation. For example, Haag et al. (1991) observed in one column experiment that the biotransformation of toluene was followed by biotransformation of *p*-xylene and, to a lesser degree, benzene, ethylbenzene, *o*-xylene, 1,3,5-trimethylbenzene, naphthalene, and chlorobenzene. The electron acceptor was not clearly identified (sulfate was present but some methane was being generated), and sorptive removal of aromatics could have been dominant.

Acton and Barker (1992) reported a forced gradient injection experiment conducted in the North Bay leachate plume. Anoxic groundwater was withdrawn, amended to contain 12 mg/L SO_4^{2-}, about 200 µg/L BTEX (*o*- and *p*-xylene only) and 1,2,4-TMB and 250 mg/L of Br^- tracer. Only toluene and perhaps *m*-xylene were interpreted to have undergone biotransformation under the sulfate-amended, but potentially methanogenic aquifer conditions. Acton and Barker (1992) estimated that it required from 6 to 11 days for injected groundwater to reach the 5 m fence. The maximum toluene concentration was only 49% of the injected concentration. This corresponds to an approximate half-life of 10 days (experiment 11, Table 18.3).

Lyngkilde and Christensen (1992) also reported declines in aromatic hydrocarbon concentrations in a methanogenic/sulfate-reducing zone from 5 to 19 m downgradient of the landfill at Vejen, Denmark. The methanogenic and sulfate-reducing zones were temporally variable, so it is not certain under which conditions biotransformation occurred. However, their Table 1 showed a minor decline in chloride concentration but a much greater decline proportionally in many aromatics within this methanogenic/sulfate-reducing zone. Taking the estimated groundwater velocity of 175 m per year, these relative attenuations, beyond the 13% attenuation of chloride attributed to dispersive dilution, were used to calculate apparent half-lives (experiment 12, Table 18.3).

Cozzarelli et al. (1995) report aromatic hydrocarbon anaerobic biotransformation under a variety of conditions in a New Jersey aquifer, but no rates are specified.

METHANOGENESIS

Methanogenesis is the microbial production of methane (CH_4) from either simple organic molecules or CO_2 plus H_2, themselves usually produced by other fermentative bacteria. Methane production is inhibited by the presence of other electron acceptors or oxidants (O_2, NO_3^-, SO_4^{2-}) and so only occurs in the most reductive environments.

A sewage sludge–derived, mixed methanogenic culture which was maintained for 5 years with periodic addition of ferulic acid was able to biotransform toluene and benzene (Grbic-Galic and Vogel, 1987). Wilson et al. (1986) reported that benzene, toluene, ethylbenzene and *o*-xylene were slowly biotransformed in methanogenic microcosms made from landfill leachate and aquifer core material. Although the results were erratic, the concentrations of these monoaromatics were reduced to less than 1% of the initial 250 to 620 μg/L concentrations within 120 weeks. Such slow rates of biotransformation are difficult to deal with in laboratory experiments where even minor oxygen leakage and/or other handling problems can seriously jeopardize the difficult and time-consuming experiments.

Wilson et al. (1992) examined a sand aquifer in Kansas contaminated by leakage of refined petroleum products, probably JP-4 fuel, from an underground pipeline. Near the pipeline an oily hydrocarbon phase was found, and total aqueous BTEX concentrations ranged from 1 to 20 mg/L. Where methane occurred in the groundwater, some BTEX compounds were significantly reduced; that is, they make up a much smaller % of the total petroleum hydrocarbons (TPH) than in samples without methane. The biotransformation of BTEX in the heavily contaminated zone was related to TPH concentrations on the assumptions that TPH would be much more conservative than BTEX since all groundwater samples came from the areas with oily phase. Preferential dissolution of the more soluble BTEX relative to TPH is unlikely to be a major source of ratio changes, since an extensive BTEX plume has not been identified at the site. The authors argue for BTEX biotransformation under anaerobic (methanogenic) conditions as the source for these changing proportions. In general, toluene appeared to be the most biotransformed, followed by the xylenes, with ethylbenzene and benzene suffering the least biotransformation. Additionally, in wells resampled after a 10-month period, methane levels increased and more extensive removal of BTEX was apparent. It was suggested that BTEX biotransformation under anaerobic, methanogenic conditions required an adaption period, following which BTEX mass removal rates were impressive. No rates of BTEX biotransformation were estimated at this site.

Another field approach was taken by Reinhard et al. (1984) and Barker et al. (1986). This involved an interpretation of the spatial distribution of solute concentrations in groundwater impacted by landfill leachate at the North Bay, Ontario landfill site. The narrow leachate plume occupies the lower portion of a sand aquifer overlying gneissic bedrock and extends almost 800 m from the landfill to a discharge zone beside a creek. The site has operated since 1962 and leachate was observed in the discharge zone before 1978. The chemical data show temporal variability, reflecting variable inputs from the landfill and fluctuating (seasonally?) groundwater flow conditions. While the "front" of the plume has long since passed through the flow system, dispersion continues to produce declining concentrations of all solutes in groundwater away from the landfill. Chloride, a conservative tracer, undergoes only dispersion, so its declining concentrations represent mixing of leachate (200 to 800 mg/L Cl) with natural groundwater (<5 mg/L Cl). The attenuation of chloride and selected aromatics is estimated by averaging the maximum concentrations at closely spaced multilevel wells sampled in July, 1985. The maximum concentration at multilevel piezometer was selected to represent the anaerobic and methanogenic core of the leachate plume, which extends at least 350 m

Table 18.2. Average Maximum Concentrations in Wells Adjacent to a 350 m Downgradient, Cocncentration Declines and Apparent Half-Life for Declines Over the Interval.

Solute	Concentration (Cl mg/L, others mg/L)		Concentration Decline Landfill to 350 m		Apparent Half-Life (days)
	At Landfill	At 350 m	(%)	Dispersion-Corrected (%)	
Chloride	330	230	30	0	—
Benzene	28	13	54	24	1500
Toluene	19	1.0	95	65	561
Ethylbenzene	60	31	48	18	2000
m- + *p*-Xylene	85	15	82	52	700
o-Xylene	24	3.7	85	55	660
1,2,4-TMB	65	17	74	44	830
Chlorobenzene	35	19	54	24	1500

downgradient of the landfill site. Table 18.2 presents these average maximum concentrations adjacent to the landfill and about 350 m downgradient. Where relative concentrations of aromatics decline more rapidly than chloride, attenuation by biotransformation in addition to dispersion is indicated. The estimated travel time along this segment is 2 years. The apparent biotransformation half-lives are listed in Table 18.3, experiment 13.

Note that these are among the longest half-lives noted in Table 18.3. The injection experiment (Table 18.3, experiment 11) produced faster biotransformation, at least for toluene, and the column experiment (Table 18.3, experiment 5) produced faster biotransformation rates for many aromatics. The low concentration of aromatics in the leachate plume could have produced the slower biotransformation, especially if the degradation was concentration dependent (first-order).

Godsy et al. (1992) monitored the concentration of selected PAHs in a groundwater plume emanating from the creosote wastes at a wood preserving plant in Pensacola, Florida. The concentration of indene decreased much faster than could be accounted for by dispersive dilution (Godsy et al., Figure 18.14). Naphthalene, methylnaphthalene isomers and acenaphthalene appeared to increase and then decrease in concentration along the groundwater flow direction, but no explanation for this complex behavior was offered. Indene biotransformation under the methanogenic aquifer conditions appears to have occurred, however.

Wilson et al. (1993) examined groundwater contaminated by leaking underground storage tanks at Sleeping Bear Dunes National Lakeshore, Michigan. Contaminants move in a complex flow system featuring a seasonally variable gradient direction usually toward the nearby (25 m) Platte River. Residence time between the tank area and two other nests of multilevel monitoring wells 9.1 and 21.3 m downgradient was estimated and the changing BTEX concentrations examined for evidence of biotransformation. The plume was anaerobic with methanogenesis the dominant redox process. In order to estimate the rate of bioattenuation or biotransformation, the behavior of BTEX was compared to 2,3-dimethylpentane, which appeared to be conservative at this site. Average 2,3-dimethylpentane concentrations in the core of the plume declined from 46 to 15 and to 6 μg/L at the tank site, 9.1 m and 21.3 m downgradient. This suggests dispersive dilution by 67% between the tank site and 175 m and by an additional 60% between 9.1 m and 21.3 m. Declines in benzene concentrations were comparable and so no biotransformation of benzene is inferred. Rapid biotransformation of toluene and significant biotransformation of ethylbenzene and all xylenes were indicated. The calculated range of approximate, apparent half-lives (first-order) are indicated in Table 18.3, experiment 14.

SUMMARY OF FIELD BIODEGRADATION RATES

Some generalizations can be drawn from the apparent half-lives listed in Table 18.3. There is little consistent difference in the calculated half-lives in different anaerobic environments, but methanogenic

Table 18.3. Approximate, Apparent Half-Lives of Aromatic Hydrocarbons—Field Studies.

Organic	Biotransformation Half-Life (per day)						
	Denitrifying			Fe/Mn Reduction		Unspecified	Sulfate-Reducing
Experiment:	**1**	**2**	**3**	**4**	**5**	**6**	**7**
Benzene	ND			ND	35	ND	ND
Toluene	13,8	1–8		24	10	95–145	58–134
Ethylbenzene	68,12	2–13	44	ND	13	200–260	
m-Xylene	54,12	1–7	44	ND	35	160–180	
o-Xylene	93,22	1-ND	46	ND	30	115–135	
p-Xylene	50,13	(a)	(a)	ND	(a)	(a)	84–303
Trimethylbenzene			36		ND	180	
Cumene					46		
Chlorobenzene					40		
Naphthalene			53		ND	140–180	16–40
	Fe/Mn-Reducing or Methanogenic					Methanogenic	
Experiment:	**8**	**9**	**10**	**11**	**12**	**13**	**14**
Benzene	39	97	29	ND	ND	800	ND
Toluene	10	4	10	10	18	300	10–30
Ethylbenzene	ND			ND	150	1000	60–280
m-Xylene		161	49		18	700	50–190
o-Xylene	10	(b)	(b)	ND	30	700	40–600
p-Xylene		(b)	(b)		(a)	(a)	70–40
Trimethylbenzene				ND		800	
Cumene	ND						
Chlorobenzene						1200	
Naphthalene							

Experiments: 1) Barbaro et al. (1992) and Barbaro (1990); 2) Hutchins and Wilson (1991); 3) Lyngkilde and Christensen (1992), 19–39 m; 4) Barbaro et al. (1992); 5) Acton & Barker (1992), column C; 6) Thierrin et al. (1992a), plume interpretation; 7) Thierrin et al. (1992b), deuterated injection exp.; 8) Cozzarelli et al. (1990); 9) Wilson et al. (1990), along flow path; 10) Wilson et al. (1990), changes over time; 11) Acton and Barker (1992), injection exp.; 12) Lyngkilde and Christensen (1992), 5–19 m; 13) This chapter, Table 18.3; 14) Wilson et al. (1993).

ND: biotransformation not found.

(a) *m*- and *p*-xylene analyzed together.

(b) Total xylenes reported as *m*-xylene.

environments might be generally the least supportive of rapid biotransformation. Toluene was usually the most rapidly biotransformed, while benzene, cumene, trimethylbenzenes, and chlorobenzene were usually the slowest to biotransform in most experiments. Xylene isomers showed no consistent biotransformation rate sequence.

The major field experiments occurred with petroleum product contamination or in landfill leachate plumes. The suggestion of substrate interactions seen in Figure 18.1 would suggest that differences in aromatic biotransformation rates might be expected when these two situations are compared. Additionally, leachate experiments used lower (100s μg/L) concentrations of aromatics compared to the 10s of mg/L of the petroleum contamination plumes. Experiments 8, 9, 10 (Table 18.3) dealt with petroleum contamination while experiments 11 and 12 were in landfill leachates, but all are under extremely reducing conditions. Benzene was biotransformed more rapidly in the petroleum plumes; toluene was biotransformed rapidly in both situations. Other aromatics showed no consistent differences.

Given the common occurrence of anaerobic, aromatic biotransformation in groundwater, why do plumes of BTEX persist for long distance and time? In part, it could be the continuing supply of BTEX from dissolution of the petroleum source. BTEX supplied at mg/L concentrations may undergo a 1000-fold attenuation and still be mapped within a plume. On the other hand, Wilson et al. (1993) describe a site where rapid, anaerobic biotransformation of BTEX appears to produce a very restricted dissolved plume of BTEX. Here, rapid biotransformation appears to offer some protection to the groundwater resource.

Clearly, the database is small and approximate, so clear trends may emerge as more and better data is collected. There certainly seems to be a potential for rapid, anaerobic biotransformation of a variety of aromatics. Apparently, this potential is being reached at some sites, while at others, it is not. The limitations of anaerobic biotransformation of aromatics at many sites needs to be identified. Availability of electron acceptors does not appear to be the limiting factor in general. Nutrients, competitive substrates, or some other factor may be limiting biotransformation rates. Perhaps such limitations can then be overcome as part of *in situ* bioremediation activities.

REFERENCES

Acton, D.W. and J.F. Barker. 1992. *In situ* biodegradation potential of aromatic hydrocarbons in anaerobic groundwaters. *J. Contam. Hydrol.* **9**:325–352.

Baedecker, M.J., D.I. Siegel, P. Bennett, and I.M. Cozzarelli. 1988. The fate and effects of crude oil in a shallow aquifer 1. The distribution of chemical species and geochemical facies. *U.S. Geol. Surv. Water-Resources Invest. Rept. 88-4220*, pp. 13–20.

Ball, H.A., M. Reinhard, and P.L. McCarty. 1991. Biotransformation of monoaromatic hydrocarbons under anoxic conditions. *In* Hinchee, R.E. and R.F. Olfenbuttel (Eds.). *In Situ Bioreclamation: Applications and Investigations for Hydrocarbon and Contaminated Site Remediation.* Butterworth-Heinemann, Boston. pp. 458–462.

Barbaro, J.R. 1990. Biotransformation of BTEX under anaerobic denitrifying conditions: a laboratory and field evaluation. MSc. Project Report, Department of Earth Sciences, University of Waterloo, Waterloo, Ontario.

Barbaro, J.R., J.F. Barker, L.A. Lemon, and C.I. Mayfield. 1992. Biotransformation of BTEX under anaerobic denitrifying conditions: Field and laboratory observations. *J. Contam. Hydrol.* **11**:245–272.

Barber, L.B., E.M. Thurman, M.P. Schroeder, and D.R. LeBlanc. 1988. Long-term fate of organic micropollutants in sewage-contaminated groundwater. *Environ. Sci. Technol.* **22**:205–211.

Barker, J.F. 1987. Volatile aromatic and chlorinated organic contaminants in groundwater at six Ontario landfills. *Poll. Res. J. Can.* **22**:33–48.

Barker, J.F., J.S. Tessman, P.E. Plotz, and M. Reinhard. 1986. The organic geochemistry of a sanitary landfill leachate plume. *J. Contam. Hydrogeol.* **1**:171–189.

Beller, H.R., M. Reinhard, and D. Grbic-Galic. 1992. Metabolic by-products of anaerobic toluene degradation by sulphate-reducing enrichment cultures. *Appl. Environ. Microbiol.* **58**:3192–3195.

Bouwer, E.J. and P.L. McCarty. 1983. Transformations of halogenated organic compounds under denitrification conditions. *Appl. Environ. Microbiol.* **45**:1295–1299.

Chiang, C.Y., J.P. Salanitro, E.Y. Chai, J.D. Colthart, and C.L. Klein. 1989. Aerobic biodegradation of benzene, toluene, and xylene in a sandy aquifer—data analysis and computer modeling. *Ground Water* **27**:823–834.

Cozzarelli, I.M., R.P. Eganhouse, and M.J. Baedecker. 1990. Transformation of monoaromatic hydrocarbons to organic acids in anoxic groundwater environments. *Environ. Geol. Water Sci.* **16**:135–141.

Cozzarelli, I.M., J.S. Herman, and M.J. Baedecker. 1995. Fate of microbial metabolism of hydrocarbons in a coastal plain aquifer: The role of electron acceptors. *Environ. Sci. Technol.* **29**:458–469.

Edwards, E. and D. Grbic-Galic. 1992. Complete mineralization of benzene by aquifer microorganisms under strictly anaerobic conditions. *Appl. Environ. Microbiol.* **58**:2663–2666.

Edwards, E.A., L.E. Wills, D. Grbic-Galic, and M. Reinhard. 1991. Anaerobic degradation of toluene and xylene—Evidence for sulphate as the terminal electron acceptor. *In* Hinchee, R.E. and R.F. Olfenbuttel (Eds.), *In Situ Bioreclamation: Applications and Investigations for Hydrocarbon and Contaminated Site Remediation.* Butterworth-Heinemann, Boston. pp. 463–471.

Edwards, E.A., L.E. Wills, M. Reinhard, and D. Grbic-Galic. 1992. Anaerobic degradation of toluene and xylene by aquifer microorganisms under sulfate-reducing conditions. *Appl. Environ. Microbiol.* **58**:794–800.

Evans, P.J., D.T. Mang, K.S. Kim, and L.Y. Young. 1991. Anaerobic degradation of toluene by a denitrifying bacterium. *Appl. Environ. Microbiol.* **57**:1139–1145.

Flyvbjerg, J., E. Arvin, B.K. Jensen, and S.K. Olsen. 1993. Microbial degradation of phenols and aromatic hydrocarbons in creosote-contaminated groundwater under nitrate-reducing conditions. *J. Contam. Hydrol.* **12**:133–50.

Gersberg, R.M., W.J. Dawsey, and H.F. Ridgway. 1991. Biodegradation of dissolved aromatic hydrocarbons in gasoline-contaminated groundwaters using denitrification. In: E.J. Calabrese and P.T. Kostecki (Eds.), *Petroleum Contaminated Soils*. Lewis Publishers, Boca Raton, FL, pp. 211–217.
Godsy, E.M., D.F. Goerlitz, and D. Grbic-Galic. 1992. Methanogenic biodegradation of creosote contaminants in natural and simulated ground-water ecosystems. *Ground Water* **30**:232–242.
Grbic-Galic, D. and T.M. Vogel. 1987. Transformation of toluene and benzene by mixed methanogenic cultures. *Appl. Environ. Microbiol.* **53**:254–260.
Haag, F., M. Reinhard, and P.L. McCarty. 1991. Degradation of toluene and p-xylene in anaerobic microcosms: evidence for sulphate as a terminal electron acceptor. *Environ. Toxicol. Chem.* **10**:1379–1389.
Hult, M.F. 1987. Movement and fate of crude oil contaminants in the subsurface environment at Bemidji, Minnesota. *In* Franks, B.J. (Ed.), U.S. Geol. Survey Program on Toxic Waste-Groundwater Contamination: Proc. Third Tech. Meeting, Pensacola, FL, March 23-27, 1987. *U.S. Geol. Surv. Open-file Rept. 87-109*, pp. C2-C5.
Hutchins, S.R. and J.T. Wilson. 1991. Laboratory and field studies on BTEX biodegradation in a fuel-contaminated aquifer under denitrifying conditions. In Hinchee, R.E. and R.F. Olfenbuttel (Eds.), *In Situ Bioreclamation: Applications and Investigations for Hydrocarbon and Contaminated Site Remediation.* Butterworth-Heinemann, Boston. pp. 157–172.
Hutchins, S.R., W.C. Downs, J.T. Wilson, G.B. Smith, D.A. Kovacs, D.D. Fine, R.H. Douglass, and D.J. Hendrix. 1991a. Effect of nitrate addition on biorestoration of fuel-contaminated aquifer: field demonstration. *Ground Water* **29**:571–580.
Hutchins, S.R., G.W. Sewell, D.A. Kovacs, and G.A. Smith. 1991b. Biodegradation of aromatic hydrocarbons by aquifer microorganisms under denitrifying conditions. *Environ. Sci. Technol.* **25**:68–76.
Kuhn, E.P., J. Zeyer, P. Eicher, and R.P. Schwarzenbach. 1988. Anaerobic degradation of alkylated benzenes in denitrifying laboratory aquifer columns. *Appl. Environ. Microbiol.* **54**:490–496.
Loveley, D.R. and D.J. Lonergan. 1990. Anaerobic oxidation of toluene, phenol, and p-cresol by the dissimilatory iron-reducing organism, GS-15. *Appl. Environ. Microbiol.* **56**:1858–1864.
Lovely, D.R., M.J. Baedecker, D.J. Lonergan, I.M. Cozzarelli, E.J.P. Phillips, and D.I. Siegel. 1989. Oxidation of aromatic contaminants coupled to microbial iron reduction. *Nature* **339**:297–299.
Lyngkilde, J. and T.H. Christensen. 1992. Fate of organic contaminants in the redox zones of a landfill leachate pollution plume (Vejen, Denmark). *J. Contam. Hydrol.* **10**:291–307.
Lyngkilde, J., J.C. Tjell, and A. Foverskov. 1988. The degradation of specific organic compounds with landfill leachate as a primary substrate. In Wolf, K., W.J. van den Brink, and F.J. Colon (Eds.), *Contaminated Soil '88,* Kluwer Academic Publishers, pp. 91–100.
Major, D.W., C.I. Mayfield, and J.F. Barker. 1988. Biotransformation of benzene by denitrification in aquifer sand. *Ground Water* **26**:8–14.
Mihelic, J.R. and R.G. Luthy. 1988. Microbial degradation of acenaphthalene and naphthalene under denitrification conditions in soil-water systems. *Appl. Environ. Microbiol.* **54**:1188–1198.
Mueller, J.G., P.J. Chapman, and P.H. Pritchard. 1989. Creosote-contaminated sites. *Environ. Sci. Technol.* **23**:1197–1201.
Reinhard, M., L.E. Wills, H.A. Ball, T. Harmon, D.W. Phipps, H.F. Ridgway, and M.P. Eisman. 1991. A field experiment for the anaerobic biotransformation of aromatic hydrocarbon compounds at Seal Beach, California. *In* Hinchee, R.E. and R.F. Olfenbuttel (ed) *In Situ Bioreclamation: Applications and Investigations for Hydrocarbon and Contaminated Site Remediation.* Butterworth-Heinemann, Boston. pp. 487–496.
Reinhard, M., N.L. Goodman, and J.F. Barker. 1984. Occurrence and distribution of hazardous organic chemicals in the leachate plumes of two sanitary landfills. *Environ. Sci. Technol.* **18**:953–961.
Schwarzenbach, R.P., W. Giger, E. Hoehn, and J.K. Schneider. 1983. Behaviour of organic compounds during infiltration of river water to groundwater. Field studies. *Environ. Sci. Technol.* **17**:472–479.
Thierrin, J., G.B. Davis, C. Barber, B.M. Patterson, F. Pribac, T.R. Power, and M. Lambert. 1992a. Natural degradation rates of BTEX compounds and naphthalene in a sulphate reducing groundwater environment. *In* Lesage, S. (Ed.), In situ bioremediation Symp. '92, Niagara-on-the-Lake, Ontario, Sept. 20–24, 1992, pp. 108–123.
Thierrin, J., G.B. Davis, C. Barber, and T.R. Power. 1992b. Use of deuterated organic compounds as groundwater tracers for determination of natural degradation rates within a contaminated zone. Presented at the 6th Internat. Symp. on Water Tracing, Karlsruhe, Germany, Sept. 21–26, 1992.
Tiedje, J.M. 1988. Ecology of denitrification and dissimilatory nitrate reduction to ammonium. *In* Zehnder, A.J.B. (Ed.), *Biology of Anaerobic Microorganisms*. John Wiley & Sons, Inc., pp. 179–244.
Wilson, B.H., J.T. Wilson, D.H. Kampbell, B.E. Bledsoe, and J.M. Armstrong. 1990. Biotransformation of monoaromatic and chlorinated hydrocarbons at an aviation gasoline spill site. *Geomicrobiology Journal* **8**:225–240.

Wilson, B.H., G.B. Smith, and J.F. Rees. 1986. Biotransformation of selected alkylbenzenes and halogenated aliphatic hydrocarbons in methanogenic aquifer material: A microcosm study. *Environ. Sci. Technol.* **20**:997–1002.

Wilson, J.T., D.H. Kampbell, and J. Armstrong. 1993. Natural bioreclamation of alkylbenzenes (BTEX) from a gasoline spill in methanogenic groundwater. *Proc. 2nd Internat. Symp. on In Situ Bioreclamation*, April 5–8, 1993, San Diego, CA.

Wilson, J.T., D.H. Kampbell, M. Cook, S.R. Hutchins, B.H. Wilson, and L.G. Kennedy. 1992. Geochemical indicators of anaerobic biodegradation of BTEX. *Proc. 85th Annual Meeting, Air and Waste Management Assoc., Kansas City, MO, June 21–26, 1992.*

Wilson, J.T., L.E. Leach, J. Michalowski, S. Vandergrift, and R. Calloway. 1989. *In-situ* bioremediation of spills from underground storage tanks: new approaches for site characterization, project design, and evaluation of performance. U.S. Environ. Protection Agency, July, 1989, EPA/600/2-89/042.

Zeyer, J., E.P. Kuhn, and R.P. Schwarzenbach. 1986. Rapid microbial mineralization of toluene and 1,3-dimethylbenzene in the absence of molecular oxygen. *Appl. Environ. Microbiol.*, 52, pp. 944–947.

Chapter 19

Air Sparging: A Primer for Application and Design

R.A. Brown

INTRODUCTION

Air sparging is rapidly becoming an important, new remedial technology. It has recently been documented as being effective for treating both petroleum hydrocarbon contamination and chlorinated solvent contamination (see References), removing as much as 10,000 lb of contaminants in as little as six months. With this level of results, there is an understandably high degree of interest in air sparging. However, there is little information currently available on air sparging.

This chapter is presented as a basic primer on the application and design of air sparging systems. The format that will be employed is a question and answer format. The topics that will be covered are a description of the technology, its benefits and limitations, how air sparging is designed, how it can be used, and what the primary issues are concerning air sparging. A keyword index follows the text to help in answering specific questions.

DESCRIPTION OF AIR SPARGING

What is Air Sparging?

Air sparging is a remedial technology based on the direct injection of air into a water-saturated soil matrix. The injected air travels horizontally and vertically through the soil matrix until it exits into unsaturated (water) soils or directly into the atmosphere.

How Does Air Sparging Differ from Older Technologies Called by the Same Name?

In the late seventies and early eighties an in-well oxygenation system was employed in *in situ* bioremediation projects. With this technique air was "sparged" into the well and the oxygenated water was allowed to migrate into the formation. The difference between this technique and air sparging is that with the old technique the air was injected into the water in the well, whereas with air sparging, the air is directly injected into the soil matrix (Figure 19.1). The well is dewatered to the depth of the sparging screen.

What are the Essential Elements of an Air Sparging System?

An air sparging system consists of four basic elements: an injector system, a compressor, a control system, and a monitoring system (Figure 19.2).

Describe the Injector (Well) System

The injector system is a conduit to carry air from the surface and inject it into the subsurface saturated soil matrix. A generalized diagram of a sparging injector (well) is pictured in Figure 19.3. The well consists of a pervious section of usually 1–2 ft connected to a solid, non-pervious pipe which extends from the pervious section to the surface. The pervious section can consist of a screen

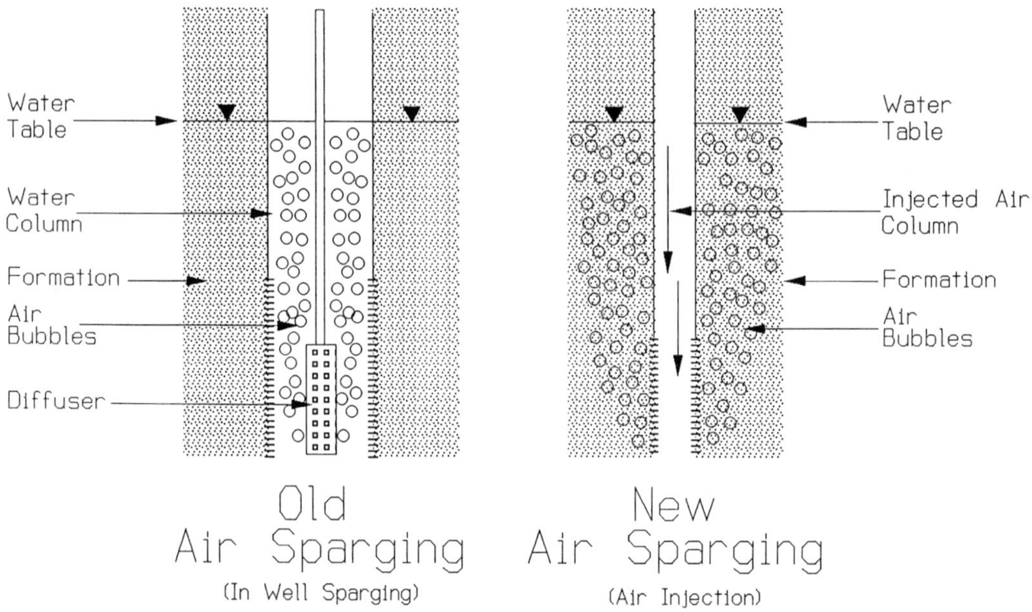

Figure 19.1. Differences between old and new air sparging technologies.

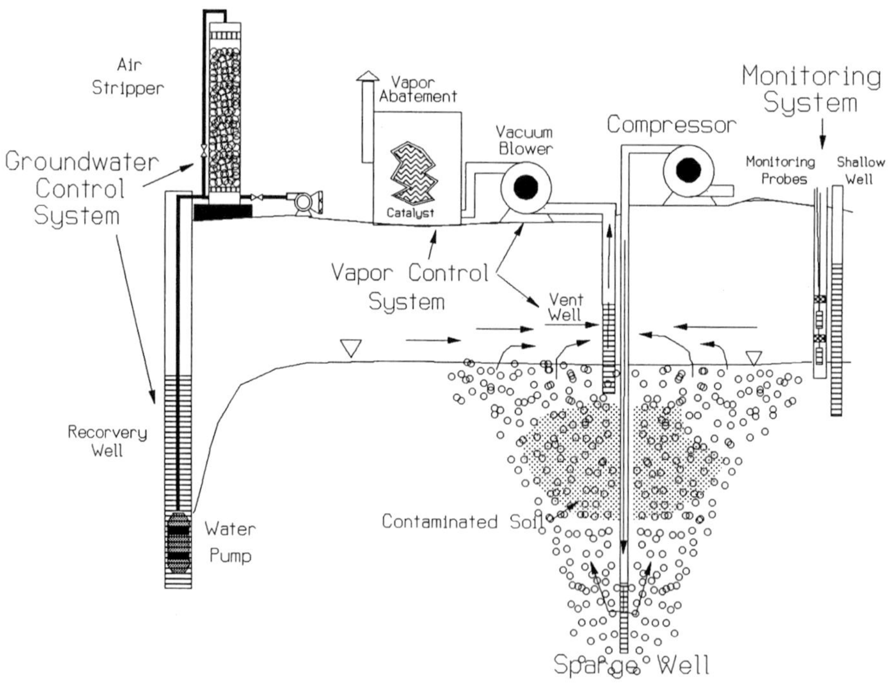

Figure 19.2. Elements of an air sparging system.

section or a diffuser (a porous pipe). The slot or hole size should be of a similar diameter to the average soil pore diameter. This allows for more efficient injection of air by producing smaller air bubbles, which have less resistance to flow (Figure 19.4).

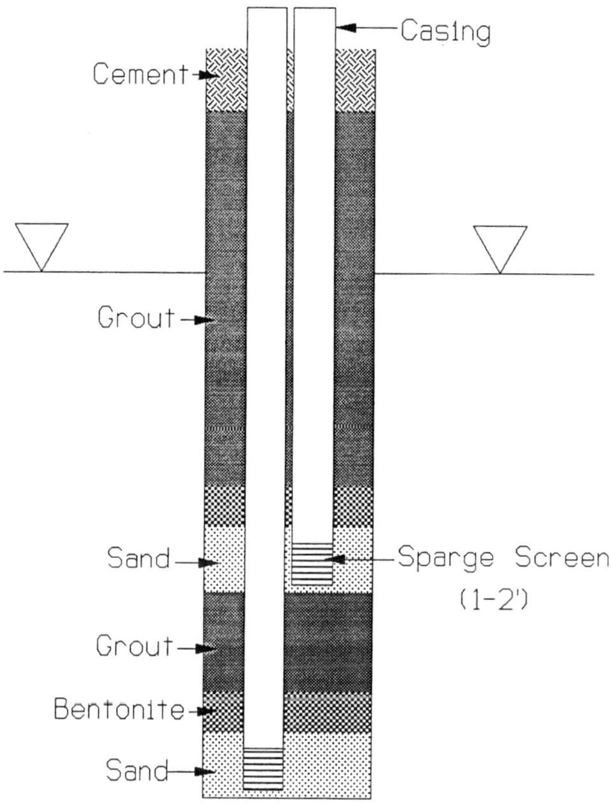

Figure 19.3. Diagram of sparge well.

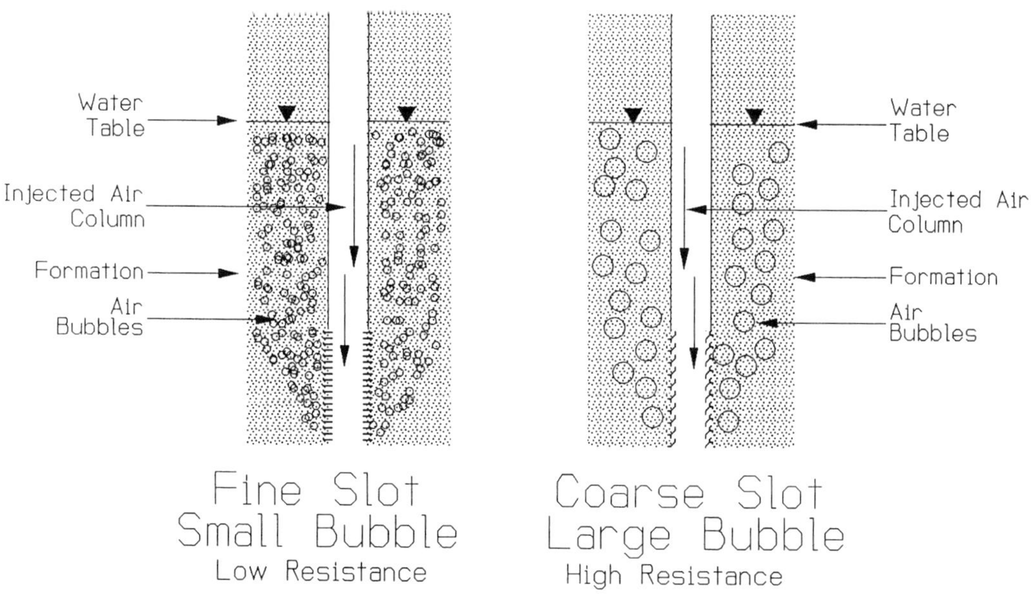

Figure 19.4. Effect of slot size on bubble formation.

The sparge well is completed by placing a sand pack around the pervious section. A 1–2 ft bentonite seal is placed above the sand pack. The well bore is then grouted to the top of the water table or the bottom of a vent or another sparge well. If there is no other well co-located in the well bore, the well bore is grouted to the surface.

If the depth of the sparge well is greater than 20 ft, multiple sparge intervals may be installed, as pictured in Figure 19.3. This nesting of sparge wells allows for greater control of the injected air.

How Does One Size a Compressor?

In sizing a compressor for a sparge system, the two most important characteristics are pressure and flow. Pressure is a function of the soil matrix and depth of the sparge points. Flow is a function of the number, location and construction of the wells. Both the pressure and flow rate for a sparge system have an optimum range. Determining these optimum ranges will be discussed later. However, once the pressure and flow requirements are determined, the compressor should be sized to supply 150% of the desired flow at 125% of the desired pressure.

What Controls May be Necessary With an Air Sparging System?

Because air sparging involves the injection of air into a groundwater aquifer, it can mobilize contaminants in the vapor state and in the dissolved state. Therefore, both vapor controls and/or groundwater controls may be necessary. Vapor controls are necessary if there are vapor receptors within the area of impact of the air sparging operation, or if there is a concern with the fate of the released (sparged) organic vapors. Groundwater controls are necessary if there are potential or impacted groundwater receptors, or if there is an established need for active containment of the groundwater plume.

How Are Organic Vapors Controlled During Air Sparging?

Organic vapors that are mobilized by the air sparging operation are controlled by the application of a soil vapor extraction (SVE) system. The SVE system captures the released vapors by creating a preferential pressure gradient towards the SVE well and away from any receptors (Figure 19.5). Once the vapors are captured, they may be treated or released to the atmosphere, away from any receptors.

How Are Dissolved Contaminants Controlled During Air Sparging?

Air sparging will strip many dissolved VOCs from groundwater, providing that there is sufficient air flow. However, if there is a concern with the migration of dissolved contaminants then a groundwater control system may be needed. The primary groundwater control system consists of a groundwater recovery well(s) and a groundwater treatment unit. A secondary groundwater control system would be barrier sparge wells which ring the treatment area and strip the groundwater of VOCs.

Since air flow does not follow the hydraulic gradient and may flow radially and vertically in any direction from the sparge well, air sparging may therefore "push" contaminants in any direction. Consequently, the groundwater controls would need to be applied in any direction there is a potential receptor, not just in the downgradient direction (Figure 19.6).

How Should an Air Sparging System be Monitored?

A number of parameters can be used to monitor the impact of an air sparging system. The best are DO (dissolved oxygen), water table elevation, pressure (or vacuum if an SVE system is also employed), and VOC concentration. The best monitoring configuration consists of a shallow monitoring well with vapor monitoring probes (Figure 19.7). The well can be used to monitor water table elevation and DO, as well as VOC levels and pressure at the water table. The probes, placed above the well

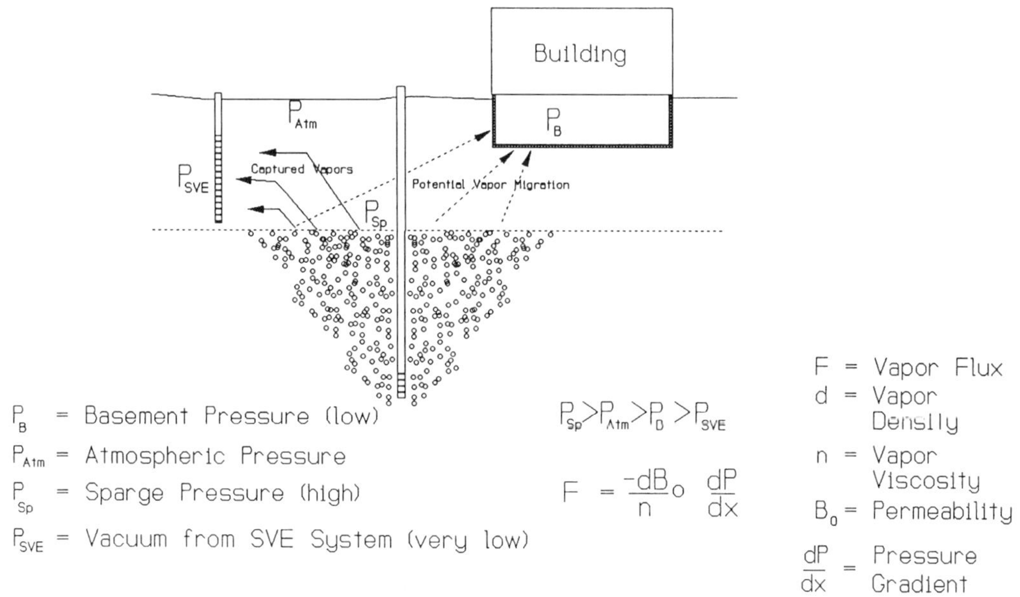

Figure 19.5. Vapor control by applying a preferential pressure gradient.

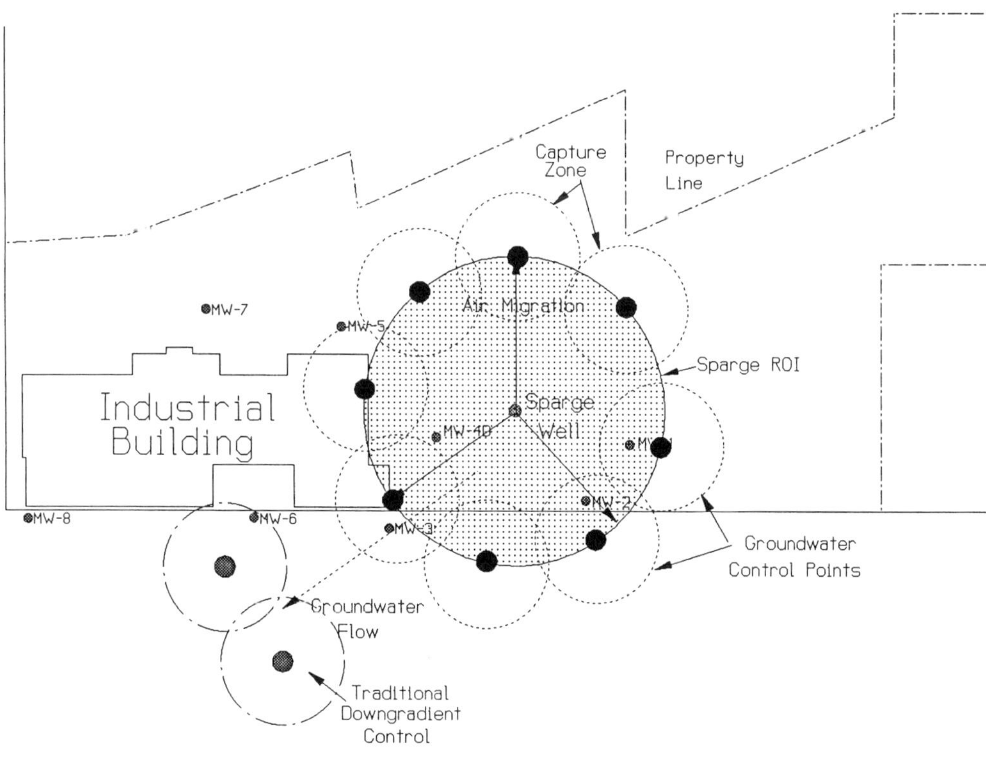

Figure 19.6. Groundwater control for air sparging compared with traditional method.

in the vadose zone, are used to monitor pressure (or vacuum if an SVE system is also employed) and VOC vapor concentration.

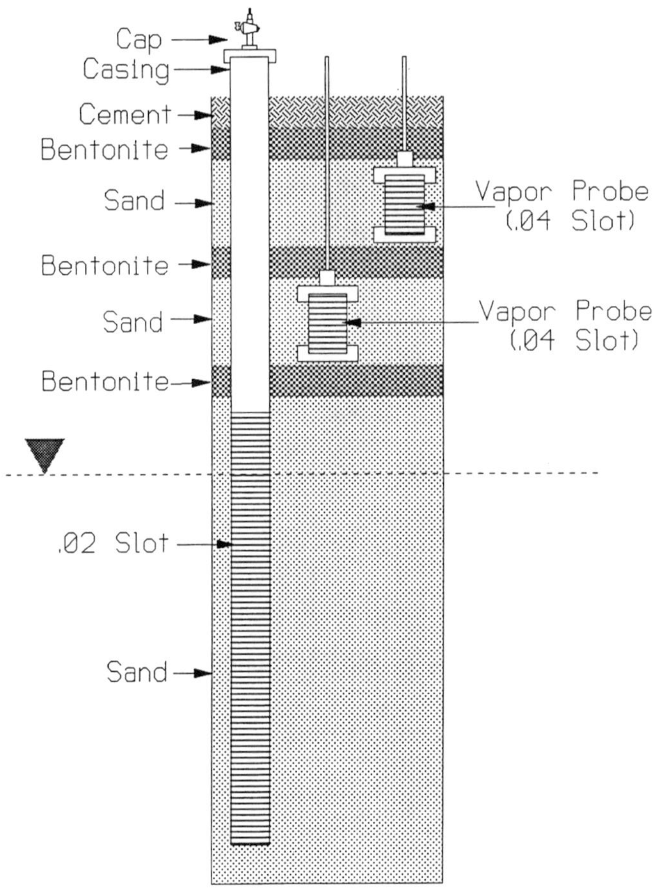

Figure 19.7. Monitoring point for sparging systems.

In addition to the above parameters, an active tracer such as helium can be used to monitor an air sparging system. The helium is injected into the air stream prior to the sparge well at a concentration of several percent. The concentration of helium that appears in wells or in vapor probes can be used to calibrate the sparging system.

BENEFITS OF AIR SPARGING:

What Are the Benefits (Effects) of Air Sparging?

There are four potential benefits that may be derived from the use of air sparging. Three of these are primary benefits:

- the volatilization and removal of adsorbed VOCs;
- the enhanced biodegradation of adsorbed and dissolved organic compounds in saturated media due to increased oxygenation; and
- the *in situ* stripping of dissolved VOCs from groundwater.

In addition to the effective treatment of dissolved and adsorbed organic compounds through volatilization, biodegradation, and stripping, air sparging provides a fourth benefit—the control of dissolved iron and manganese.

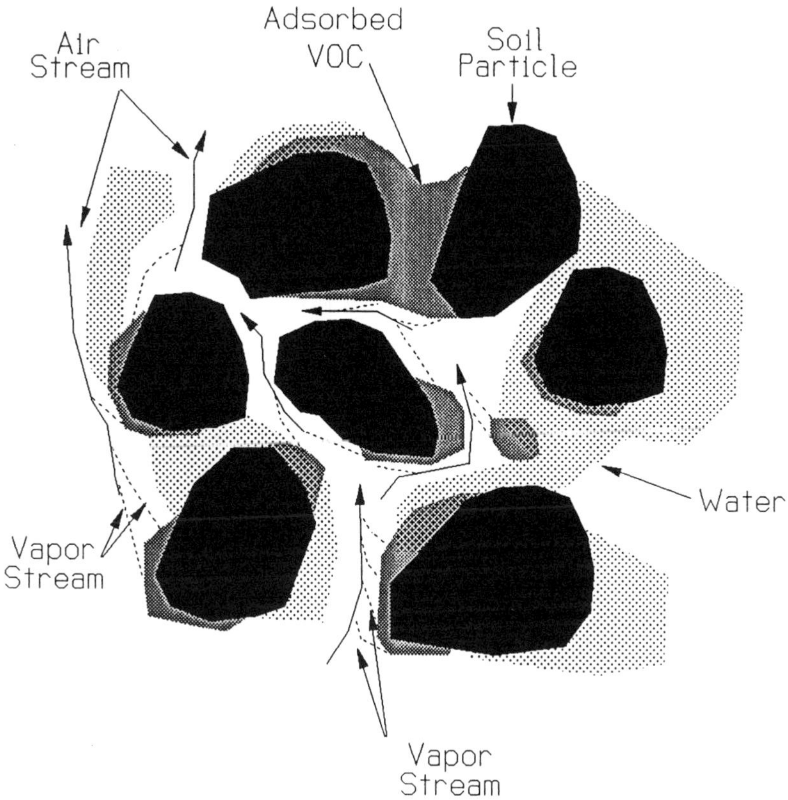

Figure 19.8. Volatilization of adsorbed VOCs during sparging.

How Does Air Sparging Remove Volatile Organic Compounds from Soil?

Volatile organic compounds (VOCs) can be removed from soils by direct volatilization. In unsaturated soils, the VOCs adsorbed to the soil are in equilibrium with their vapors. This equilibrium is defined by the vapor pressure of the VOC. If the adsorbed VOC is present above the water table it may be removed by inducing air flow by the application of a vacuum. Air passing through the soil removes the vapors from the soil pores, which causes more of the adsorbed VOC to volatilize to re-establish the vapor equilibrium.

Below the water table the volatilization of VOCs is suppressed because there is no air-filled porosity. As a result, the direct vapor phase removal of VOCs is precluded. Air sparging addresses this problem. The injection of air into the saturated soil matrix displaces water from the soil pores, causing a transient air-filled porosity or partial saturation. Adsorbed VOCs exposed to this transient air-filled porosity will evaporate into the airstream and be carried out of the saturated soil matrix (Figure 19.8), where they can be captured by a soil vapor extraction system.

What Proof Is There That Air Sparging Does, In Fact, Remove Adsorbed VOCs?

There are two methods of establishing the fact that air sparging removes VOCs from below the water table. The most effective is to measure soil concentrations below the water table before and after air sparging. The second is to measure VOC vapor concentrations in the vadose zone before and during air sparging.

While the first method is perhaps more conclusive, direct soil data for air sparging operations is not readily available. There are several reasons for this lack of data. First, it is difficult to collect and analyze contaminated soil samples from below the water table. There is always a question as to whether the VOC levels measured are adsorbed or dissolved VOCs entrained in the soil pores. Second, the concern with saturated phase contamination is usually the groundwater concentrations and, as a result, dissolved concentrations are more commonly monitored. Third, air sparging is a relatively new technology and, therefore, few complete case histories exist.

There is, however, considerable data that shows the release of VOCs from below the water table during air sparging. This data was measured in two ways:

1. concentrations of VOCs observed in the vadose zone during the injection of air; and
2. change in VOC levels in a SVE system with the addition of an air sparge system.

Both measurements show significant increases in VOCs during sparging. This data, taken for both chlorinated solvents and petroleum hydrocarbons, is summarized in Table 19.1.

How Does Air Sparging Enhance Biodegradation?

Biodegradation in the saturated zone is limited by the availability of oxygen. When a degradable organic is found in groundwater or in soils below the water table, bacteria will metabolize the organic utilizing the available DO. The groundwater often then becomes anoxic or anaerobic. This is evidenced by the by-products of anaerobic metabolism such as the accumulation of dissolved iron (Fe II) or the presence of partially dechlorinated organics. Additionally, DO measurements in areas of labile contaminants are often below 2 mg/L as opposed to a typical 5–6 mg/L for uncontaminated groundwater. Once depleted, dissolved oxygen is slow to rebound since it must diffuse into the aquifer from the surface of the water table. This process is slow because the diffusive path length is long.

Air sparging increases the availability of DO by decreasing the diffusive path length of air into the water phase (Figure 19.9). By creating a transient air-filled porosity in the soil pores by dispersing bubbles throughout the soil matrix, the diffusive path length is shortened from on the order of 100–1000 mm to 0.01–0.05 mm, a decrease of up to five orders of magnitude. With this shortening of the diffusive path length, DO supply and, therefore, biodegradation is significantly enhanced.

What Evidence Is There That Air Sparging Will Enhance Biodegradation?

The evidence is both direct and indirect. The direct evidence is the increase in CO_2 observed during air sparging. Concentrations of CO_2 of 8–12% have been observed during air sparging of petroleum hydrocarbons at depths of 12–25 ft below the water table. The indirect evidence is the increase in DO levels measured in groundwater during air sparging operations. Within 0.5 to 2 hr after sparging is initiated, DO levels in aquifers contaminated with petroleum hydrocarbons will often increase from 1–2 mg/L to saturation (>10 mg/L).

Table 19.1. Increase in VOC Levels Observed During Air Sparging Pilot Test

Sparge Flow scfm	Vent Flow scfm	VOC Conc., ppmv Before Sparging	VOC Conc., ppmv During Sparging
8	0	1670[a]	29,000
10	100	330[a]	19,000
8	0	1200[b]	4,500
25	0	890[a,b]	>10,000
15	55	100[b]	1,000

[a] Petroleum hydrocarbons.
[b] Chlorinated solvents.

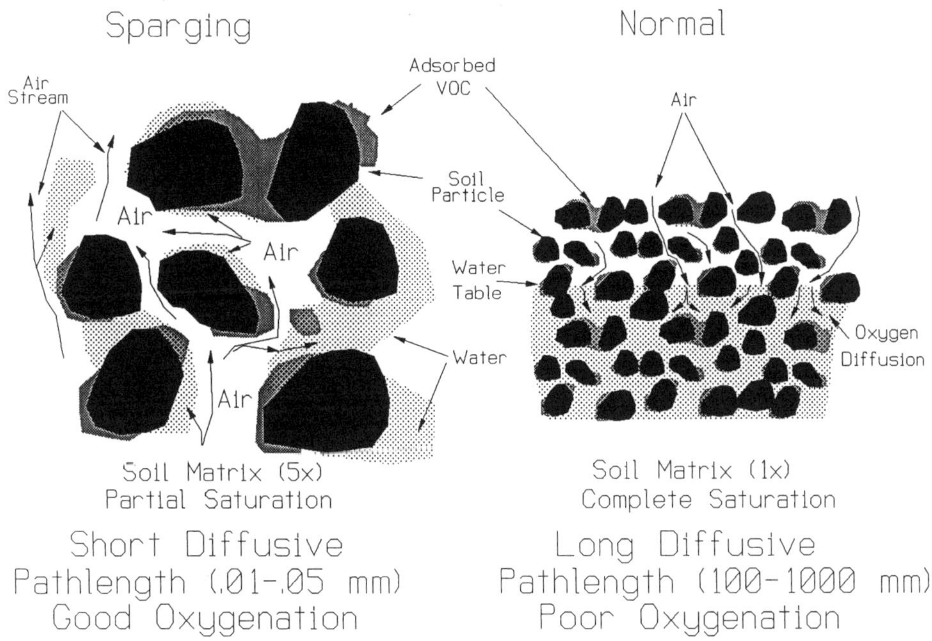

Figure 19.9. Enhanced oxygenation during sparging.

How Does Air Sparging Affect Dissolved VOCs?

Air sparging can act as a crude air stripper, with the soil matrix acting as the packing (Figure 19.10). Air traveling through the soil contacts contaminated water and strips the dissolved VOCs. The effectiveness of air sparging in treating contaminated groundwater is a function of two factors—the strippability of the VOC as measured by its Henry's Law constant, and the air-to-water ratio (volume to volume) of the sparging operation. The Henry's Law constant, K_h, must be $>10^{-5}$ for a strippable volatile constituent. Table 19.2 lists Henry's constants for some common VOCs. As can be seen, BTX components and most chlorinated solvents are readily strippable; soluble contaminants such as acetone are not. Even with a strippable contaminant, there needs to be sufficient air volume to remove the contaminant. The air-to-water ratio should be between 10:1 to 20:1. Below a 10:1 ratio, groundwater contamination may persist during air sparging. The air-to-water ratio may be estimated by dividing the volume of air added per unit time by the volume of water contained in a cone defined by the depth of the sparge point and the radius of influence.

What Evidence Is There That Air Sparging Can Strip Dissolved VOCs?

The rapid drop in groundwater concentration in downgradient monitoring wells is a good indication of a stripping operation. Table 19.3 shows a rapid drop in dissolved VOCs after only a short period of time for a couple of sites.

What Effect Does Air Sparging Have on Dissolved Metals?

Iron (Fe) and manganese (Mn) are mobilized by biological activity under anoxic conditions. They are both electron acceptors in their oxidized insoluble states (Fe^{+3}, Mn^{+4}) and are converted to the more soluble reduced state (Fe^{+2}, Mn^{+2}) by the bacterial metabolism of organic compounds.

During sparging these reduced species (Fe^{+2}, Mn^{+2}) are rapidly oxidized by oxygen to the original, less soluble Fe^{+3} and Mn^{+4}:

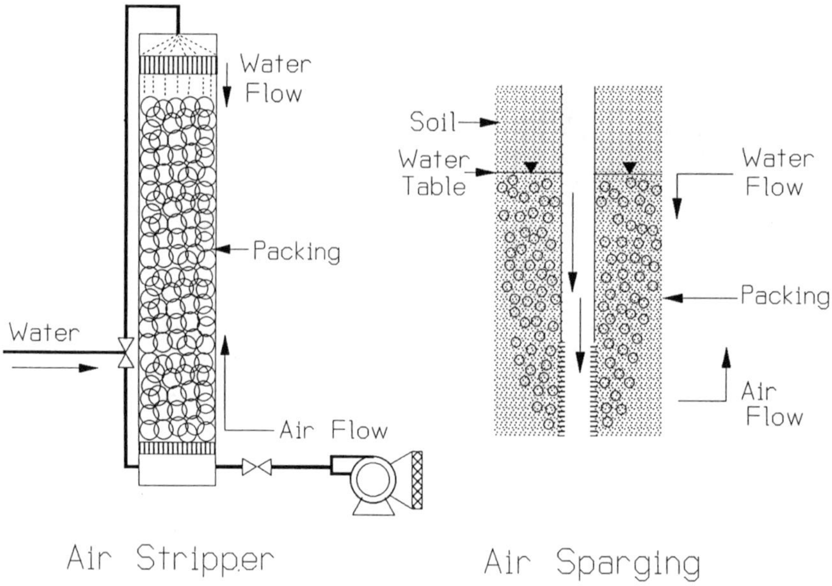

Figure 19.10. Comparison between air sparging and air stripping operations.

Table 19.2. Henry's Constant for Selected VOCs

Constituent	Henry's Constant K_h (atm-m^3-mole^1)	Vapor Pressure (mm Hg)
Naphthalene	4.1×10^{-4}	0.1
Benzene	5.6×10^{-3}	28.0
Toluene	6.3×10^{-3}	9.0
Xylene	5.7×10^{-3}	3.0
Tetrachloroethylene	1.5×10^{-2}	14.0
Trichloroethylene	9.9×10^{-3}	28.0
trans-1,2-Dichloroethene	9.4×10^{-3}	200.0
Acetone	6.8×10^{-6}	76.0

$$4Fe^{++} \rightarrow 4Fe^{+++} + 4e^- \qquad E^o = -0.77\ v$$

$$O_2 + 4H^+ + 4e^- \rightarrow 2H_2O \qquad E^o = +1.23\ v$$

$$4Fe^{++} + O_2 + 4H^+ \rightarrow 4Fe^{+++} + 2H_2O \qquad E^o = +0.46\ v$$

$$Fe^{+++} + 3OH^- \rightarrow Fe(OH)_3 \qquad K^{sp} = 4 \times 10^{-38}$$

$$2Mn^{++} + 4H_2O \rightarrow 2MnO_2 + 8H^+ + 4e^- \qquad Eo = -1.23\ v$$

$$O_2 + 4H^+ + 4e^- \rightarrow 2H_2O \qquad E^o = +1.23\ v$$

$$2Mn^{++} + O_2 + 2H_2O \rightarrow 2MnO_2 + 4H^+ \qquad E^o = +0.00\ v$$

MnO_2 (Mn^{+4}) is insoluble.

In addition to this oxidation of the soluble metals to less soluble species, air sparging prevents further biological mediated reduction by increasing the oxygen content of the groundwater and thus correcting the anoxic conditions.

Table 19.3. Change in Groundwater VOC Concentration (PPB) with Time.

Site A		
Monitoring Well #	**Before Startup**	**54 Days Sparge**
MW-1S	2,108	4
MW-3S	2,161	2
MW-4S	4,328	444
MW-5S	6,940	357
Site B		
Monitoring Well #	**Before Startup**	**2 Months Sparge**
MW-3	10,230	4,530
MW-4S	173	67
MW-5	48	18
MW-7	473	18

What Proof Is There That Air Sparging Can Reduce Dissolved Metal Levels?

There is a commercial patented technology called Vyredox which used a form of air sparging to drop out iron and manganese in an aquifer so that a water well would be protected. The Vyredox process used an array of aeration wells placed radially around a water recovery well.

LIMITATIONS OF AIR SPARGING

If Air Sparging Is Simply the Injection of Air, Why Is There a Concern with Limitations?

While air sparging appears to be a simple technology—the injection of air—it does have an inherent risk—the spread of contamination. The injected air can displace water in the soil matrix, and it can volatilize, entrain and spread volatile contaminants. Both the displaced water and the injected airstream must be adequately controlled to prevent the spread of contamination. Anything that impacts this control will limit the effectiveness of air sparging.

What are the Limitations of the Use of Air Sparging?

There are two primary types of limitations of the use of air sparging—physical and operational. Physical limitations are any conditions that restrict the flow of air through the soil matrix, such as soil permeability, geological heterogeneity, and depth. Operational limitations are conditions which impact the control of or the beneficial effect(s) of air sparging. These include pressure, flow, and depth.

How Does Permeability Affect Air Sparging?

The permeability of the soil matrix must be sufficient to allow the movement of air. This air movement must be both horizontal as well as vertical to allow the injected air to contact the area of contamination and to exit the formation so that it may be captured. Thus, one needs to consider both the horizontal as well as the vertical permeability. With respect to horizontal permeability, the matrix must be transmissive to air flow. Thus, in general, competent rock, soils having high silt or clay contents, or soils having hydraulic conductivities of $< 10^{-5}$ cm/sec will generally not be amenable to

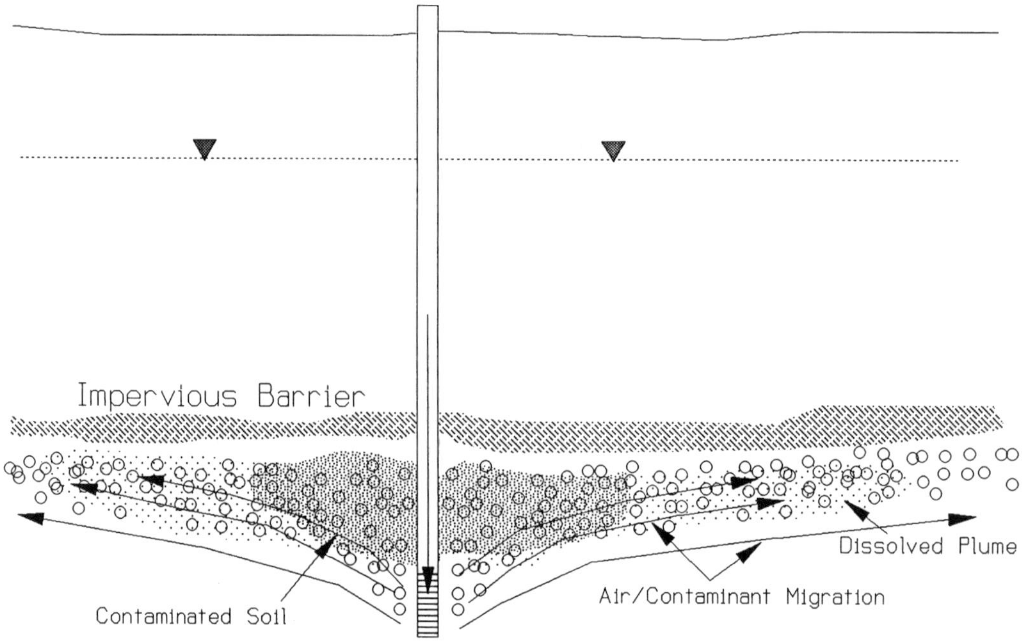

Figure 19.11. Inhibited vertical air flow due to impervious barrier.

air sparging due to the difficulty of injecting air into such matrices. Weathered rock, however, may be sufficiently permeable to allow air sparging. With respect to vertical permeability there should be no low-permeability barriers between the sparge interval and the surface of the water table. Even if the general permeability is sufficient to allow for air injection, the presence of low-permeability barriers above the sparge interval can trap the injected air and channel it and any entrained contaminants away from the treatment area (Figure 19.11), thus spreading the contamination. A sparge point *should never be set below* a low-permeability zone.

How Does One Ensure that Permeability Will Not Be an Impediment to Air Sparging?

The best insurance for preventing permeability problems is to adequately characterize the lithology of the site, both to determine the general soil conditions and to determine if there are any low-permeability barriers above the sparge interval. Ideally each sparge well should be continuously cored from the surface of the water table to ensure that the sparge interval will not be set below a low-permeability barrier.

What Is the Importance of Soil Heterogeneity in Air Sparging?

Any changes in permeability or in soil structure have the potential for trapping or channeling air flow. Air will flow preferentially through areas of high permeability. If a high-permeability layer exists above the sparge interval, air flow can be channeled (Figure 19.12).

The are two factors to consider with respect to heterogeneity. The first is the degree of heterogeneity. In general, the higher the degree of heterogeneity the higher the risk of channeling. Highly layered soils are not amenable to air sparging. The second factor is how the heterogeneity grades with depth. If layering exists, it may not impact the use of air sparging as long as the permeability increases as one approaches grade. If the opposite is true, that is, the permeability increases with depth, then the layering can potentially trap and channel the airstream.

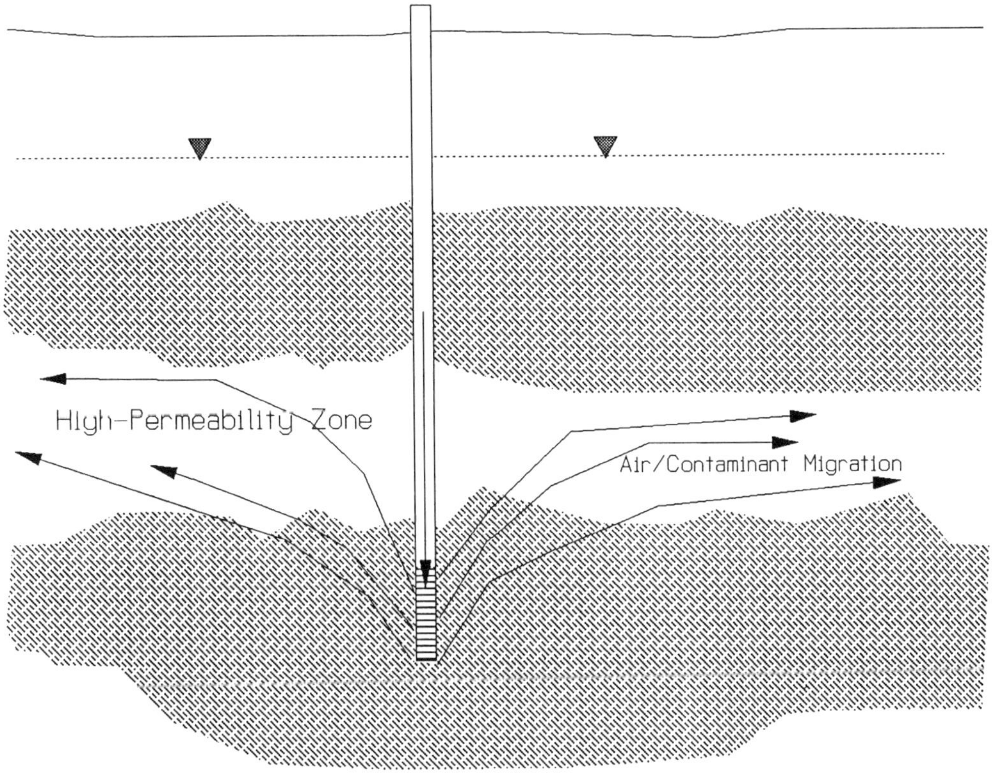

Figure 19.12. Inhibited vertical air flow due to channeling in high-permeability zone.

How Does One Determine if Heterogeneity Will Impact Air Sparging?

The first means of determining if heterogeneity is a problem is to characterize the lithology of the site. This, as with the discussion of permeability, is best done by continuously coring all sparge points below the water table. The permeability of retained soil cores can be evaluated in a laboratory either by sieve analysis or by actual air permeability testing. Layering is determined by changes in permeability.

A second method of evaluating the potential impact of heterogeneity is to conduct a tracer test. A gas tracer such as helium is injected through the sparge well and monitored with distance and orientation from the sparge point. In homogeneous soils, the dispersion of the tracer will be radial. Any channeling will cause a distortion of this radial distribution. The distortions can locate preferential pathways or barriers.

How Do Operating Conditions Affect the Use of Air Sparging?

The goal in the application of air sparging is to maximize the benefits of air sparging while maintaining control and minimizing the potential detrimental impacts. The primary operating parameters are pressure, flow rate, and depth. In addition to these parameters, another operating consideration is the existence and impact of water table mounding.

What Impact Does Pressure Have on Air Sparging?

Air pressure is what drives an air sparging system. The minimum pressure is that which will overcome the hydraulic head and displace the water column in the sparge well, allowing air to enter the soil matrix. This minimum pressure is roughly 1 psi for every 2.3 ft of water column. This pressure, termed the "breakout pressure," may be greater than 1 psi/2.3 ft due to the resistance to air flow of the soil matrix surrounding the sparge well.

There is also a maximum pressure. A common assumption is that increasing the injection pressure simply increases the radius of influence. The purported benefit is a greater treatment. The risk, however, is that when the pressure is too high, turbulent flow can occur. Under turbulent flow, the injected air will physically displace the water, rapidly spreading any contamination (Figure 19.13). This is graphically illustrated when air rotary drilling, with pressures of > 200 psi, causes "geysers" in adjacent, uncapped monitoring wells. Under high pressure, control of the air sparging system is lost. It is not true that more is always better.

The impact of injection pressure can be seen from test data. Figure 19.14 depicts the relationship between horizontal response expressed as the ratio of radius of influence to sparge depth, and injection pressure expressed as ratio of pressure to breakout pressure. As pressure is increased above the breakout pressure, air is driven horizontally from the well. There is, initially, a linear relationship between the injection pressure and the horizontal response. However, once the injection pressure is more than three times the breakout pressure, the response becomes non-linear. The force of the air becomes excessive. This is the point at which physical movement/displacement of water in the formation can occur. The design of an air sparging system, therefore, necessitates finding the optimum pressure range to provide the desired radius of influence. The maximum operating pressure, however, should never exceed three times the breakout pressure.

How Does the Air Flow Rate Affect Air Sparging?

The injection of air during air sparging provides several benefits. The benefits are a function of the air flow rate, or volume of air injected. At low flow rates the air oxygenates the water, enhancing biodegradation. At low to moderate flow rates, air sparging will enhance the partitioning of adsorbed phase contaminants into groundwater by increasing the mixing of groundwater with the contaminated soil matrix (Figure 19.15). This is the premise of an enhanced groundwater recovery process. This enhanced partitioning can be a benefit if a groundwater recovery system is operated concurrently with the air sparging system. However, if the dissolved plume is not captured, this enhanced partitioning can increase downgradient contamination as the dissolved plume is carried with the groundwater flow. At moderate to high flow rates, stripping of dissolved contaminants will occur, providing rapid treatment of contaminated groundwater. The point at which stripping occurs may be estimated by determining the air-to-water ratio. The air-to-water ratio may be estimated by dividing the volume of air added per unit time by the volume of water contained in a cone defined by the depth of the sparge point and the radius of influence. The air-to-water ratio should be between 10:1 to 20:1. Below a 10:1 ratio groundwater contamination may increase due to enhanced partitioning. At very high air flow rates, there is a risk of physically displacing the groundwater and spreading the contamination.

The impact of air sparging is dependent on the air flow rate. Whether air sparging is beneficial or detrimental depends on how the air flow rate is designed and controlled. There is an optimal operating range, where the air flow is sufficient to oxygenate groundwater, volatilize adsorbed contaminants, and strip dissolved contaminants, but not high enough to displace groundwater (Figure 19.16).

How Does Air Sparging Affect the Water Table?

When air is injected into a saturated matrix, the air displaces the water, creating a partially saturated environment. Since the air travels both horizontally and vertically, this displacement can cause an apparent mounding of the water table (Figure 19.17). When the air pressure is released, there is a rapid collapse of the water table as the entrained air degasses and the partially saturated air becomes, once again, fully saturated (Figure 19.17). Eventually, as water flows into the sparged area, the water table rebounds to the original level. The mounding and collapse of the water table are readily observed in monitoring wells in the area being sparged, as shown in Table 19.4. The observed water

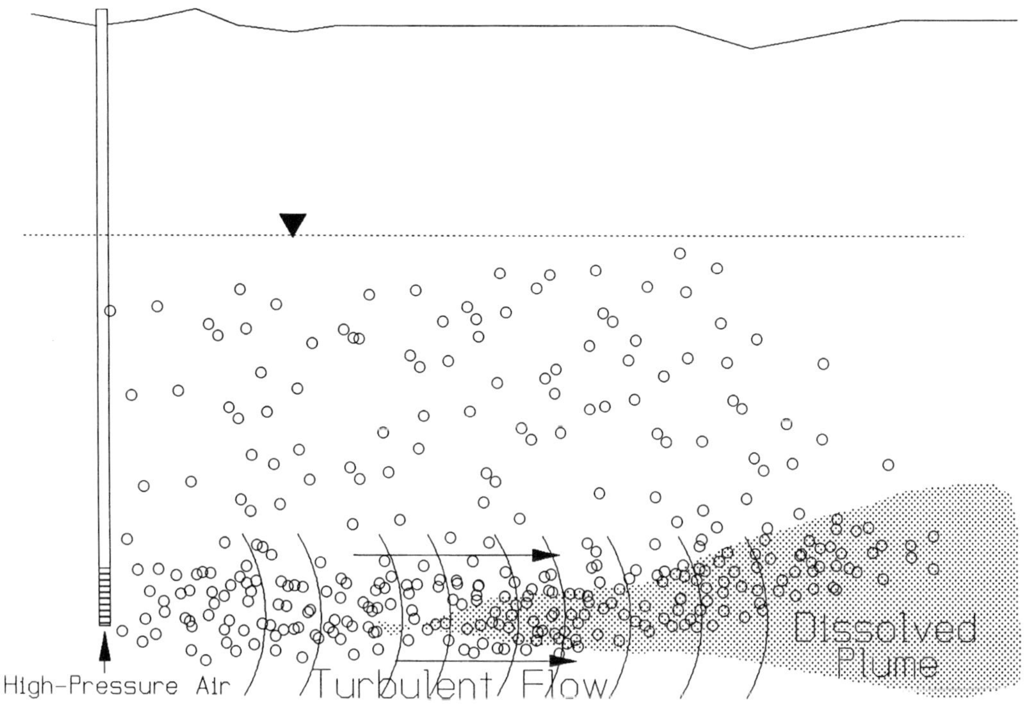

Figure 19.13. Accelerated dissolved migration due to turbulent flow caused by overpressurization.

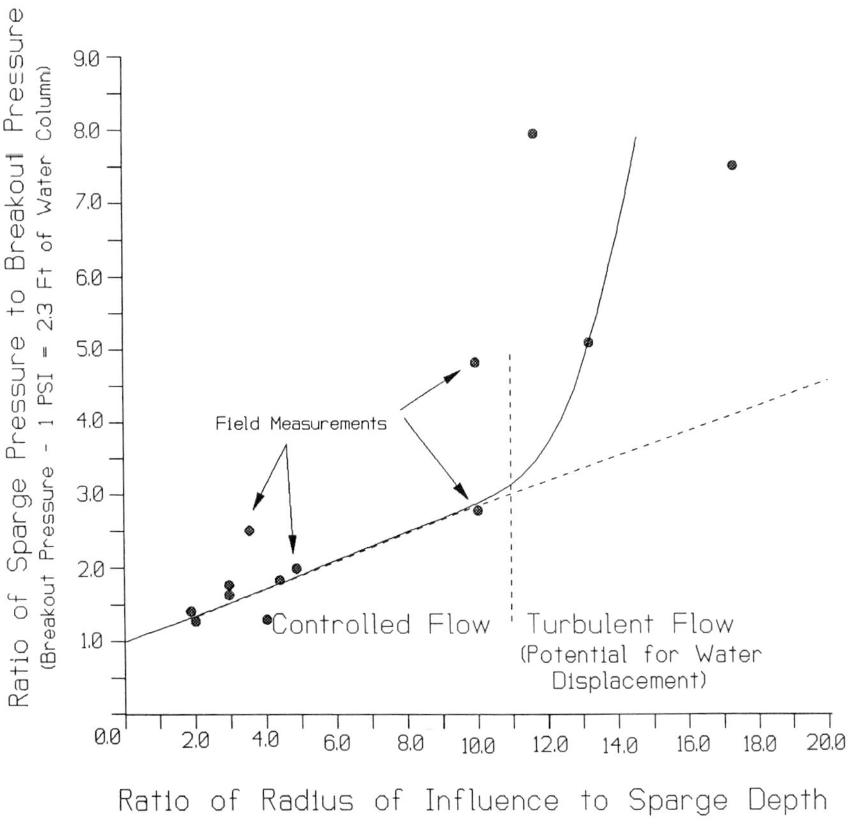

Figure 19.14. Impact of injection pressure on air sparging.

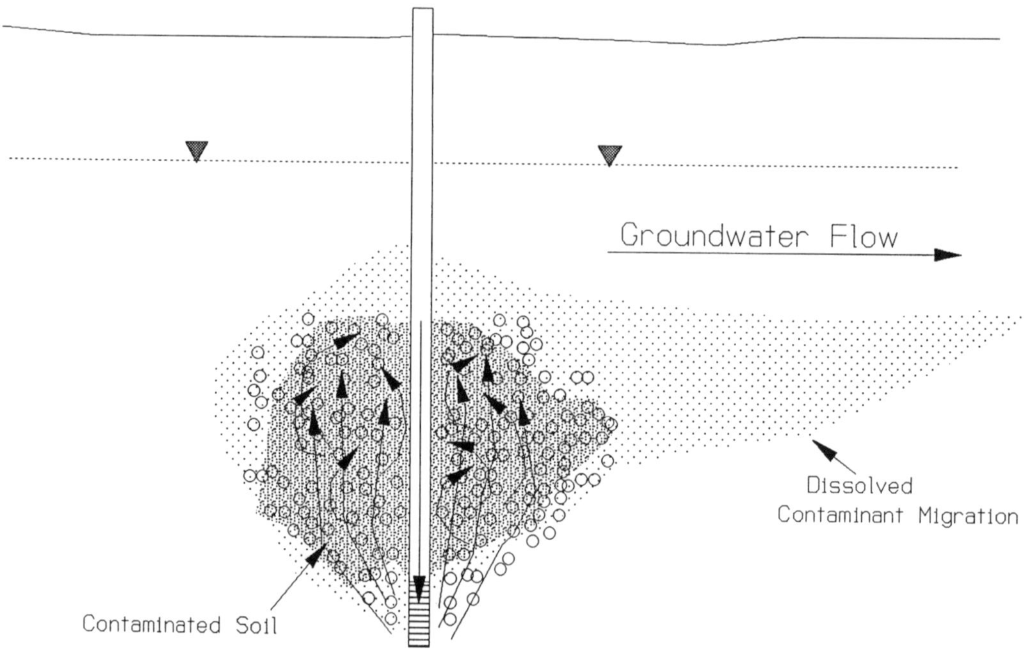

Figure 19.15. Dissolution of adsorbed phase due to mixing at low air flow rates.

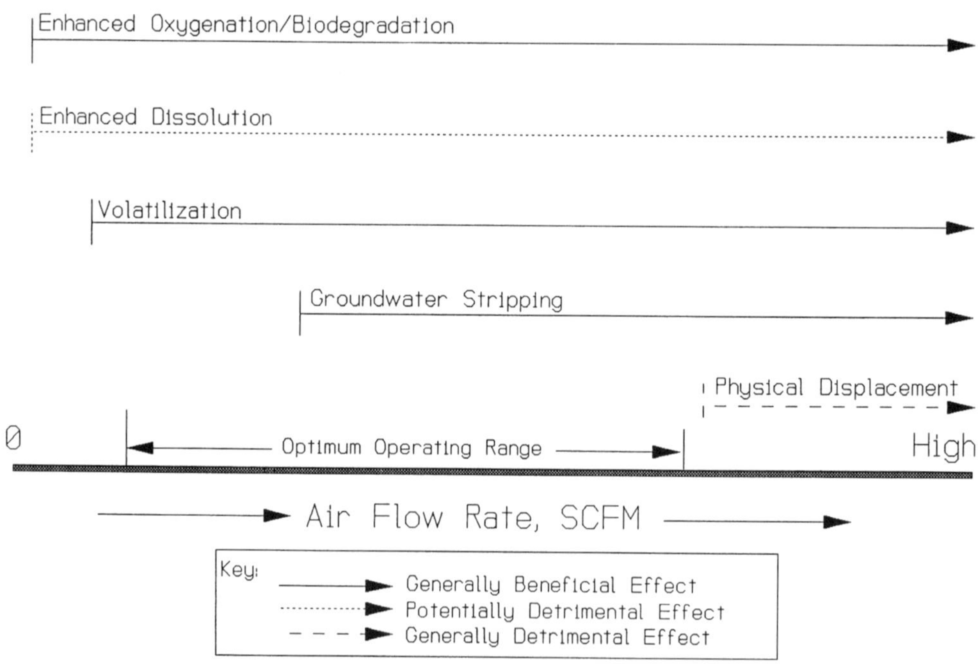

Figure 19.16. The effects of air sparging on transport/treatment as a function of air flow rate.

levels in the monitoring wells may be artificially accentuated by localized pressure effects—air pressure may support a water column in the well that exceeds the actual mounding. In addition to these monitoring well observations, air sparging has, on several occasions, actually driven water out of a formation with a shallow water table (< 3 ft).

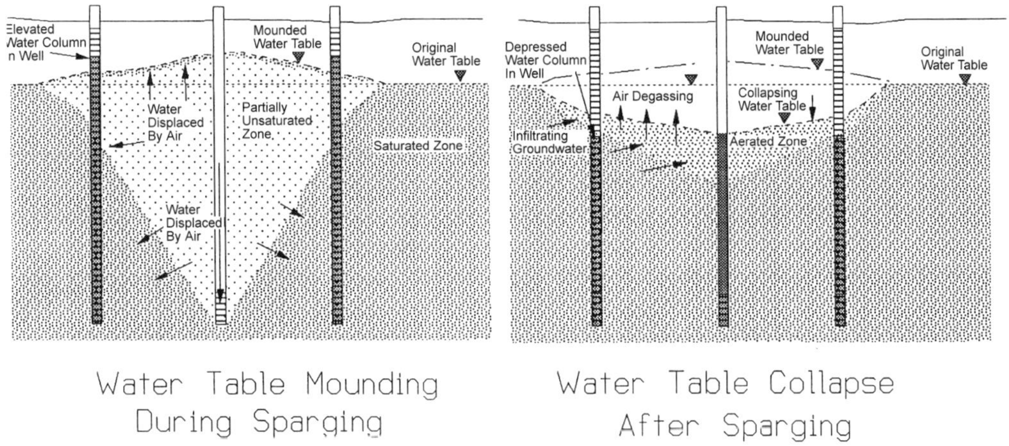

Figure 19.17. Effect of sparging on water table.

Table 19.4. Water Table Mounding and Collapse

		Water Table Elevation at:			
Well #	**Distance, ft**	**Static**	**Sparging**	**5 min After**	**10 min After**
MW-7	5	6.46	4.09	10.03	6.96
SE-19	19	6.42	6.20	6.93	6.54
S-26	29	6.71	6.55	6.96	6.77
NE-13	13	6.52	6.11	7.44	6.75

What Impact Does Water Table Mounding Have on Groundwater Flow and Contaminant Migration?

Since water table mounding occurs through the displacement of water with air, an obvious and important question is what effect does this displacement have on groundwater flow and contaminant transport. To answer this question, one needs to consider the mounding from both a short-term and long-term perspective. Short-term water table mounding causes some groundwater movement. Long term, it would appear that the water table mounding results in static groundwater flow conditions and little or no impact on contaminant transport.

When air is injected under moderate flow/pressure, water is displaced by dispersed air bubbles. The flow of water out of the sparge area is in all directions, with the distribution determined by the resistance to flow, which is a function of permeability, saturated thickness, and layering. Because the air bubbles are dispersed throughout the soil matrix, the actual volume of water displaced is small—10–25% of the saturated volume, based on a comparison of the water mounding observed with the sparging depth. Also, because the water and air bubbles are "mixed" in the soil matrix, the air strips any dissolved contaminants from the groundwater as it is displaced. Thus, the short-term impact is negligible.

Longer term, the sparge area becomes static with respect to groundwater flow. There are several reasons for this assertion. First, the hydraulic conductivity of the sparge area decreases as the water is displaced with air. Thus, groundwater flow is diverted around the sparge area. Second, the fluid density is less than static groundwater because the fluid column is actually a mixture of air and water. Thus, the true piezometric surface is less than what may be inferred from the apparent water table mounding.

Long term, a properly operated air sparging system will not cause or increase the spread of groundwater contamination. This is because, first, groundwater flow through the area is, as discussed above, static. Second, the air flow through the contaminated groundwater strips the dissolved VOCs. Third, the air stream also volatilizes and removes the adsorbed contaminants, thus reducing the long-term source of groundwater contamination. Rather than spreading groundwater contamination, a properly operated sparge system will in fact rapidly reduce groundwater contamination.

Under improper operating conditions, air sparging can potentially spread contamination. As discussed previously, with high flow or pressure the airstream physically displaces the groundwater. Rather than dispersed bubbles creating a partial saturation, high pressure/flow creates a single "bubble" which moves the entire water column, creating an unsaturated environment. Because the air does not mix with the groundwater, it does not strip the contaminants and the displaced groundwater spreads contamination.

DESIGN OF AN AIR SPARGING SYSTEM

What Is the Best Approach to Designing an Air Sparging System?

The best approach to designing an air sparging system is to use a phased approach. The first phase is to assess the applicability of air sparging to a particular site or contamination situation. The second phase is to delineate the area of contamination to be treated (horizontal and vertical delineation). The third phase is to determine the air flow properties of the site and the air flow requirements of the system. The fourth phase, the system layout, is an overlay of the air flow properties (well location and spacing) with the area of contamination. The final phase, equipment specification, is the combining of the number of wells with the air flow requirements (flow, pressure).

How Does One Determine if Air Sparging Is an Applicable Remediation Technology?

The applicability of air sparging is a function of the properties of the contaminant and of the geology/hydrogeology of the site (Table 19.5) and (Figure 19.18). The contaminant must be either volatile, biodegradable, or strippable. The site geology/hydrogeology must be conducive to the flow of air, both vertically and horizontally. These are the properties of contaminants that are ideally suited to air sparging. Contaminants outside of these values may still be amenable to air sparging however, the treatment time and effectiveness may be substantially reduced. Table 19.6 gives the treatability of a number of common contaminants.

What Type of Geological Conditions Are Suitable for Air Sparging?

Air sparging is not applicable to all geological conditions. The site conditions must allow for the unimpeded flow of air both horizontally and vertically (Figure 19.18). The important geologic characteristics are heterogeneity and permeability. Other important site conditions are depth and saturated thickness. Heterogeneity is important because layering can cause barriers or channelling, which can lead to loss of control of the injected air stream and could spread contamination. Permeability needs to be considered both horizontally and vertically. A soil with a low ratio of horizontal to vertical permeability can be of lower general permeability than a soil with a high ratio, as there is less risk of channeling. A minimum saturated thickness is necessary to prevent the injected air from "blowing out" the injection well. The following table summarizes the desirable site conditions for air sparging:

Heterogeneity	—	No impervious layers above sparge point.
	—	If layering is present, permeability increases toward grade.
Permeability	—	$>10^{-5}$ cm/sec if Horz:Vert Permeability is < 3:1.
		$>10^{-4}$ cm/sec if Horz:Vert Permeability is > 5:1.
Physical	—	Saturated thickness > 4 ft.
	—	Sparge depth < 30 ft.

Table 19.5. Contaminant Properties Suited to Sparging.

Property	Desired Value
Volatility	> 1mm Hg
Biodegradability	BOD_5 > 0.01 mg/L, or BOD_5:ThOD > 0.001
Strippability	$K_H > 10^{-5}$ (atm-m^3-mole^{-1})
Solubility[a]	< 20,000 mg/L

[a] If not biodegradable.

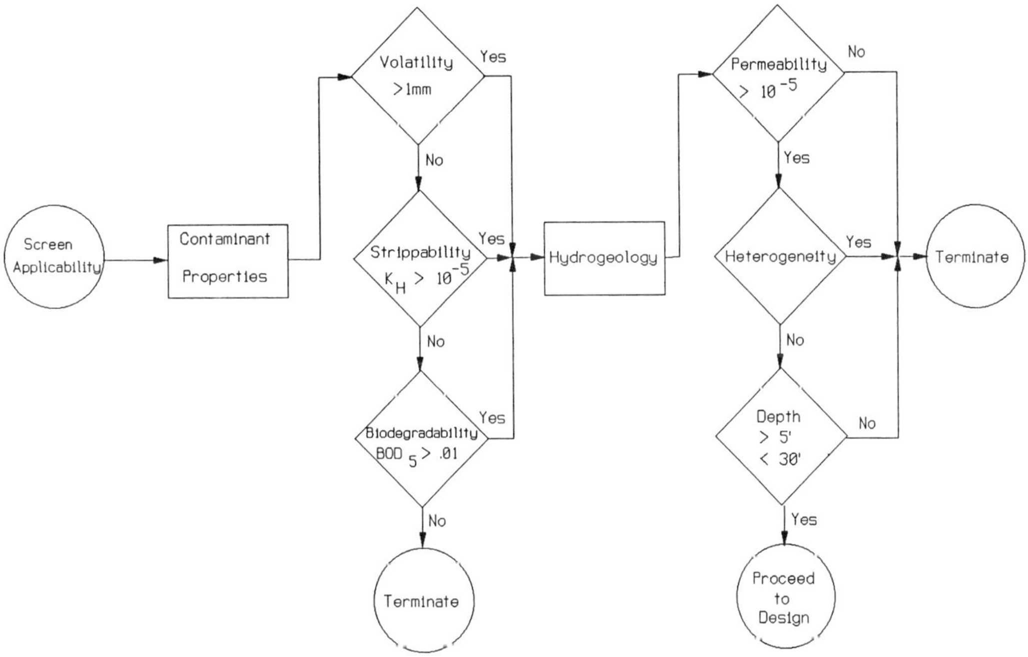

Figure 19.18. Screening procedure for determining the applicability of air sparging.

Air sparging may be operated outside of these site conditions; however, the performance may be substantially reduced and the risk of losing control and spreading contamination is substantially increased.

How is an Air Sparging System Designed?

The design process is a stepwise process (Figure 19.19). The two elements necessary to design an air sparging system are the contaminant distribution and the flow properties of the site. The contaminant distribution is the vertical and horizontal extent of contamination. This defines the area to be treated. The flow properties are the radii of influence (ROI) of the sparge and soil vapor extraction systems at a given flow/pressure. These two elements define the number and location of the sparge and vent wells. The flow rates and pressures determined during the ROI tests, combined with the number of wells, provide the basis for the equipment specifications.

What Is Necessary to Define the Contaminant Distribution?

An important part of the design process is to determine the contaminant distribution. This consists of delineating both the horizontal and vertical extent of contamination. The goal of the delineation is

Table 19.6. Treatability of Various Contaminants

Contaminant	Biodegradability	Volatility	Strippability	Solubility
		High Treatability		
Benzene	High	Moderate	High	Low
Toluene	High	Moderate	High	Low
Xylenes	High	Moderate	Moderate	Low
Gasoline	High	High	High	Low
TCE	Very Low	High	Moderate	Moderate
1,1,1-TCA	Very Low	High	Moderate	Moderate
PCE	Very Low	High	Moderate	Moderate
		Moderate Treatability		
Naphthalene	Moderate	Low	Moderate	Low
Acetone	High	High	Very Low	Very High
Diesel	High	Low	Moderate	Low
		Low Treatability		
MTBE	Low	Moderate	Low	Very High
TBA	Low	Low	Low	High
Anthracene	Low	Low	Low	Low
Benzo pyrene	Low	Low	Low	Low

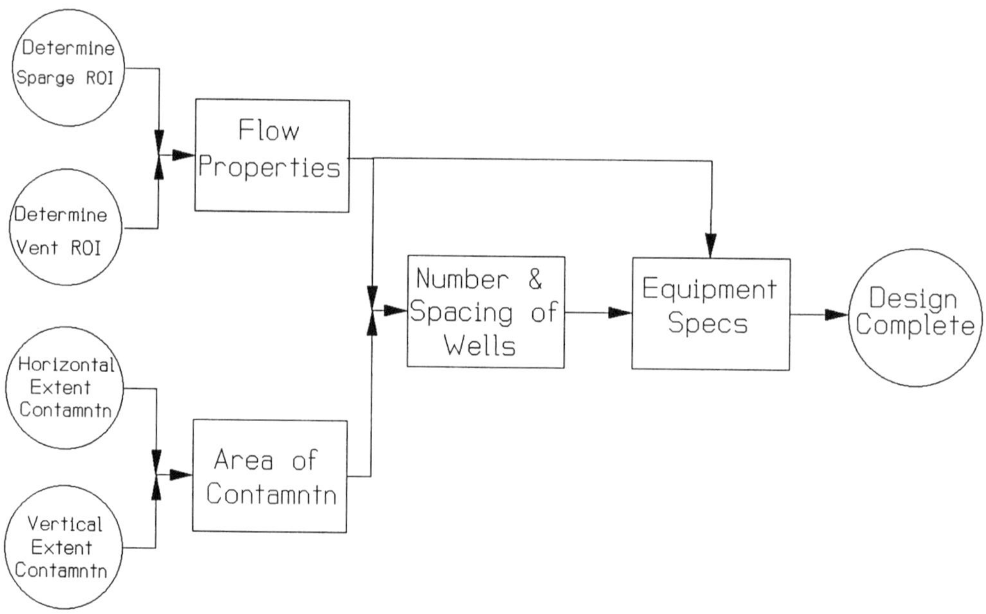

Figure 19.19. Design procedure for air sparging.

to produce two "maps" of the contamination (Figure 19.20). The first is a plan view showing the area to be treated, both soil and groundwater contamination. This "map" is derived through soil gas surveying and well sampling. The second is a geological cross section upon which the contaminant levels have been superimposed. This cross section provides a lithological profile which identifies the presence of any barriers or high-permeability channels. It also shows areas of adsorbed phase contamination. This cross section requires continuous coring and analysis (lab or field) of the cores for contaminant levels.

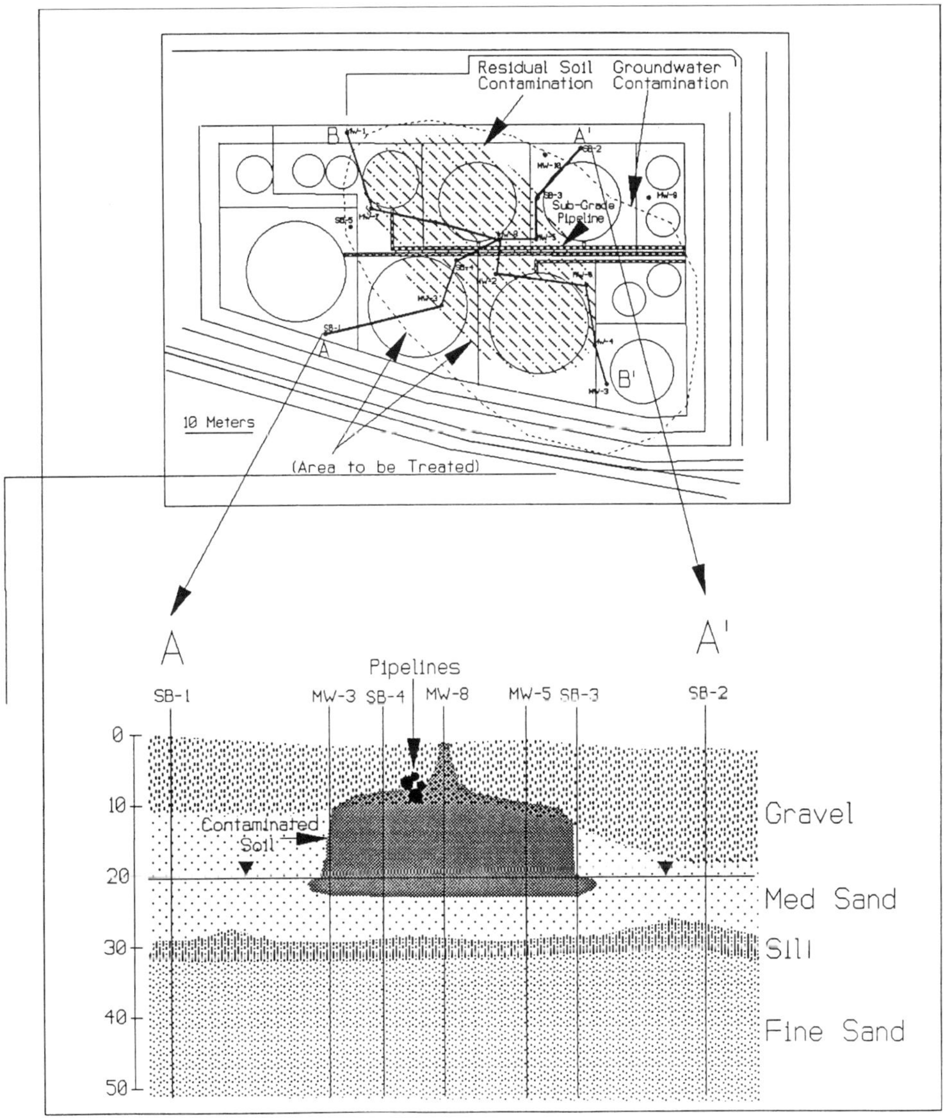

Figure 19.20. Site delineation data package.

How Does One Determine the Flow Properties of a Site?

Designing an air sparging system requires understanding the horizontal and vertical air permeability of the area to be treated. The best method is to run an onsite pilot test to determine the radius of influence for both the air sparging system and the soil vapor extraction system.

The pilot test cell consists of a combined or separate vent/sparge well(s) and a series of monitoring points placed radially around the vent/sparge well at varying distances. The monitoring points consist of a shallow monitoring well and several vapor probes separated by bentonite seals.

The pilot test is a series of short-term pump tests (2–4 hr). The protocol is to first run a sparge test where air is injected at the breakout pressure, and then 150%, 200%, and 250% of the breakout pressure, respectively. At each stage the pressure and flow are measured at the well head, and pressure, water table elevation, DO, and change in VOC concentration are measured in all probes. Carbon dioxide can also be monitored if the contaminant is biodegradable. The second test is a vent test. A vacuum is applied to the vent well at 50%, 75%, and 100% of maximum flow. Well head vacuum and flow and vacuum and VOC levels at each probe are taken at each stage. The final pump test is a combined sparge vent test. Two combinations of conditions are run—high vent, low sparge; and low vent, high sparge. The purpose of the combined test is to prove that the vent system is able to completely capture the sparged air and, therefore, maintain control of the system. Additionally, tracer tests can be run to confirm the ROIs, prove control, and check for barriers or channeling.

The data derived from these tests is used to derive a vent and sparge radius of influence either through graphical reduction using simple linear regression, or through the use of computer models. With the sparge test, the ROI should be calculated from all the parameters monitored—DO, pressure, water table elevation, and VOC concentration. Generally there is ±20% agreement in the ROIs derived with these different parameters. The ROI(s) give the spacing of the wells. The flow/pressure (vacuum) times the number of wells gives the equipment specifications. Flow/pressure values are derived from the pilot tests and from such considerations as stripping efficiency (air-to-water ratio).

These in-field pilot tests can be supplemented with laboratory tests. Sieve analyses may be run on retained cores to derive an estimate of permeability. Laboratory air permeability tests may also be run on retained cores or packed columns to determine air flow characteristics. Until more data is available on sparge systems, an in-field pilot test should always be run; laboratory tests should be used only to supplement or confirm field results.

What Other Factors Are Necessary for Design?

While the contaminant distribution and the air flow properties are the heart of the design of an air sparging system there are other factors and bases that need to be considered in the overall design. The following table lists the elements of an air sparging system and the basis for their design:

System Component	Basis for Design
Sparge Wells	
Number	Site Characterization, Pilot Test, Treatment Goals (Time and Level)
Depth	Site Characterization
Construction	General Principles, Site Characterization (Nesting)
Location	Site Construction, Site Characterization, Regulatory Requirements, Receptors
Vent Wells	
Number	Site Characterization, Pilot Test, Receptors, Treatment Goals (Time and Level)
Depth	Site Characterization
Construction	General Principles, Site Characterization (Nesting)
Location	Site Construction, Site Characterization, Regulatory Requirements, Receptors
Co-Located Wells	Site Characterization, Pilot Tests, Site Construction
Compressor	Pilot Tests, Well Design
Vacuum Pump	Pilot Tests, Well Design
Controls	
Noise	Site Location, Regulatory Requirements
Groundwater	Receptors, Regulatory Requirements
Vapors	Pilot Tests, Regulatory Requirements, Receptors
Moisture	Pilot Tests, Site Characterization
Monitoring Systems	Regulatory Requirements, Pilot Tests, Site Characterization, Treatment Goals
Recovery Wells	Receptors, Regulatory Requirements, Site Characterization

How Should an Air Sparging System Be Operated?

There is no standard way to operate an air sparging system. Different configurations and operating modes can be used to provide different short- and long-term results. There are four basic modes of

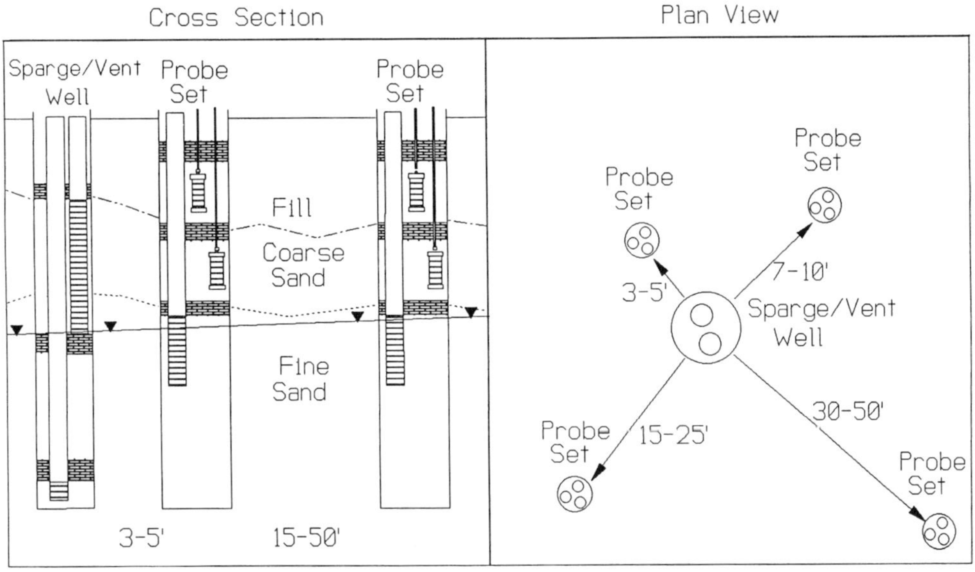

Figure 19.21. Pilot test cell schematic.

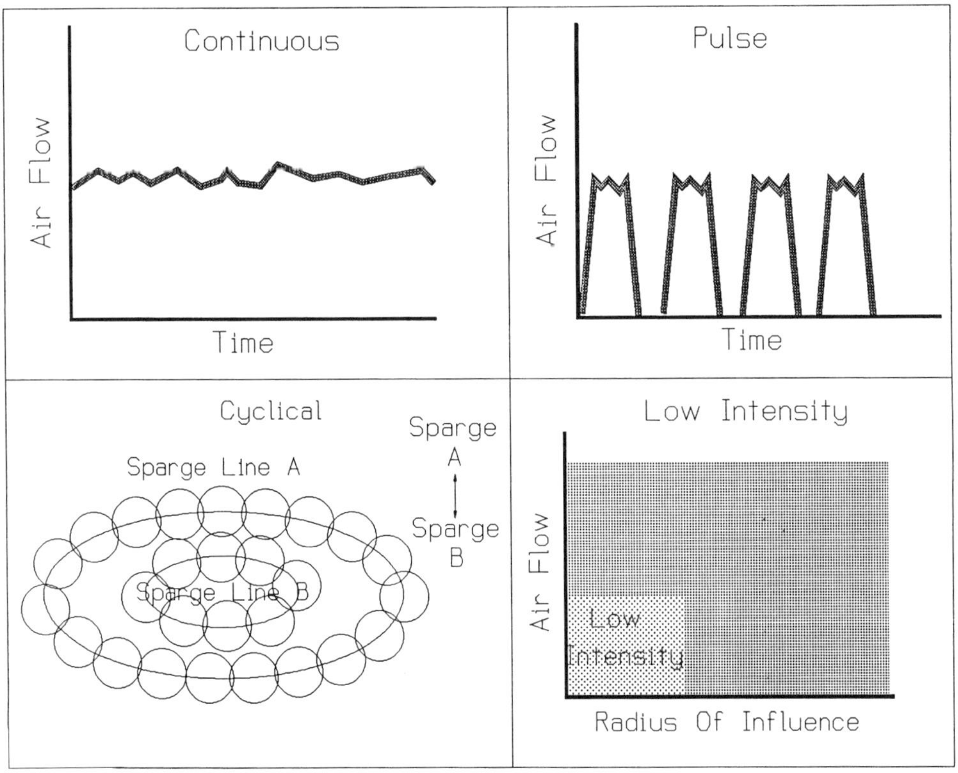

Figure 19.22. Modes of operation.

operation (Figure 19.22)—continuous, pulse, cyclical, and low-intensity. The following is a basic description of each of these modes of operation.

Mode of Operation	Description
Continuous	Air is injected continuously at a fairly constant rate through the sparge points. The length of time is determined by treatment goals and system response. Individual wells may be shut down as treatment goals are achieved. The advantage of continuous operation is that it provides the most rapid reduction in gross levels of contamination and the best long-term control of groundwater contamination. Continuous operation, however, does not necessarily attain low levels faster than other modes of operation because of diffusion limitations. Continuous operation is best applied to high-permeability, homogenous formations.
Pulse	With pulse operation, the rate of air injection is varied with time. Pulse operation is best applied to low-permeability or heterogeneous formations where diffusion limitations impact performance. The advantage of pulse operation is that the air flow pathways change, thus minimizing long-term channeling effects. Pulse operation can minimize equipment costs if the wells are sequenced. Additionally, if noise control is an issue, pulse operation can be used to minimize complaints. The disadvantage of pulse operation is that it does not provide long-term groundwater control and may require the use of groundwater recovery and treatment.
Cyclical	With cyclical operation, rings or banks of sparge wells are employed. Operation usually begins with the outermost or most downgradient ring/bank of sparge wells. The principle of operation is to strip the dissolved plume prior to treating the adsorbed phase—source area. This provides a "clean" barrier zone to any potential dissolved contaminant migration. Cyclical operation minimizes the need for groundwater control.
Low-Intensity	Low-intensity operation uses low flow and a short radius of influence. It is used to maximize biodegradation and minimize volatilization or stripping. Low-intensity operation is best used for low-volatility, low-solubility contaminants. It may also be used for polishing operations at the end of treatment to focus on small "hot spots." Low-intensity operation is an aeration or oxygenation process.

What Are the Outstanding Issues Concerning Air Sparging?

Air sparging is a relatively new technology for the United States even though it has been used for > 7 years in Europe. Consequently, there is only limited information on the application, design, operation, and effectiveness of air sparging. As a result there is a need to address some outstanding issues. These issues are:

Liability/Proof of Control	Since air sparging is an injection process, it has the potential for moving contaminants. Therefore establishing and proving control is important. The best monitoring systems for monitoring control and historical data documenting control need to be developed.
Limits of Operation	Because of the lack of operational data, there is limited information on the operational limits for air sparging—depth, permeability, heterogeneity. Two questions need to be addressed—what are the absolute limitations, and how may limitations be controlled?
Optimizing Performance	Being a new technology, there is limited information on how to optimize performance. Design and operational procedures need to be developed to maximize performance.

CONCLUSION

Air sparging is an important remedial technology. It can remove or treat contaminants through direct volatilization, biodegradation, or stripping. With air sparging, air is injected under pressure below the water table, creating a transient air-filled porosity. This enhances the biodegradation of petroleum hydrocarbon contaminants from the soil and groundwater. The net result is a rapid and significant decrease in contaminant levels.

Air sparging has two inherent dangers. First, the VOC-laden airstream can rapidly migrate through the vadose zone to low-pressure zones such as basements. This could cause a vapor hazard. To prevent this occurrence, a sparge system is operated in conjunction with a vent system. A second danger is that the injected air can mobilize groundwater contaminants rather than stripping them, causing accelerated downgradient migration. This can occur if there are vertical barriers to air migration, causing the air to be trapped, producing lateral spread. It can also occur if too much pressure is used, physically displacing the water column.

Because of these dangers, proper design is essential. This necessitates a field pilot test and careful site delineation. With proper design, the use of sparging can substantially and rapidly remediate groundwater contamination. The key to the effective use of air sparging is proper design. This entails first an understanding of the distribution of contaminants across the site. Second, the flow dynamics of air must be determined in the vadose zone and in the saturated zone. This can only be done through a properly designed pilot study. Once both the contaminant distribution and the flow dynamics are known, the number, location and type of treatment wells can be specified. This, in turn, leads to the equipment specifications. Through a careful and phased design process, air sparging can be an effective remedial system.

With effective design and careful monitoring, air sparging can be an important remedial tool. If, however, it is oversimplified and improperly designed, it can be ineffectual at best or counterproductive at worst. This chapter has been written as a basic primer for the design and use of air sparging. The goal of the chapter was to provide a basic overview of air sparging technology.

REFERENCES

Ahlfeld, D., A. Dahmani, M. Farell, and W. Ji. 1992. Detailed Field Measurements of an Air Sparging Pilot Test. *Ground Water* (submitted).

Billings, J.F., A.I. Coolely, and G.K. Billings. 1994. Microbial and Carbon Dioxide Aspects of Operating Air-Sparging Sites. *In* Air Sparging for Site Remediation, R.E. Hinchee, (Ed.), Lewis Publishers, Chelsea, MI. 56–80.

Bohler, U.J. Brauns, H. Hotzl, and M. Nahold. Air Injection and Soil Air Extraction as a Combined Method for Cleaning Contaminated Sites—Observations from Test Sites in Sediments and Solid Rocks. *In* Contaminated Soil '90, F. Arendt, M. Hinsenveld, and W.J. van den Brink (Eds.), Kluwer Academic Publishers, The Netherlands.

Boesrma, P., F.S. Petersen, P. Newman, and R. Huddleston. 1993. Use of Groundwater Sparging to Effect Hydrocarbon Biodegradation. *Petroleum Hydrocarbons and Organic Chemicals in Ground Water: Prevention, Detection and Restoration,* Proceedings, Houston, TX, 17, 557–559.

Brown, R.A. and F. Jasiulewicz. 1992. Air Sparging Used to Cut Remediation Costs. *Pollution Engineering,* July, 52–57.

Brown, R.A. and R. Fraxedas. 1991. "Air Sparging—Extending Volatilization to Contaminated Aquifers," presented at *Symposium on Soil Venting, Groundwater Technol., Inc.,* Houston, TX.

Brown, R.A., E. Henry, C. Herman, and W. Leonard. 1991. The Use of Aeration in Environmental Clean-Ups. *Petroleum Hydrocarbons and Organic Chemicals in Ground Water: Prevention, Detection and Restoration,* Proceedings, Houston, TX, 265–279.

Brown, R., C. Herman, and E. Henry. 1991. The Use of Aeration in Environmental Cleanups. *Proceedings of Haztech International Pittsburgh Waste Conference,* Pittsburgh, PA.

Brown, R.A., R.E. Payne, and P.F. Perlwitz. 1992. "Air Sparging Pilot Testing at a Site Contaminated with Gasoline," *Petroleum Hydrocarbons and Organic Chemicals in Ground Water: Prevention, Detection and Restoration,* Proceedings, Houston, TX, 17, 429–443.

Brown, R.A., R.J. Hicks, and P.M. Hicks. 1994. Use of Air Sparging for In Situ Bioremediation. *In* Air Sparging for Site Remediation, R.E. Hinchee, (Ed.), Lewis Publishers, Chelsea, MI. 38–55.

Fairbanks, P.E., L. Pennington, and J.H. Rabaideau. 1993. "Air Sparging in Reduced Permeability Sediments," *Petroleum Hydrocarbons and Organic Chemicals in Ground Water: Prevention, Detection and Restoration,* Proceedings, Houston, TX, 17, 561–574.

Felton, D.W., M.C. Leahy, L.J. Bealer, and B.A. Kline. 1992. "Case Study: Site Remediation Using Air Sparging and Soil Vapor Extraction," *Petroleum Hydrocarbons and Organic Chemicals in Ground Water: Prevention, Detection and Restoration,* Proceedings, Houston, TX, 14, 395–411.

Gudemann, H. and D. Hiller. 1988. In Situ Remediation of VOC Contaminated Soil and Groundwater by Vapor Extraction and Groundwater Aeration. *Proceedings of the Third Annual Haztech International Conference,* Cleveland, OH.

Gvirtzman, H. and S.M. Gorelick. The Concept of In-Situ Vapor Stripping for Removing VOCs from Groundwater. *In* Transport in Porous Media, Kluwer Academic Publishers, The Netherlands.

Hinchee, R.E. 1994. Air Sparging State of the Art. *In* Air Sparging for Site Remediation, R.E. Hinchee, (Ed.), Lewis Publishers, Chelsea, MI. 56–80.

Ji, W., A. Dahmani, D. Ahlfeld, J. Lin, and E. Hill. 1993. Laboratory Study of Air Sparging: Air Flow Visualization. *Ground Water Monit. Rev.,* Fall, 115–126.

Johnson, R.L., P.C. Johnson, D.B. McWhorter, R.E. Hinchee, and I. Goodman. 1993. An Overview of In Situ Air Sparging. *Ground Water Monit. Rev.,* Fall, 127–135.

Johnson, R.L. 1994. Enhancing Biodegradation with In Situ Air Sparging: A Conceptual Model. *In* Air Sparging for Site Remediation, R.E. Hinchee, (Ed.), Lewis Publishers, Chelsea, MI. 14–22.

Kaback, D.S., B.B. Looney, C.A. Eddy, and T.C. Hazen. 1989. "Innovative Ground Water and Soil Remediation: In Situ Air Stripping Using Horizontal Wells, " Presented at the Fifth Outdoor Action Conference on Aquifer Restoration, Ground Water Monitoring and Geophysical Methods, Las Vegas, NV, May 13–16.

Kampbell, D. 1993. U.S. EPA Air Sparging Demonstration at Traverse City, Michigan. *In: Environmental Restoration Symposium,* sponsored by the U.S. Air Force Center for Environmental Excellence, Brooks AFB, TX.

Leonard, W.C. and R.A. Brown. 1992. "Air Sparging: An Optimal Solution," *Petroleum Hydrocarbons and Organic Chemicals in Ground Water: Prevention, Detection and Restoration,* Proceedings, Houston, TX, 14, 349–363.

Lundegard, P.D. and G. Anderson. 1993. "Numerical Simulation of Air Sparging Performance," *Petroleum Hydrocarbons and Organic Chemicals in Ground Water: Prevention, Detection and Restoration,* Proceedings, Houston, TX, 17, 461–476.

Loden, M.E. "Air Sparging Technology Assessment," U.S. Environmental Protection Agency, Risk Reduction Engineering Laboratory/Contract No. 68-03-3409.

Marley, M.C., M.T. Walsh, and P.E. Nangeroni. 1990. "A Case Study on the Application of Air Sparging as a Complementary Technology to Vapor Extraction at a Gasoline Spill Site in Rhode Island," *Hydrocarbon Contaminated Soil Conference,* University of Massachusetts, September.

Marley, MC., D.J. Hazebrouck, and M.T. Walsh. 1991. "Air Sparging in Conjunction with Vapor Extraction for Source Removal at VOC Spill Sites," Paper presented at *6th Annual Conference on Hydrocarbon Contaminated Soils,* University of Massachusetts, Amherst, MA, September.

Marley, M.C. 1991. "Air Sparging in Conjunction with Vapor Extraction for Source Removal at VOC Spill Sites," Proceedings of Fifth National Outdoor Action Conference on Aquifer Restoration, Ground Water Monitoring, and Geophysical Methods, NWWA, May, Las Vegas, NV.

Marley, M.M., F. Li, and S. Magee. 1992. "The Application of a 3-D Model in the Design of Air Sparging Systems," *Petroleum Hydrocarbons and Organic Chemicals in Ground Water: Prevention, Detection and Restoration,* Proceedings, Houston, TX, 377–392.

Marley, M.C., D.J. Hazebrouck, and M.T. Walsh. 1992. The Application of In Situ Air Sparging as an Innovative Soils and Ground Water Remediation Technology. *Ground Water Monitoring Review,* Spring, 137–144.

Middleton, A.C. and D.H. Hiller. 1990. "In Situ Aeration of Groundwater, A Technology Overview," Presented at *Environment Canada Conference,* Montreal.

Neaville, C.C., P.C. Johnson, and R.L. Johnson. 1994. An Evaluation of In-Situ Air-Sparging Technology at a Gasoline UST Site, Hayden Island, Oregon. *In* Proceedings of the 1994 Conference on the Environmental Health of Soils, Long Beach, CA, March 30–April 1.

Ostendorf, D.W., E.E. Moyer, and E.S. Hinlein. 1993. "Petroleum Hydrocarbon Sparging from Intact Core Sleeve Samples," *Petroleum Hydrocarbons and Organic Chemicals in Ground Water: Prevention, Detection and Restoration,* Proceedings, Houston, TX, 17, 415–427.

Pankow, J.F., R.L. Johnson, and J.A. Cherry. 1993. Air Sparging in Gate Wells in Cutoff Walls and Trenches for Control of Plumes of Volatile Organic Compounds (VOCs). *Ground Water,* 31, 654–663.

Sellers, K.L. and R.P. Schreiber. 1992. "Air Sparging Model for Predicting Groundwater Cleanup Rate," *Petroleum Hydrocarbons and Organic Chemicals in Ground Water: Prevention, Detection and Restoration,* Proceedings, Houston, TX, 365–376.

Wehrle, K. 1990. In-Situ Cleaning of CHC Contaminated Sites: Model-Scale Experiments Using the Air Injection (In-Situ Stripping) Method of Granular Soils. *In* F. Arendt, et al. (Eds). *Contaminated Soil '90,* 1061–1062, Kluwer Academic Publishers, Boston.

Chapter 20

Natural Biological Attenuation: Integrative Transition Zones

T.H. Christensen, J. Lyngkilde, P.H. Nielsen, H.-J. Albrechtsen, and G. Heron

INTRODUCTION

Old landfills and waste disposal sites with no or insufficient leachate collection systems constitute a frequent risk to groundwater quality. These waste disposal sites typically contain a high amount of organic matter in terms of municipal garbage and industrial waste as well as significant fractions of chemical and hazardous waste. Because of the substantial microbial decomposition in the landfill, the leachates are strongly anaerobic and contain high concentrations of dissolved organic carbon on the order of hundreds to thousands of mg C/L. In addition, a wide spectrum of dissolved specific organic chemicals are found in the leachate, usually ranging from 100 µg/L to a few mg/L.

The presence of specific organic chemicals in the leachate is believed to constitute the major risk to the groundwater quality. A study was performed at the Vejen Landfill (Denmark) focusing on the attenuation of specific organic chemicals in the leachate plume. Since the degradation of specific organic chemicals to a large degree depends on the governing redox environment (Christensen et al., 1994), special efforts were made to describe the redox transition zones expected to be present in a leachate plume entering an aerobic aquifer.

VEJEN LANDFILL

The Vejen Landfill is located in the western part of Denmark (Figure 20.1) in a glacial outwash plain from the Weichselian Period (Late Quaternary). The landfill is partly located on the ground surface and partly dug into the ground. The upper aquifer is a 10–20 m deep sandy deposit with small random clay and silt clasts. The water table is located 3–5 m below ground surface. The landfill is located approximately 5 km from the water divide and the groundwater pore flow velocity in the area is approximately 150–200 m/yr. The area is intensively farmed and the annual precipitation is on the order of 700–900 mm. The upper part of the shallow aquifer is well-oxidized, containing free oxygen, high nitrate concentrations and some sulfate (Pedersen et al., 1991). The bottom part of the shallow aquifer is anoxic and has no nitrate.

In the period 1962–1981, the landfill received about 450,000 tons of waste, including municipal, demolition, industrial and chemical waste (Kjeldsen, 1993). The landfill has no engineered leachate collection system and a combination of various soil types forming the base of the landfill plus the presence of drainage ditches form a relatively complicated landfill geohydrology. The longitudinal transect of the aquifer, which has been studied with respect to redox zones, begins in an area primarily containing household waste and minor amounts of demolition waste. At this area of the landfill, its upstream length is approximately 100 m and the waste is placed on the ground surface to a height of 6 m. This section of the landfill was in use from 1973 to 1977. The landfill is covered by a soil cover. Leachate sampled inside the landfill contained DOC on the order of 500 mg/L, a BOD_5/COD (5 day biological oxygen demand/chemical oxygen demand) ratio of 0.08, a pH of 6.8 and levels of inorganic and organic contaminants resembling municipal landfill leachate in general (Kjeldsen, 1993).

REDOX ZONES

The redox zones in the leachate pollution plume were described by groundwater characterization, microbial bioassays and aquifer material characterization.

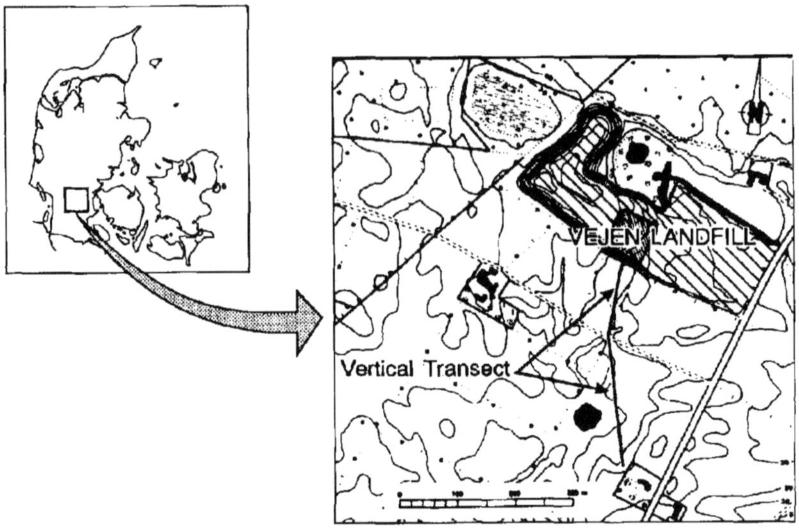

Figure 20.1. Geographical location of Vejen Landfill, Denmark. The magnified section shows the location of the vertical transect downgradient of the landfill from where groundwater samples were collected (Lyngkilde and Christensen, 1992a).

Groundwater Characterization

Groundwater samples were taken downgradient of the landfill along a 350 m transect from 1 inch (2.54 cm) iron pipes driven into the ground by a motor-driven jackhammer. The iron pipes were fitted with tips supplied with 10 cm long iron screens and Teflon® check valves. Water samples were collected at 0.5 m intervals through a Teflon® tube lowered into the iron pipe. The sampling was done by pressuring the pipe with nitrogen. The Teflon® tube was connected to a multiboard device to allow sampling of groundwater and direct analysis of oxygen, pH, redox potential, temperature and chemical conductivity in flow cells (Lyngkilde and Christensen, 1992a).

Redox-characterizing species, as well as inorganic compounds and physical parameters used for characterizing pollution levels and general groundwater quality, were quantified in approximately 200 groundwater samples. The redox-sensitive species were oxygen, nitrate, nitrite, ammonium, Mn(II), Fe(II), sulfate, sulfide and methane. The other group of parameters included chloride, calcium, dissolved organic carbon, total alkalinity, chemical conductivity, temperature, pH and redox potential. Based upon concentrations of redox-sensitive parameters, every groundwater sample collected was assigned a redox status in accordance to the criteria given in Table 20.1, thereby defining each sample as aerobic, nitrate-reducing, ferro-/manganogenic, sulfidogenic or methanogenic. For a detailed discussion on the criteria values, refer to Lyngkilde and Christensen (1992a).

The pollution plume, depicted as chloride contours, and the determined redox zones are shown in Figures 20.2 and 20.3, respectively. The chloride plume shows high concentrations on the order of 1000 mg/L close to the landfill, decreasing over the 350 m long transect to levels close to background level (35–50 mg/L). The rapid decrease of the chloride concentrations is probably due to a substantial mixing of the leachate with groundwater flowing under the clay-silt lens beneath the landfill and, downgradient of the landfill, being forced upwards due to the rise of the bottom of the upper aquifer. The redox zone map superimposes to a great extent the chloride plume, showing strongly reduced zones close to the landfill and aerobic conditions in the far downgradient part of the plume. The redox zones close to the landfill are methanogenic and sulfate-reducing. These zones are actually difficult to distinguish based on sampled groundwater because methane formation and sulfate reduction may take place at similar redox potentials and may take place in neighboring microzones not identified by groundwater samples, and also because methane detected in the groundwater may not have been generated on location but transported downgradient from methanogenic zones in the landfill. A later

Table 20.1. Criteria for Redox Parameters Used for Assigning Redox Status to Groundwater Samples (All Values in mg/L) According to Lyngkilde & Christensen (1992a).

Parameter	Aerobic	Nitrate-reducing	Manganogenic	Ferrogenic	Sulfidogenic	Methanogenic
Oxygen	> 1.0	< 1.0	< 1.0	< 1.0	< 1.0	< 1.0
Nitrate	—	—	< 0.2	< 0.2	< 0.2	< 0.2
Nitrite	< 0.1	—	< 0.1	< 0.1	< 0.1	< 0.1
Ammonium	< 1.0	—	—	—	—	—
Mn(II)	< 0.2	< 0.2	> 0.2	—	—	—
Fe(II)	< 1.5	< 1.5	< 1.5	> 1.5	—	—
Sulfate	—	—	—	—	—	< 40
Sulfide	< 0.1	< 0.1	< 0.1	< 0.1	> 0.2	—
Methane	< 1.0	< 1.0	< 1.0	< 1.0	< 1.0	> 1.0

— = no criterion is applied.

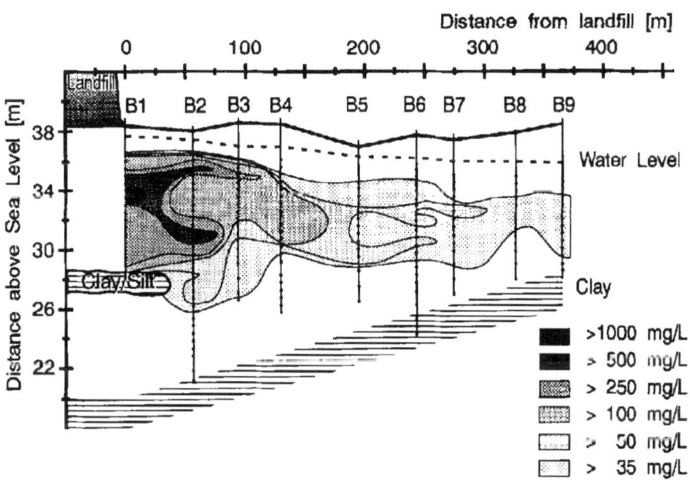

Figure 20.2. Chloride contours in a vertical transect downgradient of Vejen Landfill, Denmark. The contours indicate the pollution plume migrating from the landfill into the groundwater. The small dots on the vertical borings represent sample points (Lyngkilde and Christensen, 1992a).

resampling in the plume did not confirm the presence of sulfate-reducing conditions. At a distance of 25 m downgradient of the landfill, an extensive iron-reducing zone begins and reaches as far as 350 m downgradient from the landfill. In this zone, oxidized iron (and manganese) associated with the aquifer material is being reduced and dissolved. Between the iron-reducing zone and the aerobic groundwater zone, a nitrate-reducing zone is observed. Nitrate is present in the aquifer at substantial concentrations due to intensive agricultural activities in the area. The denitrifying zone below the plume is not caused by the plume but is a characteristic of the regional groundwater.

Microbial Bioassays

The redox zones identified in the leachate pollution plume are believed to be created by microbially mediated redox processes. To determine the presence of microbiological potential for the respective redox processes, aquifer sediments were collected and bioassays were performed in the laboratory.

Aquifer sediments were sampled from the different redox zones in aluminum pipes by means of a Waterloo-piston-sampler (Starr and Ingleton, 1992). The sediment-containing aluminum pipes were cut and sealed in the field and stored at 10°C until handled in the laboratory for the bioassays.

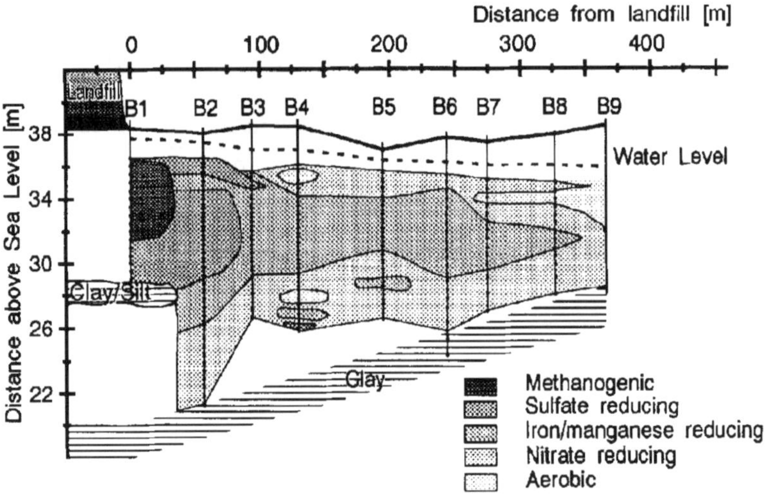

Figure 20.3. Distribution of redox zones in the groundwater downgradient of Vejen Landfill, Denmark. The small dots on the vertical borings represent individual sample points (Lyngkilde and Christensen, 1992a).

For the assays testing the potential for a range of redox-based microbial activities, 10 mL of sediment was mixed with 30 mL of sterile filtered groundwater and incubated in serum bottles with an anaerobic atmosphere (N_2-CO_2) at 10°C. Aerobic respiration was measured as $^{14}CO_2$-evolution during aerobic incubation after amendment with ^{14}C-benzoate. Denitrification was measured as accumulation of nitrous oxide (N_2O) in the headspace after addition of acetylene. Iron reduction was measured as accumulation of dissolved ferrous (Fe(II)) iron with or without amendment of amorphous ferric oxides. Sulfate reduction was measured as accumulation of $H_2{}^{35}S$ after addition of $^{35}SO_4$. Methane production was measured as accumulation of methane in the headspace. ATP and direct microscopic counting (after staining by acridine orange) were used to quantify the microbial biomass.

Microbial biomass was observed in all redox zones and ranged from 7.5×10^6 to 66×10^6 cells/gww and from 0.13 to 1.81 ng ATP/gww. Table 20.2 summarizes the results of the performed bioassays in terms of numbers of positive and negative responses. The performed bioassays confirmed that the redox zones identified from water samples did have the microbial potential to perform the relevant redox processes. However, some overlapping was observed and the redox zones therefore only reflect dominant redox conditions.

Sediment Characterization

The predominance of the iron-reducing zone would indicate that this zone and zones of lower redox level would show decreased sediment contents of iron, since reduction of insoluble ferric compounds associated with the sediment would result in dissolved ferrous ions. Part of this soluble ferrous iron would probably be transported further downgradient in the plume if not retained by ion exchange or precipitation.

Sediment samples were collected from the four dominating zones and, after freeze drying, the iron oxides were extracted by the TiCEB-method (Titanium-Citrate-EDTA-Bicarbonate) proposed by Ryan and Gschwend (1991) and quantified by atomic absorption spectrophotometry. The results are shown in Figure 20.4. The strongly reduced zones (methanogenic/sulfidogenic and iron/manganese-reducing zones) showed low contents of iron oxides on the order of 0.2 to 0.8 mg Fe/gdw sediment, while the samples from the denitrifying and aerobic zones contained 1.5 to 6 mg Fe/gdw sediment. These data indicate that a substantial part of the iron oxide pool on the sediment has been reduced and solubilized. The observed sediment characteristics support the presence of redox transition zones in the leachate pollution plume at Vejen Landfill.

Table 20.2. Summary of Results from Microbial Bioassays on Sediment from the Various Redox Zones (Number of Positive and Negative Assays).

	CH_4 Production		SO^- Reduction		Fe(III) Reduction		Denitrification		Respiration	
Zone	+	–	+	–	+	–	+	–	+	–
Aerobic	–	–	–	–	–	–	6	0	2	0
Denitrifying	–	–	–	–	0	2	7	0	2	0
Iron-reducing	1	3	0	4	4	3	6	0	–	–
Methanogenic	3	1	0	4	0	4	–	–	–	–

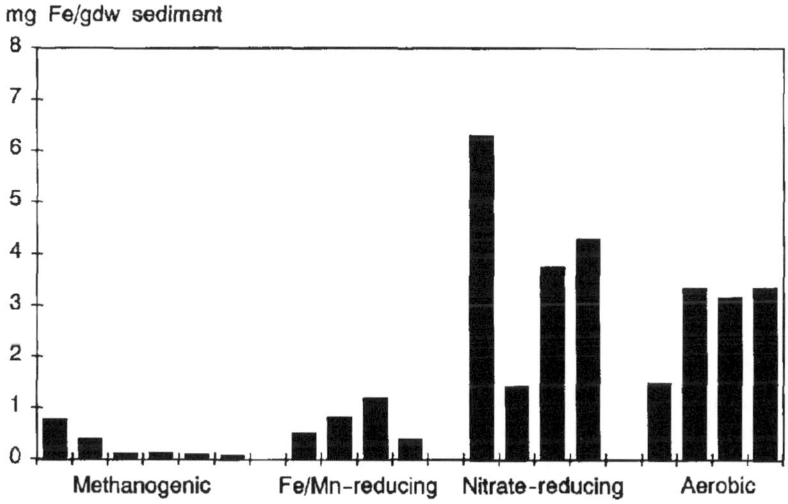

Figure 20.4. Total iron contents (TiCEB extraction, Ryan and Gschwend, 1991) of freeze-dried sediment samples from the 4 dominating redox zones downgradient of Vejen Landfill.

ORGANIC COMPOUND DISTRIBUTION IN THE PLUME

The distribution of organic compounds in the leachate plume has been determined for dissolved organic matter, expressed as nonvolatile organic carbon (NVOC) determined by standard procedures on a Dohrmann DC-80® TOC analyzer, and for specific organic chemicals quantified by a Carlo Erba QMD 1000® quadropole mass spectrometer in scanning mode after pentane extraction and acetic acid anhydride derivatization. For details on the sampling and analytical methods, refer to Lyngkilde and Christensen (1992b).

Dissolved Organic Carbon (NVOC)

The concentrations of dissolved organic carbon (determined as NVOC) varies between 300 and 500 mg C/L close to the landfill and between 10 and 20 mg C/L about 130 m downgradient of the landfill. The background level is approximately 2–5 mg C/L. The decreasing concentrations as a function of distance may be due to dilution, sorption and degradation. Sorption is believed to be very limited according to the available literature (Christensen et al., 1994); due to the relatively stationary NVOC plume observed for the first 130 m (Lyngkilde and Christensen, 1992b), sorption as an attenuation mechanism is neglected in the following discussion. Accounting for the dilution by correcting the measured NVOC values according to the chloride content of the sample revealed the trend observed in Figure 20.5.

The results in Figure 20.5 clearly indicate that degradation of dissolved organic matter takes place in the leachate plume. Comparing the location of the decrease of the dissolved organic matter with the

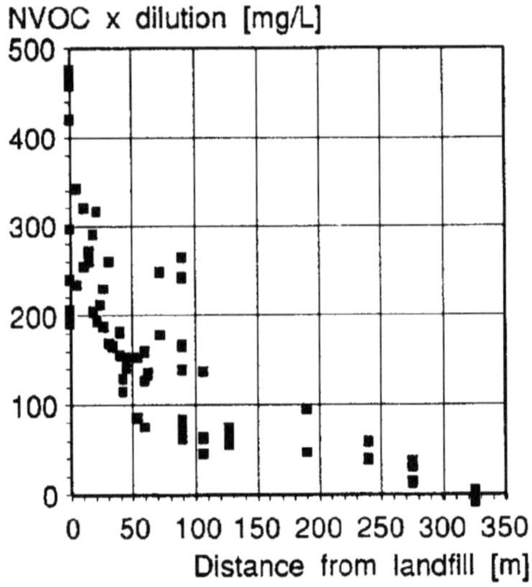

Figure 20.5. Concentration of organic matter (NVOC) in the groundwater downgradient from Vejen Landfill as a function of distance. All concentrations have been corrected for dilution according to chloride concentrations (Lyngkilde and Christensen, 1992b).

redox zone presented in Figure 20.3, it appears that some degradation takes place in the methanogenic/sulfidogenic zone but that the majority of NVOC is being degraded in the iron-reducing zone. The degradation of dissolved organic carbon in this anaerobic part of the plume is consistent with the increase of the iron content of the actual groundwater samples and with reduction of sediment-bound oxidized iron in the same part of the plume.

Specific Organic Chemicals

Immediately downgradient of the landfill, about 40 specific organic chemicals have been identified (Table 20.3), although none of them exceed 500 μg/L (Lyngkilde and Christensen, 1992b). The concentration of the compounds, as a function of the distance from the landfill, is shown in Figure 20.6, where the compounds are categorized into 6 groups. The compound concentrations were corrected for dilution according to the chloride content of the samples. The specific organic chemicals show a pattern somewhat comparable to that of the dissolved organic carbon (Figure 20.5). Interpretation of the data in Figure 20.6 with respect to the first 20 m downgradient of the landfill is somewhat difficult, but weak indications of degradation of toluene, xylenes and naphthalene in the methanogenic/sulfidogenic zone may be found (Lyngkilde and Christensen, 1992b). Degradation of all the listed compounds, excluding sorption as a significant attenuation mechanism as discussed in the following section, seems to take place in the iron-reducing zone of the leachate plume.

Lyngkilde and Christensen (1992b) observed the fate of camphor (1,7,7-trimethyl-bicyclo [2.2.1]-heptane-2-one) in the first 50 m of the plume while its oxidized derivative (1,7,7-trimethyl-bicyclo[2.2.1]-heptane-2,5-dione) was emerging in the same zone and considered this as an indication of oxidative degradation processes in this part of the leachate plume.

The herbicide Mecoprop® (MCPP,2-(4-chloro-2-methyl/phenoxy)-propanoic acid), not included in Figure 20.6, was the most significant organic trace component found in the plume and is apparently persistent within the anaerobic part of the plume. In a distance of 130 m downgradient of the landfill where no other specific organic chemical exceeds 1 μg/L, MCPP was detected at concentrations of 95 μg/L. However, in the dilute part of the plume the concentration of MCPP was decreased to below detection levels. Heron and Christensen (1992) demonstrated in laboratory experiments that MCPP was degradable under aerobic conditions in the Vejen Landfill aquifer.

Table 20.3. Concentrations of NVOC, Chloride, and Specific Organic Chemicals in the Groundwater at Three Locations Downstream from Vejen Landfill, Denmark (Lyngkilde & Christensen, 1992b).[a]

Chemical	Distance from Landfill (m) 5	19	39
NVOC (mg/L C)	346	257	82
Chloride (mg/L)	1125	984	589
Benzene	3	3	2
Toluene	39	5	1
Benzene, ethyl-	96	75	9
Xylenes, *p-/m-*	224	33	2
Xylene, *o-*	16	7	1
Benzene, C_3- isomer	27	12	<0.2
Benzene, C_3- isomer	21	13	<0.2
Benzene, C_3- isomer	48	32	<0.2
Benzene, C_3- isomer	36	24	1
Benzene, 1,3,5-trimethyl- (mesitylene)	11	9	<0.2
Benzene, 1-propenyl-	8	6	1
Benzene, C_4- isomer	3	3	4
Benzene, C_4- isomer	10	4	<0.2
Benzene, C_4- isomer	9	5	1
Benzene, C_4- isomer	6	4	<0.2
Benzene, C_4- isomer	4	5	1
Benzene, C_4- isomer	5	3	<0.2
Benzene, 1,2,3,5-tetramethyl- (isodurene)	3	2	<0.2
Benzene, 1-methyl-2-propyl-	5	3	<0.2
Naphthalene	24	10	2
Naphthalene, 2-methyl-	5	<0.2	<0.2
Naphthalene, 1-methyl-	6	1	<0.2
7-Oxabicyclo[2.2.1]-heptane, 1-methyl-4-(1-methylethyl)-	6	<0.5	<0.5
2-Oxabicyclo[2.2.2]-octane, 1,3,3-trimethyl-	8	5	1
Bicyclo[2.2.1]-heptane-2-one, 1,3,3-trimethyl- (*d*-fenchone)	126	177	21
Bicyclo[2.2.1]-heptane-2-one, 1,7,7-trimethyl- (±)- (camphor)	426	474	46
Bicyclo[2.2.1]-heptane 2,2-dimethyl-3-methylene-	72	56	2
Bicyclo[3.1.1]-heptane-2-one, 3,6,6-trimethyl-	41	6	4
Bicyclo[2.2.1]-heptane-2,5-dione, 1,7,7,-trimethyl-	12	10	9
Phenol, chloro-methyl isomer	33	30	11
Phenol, 6-chloro-2-methyl-	11	10	11

[a] The concentrations are average values from 2 to 3 samplings. All concentrations are µg/L, unless stated otherwise. The specific organic chemicals are determined by GC-MS analysis.

SORPTION OF ORGANIC CHEMICALS

Sorption was not considered to be a significant attenuation mechanism in the previous discussion on the fate of specific organic chemicals in the leachate plume. This is because, first, sorption only attenuates compounds at the compound plume front and does not, under stationary conditions, provide any retardation as to the plume distribution behind the compound migration front. The location of this front, of course, depends on the retardation of the compound in question. Second, laboratory experiments, as summarized in the following, showed very little sorption of specific organic chemicals in the Vejen Landfill aquifer.

The Vejen Landfill aquifer is very sandy and has a very low fraction of organic carbon, on the order of 0.25 mg C/gdw sediment. For such low fractions of organic matter, the traditional estimation procedures involving the octanol-water partitioning coefficient (K_{ow}) do not apply (Larsen et al., 1992c). Therefore laboratory experiments were performed to determine the retardation of specific organic chemicals by sorption onto the aquifer material.

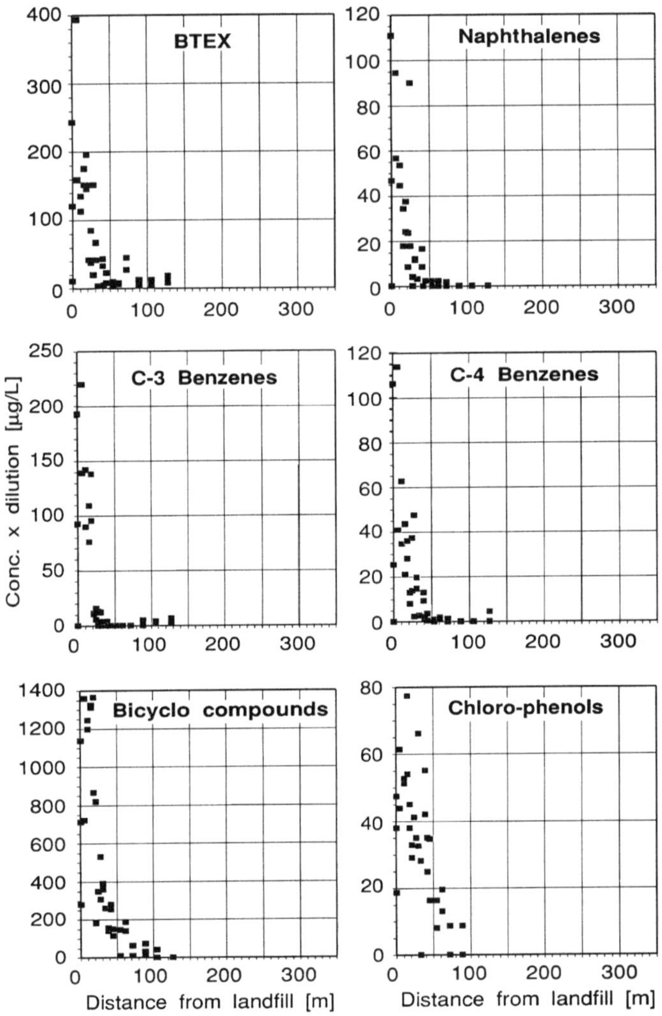

Figure 20.6. Concentrations of different groups of specific organic chemicals in the groundwater downgradient from Vejen Landfill as a function of distance. All concentrations have been corrected for dilution, according to chloride concentrations (Lyngkilde and Christensen, 1992b).

Retardation Factors

Sorptions of 12 specific organic chemicals were studied in laboratory columns at low concentrations (about 100 µg/L of each chemical). Special precautions were taken to avoid compound loss due to sorption on the experimental equipment and to volatilization (Larsen et al., 1992d). Based on breakthrough curves, as illustrated in Figure 20.7, the retardation of the specific organic chemicals was determined relative to water as depicted by the tracer breakthrough curve. The results are shown in Table 20.4.

The presented retardation factors (R) can be interpreted as the inverse of the relative solute transport velocity or the relative transport time for the compound to migrate as far as the water migrated within one time unit. Table 20.4 shows that the compounds with low K_{ow} values are retarded only a few percent, while biphenyl migrates with only a third of the velocity of water. However, since the water pore velocity is on the order of 80–120 m/yr in the plume and the plume has existed for more than 15 years, it seems unlikely that sorption is a major attenuation mechanism in the anaerobic part of the plume about 200–300 m downgradient from the landfill.

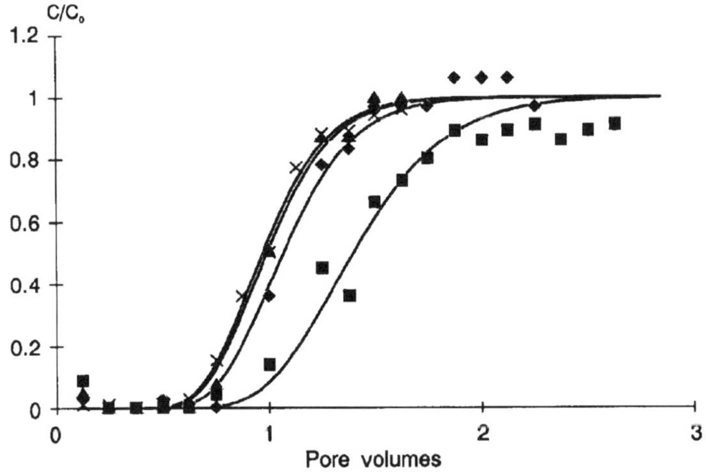

Figure 20.7. Breakthrough curves for ^3H_2O (tracer) (x) and the specific organic chemicals benzene (▲), *o*-xylene (◆) and 1,4-dichlorobenzene (■) in laboratory columns containing sediment collected from the aquifer downstream of the Vejen Landfill area (f_{oc}=0.00025).[a] After Larsen et al., 1992d.

Table 20.4. Retardation factors (R) for 12 specific organic chemicals determined in laboratory column experiments with aquifer material from the Vejen landfill area. After Larsen et al. (1992d).

	log k_{ow}	Retardation Factor[a]
Tracer (^3H_2O)		1.00
Benzene	2.15	1.02
Trichloroethylene	2.29	1.07
1,1,1-Trichloroethane	2.49	1.02
Tetrachloroethylene	2.60	1.21
Tetrachloromethane	2.62	1.10
Toluene	2.80	1.01
Indene	2.92	1.06
o-Xylene	3.12	1.10
1,4-Dichlorobenzene	3.38	1.44
1,2-Dichlorobenzene	3.38	1.43
Naphthalene	3.59	1.56
Biphenyl	4.09	2.96

[a] The retardation factor is the inverse relative solute transport velocity: $V_s/V_w = R^{-1} = (1 + \rho/E \bullet K_d)^{-1}$, where V_s is the solute velocity, V_w is the pore water velocity, R is the retardation factor, ρ is the bulk density of the aquifer material, E is the porosity and K_d the compound distribution coefficient.

Kinetic Aspects

The retardation factors presented in Table 20.4 are authentic retardation factors not influenced by kinetic effects in the laboratory columns. However, kinetic effects were observed in the experiments at high flow velocities (75 m/yr) and a bi-continuum kinetic model (Brusseau et al., 1991) was used to model the data. The Vejen aquifer material does exhibit sorption properties significantly influenced by kinetic aspects for the more hydrophobic compounds (Brusseau et al., 1991, Larsen et al., 1992a). But the model indicates that no significant kinetic effects would be observable in the Vejen Landfill leachate plume at distances exceeding approximately 50 m distance from the landfill at the actual pore flow velocities.

Effects of Leachate on Sorption

The presence of dissolved organic carbon in the leachate was hypothesized to reduce the sorption of specific organic chemicals onto the aquifer material and hence to enhance their mobility. Laboratory experiments involving leachates did show that very hydrophobic compounds would partition into the dissolved organic matter in the leachate (Larsen et al., 1992b), but the effect was only minor for the compounds found in the Vejen Landfill leachate plume (log K_{ow}<4), and affected the expected migration velocity on the order of ±20%. Larsen et al. (1992b) concluded that this effect was small in comparison with the other uncertainties in predicting the compound migration velocity.

DEGRADATION OF ORGANIC CHEMICALS

In view of the inability of dilution and sorption processes to explain the observed distribution of specific organic chemicals in the plume, it seems likely that degradation is responsible for the fate of the compounds. Although most of the degradation seems to take place in the part of the leachate plume dominated by iron reduction, very little supportive information is available in the literature as to degradability of organic chemicals under iron-reducing conditions. Loveley et al. (1989) reported on biological degradation of toluene, phenol and *p*-cresol under iron-reducing conditions by a pure culture (GS-15) isolated from a river sediment.

In order to demonstrate that degradation actually takes place in the plume, several *in situ* microcosm studies are being performed in the plume. Preliminary results of this work are presented in the following section.

In Situ Microcosms

In situ microcosms, originally developed by Gillham et al. (1990), were installed in the 4 dominating zones of the plume: methanogenic/sulfidogenic, iron-reducing, denitrifying and aerobic.

The microcosm consists of a stainless steel cylinder, open in the bottom to allow aquifer material to push into the cylinder when driven into the aquifer through a cased borehole. An illustration of an installed microcosm is shown in Figure 20.8. The microcosm is connected to the ground surface and after development by pumping, the microcosm is loaded with groundwater spiked with the specific organic chemicals and a tracer. A surplus of spiked groundwater is pumped into the microcosm to ensure an even distribution of the compounds within the microcosm volume. Samples are taken at different time intervals to follow the fate of the compounds.

The preliminary results for 1,1,1-trichloroethane, naphthalene and *p*-nitrophenol are presented in Figure 20.9. The concentration of the tracer was constant during the experiment, indicating that the microcosms were hydraulically well-functioning. The drop in concentration observed within the first 5 days for some of the compounds, in particular naphthalene, may be due to slow kinetics of the sorption process and should be interpreted with care. Highly variable patterns are observed: *p*-nitrophenol seems to degrade in all redox zones, while naphthalene seems to degrade only in aerobic and maybe in iron-reducing zones. 1,1,1-Trichloroethane seems to degrade under iron-reducing conditions and under methanogenic conditions, but not in the more oxidized zones. The results for the aerobic zone, the denitrifying zone and the methanogenic zone generally seem to be in accordance with the literature, while the data for the iron-reducing zone still is too scarce to support any definitive conclusions.

Until confirmed results are available for several specific organic chemicals, it must be concluded that the apparent degradation of a wide spectrum of organic chemicals in Vejen Landfill plume has not yet been positively demonstrated.

CONCLUSION

Based on the current investigations at the Vejen Landfill leachate plume in Denmark it can be concluded:

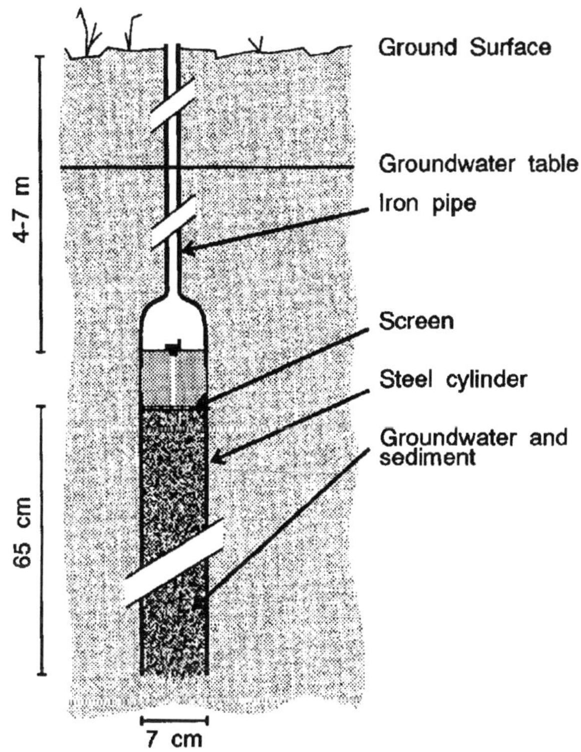

Figure 20.8. Illustration of an *in situ* microcosm installed in an aquifer.

- A thermodynamically sound sequence of redox zones have been identified in the plume based on groundwater characteristics, microbial bioassays and iron content of sediment samples. Most of the organic matter and specific organic chemicals are disappearing in the anaerobic redox zones. One exception is MCPP, which is appearing in all the anaerobic zones, but disappears in the aerobic zone.
- The iron-reducing zone seems to be the most extensive zone and provides extensive degradation of dissolved organic carbon and apparently also specific organic chemicals.
- The performed calculations and laboratory experiments show that dilution and sorption cannot account for the observed fate of the specific organic compounds in the plume.
- Preliminary results from *in situ* microcosms installed in the plume indicate the potential for degradation in all dominating redox zones, depending on the specific organic chemical. However, more observations are needed to confirm that degradation under anaerobic conditions is the major attenuation in the leachate plume.

The investigations of the integrative transition zones in the Vejen Landfill leachate plume have led to the conclusion that no enhancement of the natural remediation processes seems to be needed in this case.

ACKNOWLEDGMENTS

The results presented are part of a major research program focusing on the effects of waste disposal on groundwater. The program is funded by the Danish Technical Research Council, the Technical University of Denmark, and the Commission of the European Communities. The assistance of Anja Foverskov, Bent Skov, Laila Olsen, Mona Refstrup, and Tage V. Andersen is gratefully acknowledged.

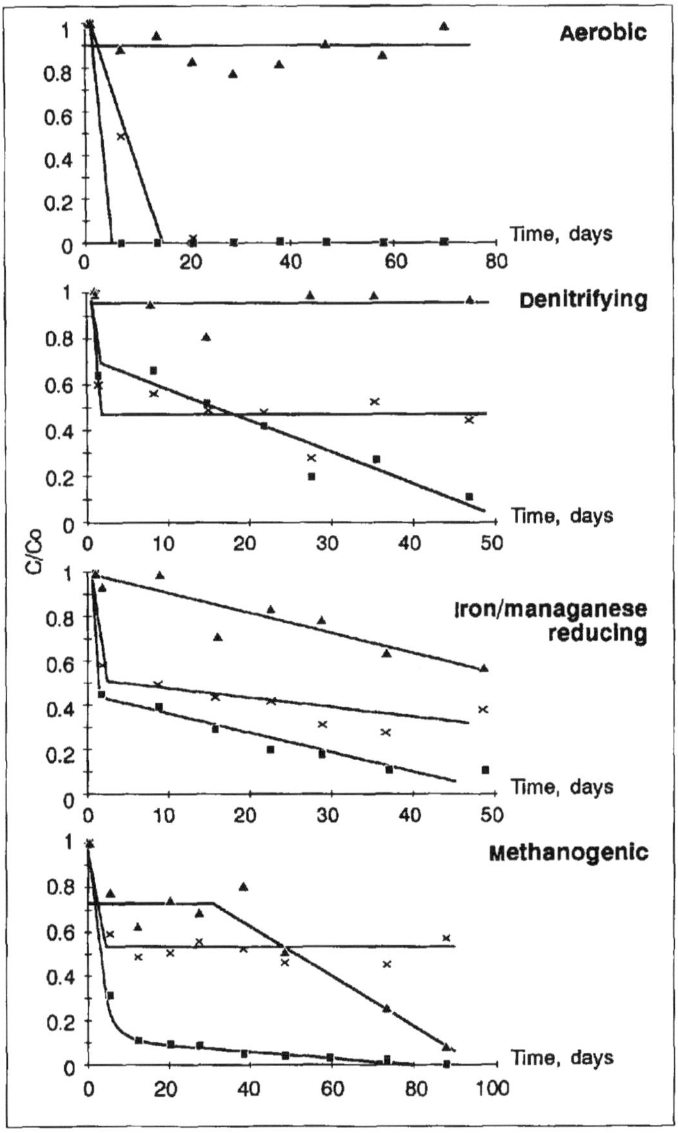

Figure 20.9. Preliminary results for 1,1,1-trichloroethane (▲), naphthalene (x) and *p*-nitrophenol (■) in *in situ* microcosms installed in the 4 dominating redox zones in the Vejen Landfill leachate pollution plume.

REFERENCES

Brusseau, M.L., T. Larsen, and T.H. Christensen. 1991. Rate-limited sorption and nonequilibrium transport of organic chemicals in low organic carbon aquifer materials. *Water Resources Research*, **27**:1137–1145.

Christensen, T.H., P. Kjeldsen, H.-J. Albrechtsen, G. Heron, P. Nielsen, P.L. Bjerg, and P.E. Holm. 1994. Attenuation of landfill leachate pollutants in aquifers. *Critical Reviews in Environmental Science and Technology,* **24**:119–202.

Gillham, R.W., R.C. Starr, and D.J. Miller. 1990. A device for in situ determination of geochemical transport parameters. 2. Biochemical reactions. *Ground Water*, **28**:858–862.

Heron, G. and T.H. Christensen. 1992. Degradation of the herbicide Mecoprop in an aerobic aquifer determined by laboratory batch studies. *Chemosphere*, **24**:547–557.

Kjeldsen, P. 1993. Groundwater pollution source characterization of an old landfill. *Journal of Hydrology,* **142**:349–371.

Larsen, T., T.H. Christensen, and M.L. Brusseau. 1992a. Predicting nonequilibrium transport of naphthalene through aquifer materials using batch determined sorption parameters. *Chemosphere*, **24**:141–153.

Larsen, T., T.H. Christensen, F.M. Pfeffer, and C.G. Enfield. 1992b. Landfill leachate effects on sorption of organic micropollutants onto aquifer materials. *Journal of Contaminant Hydrology*, **9**:307–324.

Larsen, T., P. Kjeldsen, and T.H. Christensen. 1992c. Correlation of benzene, 1,1,1-trichloroethane, and naphthalene distribution coefficients to the characteristics of aquifer materials with low organic carbon content. *Chemosphere*, **24**:979–991.

Larsen, T., P. Kjeldsen, and T.H. Christensen. 1992d. Sorption of hydrophobic hydrocarbons on three aquifer materials in a flow through system. *Chemosphere*, **24**:439–451.

Lovley, D.R., M.J. Baedecker, D.J. Lonergan, I.M. Cozzarelli, E.J.P. Phillips, and D.I. Siegel. 1989. Oxidation of aromatic contaminants coupled to microbial iron reduction. *Nature*, **339**:297–301.

Lyngkilde, J. and T.H. Christensen. 1992a. Redox zones of a landfill leachate pollution plume (Vejen, Denmark). *Journal of Contaminant Hydrology*, **10**:273–289.

Lyngkilde, J. and T.H. Christensen. 1992b. Fate of organic contaminants in the redox zones of a landfill leachate pollution plume (Vejen, Denmark). *Journal of Contaminant Hydrology*, **10**:291–307.

Pedersen, J.K., P.L. Bjerg, and T.H. Christensen. 1991. Correlation of nitrate profiles with groundwater and sediment characteristics in a shallow sandy aquifer. *Journal of Hydrology*, **124**:263–277.

Ryan, J.N. and P.M. Gschwend. 1991. Extraction of iron oxides from sediments using reductive dissolution by Titanium(III). *Clays and Clay Minerals*, **39**:509–518.

Starr, R.C. and R.A. Ingleton. 1992. A new method for collecting core samples without a drilling rig. *GWMR*, **12**:91–95.

Chapter 21

Recent Developments in Permeable *In Situ* Treatment Walls for Remediation of Contaminated Groundwater

R.W. Gillham and D.R. Burris

INTRODUCTION

The most commonly used method of groundwater remediation involves removal of the contaminated water by pumping, followed by above-ground treatment (pump-and-treat). After close to twenty years of experience with pump-and-treat remediation schemes, it is now a common view that these systems might better be viewed as a means of plume control rather than remediation. In particular, pumping times, and thus remediation costs, are commonly much greater than expected. This can be attributed to partitioning of contaminants between the solution and solid phases, geologic heterogeneity, and incomplete removal of contaminant sources, particularly immiscible phase liquids. Though considerable effort has been directed at developing more cost-effective methods for treating the discharge water, complete removal of the contaminated water from the aquifer remains a difficult and generally unresolved challenge. Mackay and Cherry (1989) discuss the limitations of pump-and-treat systems in greater detail.

Considerable effort is now being directed toward the development of *in situ* treatment methods, due to the cost-ineffective nature of pump-and-treat, as an approach to remediation. In a comprehensive review (U.S. EPA, 1990), *in situ* methods for decontamination of soil and groundwater were grouped under five categories: soil flushing; solidification and stabilization; degradation; control of volatile materials; and chemical and physical separation techniques (including permeable barriers). Burris and Cherry (1992) provide a further review of both passive and active *in situ* treatment methods.

The purpose of this chapter is to present recent developments in the use of *in situ* permeable treatment walls for remediation and containment of plumes of contaminated groundwater. Physical separation methods such as sorption or *in situ* air sparging transfer contaminants to an alternate medium, which may require subsequent treatment or disposal, and therefore do not provide an ultimate solution. The discussion will therefore be focused primarily on the use of permeable barriers to enhance chemical and biological degradation processes.

A number of chemical, photochemical and biological processes that will degrade common groundwater contaminants are known. The *in situ* application of these processes generally requires that the *in situ* geochemical/biochemical conditions be modified such that the reactions will proceed at a practical rate. An early and notable example is the injection of a hydrogen peroxide solution to enhance the aerobic degradation of the soluble mono-aromatics of petroleum products such as benzene, toluene, ethylbenzene and xylene (BTEX). To be successful, these methods require that the injected water and the resident (contaminated) water be intimately mixed. During injection, the enhancing solution tends to displace the contaminated water, with intimate mixing occurring only by dispersion at the boundary between the two solutions. Because dispersion is commonly a weak process, it is difficult to ensure that the enhancing solution reaches the contaminated zones. This situation is improved somewhat if the contaminants and enhancing agents migrate at different velocities. For example, Berry-Spark and Barker (1987) showed that nitrate contained in injected water can overtake a plume of water contaminated with BTEX. This occurs because nitrate moves at the velocity of the groundwater while the BTEX compounds are somewhat retarded as a result of partitioning between the solution and solid phases.

In spite of the potentially advantageous effects of geochemical retardation, mixing of injected and resident waters continues to be viewed as a major limitation to remediation methods that involve *in situ* chemical alteration or biostimulation. As an alternative, *in situ* permeable reaction walls could be a means of providing greater control over the treatment zone.

A simplified schematic cross section of an *in situ* reaction wall is shown in Figure 21.1. A permeable wall containing the appropriate reactants is constructed across the path of the contaminant plume and, as the contaminated water passes through the wall, the contaminants are removed/degraded, allowing uncontaminated water to emerge on the downstream side. Pump-and-treat systems are entirely active, resulting in high operating and maintenance costs. On the other hand, *in situ* treatment walls offer the potential for highly passive remediation, resulting in greater cost effectiveness. While the concept is simple and the potential advantages are significant, two major technical challenges must be addressed. The first concerns the method of installation and the second concerns the availability of suitable reactants for incorporation into the wall. A review of these topics is the primary purpose of this chapter.

It is important to realize that permeable reactive treatment walls are a relatively new concept. The earliest reference that we have found on the topic is McMurtry and Elton (1985), and to our knowledge, a wall has yet to be installed at a subsurface contamination site. In the absence of commercial performance data, this review is based on conceptual models, the results of small-scale laboratory tests and a relatively small number of small-scale field tests.

INSTALLATION

Several methods have been contemplated for the installation of reactive walls/zones, with required depth of the treatment zone, geological material, construction cost and compatibility with reactive materials being major considerations. Most methods are adaptations of those used for the construction of impermeable barriers. These will be referred to as conventional methods, and will be discussed here only briefly. Permeable reactive sections separated by impermeable sections ("funnels-and-gates"), a modification of the continuous wall concept, will also be discussed. At this time, all methods of installation are generally limited to unconsolidated surficial materials.

Conventional Methods

Open Trenches

The most direct approach to installation of a reactive wall is the excavation of a trench to an appropriate depth followed by backfilling with the reactive material. Using a backhoe or dragline, trenches of various widths can be excavated to depths well in excess of 30 m in unconsolidated geological materials. However, in most materials, if advanced to depths significantly below the water table, the walls of the trench will collapse. As an alternative, it has been proposed that sealable-joint sheet piling (Starr et al., 1992) be driven prior to excavation to support the walls. This technique allows the excavation of trenches to greater depths below the water table. Lateral earth pressures require that bracing beams be installed between the parallel rows of sheet piling as the trench is advanced. The bracing acts as an obstruction, creating difficulties in excavation. Once backfilled, the sheet piling is removed. This method has been successfully tested in a small-scale prototype test at Canadian Forces Base Borden in Ontario (O'Hannesin, 1993).

A particular disadvantage of trenching is that should the trench pass through a contaminated region, it is quite possible that the excavated material would need to be transferred to a hazardous waste disposal site. This could add substantially to the construction cost.

Polymer Slurry Trench

A method similar to that used to construct impermeable slurry walls has been used to construct deep drains and permeable curtains. In this case, the bentonite used to construct slurry walls is re-

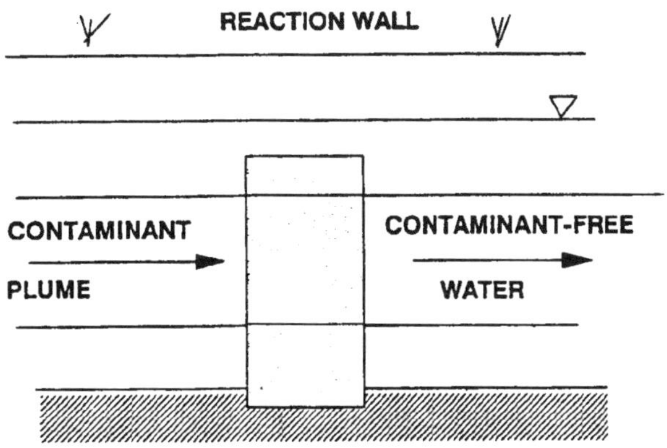

Figure 21.1. Schematic cross section through a permeable *in situ* treatment wall.

placed by a biodegradable polymer. In constructing a permeable reactive wall, the polymer forms a slurry with the reactive materials, replacing the natural geologic material as excavation proceeds. The polymer slurry maintains the walls of the trench during construction, but when construction is complete, as much of the polymer as possible is flushed out. The remainder biodegrades, leaving the trench filled with the reactive material. In principle, this method should be applicable to depths in excess of 30 m. Disposal of excavated material could add to the construction cost, and of course it would need to be established that the reactive material was chemically compatible with the polymer solution, as well as the process and products of polymer biodegradation.

Overlapping Caissons

Steel caissons can be driven to depths in excess of 30 m. The soil can then be removed using a large-diameter auger, the caisson refilled with concrete grout or a soil-bentonite mixture and the caisson removed. Using a series of overlapping caissons, a continuous impermeable wall can be constructed. A reactive wall can be constructed using a similar procedure but with the reactive material replacing the grout. As in the trench methods, disposal of contaminated material excavated from the caissons could add significantly to the cost. The method may be suitable to depths of 30 to 50 m.

Soil Mixing

Soil mixing has been used to stabilize excavations and more recently for constructing impermeable barriers. Two methods are used. One employs drilling with grout injected under high pressure as the drill stem is withdrawn. The second method uses large-diameter mixing augers in which the additives are injected through the hollow stem. These methods can be used to depths well in excess of 30 m, and the fact that little or no soil is removed can be a particular advantage. Ensuring complete mixing of the additive (reactive material) is difficult and the physical compatibility of the reactive material with the injection system may also present problems. In particular, the injection systems are normally designed for viscous fluids and therefore may not be suitable for additives, such as granular materials, that have substantially different physical properties.

Funnel-and-Gate Systems

Depending upon the nature of the reaction and the reactive material, as well as the method of installation, continuous reactive walls have several potential limitations. In particular, they can be difficult to impossible to remove, replacement may be excessively costly and, if the reactive material is costly, a continuous wall may not be competitive with conventional methods of remediation. Furthermore, active maintenance, such as periodic nutrient addition or modification of the geochemical conditions, may not be possible or practical in a continuous wall.

As an alternative, McMurtry and Elton (1985) proposed the use of impermeable sections of wall that would direct the groundwater through relatively narrow zones of reactive material. More recently this approach has been developed conceptually and explored in greater detail through mathematical modelling by Starr and Cherry (1994), who refer to the concept as the "funnel-and-gate system." Various configurations of funnel-and-gate systems are given in Figure 21.2.

The impermeable sections can be constructed by several means, including slurry walls and sealable-joint sheet piling. Sheet piling has a particular advantage in many situations in that installation does not result in excavated soil that requires disposal. The reactive section, the "gate," can have various geometries and dimensions. As suggested in Starr and Cherry (1994), the gate can contain multiple treatment zones, with each zone designed to remove a particular contaminant.

Having restricted zones of treatment provides a greater opportunity to control the treatment process, and to overcome many of the potential limitations listed previously for continuous treatment walls. A major design consideration concerns the residence time of the contaminated water within the reactive material. Effective design therefore requires that one know: 1) the residence time necessary to reach the objective concentration; and 2) the natural groundwater velocity. From this information, the ratio of the width of the funnel to the width of the gate can be selected (which determines the velocity through the gate), and knowing the velocity through the gate, the reactive zone travel distance can be determined to give the appropriate residence time. Increasing the width of the funnel to the width of the gate may reduce the amount of soil material that must be removed. However, increasing the width of the funnel also increases the velocity through the gate, which in turn forces a greater reactive zone travel distance.

A trial "funnel-and-gate" was installed by members of the Waterloo Centre for Groundwater Research at Canadian Forces Base Borden (CFB Borden) Ontario in November 1992. In this case, a 107 cm diameter steel caisson was driven through the upper sand aquifer into the aquitard at a depth of about 10 m. The sand was then removed from the caisson using a large-diameter auger and the "gate" was lowered into the caisson. The "gate" consisted of a steel pipe 76 cm in diameter with sections removed and replaced by well-screen material. Screen material was placed on both the upstream and downstream sides of the gate. The annular space between the gate and the caisson was filled with pea gravel and the interior of the gate was filled with the reactive material which, in this case, was a mixture of sand and iron grindings. The caisson was then withdrawn and the "funnel" was formed using sealable sheet piling attached to sealable joints which had previously been welded to each side of the gate. Installation of the gate took about one-half day, and installation and sealing of the sheet piling required about another day and a half. In the geological setting at Borden, and to a depth of 10 m, installation proved to be quick and moderately inexpensive. Geotechnical firms suggest that depths in excess of 30 m can be achieved using similar construction methods. The cylindrical geometry of the gate used in this test was convenient for installation and should be particularly advantageous in situations where the reactive material requires replacement on a regular basis. That is, the spent material could be removed quite effectively using the large-diameter auger.

As in the case of continuous reactive walls, the depth of the contaminant plume and the geological material are significant potential limitations to the installation of "funnels-and-gates." Further, because of the relatively low velocity of natural groundwater flow, the residence time in a continuous wall of even 1 m thickness would normally be on the order of several days to weeks. Velocities through a gate would normally be many times greater than the natural groundwater velocity, placing greater constraints on the applicability of a reaction process, in particular, the reaction rates.

REACTIVE MATERIALS

The prospect of *in situ* treatment has presented a significant challenge to the skill and imagination of geochemists, microbiologists and others involved in the broader subject area of water treatment. Numerous concepts are being developed for *in situ* treatment of a wide range of groundwater contaminants. In evaluating the ultimate utility of these concepts, several factors should be considered.

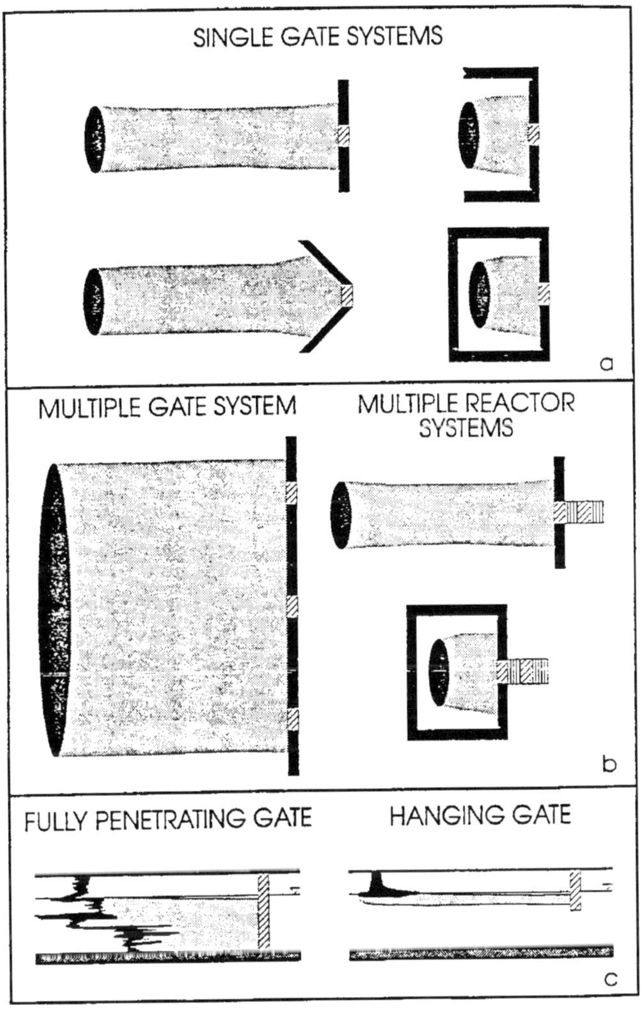

Figure 21.2. Various proposed configurations of "funnel-and-gate" systems (from Starr and Cherry, 1994).

In situ treatment normally requires large volumes of reactive materials, particularly in the case of continuous walls, and thus the cost of the reactants is a significant factor. Persistence of the reactive material is also a major factor in that frequent replacement could add substantially to the operating and maintenance costs, and could be an important consideration in selecting the method of installation. Because the capital cost of *in situ* treatment systems is relatively high, in order to be competitive, operating and maintenance costs must be low. For this reason, it is important that the treatment system be relatively passive. Frequent addition of nutrients or active modification of the natural groundwater chemistry could therefore be undesirable.

Several proposed methods of *in situ* treatment are presented in the following discussion. Because of rapid advances in this field, we do not claim completeness. Much of the work is in the conceptual or early experimental/demonstration stage, with most references coming from conference proceedings and abstracts rather than from refereed journals.

Modification of Aquifer Materials

The first concept of an *in situ* treatment wall to be presented here is a departure from the foregoing discussion in that it does not require replacement of the natural geologic material by a reactive material. Rather, a defined zone of the natural geological material, possibly in the form of a wall, is chemically modified to enhance the treatment/degradation process.

A notable example of the above concept is presented by Burris and Antworth (1990, 1992). It is proposed that when an aqueous solution of a cationic surfactant (hexadecyltrimethylammonium chloride, for example) is injected into an aquifer, the surfactant molecules would become associated with the exchange sites of the aquifer material. With an appropriate arrangement of injection wells, a continuous wall of surfactant-amended aquifer material could be created. It has also been shown that surfactant addition can increase the sorptive capacity of the aquifer material for hydrophobic organic compounds by one to three orders of magnitude. As a contaminant plume passes through the treatment zone, organic contaminants would be removed from the solution phase and concentrated on the solid phase. Though providing a temporary benefit, the concept does not rely on sorptive removal. In particular, Burris and Antworth (1990) propose that reactants that will enhance degradation processes be injected upstream of the wall. In the case of a contaminant plume emanating from a petroleum spill, for example, BTEX compounds could accumulate within the treatment zone for a period of time. The periodic addition of an oxygen-rich solution (aqueous solution of peroxide, for example) upstream of the wall should result in rapid biodegradation of the BTEX compounds within the wall. Nutrient additions or other amendments to enhance biodegradation of other types of organic contaminants are of course possible.

Though relatively passive, periodic addition of bioenhancing agents upstream of the wall would be required. Studies of the compatibility of biodegradation processes with surfactant-modified aquifer materials are continuing.

Studies of the modification of aquifer materials through the addition of cationic surfactants are also reported by Westall et al. (1992), though application in reactive treatment walls are not discussed explicitly.

Chemical Barriers for Control of Inorganic Contaminants

Various types of chemical barriers have been proposed for control of inorganic contaminants. For the most part these include redox control for removal of electroactive chemicals such as some metals, additives to precipitate certain contaminants and stimulation of biochemical removal processes.

Blowes and Ptacek (1992) proposed the use of solid phase additives to permeable reaction walls that would control the redox conditions, thus promoting the removal of electroactive metals. In a particular example, the use of various forms of reduced iron (siderite, $FeCO_3$; pyrite, FeS_2; and metallic iron Fe^0) for removal of dissolved chromium ($CrO^{2-}{}_4$) was examined. Of the three forms of iron considered, fine-grained zero-valent iron was found to be particularly effective for removal of dissolved chromium. For example, in column studies designed to simulate a small section of a reactive wall, a reactive material consisting of 50 mass % iron filings (0.5 to 1 mm in diameter) and 50 mass % quartz sand was found to reduce the influent concentration from 18 mg/L Cr to below the detection limit of 0.05 mg/L in the effluent. The columns were 15 cm in length, and after the passage of 15 pore volumes of solution, there was no detectable chromium in the effluent. More recent information shows that after the passage of 85 pore volumes, chromium was not detected in the effluent (David Blowes, personal communication).[1] Under the highly reducing conditions generated by the reduced iron, the hexavalent chromium is reduced, causing it to be removed from solution as a highly insoluble chromium hydroxide precipitate. Studies are continuing to examine the applicability of redox control for removal of other electroactive chemicals. Additional studies are in progress to examine the use of chemical additives in *in situ* treatment walls for removal of iron, sulfate and heavy metals in groundwater emanating from mine tailings impoundments (David Blowes, personal communication).

While redox control can be used to promote precipitation of certain metals, in other situations reactants can be added to promote precipitation of insoluble mineral phases or to promote precipita-

[1] Waterloo Centre for Groundwater Research, University of Waterloo, Waterloo, Ontario N2L 3G1.

tion through adjustment of pH. In a particular case study discussed in McMurtry and Elton (1985), a spill of acid waste resulted in subsurface contamination by a low pH (2) solution with high concentrations of chromium, lead and other heavy metals. As an emergency measure, a trench was excavated downslope of the spill and was filled with crushed limestone. As the contaminated water passed through the trench it was neutralized by the limestone, causing the metals to be removed from solution by precipitation reactions. After the contaminated water had passed through the trench, the limestone (containing the metal precipitates) was excavated and disposed of. No information was provided concerning the initial chemical form or oxidation state of the metals, or of the particular precipitation reactions.

Schwartz and Xu (1992) and Xu and Schwartz (1992) considered additives that could be used in a permeable treatment wall for removal of dissolved lead from groundwater. Their studies have shown that hydroxyapatite added to a reactive barrier could serve as a source of phosphate that would result in the precipitation of lead phosphate. The particular precipitate was identified as either hydroxypyromorphite or pyromorphite.

Denitrification is a biologically mediated process that results in the conversion of nitrate (NO^-_3) to nitrogen gas (N_2). If appropriate bacterial populations are present, the reaction will proceed under reducing conditions, provided there is adequate labile organic matter present to serve as an energy source for the bacteria. As a consequence, it has been proposed that a permeable layer containing a source of organic carbon could be placed beneath the tile fields of domestic septic systems for removal of nitrate from the tile effluent (Vogan, 1993). Similarly, it has been proposed that organic matter added to reactive treatment walls could enhance nitrate removal from contaminated groundwater. Laboratory and small-scale field tests reported in Vogan (1993) indicate the concept to be viable. The results show, however, that the choice of organic matter is critical in that it must release dissolved organic carbon at a rate sufficient to enhance the denitrification process, but at a rate that is sufficiently low such that the source will persist for a practicable period of time. Studies of the effectiveness of organic matter addition to permeable walls for nitrate removal are currently under investigation in several different hydrogeologic environments (Robertson, personal communication).[2]

Enhanced Biodegradation of Organic Contaminants

It is widely accepted that organic chemicals are the most serious threat to groundwater quality. Over the past ten to fifteen years, a substantial research effort has been directed toward biochemical processes for remediation of groundwater contaminated by these materials, and several of these processes are now being considered for application in reactive treatment walls. While several examples are presented here, it is safe to say that numerous other concepts or adaptations are being contemplated.

As discussed briefly in the introduction to this chapter, efforts at enhanced *in situ* bioremediation have been frustrated to a significant degree by a limited ability to cause mixing between the contaminated water and solutions containing the enhancing agents. Many of the proposed methods provide potential ways of overcoming this limitation.

Permeable trenches have been used as a means of intercepting contaminant plumes and in some cases the passage of air through the trench has been used as a means of stripping volatile contaminants from the groundwater as it passes through the trench. Pankow et al. (1993) provide a review of air sparging technologies. Though air sparging is not within the scope of this review, there is evidence that sparging can have two beneficial effects; volatile constituents are removed through transfer to the gas phase, and the supply of oxygen can enhance the biodegradation of contaminants that degrade under aerobic conditions, particularly BTEX. Results of laboratory tests of the combined benefits of sparging and the performance of a field trench installation are reported in Thomson et al. (1992). In laboratory tests with high air flow rates, removal efficiencies for BTEX and methyl tertiary-butyl ether (MTBE) of 95% were observed, and at low air flow rates, up to 50% of the removal was

[2]Waterloo Centre for Groundwater Research, University of Waterloo, Waterloo, Ontario N2L 3G1.

attributed to aerobic degradation. Though Thomson et al. (1992) describe the installation and successful performance of a field installation, no information concerning the relative importance of air stripping and biodegradation under the conditions of the field test was provided.

Kao and Borden (1992) proposed the use of a two-layer permeable reaction wall for removal of BTEX from groundwater. The upstream portion of the wall is composed of nutrient-laden (N and P) concrete briquets, followed by a layer of peat. As the contaminated water passes through the briquets, nitrate is released, which stimulates BTEX degradation through denitrification. The peat serves as a polishing medium to adsorb low concentrations of BTEX compounds that pass through the briquet layer. Laboratory column tests conducted for a period of 55 days and designed to simulate an *in situ* wall showed rapid rates of degradation for toluene and ethylbenzene, and somewhat reduced rates for the xylene isomers. Under the conditions of the test, there appeared to be little or no loss of benzene.

Bianchi-Mosquera et al. (1994) describe the use of a solid phase oxygen-release compound (ORC) to enhance degradation of BTEX compounds. The concept is similar to that of Kao and Borden (1992), except that the slow-release source of nitrate (the concrete briquets) is replaced by a slow-release (briquets or powder) source of oxygen. Bianchi-Mosquera et al. (1994) describe a field trial at CFB Borden in which permeable zones of ORC (both briquets and powder) were installed downstream of a continuous source of dissolved benzene and toluene, each at a concentration of about 4 mg/L. The results showed the ORC to be an effective slow-release source of dissolved oxygen for periods of at least 10 weeks. The results also showed the dissolved oxygen that was released to be adequate and effective in promoting the aerobic degradation of the dissolved benzene and toluene. Based on the encouraging results, the authors recommended further studies to examine the longer-term behavior of the ORC, and alternate methods for achieving contact between the ORC and contaminated groundwater. The same research team currently has a second trial in progress at the Borden site.

Devlin and Barker (1992, 1993a,b) proposed the use of an *in situ* permeable barrier to facilitate nutrient addition for the purpose of stimulating bioremediation processes. In particular, it was proposed that an *in situ* wall having a permeability substantially greater than the natural geologic material be constructed across the path of a contaminant plume. Through a series of withdrawal and injection wells, the water within the wall could be removed, amended with nutrients, and reinjected. The pulse of nutrient-amended water would then be carried downstream of the wall at the natural groundwater velocity. The periodic amendment of the water within the wall would result in a series of pulses migrating downstream from the wall. As a result of dispersion processes, the pulses would mix with the contaminated water and would also coalesce to form a continuous zone of nutrient-amended groundwater. The system would of course be only semi-passive in that periodic but regular nutrient addition would be required.

A field evaluation of the concept is currently in progress at the Borden field site. A permeable wall 1 m thick, 4 m deep and 4 m long was installed. Initial tests using nonreactive tracers showed that replacement of the water in the wall every 5 weeks resulted in pulses of water that coalesced within a travel distance of about 5 m from the wall. Using the procedure of periodic injection, the system was passive about 99% of the time. Experiments are continuing in which an aqueous source of carbon tetrachloride (CT) and trichloroethene (TCE) was injected immediately downgradient of the wall. In this case the water within the wall is periodically amended with acetate. It is proposed that the acetate will stimulate methanogenesis, which will in turn contribute to the reductive dechlorination of the CT and TCE within the aquifer, downstream of the permeable wall. Results of this test are not yet available.

The introduction of non-indigenous bacteria to aquifers for the purpose of stimulating biodegradation processes generally suffers from the same limitations discussed previously with respect to nutrient addition but with the further requirement that amendments be added to maintain the population. A study consisting of laboratory, field and modelling components is presently being developed cooperatively by the Gulf Breeze Laboratory of EPA, SBP Technologies Inc., University of West Florida, Environment Canada, and the Waterloo Centre for Groundwater Research, to investigate the feasibility of introducing bacterial populations using a funnel-and-gate system (J.A. Cherry, personal communication).[3] It is proposed that the gate consist of removable cassettes colonized with *Pseudomonas*

[3]Waterloo Centre for Groundwater Research, University of Waterloo, Waterloo, Ontario N2L 3G1.

cepacia G4 5223 PRI (referred to as "PRI"). PRI has a particular advantage in that it degrades TCE under oxidizing conditions, thus avoiding the potential formation of vinyl chloride, a problem commonly encountered under anaerobic conditions. It has the further advantage of not requiring potentially toxic stimulants such as phenol or toluene. The bacteria do require an oxidizing environment, however, and life-support chemicals including an organic substrate and nutrients must also be provided. Thus the proposed cassette installations will include slow-release sources of oxygen, nitrate and phthalate. The cassette concept is a potentially useful modification of the funnel-and-gate system since it should allow for convenient and regular replacement of the reactive materials. A further limitation to the *in situ* use of non-indigenous bacteria is the need for specific regulatory approval.

Abiotic Degradation of Organic Contaminants

The difficult requirements for maintenance of bacterial populations encourages the use of abiotic degradation processes. Unfortunately, in the absence of light, natural processes of abiotic degradation generally do not proceed at sufficient rates to be practical for removal of many organic contaminants of environmental interest. Recently, however, attention has turned to zero-valent metals as a means of enhancing the rate of abiotic degradation of halogenated organic contaminants.

It has been recognized for many decades that zero-valent metals can promote the transformation of various pure phase organic liquids. The first environmental application, to our knowledge, was the degradation of pesticides in aqueous solution (Sweeny and Fischer, 1972). Since the late 1980s, studies at the Waterloo Centre for Groundwater Research have focused on metallic iron (Fe^0) as a means of enhancing the degradation of chlorinated aliphatic organic compounds, with particular application in permeable *in situ* treatment walls (Gillham and O'Hannesin, 1992, 1994; O'Hannesin and Gillham, 1992; Gillham et al., 1993). Batch tests in which aqueous solutions of a wide range of chlorinated methanes, ethanes and ethenes were added to 100 mesh iron filings resulted in degradation rates that were three to seven orders of magnitude greater than natural abiotic rates reported in the literature. Generally the rates increased with the degree of chlorination and with increasing iron-to-solution ratio. The degradation generally appeared to be pseudo first-order, and of 14 compounds tested, only dichloromethane did not show measurable degradation. Though less chlorinated forms of the initial compound appeared as breakdown products, they could seldom account for more than 10% of the original. Furthermore, the chlorinated products of degradation subsequently degraded to non-chlorinated compounds.

Figure 21.3 shows results for a laboratory column simulation of a reactive wall. The column was packed with 15 mass % of iron filings and 85% silica sand, the flow velocity was 50 cm/day and the influent solution contained 4.69 mg/L trichloroethene (TCE). The TCE concentration profile of Figure 21.3a, showing an exponential decline in concentration with distance, as well as an exponential increase in hydrocarbon gases, was stationary in both time and position. The major constituents of the hydrocarbon gases were methane and ethane, with minor amounts of butane, butene and propane. The DCE isomers and vinyl chloride in the effluent accounted for about 8% of the original TCE. Had the column been longer, the flow rate lower or the iron content in the column greater, these also would have disappeared. In this particular test, the chloride concentration in the effluent could account for about 76% of the chlorine content of the influent TCE. Figure 21.3b includes the data of 3a, as well as the results of similar tests conducted with initial concentrations ranging from 1.27 to 61 mg/L. In this case the data are plotted as Ln relative concentration versus time. The linear and similar character of the graphs indicates that the process is pseudo first-order and independent of initial concentration. The half-life determined by fitting a first-order kinetic model to the data was about 26 hr.

Based on the encouraging laboratory results, a field trial was initiated in June 1991. Full details of the test are given in O'Hannesin (1993). A permeable treatment wall 1.6 m thick and 5.5 m wide was constructed to a depth of about 4 m below the water table at the Borden field site. Sealable sheet piling was used to facilitate installation. The wall was located about 5 m downgradient from a continuous source of dissolved TCE and PCE (Rivett et al., 1992). The plume was about 2 m in width and 1 m thick, and concentrations along its axis were about 253 and 43 mg/L of TCE and PCE, respectively.

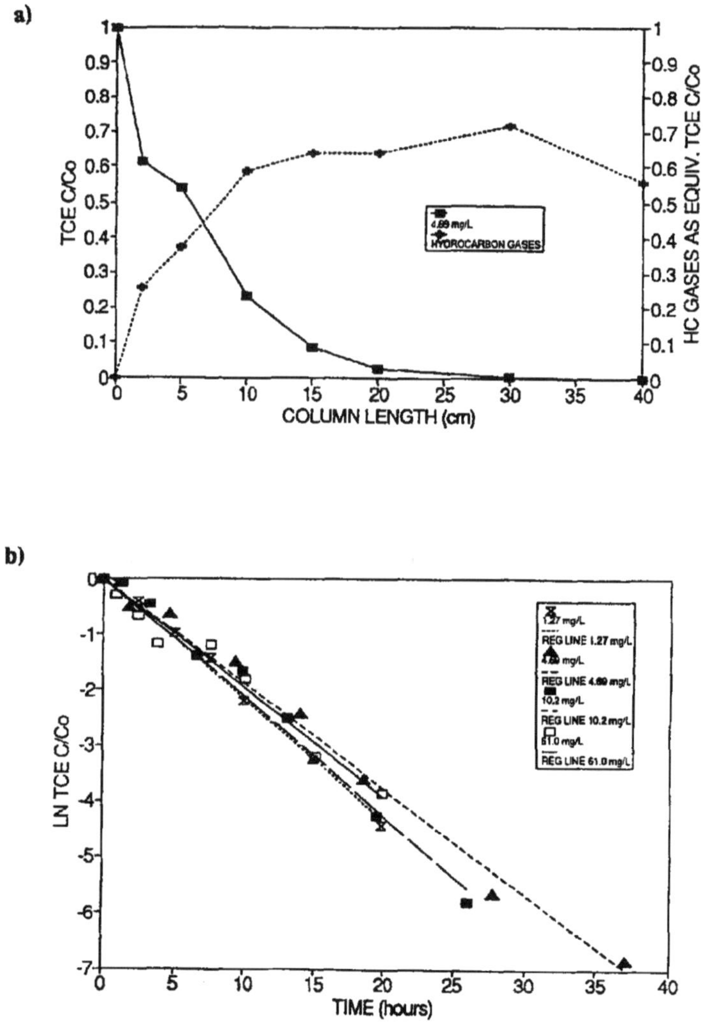

Figure 21.3. a) Column tests of TCE degradation in the presence of iron: stationary concentration profiles of TCE and breakdown products. b) Semi log plots of a relative concentration vs. time, indicating first-order behavior, with the rate constant independent of initial concentration. (After Orth and Gillham, 1996.)

The performance of the wall was monitored over a period of 474 days using a dense network of multilevel sampling wells (348 sampling points). Figure 21.4a shows the concentration of TCE, PCE and chloride across the wall on day 299. Though designed for complete remediation of the plume, the effluent concentration was about 10% of the influent. From the concentration distribution it appears that the upstream portion of the wall performed as expected, but that the rate constant declined with distance through the wall. Figure 21.4b shows that the degradation of TCE and PCE resulted in the formation of all three DCE isomers, but that these subsequently degraded with continued passage through the wall. No vinyl chloride was detected. Though variable between sampling dates, when averaged over the duration of the test, the chloride mass balance was close to 100%.

A core was taken horizontally through the wall in July 1993, 25 months after the wall was installed. From visual observation and from SEM and XRD analyses, precipitate formation could not be detected and there was no apparent change in the surfaces of the iron. Thus there is good reason to believe that the performance of the wall would persist for a substantial period of time.

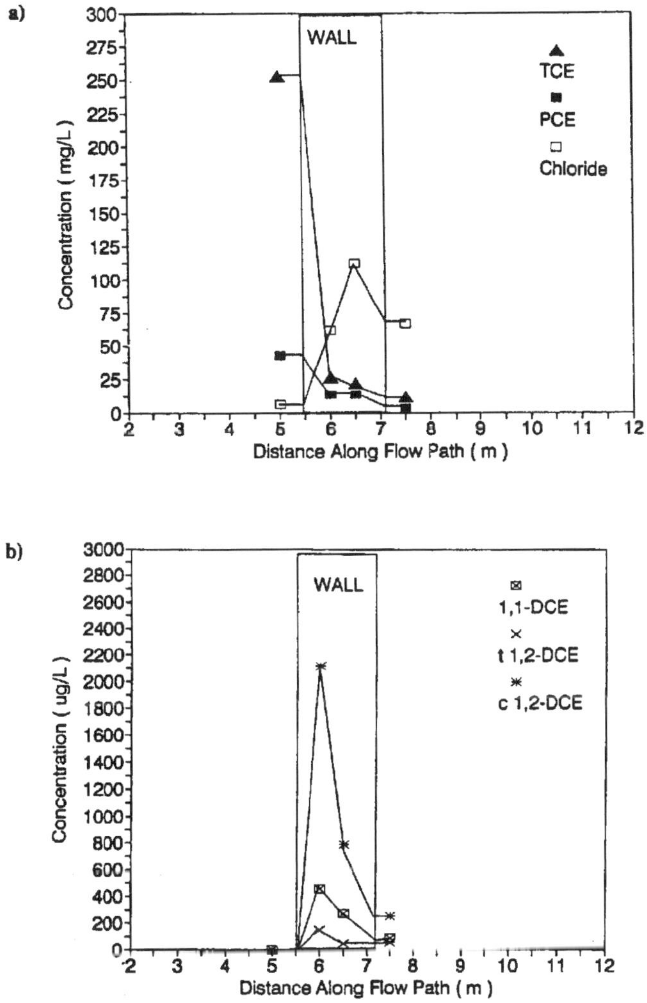

Figure 21.4. Concentrations of chlorinated compounds across the *in situ* treatment wall: a) Decline in influent concentrations of TCE and PCE, accompanied by an increase in chloride concentration. b) DCE isomers produced as a result of degradation of TCE and PCE, but subsequently degraded with distance through the wall (After O'Hannesin, 1993).

The performance of the wall was particularly encouraging in that very large amounts of TCE and PCE were degraded over the 474 day monitoring period, there was no apparent change in performance over this period, and from the analyses of the wall materials, there is good reason to expect that the performance would persist for a much longer period of time. The concentrations of both PCE and TCE in the effluent exceeded drinking water limits, apparently as a result of declining rate constants with travel distance through the wall. The reason for the apparent decline in reaction rates remains a topic of investigation.

The use of iron to enhance the degradation of chlorinated organic compounds is particularly attractive in that the process appears to be very robust, with the ability to degrade a wide range of common organic contaminants in waters with a wide range of inorganic chemical characteristics. Further, it appears to be effective at high concentrations that could be toxic to bacterial populations, as well as at very low concentrations. The reactive materials are relatively inexpensive and there is the potential for installations to be entirely passive over periods of several years and possibly decades.

Several aspects of the technology require further investigation. These include the effects of inorganic groundwater parameters on the reaction rate, persistence of the reaction over long time periods, and the range of compounds for which the process is applicable and the reaction mechanism.

OBSTACLES TO IMPLEMENTATION

In situ treatment walls are a recent concept, but based on the current level of research activity and the results of early studies, it is reasonable to conclude that several treatment methods will be developed that have commercial potential. Many of the obstacles to realizing this potential are common to all new technologies for groundwater remediation, while others are specific to the *in situ* treatment wall technologies. The very high cost of groundwater remediation schemes, coupled with their limited success, has impressed the groundwater community, regulators and industry with the difficulty of the problem. Furthermore, through enthusiasm or motivated by profit, new technologies are commonly overpromoted, falling short of expectations when applied at contaminated sites. Pervasive scepticism is therefore a major cultural impediment to the early implementation of new technologies.

In situ treatment walls require a high initial capital investment, with the benefits to be derived from greatly reduced long-term operating and maintenance costs. Implementation therefore requires a high degree of confidence on the part of the client that the benefits will indeed be realized. Unfortunately, the technologies with the greatest potential to be beneficial provide the greatest difficulty in predicting long-term performance. For example, if it is known that the reactive material must be replaced every year, then both the cost and performance should be relatively predictable, though the cost of replacement could substantially reduce the financial benefits. On the other hand, a technology that is expected to operate passively over a period of several years would provide the greatest financial benefit, but the anticipated performance could be very difficult to substantiate. Providing convincing evidence regarding long-term performance could therefore be a significant obstacle to implementation.

In situ treatment walls generally require that significant volumes of foreign materials be introduced to the subsurface. Thus, depending upon the materials used, regulatory approval could be a further obstacle to implementation.

Researchers, regulators and potential clients all have a role to play if *in situ* treatment walls are to meet their commercial potential. Researchers and purveyors of *in situ* treatment methods have a responsibility to provide thorough and impartial documentation of performance. It is important that the respective technologies be founded on basic principles or strong empirical evidence, and that possible limitations be thoroughly investigated. In order to reduce potential risk to the client, and to facilitate regulatory approval, it would be desirable if application of the technology could proceed through laboratory or prototype tests under conditions that represent, as closely as possible, the anticipated *in situ* conditions of the particular site.

Regulatory agencies have the responsibility to be informed of technological developments, and to be in a position to provide objective and technically sound evaluations of new technologies. Regulatory agencies could also support full-scale demonstrations by negotiating relaxed discharge quality requirements, should the technology not meet expectations, particularly at sites where human health is not at risk. This could substantially reduce the risk to the client of the high initial capital investment.

It is unlikely that small companies with a small number of contaminated sites will risk the necessary capital investment on technologies for which the long-term performance remains unproven. It is therefore likely that the initial clients will be large corporations or government agencies that are responsible for a large number of sites. For these organizations, the risk of applying a new technology at an individual site could be small compared to their total groundwater liability, with very large potential benefits should the technology be successful.

CONCLUSION

In situ treatment walls offer numerous potential advantages over pump-and-treat remediation schemes. In particular, they are generally passive or semi-passive, resulting in greatly reduced operating and maintenance costs. Furthermore, they are non-invasive in that major above-ground treatment works are not required. In contrast to pump-and-treat systems, which mix large amounts of uncontaminated water with the contaminated water before treatment, through proper placement of the wall, *in situ* treatment wall technologies should affect a relatively small amount of uncontaminated water. Water conservation is thus another advantage. In view of these advantages, the high level of research activity that is currently being directed toward the technology is entirely justified.

It can be said with certainty that methods suitable for installation of treatment walls now exist. Presently most of these are limited to unconsolidated surficial materials and to depths no greater than 30–50 m. Should the technology gain acceptance, installation procedures will undoubtedly be improved and refined through experience.

The technical limitation to the implementation of *in situ* treatment wall technologies concerns the treatment methods. Methods are now under development for treatment of a very wide range of inorganic and organic contaminants. Several of these show considerable promise and it is certain that many other treatment methods will be developed over the next few years. Though several methods have promise, none have yet been demonstrated at a large field scale, and few can provide, with any degree of certainty, estimates of installation and long-term operating and maintenance costs.

In addition to cultural factors such as the skepticism with which new technologies are viewed, the nature of *in situ* treatment walls mitigates against rapid implementation. In particular, because of the high initial cost, with benefits to be derived from low long-term operating and maintenance costs, few clients will be prepared to "give it a try." There is therefore a pressing need to move toward carefully designed and monitored demonstrations at contaminated sites. The confidence and experience gained from these demonstrations will be a major factor in the ultimate acceptance and implementation.

Though implementation may be slow, there is sufficient evidence to suggest that the development and implementation of *in situ* treatment wall technology is highly probable. While interest in these technologies grows, it is important to keep in mind that there are many sites, contaminants in fractured rock or at depths greater than about 30 to 50 m being notable examples, for which *in situ* reaction walls are not a viable option. Pump-and-treat or other methods of remediation and plume control will continue to play an important role in these situations.

REFERENCES

Berry-Spark, K.L. and J.F. Barker. 1987. Nitrate remediation of gasoline contaminated groundwaters. Results of controlled field experiment. Proc. of Petrol. Hydrocarbons and Organic Chemicals in Ground Water: Prevention, Detection and Restoration. NWWA/API, Houston, Nov. 17–19, 1987.

Bianchi-Mosquera, G.C., R.M. Allen-King, and D.M. Mackay. 1994. Enhanced degradation of dissolved benzene and toluene using a solid oxygen-releasing compound. *Ground Water Monitoring and Remediation,* 14(1):120–128.

Blowes, D.W. and C.J. Ptacek. 1992. Geochemical remediation of groundwater by permeable reactive walls: removal of chromate by reaction with iron-bearing solids. Proc. Subsurface Restoration Conference, June 21–24, 1992, Dallas, Texas. Available, Rice University, Dept. of Env. Sci. & Eng., P.O. Box 1892, Houston, TX, 77251.

Burris, D.R. and C.P. Antworth. 1990. Potential for subsurface in situ sorbent systems. *Groundwater Management,* 4:527.

Burris, D.R. and C.P. Antworth. 1992. In situ modification of an aquifer material by a cationic surfactant to enhance retardation of organic contaminants. *J. Contam. Hydrol.,* 4(4)325–337.

Burris, D.R. and J.A. Cherry. 1992. Emerging plume management technologies: in situ treatment zones. In proceedings of the 8th Annual Meeting of the Air and Waste Management Association. Manuscript 92-34.04.

Devlin, J.F. and J.F. Barker. 1992. In situ bioremediation of dissolved chlorinated solvents using a novel nutrient injection scheme. Abstract, Aquifer Restoration: Pump-and-Treat and the Alternatives, National Ground Water Association, Sept. 30–Oct. 2, 1992, Las Vegas Convention Center.

Devlin, J.F. and J.F. Barker. 1993a. A semi-passive injection system for in situ bioremediation. Proc., 1993 National Conference on Hydraulic Engineering and International Symposium on Engineering Hydrology, July 25–30, San Francisco, California. Sponsored by American Society of Civil Engineers, Hydraulics Division.

Devlin, J.F. and J.F. Barker. 1993b. A semi-passive injection system for in situ bioremediation. Abstract, Second International Symposium, In Situ and On-Site Bioremediation, April 5–8, San Diego, California. Sponsored by Battelle.

Gillham, R.W. and S.F. O'Hannesin. 1992. Metal-catalysed abiotic degradation of halogenated organic compounds. IAH Conference, Modern Trends in Hydrogeology, Hamilton, Ontario, May 10–13, pp 94–103.

Gillham, R.W. and S.F. O'Hannesin. 1994. Enhanced degradatioon of halogenated aliphatics by zero-valent iron. *Ground Water,* **32(6)**:958–967.

Gillham, R.W., S.F. O'Hannesin, and W.S. Orth. 1993. Metal enhanced abiotic degradation of halogenated aliphatics: laboratory tests and field trials. Proc., 1993 HazMat Central Conference, Chicago, Illinois, March 9–11, 1993.

Kao, C.-M. and R.C. Borden. 1992. Denitrification of BTEX using nutrient briquets as the nitrate source: laboratory microcosm and column studies. Proc., Subsurface Restoration Conference, June 21–24, 1992, Dallas, Texas. Available, Rice University, Dept. of Env. Sci. & Eng., P.O. Box 1892, Houston, TX 77251.

Mackay, D.M. and J.A. Cherry. 1989. Groundwater contamination: Pump-and-treat remediation. *Environ. Sci. Technol.,* 23(6):630–636.

McMurtry, D.C. and R.O. Elton. 1985. New approach to in-situ treatment of contaminated groundwaters. *Environ. Progress,* 4(3):168–170.

O'Hannesin, S.F. and R.W. Gillham. 1992. A permeable wall for in situ degradation of halogenated organic compounds. 45th Canadian Geotechnical Society Conference, Toronto, Ontario, October 25–28.

O'Hannesin, S.F. 1993. A field demonstration of a permeable reaction wall for the in situ abiotic degradation of halogenated aliphatic organic compounds. M.Sc. Thesis, University of Waterloo, Waterloo, Ontario, Canada.

Orth, W.S. and R.W. Gillham. 1996. Dechlorination of trichoroethene in aqueous solution using Fe^0. *Environ. Sci. Technol.,* **30(1)**:66–71.

Pankow, J.F., R.L. Johnson, and J.A. Cherry. 1993. Air sparging in gate wells within cutoff walls and in trenches for the control of plumes of volatile organic compounds (VOCs). *Ground Water,* 31(4):654–663.

Rivett, M.O., S. Feenstra, and J.A. Cherry. 1992. Groundwater zone transport of chlorinated solvents: A field experiment. IAH Conference, Modern Trends in Hydrogeology, Hamilton, Ontario, May 10–13.

Schwartz, F.W. and Y. Xu. 1992. Modelling the behavior of a reactive barrier system for lead. Proc., Modern Trends in Hydrogeology, Conference of the Canadian National Chapter, International Assn. of Hydrogeologists, Hamilton, Ontario, May 11–13, 1992.

Starr, R.C., J.A. Cherry, and E.S. Vales. 1992. A new type of steel sheet piling with sealed joints for groundwater pollution control. Proceedings, 45th Canadian Geotechnical Conference, Toronto, Ontario, October 26–28, 1992.

Starr, R.C. and J.A. Cherry. 1994. In situ remediation of contaminated groundwater: The funnel-and-gate system. *Ground Water,* **32(3)**:465–476.

Sweeny, K.H. and J.R. Fischer. 1972. Reductive degradation of halogenated pesticides. U.S. Patent No. 3, 640, 821; February 8.

Thomson, B.M., D.J. Findrock, and C.P. McHaley. 1992. Development of Permeable Barriers for Groundwater Remediation: Air Stripping and Biodegradation of Hydrocarbons. Proc., American Chemical Society's 4th Annual Symposium on Emerging Technologies for Hazardous Waste Management, Atlanta GA, September 23, 1992.

U.S. EPA. 1990. Handbook on in situ treatment of hazardous waste-contaminated soils. Risk Reduction Engineering Laboratory, Office of Research and Development, U.S. Environmental Protection Agency, Cincinnati, Ohio 45268. EPA/540/2-90/001, January, 1990.

Vogan, J.L. 1993. The use of emplaced denitrifying layers to promote nitrate removal from septic effluent. M.Sc. Thesis, Department of Earth Sciences, University of Waterloo, Waterloo, Ontario N2L 3G1.

Westall, J.C., J. Hatfield, H. Chen, and B.J. Brownawell. 1992. The use of cationic surfactants to modify aquifer materials to reduce the mobility of hydrophobic organic compounds—a study of equilibrium kinetics. Proc., Subsurface Restoration Conference, June 21–24, 1992, Dallas, Texas. Available, Rice University, Dept. of Env. Sci. & Eng., P.O. Box 1892, Houston, TX 77251.

Xu, Y. and F.W. Schwartz. 1992. Immobilization of lead in groundwater with a reactive barrier system. Proceedings, Subsurface Restoration Conference, June 21–24, 1992, Dallas, Texas. Available, Rice University, Dept. of Env. Sci. & Eng., P.O. Box 1892, Houston, TX 77251.

Chapter 22

Bioventing

R.E. Hinchee and B.C. Alleman

INTRODUCTION

Bioventing is a process designed to provide indigenous microorganisms with enough oxygen (O_2) to aerobically degrade target contaminants by introducing a supply of air or O_2 to O_2-deficient soils. The technology is applicable to all contaminants found in O_2-deficient soils that are more biodegradable aerobically than anaerobically. Bioventing may be applied *in situ* in either the vadose zone or the saturated zone (by means of air sparging) or *ex situ* in soil piles. Several aspects of bioventing make it an attractive technology:

- Bioventing can be used to promote *in situ* destruction of contaminants, relieving requirements for excavation.
- Bioventing limits production of waste streams that might require additional treatment.
- Bioventing systems are of a simple design and are not constrained by weather conditions.
- Costs associated with bioventing may be lower than with other conventional treatment technologies.

One of the first documented studies of bioventing resulted from an experiment designed to examine a completely different technology (Texas Research Institute, 1980; 1984). During the experiment, a large-scale model was constructed to test the effectiveness of a surfactant treatment to enhance the recovery of spilled gasoline. The experiment accounted for only 30 L of the 250 L originally spilled, and thus raised questions about the fate of the gasoline. Subsequently, a column study was conducted to determine a diffusion coefficient for soil venting. This column study evolved into a biodegradation study in which it was concluded that as much as 38% of the fuel hydrocarbon was biologically mineralized. Researchers concluded that venting would not only remove gasoline by physical means, but also could enhance microbial activity and promote biodegradation of the gasoline (Texas Research Institute, 1980; 1984).

Some of the first field-scale bioventing experiments were conducted in the Netherlands and were designed to investigate the effectiveness of bioventing for treating hydrocarbon-contaminated soils. These studies are reported in a series of papers (Anonymous, 1986; Staatsuitgeverij, 1986; van Eyk and Vreeken, 1988, 1989a, and 1989b).

Wilson and Ward (1986) suggested that using air as a carrier for O_2 could be 1,000 times more efficient than using water, especially in deep, hard-to-flood, unsaturated zones. They made the connection between soil venting and biodegradation by observing that "soil venting uses the same principle to remove volatile components of the hydrocarbon." In a general overview of the soil venting process, Bennedsen et al. (1987) concluded that soil venting provides large quantities of O_2 to the unsaturated zone, possibly stimulating aerobic degradation. They suggested that water and nutrients would also be required for significant degradation and encouraged additional investigation into this area.

Biodegradation enhanced by soil venting has been observed at several field sites. Investigators claim that significant biodegradation occurred (measured by a temperature rise) when air was supplied at a soil venting site for remediation of gasoline-contaminated soil. Investigators pumped pulses of air through a pile of excavated soil and observed a consistent rise in temperature, which they attributed to biodegradation. They claimed that the pile was cleaned up during the summer primarily by biodegradation (Conner, 1988). However, they did not control for natural volatilization from the aboveground pile, and not enough data were published to critically verify their biodegradation claim.

Ely and Heffner (1988) of the Chevron Research Company patented a bioventing process and used it at several sites. They did not provide their experimental design and data, but they did present their findings graphically. At a site contaminated by gasoline and diesel oil, they observed slightly higher removal through biodegradation than through evaporation. At a gasoline-contaminated site, results indicated that about 2/3 of the hydrocarbon removal was by volatilization and 1/3 by biodegradation. At a site containing only fuel oils, approximately 75 liters/well/day were biodegraded, while removals by volatilization were low due to low vapor pressures of the fuel oil. Ely and Heffner claimed that the process is more advantageous than strict soil venting because removal does not depend only on vapor pressure. In the examples stated in their patent, carbon dioxide (CO_2) was maintained between 6.8% and 11% and O_2 between 2.3% and 11% in vented air. The patent suggests that the addition of water and nutrients may not be acceptable because of flushing of the contaminants to the water table, but nutrient addition is claimed as part of the patent. The patent recommends flow rates between 50 and 420 m^3/min per well and states that air flows higher than those required for volatilization may be optimum for biodegradation. Although Chevron's patent is directly related to the bioventing process, other soil venting patents may also be relevant.

Researchers from the U.S. Environmental Protection Agency Robert S. Kerr Environmental Research Laboratory (RSKERL) at Traverse City, Michigan, observed a decrease in the toluene concentration in unsaturated zone soil gas, which they measured as an indicator of fuel contamination in the unsaturated zone. They assumed that advection had not occurred and attributed the toluene loss to biodegradation. The investigators concluded that because toluene concentrations decayed near the oxygenated ground surface, soil venting is an attractive remediation alternative for biodegrading light volatile hydrocarbon spills (Ostendorf and Kambell, 1989). This work was followed by a field-scale bioventing pilot study which demonstrated that providing organisms with a supply of O_2 promoted more rapid degradation of the toluene.

A thoroughly documented, full-scale bioventing study was initiated in 1988 at Hill Air Force Base (AFB) in Utah (Hinchee et al., 1991; Hinchee and Arthur, 1991; Dupont et al., 1991). This work was followed by a thoroughly documented field pilot-scale study at Tyndall AFB in Florida (Miller, 1990; Miller et al., 1991).

Bioventing has been applied for remediation of sites since the early- to mid-1980s in the form of soil venting (a.k.a. soil vacuum extraction, vacuum extraction, soil gas extraction, and *in situ* volatilization). At most if not all sites where soils are ventilated, O_2 is supplied and microbial activity is stimulated, and in many cases biodegradation is a significant contributor to remediation. In early applications of soil venting the role of biodegradation was not recognized or documented. More recently, however, soil venting vendors have begun to monitor and document biodegradation, and some are now designing and operating soil venting systems to optimize biodegradation either in addition to volatilization or to minimize volatilization.

Various forms of undocumented bioventing have been applied to a large number of sites worldwide. However, biodegradation has most probably been monitored at no more than 90% of these sites, and optimized at far fewer.

BASIC PRINCIPALS

Types of Contamination Treated With Bioventing

Bioventing is potentially applicable to any contaminant that is more readily biodegraded aerobically than anaerobically, such as most petroleum hydrocarbons (Atlas, 1986). To date, most applications have been to petroleum hydrocarbons (Hoeppel et al., 1991); however, application to polycyclic aromatic hydrocarbons (PAHs) (Lund et al., 1991; Hinchee and Ong, 1992) and to an acetone, toluene, and naphthalene mixture (Hinchee and Ong, 1992) have been reported.

In most bioventing applications, the key is biodegradability vs. volatility. If the rate of volatilization significantly exceeds the rate of biodegradation, removal takes place more through volatilization. Figure 22.1 illustrates the general relationship between a compound's physicochemical properties and its potential for bioventing.

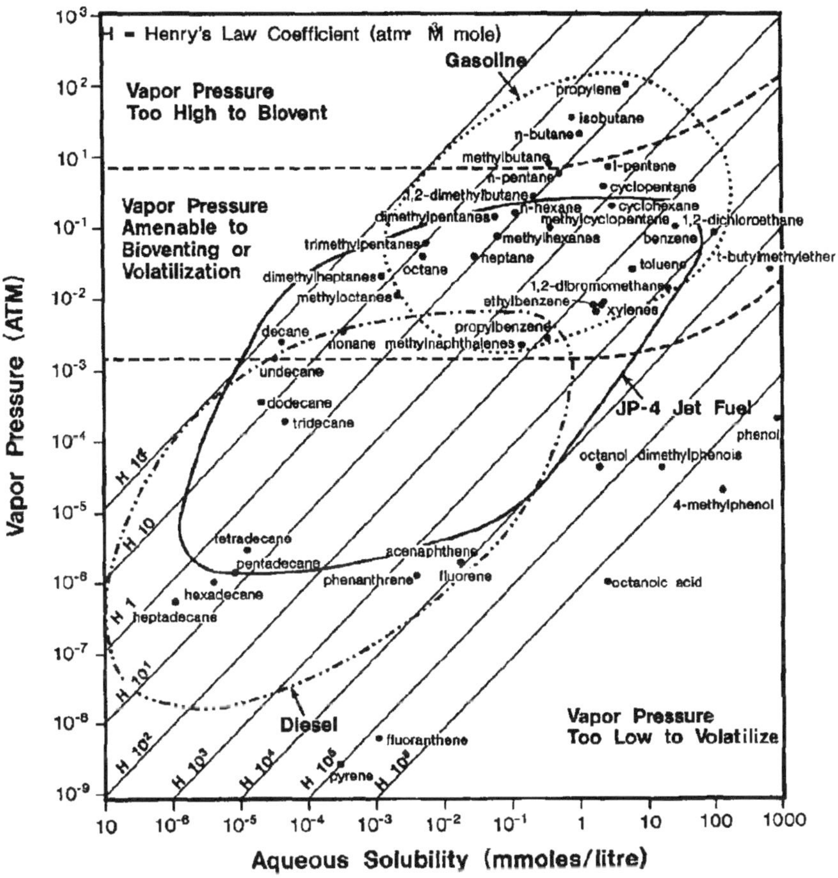

Figure 22.1. Impact of physicochemical properties on potential for bioventing

In general, low-vapor-pressure compounds below ~1 mmHg cannot be successfully removed by volatilization. Biodegradable contaminants that exhibit these low vapor pressures are potential candidates for bioventing. Higher vapor pressure compounds, above ~760 mmHg, are gases at ambient temperatures. These compounds volatilize too rapidly to be biodegraded during venting, with the typical results that the compounds are purged from the soil matrix before degradation can occur. Compounds with vapor pressures between 1 and 760 mmHg may be amenable to either volatilization or biodegradation in a bioventing system. Within this intermediate range lie many of the petroleum hydrocarbon compounds of greatest regulatory interest, such as benzene, toluene, and the xylenes. Figure 22.1 shows that various petroleum fuels are more or less amenable to bioventing. Some of the compounds found in gasoline are too volatile to easily biodegrade using a venting process. Most of the constituents of diesel fuel are sufficiently nonvolatile to preclude volatilization, whereas JP-4 jet fuel is somewhat intermediate.

Field data are lacking on the applicability of bioventing for remediation of PAH contamination. However, in the laboratory, PAH compounds have been shown to degrade much faster under aerobic conditions (Mihelcic and Luthy, 1988). The majority of PAH compounds have low volatility and are not easily purged from soils. These facts suggest that bioventing may be useful for remediating soils contaminated with PAHs.

Site Characterization Requirements

Normal site characterization data are required to implement bioventing or any other remedial technology and will not be addressed here. In addition, three site characterization tests are required to determine the applicability of bioventing at a specific site. These include a soil gas survey incorporating respiratory gas measurements, a soil gas permeability test, and an *in situ* respiration test. In addition to these tests, soil samples may be collected for nutrient analysis and microbial characterization may be desirable.

Soil gas surveys are now routine in many site characterizations. Methods for soil gas sample collection are well documented in the literature and will not be discussed here. Any method that assures collection of a soil gas sample from discrete depths is sufficient. Soil gas samples should be analyzed for contaminant hydrocarbons as well as for O_2 and CO_2. For bioventing to be successful in stimulating biodegradation, the contaminated area must be O_2 deficient. If it is not, the addition of more O_2 will have no effect and using bioventing may not be appropriate.

A full-scale bioventing design requires an estimate of the soil's permeability to fluid flow and the radius of influence of the venting wells. Onsite testing provides the most accurate estimate of the soil gas permeability, k. Onsite testing can also be used to determine the radius of influence that can be achieved for a given well configuration, air flow rate, and pressure. These data are used during system design to determine the spacing between venting wells, to size blower equipment, and to ensure that the entire site receives an adequate supply of O_2-rich air to sustain *in situ* biodegradation.

Soil gas permeability, or intrinsic permeability, can be defined as a soil's capacity for fluid flow, and varies according to grain size, soil uniformity, porosity, and moisture content. The value of k is a physical property of the soil; k does not change with different extraction/injection rates or different pressure levels.

Soil gas permeability is generally expressed in the units cm^2 or darcy (1 darcy = 1×10^{-8} cm^2). Like hydraulic conductivity, soil gas permeability may vary by more than an order of magnitude on the same site due to soil variability. The range of typical k values to be expected with different soil types is illustrated below.

Soil Type	k in Darcy
Course Sand	100–1000
Medium Sand	1–100
Fine Sand	0.1–1.0
Silts/Clays	<0.1

Source: Johnson et al. (1990)

The radius of influence (R_I) is defined as the maximum distance from the air extraction or injection well where measurable vacuum or pressure (soil gas movement) occurs. R_I not only is a function of soil properties, but also is dependent on the configuration of the venting well and extraction or injection flow rates, and is altered by soil stratification. On sites with shallow contamination, the R_I can be increased by impermeable surface barriers such as asphalt or concrete. These paved surfaces may or may not act as vapor barriers. Without a tight seal to the native soil surface, the pavement will not significantly impact soil gas flow.

Several field methods have been developed to determine soil gas permeability (see review by Sellers and Fan, 1991). One method in common use is the modified field drawdown method developed by Paul Johnson of Shell Development Company (Johnson, 1991). This method involves the injection or extraction of air at a constant rate from a single venting well while measuring the pressure/vacuum changes over time at several monitoring points in the soil away from the venting well.

An *in situ* respiration test was developed at Battelle Memorial Institute (Hinchee and Ong, 1992). The test has been used at numerous sites throughout the United States. The results are used to provide an indication of the level of biological activity and to calculate an initial estimate of the time required for remediation. The test consists of placing narrowly screened soil gas injection/withdrawal probes into soils in the unsaturated zone of a contaminated and a non-contaminated area. The soils are vented by injecting air containing an inert tracer gas for a given period of time. The apparatus for the respiration test is illustrated in Figure 22.2.

Typically, *in situ* respiration testing is conducted only in the contaminated area. However, if soil gas survey results show sub-ambient oxygen concentrations in non-contaminated areas, testing should also be done in these locations. The results will be used to determine background respiration rates resulting from biological breakdown of naturally occurring organic matter in the soil. Subtracting the

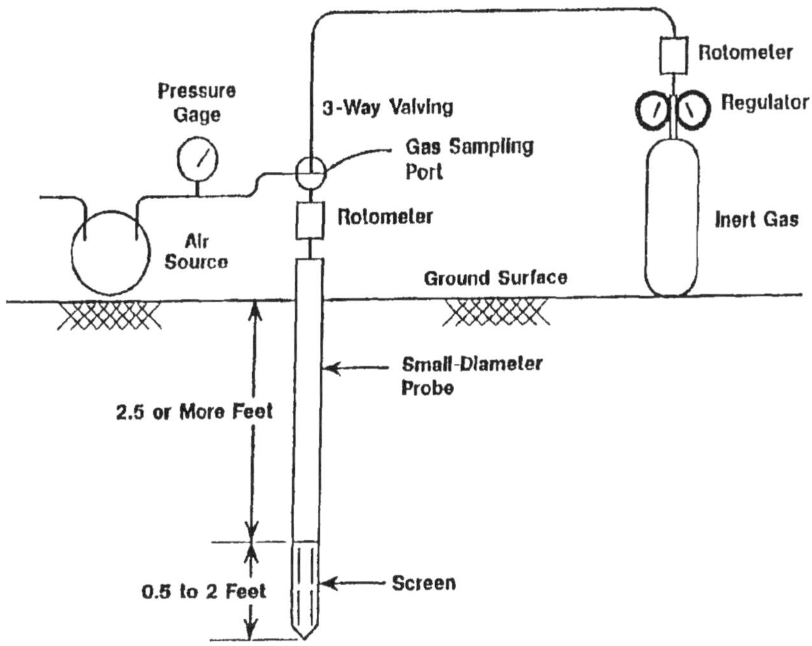

Figure 22.2. Gas injection/soil gas sampling monitoring point used by Hinchee and Ong (1992) in their *in situ* respiration studies.

background respiration rate from the rates measured in the contaminated area yields a more accurate estimate of contaminant degradation rates.

A cluster of three to four probes is usually placed in the soil of each test location. A 1 to 3% concentration of inert gas (usually helium) is added to the air, which is injected for approximately 24 hours. The air provides O_2 to the soil, whereas inert gas measurements provide data on the diffusion of O_2 from the ground surface and the surrounding soil and monitor potential leakage of the soil gas sampling system.

CO_2 and O_2 concentrations in the soil gas are measured before undertaking air and inert gas injection. Following injection, CO_2 and O_2 and inert gas concentrations are monitored over time. The monitoring points in contaminated soil at sites amenable to bioventing show a significant decline in O_2 over a 40- to 80-hr period. Figure 22.3 illustrates the average results from four such sites, along with the corresponding O_2 utilization rates in terms of percent of O_2 consumed per hour. In general, little or no O_2 utilization was measured in the uncontaminated background monitoring point.

The biodegradation rates measured by the *in situ* respiration test appear to be representative of those for a full-scale bioventing system. Miller (1990) conducted a 9-month bioventing pilot project at Tyndall AFB at the same time Hinchee and Ong (1992) were conducting their *in situ* respiration test. The O_2 utilization rates (Miller, 1990), measured from nearby active treatment areas, were virtually identical to those measured in the *in situ* respiration test.

CO_2 production proved to be a less useful measure of biodegradation than O_2 disappearance. The biodegradation rate in milligrams of hexane-equivalent/kilograms of soil per day based on CO_2 appearance is usually less than can be accounted for by the O_2 disappearance. The Tyndall AFB site was an exception. That site had low-alkalinity soils and low-pH quartz sands, and CO_2 production actually resulted in a slightly higher estimate of biodegradation (Miller, 1990). In the case of the higher-pH and higher-alkalinity soils at Fallon Naval Air Station (NAS) and Eielson AFB, little or no gaseous CO_2 production was measured (Hinchee and Ong, 1992). This could be due to the formation of carbonates from the gaseous evolution of CO_2 produced by biodegradation at these sites. A similar problem was encountered by van Eyk and Vreeken (1988) in their attempt to use CO_2 evolution to quantify biodegradation associated with soil venting. As a rule, O_2 utilization is a more reliable measure of bioventing-induced biodegradation than is CO_2 consumption.

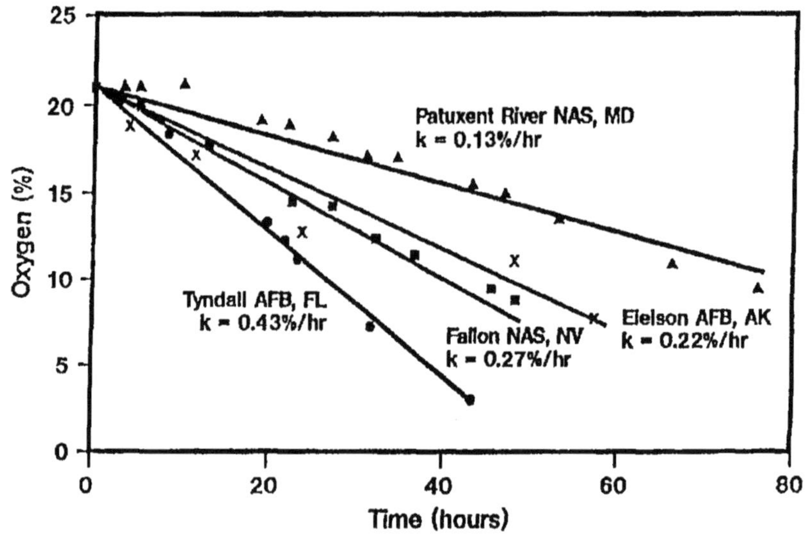

Figure 22.3. Average oxygen utilization rates measured at four test sites.

Impacts of Site Characteristics on Applicability

Assuming that the contaminants present are amenable to bioventing, gas permeability is probably the most important site characteristic. Soils must be sufficiently permeable to allow movement of enough gas to provide adequate O_2 for biodegradation. Gas permeability is a function of both soil structure and grain size, as well as of soil moisture content. Even in a coarse sand, if soil moisture content is high, adequate gas flow may not be possible. The site must be sufficiently permeable to allow an approximate minimum of 1 soil gas exchange per week. Typically, permeability in excess of 1 darcy is adequate. When the permeability falls below ~0.1 darcy, gas flow is primarily through either secondary porosity (such as fractures) or any more permeable strata that may be present (such as thin sand lenses).

The feasibility of bioventing in these low-permeability soils is a function of the distribution of flow paths and diffusion of air to and from the flow paths within the contaminated area. In a soil with reasonably good diffusion, a maximum separation of 2 to 4 feet between the flow path and the contaminant may still result in treatment. Gas permeability is a very site-specific characteristic. Bioventing has been used successfully to provide oxygen and support biodegradation of contaminants in some low-permeability soils including silty clay sites at Fallon NAS in Nevada (unpublished data) and at Eielson AFB in Alaska (Leeson et al., 1992). At a clay site on Tinker AFB in Oklahoma, bioventing has met with less success.

In addition to gas permeability, hydraulic permeability may be important if it is necessary to add nutrients or moisture or if it is necessary to dewater a site.

Another important site characteristic is contaminant distribution. Bioventing is primarily a vadose zone treatment process. The vadose zone may be extended through dewatering, but contamination below the water table cannot be treated.

Treatability Studies (Field and Laboratory)

Laboratory experiments are sometimes conducted prior to or simultaneous with a bioventing application (Hinchee and Arthur, 1991). These experiments provide data on the biodegradability of the contamination and the need for moisture or nutrient addition. Laboratory studies include both column and batch microcosm configurations. Although valuable data can be obtained through laboratory experimentation, the extrapolation of the data to the field is often difficult, resulting in the need for field verification of laboratory findings.

Field treatability studies are generally limited to *in situ* respiration tests or single-vent pilot-scale bioventing systems. When contamination is at shallow depths, push-tube respiration tests can be conducted. If higher respiration rates result in contaminated soil versus a non-contaminated background area, degradation of the contamination is assumed to be exerting the increased O_2 demand. If the contamination is at a significant depth, single-vent well systems can be installed, and either an *in situ* respiration test or a short-term bioventing test can be conducted. Conducting an *in situ* respiration test requires installation of vent well and monitoring systems in a background area to measure background respiration. Conducting a short-term bioventing test requires analysis of samples for contaminant concentrations before and after venting, and success is based on the decrease in these concentrations.

Process Monitoring

Bioventing process monitoring requires recording measurements of both the equipment operation and the system performance. Equipment operation is based on measurements of flow rates and pressures. Decreases in either or both of these could indicate equipment deterioration and signal the need for maintenance. Measurements of contamination concentrations, subsurface temperatures, and O_2 and CO_2 concentrations are made continuously to monitor system performance.

In situ respiration tests, conducted during periodic system shutdowns, provide an indirect measurement of contaminant degradation rates. It is important to conduct these tests to update the degradation rates and to monitor the remediation progress. Degradation rates can change over time as the dynamics of the microbial population change and/or changes occur in the contaminant composition occur.

Field Applications and Configurations

The use of bioventing to enhance biodegradation is based on a design to keep air flow through hydrocarbon-contaminated soils at rates and configurations that will (1) ensure adequate oxygenation for aerobic biodegradation, and (2) minimize or eliminate the production of a hydrocarbon-contaminated offgas. The addition of nutrients and moisture may be desirable to increase biodegradation rates; however, field research to date indicates that these additions may not always be needed (Dupont et al., 1991; Miller et al., 1991). If necessary, nutrient and moisture addition could take any of a variety of configurations. Dewatering may at times be necessary, depending on the distribution of contaminants relative to the water table. An important feature of good bioventing application is the use of narrowly screened soil gas monitoring points to sample gas in short, vertical sections of the soil. These points are utilized to monitor local O_2 concentrations, as the O_2 levels in the vent well do not represent local conditions.

Typically, a soil venting system is installed to draw air from a vent well in the area of greatest contamination. This configuration allows straightforward monitoring of the offgases. However, its disadvantage is that hydrocarbon offgas concentrations are maximized and could require permitting and treatment. Furthermore, all of the capillary fringe contamination may not be treated.

Figure 22.4 shows a bioventing system that involves air injection only. Although this is the lowest cost configuration, careful consideration must be given to the fate of injected air. The objective is for most, if not all, of the hydrocarbons to be degraded, and for CO_2 to be emitted at some distance from the injection point. If a building or subsurface structure were to exist within the radius of influence of the well, hydrocarbon vapors might be forced into that structure. Thus, protection of subsurface structures may be required. A bioventing system with this configuration was installed at Hill AFB in 1991 and is currently under study by U.S. EPA Risk Reduction Engineering Laboratory (RREL) and the U.S. Air Force.

Figure 22.5 is an illustration of a configuration in which air is injected (the injection may also be by passive well) into the contaminated zone and withdrawn from clean soils. This configuration allows the more volatile hydrocarbons to degrade before the air is withdrawn, thereby eliminating contaminated offgases. This configuration typically does not require air emission permitting, although

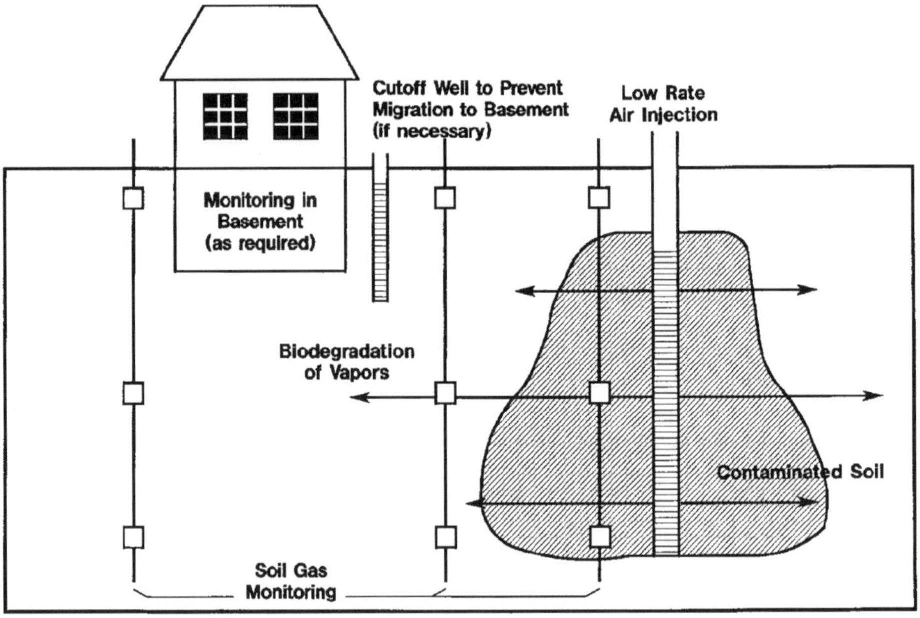

Figure 22.4. Conceptual layout of bioventing process with air injection only.

site-specific exceptions may apply. Currently, work is being conducted at a gasoline site near Atlanta, Georgia, using a configuration similar to the site shown in Figure 22.5.

Figure 22.6 illustrates a configuration that may alleviate the threat to subsurface structures while achieving the same basic effect as air injection alone. In this configuration, soil gas is extracted near the structure of concern and reinjected at a safe distance. If necessary, makeup air can be added before injection. A system with a configuration similar to this is being implemented at a site on Eglin AFB in Florida.

Figure 22.7 illustrates a conventional soil venting configuration at sites where hydrocarbon emissions to the atmosphere are not a problem. This may be the preferred configuration. Dewatering, and nutrient and moisture additions are also illustrated. Dewatering will allow more effective treatment of deeper soils. The optimal configuration for any given site will, of course, depend on site-specific conditions and remedial objectives.

STATE OF THE PRACTICE

Summary of Reported Case Histories

Hill AFB Site

A spill of approximately 100,000 L of JP-4 jet fuel occurred when an automatic overflow device failed at Hill AFB in Ogden, Utah. Contamination was limited to the upper 20 m of a delta outwash of the Weber River. This surficial formation extends from the surface to a depth of approximately 20 m and is composed of mixed sand and gravel with occasional clay stringers. Depth to regional groundwater is approximately 200 m; however, water occasionally may be found in discontinuous perched zones. Soil moisture averaged less than 6% in the contaminated soils.

The collected soil samples had JP-4 fuel concentrations up to 20,000 mg/kg, with an average concentration of approximately 400 mg/kg (Oak Ridge National Laboratory, 1989). Contaminants were unevenly distributed to depths of 20 m. Vent wells were drilled to approximately 20 m below the ground surface and were screened from 3 to 18 m below the surface. A background vent was

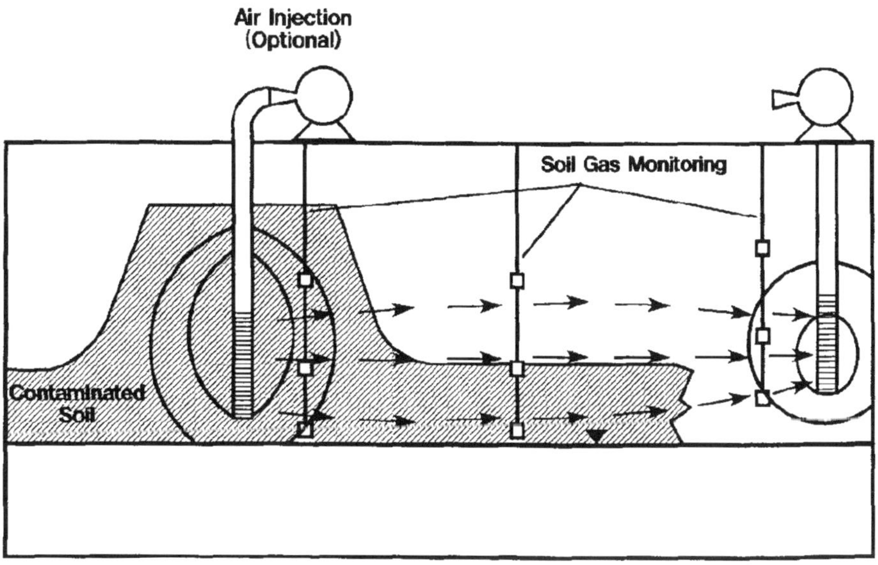

Figure 22.5. Conceptual layout of bioventing process with air withdrawn from clean soil.

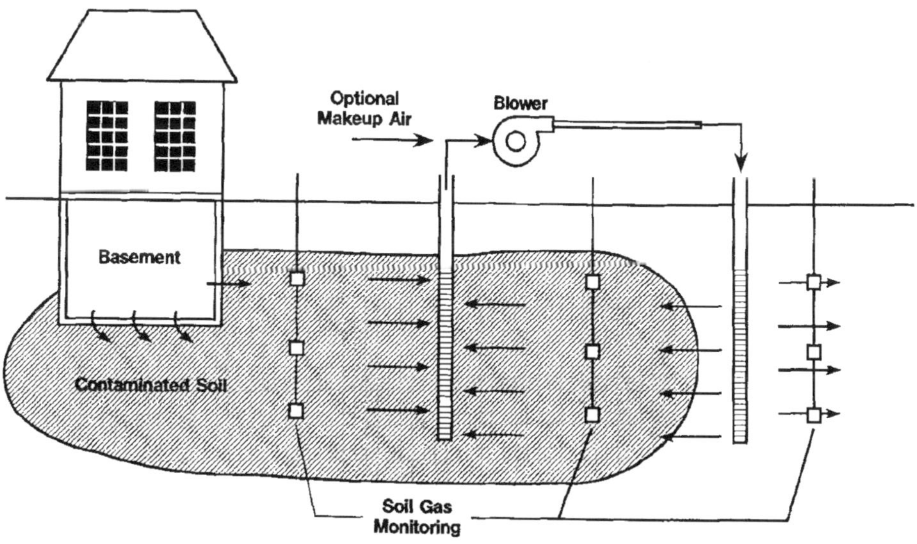

Figure 22.6. Conceptual layout of bioventing process with soil gas reinjection.

installed in an uncontaminated location in the same geological formation approximately 200 m north of the site.

Venting was initiated in December 1988 by air extraction at a rate of ~40 m^3/hr. The offgas was treated by catalytic incineration, and it was initially necessary to dilute the highly concentrated gas to remain below explosive limits and within the incinerator's hydrocarbon operating limits. The venting rate was gradually increased to ~2,500 m^3/hr as hydrocarbon concentration levels dropped. During the period between December 1988 and November 1990, more than 1.0×10^6 m^3 of soil gas was extracted from the site. In November 1989, ventilation rates were reduced to between ~500 and 1000 m^3/hr to provide aeration for bioremediation while reducing offgas generation. This change allowed removal of the catalytic incinerator, saving ~$6,000 per month.

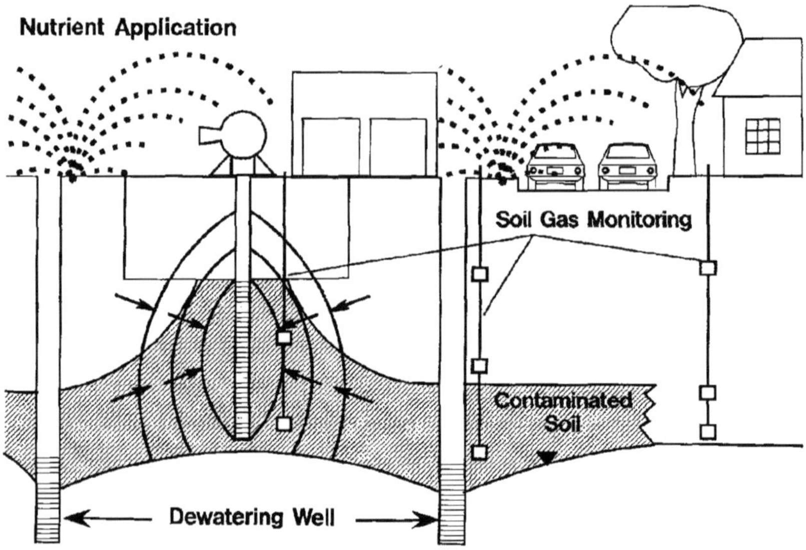

Figure 22.7. Conceptual layout of bioventing process with air injection into contaminated soil, coupled with dewatering and nutrient application.

During extraction, O_2 and hydrocarbon concentrations in the offgas were measured. To quantify the extent of biodegradation at the site, the O_2 was converted to an equivalent basis. This was based on the stoichiometric O_2 requirement for hexane mineralization. JP-4 hydrocarbon concentrations were determined based on direct readings of a total hydrocarbon analyzer calibrated to hexane. Based on these calculations, the mass of the JP-4 fuel as carbon removed was ~50,000 kg volatilized and 40,000 kg biodegraded. Figures 22.8 and 22.9 illustrate these results.

Hinchee and Arthur (1991) conducted bench-scale studies using soils from this site and found that, in the laboratory, both moisture and nutrients became limiting upon achievement of aerobic conditions. This led to the addition of first moisture and then nutrients in the field. The results of these field additions are shown in Figure 22.8. Moisture addition clearly stimulated biodegradation; nutrient addition did not.

The failure to observe an effect of nutrient addition could be explained by a number of factors, including:

- The nutrients failed to move in the soils; this is a problem particularly for ammonia and phosphorus (see Aggarwal et al., 1991).
- Remediation of the site was entering its final phase, and the nutrient addition may have been too late to result in an observed change.
- Nutrients simply may not have been limiting.

Tyndall AFB Site

As a followup to the Hill AFB research, a more controlled study was designed at Tyndall AFB. The experimental area in this study was located at a site where past JP-4 fuel storage had resulted in contaminated soils. The nature and volume of fuel spilled or leaked were unknown. The site soils are a fine- to medium-grained quartz sand. The depth to groundwater is 0.5 to 1.5 m.

Four test cells were constructed to allow control of gas flow, water flow, and nutrient addition. Test cells V1 and V2 were installed in the hydrocarbon-contaminated zone; the other two were installed in uncontaminated soils. Initial site characterization indicated the mean soil hydrocarbon levels were 5,100 and 7,700 mg of hexane equivalent/kg in treatment plots V1 and V2, respectively. The contaminated area was dewatered, and hydraulic control was maintained to keep the depth to water at

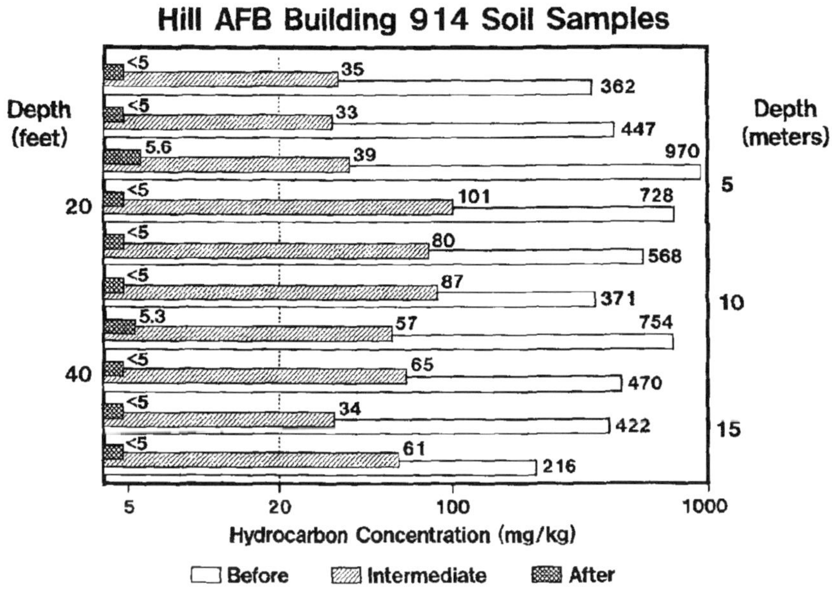

Figure 22.8. Cumulative hydrocarbon removal from the Hill AFB Building 914 soil venting site.

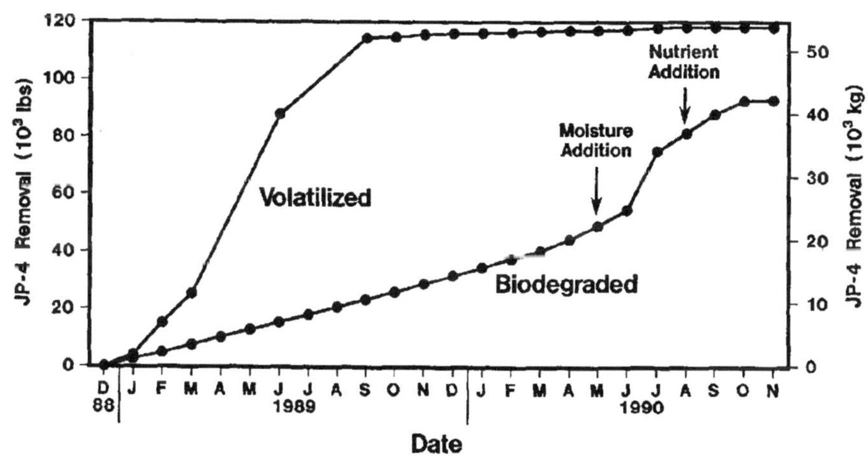

Figure 22.9. Results of soil analysis at Hill AFB before and after venting. (Each bar represents the average of 14 or more samples.)

~2 m. This exposed more of the contaminated soil to aeration. During normal operation, air-flow rates were maintained at approximately one air-filled void volume per day.

Biodegradation and volatilization rates were much higher at the Tyndall AFB site than those observed at Hill AFB; these higher rates were likely due to higher average levels of contamination, warmer temperatures, and the presence of moisture. After 200 days of aeration, an average hydrocarbon reduction of ~2,900 mg/kg was observed. This average represents a reduction in total hydrocarbons of approximately 40%.

The study was terminated because the process monitoring objectives had been met; biodegradation was still vigorous. Although the total petroleum hydrocarbons had been reduced by only 40%, the low-molecular-weight aromatics—benzene, toluene, ethylbenzene, and xylenes (BTEX)—were reduced by more than 90% (see Figure 22.10). It appears that the bioventing process more rapidly removes the BTEX compounds than the other JP-4 fuel constituents.

Another important observation of this study was the effect of temperature on the biodegradation rate. Miller (1990) found that the van Hoff-Arrhenius equation provided an excellent model of temperature effects. In the Tyndall AFB study, soil temperature varied by only ~7°C, yet biodegradation rates were approximately twice as high at 25°C than at 18°C.

In the Tyndall AFB study, the effects of moisture and nutrients were observed in a field test. Two side-by-side plots received identical treatment, except that one (V2) received both moisture and nutrients from the outset of the study while the other plot (V1) received neither for 8 weeks, then moisture only for 14 weeks, followed by both moisture and nutrients for 7 weeks. As illustrated in Figure 22.11, no significant effect of moisture or nutrients was observed. The lack of moisture effect contrasts with the Hill AFB findings, but is most likely the result of contrasting climatic and hydrogeologic conditions. Hill AFB is located on a high-elevation desert with a very deep water table. Tyndall AFB is located in a moist subtropical environment, and at the site studied, the water table was maintained at a depth of approximately 2 m.

The nutrient findings support field observations at Hill AFB that the addition of nutrients does not stimulate biodegradation. Based on acetylene reduction studies, Miller (1990) speculates that adequate nitrogen was present due to nitrogen fixation. Both the Hill and Tyndall AFB sites were contaminated for several years before the bioventing studies, and both sites were anaerobic. It is possible that nitrogen fixation, which is maximized under these conditions, provided the required nutrients. In any case, these findings show that nutrient addition is not always required.

In the Tyndall study, a careful evaluation of the relationships between air flow rates and both biodegradation and volatilization was made. It was found that extracting air at the optimal rate for biodegradation resulted in 90% removal by biodegradation and 10% removal by volatilization. It was also found that passing the 10% volatilized contaminants through clean soil resulted in complete biodegradation of the volatilized contaminants.

Applicability and Limitations

The primary problems encountered with bioventing are:

- Accurate estimate of emissions—One key variable in bioventing cost and design is the need or lack of need for offgas treatment. To permit an emission with or without offgas treatment, regulators typically require an estimate of emission rate.
- Time required for remediation—Bioventing that depends primarily on biodegradation is a slow process requiring 2 or more years for remediation. At many sites this can be a problem.
- Determining effectiveness on non-petroleum hydrocarbons—Many sites are contaminated with a mixture of chemical wastes, and little is known of the effectiveness of bioventing on non-petroleum hydrocarbons.
- Regulatory acceptance—With this technology, as with many emerging technologies, obtaining regulatory acceptance can be difficult.

Permitting Problems

The primary obstacle to regulatory acceptance is demonstrating potential effectiveness. Many regulators are unfamiliar with the technology and are skeptical of accepting a remedial approach that may require 2 to 5 years to show results. In general, after the responsible regulator(s) agree to allow the use of bioventing, permitting is not difficult.

Reported Costs

As with many emerging technologies, not much published experience exists to precisely determine cost; however, some general guidelines are available. The basic cost of soil venting equipment has been estimated to be as low as \$10 per yd^3. Offgas treatment costs can run \$30 to \$50/yd^3 (Long, 1992). The key to the cost of bioventing is the monitoring costs, and these are very site specific. Due

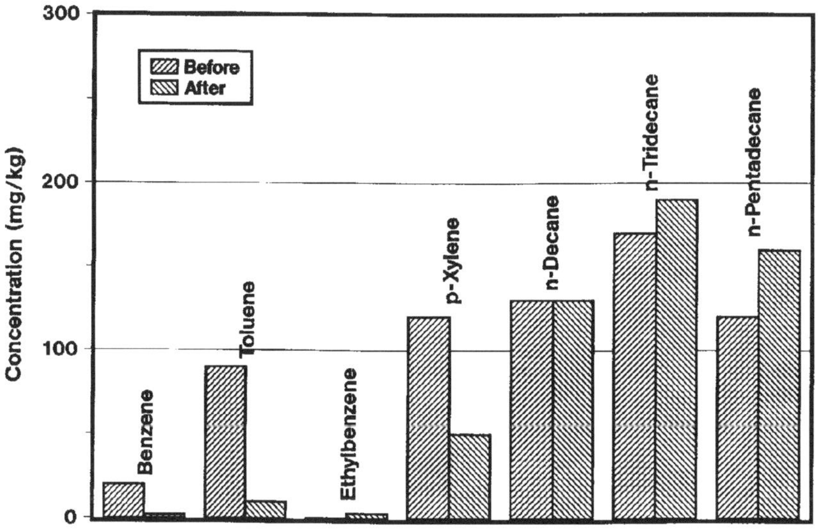

Figure 22.10. Results of soil analysis from Plot V2 at Tyndall AFB before and after venting. (Each bar represents the average of 21 or more soil samples.)

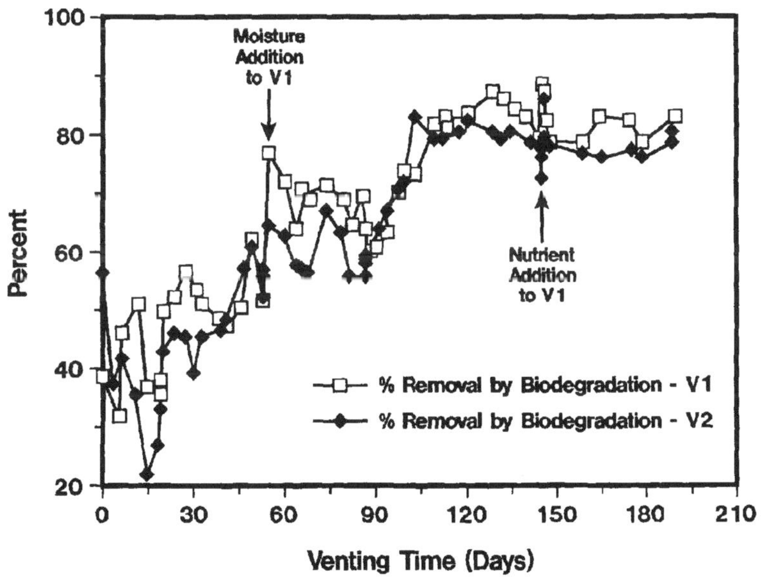

Figure 22.11. Cumulative percent hydrocarbon removal at Tyndall AFB for sites V1 and V2.

to the relatively long time period required for biodegradation, a bioventing site optimized for biodegradation as opposed to volatilization will incur higher monitoring costs. On some sites this may be offset by not having gas treatment costs, thus reducing total remediation costs.

SUMMARY AND CONCLUSIONS

Bioventing has been performed and monitored at several field sites contaminated with middle distillate fuels, mainly JP-4 jet fuel. Yet the effects of environmental variables on bioventing treatment rates are not well understood. *In situ* respirometry at additional sites with drastically different geologic conditions has further defined environmental limitations and site-specific factors that are pertinent to successful bioventing. However, the relationships between respirometric data and actual bioventing treatment rates have not been clearly determined. Additional field respirometry and closely monitored field pilot bioventing studies at the same sites are needed to determine what types of contaminants can be successfully treated *in situ* by bioventing and what the environmental limitations are. Studies are also needed on a wide variety of contaminants. Studies to date clearly show that many preconceptions regarding the factors that control bioventing rates can be wrong. For example, active respiration at a subarctic site at Eielson AFB near Fairbanks, Alaska, suggests that good hydrocarbon degradation can occur *in situ* at locations that are continually subjected to a cold environment. Failure to accelerate biodegradation rates by adding nitrogen fertilizer to biovented soils that contain low nitrogen levels indicates that nutrient addition at some sites may not be required. Also, fine-grained, moist, clayey soils have been readily aerated and showed aerobic respiration, indicating that bioventing is feasible at times in soils having low permeabilities. Other low-permeability sites have not proven amenable to bioventing. Better site diagnostics are needed to identify sites with biodegradation potential.

Vapor phase biodegradation occurs and can take place *in situ.* The question of how soil sorption and partitioning of volatile organic compounds into soil air affect biodegradation rates was addressed earlier by McCarty (1987). This question needs further attention as the movement of the vapor phase in soils is complex and dependent on changing soil environmental conditions.

Bioventing rates need to be determined under varying vapor extraction rates, because an important purpose for bioventing is to biodegrade the vapor within the soil profile. The minimal soil aeration levels that provide for high degradation rates must be determined under different soil conditions. Interaction of the vapor phase with soil particles and microorganisms in the uncontaminated soil profile needs further research in both the laboratory and in the field.

REFERENCES

Aggarwal, P.K., J.L. Means, and R.E. Hinchee. 1991. Formulation of Nutrient Solutions for In Situ Bioremediation. *In* Hinchee, R.E. and R.F. Olfenbuttel, eds., *In Situ and On-Site Bioreclamation*, Butterworth-Heinemann, Stoneham, MA, pp. 51–66.

Anonymous. 1986. In Situ Reclamation of Petroleum Contaminated Sub-soil by Subsurface Venting and Enhanced Biodegradation. *Research Disclosure*, No. 26233, 92–93.

Atlas, R.M. 1986. Microbial Degradation of Petroleum Hydrocarbons: An Environmental Perspective. *Microbiol. Rev.* **45**:180–209.

Bennedsen, M.B., J.P. Scott, and J.D. Hartley. 1987. Use of Vapor Extraction Systems for In Situ Removal of Volatile Organic Compounds from Soil. *In* Proceedings of National Conference on Hazardous Wastes and Hazardous Materials, Washington, DC, 92–95.

Conner, J.S. 1988. Case Study of Soil Venting. *Poll. Eng.* **7**:74–78.

Dupont, R.R., W. Doucette, and R.E. Hinchee. 1991. Assessment of In Situ Bioremediation Potential and the Application of Bioventing at a Fuel-Contaminated Site. *In* Hinchee, R.E. and R.F. Olfenbuttel, eds., *In Situ and On-Site Bioreclamation*, Butterworth-Heinemann, Stoneham, MA, pp. 262–282.

Ely, D.L. and D.A. Heffner. 1988. Process for In-Situ Biodegradation of Hydrocarbon Contaminated Soil, U.S. Patent Number 4,765,902.

Hinchee, R.E. and M. Arthur. 1991. Bench-Scale Studies of the Soil Aeration Process for Bioremediation of Petroleum Hydrocarbons. *J. Appl. Biochem. Biotech.* **28/29**:901–906.

Hinchee, R.E., D.C. Downey, R.R. Dupont, P. Aggarwal, and R.N. Miller. 1991. Enhancing Biodegradation of Petroleum Hydrocarbon through Soil Venting. *J. Hazardous Materials.* **27**:315–325.

Hinchee, R.E. and S.K. Ong. 1992. A Rapid In Situ Respiration Test for Measuring Aerobic Biodegradation Rates of Hydrocarbons in Soil. Accepted for Publication in the *Journal of the Air & Waste Management Association.* 21 pp.

Hoeppel, R.E., R.E. Hinchee, and M.R. Arthur. 1991. Bioventing Soils Contaminated with Petroleum Hydrocarbons. *J. Industrial Microbiology.* **8**:141–146.

Hogg, D.S., R.J. Burden, and P.J. Riddell. 1992. In Situ Vadose Zone Bioremediation of Soil Contaminated with Non-Volatile Hydrocarbon. Presented at HMCRI Conference, February 4, San Francisco, CA.

Johnson, P.C., M.W. Kemblowski, and J.D. Colthart. 1990. Quantitative Analysis for the Cleanup of Hydrocarbon-Contaminated Soils by In-Situ Soil Venting. *Ground Water.* **28**:413–429.

Johnson, P.C. 1991. *HyperVentilate Users Manual*, Shell Development, Houston, TX, 35 pp.

Leeson, A., R.E. Hinchee, J. Kittle, G. Sayles, C.M. Vogel, and R.N. Miller. 1992. Optimizing Bioventing in Shallow Vadose Zones and Cold Climates. Submitted for Publication in the *Proceedings of the In-Situ and On-Site Bioremediation Symposium*, Niagara on the Lake, Canada.

Long, G. 1992. Bioventing and Vapor Extraction: Innovative Technologies for Contaminated Site Remediation. *J. Air and Waste Mgt. Assoc.* **43**:345–348.

Lund, N.-Ch., J. Swinianski, G. Gadehus, and D. Maier. 1991. Laboratory and Field Tests for a Biological In Situ Remediation of a Coke Oven Plant. In Hinchee, R.E. and R.F. Olfenbuttel eds., *In Situ and On-Site Bioreclamation*, Butterworth-Heinemann, Stoneham, MA, pp. 396–412.

McCarty, P.L. 1987. Bioengineering Issues Related to In Situ Remediation of Contaminated Soils and Groundwater. In: *Proc. Conf. on Reducing Risk from Environmental Chemicals Through Biotechnology.* Seattle, WA, July.

Mihelcic, J.R. and R.G. Luthy. 1988. Degradation of Polycyclic Aromatic Hydrocarbon Compounds under Various Redox Conditions in Soil-Water Systems. *Appl. Environ. Microbiol.* **54**:1182–1187.

Miller, R.N. 1990. A Field Scale Investigation of Enhanced Petroleum Hydrocarbon Biodegradation in the Vadose Zone Combining Soil Venting as an Oxygen Source with Moisture and Nutrient Additions. Ph.D. thesis. Utah State University, Logan, Utah.

Miller, R.N., R.E. Hinchee, and C. Vogel. 1991. A Field-Scale Investigation of Petroleum Hydrocarbon Biodegradation in the Vadose Zone Enhanced by Soil Venting at Tyndall AFB, Florida. In Hinchee, R.E. and R.F. Olfenbuttel, eds., *In Situ and On-Site Bioreclamation*, Vol. 1., Butterworth Publishers, Stoneham, MA, pp. 283–302.

Oak Ridge National Laboratory. 1989. Soil Characteristics: Data Summary, Hill Air Force Base Building 914 Fuel Spill Soil Venting Project, an unpublished report to the U.S. Air Force.

Ostendorf, D.W. and D.H. Kambell. 1989. Vertical Profiles and Near Surface Traps for Field Measurement of Volatile Pollution in the Subsurface Environment. In: *Proceedings of NWWA Conference on New Techniques for Quantifying the Physical and Chemical Properties of Heterogeneous Aquifers*, Dallas, TX.

Sellers, K. and C.Y. Fan. 1991. Soil Vapor Extraction: Air Permeability Testing and Estimation Methods. In: *Proceedings of the 17th RREL Hazardous Waste Research Symposium*, EPA/600/9-91/002, April.

Staatsuitgeverij. 1986. Proceedings of a Workshop, 20–21 March, 1986. *Bodembeschermingsreeks* No. 9; *Biotechnologische Bodemsanering*, pp. 31–33. Rapportnr. 851105002, ISBN 90-12-054133, Ordernr. 250-151 59; Den Haag, The Netherlands.

Texas Research Institute. 1980. *Laboratory Scale Gasoline Spill and Venting Experiment.* American Petroleum Institute, Interim Report No. 7743-5:JST.

Texas Research Institute. 1984. *Forced Venting to Remove Gasoline Vapor from a Large-Scale Model Aquifer*, American Petroleum Institute, Final Report No. 82101-F:TAV.

van Eyk, J. and C. Vreeken. 1988. Venting-Mediated Removal of Petrol from Subsurface Soil Strata as a Result of Stimulated Evaporation and Enhanced Biodegradation. *Med. Fac. Landbouww. Riiksuniv. Gent.* **53**:1873–1884.

van Eyk, J. and C. Vreeken. 1989a. Model of Petroleum Mineralization Response to Soil Aeration to Aid in Site-Specific, In Situ Biological Remediation. In Jousma et al., eds., *Groundwater Contamination: Use of Models in Decision-Making, Proceedings of an International Conference on Groundwater Contamination*, Kluwer, Boston/London, pp. 365–371.

van Eyk, J. and C. Vreeken. 1989b. Venting-Mediated Removal of Diesel Oil from Subsurface Soil Strata as a Result of Stimulated Evaporation and Enhanced Biodegradation. In: *Hazardous Waste and Contaminated Sites, Envirotech Vienna*, Vol. 2, Session 3, ISBN 389432-009-5, Westarp Wiss., Essen, pp. 475–485.

Wilson, J.T. and C.H. Ward. 1986. Opportunities for Bioremedation of Aquifers Contaminated with Petroleum Hydrocarbons. *J. Ind. Microbiol.* **27**:109–116.

Chapter 23

Aerobic Cometabolism of Chlorinated Aliphatic Hydrocarbons

P.L. McCarty

INTRODUCTION

Chlorinated aliphatic hydrocarbons (CAHs), such as tetrachloroethene (PCE), trichloroethene (TCE), and 1,1,1-trichloroethane (TCA), are widely used cleaning solvents, and as such have often been discarded with little consideration for the environmental consequences. As a result, they and some of their degradation products, such as 1,1-dichloroethane (1,1-DCA), 1,1-dichloroethene (1,1-DCE), *cis*- and *trans*-1,2-dichloroethene (*c*-DCE, *t*-DCE), and vinyl chloride (VC) are among the most prevalent hazardous organic compounds found present in soils and the subsurface environment (Vogel et al., 1987). Industrial production of some CAHs, such as 1,2-dichloroethane (1,2-DCA) and VC, is much greater than for the above chlorinated solvents, but since they are largely consumed in the manufacture of other useful products, groundwater contamination from this source is much less than from the waste solvents and the transformation products formed from them. Many CAHs are known or suspected carcinogens, and so methods for their removal from the subsurface are being sought.

Just over one decade ago, most CAHs were considered to be non-biodegradable. What appeared to be transformation products of chlorinated solvents then started to be found in groundwaters, and this led to expanded efforts to determine the chemical and biological processes responsible. It was found that most of the CAHs can in fact be transformed by biological processes, but at times, the microorganisms responsible cannot obtain energy for growth from the transformations. Such transformations are brought about by cometabolism, that is, through interactions of the CAHs with biological molecules produced for other purposes. There are now widespread efforts to take advantage of cometabolism for the transformation of CAHs in groundwater.

Biotransformation of CAHs can occur aerobically or anaerobically (Vogel et al., 1987). Anaerobic cometabolic processes generally result in reductive dehalogenation of the compounds, a process in which a halogen atom is replaced with a hydrogen atom. The anaerobic transformation process occurs widely in contaminated groundwaters, leading to formation of 1,2-DCE and VC from PCE and TCE. Further reduction of VC has also been observed, leading to the formation of ethene. Aerobic transformation of CAHs can also occur, generally through cometabolism and leading to oxidized products. In general, the anaerobic reductive process is faster with the more highly halogenated compounds, while the aerobic process is faster with the less halogenated compounds. This suggests that degradation of highly chlorinated CAHs may benefit from treatment first by anaerobic processes to remove some of the halogens from the molecule, followed by aerobic cometabolism in order to achieve complete oxidation of the intermediate compounds so formed. CAHs can also be transformed abiotically (Vogel et al., 1987). Such transformations generally involve substitution (replacement of halogen with a hydroxyl), elimination (removal of two halogens from adjoining carbons), or reductions (replacement of halogen with hydrogen). This chapter does not emphasize abiotic or anaerobic transformation processes. It does emphasize primarily the aerobic cometabolic transformation processes.

AEROBIC COMETABOLISM OF CAHs

Primary Substrates and Cometabolism

Organic compounds can be biotransformed by microorganisms through two basically different processes: (1) use as a primary substrate, and (2) cometabolism. In the first process biodegradation occurs when the organism consumes the organic compound as a primary substrate to satisfy its energy and organic carbon needs. This is the usual process for organic decomposition in nature, and the process generally captured in the vast majority of biological treatment processes designed for municipal and industrial wastewaters. Knowledge about organism growth and kinetics with primary substrate utilization is quite extensive.

CAH Usage as Primary Substrate

Some few of the CAHs have been shown to serve as primary substrates for energy and growth by some microorganisms (Table 23.1). These same compounds can be transformed through cometabolism as well. Since use as a primary substrate is much more efficient, knowledge of when this might occur can be beneficial. Among the C_1 compounds, dichloromethane (DM) can be used as a primary substrate under both aerobic and anaerobic conditions. DM can be completely mineralized while serving as a primary substrate under anaerobic conditions by municipal digesting sludge microorganisms (Freedman and Gossett, 1991; Klecka and Gonsior, 1984; Rittmann and McCarty, 1980). Pure cultures of the genera *Pseudomonas* and *Hyphomicrobium* have been isolated which can grow aerobically on DM as the sole carbon and energy source (Brunner et al., 1980; Kohler-Staub et al., 1986; LaPat-Polasko et al., 1984; Stucki et al., 1981).

The two-carbon saturated CAH, 1,2-dichloroethane (1,2-DCA), can be also used as a primary energy source under aerobic conditions (Janssen et al., 1985; Stucki and Leisinger, 1983). One unsaturated two-carbon CAH, vinyl chloride, also has been shown to be available as a primary substrate for energy and growth under aerobic conditions (Hartmans and de Bont, 1992; Hartmans et al., 1985). These few exceptions noted to date indicate that only the less halogenated one- and two-carbon CAHs tend to be used as primary substrates for energy and growth, and that the organisms that are capable of doing this are not necessarily widespread in the environment. The biological transformation of most of the CAHs depends upon cometabolism.

Aerobic Cometabolic Transformations of CAHs

Cometabolism is the fortuitous transformation of an organic compound by enzymes or other biomolecules produced by organisms for other purposes (Jensen, 1963; Leadbetter and Foster, 1959).

Table 23.1. Potential for CAH Biotransformation Under Aerobic or Anaerobic Conditions (McCarty and Semprini, 1994).

	Primary Substrate		Aerobic Cometabolism	Anaerobic Reductive Dehalogenation	
Compound	Aerobic Potential	Anaerobic Potential	Potential[a]	Potential[a]	CAH Product
CCl_4			0	XXXX	$CHCl_3$
$CHCl_3$			X	XX	CH_2Cl_2
CH_2Cl_2	Yes	Yes	XXX		
CH_3CCl_3			X	XXXX	CH_3CHCl_2
CH_3CHCl_2			X	XX	CH_3CH_2Cl
CH_2ClCH_2Cl	Yes		X	X	CH_3CH_2Cl
CH_3CH_2Cl	Yes		XX	[b]	
$CCl_2{=}CCl_2$			0	XXX	$CHCl{=}CCl_2$
$CHCl{=}CCl_2$			XX	XXX	$CHCl{=}CHCl$
$CHCl{=}CHCl$			XXX	XX	$CH_2{=}CHCl$
$CH_2{=}CCl_2$			X	XX	$CH_2{=}CHCl$
$CH_2{=}CHCl$	Yes		XXXX	X	

[a] 0—very small if any potential; X—some potential; XX—fair potential; XXX—good potential; XXXX—excellent potential.
[b] Readily hydrolyzed abiotically, with half-life on order of one month.

Here, the organisms obtain no obvious or direct benefit from the transformation. Indeed, it may be harmful to them. Cometabolism is also a natural process, but it has not been used extensively for treatment of organic wastes, and knowledge of the process and its practical application are by comparison quite limited. Because of the potential usefulness of biotransformation of CAHs, there is much ongoing research to learn how cometabolism might be applied. For cometabolism to occur, an active population of microorganisms producing the cometabolizing biomolecules must be present. This means that the appropriate primary substrates for growth and maintenance of these organisms must also be present. This is an aspect that adds greater complexity and cost to cometabolic biotransformations.

Some few CAHs, particularly halomethanes such as chloromethane, bromomethane, dichloromethane, and chloroform were shown earlier to be biodegradable by microorganisms through cometabolism (Dalton and Stirling, 1982; Stirling and Dalton, 1979; Stirling and Dalton, 1985). Knowledge that a broader range of CAHs can be oxidized aerobically through cometabolism is more recent. Wilson and Wilson (1985) showed for the first time that TCE may be susceptible to aerobic degradation through use of soil microbial communities fed natural gas. The processes involved are illustrated by the following equations for TCE cometabolism by methanotrophic bacteria, organisms that oxidize methane for energy and growth:

Methane Oxidation

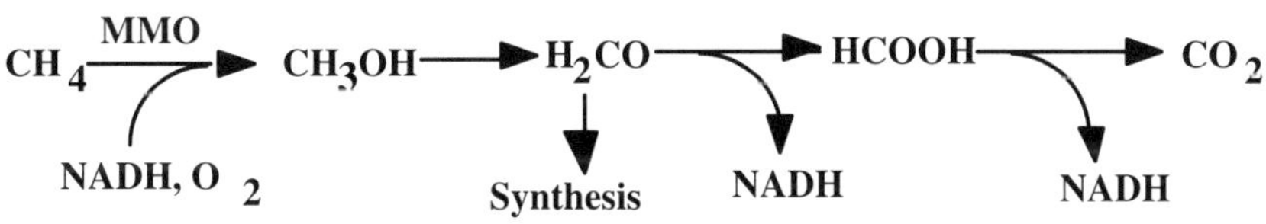

(23.1)

TCE Epoxidation

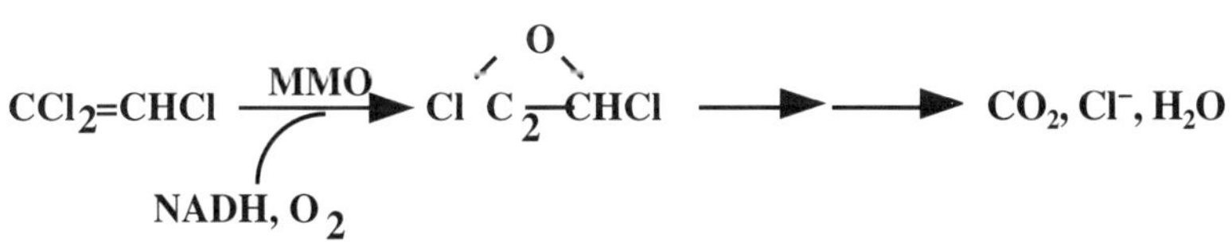

(23.2)

Methanotrophs use an oxygenase (methane monooxygenase or MMO) to catalyze the oxidation of methane to methanol. This requires energy or reducing power in the form of NADH or NADPH. In subsequent steps, excess NADH or NADPH is regenerated, leading to energy that is available for organism growth as well. Compounds in the oxidative pathway (methanol, formaldehyde, and formate) can all serve to provide the necessary reducing power for MMO activity. MMO can also oxidize TCE fortuitously (Equation 23.2) to form TCE epoxide (Fox et al., 1990; Little et al., 1987), an unstable compound that chemically undergoes decomposition to yield a variety of products, including carbon monoxide, formic acid, glyoxylic acid, and a range of chlorinated acids (Miller and Guengerich, 1983). In mixed cultures, as occurs in nature, cooperation between the TCE oxidizers and other bacteria occurs, and TCE is further mineralized to carbon dioxide, water, and chloride (Fogel et al., 1986; Henry and Grbic-Galic, 1991; Henson et al., 1989; Roberts et al., 1989; Uchiyama et al., 1992).

Since the report of Wilson and Wilson (1985) with TCE cometabolism, much scientific research addressing this phenomenon has been performed. Generally, aerobic cometabolic transformations of CAHs are catalyzed by monooxygenases or dioxygenases, the function of which is to oxidize compounds by directly inserting either one or two oxygen atoms, respectively, from molecular oxygen into the compound as illustrated in Equations 23.1 and 23.2. Such oxidation requires energy, which is generally supplied to the enzyme in the form of NADH or NADPH through a complex enzyme

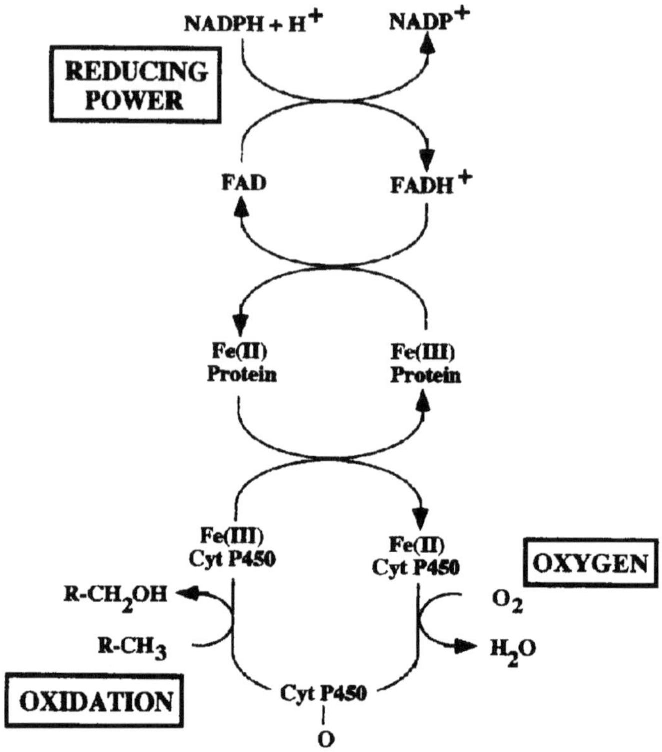

Figure 23.1. Typical monooxygenase electron transport chain based upon cytochrome P450 (after Lehninger, 1970).

system as illustrated in Figure 23.1 for the hydroxylation of a methyl group by cytochrome P_{450} (Lehninger, 1970). Because of the complexity of the enzyme system and the need for energy to keep the enzyme active, intact microorganisms would generally be required in a practical cometabolic system for CAH oxidation.

Any organism with an active mono- or di-oxygenase has the potential for cometabolic transformation of CAHs. Among the groups of bacteria that have so far been demonstrated to be capable of transforming TCE and other CAHs in this manner are not only the methane oxidizers (Alvarez-Cohen and McCarty, 1991d; Fogel et al., 1986; Henry and Grbic-Galic, 1990; Henry and Grbic-Galic, 1991; Little et al., 1987; Oldenhuis et al., 1991; Oldenhuis et al., 1989; Tsien et al., 1989); but also propane oxidizers (Wackett et al., 1989), ethylene and propene oxidizers (Ensign et al., 1992; Roberts et al., 1989; Hartmans and de Bont, 1992); toluene, phenol, or cresol oxidizers (Folsom et al., 1990; Harker and Kim, 1990; Nelson et al., 1987; Nelson et al., 1986; Nelson et al., 1988; Wackett and Gibson, 1988); ammonia oxidizers (Arciero et al., 1989; Rasche et al., 1991); isoprene oxidizers (Ewers et al., 1991), and vinyl chloride oxidizers (Hartmans and de Bont, 1992).

While many different microorganisms are capable of CAH transformation, there is currently insufficient information on the relative advantages and disadvantages of the different microbial groups to definitively recommend one over the other, but each may have its place. Most research to date has been conducted with the methane oxidizers and the group of bacteria containing toluene mono- or dioxygenase, either one or both of which can be induced with primary substrates such as toluene, phenol or cresol.

The mono- and di-oxygenases for the above organisms are often non-specific and fortuitously initiate oxidation of a variety of compounds including most of the CAHs (Table 23.1). The exceptions are highly chlorinated compounds such as CT and PCE. However, the relative susceptibility of the various CAHs to oxidation by the different oxygenase systems varies considerably from one culture to the next and with the particular oxygenase system as well. For example, the toluene-induced organism studied by Wackett and Gibson (1988) transformed TCE and *c*-DCE best and VC not at all, while the isoprene oxidizing microorganism reported by Ewers et al. (1991) oxidized VC the fastest and TCE the slowest. Also, reported susceptibilities of different CAHs to methane oxidizing bacteria varies widely. There are reports of TCA transformation by such organisms by some researchers (Arvin, 1991; Strand et al., 1990), but not by others (Henson et al., 1989; Semprini et al., 1991). Oldenhuis et al. (1989) found partial TCA transformation occurred with the presence of soluble MMO but not particulate MMO; however, even with soluble MMO the rates were quite low compared to that for other CAHs.

In general, oxygenases act on unsaturated CAHs such as TCE, by adding oxygen across the double bond to form an epoxide (Equation 23.2). With saturated CAHs such as CF or TCA, a hydroxyl group is generally substituted for one of the hydrogen atoms in the CAH molecule. Frequently the resulting products from CAH oxidation are chemically unstable and decompose as described above for TCE, yielding products that are further metabolized by other microorganisms present in a mixed culture, thereby effecting complete mineralization as already indicated for TCE.

Rates of CAH Transformation

Overview

As noted above the oxygenases reported to be capable of CAH cometabolism are those associated with oxidation of methane, ethene, ethane, propane, butane, isoprene, toluene, phenol, and ammonia. Rates of oxidation of different CAHs by different oxygenases vary markedly, with no clear predictable pattern. Transformation rates are often expressed using Michaelis-Menton (MM) kinetics (Bailey and Ollis, 1986):

$$-\frac{dS}{dt} = \frac{kXs}{S + K_s} \tag{23.3}$$

where

S (mg/L)	=	the concentration of substrate being transformed by the enzyme;
t (d)	=	time;
k(mg substrate/mg cells-d)	=	the maximum rate of substrate utilization by bacterial cells;
X(mg/L)	=	the cell concentration; and
K_s(mg/L)	=	the half-velocity coefficient, or substrate concentration at which the rate is half the maximum value.

With MM kinetics, the enzyme rather than cell concentration would be used, but from a practical sense with the concentration of whole cells being the usual measure available, it is used here instead.

There have been several reports of CAH transformation rates by pure and enriched cultures, especially with TCE, the most prevalent CAH in groundwater. Different units of expression are frequently used, and so it is sometimes difficult to compare results from one study to the next. Table 23.2 contains a listing of maximum rates (k) reported in the presence of high CAH concentrations where the rates are not substrate limited. Much of the data summarized in the table was originally reported as rate per unit mass of cell protein. For the purpose of Table 23.2, these rates were multiplied by 0.5 on the assumption that the protein content of the cells represented 50% of the cell dry weight. Generally, the rates listed represent the initial values, which tend to be the highest. In some cases, the rates were measured in the presence of an external source of non-competing reducing power, such as formate. Such rates tend to be higher than without external reducing power, suggesting that rates may often be limited by a deficiency of NADH or NADPH. With mixed cultures, the active microorganisms may represent only a small fraction of the total microbial population, and thus the reported rates per unit cell mass would tend to be lower than with pure cultures.

Table 23.2. Maximum Reported Rates of CAH Cometabolism with Different Primary Substrates (g CAH/g Cells-Day).

Reference	With Formate?	PCE	TCE	*c*-DCE	*t*-DCE	1,1-DCE	VC	TCA	CF	Organism
Ammonia										
Arciero et al. (1989)	No		0.8							*N. europaea*
Ethene										
Hartmans and de Bont (1992)	No						1.7			*Mycobacterium* strain E3
Isoprene										
Ewers et al. (1991)	No	0.00	0.03	0.34	0.10	1.1	0.8			
Methane										
Alvarez-Cohen and McCarty (1991b)	No		0.5						0.4	Mixed culture
Alvarez-Cohen and McCarty (1991b)	Yes		8						1.5	Mixed culture
Brusseau et al. (1990)	No		6							*M. trichosporium* OB3b
Brusseau et al. (1990)	Yes		19							*M. trichosporium* OB3b
DiSpirito (1992)	Yes		0.06							*Methylocystis parvis* OBBP
DiSpirito (1992)	Yes		0.03							*Methylomonas* sp. A45
Dolan and McCarty (1994)	No		0.5	0.4	0.4	0.1	0.1			Mixed culture
Dolan and McCarty (1994)	Yes		0.9	3	1.4	0.1	0.5			Mixed culture
Fogel et al. (1986)	No		0.004	0.013	0.013	0.008	0.024			Mixed culture
McFarland et al. (1992)	No		0.02							Mixed culture
McFarland et al. (1992)	Yes		0.03							Mixed culture
Oldenhuis et al. (1991)	Yes		55	25	46	0.8		5	94	*M. trichosporium* OB3b
Strand et al. (1990)	No		0.03					0.005		Mixed culture

Phenol								
Folsom et al. (1990)	No		0.8					*P. cepacia* G4
Propane								
Wackett et al. (1989)	No		0.06					*M. vaccae* JOB5
Phelps et al. (1990)	No		0.08					Mixed culture
Propene								
Hartmans and de Bont (1992)	No						1.5	*Xanthobacter* strain Py2
Toluene								
Wackett and Gibson (1988)	No	0.00	0.17				0.00	*P. putida* F1
Vinyl Chloride								
Hartmans and de Bont (1992)	No		<0.2	4.2	3.5	1.4	5	*M. aurum* L1

With some organisms, i.e., the methane-oxidizers, it is suggested that maximum rates of TCE transformation require a soluble form of MMO, which is induced only with a copper-deficient medium (Patel et al., 1979). The increased rates of CAH transformation resulting from copper deficiency have been demonstrated (Henry and Grbic-Galic, 1990; Oldenhuis et al., 1989). The very high TCE transformation rates listed in Table 23.2 (>1 g TCE/g cells-d) were all obtained with copper deficiency when soluble MMO was expressed. However, the particulate form of MMO can also oxidize TCE, at least with some microorganisms (DiSpirito et al., 1992; Henry and Grbic-Galic, 1990; Oldenhuis et al., 1989), although the rates are much lower than with the soluble form. Since copper concentration may be difficult to control in the field, the lower rates listed for methanotrophs may be more typical of field conditions.

There are various reports concerning relative rates of CAH transformation by organisms with different oxygenase systems, although frequently the rates have not been reported in a manner to provide a quantitative comparison for inclusion in Table 23.2. Nevertheless, the comparative evaluations reported are useful. Table 23.3 summarizes some of this information. First, the data indicate that PCE is not oxidized by aerobic cometabolism. This comparative evaluation, as well as other reports, indicate CT is also not transformed. TCA is frequently indicated not to be oxidized, although there are some reports to the contrary (Oldenhuis et al., 1989; Strand et al., 1990). The possibility of TCA oxidation may be culture dependent. The listing in Table 23.3 also indicates that VC is one of the most rapidly degraded CAHs and 1,1,-DCE the slowest. In the case of the latter compound, transformation product toxicity (TPT) may be a major factor affecting observed rates. This phenomenon is discussed later. Perhaps the major conclusion that can be drawn from these data is that in general, the less highly halogenated compounds are oxidized more rapidly than are more halogenated ones. The other is that the relative rates are oxygenase and organism specific. For example, in a comparative evaluation, *t*-DCE was transformed rapidly by methane-oxidizing mixed cultures, while *c*-DCE was transformed slowly, but with a phenol-oxidizing culture, just the opposite result was found (Hopkins et al., 1993).

The different factors that can affect CAH transformation rates make direct comparisons between different organisms and different oxygenase systems somewhat difficult. Also, while the maximum transformation rates listed in Table 23.2 and relative rates in Table 23.3 may be useful, there are other factors of perhaps more importance to the practical application of aerobic CAH cometabolism. Included are: (1) competitive inhibition, (2) reducing power availability, (3) transformation product toxicity, and (4) compound toxicity.

Competitive Inhibition

The significance of competitive inhibition is illustrated by the example of TCE epoxidation by MMO suggested by Equations 23.1 and 23.2. Since both methane and TCE are oxidized by the same enzyme, they act as competitive substrates to one another. That is, the presence of methane reduces the rate at which TCE is oxidized and vice versa. This presents somewhat of a problem, as methane is required to induce MMO, and is generally also used as the source for the required reducing power. When both the primary substrate oxidized by the oxygenase and the CAH to be transformed are present together, then competitive inhibition kinetics are likely to apply as given in the following modifications of Equation 23.3:

$$-\frac{dS_1}{dt} = \frac{kXS_1}{S_1 + K_{s1}\left(1 + \frac{S_2}{K_{s2}}\right)} \tag{23.4}$$

$$-\frac{dS_2}{dt} = \frac{kXS_2}{S_2 + K_{s2}\left(1 + \frac{S_1}{K_{s1}}\right)} \tag{23.5}$$

Table 23.3. Relative Rates of CAH Transformation by Different Cultures[a].

Reference	With Formate?	PCE	TCE	*c*-DCE	*t*-DCE	1,1-DCE	VC	TCA	CT	Organism
Methane										
Oldenhuis et al. (1991)	Yes	N	1	1	1	5		4	N	*M. trichosporium* OB3b
Henson et al. (1989)	No	N	3	2	1			N		Mixed culture
Semprini et al. (1990)	No		4	3	2		1	N		Mixed culture
Fogel et al. (1986)	No	N	4	2	2	3	1			Mixed culture
Dolan and McCarty (1994)	No		3	1	2	5	4			Mixed culture
Isoprene										
Ewers et al. (1991)	No	N	5	3	4	5	1			
Propane										
Wackett et al. (1989)	No	N	4	2	3	5	1			*M. vaccae* JOB5
Propene										
Ensign et al. (1992)	No	N	4	3	2	5	1			*Xanthobacter* strain Py2
Vinyl Chloride										
Hartmans and de Bont (1992)	No		N	2	3	4	1			*M. aurum* L1
Toluene										
Wackett and Gibson (1988)	No	N	1	1	4	3	N			*P. putida* F1
Phenol										
Hopkins et al. (1993)	No		3	1	2			N		Mixed culture

[a] Numbers represent relative order of transformation rates for compounds studied, with 1 representing compound with the highest rate. N indicates compound was not transformed.

where

Subscript 1 refers to the primary energy-generating substrate,
Subscript 2 refers to the CAH being cometabolized

In the case of Equation 23.4 for primary substrate utilization, it is clear that when the CAH concentration (S_2) increases, the rate of primary substrate concentration usage decreases. A similar conclusion for the effect of primary substrate concentration on CAH transformation applies with Equation 23.5. Competitive inhibition of this type has been reported frequently (Semprini et al., 1991; Hou et al., 1979; Patel et al., 1982; Strand et al., 1988; Folsom et al., 1990; Oldenhuis et al., 1991; Leahy et al., 1989).

One method suggested to circumvent the problem of competitive inhibition is to pulse feed the primary substrate just frequently enough for enzyme induction and to supply reducing power, but in a manner to avoid having both primary substrate and CAH present at the same time (Hou, 1984; Semprini et al., 1991). Another approach is to provide primary substrate on occasion only to induce the oxygenase, but generally to use another, non-competitive substrate for reducing power. For example, with MMO, Equation 23.1 indicates that methanol or formate might be used to supply reducing power for methanotrophs and would not compete with CAH for the MMO. Such an approach was illustrated in a field study of *in situ* bioremediation of CAHs by methanotrophs (Semprini et al., 1991). Here, formate was added to an aquifer following methanotrophic growth by methane addition. Formate reduced the effects of competitive inhibition. However, the effect lasted no more than a few days. Methane addition was then again required. Another concern in a practical sense is that if formate addition were continued, a population of formate-using microorganisms might outcompete the methanotrophs, leading to a possible loss in MMO activity from the system.

Another approach that can be used for surface treatment of CAH-contaminated water is the use of staged reactors. Here, primary substrate is supplied to a growth reactor where oxygenase-containing microorganisms are produced. The active microorganisms are then introduced into a second reactor through which the CAH-containing water passes and there CAH decomposition is effected (Alvarez-Cohen and McCarty, 1991c; Folsom and Chapman, 1991; McFarland et al., 1992). This approach requires either that the microorganisms have sufficient internal reserves of reducing power to sustain them sufficiently long in the absence of a primary energy substrate, or that a non-competitive substrate also be added to the second reactor as an external reducing power source (McFarland et al., 1992). Reducing power availability thus becomes an important consideration for any approach that attempts to avoid competitive inhibition.

Reducing Power

One approach to circumvent the problems with competitive inhibition noted above is to take advantage of the reducing power sometimes stored in methanotrophic cells, perhaps in the form of polyhydroxybutyrate (PHB) (Henry and Grbic-Galic, 1991). Freshly grown cells with PHB storage can then be mixed with TCE-containing water, and in the "resting" state will epoxidate TCE in the absence of methane. Thus, competitive inhibition can be avoided. This effect is the basis for the pulsing approach or the use of two-stage reactors noted in the previous section. PHB storage is a well-known phenomenon, and generally occurs as a result of a deficiency in a required nutrient for growth, such as nitrogen (Nichols and White, 1989).

Problems with relying on internal reducing power alone are that some cells do not store reducing power, and others have only a limited capacity to do so. The extent of reducing-power storage appears to be a function of the species and growth conditions. The rate of CAH transformation and the extent of TCE transformation often are lower when internal reducing power rather than an external source is used (Alvarez-Cohen and McCarty, 1991d). These factors need to be considered when taking advantage of internal reducing power.

Transformation Product Toxicity (TPT)

A major factor affecting the economics of cometabolic CAH degradation is transformation product toxicity. This phenomenon was first suggested in a report by Wackett and Gibson (1988) on TCE transformation by toluene-fed *Pseudomonas putida* F1. The rapid decline in TCE transformation rate by resting cells was thought likely to result from toxic intermediate products formed. They referenced earlier work by Miller and Guengerich (1982), who reported that the oxidation of TCE by rat liver cytochrome P-450 monooxygenase yielded an epoxide that could directly alkylate cellular nucleophiles, rearrange to form a reactive acyl chloride, or undergo spontaneous hydration and HCl elimination to yield a different acyl chloride. Subsequently, Green and Dalton (1989) found a similar effect in TCE and 1,1-DCE transformation with purified methane monooxygenase from *Methylococcus capsulatus* (Bath). Here transformation of either TCE or 1,1-DCE resulted in the inactivation of MMO, with 1,1-DCE causing the greatest inactivation. Intermediate products reported from TCE transformation were TCE oxide, formate, carbon monoxide, and HCl, and from 1,1-DCE were formate, formaldehyde, and monochloroacetic acid. From the latter it was believed that the inactivation resulted from monochloroacetic acid, which would cause alkylation of the heme group. Green and Dalton (1989) also suggested MMO acted similarly to cytochrome P-450 in its transformations.

Subsequent studies indicated that TCE cometabolism not only results in the oxygenase inactivation noted above, but also causes interactions with other cellular components, leading to death of the microorganism. This effect has been observed with methane (Alvarez-Cohen and McCarty, 1991b; Alvarez-Cohen and McCarty, 1991d; Fox et al., 1990; Henry and Grbic-Galic, 1991; Oldenhuis et al., 1991), phenol (Wackett and Householder, 1989), propylene (Ensign et al., 1992), vinyl chloride (Hartmans and de Bont, 1992), and ammonia (Rasche et al., 1991) oxidizing microorganisms.

One of the most significant practical problems with TPT is the effect it has on the quantity of primary substrate needed to degrade a given amount of CAH. Since TPT tends to kill cells, additional primary substrate must be added to produce cells for replacement. This additional primary substrate has a cost, and in some cases this may be the greatest cost component of the treatment system. The transformation capacity (T_c) is one measure of the effect of TPT. It represents the mass of a given CAH that can be transformed by a given mass of resting cells (g CAH/g cells). The transformation yield (T_y) is another useful measure, and represents the mass of CAH transformed per unit mass of primary substrate consumed. T_c and T_y are related as follows:

$$T_y = T_c / Y_{net} \tag{23.6}$$

where Y_{net} is the net yield of cells from a given reactor, or net grams of cells produced per gram of substrate consumed. Table 23.4 lists T_c values, and Table 23.5 lists T_y values for different primary substrates and CAHs. Unfortunately, experimental data from which these coefficients can be derived have generally not been reported, and thus the range of values listed is quite limited.

The few coefficients listed in Tables 23.3 and 23.4 indicate that the addition of an external source of reducing power such as formate can increase T_c and T_y considerably, but such additions have their cost as well. In the absence of such additions, the maximum value for T_y with TCE and methanotrophs is 0.032 g TCE/g methane. This means that the transformation of 1 kg TCE requires a minimum of 30 kg methane. To illustrate the problem in another way, transformation of 1 mol TCE requires a minimum of about 250 mol of methane. Obviously, overall methane turnover by MMO is much greater than TCE turnover by MMO, even in the best of circumstances. In order for biodegradation of CAHs to be cost-effective, methods for increasing T_y are needed.

Possible methods for increasing T_y or otherwise reducing the TPT problem are suggested by newer reports. Rasche et al. (1991) indicated that ammonia-oxidizing bacteria have a repair mechanism that can reduce the effects of TPT, thus reducing the amount of ammonia to grow replacement cells. Ewers et al. (1991) suggested the use of isoprene as a primary substrate, as bacteria which oxidize isoprene harbor an epoxide-detoxifying enzyme so that they are less sensitive to suicide inactivation by CAH transformation. Hartmans and de Bont (1992) found chlorooxirane, the epoxide transformation intermediate, caused inactivation of cells degrading VC. However, they also found that the presence of dead cells reduced the toxic effect, from which they concluded the toxic metabo-

Table 23.4. Estimated Transformation Capacities (T_c) for CAH-Transforming Cultures (g CAH/g cells).

Reference	With Formate?	TCE	*c*-DCE	*t*-DCE	1,1-DCE	VC	CF	Culture
Methane								
Alvarez-Cohen and McCarty (1991a, 1991b)	No	0.036					0.007	Mixed culture
Alvarez-Cohen and McCarty (1991a, 1991b)	Yes	0.08					0.015	Mixed culture
Dolan and McCarty (1994)	No	0.09	0.09	0.09	0.01	0.08		Mixed culture
Dolan and McCarty (1994)	Yes	0.11	0.35	0.47	0.01	0.14		Mixed culture
Nelson and Jewell (1993)	No					0.036		Mixed culture
Ethene								
Hartmans and de Bont (1992)	No					0.08		

lite of VC degradation accumulated extracellularly. This suggests that the presence of other cells or other toxic product deactivators in the media may have a beneficial effect in reducing TPT, a concept that might be exploited in a reactor system. More information on TPT effects for different CAHs, primary substrates, and organisms, and methods to overcome these effects, would be beneficial.

CAH Toxicity

CAHs tend to be relatively toxic to microorganisms, and this can have an impact on the design of reactors as well as on the upper concentrations of CAHs that can be treated biologically. Evaluation of the direct toxicity of CAHs, per se, to microorganisms is made complex because they also cause competitive inhibition and transformation product toxicity. These three separate forms of inhibition need to be distinguished from one another because they impact reaction kinetics and reactor design differently. There are ways by which this toxicity can be assayed separately from the other forms of inhibition (Alvarez-Cohen and McCarty, 1991d), but in many reports, such distinctions are often not made. Thus, when toxic levels of a CAH are reported, it is not clear whether the inhibitory effect noted was indeed due to direct toxic action of the compound itself or some other factor. With direct toxicity, the level of inhibition increases with increase in the concentration of the toxicant. The control strategy here would be either to reduce the concentration of toxicant to which the organism is exposed, or to use a culture that can adapt to higher levels of the toxicant. Other strategies may be more appropriate with competitive inhibition or TPT.

With methane-oxidizing bacteria, Alvarez-Cohen and McCarty (1991b) found that CF and TCE were not directly inhibiitory to a mixed culture at the maximum concentrations evaluated of 60 mg/L and 36 mg/L, respectively. A range of CAHs, including TCE, *c*-DCE, *t*-DCE, 1,1-DCE, CF, and TCA, were reported to be nontoxic at the highest concentrations evaluated with a pure culture, which were between 17 and 27 mg/L (Oldenhuis et al., 1989). However, with another mixed culture, Strand et al. (1990) reported only 7.7 mg/L TCE inhibited methane use rate, and Broholm et al. (1990) found 13 mg/L inhibited methane use rate. In the latter case, the inhibition was suggested to be caused by competitive inhibition, rather than by direct toxicity. Janssen et al. (1987) reported 6 mg/L TCE inhibited growth of a methanotrophic consortium with methane but not with methanol as a substrate. Indeed, they found this to be true with CF at 8 mg/L and TCA at 4 mg/L. No inhibition was found with *t*-DCE at 4.5 mg/L or PCE at 5.5 mg/L. Based upon these results, they suggested that the toxicity may

Table 23.5. Estimated Transformation Yields (T_y for CAH-Transforming Cultures (g CAH/g Primary Substrate).

Reference	With Formate?	TCE	*c*-DCE	*t*-DCE	1,1-DCE	VC	TCA	CF	Culture
Methane									
Alvarez-Cohen and McCarty (1991a, 1991b)	No	0.013						0.002	Mixed culture
Alvarez-Cohen and McCarty (1991a, 1991b)	Yes	0.026						0.005	Mixed culture
Arvin (1991)	No	0.034	0.15	0.24			0.034		Mixed culture
Dolan and McCarty (1994)	No	0.032	0.031	0.031	0.003	0.028			Mixed culture
Dolan and McCarty (1994)	Yes	0.039	0.12	0.16	0.003	0.049			Mixed culture
Phelps et al. (1990)	No	0.06							Mixed culture
Propane									
Phelps et al. (1990)	No	0.04							Mixed culture
Wilcox et al. (1992)	No	0.005							Mixed culture

be the result of the inhibition of some specific step in the assimilation of methane, rather than to a general inhibitory effect of the CAHs themselves. They suggested even then that the inhibitory CAHs may act as suicide substrates, an observation that is consistent with the TPT noted above. These observations provide further illustration of the caution that is needed in drawing general conclusions about toxicity from reported data. In addition, contrary to the Janssen et al. (1987) report, Broholm et al. (1990) found that TCA was not inhibitory to methane utilization at concentrations up to about 100 mg/L.

With organisms using other primary substrates as oxygenase inducers, concentrations in the range of 20 mg/L also do not appear inhibitory, although there are few reports. With propane oxidizers Wackett et al. (1989) studied concentrations in the range of 2 mg/L for several CAHs and reported no toxicity; with toluene oxidizers, Wackett and Gibson (1988) found no TCE toxicity at 11 mg/L, but reported 43 mg/L to be toxic; with several different primary substrates, including trypticase/yeast extract, propane, and methane, Fliermans et al. (1988) found no TCE toxicity at 50 mg/L, but did note some at 150 mg/L; with phenol Folsom et al. (1990) found TCE at 24 mg/L to increase the lag phase, but little significant toxicity was found; and with ammonia, Rasche et al. (1991) studied TCE at 17 mg/L and did not report toxicity. Thus, from the various reports, CAH concentrations in the range of 20 mg/L do not appear generally to be directly toxic to microorganisms with potential for CAH aerobic cometabolism. It is not yet clear where the toxic threshold may be, but it could be 100 mg/L or higher, and it is likely to be organism dependent.

KINETICS OF CAH TRANSFORMATION

The above discussion on maximum reaction rates, competitive inhibition, reducing power, and transformation product toxicity raises questions about the availability of predictive models for the cometabolic process that consider all these important factors in a way that would be useful for engineering design. A comprehensive model that places all these relationships into proper perspective is sorely needed, as are empirical data from which appropriate coefficients can be derived. One approach that may be useful for a two-stage reactor or other system that depends either upon resting cells, or else cells supplied with a non-competitive source of reducing power, was given by Alvarez-Cohen and McCarty (1991a). Here, TPT was considered to reduce cell activity in direct proportion to the quantity of CAH transformed, leading to a relationship between T_c and such reduction:

$$\frac{1}{T_c} = \frac{dX}{dS} = \frac{\text{cell mass inactivated}}{\text{mass of CAH transformed}} \tag{23.7}$$

Equation 23.7 can be used together with Equation 23.3 to evaluate the effect of T_c on CAH transformation with time. For example, considering a batch or ideal plug-flow reactor in which an initial CAH concentration (S_o) and initial active bacterial concentration (X_o) are present in the absence of a growth-producing primary substrate, the rate of CAH transformation is given by Alvarez-Cohen and McCarty (1991a) as:

$$-\frac{dS}{dt} = \frac{kS\left[X_o - \frac{1}{T_c}\left(S_o - S\right)\right]}{S + K_S} \tag{23.8}$$

This equation can be integrated to determine the CAH concentration (S) at any time t (Alvarez-Cohen and McCarty, 1991a). Equation 23.8 has been applied to the analysis of a two-stage reactor system in which active cells are grown in the first stage, or growth reactor, and then transferred to the second stage, or treatment reactor, where treatment of CAH-contaminated water occurs (Alvarez-Cohen and McCarty, 1991c). If a competitive primary growth substrate is also present with the contaminants in a single reactor, then the kinetics become more complicated. An adequate mathematical formulation for this problem is not yet (1994) available.

Field Studies

There is no well-documented full-scale experience with *in situ* aerobic cometabolic transformation of CAHs upon which to base full-scale design. However, limited small field-scale pilot studies have been conducted in order to determine the effectiveness of certain approaches to remediation. One such effort took place at the Moffett Naval Air Base (Moffett Field) pilot facility in Mountain View, California, to evaluate the capacity of native microorganisms, i.e., bacteria indigenous to the groundwater zone, to cometabolically degrade CAHs.

The Moffett field studies were conducted in a small sand/gravel confined aquifer of 1.5 m total height located about 4.5 m below ground surface under conditions typical of groundwater contamination by CAHs. Extracted groundwater was amended with chemicals and injected into the aquifer at a flow rate of about 1.5 L/min. Groundwater was removed at a rate of about 10 L/min from an extraction well located 6 m away in order to create a regional flow from the injection to the extraction well. Three monitoring wells were located between the injection and extraction wells in order to obtain samples for evaluating the effectiveness of *in situ* treatment. Two aerobic systems with oxygen as the electron acceptor have been tested: (1) methanotrophs that use methane as a primary substrate and cometabolically transform CAHs with methane monooxygenase (MMO); and (2) phenol utilizers in which toluene oxygenase (TO) serves as the cometabolizing enzyme. Target contaminants in these studies were TCE, *c*-DCE, *t*-DCE, and VC, initially in the concentration range of 40 to 120 μg/L.

The experimental approach taken was similar to that proposed for bioremediation in the field (McCarty et al., 1991). Extracted groundwater from the treatment zone would here be amended with the growth substrate and oxygen, and reinjected to stimulate indigenous growth. CAHs in the extracted groundwater would be reinjected into the biostimulated zone. Bioremediation conducted in this manner would promote the degradation of in-place contaminants as well as contaminants that are extracted and reinjected, thus obviating the need for aboveground treatment.

The Moffett Field experiments conducted to simulate the above approach were performed as a series of stimulus-response tests. The stimulus was the injection of the compounds of interest and the response was their concentration history at monitoring locations. The tests included bromide tracer tests to study advection and dispersion, transport tests with CAHs to study the retardation process due to sorption and to evaluate whether their transformation occurred in the absence of active biostimulation, and biostimulation and biotransformation tests to evaluate CAH transformation following the introduction of the primary substrates and oxygen.

Results with Methane Utilizers

Detailed discussions of the experimental methodology and results of the methanotrophic studies have been published (Roberts et al., 1990; Semprini and McCarty, 1991; Semprini et al., 1990). Indigenous methanotrophs were stimulated in three successive field seasons through the addition of groundwater supplemented with methane (5 to 7 mg/L) and oxygen (20 to 25 mg/L), which were introduced into the test zone in alternating pulses without any other supplementary nutrients (N or P).

Figure 23.2 shows the concentration history of methane and oxygen at the S_2 observation well during the initial biostimulation experiment along with model simulations (Semprini and McCarty, 1991). The S_2 observation well was located 2.2 meters and a flow time of about one day downgradient from the injection well. During the period of 200 to 430 hr, methane and oxygen concentrations decreased, indicating the growth of methane utilizers. In order to control the clogging of the injection well and borehole interface, the alternate pulse injection of methane- and oxygen-containing groundwater was initiated at 430 hr, with a pulse cycle time of 4 and 8 hr, respectively. The model simulations, represented by the solid line, matched the field observations using a reasonable set of biological and transport input parameters. Simulation modeling supported the conclusions that methanotrophic bacteria were stimulated in the test zone, that biofouling of the near well-bore region was limited by the pulsing methodology, and that these processes can be simulated when appropriate rate and transport equations are used.

Figure 23.3 shows the response at the S_2 well of the target contaminant compounds in the third season, and model simulations (Semprini and McCarty, 1992). Transformation of the organic target compounds ensued immediately following the introduction of methane at time zero, increasing with time as the bacterial population grew. Significant transformation of VC and *t*-DCE were observed, followed by *c*-DCE and TCE (not shown). TCA was a groundwater contaminant at the field site, and its possible transformation was followed as well. No transformation of TCA was found, with its concentration of about 100 μg/L. Competitive inhibition of VC and *t*-DCE transformation by methane was indicated in response to the dynamic pulsing of methane and oxygen that was initiated at 20 hr. In order to effectively simulate the field observations, both competitive inhibition kinetics and rate-limited sorption and desorption were required in the biotransformation model, thus reinforcing the conclusion that these processes were occurring, and need to be considered with *in situ* remediation. The rate of TCE transformation corresponded with the low values listed in Table 23.2, which suggests that the particulate form of MMO was operative in this system. The simulations also indicated that the overall rate of decrease in VC concentration may have been limited by the rate of its desorption from the aquifer solids. Thus, desorption may significantly increase the time required for cleanup by enhanced microbial processes.

The comparison of the cometabolic rate parameters obtained from the modeling exercise indicated that VC and *t*-DCE were transformed at rates similar to that of methane, the primary substrate added for growth, while *c*-DCE and TCE were transformed at rates one to two orders of magnitude slower that methane. The simulations indicate that *t*-DCE concentration decreased more slowly than VC since it was more strongly sorbed (higher K_d), and thus a greater contaminant mass had to be degraded. The order of magnitude difference in rates found for *c*-DCE and *t*-DCE illustrates that a small change in chemical structure can have a large effect on the cometabolic transformation rate.

The specific conclusions from this study were:

1) the stimulation of indigenous methanotrophs can be accomplished through methane and DO addition;

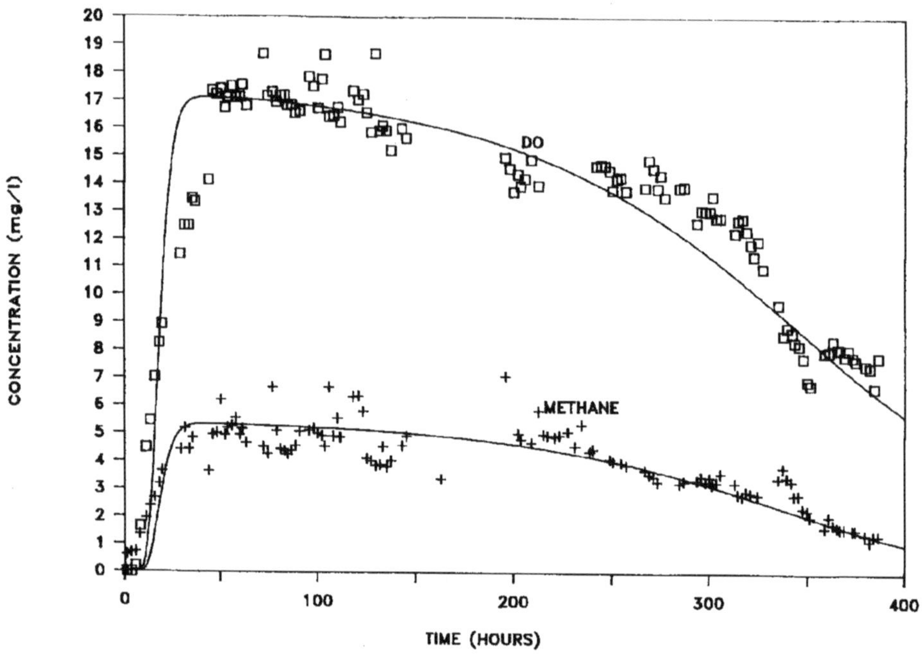

Figure 23.2. Concentration history of methane and oxygen at the S_2 monitoring well following the initial injection at Moffett Field, and model simulations (after Semprini and McCarty, 1991).

2) the rates and extents of transformation of CAHs are compound specific;
3) the average percentage transformations achieved in a 2-m biostimulated zone were: TCE, 20%; *c*-DCE, 50%; *t*-DCE, 90%; and vinyl chloride, 95%;
4) the cometabolic transformation was strongly tied to methane utilization; upon stopping methane addition, transformation rapidly ceased;
5) the cometabolic transformation was competitively inhibited by methane, an effect that reduced the transformation rate;
6) only a temporary enhancement of the cometabolic transformation could be achieved by substituting formate (a noncompetitive substrate) for methane;
7) the rate of transformation was limited by the rate of desorption from the aquifer solids, especially for the more rapidly degraded VC and *t*-DCE;
9) the results agreed with those obtained in laboratory soil microcosm studies, performed under conditions that mimicked the field tests.

Results with Phenol Utilizers

In the methanotrophic studies, limited degradation of TCE and *c*-DCE was achieved. The objective of this study was to evaluate TO as an alternative for *in situ* biodegradation of TCE, *c*-DCE and *t*-DCE at Moffett Field. This was accomplished through the introduction of phenol as a primary growth substrate and oxygen as an electron acceptor (Hopkins et al., 1993). The evaluation was performed at the same site as the methanotrophic study, using the same experimental methodology, and at a similar contaminant concentration range, permitting a direct comparison of the TO system with the MMO system. Active biostimulation was initiated through the pulsing of phenol and oxygen at time-averaged concentrations ranging from 6 to 12 mg/L, and 32 mg/L, respectively.

The concentration responses of TCE and *c*-DCE at the monitoring well located 2 m from the injection well are shown in Figure 23.4 (Hopkins et al., 1993). Decreases in *c*-DCE and TCE concentrations were associated with decreases in DO, indicating cometabolic transformations resulted from biostimulation. Significant degradations of *c*-DCE, and TCE were observed, with *c*-DCE being more

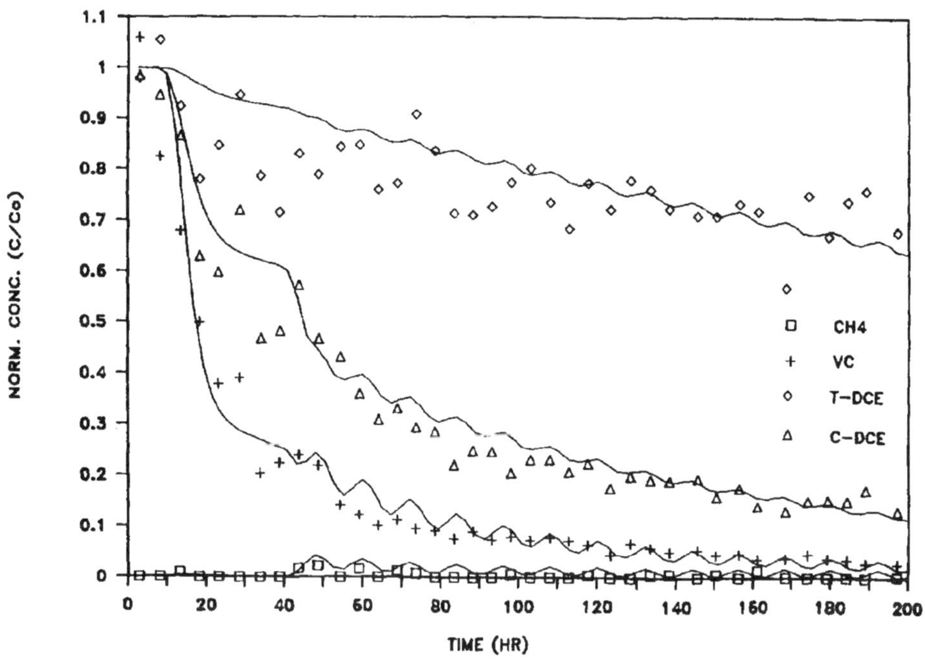

Figure 23.3. Concentration history of chlorinated aliphatic hydrocarbons and methane at the S_2 monitoring well following biostimulation with methane and oxygen at Moffett Field, and model simulations (after Semprini and McCarty, 1992).

rapidly degraded than TCE. The *c*-DCE concentration decreased by approximately 60 to 70% and TCE by 20 to 30%, during the period of 6 mg/L phenol addition. Doubling the phenol injection concentration 500 hr after the study was initiated resulted in a greater transformation of both TCE and *c*-DCE, with 85 to 90%, and over 90%, transformed, respectively. Here, *t*-DCE (not shown) was the least transformed of the three compounds studied. As with the methanotrophic study, no significant transformation of the TCA groundwater contaminant was observed.

The specific conclusions from this study were:

1) the stimulation of indigenous phenol-utilizers was accomplished through phenol and oxygen addition;
2) the enhanced population effectively degraded TCE and *c*-DCE, but was less effective in degrading *t*-DCE;
3) transformations achieved in a 2-m biostimulated zone were: TCE, 85%; and *c*-DCE over 90%;
4) the cometabolic transformation was competitively inhibited by phenol;
5) the cometabolic transformation was strongly tied to the amount of phenol utilized;
6) the results agreed with microcosm studies that were performed under similar conditions to the field study.

Comparison Between the Methane and Phenol Studies

The pilot-scale studies both demonstrated *in situ* biodegradation of CAHs could be achieved. The degradation was shown to be compound specific in both cases. The phenol utilizers more effectively degraded TCE and *c*-DCE, while the methane utilizers more effectively degraded *t*-DCE, on a percentage basis. Both studies showed that cometabolic transformation was closely associated with primary substrate utilization, and that the primary substrate also competitively inhibited the transformation, slowing the transformation rates. The concentrations of the CAHs were relatively low (≤ 100 μg/L), thus the results should not be extrapolated to higher concentrations. There is a need for studies over

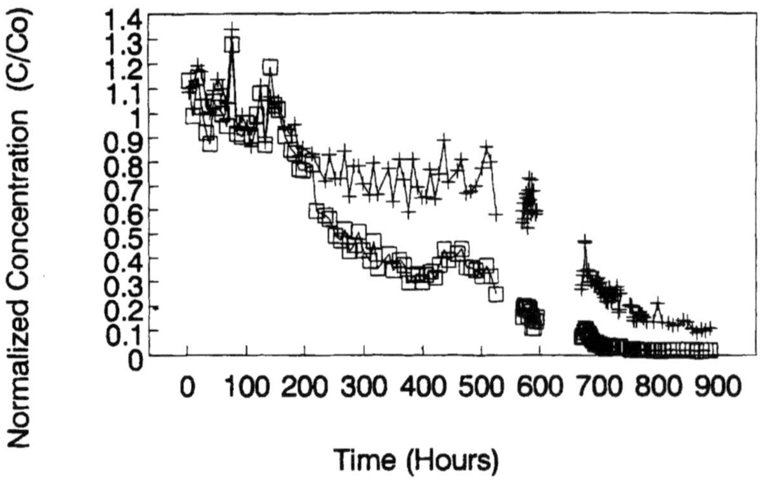

Figure 23.4. Concentration history of TCE (+) and *c*-DCE (□) at the SSE_2 monitoring well following biostimulation with phenol and oxygen at Moffett Field (after Hopkins et al., 1993).

a range of contaminant concentrations. Future studies need to explore CAH concentration effects on degradation efficiency, and on other primary substrates that may stimulate more effective oxygenase systems.

VC cometabolic decomposition was evaluated at Moffett field in the case of methane addition only, where over 90% removal was obtained. Because of the high efficiency of VC removal, the use of methanotrophic bacteria for removal of high concentrations of VC at a Superfund site in St. Joseph, Michigan, is under active consideration (McCarty et al., 1991).

Introduction of Chemicals into Groundwater

One of the most significant challenges for *in situ* bioremediation of CAHs is how to introduce chemicals needed by microorganisms for growth into the subsurface environment and mix them with the contaminants to be degraded. Unless a suitable primary substrate for stimulating cometabolic biotransformation of CAHs is already present in the aquifer, one will need to be added, as will oxygen that is required for its oxidation. For example, if methanotrophic cometabolism is desired, then methane and oxygen both must be added. These gases are of limited solubility in water and so added concentrations together with concentrations of other gaseous components such as molecular nitrogen collectively must be below the saturation partial pressure in the aquifer, which may not be much higher than 1 atm in shallow groundwaters. A possible procedure is to mix the chemicals of interest with water and introduce them into the groundwater. The introduced water could push away the native groundwater containing the contaminants, so that the required mixing between the contaminants and the introduced chemicals will not occur. However, if the contaminants sorb somewhat to the soils, then a benefit will be obtained, as the sorbed contaminants will desorb into the introduced water, thus bringing the contaminants and introduced chemicals together.

Another engineering consideration is the problem of excessive microbial growth near the point where primary substrate is introduced into the groundwater. Microorganisms will tend to grow near the point of primary substrate introduction where concentrations are the highest. In order to avoid such clogging, a strategy is needed that will make growth difficult near the point of injection. The periodic introduction of inhibitory chemicals such as chlorine or ozone may be required, or a strategy such as pulsing of the primary substrate and oxygen, as performed in the Moffett Field experiments described previously, so that they do not occur together near the injection point, are possibilities. Full-scale experience with such strategies is limited and so documented guidelines for application are not available.

One potential method for accomplishing the mixing of nutrients with groundwater contaminants is to use a pump-and-treat extraction and injection system with bioremediation added (McCarty et al., 1991). The Moffett Field experiments simulated this approach. Here, all of the biological treatment is carried out in the aquifer itself. Groundwater is extracted through a series of wells spanning the contaminant plume in a direction perpendicular to that of groundwater flow. At the surface, chemicals such as methane and oxygen are added to the extracted groundwater, either together or in alternating pulses, depending upon the extent to which a stimulated methanotrophic population has been developed.

The alternating pulses are used to distribute the microbial growth throughout the test zone. The groundwater containing the appropriate primary substrates and electron acceptors (i.e., methane and oxygen) and extracted contaminants is reinjected into the treatment zone, possibly through a series of wells distributed parallel to the extraction wells. Other configurations are also possible. In the subsurface biotreatment zone, both the in-place contaminants and the reinjected contaminants are biologically degraded. Another alternative is to use a combination of above-ground treatment and *in situ* treatment. In this case the contaminants would be removed at the surface, and only the required amendments would be added to the extracted groundwater prior to reinjection.

Another possible system for delivering the needed chemicals is subsurface groundwater recirculation (Herrling, 1991; MacDonald and Kitanidis, 1993). This eliminates the need to pump contaminated groundwater to the surface. This mixing method, which is under development, uses a subsurface recirculation unit with an upper and lower screen, and a pump to induce flow through the unit and through the porous formation. For methanotrophic treatment, methane and oxygen would be introduced directly into the groundwater recirculating through the well. This method would eliminate pumping the contaminated groundwater to the surface, surface treatment, and subsequent reinjection. One possibility is that several recirculation units could span across a plume, making a biologically reactive barrier through which groundwater would flow and be treated.

Still another approach is that being used at the Savannah River Laboratory of the U.S. Department of Energy. Here, horizontal wells have been introduced into the vadose and saturated zone of a contaminated aquifer (Hazen, 1992). Methane is mixed with air and introduced directly into the horizontal well in the saturated zone, and bubbles up into the aquifer through perforations in the well. The injected gas moves upward through the aquifer and into the vadose zone, where methane oxidation can occur as well. While conducted at full scale, the effectiveness of this system yet must be evaluated. Certainly other approaches are also possible. The effectiveness and relative advantages and disadvantages of the different possible procedures attempted and contemplated still must be determined.

SUMMARY

Most CAHs cannot be used as primary substrates for energy and growth of bacteria, but may be degraded through aerobic cometabolism. Exceptions to date are PCE and CT. Aerobic cometabolism here is carried out by monooxygenases and dioxygenases that are generally used by microorganisms to effect the initial step in the oxidation of hydrocarbons and other compounds that are used as primary substrates for energy and growth. Any oxygenase system has the potential for initiating CAH degradation. Those that have been explored to date are various alkane and alkene oxygenases, toluene monooxygenase and dioxygenase, and ammonia monooxygenase. Methane monooxygenase and toluene oxygenases have to date been the most studied, and provide good potential, at least for some of the CAHs.

The susceptibility of CAHs to aerobic cometabolic transformation varies widely with both the CAH and the oxygenase system, thus to date, one system cannot be recommended in general over others. A broader search to evaluate more oxygenase systems is desirable. This should be performed in a systematic way so that results can be compared in a meaningful manner.

A great need is to learn more about the kinetics of CAH aerobic cometabolism. CAH transformation requires a source of reducing power, which can be supplied either internally or externally. If an external source that is also oxidized by the oxygenase is used, then competitive inhibition in the CAH transformation rate is likely. If non-competitive sources of external reducing power are used, then there is the danger in a mixed-culture system that other non-oxygenase-containing organisms may come into dominance, or that the required oxygenase will not be induced.

A major problem with CAH cometabolism is that the intermediate products formed are generally toxic both to the oxygenase and to cell structure. Thus, CAHs tend to be suicide substrates for the microorganisms. This means that more primary growth substrate is required to produce replacement cells, and thus the cost of treatment may be high. Measures of this requirement are the CAH transformation capacity and transformation yields of cells. Limited information is available for these two important measures to permit making an economic analysis of the relative suitability of different oxygenase systems for CAH degradation. Suitable models that incorporate the effects of competitive inhibition, internal and external sources of reducing power, and CAH transformation product toxicity on the rate of CAH transformation and primary substrate requirements are lacking. Also, proven methods to introduce primary substrates and nutrients into the subsurface environment and adequately mix them with the CAHs to be transformed are needed. These are important areas of research that are likely to lead to better ways to exploit aerobic biodegradation of CAHs through cometabolism.

ACKNOWLEDGMENTS

Research support from the Gas Research Institute through a subcontract with Radian Corporation and from the U.S. Department of Energy is gratefully acknowledged. Research described in this chapter was also supported by the U.S. Environmental Protection Agency Biosystems Program and the Robert S. Kerr Environmental Research Agency through the Western Region Hazardous Substance Research Center under Agreement R-815738, but has not been subjected to Agency review and therefore does not necessarily reflect the views of the Agency, and no official endorsement should be inferred.

REFERENCES

Alvarez-Cohen, L. and P.L. McCarty. 1991a. A Cometabolic Biotransformation Model for Halogenated Aliphatic Compounds Exhibiting Product Toxicity. *Environ. Sci. Technol.*, **25**(8):1381–1387.

Alvarez-Cohen, L. and P.L. McCarty. 1991b. Product Toxicity and Cometabolic Competitive Inhibition Modeling of Chloroform and Trichloroethylene by Methanotrophic Resting Cells. *Appl. Environ. Microbiol.*, **57**(4):1031–1037.

Alvarez-Cohen, L.M. and P.L. McCarty. 1991c. Two-Stage Dispersed-Growth Treatment of Halogenated Aliphatic Compounds by Cometabolism. *Environ. Sci. Technol.*, **25**(8):1387–1392.

Alvarez-Cohen, L.M. and P.L. McCarty. 1991d. Effects of Toxicity, Aeration, and Reductant Supply on Trichloroethylene Transformation by a Mixed Methanotrophic Culture. *Appl. Environ. Microbiol.*, **57**(1):228–235.

Arciero, D., T. Vannelli, M. Logan, and A.B. Hooper. 1989. Degradation of trichloroethylene by the ammonia-oxidizing bacterium *Nitrosomonas europaea*. *Biochem. Biophys. Res. Commun.*, **159**:640–643.

Arvin. 1991. Biodegradation Kinetics of Chlorinated Aliphatic Hydrocarbons with Methane Oxidizing Bacteria in an Aerobic Fixed Biofilm Reactor. *Water Res.*, **25**(7):873–881.

Bailey, J.E. and D.F. Ollis. 1986. *Biochemical Engineering Fundamentals* (2nd ed.). McGraw-Hill, Inc., New York.

Broholm, K., B.K. Jensen, T.H. Christensen, and L. Olsen. 1990. Toxicity of 1,1,1-Trichloroethane and Trichloroethylene on a Mixed Culture of Methane-Oxidizing Bacteria. *Appl. Environ. Microbiol.*, **56**(8):2488–2493.

Brunner, W., D. Staub, and T. Leisinger. 1980. Bacterial Degradation of Dichloromethane. *Appl. Environ. Microbiol.*, **40**(5):950–958.

Brusseau, G.A., H.C. Tsien, R.S. Hanson, and L.P. Wackett. 1990. Optimization of Trichloroethlyene Oxidation by Methanotrophs and the Use of a Colorimetric Assay to Detect Soluble Methane Monooxygenase Activity. *Biodegradation*, **1**:19–29.

Dalton, H. and D.I. Stirling. 1982. Co-metabolism. *Phil. Trans. R. Soc. Lond. B*, **297**:481–496.

DiSpirito, A.A. et al. 1992. Trichloroethylene Oxidation by the Membrane-Associated Methane Monooxygenase in Type I, Type II and Type X Methanotrophs. *Biodegradation*, **2**(3):151–164.

Dolan, M. and P.L. McCarty. 1994. Factors Affecting Transformation of Chlorinated Aliphatic Hydrocarbons by Methanotrophs. In Norris, R.D. et al. *Handbook of Bioremediation,* Lewis Publishers, Boca Raton, FL, pp. 303–308.

Ensign, S.A., M.R. Hyman, and D.J. Arp. 1992. Cometabolic Degradation of Chlorinated Alkenes by Alkene Monooxygenase in a Propylene-Grown *Xanthobacter* Strain. *Appl. Environ. Microbiol.*, **58**(9):3038–3046.

Ewers, J., W. Clemens, and H.J. Knackmuss (eds.) 1991. Biodegradation of Chloroethenes Using Isoprene as Co-Substrate. *International Symposium on Environmental Biotechnology*, Vol 1. Royal Flemish Society of Engineers, Ostend, Belgium, pp. 77–83.

Fliermans, C.B. et al. 1988. Mineralization of trichloroethylene by heterotrophic enrichment cultures. *Appl. Environ. Microbiol.*, **54**(7):1709–1714.

Fogel, M.M., A.R. Taddeo, and S. Fogel. 1986. Biodegradation of Chlorinated Ethenes by a Methane-Utilizing Mixed Culture. *Appl. Environ. Microbiol.*, **51**(4):720–724.

Folsom, B.R. and P.J. Chapman. 1991. Performance Characterization of a Model Bioreactor for the Biodegradation of Trichloroethylene by *Pseudomonas cepacia* G4. *Appl. Environ. Microbiol.*, **57**(6):1602–1608.

Folsom, B.R., P.J. Chapman, and P.H. Pritchard. 1990. Phenol and Trichloroethylene Degradation by *Pseudomonas cepacia* G4: Kinetics and Interactions between Substrates. *Appl. Environ. Microbiol.*, **56**(5):1279–1285.

Fox, B.G., J.G. Borneman, L.P. Wackett, and J.D. Lipscomb. 1990. Haloalkene Oxidation by the Soluble Methane Monooxygenase from *Methylosinus trichosporium* OB3b: Mechanistic and Environmental Implications. *Biochemistry*, **29**(27):6419–6427.

Freedman, D.L. and J.M. Gossett. 1991. Biodegradation of Dichloromethane and Its Utilization and Growth Substrate under Methanogenic Conditions. *Appl. Environ. Microbiol.*, **57**(10):2847–2857.

Green, J. and H. Dalton. 1989. Substrate Specificity of Soluble Methane Monooxygenase. *J. Biol. Chem.*, **264**(30):17698–17703.

Harker, A.R. and Y. Kim. 1990. Trichloroethylene Degradation by Two Independent Aromatic-Degrading Pathways in *Alcaligenes eutrophus* JMP134. *Appl. Environ. Microbiol.*, **56**(4):1179–1181.

Hartmans, S. and J.A.M. de Bont. 1992. Aerobic Vinyl Chloride Metabolism in *Mycobacterium aurum* L1. *Appl. Environ. Microbiol.*, **58**(4):1220–1226.

Hartmans, S., J.A.M. de Bont, J. Tramper, and K.C.A.M. Luyben. 1985. Bacterial Degradation of Vinyl Chloride. *Biotechnol. Lett.*, **7**(6):383–388.

Hazen, T.C. 1992. *Test Plan for In Situ Bioremediation Demonstration of the Savannah River Integrated Demonstration Project DOE/OTD TTP No. SR 0566-01 (U)* No. WSRC-RD-91-23). Westinghouse Savannah River Company.

Henry, S.H. and D. Grbic-Galic. 1990. Effect of mineral media on trichloroethylene oxidation by aquifer methanotrophs. *Microb. Ecol.*, **20**:151–169.

Henry, S.M. and D. Grbic-Galic. 1991. Influence of Endogenous and Exogenous Electron Donors and Trichloroethylene Oxidation Toxicity on Trichloroethylene Oxidation by Methanotrophic Cultures from a Groundwater Aquifer. *Appl. Environ. Microbiol.*, **57**(1):236–244.

Henson, J., M. Yates, and J. Cochran. 1989. Metabolism of chlorinated methanes, ethanes, and ethylenes by a mixed bacterial culture growing on methane. *Journal of Industrial Microbiology*, **4**:29–35.

Herrling, B. 1991. Hydraulic Circulation System for *In Situ* Bioreclamation and/or In Situ Remediation of Strippable Contamination. In R.E. Hinchee and R.F. Olfenbuttel (eds.), *In Situ Bioreclamation*, Butterworth-Heinemann, Boston, pp. 173–195.

Hopkins, G.D., L. Semprini, and P.L. McCarty. 1993. Microcosms and In-Situ Field Studies of Enhanced Biotransformation of Trichloroethylene by Phenol-Utilizing Microorganisms. *Appl. Environ. Microbiol.* **59**:2277–2285.

Hou, C. 1984. Propylene oxide production from propylene by immobilized whole cells of *Methylosinus* sp. CRL 31 in a gas solid bioreactor. *Appl. Microbiol. Biotechnol.*, **19**:1–4.

Hou, C.T. et al. 1979. Microbial Oxidation of Gaseous Hydrocarbons: Production of Methyl Ketones from Their Corresponding Secondary Alcohols by Methane- and Methanol-Grown Microbes. *Appl. Environ. Microbiol.*, **38**(1):135–142.

Janssen, D.B., G. Grobben, and B. Witholt. 1987. Toxicity of Chlorinated Aliphatic Hydrocarbons and Degradation by Methanotrophic Consortia. In *Proc. 4th European Congress on Biotechnology*, 3 pp. 515–515.

Janssen, D.B., A. Scheper, L. Dijkhuizen, and B. Witholt. 1985. Degradation of Halogenated Aliphatic Compounds by *Xanthobacter autotrophicus* GJ10. *Appl. Environ. Microbiol.*, **49**(3):673–677.

Jensen, H.L. 1963. Carbon nutrition of some microorganisms decomposing halogen-substituted aliphatic acids. *Acta Agr. Scand.*, **14**:404–414.

Klecka, G.M. and S.J. Gonsior. 1984. Reductive Dechlorination of Chlorinated Methanes and Ethanes by Reduced Iron (II) Porphyrins. *Chemosphere*, **13**(3):391–402.

Kohler-Staub, D. et al. 1986. Evidence for Identical Dichloromethane Dehalogenases in Different Methylotrophic Bacteria. *J. Gen. Microbiol.*, **132**:2837–2843.

LaPat-Polasko, L.T., P.L. McCarty, and A.J.B. Zehnder. 1984. Secondary Substrate Utilization of Methylene Chloride by an Isolated Strain of *Pseudomonas* sp. *Appl. Environ. Microbiol.*, **47**:825–830.

Leadbetter, E.R. and J.W. Foster. 1959. Oxidation products formed from gaseous alkanes by the bacterium *Pseudomonas methanica. Arch. Biochem. Biophys.*, **82**:491–492.

Leahy, M.C., M. Findlay, and S. Fogel. 1989. Biodegradation of Chlorinated Aliphatics by a Methanotrophic Consortium in a Biological Reactor. In *Biotreatment*, Hazardous Materials Control Research Institute, Washington, DC, pp. 3–9.

Lehninger, A.L. 1970. *Biochemistry*. Worth Publishers, Inc., New York.

Little, C.D. et al. 1987. Trichloroethylene Biodegradation by a Methane-Oxidizing Bacterium. *Appl. Environ. Microbiol.*, **54**(4):951–956.

MacDonald, T.R. and P.K. Kitanidis. 1993. Modeling the Free-Surface of an Unconfined Aquifer Near a Recirculation Well. *Ground Water,* **31**(5):774–788.

McCarty, P.L. and L. Semprini. 1994. Ground-Water Treatment for Chlorinated Solvents. In Norris, R.D. et al. *Handbook of Bioremediation,* Lewis Publishers, Boca Raton, FL, pp. 87–116.

McCarty, P.L. et al. 1991. *In Situ* Methanotropohic Bioremediation for Contaminated Groundwater at St. Joseph, Michigan. In R.E. Hinchee and R.G. Olfenbuttel ed., *On-Site Bioreclamation Processes for Xenobiotic and Hydrocarbon Treatment*, Butterworth-Heinemann, Boston, pp. 16–40.

McFarland, M.J., C.M. Vogel, and J.C. Spain. 1992. Methanotrophic Cometabolism of Trichloroethylene (TCE) in a Two Stage Bioreactor System. *Water Res.*, **26**(2):259–265.

Miller, R.E. and F.P. Guengerich. 1982. Oxidation of Trichloroethylene by Liver Microsomal Cytochrome P-450: Evidence for Chlorine Migration in a Transition State Not Involving Trichloroethylene Oxide. *Biochemistry*, **21**:1090–1097.

Miller, R.E. and F.P. Guengerich. 1983. Metabolism of Trichloroethylene in Isolated Hepatocytes, Microsomes, and Reconstituted Enzyme Systems Containing Cytochrome P-450. *Cancer Res.*, **43**:1145–1152.

Nelson, M.J.K., S.O. Montgomery, W.R. Mahaffey, and P.H. Pritchard. 1987. Biodegradation of Trichloroethylene and Involvement of an Aromatic Biodegradative Pathway. *Appl. Environ. Microbiol.*, **53**(5):949–954.

Nelson, M.J.K., S.O. Montgomery, E.J. O'Neill, and P.H. Pritchard. 1986. Aerobic metabolism of trichloroethylene by a bacterial isolate. *Appl. Environ. Microbiol.*, **52**(2):383–384.

Nelson, M.J.K., S.O. Montgomery, and P.H. Pritchard. 1988. Trichloroethylene Metabolism by Microorganisms That Degrade Aromatic Compounds. *Appl. Environ. Microbiol.*, **54**(2):604–606.

Nelson, Y. and W.J. Jewell. 1993. Vinyl Chloride Biodegradation with Methanotrophic Attached Films. *Jour. Environ. Engineering*, **119**(5):890–907.

Nichols, P.D. and D.C. White. 1989. Accumulation of poly-b-hydroxybutyrate in a methane-enriched, halogenated hydrocarbon-degrading soil column; implications for microbial community structure and nutritional status. *Hydrobiologia*, **176/177**:369–377.

Oldenhuis, R., J.Y. Oedzes, J.J. van der Waarde, and D.B. Janssen. 1991. Kinetics of Chlorinated Hydrocarbon Degradation by *Methylosinus trichosporium* OB3B and Toxicity of Trichloroethylene. *Appl. Environ. Microbiol.*, **57**(1):7–14.

Oldenhuis, R., R.L.J. M. Vink, D.B. Janssen, and B. Witholt. 1989. Degradation of Chlorinated Aliphatic Hydrocarbons by *Methylosinus trichosporium* OB3b Expressing Soluble Methane Monooxygenase. *Appl. Environ. Microbiol.*, **55**(11):2819–2826.

Patel, R.N., C.T. Hou, A.I. Laskin, and A. Felix. 1982. Microbial Oxidation of Hydrocarbons: Properties of a Soluble Methane Monooxygenase from a Facultative Methane-Utilizing Organism, *Methylobacterium* sp. Strain CRL-26. *Appl. Environ. Microbiol.*, **44**(5):1130–1137.

Patel, R.N. et al. 1979. Microbial Oxidation of Gaseous Hydrocarbons. II. Hydroxylation of Alkanes and Epoxidation of Alkenes by Cell-Free Particulate Fractions of Methane-Utilizing Bacteria. *J. Bacteriol.*, **139**(2):675–679.

Phelps, T.J. et al. 1990. Biodegradation of trichloroethylene in continuous-recycle expanded-bed bioreactors. *Appl. Environ. Microbiol.*, **56**(6):1702–1709.

Rasche, M.E., M.R. Hyman, and D.J. Arp. 1991. Factors Limiting Aliphatic Chlorocarbon Degradation by *Nitrosomonas europaea*: Cometabolic Inactivation of Ammonia Monooxygenase and Substrate Specificity. *Appl. Environ. Microbiol.*, **57**(10):2986–2994.

Rittmann, B.E. and P.L. McCarty. 1980. Utilization of Dichloromethane by Suspended and Fixed Film Bacteria. *Appl. Environ. Microbiol.*, **39**:1225–1226.

Roberts, P.V., G.D. Hopkins, D.M. Mackay, and L. Semprini. 1990. A Field Evaluation of In-Situ Biodegradation of Chlorinated Ethenes: Part 1, Methodology and Field Site Characterization. *Ground Water*, **28**(4):591–604.

Roberts, P.V. et al. 1989. *In-Situ Aquifer Restoration of Chlorinated Aliphatics by Methanotrophic Bacteria.* EPA/600/2-89/033). U.S. EPA Center for Environmental Information, Cincinnati, OH.

Semprini, L. et al. 1991. A Field Evaluation of In-Situ Biodegradation of Chlorinated Ethenes: Part 3, Studies of Competitive Inhibition. *Ground Water*, **29**(2):239–250.

Semprini, L. and P.L. McCarty. 1991. Comparison Between Model Simulations and Field Results for In-Situ Biorestoration of Chlorinated Aliphatics: Part 1. Biostimulation of Methanotrophic Bacteria., *Ground Water,* **29**(3):365–374.

Semprini, L. and P.L. McCarty. 1992. Comparison Between Model Simulations and Field Results for In-Situ Biorestoration of Chlorinated Aliphatics: Part 2. Cometabolic Transformations. *Ground Water*, **30**(1):37–44.

Semprini, L., P.V. Roberts, G.D. Hopkins, and P.L. McCarty. 1990. A Field Evaluation of In-Situ Biodegradation of Chlorinated Ethenes: Part 2. Results of Biostimulation and Biotransformation Experiments. *Ground Water*, **28**:715–727.

Stirling, D.I. and H. Dalton. 1979. The Fortuitous Oxidation and Cometabolism of Various Carbon Compounds by Whole-Cell Suspensions of *Methylococcus capsulatus* (Bath). *FEMS Microbiol. Lett.*, **5**:315–318.

Stirling, D.I. and H. Dalton. 1985. Aerobic Metabolism of Methane and Methanol. In M. Moo-Young, A.T. Bull, and H. Dalton, eds., *Comprehensive Biotechnology*, Pergamon Press, Oxford, pp. 379–408.

Strand, S.E., M.D. Bjelland, and H.D. Stensel. 1990. Kinetics of chlorinated hydrocarbon degradation by suspended cultures of methane-oxidizing bacteria. *Research Journal Water Pollution Control Federation*, **62**(2):124–129.

Strand, S.E., R.M. Seamons, M.D. Bjelland, and H.D. Stensel. 1988. Kinetics of Methane-Oxidizing Biofilms for Degradation of Toxic Organics. *Water Science Tech.*, **20**:167–173.

Stucki, G., R. Galli, H.R. Ebersold, and T. Leisinger. 1981. Dehalogenation of Dichloromethane by Cell Extracts of *Hyphomicrobium* DM2. *Arch. Microbiol.*, **130**:366–371.

Stucki, G. and T. Leisinger. 1983. Bacterial degradation of 2-chloroethanol proceeds via 2-chloroacetic acid. *FEMS Microbiol. Lett.*, **16**:123–126.

Tsien, H.C., G.A. Brusseau, R.S. Hanson, and L.P. Wackett. 1989. Biodegradation of Trichloroethylene by *Methylosinus trichosporium* OB3b. *Appl. Environ. Microbiol.*, **55**(12):3155–3161.

Uchiyama, H., T. Nakajima, O. Yagi, and T. Nakahara. 1992. Role of Heterotrophic Bacteria in Complete Mineralization of Trichloroethylene by *Methylocystis* sp. Strain M. *Appl. Environ. Microbiol.*, **58**(9):3067–3071.

Vogel, T.M., C.S. Criddle, and P.L. McCarty. 1987. Transformations of Halogenated Aliphatic Compounds. *Environ. Sci. Technol.,* **21**:722–736.

Wackett, L.P., G.A. Brusseau, S.R. Householder, and R.S. Hanson. 1989. Survey of Microbial Oxygenases: Trichloroethylene Degradation by Propane-Oxidizing Bacteria. *Appl. Environ. Microbiol.*, **55**(11):2960–2964.

Wackett, L.P. and D.T. Gibson. 1988. Degradation of trichloroethylene by toluene dioxygenase in whole-cell studies with *Pseudomonas putida* F1. *Appl. Environ. Microbiol.*, **54**(7):1703–1708.

Wackett, L.P. and S.R. Householder. 1989. Toxicity of Trichloroethylene to *Pseudomonas putida* F1 is Mediated by Toluene Dioxygenase. *Appl. Environ. Microbiol.*, **55**(10):2723–2725.

Wilcox, D.W., R.L. Autenrieth, and J.S. Bonner. 1992. Biodegradation of Vapor Phase Trichloroethylene by *Mycobacterium vaccae* JOB5 in the Presence of Propane. In *Water Environment Federation*, AC92-006-004. Water Environment Federation, New Orleans, pp. 29–39.

Wilson, J.T. and B.H. Wilson. 1985. Biotransformation of Trichloroethylene. *Appl. Environ. Microbiol.*, **49**(1):242–243.

Chapter 24

Natural Chemical Attenuation of Halogenated Hydrocarbon Compounds via Dehalogenation

M. Reinhard, G.P. Curtis, and J.E. Barbash

INTRODUCTION

The chemistry of haloaliphatic compounds (HACs) in the environment is diverse and, in many cases, complex. The chemical and physical properties of HACs span a wide range and thus their biological, chemical and physical behavior vary widely. The abiotic transformation of HACs depends on structural features of the substrate (e.g., degree of unsaturation, halogenation, steric hindrance), on the solution conditions, such as pH, redox conditions, temperature and the presence of different nucleophiles (Haag and Mill, 1988b, Mabey and Mill, 1978; Vogel et al., 1987a,b; Barbash and Reinhard, 1992a,b) as well as on the presence of solids (Kriegman-King and Reinhard, 1991; Kriegman-King and Reinhard, 1992; Kriegman-King and Reinhard, 1994).

Because of the influence of biological activity on chemical conditions and vice versa, it can be difficult to unambiguously attribute HAC disappearance to chemical or biological factors. Biological and chemical transformations may also occur sequentially and competitively. For instance, 1,1,1-trichloroethane may be transformed by different, parallel pathways into 1,1-dichloroethylene via dehydrohalogenation, acetic acid via hydrolysis and 1,1-dichloroethane via biological reduction (Vogel and McCarty, 1987). The purpose of this chapter is to review the chemical and environmental factors that affect the transformation of HACs under groundwater conditions.

Under the relatively mild chemical conditions typically found in groundwater (pH 4 to 9, temperature $\leq 25°C$, ionic strength $I < 0.01M$), the carbon-halogen bond (R-X, X = Cl, Br, I) may be cleaved abiotically by one of the following reactions: dissociative electron transfer (ET) (reduction), or nucleophilic substitution or dehydrohalogenation (elimination of hydrogen halide). HACs are relatively resistant to abiotic oxidation, especially the highly halogenated ones. In contrast, with increasing extent of halogenation, HAC reduction becomes more likely (Vogel et al., 1987). For some HACs, biologically mediated oxidation (hydroxylation or epoxidation) is observed (refer to other chapters in this book).

REDUCTIONS

HAC reductions may be initiated by the transfer of a single electron. Transfer of the first electron to the R-X bond is often rate limiting. It leads to bond dissociation and the formation of an extremely reactive radical (R•) (Eberson, 1987). The radical may react further in a number of ways as discussed below. Oxidation of the donor, D, and reduction of RX may be written as two half-reactions:

$$D \rightarrow D^+ + e \tag{24.1}$$

$$RX + e \rightarrow R\bullet + X^- \tag{24.2}$$

An example is the reduction of carbon tetrachloride by a single ET step which yields the extremely reactive trichloromethyl radical and Cl^-:

$$CCl_4 + e \rightarrow CCl_3\bullet + Cl^- \tag{24.3}$$

Estimation of Reaction Rates

Rates and mechanisms of alkyl radical formation have been intensively investigated and interpreted in terms of the Marcus theory (Eberson, 1982; Eberson, 1987). The Marcus theory was originally developed for predicting the rates of outer-sphere ET reactions between an electron donor and an electron acceptor on the basis of thermodynamic parameters. Reactions are termed outer-sphere when the inner coordination shells of the participating metal ions are intact in the transition state, and no ligand-to-metal bonds are broken or formed in the transition state. Most metal reagents, however, reduce alkyl halides by an inner-sphere process. For example, the reduction of alkyl halides by iron porphyrins occurs by an axial inner-sphere mechanism (Wade and Castro, 1973). However, it has been found that in certain cases where the HAC reduction proceeds by an inner-sphere mechanism, the Marcus theory is still a useful tool for interpreting and correlating kinetic data (Eberson, 1987).

In applying the Marcus theory to alkyl halide reduction, Eberson (1987) assumed that the transition state for dissociative ET is formulated as two mesomeric forms—one in which the C-X bond is stretched to the point of breaking, and a second one consisting of an alkyl radical paired with a halide ion (Eberson, 1987):

$$RX + D \leftrightarrow [R\bullet X\ D \leftrightarrow R\bullet\ X^- + D^+]^{\#} \rightarrow R\bullet + X^- + D^+ \qquad (24.4)$$

The Marcus equation relates the rate of alkyl radical formation to the thermodynamic parameters $\Delta G^{\circ\prime}$, the standard free energy change of the single ET step, and the reorganization energy, λ, viz:

$$k_{obs} = 10^{11} \times \exp \frac{-\lambda\{1 + (\Delta G^{\circ\prime}/\lambda)\}^2}{4 \times RT} \qquad (24.5)$$

where R is the gas constant and T is the absolute temperature. $\Delta G^{\circ\prime}$ is the standard free energy change of the single ET indicated in Equation 24.2, in solution, adjusted for interactions between the reactant molecule and the solvent. These adjustments typically amount to less in 1 kcal/mol in aqueous solutions. The parameter λ accounts for the energy required to distort solvent molecules while forming the transition state. The prefix factor of 10^{11} ($M^{-1}s^{-1}$) equals the assumed frequency factor for a bimolecular reaction. The standard free energy of the reaction is related to the standard reduction potentials of the oxidant and reductants, E_{ox} and E_{red}, respectively, through the Nernst equation:

$$\Delta G^\circ = -nF(E_{ox} - E_{red}) \qquad (24.6)$$

where

n is the number of electrons transferred (n=1 for single ET); and
F is Faraday's constant.

The reduction potentials of HACs are difficult or impossible to measure at an electrode because the reduction tends to be irreversible. Estimation methods are therefore often used. Curtis (1991) estimated the one-electron reduction potentials for some HACs based on an approach reported by Eberson (1982). These estimated reduction potentials of selected HACs and the relative half-lives (chloroform=1) for their reaction with a hypothetical electron donor with E_o of –0.1V are indicated in Table 24.1. The data shown in Table 24.1 clearly indicate that the E_o values and the half-lives are influenced by the position, number and type of halogen substitution. Equation 24.5 indicates that the absolute rates increase with decreasing E° of the hypothetical electron donor while the ranking of the HACs remains the same.

The calculated one-electron reduction potentials range from 0.36 V for hexachloroethane to –0.88 V for methyl chloride, and, in general, decrease with decreasing extent of halogenation. The calculated half-lives follow the opposite trend. They increase by approximately eleven orders of magnitude from hexachloroethane to methyl chloride. In biological and chemical systems, qualita-

Table 24.1. One-Electron Reduction Potentials (E_o) of Selected HACs and Normalized Half-Lives for Reactions with a Hypothetical Electron Donor (E_o = –0.1V, 25°C, λ = 50 kcal/mol).

R-X	R•	E_o (V)	rel. $t_{1/2}$ ($CHCl_3$=1)
$Cl_3C\text{-}Cl_3\mathbf{C}$	$Cl_3C\text{-}Cl_2\mathbf{C}$•	0.36	2×10^{-5}
$Cl_2HC\text{-}Cl_3\mathbf{C}$	$Cl_2HC\text{-}Cl_2\mathbf{C}$•	0.32	5×10^{-5}
Br_4C	Br_3C•	0.26	1×10^{-4}
Cl_4C	Cl_3C•	0.13	1×10^{-3}
$ClH_2C\text{-}Cl_3\mathbf{C}$	$ClH_2C\text{-}Cl_2\mathbf{C}$•	0.09	2×10^{-3}
$Cl_2HC\text{-}Cl_2H\mathbf{C}$	$Cl_2HC\text{-}ClH\mathbf{C}$•	0.05	5×10^{-3}
Br_3HC	Br_2HC•	–0.06	3×10^{-2}
Cl_3HC	HCl_2C•	–0.23	1
$H_3C\text{-}Cl_3\mathbf{C}$	$H_3C\text{-}Cl_2\mathbf{C}$•	–0.23	1
$H_3C\text{-}Cl_2H\mathbf{C}$	$H_3C\text{-}ClH\mathbf{C}$•	–0.40	40
Cl_2H_2C	H_2ClC•	–0.50	400
Br_2H_2C	BrH_2C•	–0.51	500
BrH_3C	H_3C•	–0.54	900
$ClH_2C\text{-}ClH_2\mathbf{C}$	$ClH_2C\text{-}H_2\mathbf{C}$•	–0.57	2×10^{2}
$BrH_2C\text{-}BrH_2\mathbf{C}$	$BrH_2C\text{-}H_2\mathbf{C}$•	–0.70	5×10^{4}
$H_3C\text{-}ClH_2\mathbf{C}$	$H_3CH_2\mathbf{C}$•	–0.83	1×10^{6}
ClH_3C	H_3C•	–0.88	4×10^{6}

Reference: modified from Curtis (1991) using data of Taylor et al., 1991.

tively the same trend is observed (Vogel et al., 1987, Gantzer and Wackett, 1991). The model is sensitive to both the distribution within the molecule and the types of halogen present. For example, the substituents on the R′-CCl_2• radical strongly affect E_o. For R′ = CCl_3, E_o is higher than for other substituents that attract electrons less strongly. The estimated relative half-lives increase by 4 orders of magnitude as R′ is changed from CCl_3 to H. Interestingly, brominated and chlorinated analogs do not always follow the same ranking. Tetrabromomethane has a higher E_o than carbon tetrachloride and thus should accept an electron more readily, but the situation is reversed in the case of bromoform and chloroform. Table 24.1 also illustrates that the half-life changes by approximately a factor of ten for each change of 0.1 V in the reduction potential. The computed half-lives do not reflect real conditions but are a useful basis for comparing relative rates in different systems and classifying HACs according to their tendency to accept electrons.

There are few data available in the literature to quantitatively evaluate the predicted rates. The calculated E_o of carbon tetrachloride and bromoform are 0.23 V and 0.42 V lower, respectively, than that of hexachloroethane. According to the model, carbon tetrachloride and bromoform should be reduced approximately one and three orders of magnitude slower, respectively, than hexachloroethane. Curtis (1991) found qualitative agreement with this prediction; carbon tetrachloride was reduced 2 to 6 times slower than hexachloroethane and bromoform reacted more than 100 times more slowly than hexachloroethane. The systems studied included: (1) humic acid and hematin (iron porphyrin) solutions in the presence of excess HS^- and Fe^{2+} (added as relatively inert reductants to reduce the electron transfer (ET) agents) and (2) a reduced anthrahydroquinone-2,6-disulfonate solution. Gantzer and Wackett (1991) found carbon tetrachloride to react faster than the halogenated ethylenes.

Environmental Electron Donors

The electron donors present in aquifers and soils that reduce organic compounds are poorly characterized, largely because the composition of sediments is so complex. Furthermore, ET can be mediated by catalysts (ET mediators). Curtis (1991) fractionated a sandy aquifer using a magnetic separator and found measurable levels of reducing capacity distributed among different mineral fractions including biotite, ilmenite, and magnetite. However, most of the reactivity was associated with the quartz-feldspar-carbonate fraction which also contained the greatest fraction of organic carbon. In subsequent investigations, he demonstrated that humic acid extracted from the Borden aquifer material could reductively transform hexachloroethane and carbon tetrachloride when a bulk electron donor such as HS^- or Fe^{2+} was present. Carreira and Wolfe (1992), in an attempt to characterize the ET mediators present in sediment, have isolated a component highly efficient at reducing 2,4,6-trinitrotoluene. They noted that this component exhibited properties of an enzyme, contained two iron atoms and had a maximum mass of about 300,000 daltons. In their review, Mohn and Tiedje (1992) note that transition metal complexes released from decomposing organisms could affect the environmental fate of HACs.

Table 24.2 lists the reduction potentials of selected environmental electron donors, including some minerals, natural humic and fulvic acids, and metal complexes that have been used in studies of HAC reduction. Dehalogenation studies have been conducted with each of the electron donors listed. In general, it has been found that the electron donors with the lowest reduction potential react rapidly with the broadest array of substrates. For example, iron porphyrins and corrinoids reduce chlorinated methanes, ethanes, ethenes and some chlorobenzenes (Wade and Castro, 1973; Klecka and Gonsior, 1984; Gantzer and Wackett, 1991). Kray and Castro (1964) reported similarly rapid reduction of vicinal dihalides by Cr^{2+} (in 1:1 dimethylformamide-water). Reductions involving ferrous iron are relatively slow, even for hexachloroethane and carbon tetrachloride (Curtis, 1991). In the presence of HS^- the transformation rates were found to be approximately a factor of 3 to 4 faster than in the presence of Fe^{2+}. Pyrite and biotite (a sheet silicate containing small amounts of reduced iron) were found to reduce hexachloroethane and carbon tetrachloride (Kriegman-King and Reinhard, 1991, 1992, 1994). However, since it involves a heterogeneous reaction, it is difficult to rank the biotite system among the homogeneous solution systems. Humic acid has been shown to promote the reduction of hexachloroethane and carbon tetrachloride by the bulk electron donors HS^- and Fe^{2+}. In general, the observed trends agree with the Marcus model, which predicts (i) an increase in reduction rate with decreasing reduction potential of the electron donor (except in the so-called "inverted" region (Eberson, 1987)), and (ii) an increase in reduction rate with increasing reduction potential of the substrate.

Reactions Involving Radical Intermediates

As stated earlier, the rate-determining step in HAC reduction is typically the transfer of the initial electron. The products of the reaction depend on the reactions of the alkyl radical formed by this initial step. These reactions include hydrogenolysis, dihalo-elimination and radical coupling (dimerization).

Hydrogenolysis involves accepting of a second electron and a proton

$$R\bullet + e + H^+ \rightarrow RH \tag{24.7}$$

or the abstraction of a hydrogen radical, H•, from another molecule, viz.:

$$R\bullet + R'H \rightarrow RH + R'\bullet \tag{24.8}$$

Dihalo-elimination. If a halogen is present on the β-carbon, this second halogen may also be lost, leading to the formation of an alkene

$$Cl_3CCCl_2\bullet + e \rightarrow Cl_2C{=}CCl_2 + Cl^- \tag{24.9}$$

Dimerization. The radical may combine with a second radical to form a coupling product:

$$R\bullet + R'\bullet \rightarrow R\text{-}R' \tag{24.10}$$

Table 24.2. Reduction Potentials of Selected Environmental Electron Donors.

Half-Reaction[1]	E°(V)	E′(V)[2]	Ref.
$Fe^{3+} + e = Fe^{2+}$	0.77	0.05 [3]	a
$Humic^{(ox)} + 1/2H^+ + e = Humic^{(red)}$	0.7	0.5	b[4]
$\{Fe^{3+}\}_{Biotite} + 1/Z\ M^{Z+} + e = \{Fe^{2+}, 1/ZM^{Z+}\}_{Biotite}$	0.33	n.d.	c[5]
$1/2\ A(=O)_2 + H^+ + e = 1/2\ A(-OH)_2$	0.23	–0.18	d[6]
$1/64\ S_8 + 1/4H^+ + e = 1/8\ H_2S$	0.14	–0.27	a
Fe^{3+}-porph + e = Fe^{2+}-porph	–0.20	n.d.	e
$Cr^{3+} + e = Cr^{2+}$	–0.41	n.d.	a

References: a. Bratsch, 1989; b. Szilagyi, 1973, 1971; c. White and Yee, 1985; d. Clark, 1960; e. Vogel et al., 1987.

Conditions: 1. All species are in aqueous solution; 2. E′ is the reduction potential calculated from the Nernst equation for "typical natural water": pH=7, activities of Cl^- and HCO_3^- = 10^{-3} M, of Fe^{2+}_{aq} and Mn^{2+}_{aq} = 10^{-6}M; 3. Includes the precipitation of amorphous iron hydroxide; $K_{sp}=10^{-37}$; 4. Stoichiometry and E′ estimated from Eh-pH measurement by Szilagyi, 1973; 5. M^{Z+} is an exchangeable cation necessary to maintain charge balance for the mineral. The braces {} indicate that the iron is part of the mineral; and 6. $A(=O)_2$: Anthraquinone-2,6-disulfonate; $A(-OH)_2$: Anthrahydroquinone-2,6-disulfonate.

Which of these reactions predominates depends on the nature of the alkyl radical and on the solution composition. For example, hexachloroethane has been found in several studies to be reduced almost exclusively to tetrachloroethylene, the dihalo-elimination product (Jafvert and Wolfe, 1987, Curtis, 1991). In contrast, the reduction of carbon tetrachloride can lead to several different products, including chloroform, carbon monoxide, formate and cell-bound products (Criddle and McCarty, 1991). Gantzer and Wackett (1991) found low yields of hydrogenolysis products when chlorinated ethylenes were reacted with porphyrin complexes. For example, the vinyl chloride yield was only 9% when reduction of *cis*-dichloroethylene was mediated with vitamin B_{12}. The yield was dependent on the type of catalyst used. They suspected that the low yield was due to radical coupling which competed with hydrogenolysis.

In summary, relative rates of reduction reactions can be estimated by using the Marcus theory together with estimated reduction potentials. Such estimations are useful in assessing the relative trends among a set of compounds. Prediction of the actual rate of a reduction reaction in the environment is beyond our current capabilities largely because the electron donors and their reduction potentials remain poorly understood. In systems as complex and heterogeneous as those encountered in the subsurface, the identification of electron donors that reduce HACs remains elusive.

Laboratory data indicate that the inorganic species Fe^{2+} and HS^- reduce HACs such as carbon tetrachloride and hexachloroethane. Moreover, the rate of these reductions is promoted by the presence of natural materials including the surfaces of mineral and natural organic matter. Laboratory data also indicate that many HACs are rapidly transformed by reduced-iron porphyrins and enzymatic co-factors. The significance of reduced porphyrin complexes and enzymatic co-factors as ET mediators in the environment outside cells is still largely unknown. The products of reduction reactions can be difficult to predict because they depend on the intermediate radical and the solution composition. In contrast to this difficulty, substantial progress has been achieved in quantitatively understanding nucleophilic substitution and dehydrohalogenation reactions, which are discussed in the next section.

NUCLEOPHILIC SUBSTITUTION (INCLUDING HYDROLYSIS) AND DEHYDROHALOGENATION

The nucleophilic substitution and dehydrohalogenation (elimination) reactions of most HACs encountered in surface or groundwaters are bimolecular processes (S_N2 and E2). Both types of reaction involve the rate-limiting attack of a Lewis base (a species with an unshared electron pair), either on a saturated carbon to form the substitution product, or a hydrogen attached to the β-carbon leading to the olefinic dehydrohalogenation product (March, 1985). When attacking the carbon, the Lewis base is acting as a nucleophile, Nu, and, when attacking the hydrogen, it is acting as a Brønsted base. Nu can be charged or neutral and, as a Brønsted base, accept a proton from water.

Whether a given Lewis acid reacts with a particular substrate via substitution or dehydrohalogenation (or both) depends on the chemical properties of both Nu (or Nu^-) and the substrate. The dual function of the Lewis base is illustrated below for a 1,2-dihaloethane (Z represents other halogen substituents):

Substitution

$$CH_2Z-CH_2X+Nu^- \xrightarrow{k_n} CH_2Z-CH_2Nu+X^- \quad (24.11)$$

Dehydrohalogenation

$$\overset{\beta}{CH_2Z}-\overset{\alpha}{CH_2X}+Nu \xrightarrow{k_e} HZC=CH_2+X^-+Nu+H^+ \quad (24.12)$$

Note that instead of HX, HZ could be removed in Reaction 24.12, leading to $CH_2 = CHX$.

Both OH^- and H_2O are Lewis bases capable of effecting the dehalogenation of a haloaliphatic compound via hydrolysis (i.e., replacement of the halogen with a hydroxyl group via nucleophilic substitution) or dehydrohalogenation. In many cases, hydrolysis and dehydrohalogenation occur simultaneously or sequentially. Examples are indicated below and include 1,2-dibromoethane (Weintraub et al., 1986; Haag and Mill, 1988a), 1,2-dichloroethane (Barbash and Reinhard, 1989a), 1,1,1-trichloroethane (Haag and Mill, 1988a), 1-bromo-2-chloroethane (Barbash and Reinhard, 1992b, Barbash 1993) and 1,2-dibromo-3-propane (DBCP) (Burlinson et al., 1982; Deeley et al., 1991).

A general form of the rate law for the disappearance of a haloaliphatic substrate (RX) from aqueous solution via S_N2 and E2 processes due to attack by Nu may be written as:

$$-d[RX]/dt = k_{Nu}^{obs}[RX] \quad (24.13)$$

where

k_{Nu}^{obs} = the sum of several terms accounting for the competing reactions.

$$k_{Nu}^{obs} = \sum_{i=1}^{n} k_i^s[Nu_i] + \sum_{j=1}^{m} k_j^e[Nu_j] \quad (24.14)$$

where

k_i^s = the second-order rate constant for nucleophilic substitution by Nu_i, the ith of n nucleophiles;
k_j^e = the second-order rate constant for dehydrohalogenation catalyzed by Nu_j, the jth of m nucleophiles capable of effecting dehydrohalogenation; and
[Nu] = the nucleophile concentration.

The superscripts "obs," "s" and "e" denote rate constants for the observed overall disappearance, substitution and dehydrohalogenation of the substrate, respectively.

For hydrolysis reactions, the superscript "h" may be used instead of "s." If water is the only nucleophile and/or base which is present in kinetically significant concentration, k^{obs} is given by

$$k^{obs} = k^{h}_{H_2O} + k^{e}_{H_2O} \quad (24.15)$$

where $k^{h}_{H_2O}$ and $k^{e}_{H_2O}$ are pseudo-first-order rate constants for reactions with water. If the reactions with both OH^- and water are significant and hydrolysis and dehydrohalogenation occur simultaneously, k^{obs} is given as:

$$k^{obs} = k^{h}_{H_2O} + k^{e}_{H_2O} + \left\{k^{h}_{OH^-} + k^{e}_{OH^-}\right\}\left[OH^-\right] \quad (24.16)$$

The half-life in water represents an upper estimate of a compound's half-life in aqueous solution, assuming that partitioning into hydrophobic solids or the gas phase does not remove the HAC from the aqueous environment. The presence of other nucleophiles—either naturally occurring or added in laboratory systems for pH or redox control—can shorten the half-lives significantly, or even shift reaction pathways, due to competitive nucleophilic substitution reactions or catalytic effects (Barbash and Reinhard, 1989a, 1992a).

The reactivities of nucleophiles and substrates have been quantitatively characterized using the Swain-Scott (1953) equation. These authors have correlated the reactivities of several nucleophiles using the relationship:

$$\log(k_i/k_o) = s\ n \quad (24.17)$$

where

k_i	=	the reaction rate with nucleophile "i";
k_o	=	the reaction rate with water; and
"s" and "n"	=	empirical constants for the substrate and the nucleophile, respectively.

By definition, n = 0.00 for water as the nucleophile, and, in the study of Swain and Scott (1953), s = 1 for methyl bromide as the substrate. Table 24.3 lists "n" values of environmentally relevant nucleophiles compiled by Koskikallio (1972) and Haag and Mill (1988a), and based on a variety of different standard substrates.

The relative sensitivity of a substrate to base-catalyzed or neutral reactions in water is indicated by the pH value at which the base-catalyzed and the neutral reactions are equally fast (Mabey and Mill, 1978). This pH is denoted as I_{NB} and obtained by equating the water- and OH^--promoted terms in Equation 24.16. By expressing $[OH^-]$ in terms of the ionization product of water, K_w, and setting the activity coefficients for H^+ and OH^- to unity, solving for I_{NB} yields:

$$I_{NB} = -\log \frac{KW\left(k^{h}_{OH^-} + k^{e}_{OH^-}\right)}{k^{h}_{H_2O} + k^{e}_{H_2O}} \quad (24.18)$$

As indicated in Table 24.4, for most HACs which react predominantly by hydrolysis, the I_{NB} is above 10, with the exception of chloroform. For the HACs which react via elimination, the I_{NB} is generally lower, ranging from 4.5 to 8.1. Acid catalysis of HAC hydrolysis does not seem to occur (Mabey and Mill, 1978).

Table 24.4 combines data reported by Jeffers et al. (1989) with those compiled by Mabey and Mill (1978) and several other authors. The HACs are arranged in order of decreasing reduction potential and, therefore, in order of decreasing tendency to accept an electron. Exceptions are tetrachloroethylene, trichloroethylene and DBCP, which have not yet been evaluated with respect to their reduction potential (Table 24.4a).

Table 24.3. Nucleophilic Constants for Environmentally Relevant Nucleophiles.

Nucleophile	n (av)[a]	n-hexBr[b]
S_n^{2-} (n=4,5)		7.2
SO_3^{2-}	5.34	5.4
HS^-		5.4
$S_2O_3^{2-}$	5.18	5.1
HO^-	3.59	3.3
NO_2^-	3.2	3.3
Br^-	2.79	
HCO_3^{2-}	2.23	1.9
SO_4^{2-}	1.88	
Cl^-	1.82	
$NO3^-$	0.99	
H_2O	0.00	0.00

[a] Averaged values from three different reference substrates (methyl perchlorate, methyl benzenesulfonate and methyl nitrate), (Koskikallio, 1972).
[b] n-Hexyl bromide used as reference substrate (Haag, 1988a).

The data indicate that half-lives in water decrease with decreasing extent of halogenation, except in the cases where dehydrohalogenation is possible, as noted previously (Vogel et al., 1987, Jeffers et al., 1989). Hexachloroethane, carbon tetrachloride, chloroform, bromoform and the alkenes PCE and TCE have half-lives greater than one hundred years (at 25°C) and, for practical purposes, may be considered to be inert in water under most environmental conditions. For carbon tetrachloride, two different rate laws are reported, but half-lives are relatively long in both cases. Of this group, hexachloroethane, carbon tetrachloride and, to a lesser extent, bromoform have relatively high estimated one-electron reduction potentials (Table 24.1). Therefore, these compounds are more likely to transform via dissociative ET, assuming a suitable electron donor is present. Chloroform, on the other hand, is relatively inert in reactions with water and nonreactive with reductants (Curtis, 1991).

Most HACs with bromine or chlorine atoms in vicinal (α,β) or geminal (both α) positions hydrolyze within a relatively narrow time window [half-lives between 19 yr (DBCP) and 72 yr (1,2-dichloroethane)]. 1,1,1-Trichloroethane and 1,2-dibromoethane, with half-lives of 1.1 yr and 4.6 yr, respectively, are exceptions. Dehydrohalogenation is relatively fast for several HACs (e.g., pentachloro- and tetrachloroethane), but the olefinic products formed are both stable and undesirable. The R-X compounds with R = methyl or ethyl hydrolyze relatively rapidly, with half-lives significantly less than a year. An exception is ethyl bromide, which is reported to have a half-life of 4.4 yr (Mabey and Mill, 1978). As noted earlier, both the overall rates of dehalogenation and the yields of alkene products may increase in the presence of pH buffers (Burlinson et al., 1982; Deeley et al., 1991; Barbash and Reinhard, 1989a, 1992a).

The Influence of Temperature

Because in most cases rates at ambient temperatures are slow and thus difficult to measure, rate constants are typically obtained at high temperatures (> 50°C). The Arrhenius equation is then used to extrapolate to ambient temperatures. The Arrhenius equation is

$$\log_{10}(k) = \log_{10}(A) - \frac{E_a}{2.303 \times RT} \tag{24.19}$$

where A and E_a are the pre-exponential factor and activation energy, respectively, R is the universal gas constant, and T is temperature (expressed in degrees Kelvin). If E_a is determined from Arrhenius

Table 24.4. Neutral and Basic Hydrolysis Constants, I_{NB}, and $t_{1/2}$ of Selected HACs at 25°C.

Substrate	$k_{H_2O}^{obs}$ min^{-1}	k_{OH}^{obs} $min^{-1}M^{-1}$	I_{NB} pH	$t_{1/2}$ (pH7)	E[1)]	Ref.[2)]
CCl_3-CCl_3	0.00	7.2×10^{-9}		1.8×10^9yr		a
CCl_2H-CCl_3	5.9×10^{-7}	1.3×10^3	4.7	4 d	ma	a
CCl_2H-CCl_3	0.0	1.6×10^3		3 d	ma	b
CCl_4	3.3×10^{-8}	0.0		40 yr		a
CCl_4	sec. order			7,000 yr		c
$CHBr_3$	1.9×10^{-9}			690yr		c
$CClH_2$-CCl_3	2.6×10^{-8}	2.1×10^{-2}	8.1	47 yr	ma	a
$CHCl_3$	1.9×10^{-10}	5.2×10^{-3}	6.6	1,800 yr		a
CCl_3-CH_3	1.2×10^{-6}	0.0		1.1 yr	mi	a
$CHCl_2$-$CHCl_2$	9.7×10^{-9}	32	4.5	150 d	ma[d)]	a,d
CH_3-$CHCl_2$	2.1×10^{-8}	7.2×10^{-7}	12.5	61 yr	mi	a
CH_2BrCl	3×10^{-8}			44 yr		c
CH_2Cl_2	1.9×10^{-9}	1.2×10^{-6}	11.2	68 yr		c,e
CH_2Br_2	7.2×10^{-9}			18 yr		c
CH_3Br	2.4×10^{-5}	8.5×10^{-3}	11.5	20 d		c
CH_2Cl-CH_2Cl	2.1×10^{-8}			63 yr	tr	f
CH_2Cl-CH_2Cl	1.8×10^{-8}	1.0×10^{-4}	10.2	72 yr		a
CH_2Br-CH_2Br	2.8×10^{-7}	0.0		4.6 yr	tr	f
CH_2Br-CH_3	3.0×10^{-7}			4.4 yr		c
CH_3Cl	1.4×10^{-6}	3.7×10^{-4}	11.6	34 d		c

Table 24.4a. Compounds Not Ranked in Order of Estimated Reduction Potential.

DBCP	5.5×10^{-9}	0.34	6.2	33 yr	ma	g
DBCP	0	0.7		19 yr	ma	h
TCE	0	1.1×10^{-5}		$1.2\ 10^6$		a
PCE	0	1.4×10^{-8}		$9.6\ 10^8$		a

1) *Elimination yields:* ma: major (>50%); mi: minor (10–50%); tr: trace (<10%). Yield may depend on system studied (see text).
2) *References:* a. Jeffers et al., 1989; b. Roberts and Gschwend, 1991; c. Mabey and Mill, 1978; d. Haag and Mill, 1988b; e. Fells and Moelwyn-Hughes, 1958; f. Barbash and Reinhard, 1989a; g. Burlinson et al., 1982; h. Deeley et al., 1991.

plots where k is the overall or observed reaction rate and where several reactions occur competitively, the E_a is termed the "apparent" E_a value. A change of E_a with temperature could indicate a shift in relative importance among competing mechanisms.

In most reports of HAC transformation in water, E_a was found to be constant within the experimental error over the relatively narrow temperature ranges investigated. Few studies have attempted to verify whether the extrapolated values agree with those measured at ambient temperature. Deeley et al. (1991) have studied DBCP transformation over the temperature range from 21°C to 55°C. They used γ-irradiated soil samples to prevent biological transformation, and flame-sealed glass ampules to avoid losses from the system. The overall rate calculated from DBCP disappearance at 21°C was faster than that predicted from high temperature data, and E_a appeared to decrease with decreasing temperature. A possible explanation for this observation was that a different mechanism predominated at the lower temperatures (Deeley et al., 1991).

Activation energies for substitution reactions are dependent on the type of nucleophile involved. Barbash and Reinhard (1989a) have determined E_a values for the reactions involving the substrates EDB and 1,2-DCA, and the nucleophiles HS^- and H_2O. E_a for the reaction with HS^- were significantly smaller than for reaction with H_2O for both substrates. E_a values for the reaction of HS^- with 1,2-dichloroethane and 1,2-dibromoethane were 22 kcal/mol and 20 kcal/mol, respectively, compared to the E_a for the overall reaction with water, which was 26 kcal/mol for both substrates. The lower E_a values would be expected for softer nucleophiles reacting with the relatively soft sp^3-carbon compound to halogen (Pearson, 1963).

Homogeneous Solution Reactions Involving Bisulfide

Reactions of HACs with reduced sulfur nucleophiles have been summarized by Barbash and Reinhard (1989b). Because HS^- is the most abundant sulfur nucleophile in hypoxic environments, most of the recent studies of nucleophilic substitutions of HACs have focused on reactions involving this anion (Schwarzenbach et al., 1985; Haag and Mill, 1988a; Barbash and Reinhard, 1989a and 1992b; Roberts and Gschwend, 1992).

The importance of nucleophilic substitution reactions relative to reactions with water (hydrolysis and dehydrohalogenation) can be evaluated by defining a critical Nu concentration, $[Nu]_c$, at which reactions with Nu and water are equally fast (Barbash and Reinhard, 1989a,b):

$$K_{Nu}\,[Nu]_c = k_{H_2O}^{obs} \tag{24.20}$$

Since both k_{Nu} and $k_{H_2O}^{ov}$ are substrate- and temperature-dependent, $[Nu]_c$ is also substrate- and temperature-dependent. In Table 24.5, $[HS^-]_c$ values are compiled for different substrates. It is evident that for the 1,2-dihaloethanes and dihalomethanes listed, reactions with HS^- may be significant in hypoxic groundwaters containing dissolved sulfide, in which the total sulfide concentrations are typically within the range of $1\text{–}10^3$ μM (Barbash and Reinhard, 1989b). These data also indicate that the difference in nucleophilicity between HS^- and H_2O becomes more pronounced when the leaving group on the substrate is substituted with a better one (in this case, chlorine replaced by bromine).

The bisulfide anion is a soft Lewis base (Pearson, 1963) with a pK_i of 7.01 at 25°C (Millero, 1986), reacting almost exclusively via substitution with HACs (Schwarzenbach et al., 1985; Barbash and Reinhard, 1989a and 1992b). This displacement gives rise to the corresponding thiol, as illustrated by Reaction 21:

$$RX + HS^- \rightarrow RSH + X^- \tag{24.21}$$

Thiolate anions are more powerful nucleophiles than HS^- (Barbash and Reinhard, 1989b). Thus, RS^- may also displace a halogen from a halo-aliphatic compound, giving rise to a thioether, e.g.,

$$RS^- + R'X \rightarrow RSR' + X^- \tag{24.22}$$

Other reactions in which the thiolate anion may engage include oxidations (yielding disulfides) or, if another good leaving group is present on the molecule, intramolecular displacement (giving rise to a cyclic thioether). Various combinations of substitution and oxidation reactions may generate complicated assemblages of thiols, thioethers and polysulfides (Schwarzenbach et al., 1985; Weintraub et al., 1986). Oxidation of HS^- may lead to polysulfides, which are also more powerful nucleophiles than HS^-, as was demonstrated by Haag and Mill (1988a) (see Table 24.3). The reactions of dihalomethanes with HS^- as the nucleophile were studied in detail by Roberts et al. (1992). Products formed included several di- and poly(thiomethylene) compounds, but no cyclic products.

The Effect of Solids on HAC Reactions

The effect of solids on HAC transformation rates has been studied in a number of cases. Haag and Mill (1988b) have measured the half-lives with and without aquifer solids present and have concluded that for 1,1,2,2-tetrachloroethane, 1,1,1-trichloroethane and 1,2-dibromoethane the half-lives and products were not significantly different with and without subsurface sediment present. Similarly for 1,2-dibromo-3-chloroethane (DBCP), Deeley et al. (1991) found no effect of the subsurface sediments. However, they observed that more hydrolysis products appeared in the groundwater than in phosphate-buffered controls.

Solids appear to have an effect under anaerobic conditions where HS^- is present, however. The transformation of carbon tetrachloride by $HS^-(aq)$ was found to be more rapid in the presence of the

Table 24.5. Nucleophilic Rate Constants of HS^- and $[Nu]_c$ Values for Some HACs at 25°C.

Substrate	k_{HS^-} [a] $M^{-1}min^{-1}$	$[Nu]_c$ μM	Ref.
CH_2Cl_2	1.1×10^{-3}	17	a
CH_2BrCl	4.4×10^{-3}	7	
CH_2Br_2	3.15×10^{-3}	2	a
$CH_2Cl\text{-}CH_2Cl$	3.5×10^{-4}	100	b
$CH_2Br\text{-}CH_2Br$	2.4×10^{-2}	20	b

[a] Defined as in Equation 24.21: k_{HS^-} $[HS^-] = k$.

References: a. Roberts et al. 1992; b. Barbash and Reinhard, 1989a.

minerals biotite, vermiculite and pyrite than in homogeneous systems (Kriegman-King and Reinhard, 1991, 1992, 1994). In the biotite systems, approximately 80–85% of the carbon tetrachloride was transformed to CO_2 via the measured intermediate CS_2, whereas chloroform comprised 5–15% of the products. At 25°C, the half-lives of CCl_4 in the presence of 1 mM HS^- were calculated to be 2600, 160, and 50 days for the homogenous, vermiculite and biotite systems, respectively.

SUMMARY AND CONCLUSIONS

Under groundwater conditions, HACs are transformed via three basic processes: nucleophilic substitution (including hydrolysis), dehydrohalogenation, and reduction. The importance of these three processes depends on the substrate structure and the geochemical conditions. Important structural features and geochemical characteristics that affect half-lives include the extent of halogenation and the presence of hydrogen atoms, the redox potential of the solution, solution pH and the presence of other reactive species, such as reduced sulfur nucleophiles.

Theoretical rate data for reaction via ET have been developed by Curtis (1991) and Curtis and Reinhard (1992) based on a method presented by Eberson (1987) incorporating the Nernst, Marcus and the Eyring equations. The approach appears useful for rank-ordering HACs with respect to their reactivity in redox reactions. The theoretical reduction rates were compared with measured hydrolysis and dehydrohalogenation data. In accordance with previous analyses (e.g., Vogel et al., 1987) the principal factor affecting the HAC transformation rates was found to be the average oxidation number of HAC compounds, a number which increases with the degree of halogenation. Highly halogenated HACs are readily reduced unless they contain a hydrogen that makes the compound susceptible to dehydrohalogenation. For many HACs that are susceptible to slow hydrolysis, nucleophilic substitution by reduced sulfur nucleophiles—if the latter are present—can be a significant alternate pathway, leading to substantially shorter half-lives. However, the sulfur products formed could be of potential environmental concern. Other examples of transformations yielding products of potential concern include the formation of 1,1-DCE from 1,1,1-TCA, which is resistant to attack by HS^- (Haag and Mill, 1988a; Barbash and Reinhard, 1992b) and the formation of vinyl halides from the dehalogenation of dihaloethanes (Barbash and Reinhard, 1989a; 1992a and b; Barbash, 1993).

Finally, it should be noted that the rates of many of these reactions, even though they are slow and difficult to measure under ambient conditions, are significant when compared with the duration of cleanup efforts at groundwater contamination sites, which usually take several years. Inadequacies in the currently available database on abiotic contaminant transformations often prevent investigators from accurately assessing the significance of abiotic processes at a given site. One consequence of these data gaps may be the design of inappropriate strategies for aquifer remediation if the time scales of the transformations are substantially shorter than that of the cleanup effort. In such cases, an inadequate understanding of abiotic transformation mechanisms may present difficulties if the products of these reactions are considerably less amenable than the original contaminant to removal using the treatment system employed.

ACKNOWLEDGMENTS

Work described in this chapter has been supported in part by the Robert S. Kerr Environmental Research Laboratory of the U.S. Environmental Protection Agency, Ada, Oklahoma, through CR-816776 (Stephen R. Hutchins, Project Officer). Although the research described in this chapter has been funded by the U.S. Environmental Protection Agency, it does not necessarily reflect the views of this agency, nor should any official endorsements be inferred.

REFERENCES

Barbash, J.E. and M. Reinhard. 1989a. Abiotic Dehalogenation of 1,2-Dichloroethane and 1,2-Dibromoethane in Aqueous Solution Containing Hydrogen Sulfide. *Environ. Sci. Technol.,* **23**:1349–1358.

Barbash, J.E. and M. Reinhard. 1989b. Reactivity of Sulfur Nucleophiles Toward Halogenated Organic Compounds in Natural Waters. In Saltzman, E.S. and W.J. Cooper, eds., *Biogenic Sulfur in the Environment;* ACS Symposium Series 39; American Chemical Society: Washington, DC, pp. 101–138.

Barbash, J.E. and M. Reinhard. 1992a. The Influence of pH Buffers and Nitrate Concentration on the Rate and Pathways of Abiotic Transformation of 1,2-Dibromoethane (EDB) in Aqueous Solution. Preprint Extended Abstract, presented before the Division of Environmental Chemistry, American Chemical Society, San Francisco, CA, pp. 674–677.

Barbash, J.E. and M. Reinhard. 1992b. Abiotic Reactions of Halogenated Ethanes and Ethylenes in Buffered Aqueous Solutions Containing Hydrogen Sulfide. Preprint Extended Abstract, presented before the Division of Environmental Chemistry, American Chemical Society, San Francisco CA, April 5–10, 1992, pp. 670–673.

Barbash, J.E. 1993. Abiotic Reactions of Halogenated Ethanes and Ethylenes with Sulfide, Nitrate and pH-Buffer Anions in Aqueous Solution. Ph.D. thesis, Stanford University, Stanford, CA.

Bratsch, S.G. 1989. Standard Electrode Potentials and Coefficients in Water at 298.15K. *J. Phys. Chem. Ref. Data,* **18**:1–21.

Burlinson, N.E., L.A. Lee, and D.H. Rosenblatt. 1982. Kinetics and Products of Hydrolysis of 1,2-Dibromo-3-chloropropane. *Environ. Sci. Technol.,* **16**:627–632.

Carreira, L.H. and N.L. Wolfe. 1992. Isolation and Characterization of a Sediment Component that Reduces 2,4,6-Trinitritoluene. Preprint Extended Abstract, presented before the Division of Environmental Chemistry, American Chemical Society, San Francisco, CA, pp. 511–512.

Clark, W.M. 1960. *Oxidation-Reduction Potentials of Organic Systems*, The Williams and Wilkins Co., Baltimore, MD.

Criddle, C.S. and P.L. McCarty. 1991. Electrolytic Model System for Reductive Dehalogenation in Aqueous Solution. *Environ. Sci. Technol.,* **25**:973–978.

Curtis, G.P. and M. Reinhard. 1992. Reductive Dehalogenation of Hexachloroethane, Carbon Tetrachloride, and Bromoform by Anthrahydroquinone Disulfonate and Humic Acid. Preprint Extended Abstract, presented before the Div. of Environ. Chem., American Chem. Soc., San Francisco, CA, pp. 499–501.

Curtis, G.P. 1991. Reductive Dehalogenation of Hexachloroethane and Carbon Tetrachloride by Aquifer Sand and Humic Acid. Ph.D. thesis, Stanford University, Stanford, CA.

Deeley, G.M., M. Reinhard, and S.M. Stearns. 1991. Transformation and Sorption of 1,2-Dibromo-3-chloropropane in Subsurface Samples Collected at Fresno, CA. *J. Environ. Qual.* **20**:547–556.

Eberson, L. 1982. Electron Transfer Reactions in Organic Chemistry II. An Analysis of Alkyl Halide Reduction by Electron Transfer Reagents on the Basis of the Marcus Theory. *Acta Chem. Scand.,* B **36**:533–543.

Eberson, L. 1985. The Marcus Theory of Electron Transfer, a Sorting Device for Toxic Compounds. In *Advances in Free Radical Biology & Medicine*, Vol. 1, Pergamon Press, New York, NY, pp. 19–90.

Eberson, L. 1987. *Electron Transfer Reactions in Organic Chemistry*. Springer, Berlin.

Fells, I. and E.A. Moelwyn-Hughes. 1958. The Kinetics of the Hydrolysis of Methylene Dichloride, *J. Chem. Soc.* (B) 1326.

Gantzer, C.J. and L.P. Wackett. 1991. Reductive Dechlorination Catalyzed by Bacterial Transition-Metal Coenzymes. *Environ. Sci. Technol.,* **25**:715–722.

Haag, W.R. and T. Mill. 1988a. Some Reactions of Naturally Occurring Nucleophiles with Haloalkanes in Water. *Environ. Tox Chem.,* **7**:917–924.

Haag, W.R. and T. Mill. 1988b. Effect of Subsurface Sediments on Hydrolysis Reactions. *Environ. Sci. Technol.,* **22**:658–663.

Jafvert, C.T. and N.L. Wolfe. 1987. Degradation of Selected Halogenated Ethanes in Anoxic Sediment Water Systems. *Environ. Tox. Chem.,* **6**:827–837.

Jeffers, P.M., L.M. Ward, L.M. Woytowitch, and N.L. Wolfe. 1989. Homogenous Hydrolysis Rate Constants for Selected Chlorinated Methanes, Ethanes and Propanes. *Environ. Sci. Technol.,* **23**:965–969.

Klecka, G.M. and S.J. Gonsior. 1984. Reductive Dechlorination of Chlorinated Methanes and Ethanes by Reduced Iron(II) Porphyrins. *Chemos.,* **13**:391–402.

Koskikallio, J. 1972. Nucleophilic Reactivity Part VI. An Ion-Pair Mechanism for Reactions of Methylsulphonate with Nucleophiles in Water. *Acta Chem. Scand.,* **26**:1201–1208

Kray, W.C., Jr. and C.E. Castro. 1964. The Cleavage of Bonds by Low Valent Transition Metal Ions. The Homogeneous Dehalogenation of Vicinal Dihalides by Chromous Sulfate. *J. Am. Chem. Soc.,* **86**:4603–4608.

Krone, U.E., R.T. Thauer, and H.P.C. Hogenkamp. 1989. Reductive Dehalogenation of Chlorinated C_1-Hydrocarbons Mediated by Corrinoids. *Biochemistry,* **28**:4908–4914.

Kriegman-King, M.R. and M. Reinhard. 1991. Reduction of Hexachloroethane and Carbon Tetrachloride at Surfaces of Biotite, Vermiculite, Pyrite and Marcasite. In R.A. Baker, ed. *Organic Substances and Sediments in Water*, Vol. 2, Lewis Publishers, Chelsea, MI, pp. 349–364.

Kriegman-King, M.R. and M. Reinhard. 1992. Transformation of Carbon Tetrachloride in the Presence of Sulfide, Biotite, and Vermiculite. *Environ. Sci. Technol.,* **26**:2198–2206.

Kriegman-King, M.R. and M. Reinhard. 1994. Transformation of Carbon Tetrachloride by Pyrite in Aqueous Solution. *Environ. Sci. Technol.,* **28**:692–700.

Mabey, W. and T. Mill. 1978. Crit. Rev. of Hydrolysis of Organic Compounds in Water under Environmental Conditions. *J. Phys. Chem. Ref. Data,* **7**:383–415.

March, J. 1985. *Advanced Organic Chemistry*, John Wiley & Sons, Inc., New York, NY.

Millero, F. 1986. The Thermodynamics and Kinetics of the Hydrogen Sulfide System in Natural Waters. *Mar. Chem.,* **18**:121–147.

Mohn, W.W. and J.M. Tiedje. 1992. Microbial Reductive Dehalogenation. *Microbiol. Rev.* **56**:482–507.

Pearson, R.G. 1963. Hard/Soft Acids and Bases *J. Am. Chem. Soc.,* **85**:3533–3539.

Roberts, L.A. and P.M. Gschwend. 1991. Mechanism of Pentachloroethane Dehydrochlorination to Tetrachloroethane. *Environ. Sci. Technol.,* **25**:76–86.

Roberts, L.A., P.N. Sanborn, and P.M. Gschwend. 1992. Nucleophilic Substitution Reactions of Dihalomethanes with Hydrogen Sulfide Species. *Environ. Sci. Technol.,* **26:**2263–2274.

Schwarzenbach, R.P., W. Giger, C. Schaffner, and O. Wanner. 1985. Groundwater Contamination by Volatile Halogenated Alkanes: Abiotic Formation of Volatile Sulfur Compounds Under Anaerobic Conditions. *Environ. Sci. Technol.,* **19**:322–327.

Skogerboe, R.K. and S.A. Wilson. 1981. Reduction of Ionic Species by Fulvic Acid. *Anal. Chem.,* **53**:228–232.

Stumm, W. and J.J. Morgan. 1981 *Aquatic Chemistry*, Wiley, New York, NY.

Swain, C.G. and C.B. Scott. 1953. Quantitative Correlation of Relative Reactivities. Comparison of Hydroxide Ion with Other Nucleophilic Reagents toward Alkyl Halides, Esters, Epoxides and Acylhalides *J. Amer. Chem. Soc.,* **75**:141–147.

Szilagyi, M. 1971. Reduction of Fe^{3+} Ion by Humic Acid Preparations. *Soil Sci.,* **111**:233–235.

Szilagyi, M. 1973. The Redox Properties and the Determination of the Normal Potential of the Peat-water System. *Soil Sci.,* **115**:434–437.

Taylor, P.H., B. Dellinger, and D.A. Tirey. 1991. Oxidative Pyrolysis of CH_2Cl_2, $CHCl_3$, and CCl_4—I. Incineration Implications. *Int. J. Chem. Kin.,* **23**:1051–1074.

Vogel, T.M. and P.L. McCarty. 1987a. Abiotic and Biotic Transformation of 1,1,1-Trichloroethane Under Methanogenic Conditions. *Environ. Sci. Technol.,* **21**:1208–1213.

Vogel, T.M., C.S. Criddle, and P.L. McCarty. 1987b. Transformation of Halogenated Aliphatic Compounds. *Environ. Sci. Technol.,* **21**:722–736.

Wade, R.S. and C.E. Castro. 1973. Oxidation of Heme Proteins by Alkyl Halides, *J. Am. Chem. Soc.,* **95**:226–231.

Weintraub, R.A., G.W. Jex, and H.A. Moye. 1986. In W.Y. Garner, R.C. Honeycutt, and H.N. Nigg, eds. ACS Symposium Series No. 315, American Chemical Soc. Washington, DC, pp. 294–310.

White, A.F. and A. Yee. 1985. Aqueous Oxidation-Reduction Kinetics Associated With Coupled Electron-Cation Transfer from Iron-Containing Silicates. *Geochem. Cosmochim. Acta,* **49**:1263–1275.

Wilson, S.A. and J.H. Weber. 1979. An EPR Study of the Reduction of Vanadium (V) to Vanadium (IV) by Fulvic Acid. *Chem. Geol.,* **26**:345–354.

Chapter 25

Natural Aerobic Biological Attenuation

H.S. Rifai

INTRODUCTION

Natural aerobic biodegradation is the process by which indigenous microorganisms break down chemicals present in the soil or groundwater, using oxygen as an electron acceptor, into other products and sometimes to carbon dioxide and water. In this chapter, we review the current state-of-the-art of quantifying aerobic biodegradation in laboratory and field demonstrations. In addition, we discuss proposed field protocols and modeling tools for assessing passive aerobic biodegradation as a remediation technology for hazardous waste sites.

Generally, there are six basic requirements for biodegradation:

1. *The presence of the appropriate organisms*—microorganisms in groundwater aquifers are almost invariably bacteria. It is preferable to have indigenous microorganisms at a field site which are capable of degrading the contaminants at the site. Although there has been much work done on inocculating aquifers with natural or genetically engineered organisms, the success of these experiments has not been fully established.
2. *Energy source*—organic carbon is required as an energy source and is used by the organisms for cell maintenance and growth. The organic carbon is transformed into inorganic carbon, energy and electrons.
3. *Carbon source*—about 50% of the dry weight of bacteria is carbon. Organic chemicals serve as both carbon and energy sources. As a carbon source, organic carbon is used in conjunction with energy to generate new cells.
4. *Electron acceptor*—some chemical must accept the electrons released from the energy source. Typical acceptors include O, NO, SO, and CO.
5. *Nutrients*—the required nutrients include: nitrogen, phosphorus, calcium, magnesium, iron and trace elements. Nitrogen and phosphorus are needed in largest quantities. Elemental ratios of nitrogen and phosphorus in cell dry mass are given by CHONP, so that nitrogen is approximately 12% and phosphorus is 2–3% of dry mass.
6. *Acceptable environmental conditions*—such as pH, the presence of toxic chemicals, and temperature.

The dilemma that scientists and engineers are faced with today is how to verify and quantify natural aerobic biodegradation at a field site. The current approach involves conducting a series of laboratory studies of biodegradation and site-specific field characterization studies. The interest in understanding biodegradation has also prompted researchers to develop models to simulate the process. The implementation of this approach has differed from one site to the next, and has not been adequately developed as a standard procedure.

Biodegradation of Organics in the Subsurface

The most aerobically biodegradable compounds in the subsurface have been petroleum hydrocarbons such as gasoline, crude oil, heating oil, fuel oil, lube oil waste, and aviation gas (Lee et al., 1987). In particular, benzene, toluene, xylene, and ethylbenzene are very degradable. Other compounds such as alcohols (isopropanol, methanol, ethanol), ketones (acetone, methyl ethyl ketone) and glycols (ethylene glycol) are also aerobically biodegradable. Recent studies have expanded the list of aerobically degraded compounds to include methylated benzenes, chlorinated benzenes (Kuhn et al., 1985), chlorinated phenols (Suflita and Miller, 1985), methylene chloride (Jhaveri and Mazzacca,

1983), naphthalene, methylnaphthalenes, dibenzofuran, fluorene, and phenanthrene (Wilson et al., 1985; Lee and Ward, 1985).

LABORATORY TESTING

Generally, in a laboratory testing program for a field site, core samples of subsurface solids and groundwater samples are aseptically collected from contaminated areas. The objectives of the laboratory testing are to determine the chemistry of soils and groundwater, to verify the presence of microorganisms, and to measure the response of native microorganisms to stimulation by the addition of contaminants and/or nutrients.

Microbial investigations of the subsurface have revealed that all aquifers examined thus far support a microbial population. Such studies include direct microscopic, cultivation, metabolic, and biochemical evidence for microorganisms in aseptically obtained aquifer material. Typical microbial numbers range from 1×10^5 to 1×10^7 cells per gram dry weight (cells/gdw) or cells/mL (Ghiorse and Balkwill, 1983; Wilson et al., 1983; Balkwill and Ghiorse, 1985; and Beeman and Suflita, 1987). Relatively high numbers of microorganisms have been detected in both contaminated and pristine aquifers of varying depth and geological composition. The picture that emerges from microbiological studies is that subsurface microorganisms tend to be small, capable of response to the influx of nutrients, and primarily attached to solid surfaces (Suflita, 1989).

Sampling Techniques

To make valid interpretations of the data resulting from subsurface investigations, the samples analyzed must not be contaminated with nonindigenous microorganisms. Potential sources of contamination include organisms present on drilling machinery, in surface soil layers, and in drilling muds. Most current sampling efforts rely on the recovery and subsequent removal of the outer few centimeters and the top and bottom portions of aquifer cores because of possible contamination by nonindigenous bacteria. Only the center portions of an aquifer core are used for analysis.

Ideally, the dissection process of the core occurs as soon as possible after the core is removed from the ground. In the field, a sterilized paring device is used in the dissection process (McNabb and Mallard, 1984). The paring device has an inner diameter that is smaller than the diameter of the core itself. The aquifer material is extruded out of the core barrel used for sampling.

Methods for Detection, Enumeration, and Determination of Metabolic Activity

The main questions that need to be answered about subsurface microorganisms in relation to biodegradation of contaminated soils and groundwater can be summarized as follows:

1. Are the microorganisms metabolically active?
2. How diverse is their metabolism?
3. What factors serve to stimulate and/or limit their growth and activity?
4. Can we take advantage of their metabolism for aquifer remediation?

There is no lack of methods that might be applied to detect microorganisms and estimate their biomass and metabolic activities in subsurface samples. The applicable technologies include: microscopic methods, cultural methods, biochemical indicators, radioscopic methods, and microcosms. The presence of microorganisms in subsurface samples and some of their distinguishing morphological features can be established by direct light and electron microscopy. Generally, the water or sediment sample can be stained with a fluorescent dye to improve the visibility of the cells in the light microscope.

Standard cultural procedures which include plate counts, most probable number counts, and enrichment culture procedures, are more sensitive than microscopic methods of detection. Biochemical indicators of metabolic activity and detection of specific molecular markers in natural samples such as ATP (adenosine triphosphate) provide very sensitive estimates for biomass and activity. These methods, however, cannot show directly the presence or activity of cells, and therefore results should be interpreted cautiously. Ideally, biochemical indicator methods should be supported by direct microscopic and cultural methods applied to the same samples.

Radioisotope-labeled substrate transformation and uptake methods provide another sensitive means of measuring metabolic activity and growth of microorganisms in subsurface samples. Ideally, the rate of labeled substrate transformation should be compared to total direct counts, plate counts, or another biomass indicator to normalize the rates to population density (Ghiorse and Wilson, 1988).

Microcosms

Pritchard and Bourquin (1984) defined a microcosm study as "an attempt to bring an intact, minimally disturbed piece of an ecosystem into the laboratory for study in its natural state. It is the establishment of a physical model or simulation of part of the ecosystem in the laboratory within definable physical and chemical boundaries under a controlled set of experimental conditions." Microcosms have been used extensively to determine the biodegradability of organic contaminants at the laboratory scale and under the influence of the site-specific physical, chemical and hydrological characteristics.

Microcosms range from simple batch incubation systems to large and complex flow-through devices. Figure 25.1 shows three different types of microcosms designed by Dunlap et al. (1972), Bengtsson (1981) and Wilson et al. (1981). Microcosms are used to: (1) identify biodegradable pollutants; (2) determine the metabolic pathways of biotic or abiotic transformations; and (3) relate the decay of a particular pollutant, if possible, in a microcosm to the rate of biotransformation *in situ* without relying on indirect measures of microbial biodegradation for rate prediction (Suflita, 1989).

Microcosms offer several experimental advantages. For example, results from microcosm studies are replicable and allow appropriate controls to be employed. Furthermore, microcosms provide a time-efficient method for evaluating biodegradation potential at a field site. Conversely, microcosms suffer from some limitations, mainly that microcosms can disturb the normal structure of an ecosystem and generally possess high surface-to-volume ratios. These disturbances may influence biodegradation estimates and prevent direct extrapolation of microcosm-determined biodegradation rates to the field scale.

FIELD DEMONSTRATIONS

A number of field studies of natural attenuation have been presented in the general literature over the past few years. These studies are reviewed to illustrate the type of field measurements that could be made to support their laboratory counterparts.

The United Creosoting Company (UCC) Site in Conroe, Texas

An injection-production test was conducted at the United Creosoting Company (UCC) Superfund site in Conroe, Texas (Borden and Bedient, 1987). The aim of the field test was to obtain an accurate evaluation of the effects of adsorption and biotransformation on the fate and transport of the organic contaminants present at the site. The United Creosoting Company (UCC) site was operated as a wood preserving facility from 1946 to 1972.

Wastes generated by the wood creosoting operation were disposed of in two unlined ponds (see Figure 25.2). These wastes were composed of predominantly polycyclic aromatic hydrocarbons which originated in the creosote and pentachlorophenol (PCP). In 1972, the facility was closed and the land sold off for use by a small commercial operation and later subdivided for a housing development. Monitoring of the site has shown elevated levels of organic contaminants in the soil and groundwater. In addition to the organic wastes, elevated levels of chloride are present in the groundwater.

The geology at the UCC site is heterogeneous with numerous changes between relatively clean well sorted sand and sandy clay. Two separate water-bearing zones have been identified: 1) a lower-

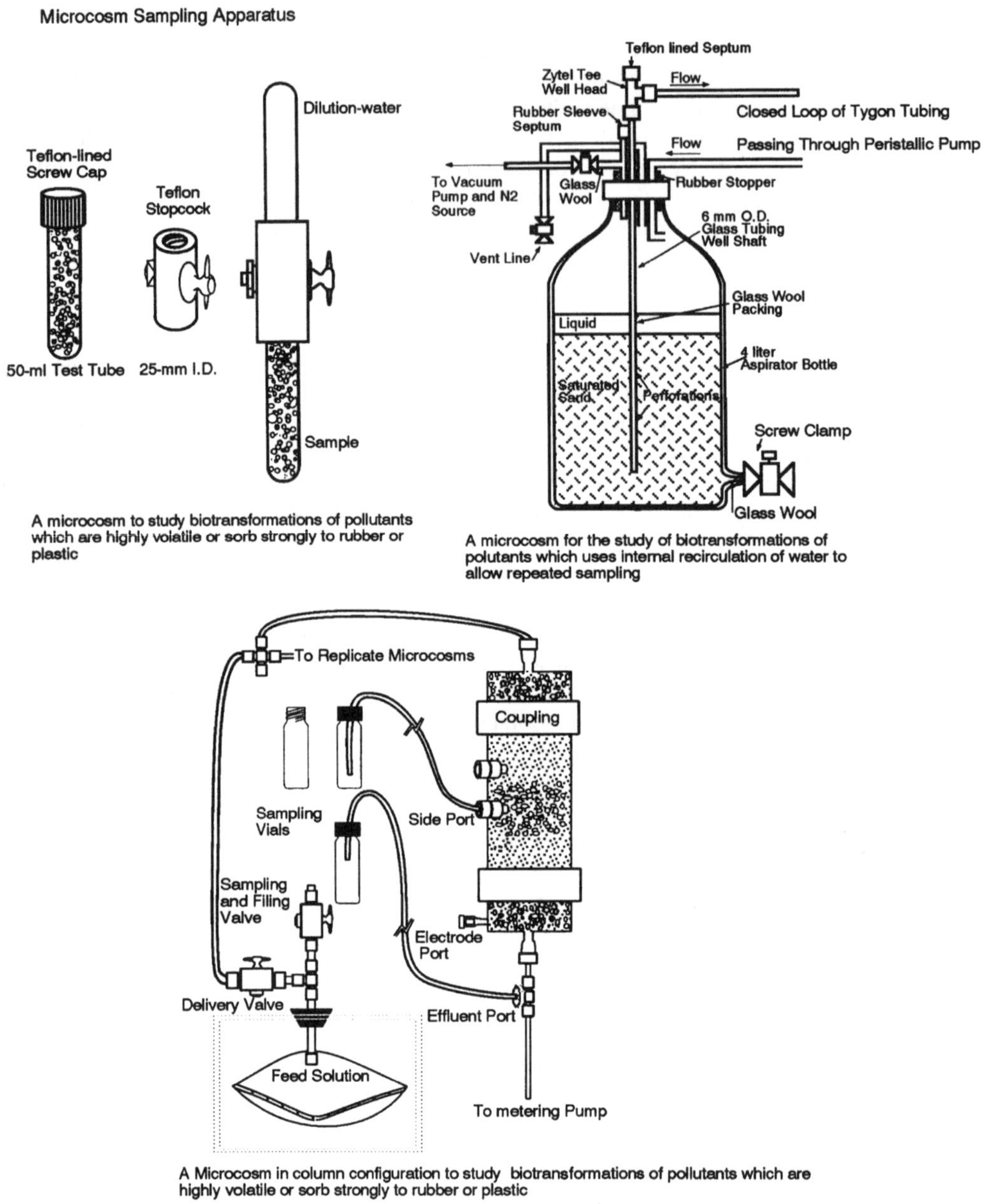

Figure 25.1. Schematics of microcosms.

permeability unconfined zone, and 2) a slightly higher-permeability semi-confined zone. These two zones are separated by a thin semi-confining clay layer. The injection-production test (Figure 25.3) was performed in the semi-confined zone approximately 75 m south of the former ponds (Figure 25.2). This zone is composed of medium fine clayey sand with a permeability of approximately 1 m/d. The piezometric gradient in the area is low, resulting in groundwater velocities in the range of 5 m/yr.

Groundwater was produced from the two outside wells (RU28 and RU29) and injected into the center well (RU1) for seven days. A nonreactive tracer, chloride, and naphthalene and paradichlorobenzene (*p*-DCB) were added to the injection stream for the first 24 hours. The two organics have approximately equal affinities for adsorption (log octanol-water partition coefficient, K=3.37 for naphthalene and 3.39 for *p*-DCB) but show considerably different affinities for biodegradation.

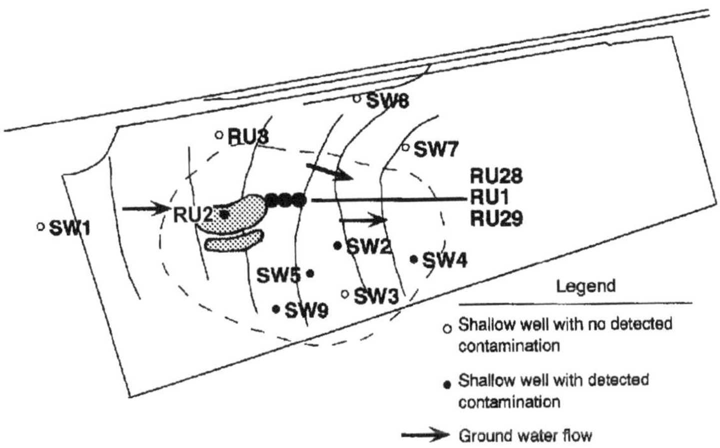

Figure 25.2. United Creosoting Co. site map. (Reference: Borden and Bedient, 1987.)

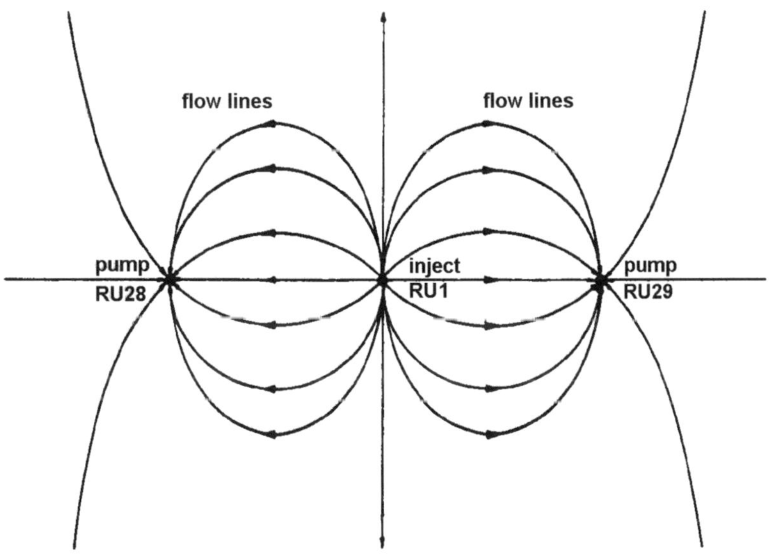

Figure 25.3. Schematic of injection-production test. (Reference: Borden and Bedient, 1987.)

Breakthrough curves for chloride, *p*-DCB, and naphthalene in wells RU28 and RU29 were compared to obtain estimates of retardation and biotransformation. Normalized concentration distributions for chloride, naphthalene and *p*-DCB in RU28 are shown in Figure 25.4. The data in Figure 25.4 show that the breakthrough curves for naphthalene and *p*-DCB have the same general appearance as the chloride curve, although the area under the curves is significantly less. The observed retardation factors and fractional recoveries for naphthalene and *p*-DCB are shown in Table 25.1.

The results in Table 25.1 indicated that there was essentially very little retardation of the organic compounds (the aquifer has very low organic carbon content which is below the analytical detection limit of 0.01% by weight), yet there was a significant loss of the organic compounds from solution. The observed loss of mass (65% for naphthalene and 48% for *p*-DCB) was attributed by Borden and Bedient (1987) to biotransformation in the aquifer. The extent of biodegradation was expected to be greater for naphthalene than for *p*-DCB since the microbial population had been adapted to naphthalene but not to *p*-DCB.

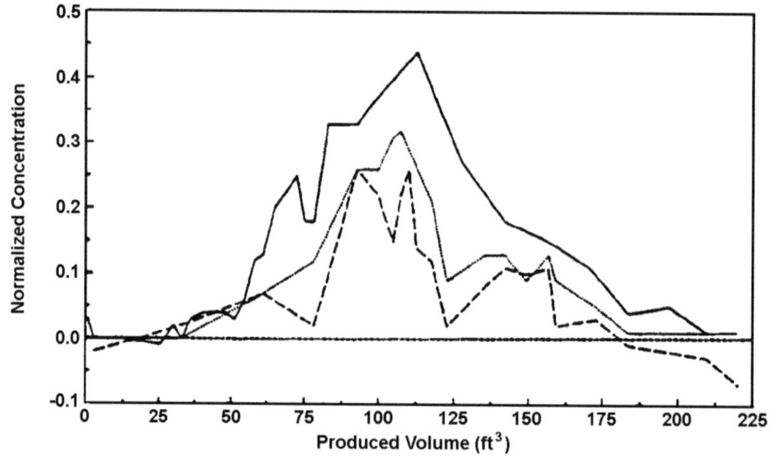

Figure 25.4. Normalized concentration distributions for chloride (solid line) *p*-DCB (short dash) and naphthalene (long dash) in RU28. (Reference: Borden and Bedient, 1987.)

Table 25.1. Summary of Normalized Data for Well RU29.

Moments[a]	Chloride	Conductivity	Naphthalene	*p*-DCB
Eo	31.8	36.49	11.1	16.68
Ev	3650	4287	1312	1848
μ	114.8	117.5	118.2	110.8
R	1	1.02	1.03	0.97
Fract. Recovery	1	1.15	0.35	0.52

[a] Eo and Ev are statistical moments of the data, μ is the ratio of Eo and Ev and R is the retardation coefficient. Fractional recovery is defined as the ratio of Eo for the organic divided by Eo for chloride.

Source: Borden and Bedient, 1987.

Borden Landfill in Canada

Barker et al. (1987) introduced 7.6 mg/L total BTX below the water table (in addition to chloride as a tracer) into the sandy aquifer at the Borden landfill. Average concentrations of benzene, toluene, *p*-xylene, *m*-xylene and *o*-xylene were 2360, 1750, 1080, 1090, and 1290 μg/L, respectively. The migration of the contaminants was monitored using a dense sampling network. The horizontal aspect of the plume, produced by using vertically integrated concentrations of each constituent at each multilevel piezometer, is shown in Figure 25.5. Chloride behaves conservatively and illustrates the effects of advection and dispersion. Although the organics follow the same path, their centers of mass are delayed. This is mainly due to sorption at the site.

The apparent decline in total mass is shown in Figure 25.6. All the injected mass of BTX was lost within 434 days due to biodegradation. The rates of mass loss appear to be similar for all the monoaromatics; benzene was the only component to persist beyond 270 days. Laboratory microcosm studies confirmed the conclusions regarding biodegradation drawn from the field experiment.

Gas Plant Facility in Michigan

Soluble hydrocarbon and dissolved oxygen (DO) were characterized in a shallow aquifer beneath a gas plant facility in Michigan by Chiang et al. (1989). The distributions of benzene, toluene, and xylene (BTX) in the aquifer had been monitored in 42 wells for a period of three years. A general site plan including the locations of the wells is shown in Figure 25.7.

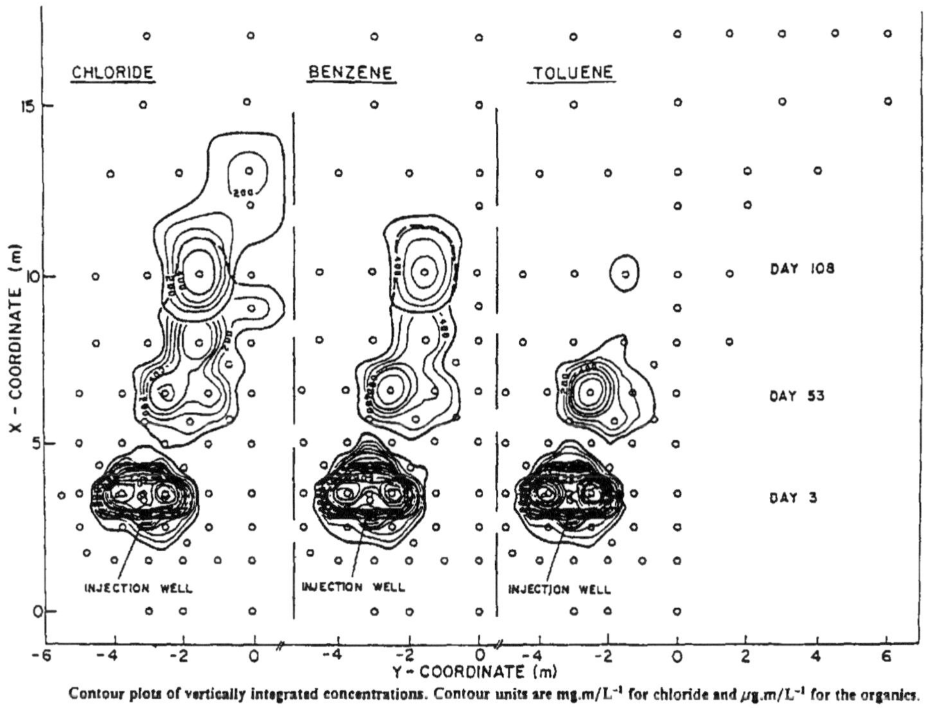

Figure 25.5. Chloride, benzene and toluene contours. (Reference: Barker et al., 1987.)

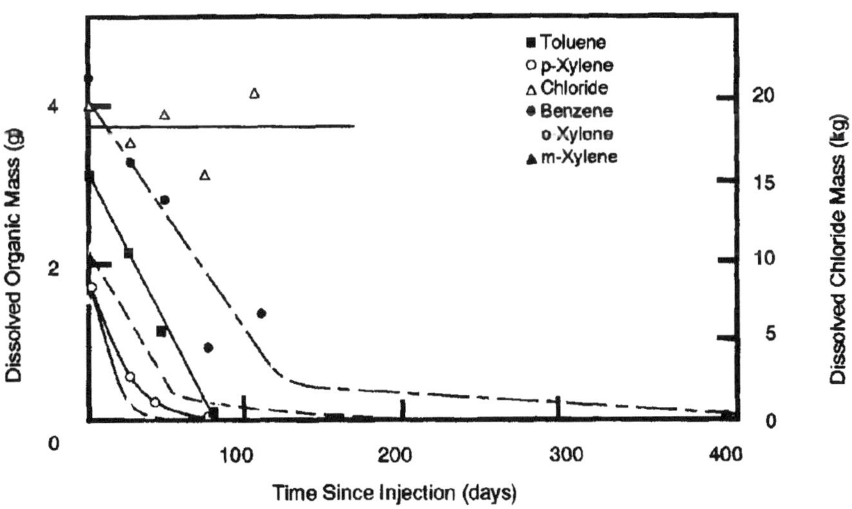

Figure 25.6. Masses of chloride and organics remaining in the groundwater. (Reference: Barker et al., 1987.)

Results from the three-year sampling period showed a significant reduction in total benzene mass with time in groundwater (Figure 25.8). The plume sampled in 1984 contained an approximate total mass of 9.83 kg, while the plume sampled in 1985 and 1986 contained 5.66 kg and 2.25 kg, respectively. Chiang et al. (1989) determined the attenuation rates of the soluble benzene, and determined the effects of DO on the biodegradation of BTX through a combination of material balance, statistical analyses, soluble transport modeling, and laboratory microcosm experiments.

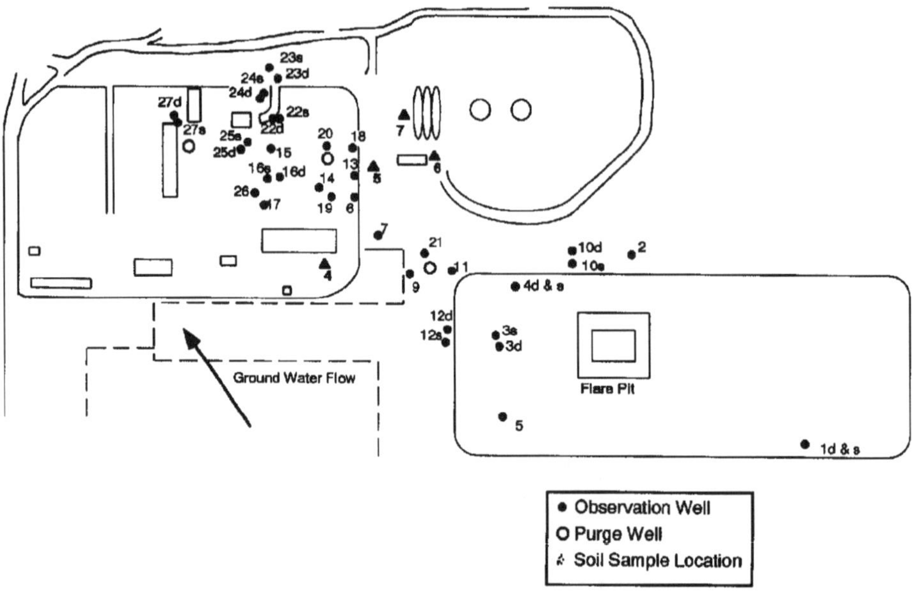

Figure 25.7. General plan of gas plant facility. (Reference: Chiang et al., 1989.)

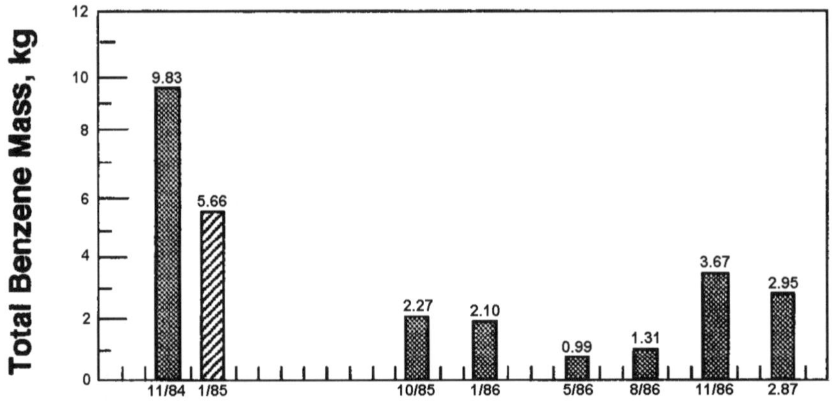

Figure 25.8. Calculated total soluble benzene mass in aquifer vs. time, November 1986 to February 1987. (Reference: Chiang et al., 1989.)

Dissolved oxygen concentrations in the groundwater were measured by the Winkler titration method and with a field DO probe on two occasions (February and July 1987). Figure 25.9 shows the total BTX and DO concentration distributions of the February and July analyses, respectively. It can be seen from Figure 25.9 that the DO concentrations are high at low BTX concentrations and vice versa. The DO concentrations increase from their low values of < 1.0 ppm inside the 1000 ppb BTX contour to higher concentrations of > 1.0 ppm outside the 100 ppb BTX contour line.

Cliffs-Dow Superfund Site

Groundwater at the Cliffs-Dow site is contaminated with low levels of phenolic and polycyclic compounds. Based on the analysis of samples obtained from monitoring wells, Klecka et al. (1990) found that the levels of organic contaminants are reduced to near or below the detection limit within

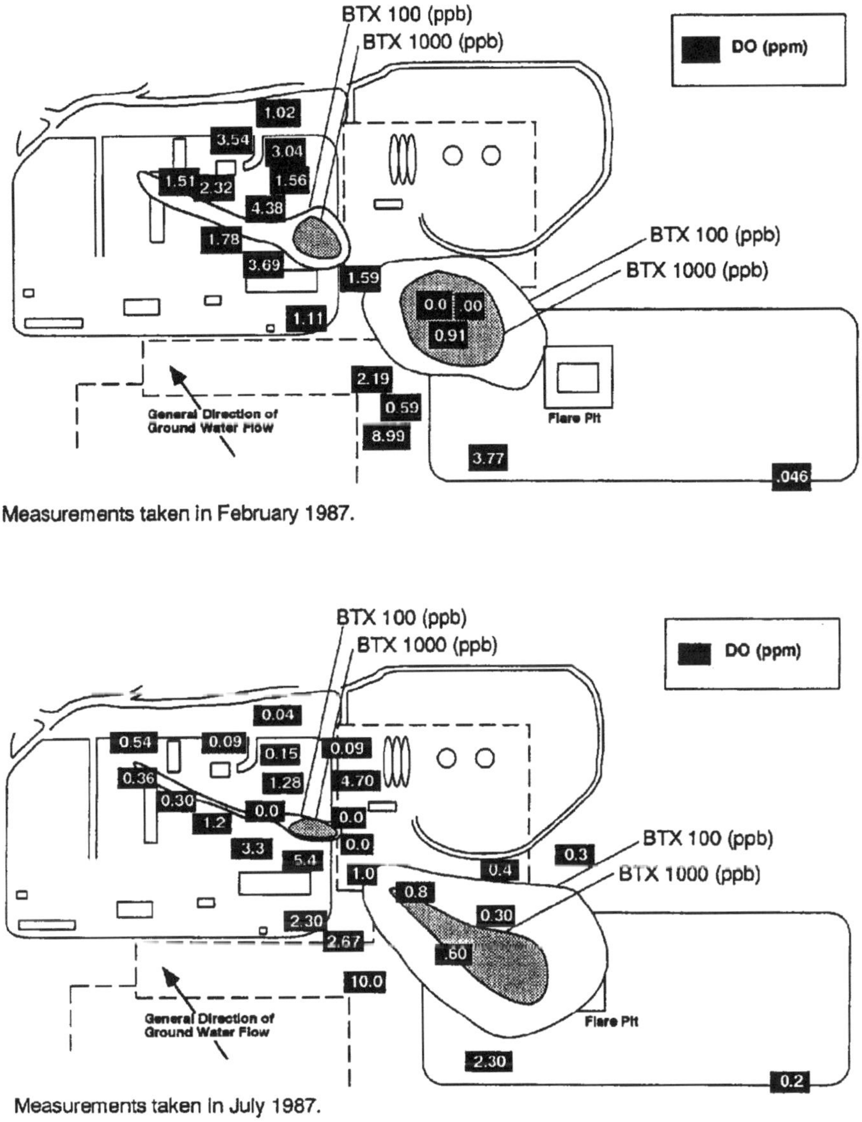

Figure 25.9. BTX and dissolved oxygen in groundwater. (Reference: Chiang et al., 1989).

a distance of 100 m downgradient from the source. Further analyses of the groundwater chemistry were used to verify that biodegradation was occurring at the site and causing the disappearance of the contaminants.

Figure 25.10 shows the location of five soil borings and 20 groundwater monitoring wells at the site. The aquifer sediments at the site consist of mostly coarse sands and gravels. The hydraulic conductivity ranges between 3.5× 10 to 4.6× 10 cm/sec. The principal contaminants found at the site near the source area include phenol, several methyl-substituted phenols, and naphthalene at concentrations ranging from 220 μg/L to 860 μg/L. Results of the characterization of the groundwater are shown in Table 25.2.

Klecka et al. (1990) attributed the reduction of dissolved oxygen from 1.2 mg/L at the upgradient well 85-1 to less than 0.1 mg/L at wells B-3A and 85-3 to biodegradation. This conclusion, Klecka argues, is supported by the increase in total inorganic carbon at well B-3A. Microcosm studies were used by Klecka et al. (1990) to simulate the aerobic biodegradation of the phenols at the site.

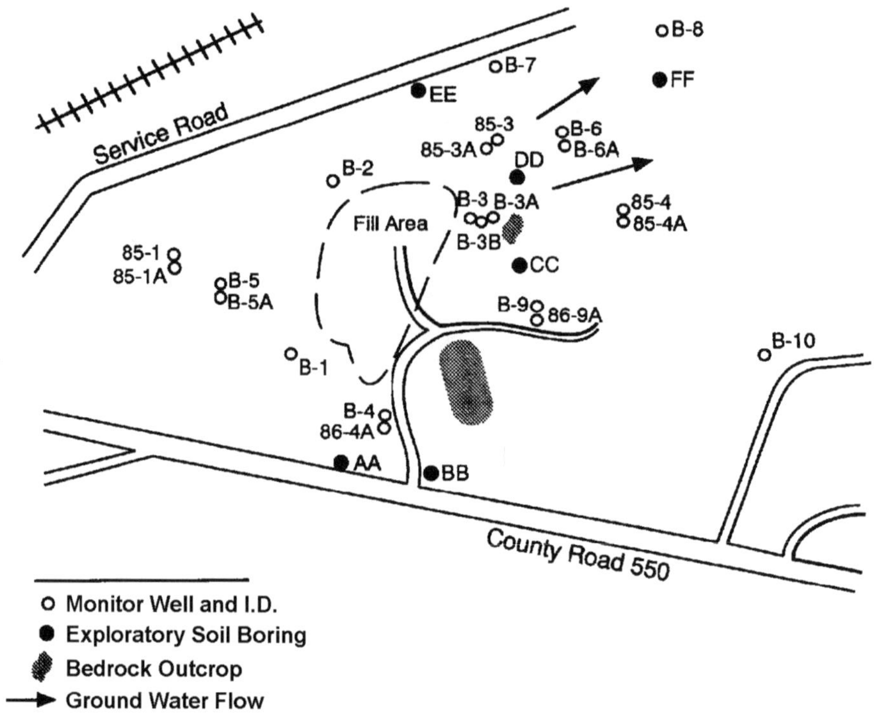

Figure 25.10. Site map showing locations of groundwater monitoring well and exploratory soil borings. (Reference: Klecka et al., 1990).

Table 25.2. Characteristics of Ground Water from the Cliffs-Dow Site.

Parameter	Well 85-1	Well B-3A	Well 85-3	Well B-6
Temperature (°C)[a]	—	9.2	8.0	8.5
Dissolved oxygen (ppm)[a]	1.2	<0.1	0.1[b]	1.6
Total organic C (ppm)	3.2	131.7	2.6	2.8
Total inorganic C (ppm)	39.2	381.0	48.0	17.4
Total phenolics (ppb)	—	47[c]	<4	—
Total polycyclics (ppb)	—	<40	<4	—
pH	6.1	6.4	5.5	6.1
Conductivity (μmhos)	220	1180	335	65
Phosphate (ppm)	<1	<1	<1	<1
Nitrate (ppm)	1.7	<0.1	0.1	<0.1
Sulfate (ppm)	48	7	124	14
Chloride (ppm)	28	14	70	3

[a] Temperature and dissolved oxygen analyzed during sampling.
[b] Degassing of sample noted during sampling.
[c] Phenolic contaminant identified as 2,4-dimethylphenol.
Source: Klecka et al., 1990.

AEROBIC BIODEGRADATION MODELS

The problem of quantifying biodegradation in the subsurface can be addressed by using models which combine physical, chemical and biological processes. Developing such models is not simple, however, due to the complex nature of microbial kinetics, the limitations of computer resources, the lack of field data on biodegradation, and the need for robust numerical schemes that can simulate the physical, chemical, and biological processes accurately.

The main expressions that have been utilized for modeling aerobic biodegradation include:

1. Monod Kinetics
2. First-Order Decay Kinetics
3. An Instantaneous Reaction Assumption

The reduction of contaminant concentrations using Monod kinetics (Monod, 1942) can be expressed as:

$$\Delta C = M_t\, \mu_{max}\left(\frac{C}{K_c + C}\right)\Delta t \tag{25.1}$$

where

C = the contaminant concentration;
M_t = the total microbial concentration;
μ_{max} = the maximum contaminant utilization rate per unit mass microorganisms;
K_c = the contaminant half saturation constant; and
Δt = the time interval being considered.

Incorporating Equation 25.1 into the one-dimensional transport equation, for example, results in:

$$\frac{\partial C}{\partial t} = D_x \frac{\partial^2 C}{\partial x^2} - v\frac{\partial C}{\partial x} - M_t \mu_{max}\left(\frac{C}{K_c + C}\right) \tag{25.2}$$

where

v = the seepage velocity; and
D_x = is the dispersion coefficient.

For aerobic biodegradation, and assuming that oxygen and the contaminant are the only substrates required for growth, the change in contaminant and oxygen concentrations due to biodegradation is given by (Borden and Bedient, 1986):

$$\Delta C = M_t\, \mu_{max}\left(\frac{C}{K_c + C}\right)\left(\frac{O}{K_o + O}\right) \tag{25.3}$$

$$\Delta O = M_t\, \mu_{max} F\left(\frac{C}{K_c + C}\right)\left(\frac{O}{K_o + O}\right)\Delta t \tag{25.4}$$

where

O = the oxygen concentration;
K_o = the oxygen half saturation constant;
F = the ratio of oxygen to contaminant consumed; and
All other parameters being as defined previously.

Incorporating Equations 25.3 and 25.4 into the transport equation results in a system of partial differential equations (Borden and Bedient, 1986):

$$\frac{\partial C}{\partial t} = \frac{1}{R_c} \bullet \nabla \bullet (D\nabla C - vC) - \frac{M_t \bullet M_{max}}{R_c} \left(\frac{C}{K_c + C} \right) \left(\frac{O}{K_o + O} \right) \tag{25.5}$$

$$\frac{\partial O}{\partial t} = \nabla \bullet (D\nabla O - vO) - M_t \bullet M_{max} \bullet F \left(\frac{C}{K_c + C} \right) \left(\frac{O}{K_o + O} \right) \tag{25.6}$$

$$\frac{\partial M_s}{\partial t} = \frac{1}{R_m} \nabla \bullet (D\nabla M_s - vM_s) + M_s \, \mu_{max} \, Y \left(\frac{C}{K_c + C} \right) \left(\frac{O}{K_o + O} \right) + \frac{k_c Y \, (OC)}{R_m} - bM_s \tag{25.7}$$

where

R_c	=	the retardation coefficient for the contaminant;
M_s and M_t	=	the concentration of microbes in solution and the total microbial concentration, respectively ($M_t = R_m \bullet M_s$ where R_m is the microbial retardation factor);
Y	=	the microbial yield coefficient;
OC	=	the natural organic carbon concentration; and
b	=	the microbial decay rate.

First-order kinetics represent an exponential decay and are a simplification of Monod kinetics. When $C << K_c$, the Monod expression

$$\mu = \mu_{max} \left(\frac{C}{K_c + C} \right)$$

reduces to:

$$\mu = \frac{\mu_{max}}{K_c} \bullet C \tag{25.8}$$

and the reaction approaches first-order with

$$\frac{\mu_{max}}{K_c} = k = \text{first - order rate constant}$$

The biodegraded concentration can be estimated using:

$$C = C_0 \bullet e^{-kt} \tag{25.9}$$

where

C is the biodegraded concentration of the chemical;
C_0 is the starting concentration; and
k is the rate of decrease of the chemical.

First-order rate constants are particularly useful since they allow half-lives for chemicals to be estimated:

$$t_{1/2} = \frac{0.693}{k} \tag{25.10}$$

The instantaneous reaction model, a model first proposed by Borden and Bedient (1986), was used in BIOPLUME II. The main assumption in the model is that microbial biodegradation kinetics are fast in comparison with the transport of oxygen, and that the growth of microorganisms and utilization of oxygen and organics in the subsurface can be simulated as an instantaneous reaction between the organic contaminant and oxygen. From a practical standpoint, the instantaneous reaction model assumes that the rate of utilization of the contaminant and oxygen by the microorganisms is very high, and that the time required to mineralize the contaminant is very small, or almost instantaneous. Biodegradation is calculated from the relation:

$$\Delta C_R = -\frac{O}{F} \tag{25.11}$$

where ΔC_R is the change in contaminant concentration due to biodegradation.

Many biodegradation models have been developed in recent years, most of which utilize some form of the three expressions presented earlier (see Table 25.3). Molz et al. (1986) and Widdowson et al. (1987) presented one-dimensional and two-dimensional models for aerobic biodegradation of organic contaminants in groundwater coupled with advective and dispersive transport. Their results indicate that biodegradation would be expected to have a major effect on contaminant transport when proper conditions for growth exist. Simulations of two-dimensional transport suggested that under aerobic conditions, microbial degradation reduces the substrate concentration profile along longitudinal sections of the plume and retards the lateral spread of the plume. Anaerobic conditions developed in the plume center due to microbial consumption and limited oxygen diffusion into the plume interior. Widdowson et al. (1988) extended their 1986 and 1987 studies to simulate oxygen- and/or nitrate-based respiration.

Srinivasan and Mercer (1988) presented a one-dimensional, finite difference model, BIO1D, for simulating biodegradation and sorption processes in saturated porous media. The model formulation allowed for accommodating a variety of boundary conditions and process theories. Aerobic biodegradation was modeled using a modified Monod function. The model presented is an extension of that presented by Borden and Bedient (1986).

The BIOPLUME II model was developed by modifying an existing two-dimensional transport model developed by the USGS and known as the Method of Characteristics (MOC) model (Konikow and Bredehoeft, 1978). Two governing equations are solved in BIOPLUME II: the groundwater flow equation and the transport equation. The basic concept applied to modify the USGS MOC model and to develop the BIOPLUME II model includes the use of a dual-particle mover procedure to simulate the transport of oxygen and contaminants in the subsurface. The transport equation is solved twice at every time step to calculate the oxygen and contaminant distributions. The two plumes are combined using the principle of superimposition to simulate the instantaneous reaction between oxygen and the contaminants.

Figure 25.11 is a conceptual schematic of the BIOPLUME II model. At the top of the figure, a plan view of the contaminant and oxygen plumes with and without biodegradation is shown. After the two plumes are superimposed, the contaminant plume is reduced in size and concentrations. The dissolved oxygen is depleted in zones of high contaminant concentrations and reduced in zones of relatively moderate contaminant concentrations. The bottom schematics in Figure 25.11 present transects down the plume centerline and help to illustrate the distributions of contaminant and oxygen concentration with and without biodegradation. It is noted that field data have verified the correlation between oxygen and contaminant concentrations at sites.

Many of the aerobic models developed to date have been used to simulate aerobic biodegradation at field sites. Borden et al. (1986) applied the first version of the BIOPLUME model to simulate biodegradation at the Conroe Superfund site in Texas. Oxygen exchange with the unsaturated zone

Table 25.3. Aerobic Biodegradation Models

Model Name	Author
None	Molz et al., 1986
BIOPLUME	Borden et al., 1986
None	Domenico, 1987
BIO1D	Srinivasan and Mercer, 1988
None	Widdowson et al., 1988
None	Celia et al., 1989
BIOPLUME II	Rifai et al., 1988
None	MacQuarrie et al., 1990
BIOPLUS	Wheeler et al., 1987
ULTRA	Tucker et al., 1986
None	Odencrantz et al., 1990

was simulated as a first-order decay in hydrocarbon concentration. The loss of hydrocarbon due to horizontal mixing with oxygenated groundwater and resulting biodegradation were simulated by generating oxygen and hydrocarbon distributions independently and then combining by superimposition. Simulated oxygen and hydrocarbon concentrations closely matched the observed values.

Rifai et al. (1988) applied the BIOPLUME II model to the jet fuel spill at the Traverse City field site in Michigan. Rifai et al. (1988) simulated the natural aerobic biodegradation at the site and were able to match the observed laboratory and field rates of biodegradation of 1% per day. Chiang et al. (1989) evaluated a first-order decay biodegradation approach and the BIOPLUME II model for simulating biodegration at the gas plant facility. Using the model and assuming first-order decay, several simulations were made to match the observed benzene concentration distribution of 1/22/85 by setting the observed concentration distribution of 11/1/84 as the initial condition. The variables involved included the distribution of the leakage/spill rates between the flare pit and the slop oil tanks and macrodispersivities of the aquifer. The BIOPLUME II model was used to simulate the July 1987 data and set the observed concentration distribution of February 1987 as an initial condition.

The migration of organic constituents at the Cliffs-Dow Superfund Site was simulated by Klecka et al. (1990) using the BIO1D model and assuming a first-order decay expression. Half-lives for the contaminants at the site were estimated from the results of soil microcosm experiments based on the time required for 50% disappearances of the parent compound. Model simulations performed using a half-life of two days indicated that the levels of the phenolic components were reduced by greater than 99% within a distance of 30 m downgradient of the source. When the half-life was increased by a factor of 10, the concentrations were reduced to a similar extent within 75 m. Because of the dominance of biodegradation, increases in groundwater velocity from 0.2 to 0.46 m/day (the range of velocities measured at the site), had minor effects on the level of attenuation predicted with the model.

IMPLICATIONS FOR REMEDIATION

As we come to grips with the almost impossible task of actively remediating all the contaminated sites in the United States, due to economic or technical feasibility considerations, it might be possible to view natural aerobic biodegradation as a viable attenuation technology. The technology might be used for sites which are associated with lower risks to human health and the environment and for sites which cannot attain cleanup standards using other technologies. Some of the requirements for establishing technical feasibility of passive aerobic remediation include:

- *Developing a standard field test(s) of biodegradation:* A detailed field protocol for measuring biodegradation at the field scale is needed. The protocol needs to identify which parameters need to be measured and the frequency and spatial distribution of these measurements.

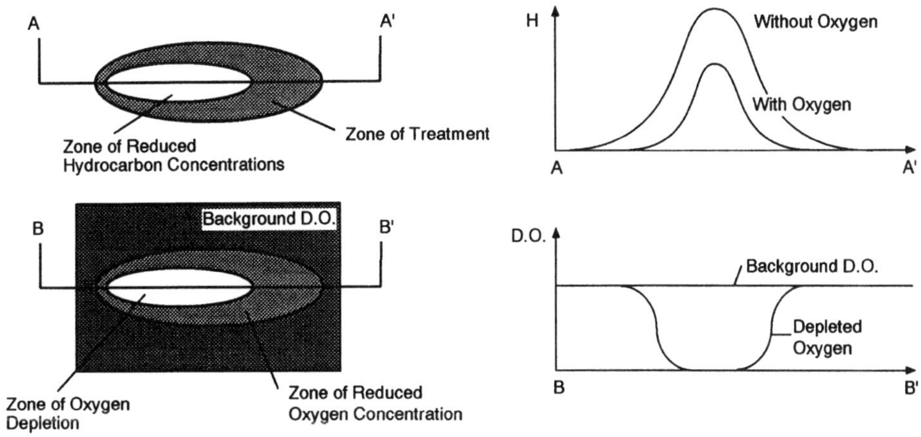

Figure 25.11. Principle of superimposition in BIOPLUME II.

- *Determining the rate of biodegradation at the field scale:* It is necessary to quantify the rate of disappearance of contaminants due to biodegradation and hence the time required for attenuation to acceptable levels. The field rate of biodegradation may be determined from some of the parameter measurements carried out at the site or inferred from standardized laboratory microcosm studies.
- *Establishing source control, if necessary:* Some field situations where a significant amount of free product or residual remains in the soil/groundwater system may require implementation of source control measures.

PROPOSED FIELD PROTOCOL

Research aimed at developing a field protocol for assessing aerobic biodegradation of BTX at field sites has been initiated over the past year at Rice University and North Carolina State University. The conceptual framework for these projects involves analyzing the soil and groundwater chemistry at two gasoline spill sites undergoing aerobic biodegradation as well as completing modeling studies to predict future site conditions. The main parameters which are being monitored at the two sites are listed in Table 25.4. The objective of the sampling program is to determine which parameters could be used as indicators of biodegradation, and as such provide some idea on the rate of the reactions.

CONCLUSIONS

A significant amount of research has been conducted in the laboratory and under field conditions to verify the occurrence of biodegradation and quantify the rates of associated contaminant attenuation. These studies seem to indicate that natural aerobic biodegradation might be a viable remediation technology at some sites. Research is necessary, however, to develop a field protocol for assessing the attenuation resulting from biodegradation processes. The protocol needs to provide data that would allow researchers to estimate the time required for cleanup and thereby study the technical feasibility of natural biodegradation as a remedial alternative for a given site.

ACKNOWLEDGMENTS

Research support from the Shell Chair in Environmental Science and the American Petroleum Institute is gratefully acknowledged.

Table 25.4. Proposed Field Sampling.

Media	Parameter
Soil	Microbial counts/activity TOC BTX
Ground Water	Temperature Dissolved oxygen Carbon dioxide Conductance pH, TDS Redox potential Ca, Mg, Na, Mn, Fe, SO_4, Cl Total alkalinity BOD, COD, TOC, BTX NO_3, NO_2, NH_4, PO_4
Soil Vapor	O_2, CO_2, CH_4, H_2S

REFERENCES

Balkwill, D.L. and W.C. Ghiorse. 1985. Characterization of Subsurface Bacteria Associated with Two Shallow Aquifers in Oklahoma. *Appl. Environ. Microbiol.,* **3**:580–588.

Barker, J.F., G.C. Patrick, and D. Major. 1987. Natural Attenuation of Aromatic Hydrocarbons in a Shallow Sand Aquifer. *Ground Water Monitoring Review,* **7**(1):64–71.

Beeman, R.E. and J.M. Suflita. 1987. Microbial Ecology of a Shallow Unconfined Ground Water Aquifer Polluted by Municipal Landfill Leachate. *Microb. Ecol.,* **14**:55–66.

Bengtsson, S.G.L. 1981. Presented at the First International Conference on Ground Water Quality Research, October 7–10, Rice University, Houston, Texas.

Borden, R.C. and P.B. Bedient. 1986. Transport of Dissolved Hydrocarbons Influenced By Oxygen-Limited Biodegradation: 1. Theoretical Development. *Water Resources Research,* **13**:1973–1982.

Borden, R.C. and P.B. Bedient. 1987. In Situ Measurement of Adsorption and Biotransformation at a Hazardous Waste Site. Water Resources Bulletin, American Water Resources Association, **4**:629–636.

Borden, R.C., P.B. Bedient, M.D. Lee, C.H. Ward, and J.T. Wilson. 1986. Transport of Dissolved Hydrocarbons Influenced by Oxygen-Limited Biodegradation: 2. Field Application. *Water Resources Research,* **13**:1983–1990.

Celia, M.A., J.S. Kindred, and I. Herrera. 1989. Contaminant Transport and Biodegradation, 1. A Numerical Model for Reactive Transport in Porous Media. *Water Resources Research,* **25**(6):1141–1148.

Chiang, C.Y., J.P. Salanitro, E.Y. Chai, J.D. Colthart, and C.L. Klein. 1989. Aerobic Biodegradation of Benzene, Toluene and Xylene in a Sandy Aquifer—Data Analysis and Computer Modeling. *Ground Water,* **6**:823–834.

Domenico, P.A. 1987. An Analytical Model for Multidimensional Transport of a Decaying Contaminant Species. *J. Hydrology,* **91**:49–58.

Dunlap, W.J., R.L. Cosby, J.F. McNabb, B.E. Bledsoe, and M.R. Scalf. 1972. *Ground Water,* **10**:107.

Ghiorse, W.C. and D.L. Balkwill. 1983. Enumeration and Morphological Characterization of Bacteria Indigenous to Subsurface Environments. *Dev. Ind. Microbiol.,* 213–225.

Ghiorse, W.C. and J.T. Wilson. 1988. Microbial Ecology of the Terrestrial Subsurface, *Adv. Appl. Microbiol.,* **33**:107–172.

Jhaveri, V. and A.J. Mazzacca. 1983. "Bioreclamation of Ground and Groundwater. A Case History," Proc. 4th National Conference on Management of Uncontrolled Hazardous Waste Sites, Washington, DC, 242 pp.

Klecka, G.M., J.W. Davis, D.R. Gray, and S.S. Madsen. 1990. Natural Bioremediation of Organic Contaminants in Ground Water: Cliffs-Dow Superfund Site. *Ground Water,* **4**:534–543.

Konikow, L.F. and J.D. Bredehoeft. 1978. "Computer Model of Two-Dimensional Solute Transport and Dispersion in Ground Water, Automated Data Processing and Computations," Techniques of Water Resources Investigations of the U.S.G.S., Washington, DC, 100 pp.

Kuhn, E.P., P.J. Colberg, J.L. Schnoor, O. Wanner, A.J.B. Zehnder, and R.P. Schwarzenbach. 1985. Microbial Transformations of Substituted Benzenes during Infiltration of River Water to Groundwater: Laboratory Column Studies. *Environ. Sci. Technol.*, **19**(10):961–967.

Lee, M.D. and C. H. Ward. 1985. "Microbial Ecology of a Hazardous Waste Disposal Site: Enhancement of Biodegradation," Proc. Second International Conference on Ground Water Quality, OSU University Printing Services, Stillwater, OK, 25–27.

Lee, M.D., V.W. Jamison, and R.L. Raymond. 1987. "Applicability of In Situ Bioreclamation as a Remedial Action Alternative," Proc. of Petroleum Hydrocarbons and Organic Chemicals in Ground Water: Prevention, Detection and Restoration, Houston, TX, National Water Well Association, 167–185.

MacQuarrie, K.T.B., E.A. Sudicky, and E.O. Frind. 1990. Simulation of Biodegradable Organic Contaminants in Ground Water, 1. Numerical Formulation in Principal Directions. *Water Resources Research*, **26**(2): 207–222.

McNabb, J.F. and G.E. Mallard. 1984. Microbiological Sampling in the Assessment of Groundwater Pollution. In *Groundwater Pollution Microbiology,* G. Bitton and C.P. Gerba (Eds.), 89–115, John Wiley & Sons, Inc., NY.

Molz, F. J., M.A. Widdowson, and L.D. Benfield. 1986. Simulation of Microbial Growth Dynamics Coupled to Nutrient and Oxygen Transport in Porous Media. *Water Resources Research*, **22**(8):1207–1216.

Monod, J. 1942. "Recherches sur la Croissance des Cultures Bacteriennes," Herman & Cie., Paris.

Pritchard, P.H. and A.W. Bourquin. 1984. "The Use of Microcosms for Evaluation of Interactions Between Pollutants and Microorganisms," In: *Adv. Microbial Ecology,* K.C. Marchall (Ed.), Plenum Publishing Corporation, **7**:133–215.

Rifai, H.S., P.B. Bedient, J.T. Wilson, K.M. Miller, and J.M. Armstrong. 1988. Biodegradation Modeling at Aviation Fuel Spill Site. *J. Env. Eng.*, **5**:1007–1029.

Semprini, L. and P.L. McCarty. 1991. Comparison Between Model Simulations and Field Results for In-Situ Biorestoration of Chlorinated Aliphatics: Part 1. Biostimulation of Methanotrophic Bacteria. *Groundwater*, **29**(3):365–374.

Srinivasan, P. and J.W. Mercer. 1988. Simulation of Biodegradation and Sorption Processes in Ground Water. *Ground Water*, **4**:475–487.

Suflita, J.M. 1989. "Microbiological Principles Influencing the Biorestoration of Aquifers." EPA/625/4-89/019, 85–99.

Suflita, J.M. and G.D. Miller. 1985. The Microbial Metabolism of Chlorophenolic Compounds in Groundwater Aquifers. *Env. Toxicol. Chem.*, **4**:751–758.

Tucker, W.A., C.T. Huang, J.M. Bral, and R.E. Dickinson. 1986. "Development and Validation of the Underground Leak Transport Assessment Model (ULTRA)," in Proceedings of the Petroleum Hydrocarbons and Organic Chemicals in Ground Water: Prevention, Detection and Restoration, November 12–14.

Odencrantz, J.E., A.J. Valocchi, and B.E. Rittmann. 1990. "Modeling Two-Dimensional Solute Transport with Different Biodegradation Kinetics," in *Proceedings* of the Petroleum Hydrocarbons and Organic Chemicals in Ground Water: Prevention, Detection, and Restoration, Oct 31–Nov 2, 1990, Houston, TX, 355–368.

Wheeler, M.F., C.N. Dawson, P.B. Bedient, C.Y. Chiang, R C. Borden, and H.S. Rifai. 1987. "Numerical Simulation of Microbial Biodegradation of Hydrocarbons in Ground Water," in Proceedings of the Solving Ground Water Problems With Models Conference and Exposition; February 10–12, 1987; Denver, CO.

Widdowson, M.A., F.J. Molz, and L.D. Benfield. 1987. "Development and Application of a Model for Simulating Microbial Growth Dynamics Coupled to Nutrient and Oxygen Transport in Porous Media," in *Proc.* AGWSE/IGWMCH Conference on Solving Ground Water Problems with Models, Denver, CO, National Water Well Association, Dublin, OH, 28–51.

Widdowson, M.A., F.J. Molz, and L.D. Benefield. 1988. A Numerical Transport Model for Oxygen- and Nitrate-Based Respiration Linked to Substrate and Nutrient Availability in Porous Media. *Water Resources Research*, **9**:1553–1565.

Wilson, J.T., M.J. Noonan, and J.F. McNabb. 1981. Presented at the First International Conference on Ground Water Quality Research, October 7–10, Rice University, Houston, TX.

Wilson, J.T., J.F. McNabb, J.W. Cochran, T.H. Wang, M.B. Tomson, and P.B. Bedient. 1985. Influence of Microbial Adaption on the Fate of Organic Pollutants in Ground Water. *Environ. Toxicol. Chem.*, **4**:721–726.

Wilson, J.T., J.F. McNabb, D.L. Balkwill, and W.C. Ghiorse. 1983. Enumeration and Characterization of Bacteria Indigenous to a Shallow Water-Table Aquifer. *Ground Water,* **2**:134–142.

Chapter 26

In Situ Transformation of Halogenated Aliphatic Compounds Under Anaerobic Conditions

L. Semprini

INTRODUCTION

Halogenated aliphatic compounds (HACs) are among the most prevalent hazardous chemical contaminants found in municipal and industrial wastewaters, landfills and landfill leachates, industrial sludges, waste disposal sites, and groundwater. Several of these chemicals have been widely used as solvents for household, commercial, and industrial cleaning and degreasing operations. Included here are chlorinated solvents such as carbon tetrachloride (CT), tetrachloroethene (PCE), trichloroethene (TCE), 1,1,1-trichloroethane (TCA), and dichloromethane (DM). In the past, used solvents were often discharged indiscriminately into the environment. A characteristic of these man-made solvents is their relative resistance to breakdown by natural microorganisms, and hence they have tended to persist and accumulate in the environment. While efforts are now being made to clean with other less harmful and less environmentally persistent chemicals, or to clean and recycle used chlorinated solvents, there is a legacy of chlorinated solvent–contaminated environments that need to be remediated.

This chapter will consider reductive biotransformations of HACs under anaerobic conditions, with an emphasis on CAHs (chlorinated aliphatic hydrocarbons). Reductive processes are of interest for several reasons. Anaerobic conditions are often associated with subsurface contamination, especially in the saturated zone. Thus, there is the potential for fortuitous transformation of HACs to occur. Secondly, enhanced *in situ* transformation of HACs is a potential remediation process for converting the HACs to non-toxic end products, or transforming them to less halogenated forms that might be more easily treated in situ by aerobic cometabolic processes, or more easily removed from the subsurface and treated aboveground. Along with a review of the reductive processes, results from microcosm studies and field studies, where fortuitous and enhanced transformations have been observed, will be presented.

ANAEROBIC TRANSFORMATIONS

Transformation Reactions

The potential for anaerobic biological transformations of HACs in the subsurface was demonstrated in 1981 (Bouwer et al., 1981). Subsequently, HACs in general have been found to be transformed under a variety of environmental conditions in the absence of oxygen. Some transformations occur chemically (abiotic), as well as through the direct action of microorganisms (biotic).

Vogel et al. (1987) reviewed degradation mechanisms of C1 and C2 halogenated solvents. These mechanisms included biotic and abiotic steps. In some cases combinations of both biotic and abiotic reactions involve several dehalogenation mechanisms. Reductive dehalogenations (Table 26.1) are one of the mechanisms for dehalogenation that occurs especially (but not exclusively) in anaerobic environments. In the hydrogenolysis reaction (I), a hydrogen replaces the halogen substitute on the C1 or C2 molecule, and two electrons are transferred. The example shown is the reduction of CT to chloroform (CF). In vicinal reduction (II) (dihalo-elimination reaction), two electrons are transferred and two halogen substitutes are removed from adjacent carbon atoms with the formation of an addi-

Table 26.1. Selected Transformation Reactions of Halogenated Aliphatic Compounds Likely to Occur Under Anaerobic Conditions (after Vogel et al., 1987).

Reaction Type	Example
Reductions	
I) Hydrogenolysis	$CCl_4 + H^+ + 2e^- \rightarrow CCl_3H + Cl^-$
II) Dihalo-elimination	$CCl_3CCl_3 + 2e^- \rightarrow CCl_2 = CCl_2 + 2Cl^-$
Substitution	
III) Dehydrohalogenation	$CH_3CCl_3 \rightarrow CH_2 = CCl_2 + HCl$

tional bond between the carbon atoms. The example shown is the reduction of hexachloroethane (HCE) to PCE. The abiotic dehydrohalogenation reaction (III) results in the elimination of a halide and can be combined with hydrolysis. The example shown is the abiotic transformation of TCA to 1,1-dichloroethene (1,1-DCE).

Biotic reductive processes of HACs appear to be occurring mostly through cometabolism. Here the HACs are not used for energy or growth by the microorganisms, but are transformed fortuitously by enzymes or cofactors being used for other normal metabolic processes. Cometabolism generally occurs when other chemicals are also present that can be used for energy and growth. This helps explain why transformations occur at some sites but not at others. Under anaerobic conditions halogenated compounds can undergo reductive dehalogenation (Bouwer and McCarty, 1982; Suflita et al., 1982), a partial transformation which sometimes detoxifies but in other cases activates the pollutant. Here the HACs may be acting fortuitously as electron acceptors, but the active microorganisms are not getting any benefit from the transformation.

In some cases microorganisms may be utilizing the halogenated compounds as electron acceptors, generating a useful form of energy when the HAC is coupled to a suitable electron donor. It has been suggested that PCBs may represent such electron acceptors and that the microorganisms, which reductively dechlorinate PCBs in anoxic environments, may be at an advantage relative to other members of the microflora (Brown et al., 1987). For some other chlorinated compounds, such as 3-chlorobenzoic acid, it has been proven that the dehalogenating microbial communities indeed obtain energy from this dechlorination (Dolfing and Tiedje, 1987). Holliger (1992) recently isolated a bacteria that uses reduced PCE and TCE as electron acceptors of choice.

Redox Conditions

Redox conditions in the subsurface are often regulated by microbial processes. Common anaerobic electron acceptors and the associated microbial process, in the order of their redox potential, are: nitrate (denitrification); Mn(IV) (manganese reduction); Fe(III) (iron reduction); sulfate (sulfate reduction); and carbon dioxide (methanogenesis). Many of the CAHs are highly oxidized and have redox potentials above those of common electron acceptors (Figure 26.1). For example, the reduction of HCE to PCE has a redox potential greater than O_2 to H_2O. Thus, it is anticipated that HCE might be easily reduced, even under aerobic conditions. The redox potentials generally decrease as the HAC becomes less halogenated. Reductions likely to occur under sulfate-reducing and methanogenic conditions are as follows, from higher to lower potential: CT to CF; PCE to TCE; TCA to 1,1-dichloroethane (1,1-DCA); TCE to DCE; and DCE to vinyl chloride (VC).

Transformation Pathways

HACs in general have been found to transform under a variety of environmental conditions in the absence of oxygen. Figure 26.2 illustrates the major biotic and abiotic pathways that CAHs may undergo at anaerobic contaminated sites. Three groups of CAHs are presented here: C1 compounds (halomethanes), with CT representing one parent compound; C2 single-bonded compounds (haloethane), with TCA as a parent compound; and C2 double-bonded compounds (haloethenes),

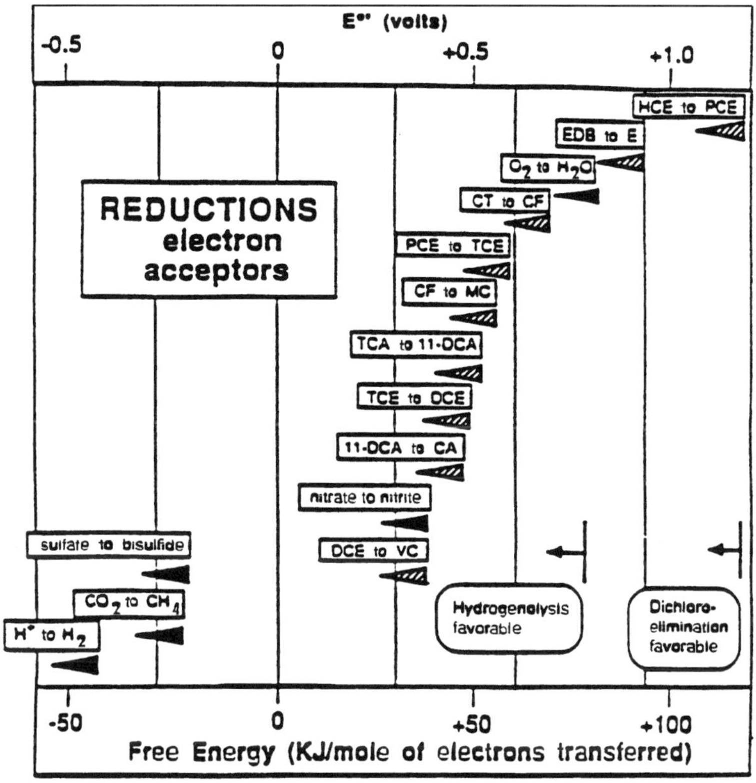

Figure 26.1. Half-reaction reduction potentials for reducing and oxidizing agents (reprinted from Vogel et al., 1987 with permission from the American Chemical Society).

with PCE as a parent compound. The transformation of these groups will be discussed in the following.

Halomethanes (CT, CF, CFC-11)

The transformation of CT (Figure 26.2) shows CF, DM, and CO_2 as potential transformation products. Criddle and McCarty (1991) present parallel pathways for the transformation conversion of CT to these products, as well as carbon monoxide (CO), formic acid, and HCE. While the production of CF and DM result in reductions of the carbon, the transformation to CO_2 results in no net change in the formal oxidation state of the carbon. CT has been assimilated into cell mass, mineralized to CO_2, and transformed to CF (Bouwer and McCarty, 1985; Bouwer and Wright, 1988; Bae et al., 1990) under conditions ranging from denitrifying to methanogenic. Some pure cultures have transformed CT to CF, quantitatively under both sulfate-reducing and methane-producing conditions (Egli et al., 1987; 1988). An acetogen, *Acetobacterium woodii*, was isolated that converted CT to DM and CO_2 (Egli et al., 1988). A *Clostridium* sp., identified from the effluent of an anaerobic reactor, biotransformed CT to CF as an intermediate with further transformation to DM and unidentified products (Gälli and McCarty, 1989a). A pure denitrifying culture was isolated, *Pseudomonas* sp. strain KC, that degraded CT, 50% to CO_2 and about 40% to nonvolatile compounds, with minimal CF production (Criddle et al., 1990a). CT was also transformed by *Escherichia coli* under fermenting conditions (Criddle et al., 1990b). The extent of reported mineralization of CT to CO_2 varies, ranging from 10 to 99%. The abiotic transformation of CT has also been observed in the presence of sulfide minerals (Kriegman and Reinhard, 1991) and by humic acids in the presence of sulfide minerals (Curtis, 1991).

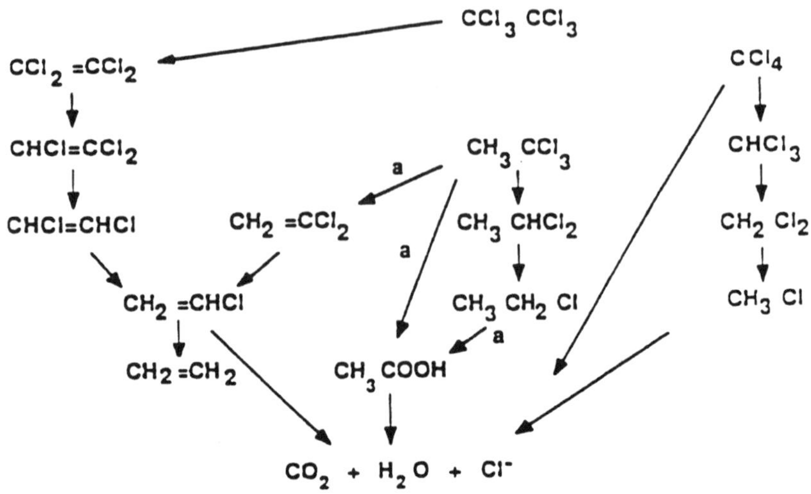

Figure 26.2. Abiotic and biotic transformation pathways of selected chlorinated aliphatic hydrocarbons under anaerobic conditions. Arrows with "a" indicate abiotic transformations; other arrows represent biotic transformations (after Vogel et al., 1987).

Trichlorofluoromethane (CFC-11) is another halomethane of environmental interest. Wolf et al. (1978) demonstrated that CFC-11 is reductively changed to dichlorofluoromethane (CFC-12) in rat liver. Disappearance of CFC-11 and (CFC-12) in soils (Khalil and Rasmussen, 1989), rice fields (Khalil et al., 1990), and in methanogenic freshwater sediments (Loveley and Woodward, 1990) have been reported; the latter study indicated the removal process may have been biotic. Here CFC-11 would be expected to be transformed more slowly than the fully chlorinated CT, since the fluoride-carbon bond would be more resistant to conversion (McCarty and Reinhard, 1993).

Haloethanes (TCA, 1,1-DCA, HCE, CFC-113)

TCA may be transformed both biotically and abiotically (Figure 26.2). The abiotic reaction results in the formation of 1,1-DCE and acetic acid. The rates are relatively slow, with a half-life for TCA on the order of one year (Vogel et al., 1987; Cline and Delfino, 1989; Jeffers et al., 1989) at 25°C. Based on Arrhenius relationships, the half-lives would be considerably longer at lower aquifer temperatures (Vogel et al., 1987).

TCA can be reduced under anaerobic conditions, and most rapidly under methanogenic conditions (Vogel and McCarty, 1985; 1987). TCA may be biologically transformed into 1,1-DCA, which can be further reduced to chloroethane (CA). CA is relatively stable biologically, but abiotically can be transformed into ethanol and chloride, thus rendering it relatively nontoxic. TCA can also be transformed under less reducing conditions of fermentation of amino acids (Gälli and McCarty, 1989a; 1989b). Here about 30–40% of the TCA was observed as 1,1-DCA while the remainder appeared as different products, including acetic acid. Criddle and McCarty (1991) discuss parallel pathways of TCA transformation where reduction could be followed by oxidation. They indicate that this might explain the observation of acetate formation from TCA by microorganisms. Thus, when TCA is discharged to the subsurface, a variety of abiotic and biotic transformation products may be found there in later years.

HCE is a fully saturated haloethane. Its reduction to PCE has a higher redox potential than O_2 to H_2O. Thus, its microbial reduction might even be expected under aerobic conditions in an aquifer. Biological transformations of HCE to PCE have been indicated by Criddle et al., (1986). Abiotic transformation to PCE has also been observed (Curtis, 1991; Kriegman and Reinhard, 1991).

1,1,2-trichloro-1,2,2-trifluoroethane (CFC-113) is another fully saturated haloethane of environmental interest. It can be reductively transformed by abiotic and biotic processes (Lesage et al., 1990; 1992). CFC-113 would be expected to be transformed more slowly than fully chlorinated HCE.

Haloethenes (PCE, TCE, DCE, and VC)

The sequential reduction of the double-bonded CAHs, including PCE and TCE, is shown in Figure 26.2. These transformations have been observed under methanogenic conditions (Parsons et al., 1984; Vogel and McCarty, 1985; Barrio-Lage et al., 1986; Fathepure et al., 1987). VC can also go even further when reduced biologically to ethene (Freedman and Gossett, 1989; DiStefano et al., 1991), and ethane (Belay and Daniels, 1987), which are essentially harmless. Käster (1991a) observed the partial transformation of PCE and TCE to *c*-DCE in cultures in the transition from aerobic to anaerobic conditions, accompanied by the release of sulfide and a decrease in the redox potential from 0 to –150 mV. These subsurface transformations are of particular concern due to the production of VC, which is a known carcinogen.

Microbial Systems—Undefined Cultures, Pure Cultures, Cell-Free Extracts

Mohn and Tiedje (1992) reviewed studies with undefined cultures and pure cultures that carry out reductive dehalogenation for many environmental contaminants of interest, including halomethanes, haloethanes, and haloethenes. Transformations have been observed in undefined cultures that have been inoculated from soil, sewage sludge, subsoil, marl, and muck. Pure cultures of denitrifiers, sulfate-reducers, methanogens, acetogens, and fermenters have been isolated that carry out the reduction of CAHs and HACs of interest.

Mohn and Tiedje (1992) also reviewed ecological and physiological principles that are beginning to emerge on the reductive process, including biological activity, acclimation, enrichment, specificity, other donor acceptors, nutrients, temperature, and substrate availability and specificity, products of transformations, and inhibition. The reader is referred to this review for a detailed discussion of these topics.

Reduced compounds found in microbes can dehalogenate HACs. Mohn and Tiedje (1992) also reviewed recent studies of HAC reduction by hematin, vitamin B_{12}, F_{430}, P-450, corrinoids, and ferroheme. In cell-free studies reduced metal-organic compounds have been shown to dehalogenate HACs (Gantzer and Wackett, 1991). The metal-organic compound becomes oxidized in these reactions and must be reduced to react again. Metabolic processes of the cell could reduce the metal-organic compound(s), with the electron donor (growth substrate) driving the reaction.

Rates of Reaction

Vogel et al. (1987) reviewed the role of redox conditions on the degree and rates of dehalogenation. Generally, as the number of halogenated substitutes decreases, removal of additional halogen substitutes by reduction becomes energetically and kinetically more difficult. For example, CT is transformed more rapidly to CF than CF to DM (Bouwer and McCarty, 1985; Bouwer and Wright, 1988; Gälli and McCarty, 1989a). Under methanogenic conditions Bouwer and McCarty (1985) observed CT to be transformed three times faster than CF. Under fermentation conditions CT was transformed 10 times faster than CF (Gälli and McCarty, 1989a). Vogel (1988) found VC to be transformed at a rate over two orders of magnitude lower that TCE. This trend has been also observed in cell-free reactions using vitamin B_{12}, coenzyme F_{430}, and hematin, where the rates of reactions of PCE > TCE > DCE > VC > ethene were correlated with the standard reduction potentials. In these enzyme studies PCE was reduced 10,000 times faster than VC (Gantzer and Wackett, 1991).

Reduction reaction rates are usually faster the lower the redox state, as discussed by Vogel et al. (1987) and Criddle and McCarty (1991). In a column study Bouwer and Wright (1988) observed faster rates of transformation of CT, TCA, and HCE under methanogenic compared to sulfate-reducing conditions, which was generally faster than denitrifying conditions. In electrolytic reductions, as conditions were made more reducing, Criddle and McCarty (1991) observed CT reduction beyond CF to DM. Käster (1991b) indicated that PCE and TCE reduction to *c*-DCE in aquifers, but not to VC, may be tied to redox conditions being higher than methanogenic conditions.

Few CAHs have been shown to serve as primary substrates for energy and growth. DM can be used as a primary substrate under both aerobic and anaerobic conditions. DM can be completely mineralized, while serving as a primary substrate, under anaerobic conditions by municipal sludge-digesting microorganisms (Rittmann and McCarty, 1980; Klecka, 1982; Freedman and Gossett, 1991).

HACs as Electron Acceptors

Until recently the direct use of HACs as electron acceptors for energy and growth had not been demonstrated. The HACs were thought to be transformed cometabolically as electron acceptors. Recent studies have indicated that microbes may obtain benefit from using HACs as electron acceptors. Holliger (1992) isolated a bacterial PER-K23 that reduced PCE and TCE to *c*-DCE while being fed H_2/formate as electron donors. PCE and TCE were the electron acceptors of choice, while other electron acceptors such as nitrate, sulfate, and carbon dioxide did not support growth.

DiStefano et al. (1991) found an anaerobic mixed culture grown on methanol degraded 55 mg/L of PCE to VC and ethene. Approximately 70% of the methanol was converted to acetate, while about 30% was associated with dechlorination reactions. Thus, a large fraction of the electron donor participated in the dechlorination reactions.

The ability of microbes to derive energy while using the CAH as the electron acceptor has implications on *in situ* treatment processes. Direct use of CAH as the electron acceptor would result in efficient use of the electron donors that need to be added to drive the reactions. It is anticipated that when used as electron acceptors, high concentrations of CAHs (>> 1 mg/L) might be treated.

Electron Donors

Reductive dehalogenation depends on the availability of electron donors (Fathepure and Boyd, 1988; Baek and Jaffe, 1989; Freedman and Gossett, 1989). At contamination sites, along with HACs, other co-contaminants, such as fuels, alcohols, ketones, organic acids, and unidentified forms of COD, are often present that can serve as electron donors and drive the anaerobic transformations. Sewell and Gibson (1991), for instance, recently found that toluene (a component of gasoline) could serve as an electron donor for the reduction of PCE to DCE.

For enhanced *in situ* transformation, electron donors may have to be added. Electron donors that have been used in laboratory and field studies are alcohols (methanol, ethanol), sugars (glucose), organic acids (acetate, lactate, pyruvate, benzoate), toluene, or H_2. In a recent study DiStefano et al. (1992) showed hydrogen served as the electron donor for PCE dechlorination to VC and ethene in anaerobic enrichment cultures temporally. Sustained dechlorination for extended periods was obtained when filtered supernatant from a methanol-fed culture was added along with hydrogen. They indicate that methanol and other substrates shown to support dechlorination in pure and mixed cultures may be serving to create a H_2 pool, and thus electron donors that create large hydrogen pools should be selected for bioremediation or methods to use H_2 directly should be devised. More research is needed to determine the role of the electron donors in achieving effective *in situ* transformation.

FIELD STUDIES OF ANAEROBIC TRANSFORMATIONS

Fortuitous Transformations

Fortuitous transformations of HACs have been observed in several field studies (Table 26.2). Decreases with time in concentration of the contaminants, as well as the observation of anticipated products, have indicated transformations. Reductions have occurred under conditions ranging from aerobic to methanogenic. Only in a few cases has the electron donor driving the reaction been identified.

Figure 26.3 shows the decrease in concentration of HACs observed during the Palo Alto, California, groundwater recharge study under the prevailing anaerobic conditions (Roberts et al., 1982). The concentrations of the trihalomethane group of compounds (e.g., CF [$CHCl_3$] and bromoform [$CHBr_3$])

Table 26.2. Fortuitous Transformation of HAC in Aquifers.

Compounds Degraded	Products Detected	Electron Donor	Redox State	Study
CF, $CHBr_3$ TCA, PCE	none	unknown	anaerobic	1
HCE, $CHBr_3$	PCE	unknown	aerobic	2
PCE	TCE, *c*-DCE VC, ethene, ethane	methanol acetic acid	methanogenic	3
TCE	*c*-DCE, VC	unknown methanogenic?		4, 5
TCE	*c*-DCE, VC ethene	COD unidentified	methanogenic	6
PCE	*c*-DCE, TCE	toluene fuel components partially oxidized fuel	methanogenic	7
PCE TCE	*c*-DCE	COD (mixed organics)	denitrifying sulfate-reducing?	8
CT	CF	unknown	denitrifying	9

(1) Roberts et al. (1982); (2) Roberts et al. (1986); (3) Major et al. (1991); (4) Wilson et al. (1991); (5) Ehlke et al. (1991); (6) McCarty and Wilson (1992); (7) Sewell et al. (1990); (8) Käster (1991b); (9) Brouns et al. (1991).

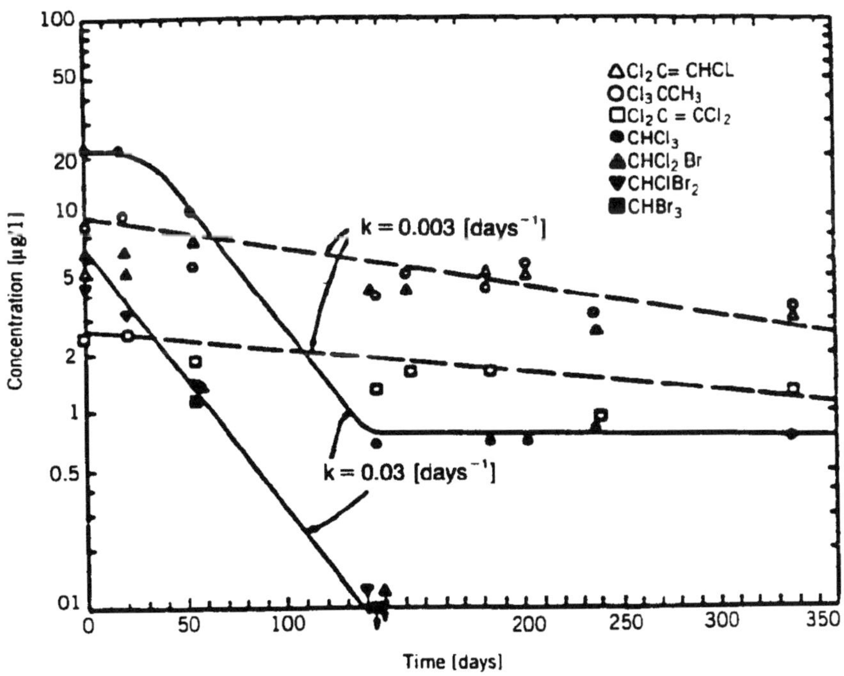

Figure 26.3. Decrease in concentration with time of halogenated aliphatic compounds during the Palo Alto recharge study (reprinted from Roberts et al., 1982).

declined much more rapidly (half-life of ~1 month) than those of the two-carbon compounds such as TCA (half-life of ~8 months). It is not known whether the transformations were mediated by microorganisms or catalyses through abiotic reactions. Here organic compounds present in treated wastewater may have served as a carbon and energy source (Rittmann et al., 1980).

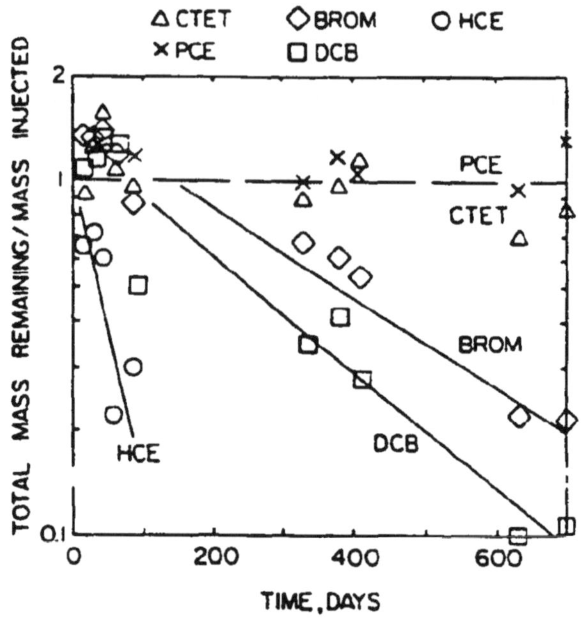

Figure 26.4. Decrease with time in total mass (sorbed plus solution) of chemicals during the Borden transport experiment (from Roberts et al., 1986).

A natural-gradient field experiment at the Borden field site, Canada (Roberts et al., 1986), provides another example of observations of transformations under field conditions (Figure 26.4). In this aerobic aquifer HCE was reduced to PCE. In supporting laboratory studies, Criddle et al. (1986) demonstrated that HCE could be reduced to PCE, possibly by microbial action, under aerobic conditions in the presence of aquifer material. Curtis (1991) demonstrated the abiotic reduction of HCE to PCE in the presence of Borden aquifer solids. Bromoform transformation was also observed in the Borden field tests; however, in laboratory studies transformation did not occur in the presence of oxygen. But transformation did occur under denitrifying conditions in laboratory studies (Bouwer and McCarty, 1983). Carbon tetrachloride and PCE were not significantly degraded in the field tests, consistent with aerobic laboratory studies. McCarty et al. (1992) suggest that micro-anoxic zones in the field were perhaps responsible for the slow disappearance of bromoform, or the transformations may have been abiotic, as indicated by Curtis (1991).

Major et al. (1991) showed both field and laboratory evidence for the *in situ* biotransformation of PCE to ethene and ethane at a chemical transfer facility site in North Toronto. At this site several potential electron donors, including methanol and acetate, were co-contaminants along with PCE. Field evidence and supporting laboratory microcosm studies indicated that anaerobic transformations were occurring. Methane was detected throughout the site and sulfate concentrations were reduced, indicating methanogenic conditions probably existed. TCE, *c*-DCE, and VC were observed as chlorinated transformation products of PCE. Ethene and ethane (only in association with ethene) were also detected, indicating complete dechlorination of the CAHs. Methanol and acetic acid were also detected in the aquifer where the transformations were suspected. Chloride levels were evaluated above background where transformations had occurred or were occurring, which may have resulted from the dechlorination of the CAHs.

Population enumeration showed that viable methanogenic populations and sulfate-reducing populations were present in subsurface sediments (Major et al., 1991). Microbially active microcosms were stimulated from subsurface sediments fed site groundwater amended with methanol, acetate, methanol/acetate, and PCE. The methanol/acetate-amended microcosm produced the most ethene. *c*-DCE was the major DCE isomer observed, consistent with the field observations (Figure 26.5). PCE was degraded rapidly with a transient buildup and decreases of TCE, *c*-DCE, and VC. VC transformation appears to increase after *c*-DCE is removed. It may be here that *c*-DCE was inhibiting the

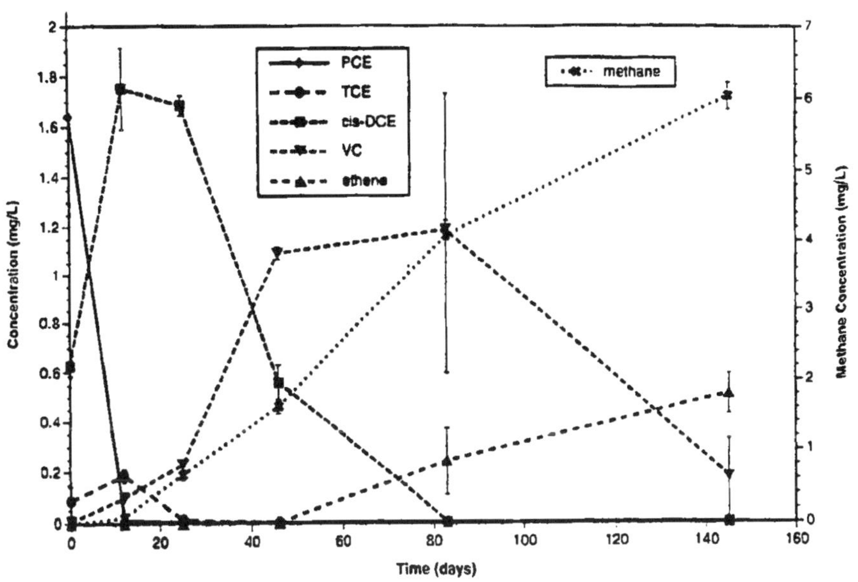

Figure 26.5. Transformation of PCE and the production or transformation of TCE, *c*-DCE, and VC, in a methanogenic microcosm fed methanol + acetate (reprinted from Major et al., 1991).

transformation of VC, as was recently observed by Zinder et al. (1992), in anaerobic cultures fed methanol. Production of ethene is shown, indicating the complete dechlorination of the CAHs.

Wilson et al. (1991) and Ehlke et al. (1991) conducted field and microcosm studies on groundwater contamination at the Picatinny Arsenal, New Jersey. TCE is the major contaminant at that site, while TCA and PCE are minor contaminants. *c*-DCE and VC, which were not original contaminants, have accumulated in the groundwater, indicating reductive dechlorination has occurred. The source of electron donor in the aquifer was not reported. Microcosms were constructed with samples obtained from within and outside the contaminant plume and were fed different primary substrates. Fastest removal rates were obtained from sites within the plume with first-order TCE transformation rates ranging between 0.012 to 0.02 per week, and *c*-DCE rates from 0.01 to 0.18 per week. The addition of electron donors either had no effect or an inhibitory effect on transformation rates. These rates agreed with estimates made for the field based on concentration measurements and possible residence times.

McCarty et al. (1991) reported results from a contaminated aquifer in St. Joseph, Michigan where *in situ* bioremediation using aerobic treatment was being considered. Evidence of anaerobic transformation of TCE was indicated, with DCE and VC observed in monitoring well samples (Figure 26.6). DCE and VC migrated ahead of TCE, which may have resulted from their being less retarded than TCE, and TCE being effectively anaerobically transformed and not transported as far. McCarty and Wilson (1992) presented half-reactions for the reduction of TCE to *c*-DCE, VC, and ethene, and the equivalent decrease in COD to drive the reactions (Table 26.3). The formation of methane is shown to require the greatest COD decrease per gram of product formed. An efficient use of an electron donor (in this case COD) would limit the production of methane, using CO_2 as an electron acceptor, and would favor the use of the CAHs directly as electron acceptors.

Evidence for complete dechlorination to ethene was provided in detailed sampling of a portion of the plume noted by the three transects in Figure 26.7. COD concentration reductions as a potential electron donor driving the anaerobic reactions were noted to coincide with DCE and VC observations (Figures 26.6 and 26.7). *c*-DCE was the predominant DCE isomer observed. Based on the detailed aqueous measurements at selected locations, 8 to 36 mg/L of COD reduction was associated with dechlorination reactions, while 18 to 47 mg/L were associated with methane formation, which represents a significant amount of the COD reduction, as indicated in Figure 26.7 (McCarty and Wilson,

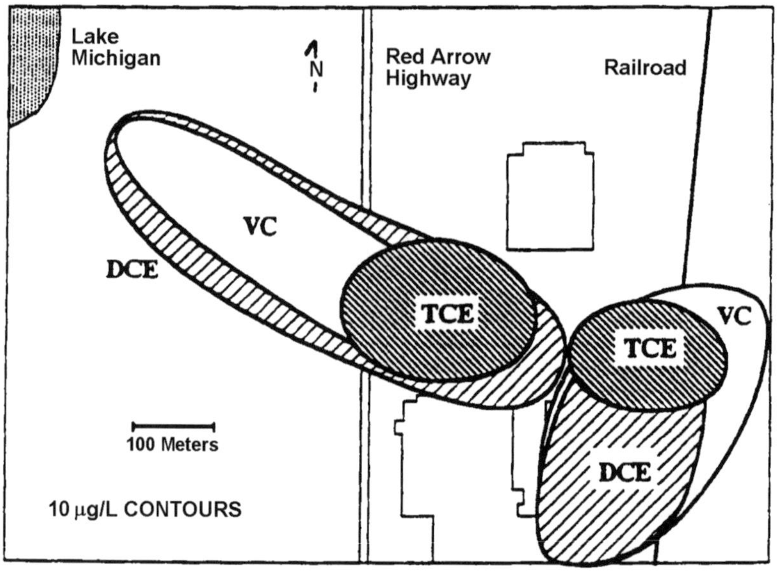

Figure 26.6. Chlorinated aliphatic hydrocarbons: 10 µg/L contours at St. Joseph, Michigan, site (from McCarty and Wilson, 1992).

Table 26.3. Half-Reactions Indicating Electron Equivalents of Change and Associated Equivalent COD Decrease Associated with the Change (from McCarty and Wilson, 1992).

Transform. Product	Mol. Wt.	Half-Reaction	Equiv. COD Decrease g COD/g Product[a]
Methane	16	$CO_2 + 8H^+ + 8e^- \rightarrow CH_4 + 2H_20$	4
DCE	97	$CHCl = CCl_2 + H^+ + 2e^- \rightarrow CHCl = CHCl + Cl^-$	0.16
VC	62.5	$CHCl = CCl_2 + 2H^+ + 4e^- \rightarrow CH_2 = CHCl + 2Cl^-$	0.51
Ethene	28	$CHCl = CCl_2 + 3H^+ + 6e^- \rightarrow CH_2 = CH_2 + 3Cl^-$	1.71

[a] From $[O_2 + 4H^+ + 4e^- \rightarrow 2H_20]$, the half-reaction for oxygen, one electron equivalent of COD equals one-fourth of molecular oxygen, or 8 g; thus Equiv. COD Decrease = 8n/mol. wt.), where n is the number of electrons in the half-reaction.

1992). Flux estimates indicate approximately 20% of the CAHs have been transformed to ethene (Semprini et al., 1992).

Anaerobic transformation of PCE to TCE and *c*-DCE has been indicated at the U.S. Coast Guard Air Station at Traverse City, Michigan (Sewell et al., 1990), where co-contamination of PCE with gasoline exists. Field observations indicate that the fuel components may have served as electron donors, with PCE and toluene concentration decreases associated with the observations of TCE, *c*-DCE, and methane. Sewell and Gibson (1991) demonstrated in laboratory anaerobic microcosm studies, using site subsurface materials, that toluene could serve as an electron donor to drive the reduction of PCE to *c*-DCE.

Käster (1991a,b) noted that in contaminated aquifers in Europe transformation of PCE and TCE to *c*-DCE was observed in aquifers that were anaerobic, but not methanogenic. He indicated that further reduction to VC was probably limited in aquifers where lower reduction potentials of methanogenesis were not achieved. This may help explain why VC is observed at some sites, while not at others.

At the U.S. Department of Energy's (DOE) Hanford Site an aquifer has been contaminated with CT and nitrate (Brouns et al., 1991). The observations of CF along with CT have indicated that biological transformations are probably occurring in the aquifer. Upon the addition of acetate as an

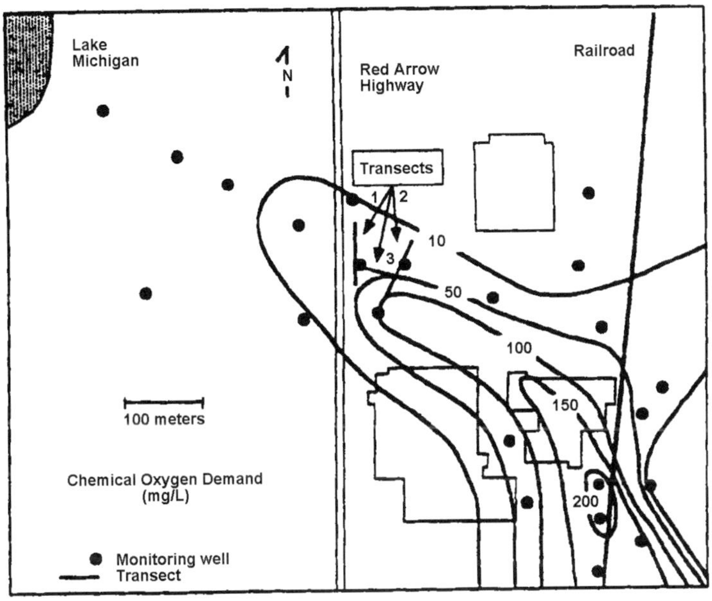

Figure 26.7. COD contours from the St. Joseph, Michigan, site (from McCarty and Wilson, 1992).

electron donor to laboratory microcosms, denitrification was observed, along with CT transformation and some formation of CF.

The fortuitous transformation of CAHs anaerobically has been documented in several field studies discussed here and has been implicated at many other contaminated sites. The development of methanogenic conditions has been observed at sites where complete mineralization of PCE and TCE to ethene has been observed. For HCE and CT, which are fully chlorinated, reductions have occurred under much less reducing conditions.

Most often the electron donors driving the reactions have not been identified. Methanol and acetic acid were observed at the site studied by Major et al. (1991) and were suspected as potential donors. Microcosm studies confirmed this possibility. Sewell et al. (1990) found fuel components may drive the anaerobic reductions, which is important where co-contamination of HACs with fuels exists.

c-DCE has been the predominant DCE isomer observed at sites of PCE and TCE transformation, consistent with laboratory culture and microcosm studies. VC is often observed as a transformation intermediate. Complete transformation of VC to ethene has been observed both in the field and in laboratory microcosms, under methanogenic conditions.

Microcosm studies using subsurface solids have been shown to be a valuable method for determining if organisms are present that can carry out the transformations. Microcosms may also be used to determine expected transformation products and to evaluate potential electron donor additions.

The observations of fortuitous transformations in the field are encouraging and suggest that application of electron donors and other nutrients in a controlled manner has promise for enhanced transformations. The observations indicate that significant fortuitous transformations may be achieved at sites without active enhancement. A better understanding of the processes will aid in taking advantage of these fortuitous transformations for site remediation.

ENHANCED *IN SITU* ANAEROBIC TRANSFORMATION OF HACS

Results of a Pilot-Scale Study

A pilot-scale study of enhanced *in situ* transformation of CT and other HACs under anaerobic conditions was performed at the Moffett Field Naval Air Station Test Facility (Semprini et al., 1991b;

1992). The goals of this field evaluation were to determine whether reductive transformation of CT could be accomplished under the mildly reducing conditions of denitrification, what factors affect the rates and extents of transformation, and what transformation intermediates might be formed. In addition the disappearance of other contaminants present in the site groundwater, including TCA and two chlorofluorocarbons, CFC-11, and CFC-113, was followed.

The work assessed under field conditions the capacity of indigenous organisms to metabolize the target HACs under anoxic conditions created by microbial growth. Since nitrate was present in groundwater at the site, the ability of organisms growing under denitrifying conditions was evaluated.

The test site consists of a layer of sand and gravel, approximately 5 m below the surface and 1.2 m thick, well confined above and below by a silty clay layer of low permeability (Figure 26.8). The transmissivity of the test zone is high, approximately 100 m^2/day. The dissolved oxygen concentration was below detection (0.2 mg/L), while nitrate and sulfate, two potential electron acceptors, were present at concentrations of 25 mg/L (as nitrate) and 700 mg/L (as sulfate). The groundwater was contaminated with TCA (50 µg/L), CFC-113 (6 µg/L), and CFC-11 (3 µg/L). The target compound, CT, was not present and therefore was continuously added in a controlled manner to the injected water. Nitrate present in the subsurface groundwater served as a source of N. Phosphorus concentrations in the groundwater were low (0.1 mg/L); however, phosphate minerals that are present probably provided a source of phosphate (Roberts et al., 1989).

The experimental methodology entailed creating a flow field dominated by pumping from an extraction well (P), while introducing solutes in known amounts at the SI injection well, (Figure 26.8). Evidence of transformation was then assessed by quantitative examination of the HACs' concentration histories at the S1, S2, and S3 monitoring wells located 1, 2.2, and 3.8 m from the injection well, respectively. A specially designed automated data acquisition and control system provided continuous records of high-accuracy data over sustained periods, which enabled mass balances to be made with relative errors of only a few percent (Hopkins et al., 1988).

Groundwater was injected at a rate 1.5 L/min into the SI well and extracted at a rate of 10.0 L/min into well P to ensure that the injected fluid completely permeated the regions around the S1 and S2 monitoring wells, and that the extraction well recovered more than 95% of the injected fluid (Roberts at al., 1990). The injected fluid was the extracted groundwater, amended with the chemicals of interest. Since the injected fluid represented only 15% of the extracted groundwater, the nitrate, sulfate,

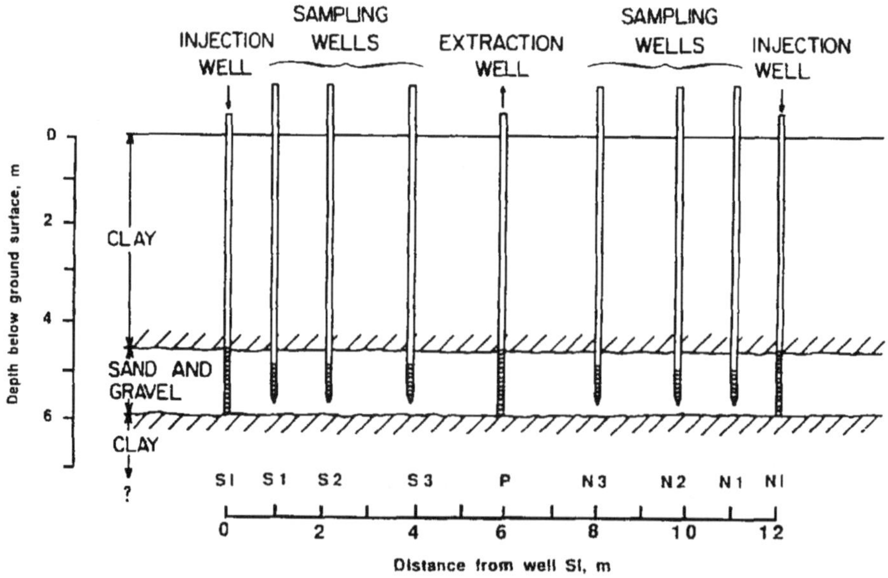

Figure 26.8. Schematic of the test zone at the Moffett Field bioremediation test facility (reprinted from Roberts et al., 1990 with permission from Ground Water Publishing Company).

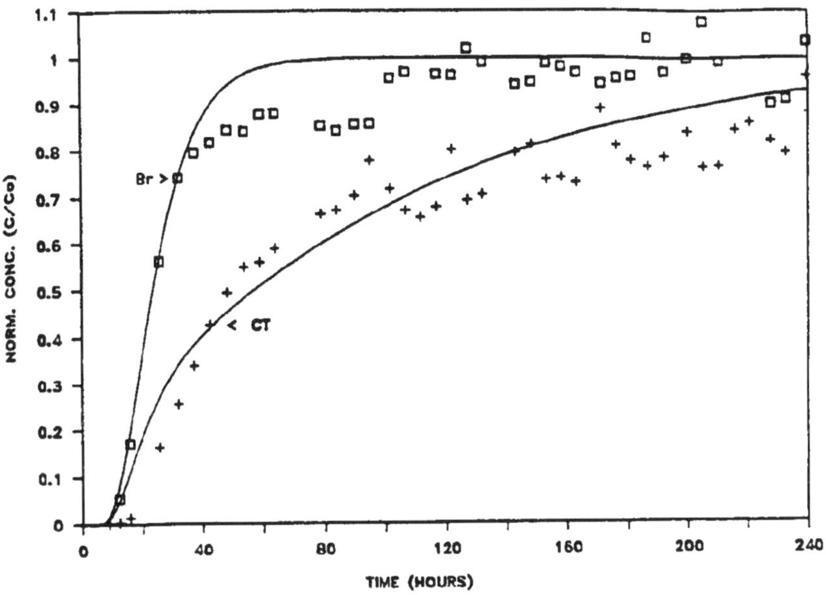

Figure 26.9. Bromide and CT breakthrough at the S2 well and corresponding simulation using a nonequilibrium sorption model (from Semprini et al., 1991b).

TCA, CFC-11, and CFC-113 present in the native groundwater were not added, but were reinjected, while CT and acetate, which were not present, were added to the injected fluid.

Initial experiments served to study CT transport in the absence of biostimulation. CT was continuously injected at a concentration of 70 µg/L in the presence of nitrate. Figure 26.9 shows the breakthrough at the S2 observation well along with model simulations, using a 1-D advection/dispersion transport model with rate-limited sorption (Semprini et al., 1991b). The breakthrough CT is shown to be retarded compared to the bromide tracer, with retardation factors ranging from 1.5 to 2.0. The CT concentration rapidly approached the injection concentration, indicating minimal transformation and sorption with prolonged injection. A minor amount of CF production was observed soon after CT addition, with the maximum CF concentration representing 3 to 4% of the CT added. Thus, minor CT transformation was observed before biostimulation through acetate addition.

Biostimulation of the test zone was initiated through acetate addition as a growth substrate. Clogging of the injection well and borehole was effectively controlled by adding the acetate in a high-concentration pulse (320 to 600 mg/L) for a period of 1 hr in a 13-hr pulse cycle, yielding a time-averaged concentration of 25–46 mg/L. Nitrate utilization commenced immediately after the introduction of acetate and was complete within 100 hr (Figure 26.10), while acetate utilization also commenced immediately but was expressed somewhat more slowly and less completely because of the stoichiometric excess applied. The onset of CT transformation was observed after approximately 350 hours (Figures 26.10 and 26.14), after which the CT concentrations at the monitoring wells gradually declined, more rapidly at the more distant (S2) well (Figure 26.14) than at the nearer (S1) well (Figure 26.10). Chloroform (CF) appeared as an intermediate product of the CT transformation at all of the sampling points, in an amount corresponding to approximately one-half to two-thirds of the CT that disappeared.

The response indicated that the most rapid rates of CT transformation did not occur in the first meter of transport where most of the nitrate and acetate were consumed, but in further zones where significantly less acetate and nitrate were consumed. The denitrifying population stimulated near the injection well did not participate in the transformation process to the same extent as microbes stimulated farther away. CT transformation by denitrifiers may have been strongly inhibited by nitrate. Another possibility is that a secondary population, living on acetate or decay products of denitrifiers,

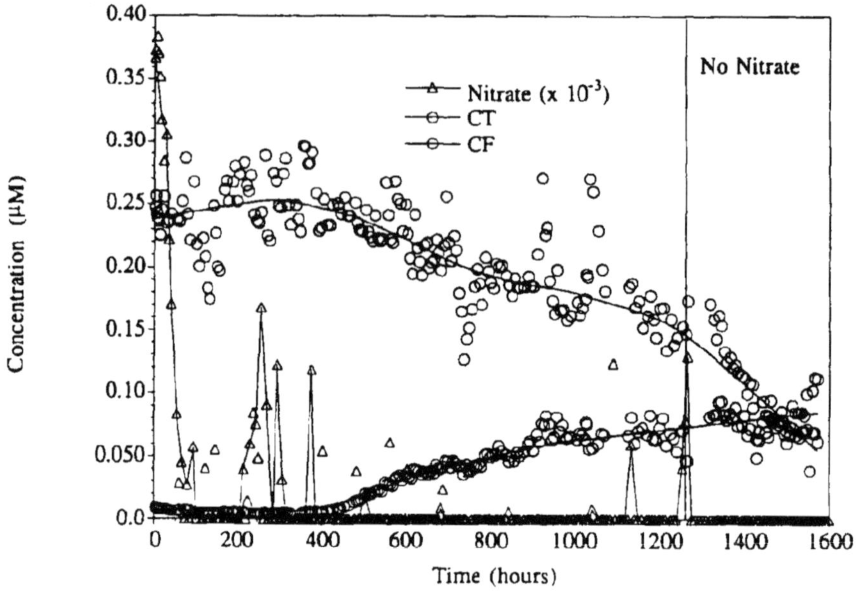

Figure 26.10. Nitrate, CT, CF response at the S1 well due to anoxic biostimulation of the Moffett Field test zone.

grew slowly and was responsible for the transformation. The growth of the secondary population and transformation of CT may have been inhibited by the presence of nitrate.

To study the effect that nitrate had on biotransformation, nitrate was removed from the reinjected fluid using a surface reactor fed acetate. Upon removing nitrate, CT rapidly decreased at the S1 well while CF increased to a lesser extent (Figure 26.10). Of the CT transformed, 55 to 67% appeared as CF before nitrate was eliminated, whereas only 30 to 40% was observed as CF after nitrate was terminated. CF was the main chlorinated intermediate found. DM and chloromethane, possible intermediate products of CT transformation, were not detected at the detection limit of 1 µg/L.

There was no direct evidence for the stimulation of sulfate-reducing bacteria or methanogenic bacteria when nitrate was completely removed. Neither sulfide nor methane was detected in groundwater extracted from the test zone. Reactions with test zone minerals may have scavenged sulfide from the groundwater. Substantial acetate utilization persisted in the absence of nitrate feed (8 mg/L), suggesting that sulfate may have served as an electron acceptor.

The background contaminants, including TCA, CFC-11, and CFC-113, were also partially transformed under the influence of anoxic biostimulation. The response of the HACs was similar to that of CT (Figure 26.11), indicating similar process(es) are acting to degrade the contaminants but at different rates. The ranked rates of transformation are as follows: CT > CFC-11 > CFC-113 > TCA. All HACs (except TCA) showed a significant increase in the semi-log slope after nitrate was removed, indicating accelerated rates of transformation. Similar changes in slopes occurred both for CT and CFC-11, with a greater change observed at the S1 well, since most of the nitrate was present between the injection and S1 well before being removed.

The maximum extents of transformation in the 3-m biostimulated zone were estimated by comparisons with bromide as a conservative tracer. The 95% confidence intervals of the percent transformation are as follows: TCA, 8–16%; CFC-113, 10–30%; CFC-11, 69–75%, and CT, 95–97%. As indicated in Figure 26.11, steady-state conditions had not been achieved by the end of the experiments; thus these transformation percentages are considered conservative estimates.

The products of CFC-11 and CFC-113 were not detected, probably due to the low concentration of the parent compound. A probable intermediate of CFC-11 is CFC-12 (Wolf et al., 1978; McCarty and Reinhard, 1993). Probable intermediates of CFC-113 transformation are HCFC-123a (1,2-dichloro-1,2,2-trifluoroethane) and CFC-1113 (chlorotrifluoroethene) (Lesage et al., 1990; 1992). 1,1-DCA is

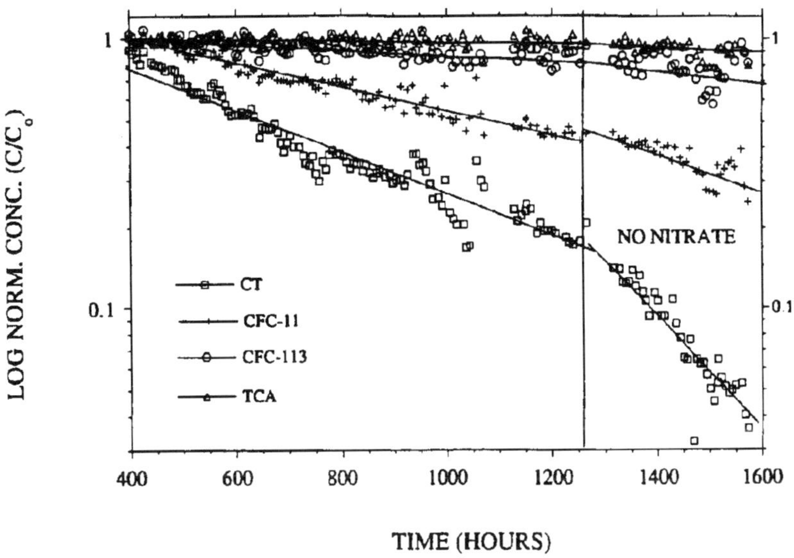

Figure 26.11. Semi-log response of CT, CFC-11, CFC-113, and TCA due to the anoxic biostimulation of the Moffett Field test zone (reprinted from Semprini et al., 1992).

an expected intermediate of reductive TCA transformation. 1,1-DCA was present as a background contaminant in the test zone and DCA increases were statistically insignificant at the 95% confidence interval.

The extent of complete mineralization of CT to CO_2 and chloride was not determined in the field study. However, supporting laboratory microcosm studies with test zone core materials and with ^{14}C-radio-labeled CT conducted under similar conditions of the field tests (Lang et al., 1990) confirmed that a significant amount of CT was completely mineralized to CO_2, while CF production was similar to that observed in the field and represented 30 to 40% of the CT transformed.

Modeling of the Pilot-Scale Study

A non-steady-state model, developed for simulating the biostimulation and biotransformation tests, proved useful in interpreting the results of the field experiments. Details of the model are provided by Semprini et al. (1991a). The model accounts for the basic phenomena of microbial growth, electron donor and electron acceptor utilization, biotransformation of the HACs, and the formation and possible transformation of intermediate products. The model simulates the growth and metabolism of two microbial populations: a denitrifying population and a second population that utilizes the respiration products of denitrifiers as depicted in Figure 26.12. In the simulations either population could have transformed the HACs. Model simulations, however, indicated that the second population transformed the HACs.

Simulations of acetate and nitrate response at the S2 well are shown in Figure 26.13. Kinetic parameters used in the simulations are those reported by Rittmann et al. (1988) for a denitrifying population. The initial population in the test zone was unknown and used a fitting parameter. The model successfully simulates the uptake of acetate and nitrate observed along with the attenuation in both acetate and nitrate pulses due to transport and biological uptake. However, simulation of CT transformation due to stimulated denitrifiers alone did not match the field observations, with more rapid CT transformation being predicted than was observed in the field.

A simulation of the CT and CF response at the S2 well is shown in Figure 26.14. The simulation shows the decrease in CT and the increase in CF due to stimulation of the secondary population. The simulation indicates that the CF itself was transformed, causing a plateau in concentration at a later

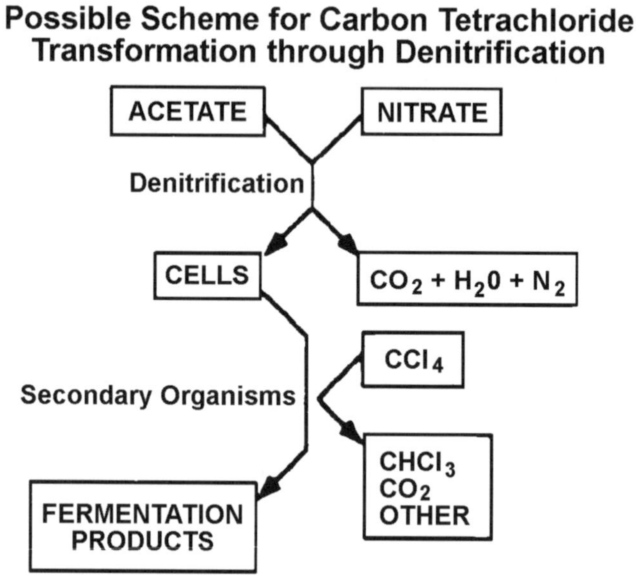

Figure 26.12. Conceptual two-population model for CT transformation that was included in model simulations of the Moffett Field biostimulation tests.

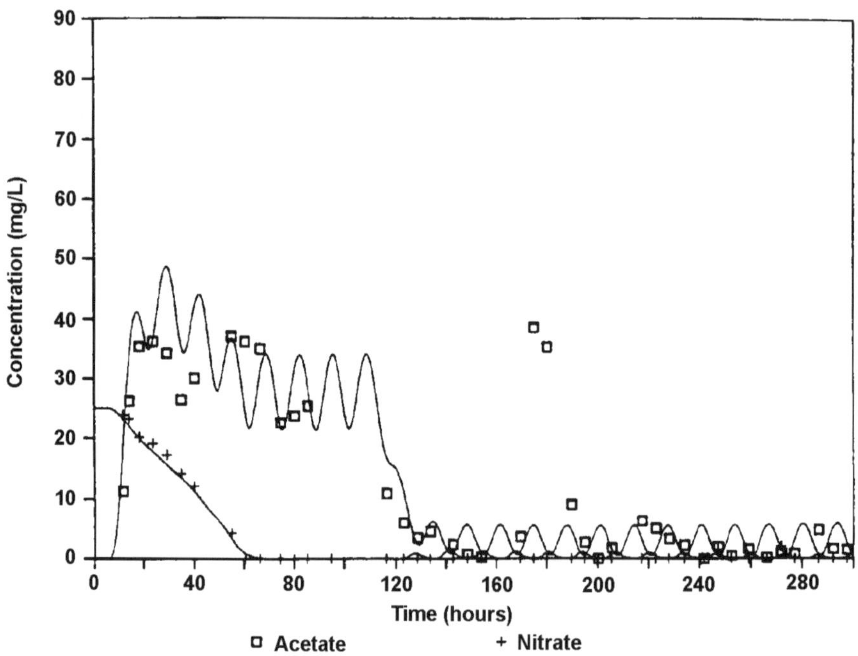

Figure 26.13. Simulations of acetate and nitrate response at the S2 well due to biostimulation.

time, followed by a slight decrease consistent with the field observations. The simulations demonstrate the importance of modeling both the parent compound and the intermediates.

The response of CFC-11, CFC-113, and TCA was also simulated, with the rate of the cometabolic transformation being the main fitting parameter. The values of the first-order rate coefficients (k/K_s) determined from the model fit the field observations given in Table 26.4. The coefficients are within

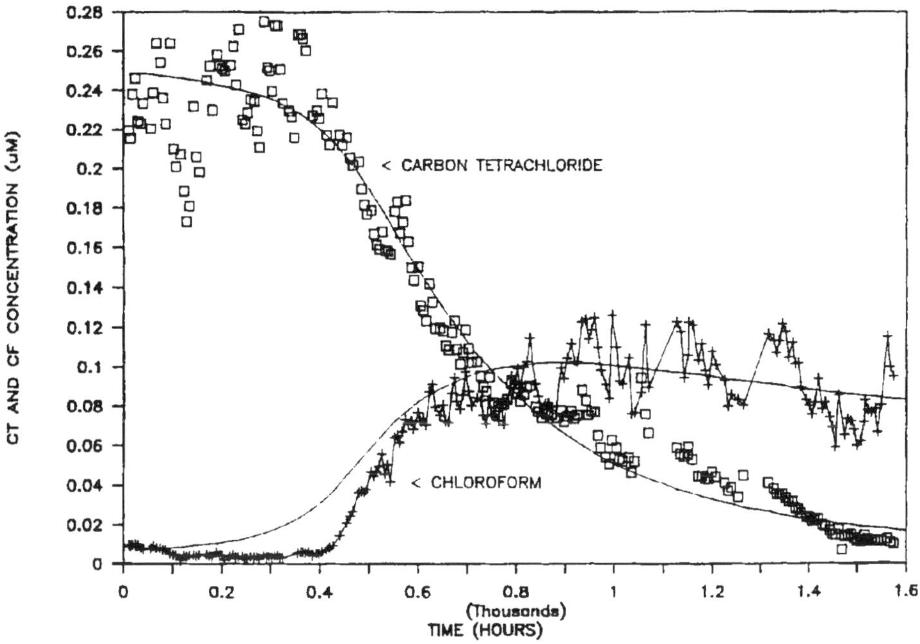

Figure 26.14. Simulations of CT and CF response at the S2 well due to biostimulation.

Table 26.4. Values of k/K_s for this study (from Semprini et al., 1991b)

Compound	Field Study $\frac{k}{K_s}$ $\left(\frac{L}{mg-cells \cdot d}\right)$	Gälli and McCarty (1989a) $\frac{k}{K_s}$[a] $\left(\frac{L}{mg-cells \cdot d}\right)$	Bouwer and Wright (1988) $\frac{k}{K_s}$[b] $\left(\frac{L}{mg-cells \cdot d}\right)$
CT	0.40	0.12	0.20
CFC-11	0.16	—	—
CF	0.08	0.012	—
CFC-113	0.04	—	—
TCA	0.01	0.007	0.005

[a]Reported rates on protein basis, converted to cell based on protein being 65% of the cell mass.
[b]Under sulfate-reducing conditions.

range of those reported for microbial transformations under sulfate-reducing conditions and by a pure *Clostridium* culture. The rate coefficient for CT is 40 times greater than for TCA, consistent with the results of Bouwer and Wright (1988). CF transformation is five times slower than CT transformation, while Gälli and McCarty (1989a) reported a rate for a *Clostridium* sp. culture that was ten times slower. The simulation indicates that the rate coefficient for CFC-11 was less than half that of CT. This most likely results from the substitution of one fluorine atom for a chlorine atom. Both the absolute rates and relative rate coefficients from laboratory studies and the field study appear reasonable, which is encouraging.

SUMMARY

Reductive transformation of HACs by microorganisms is occurring fortuitously at contaminated sites and is a potential process for enhanced *in situ* remediation. The pathways presented suggest that it is possible to render harmless essentially any chlorinated aliphatic compound under anaerobic conditions. Recent laboratory studies and field observations indicate complete transformation of VC to ethene, which is encouraging. More research is needed to determine how complete transformation can be effectively achieved in the field.

Several problems hinder this potential approach to *in situ* bioremediation. First, the biotic transformations generally involve cometabolism, meaning that other organic compounds must be present to serve as primary substrates for organism growth. In the enhanced case presented here, acetate had to be added. Methods are needed for effective addition and mixing of nutrients in the subsurface. One advantage of the anaerobic processes, however, over an aerobic cometabolic process is that addition of an electron acceptor, such as oxygen, is not also necessary. Recent laboratory studies have indicated that the electron acceptor for growth might even be the HACs. This would result in effective use of the electron donors that are added for enhanced remediation. Another advantage is that anaerobes tend to grow more slowly than aerobes. Thus it may be possible to grow anaerobes over greater areas in the subsurface using fewer wells to add electron donors.

Another problem is that the rates of anaerobic transformation are much greater for the highly chlorinated compounds than for the less chlorinated compounds, so that the latter persist longer in the environment. *c*-DCE and VC are observed in field studies of fortuitous transformations of PCE and TCE, while the formation of CF occurred in the enhanced *in situ* transformation of CT. Thus partial transformation of HACs to nontoxic end products may be achieved, while some toxic end products needing further treatment may remain. Here it may be possible to use an anaerobic process followed by an aerobic cometabolic process which works well at degrading less halogenated products of anaerobic transformation, as described by McCarty in Chapter 23 of this volume on aerobic cometabolic treatment.

Some of the anaerobic transformation products are more hazardous than the parent compounds. Examples are TCE transformation to VC, and TCA transformation to 1,1-DCE abiotically. Here more information is needed to determine if transformation can be promoted without the formation of toxic products.

Reaction rates tend to be greater under highly reducing conditions associated with methane formation than under the less reducing conditions associated with denitrification. The latter is the main anaerobic process occurring when excess nitrates are present. Reductive transformation rates are somewhat intermediate between the two under conditions favoring sulfate reduction (sulfate present, but no nitrate). In practice, problems might arise with anaerobic transformation if high nitrate or sulfate concentrations are present in the groundwater. The utilization of these electron acceptors might first be required before effective transformation is observed. In the enhanced *in situ* study, accelerated CT transformation was observed after nitrate was completely removed from the injected fluid. The amount of acetate needed here to achieve transformation might have been less if less nitrate were present in the aquifer. Creation of methanogenic conditions may require the addition of large amounts of electron donor to first utilize more energetically favorable electron acceptors present.

When the proper environmental conditions are present, microorganisms that can bring about the transformations through cometabolism must also be present. Thus, with anaerobic conditions one cannot count upon sufficiently high rates and complete transformation to harmless products to occur in groundwater, unless all the right conditions are present. Here microcosm studies conducted with aquifer solids and groundwater can help determine if the required microorganisms are present and whether appropriate nutrient needs are met for effective transformation. Studies reported here indicate that microcosms are a valuable means for evaluating the potential for *in situ* remediation.

Transport of the HACs must also be considered. The less halogenated HACs would tend to be less hydrophobic, and thus less strongly sorbed onto the aquifer solids. Where fortuitous transformations are occurring the less chlorinated, and possibly more toxic products, may be transported more rapidly away from the parent compounds. Contamination of greater areal extents may result, as was indicated at the St. Joseph site. In enhanced transformation the products formed might be more easily removed from the aquifer than the more strongly sorbed parent compounds. Here, anaerobic treatment in the aquifer might be considered, if parent compounds are especially difficult to remove. The less halogenated products might also be more easily destroyed by biological or physical/chemical processes at the surface.

Water quality aspects must also be considered. Water quality might deteriorate in an aquifer that is driven anaerobic. The creation of anaerobic conditions may result in the increase in metals, both in inorganic and chelated forms. Fermentation reactions may lead to taste or odor problems in the water. An increase in carbon content of the water might also result, which may potentially cause problems when the groundwater is chlorinated for disinfection.

The modeling of anaerobic transformation processes is a valuable tool, aiding in the interpretation of field tests. Modeling will also serve as a tool in the design of enhanced anaerobic transformation processes by helping to identify nutrient delivery schemes that might be applied to achieve effective transformation. Modeling might also serve to determine what products may persist in an enhanced remediation scheme or to determine the ultimate result of fortuitous transformations.

ACKNOWLEDGMENTS

This chapter was written with support from the U.S. Environmental Protection Agency through its sponsored Western Region Hazardous Substance Research Center under Agreement CR-815738. Research at the Moffett Field site described in this chapter was supported by the U.S. Environmental Protection Agency Biosystems Program and the Robert S. Kerr Environmental Research Laboratory under Agreement CR-815816, but has not been subjected to Agency review and therefore does not necessarily reflect the views of the Agency, and no official endorsement should be inferred.

REFERENCES

Bae, W., J.E. Odencrantz, B.E. Rittmann, and A.J. Valocchi. 1990. Transformation Kinetics of Tracer-Level Halogenated Organic Contaminants in a Biologically Active Zone (BAZ) Induced by Nitrate Injection. *J. Contam. Hydrol.*, **6**:53–68.

Baek, N.M. and P.R. Jaffe. 1989. The Degradation of Trichloroethylene in Mixed Methanogenic Cultures. *J. Environ. Quality*, **18**:515–518.

Barrio-Lage, G., F.Z. Parsons, R.S. Nassar, and P.A. Lorenzo. 1986. Sequential Dehalogenation of Chlorinated Ethenes. *Environ. Sci. Technol.*, **20**:96–99.

Belay, N. and L. Daniels. 1987. Production of Ethane, Ethylene, and Acetylene from Halogenated Hydrocarbons by Methanogenic Bacteria. *Appl. Environ. Microbiol.*, **53**:1604–1610.

Bouwer, E.J. and P.L. McCarty. 1982. Transformations of 1- and 2-Carbon Halogenated Aliphatic Organic Compounds Under Methanogenic Conditions. *Appl. Environ. Microbiol.*, **45**:1286–1294.

Bouwer, E.J. and P.L. McCarty. 1983. Transformation of Halogenated Organic Compounds under Denitrification Conditions. *Appl. Environ. Microbiol.*, **45**:1295–1299.

Bouwer, E.J. and P.L. McCarty. 1985. Utilization Rates of Tracer Halogenated Organic Compounds in Acetate Grown Biofilms. *Biotech. Bioeng.*, **27**:1564–1571.

Bouwer, E.J. and J.P. Wright. 1988. Transformation of Trace Halogenated Aliphatics in Anoxic Biofilm Columns. *J. Contam. Hydrol.*, **2**:155–169.

Bouwer, E.J., B.E. Rittmann, and P.L. McCarty. 1981. Anaerobic Degradation of Halogenated 1- and 2-Carbon Organic Compounds. *Environ. Sci. Technol.*, **15**:596–599.

Brouns, T.M. et al. 1991. Bioremediation of Hanford Groundwater. PNL-SA-19646, Pacific Northwest Laboratory, Richland, WA.

Brown, J.F., Jr., R.E. Wagner, H. Feng, D.L. Bedard, M.J. Brennan, J.C. Carnahan, and R.J. May. 1987. Environmental Dechlorination of PCBs. *Environ. Toxicol. Chem.*, **6**:579–593.

Cline, P.V. and J.J. Delfino. 1989. Transformation Kinetics of 1,1,1-Trichloroethane to the Stable Product 1,1-Dichloroethene. In *Biohazards of Drinking Water Treatment*, Lewis Publishers, Inc., Chelsea, MI, pp. 47–56.

Criddle, C.S. and P.L. McCarty. 1991. Electrolytic Model System for Reductive De-halogenation in Aqueous Environments. *Environ. Sci. Technol.*, **25**:973–978.

Criddle, C.S., C. Elliott, P.L. McCarty, and J.F. Barker. 1986. Reduction of Hexachloroethane to Tetrachloroethylene in Groundwater. *J. Contam. Hydrol.*, **1**(1/2):133–142.

Criddle, C.S, J. T. Dewitt, D. Grbi-Galiand, and P.L. McCarty. 1990a. Transformation of Carbon Tetrachloride by *Pseudomonas* sp. Strain KC under Denitrifying Conditions. *Appl. Environ. Microbiol.*, **56**:3240–3246.

Criddle, C.S., J.T. Dewitt, and P.L. McCarty. 1990b. Reductive Dehalogenation of Carbon Tetrachloride by *Escherichia coli* K-12. *Appl. Environ. Microbiol.*, **56**:3247–3254.

Curtis, G.P. 1991. Reductive Dehalogenation of Hexachloroethane and Carbon Tetrachloroethane by Aquifer Sand and Humic Acid. Ph.D. Dissertation, Stanford University, Stanford, CA.

DiStefano, T.D., J.M. Gossett, and S.H. Zinder. 1991. Reductive Dechlorination of High Concentrations of Tetrachloroethane to Ethane by Anaerobic Enrichment Culture in the Absence of Methanogenesis. *Appl. Environ. Microbiol.*, **57**:2287–2292.

DiStefano, T.M., J.M. Gossett, and S.H. Zinder. 1992. Hydrogen as an Electron Donor for Dehalogenation of Tetrachloroethane by an Anaerobic Mixed Culture. *Appl. Environ. Microbiol.*, **58**:3622–3629.

Dolfing, J. and J.M. Tiedje. 1987. Growth Yield Increase Linked to Reductive Dechlorination in a Defined 3-Chlorobenzoate Degrading Methanogenic Coculture. *Arch. Microbiol.*, **149**:102–105.

Egli, C., R. Scholtz, A.M. Cook, and T. Liesinger. 1987. Anaerobic Dechlorination of Tetrachloromethane and 1,2-Dichloroethane to Degradable Products by Pure Cultures of *Desulfobacterium* sp. and *Methanobacterium* sp. *FEMS Microbiol. Lett.*, **43**:257–261.

Egli, C., T. Tsuchan, R. Scholtz, A.M. Cook, and T. Lieslinger. 1988. Transformation of Tetrachloromethane to Dichloromethane and Carbon Dioxide by *Acetobacterium Woodi. Appl. Environ. Microbiol.*, **55**:2819–2824.

Ehlke, T.A. et al. 1991. Biotransformation of cis-1,2-dichloroethylene in Aquifer Material from Picatinny Arsenal, Morris County, NJ. In U.S. Geological Survey, Water-Resources Investigations Report, 91-4034, Reston, VA, pp. 704–707.

Fathepure, B.Z. and S.A. Boyd. 1988. Dependence of Tetrachloroethylene Dechlorination on Methanogenic Substrate Consumption by *Methanosarcina* sp. Strain DCM. *Appl. Environ. Microbiol.*, **54**:2976–2980.

Fathepure, B.Z., J.P. Nengu, and S.A. Boyd. 1987. Anaerobic Bacteria That Dechlorinate Perchloroethene. *Appl. Environ. Microbiol.*, **53**:2671–2674.

Freedman, D.L. and J.M. Gossett. 1989. Biological Reductive Dechlorination of Tetrachloroethylene and Trichloroethylene to Ethylene under Methanogenic Conditions. *Appl. Environ. Microbiol.*, **55**:2144–2151.

Freedman, D.L. and J.M. Gossett. 1991. Biodegradation of Dichloromethane and Its Utilization and Growth Substrate under Methanogenic Conditions. *Appl. Environ. Microbiol.*, **57**(10):2847–2857.

Gälli, R. and P.L. McCarty. 1989a. Biotransformation of 1,1,1-Trichloroethane, Trichloromethane, and Tetrachloromethane by *Clostridium* sp. *Appl. Environ. Microbiol.*, **55**:837–844.

Gälli, R. and P.L. McCarty. 1989b. Kinetics of Biotransformation of 1,1,1-Trichloroethane by *Clostridium* sp. Strain TCA IIB. *Appl. Environ. Microbiol.*, **55**:845–851.

Gantzer, C.J. and L.P. Wackett. 1991. Reductive Dechlorination Catalyzed by Bacterial Transition-Metal Coenzymes. *Environ. Sci. Technol.*, **25**:715–722.

Holliger, C. 1992. Reductive Dehalogenation by Anaerobic Bacteria. Ph.D. Thesis. Dept. of Microbiology, Agricultural University, Wageningen, The Netherlands.

Hopkins, G.D., L. Semprini, P.V. Roberts, and D.M. Mackay. 1988. An Automated Data Acquisition System for Assessing In-Situ Biodegradation of Chlorinated Aliphatic Compounds. *Proc. of the Second Outdoor Conference on Groundwater Monitoring and Aquifer Restoration*, NWWA, Las Vegas, NV, May 23–26, Vol. 1., pp. 201–203.

Jeffers, P.M., L.M. Ward, L.M. Woytowitch, and N.L. Wolfe. 1989. Homogeneous Hydrolysis Rate Constants for Selected Chlorinated Methanes, Ethanes, Ethenes, and Propanes. *Environ. Sci. Technol.*, **23**(8):965–969.

Käster, M. 1991a. Reductive Dechlorination of Tri- and Tetrachloroethylenes Depends on Transition from Aerobic to Anaerobic Conditions. *Appl. Environ. Microbiol.*, **57**:2039–2046.

Käster, M. 1991b. Reductive Dechlorination of Tri- and Tetrachloroethylene by Nonmethanogenic Enrichment Cultures. In R.E. Hinchee and R.G. Olfenbuttel, eds., *On-Site Bioreclamation Processes for Xenobiotic and Hydrocarbon Treatment*, Butterworth-Heinemann, Boston, MA, pp. 134–146.

Khalil, M.A.K. and R.A. Rasmussen. 1989. The Potential of Soil as Sink of Chlorofluorocarbons and Other Man-Made Chlorocarbons. *Geophys. Res. Let.* **16**:679–682.

Khalil, M.A.K., R.A. Rasmussen, M.X. Wang, and L. Ren. 1990. Emissions of Trace Gases from Chinese Rice Fields and Biogas Generators: CH_4, N_2O, CO, O_2, Chlorocarbons, and Hydrocarbons. *Chemosphere,* **20**:207–226.

Klecka, G.M. 1982. Fate and Effects of Methylene Chloride in Activated Sludge. *Appl. Environ. Microbiol.* **44**:701–707.

Kriegman, M.R. and M. Reinhard. 1991. Reduction of Hexachloroethane and Carbon Tetrachloride at Surfaces of Biotite, Vermiculite, Pyrite, and Marcasite. In R.A. Baker, ed., *Organic Substances and Sediments in Water*, ACS Symposium Series, American Chemical Society, Washington, DC, pp. 349–364.

Lang, M., L. Semprini, P.V. Roberts, P.L. McCarty, and D.B. Janssen. 1990. Biotransformation of Carbon Tetrachloride in Batch-Exchanged Laboratory Soil Columns under Anoxic Conditions. *EOS, Trans. Amer. Geophys. Union*, **71**(43):1323.

Lesage, S., R.E. Jackson, M.W. Priddle, and P.G. Riemann. 1990. Occurrence and Fate of Organic Solvent Residues in Anoxic Groundwater at the Gloucester Landfill, Canada. *Environ. Sci. Technol.*, **24**:559–565.

Lesage, S., S. Brown, and K.R. Hosler. 1992. Degradation of Chlorofluorocarbon-113 Under Anaerobic Conditions. *Chemosphere*, **24**:1225–1243.

Loveley, D.R. and J.C. Woodward. 1990. Consumption of Freons F-11 and F-12 in Methane Producing Aquatic Sediments. *EOS, Trans. Amer. Geophys. Union*, **71**:1236.

Major, D.W., E.W. Hodgins, and B.J. Butler. 1991. Field and Laboratory Evidence of In-Situ Biotransformation of Tetrachloroethene to Ethene and Ethane at a Chemical Transfer Facility in North Toronto. In R.E. Hinchee and R.G. Olfenbuttel, eds., *On-Site Bioreclamation Processes for Xenobiotic and Hydrocarbon Treatment*, Butterworth-Heinemann, Boston, pp. 113–133.

McCarty, P.L. and J.T. Wilson. 1992. Natural Anaerobic Treatment of a TCE Plume at St. Joseph MI NPL Site. In Bioremediation of Hazardous Wastes. EPA/600/R-92/126, U.S. EPA ORD, Washington, DC, pp. 47–66.

McCarty, P.L and M. Reinhard. 1993. Biological and Chemical Transformations of Halogenated Aliphatic Compounds in Aquatic and Terrestrial Environments. In R.S. Oremland, ed., *The Biogeochemistry of Global Change: Radiative Trace Gases*, Chapman & Hall, Inc., NY, pp. 839–852.

McCarty, P.L. et al. 1991. In Situ Methanotrophic Bioremediation for Contaminated Groundwater at St. Joseph, Michigan. In R.E. Hinchee and R.G. Olfenbuttel, eds., *On-Site Bioreclamation Processes for Xenobiotic and Hydrocarbon Treatment*, Butterworth-Heinemann, Boston, pp. 16–40.

McCarty, P.L., P.V. Roberts, M. Reinhard, and G. Hopkins. 1992. Movement and Transformations of Halogenated Aliphatic Compounds in Natural Systems. In J.L. Schnoor, ed., *Fate of Pesticides and Chemicals in the Environment*, John Wiley & Sons, Inc., NY, pp 191–209.

Mohn, W.W. and J.M. Tiedje. 1992. Microbial Dehalogenation. *Microbial Reviews*, **56**:482–507.

Parsons, F., P.R. Wood, and J. DeMarco. 1984. Transformations of Tetrachloroethene and Trichlorothene in Microcosms and Groundwater. *J. Amer. Water Works. Assoc.*, **76**:56–59.

Rittmann, B.E. and P.L. McCarty. 1980. Utilization of Dichloromethane by Suspended and Fixed-Film Bacteria. *Appl. Environ. Microbiol.*, **39**:1225–1226.

Rittmann, B.E., P.L. McCarty, and P.V. Roberts. 1980. Trace-Organics Biodegradation in Aquifer Recharge. *Ground Water*, **18**(3):236–243.

Rittmann, B.E., A.J. Valocchi, J.E. Odencrantz, and W. Bae. 1988. In-Situ Bioreclamation of Contaminated Groundwater. Illinois Hazardous Waste Research and Information Center, HWRIC RR 031.121.

Roberts, P.V., J. Schreiner, and G.D. Hopkins. 1982. Field Study of Organic Water Quality Changes During Groundwater Recharge in the Palo Alto Baylands. *Water Research*, **16**:1025–1035.

Roberts, P.V., M.N. Goltz, and D.M. Mackay. 1986. A Natural-Gradient Experiment on Solute Transport in a Sand Aquifer III. Retardation Estimates and Mass Balances for Organic Solutes. *Water Resources Research*, **22**(13):2047–2058.

Roberts, P.V., L. Semprini, G.D. Hopkins, D. Grbi-Gali, P L. McCarty, and M. Reinhard. 1989. In-Situ Aquifer Restoration of Chlorinated Aliphatics by Methanotrophic Bacteria. EPA/600/2-89/033, R.S. Kerr Environmental Research Lab.

Roberts, P.V., G.D. Hopkins, D.M. Mackay, and L. Semprini. 1990. A Field Evaluation of In-Situ Biodegradation of Chlorinated Ethenes: Part 1, Methodology and Field Site Characterization. *Ground Water*, **28**:591–604.

Semprini, L., G.D. Hopkins, D.B. Janssen, M. Lang, P.V. Roberts, and P.L. McCarty. 1991a. In-Situ Biotransformation of Carbon Tetrachloride Under Anoxic Conditions, U.S. EPA ORD, EPA/600/2-90/060.

Semprini, L., G.D. Hopkins, P.V. Roberts, and P.L. McCarty. 1991b. In-Situ Biotransformation of Carbon Tetrachloride, Freon-113, Freon-11, and 1,1,1-TCA under Anoxic Conditions. In R.E. Hinchee and R.G. Olfenbuttel, eds., *On-Site Bioreclamation Processes for Xenobiotic and Hydrocarbon Treatment*, Butterworth-Heinemann, Boston, pp. 41–58.

Semprini, L. et al. 1992. Detailed Chemical Characterization of Chlorinated Aliphatic Hydrocarbons at St. Joseph, Michigan, NPL Site. *EOS, Trans. Amer. Geophys. Union*, **73**(43):243.

Semprini, L., G.D. Hopkins, P.L. McCarty, and P.V. Roberts. 1992. In-Situ Transformation of Carbon Tetrachloride and Other Halogenated Compounds Resulting from Biostimulation Under Anoxic Conditions. *Environ. Sci. Technol.*, **26**:2454–2461.

Sewell, G.W. and S.A. Gibson. 1991. Stimulation of the Reductive Dechlorination of Tetrachloroethene in Anaerobic Aquifer Microcosms by the Addition of Toluene. *Environ. Sci. Technol.*, **25**:982–984.

Sewell, G.W., S.A. Gibson, and H.H. Russell. 1990. Anaerobic In-Situ Treatment of Chlorinated Ethenes. WPCF Annual Conference, Washington, DC.

Suflita, J.M., A. Horowitz, D.R. Shelton, and J.M. Tiedje. 1982. Dehalogenation: A Novel Pathway for the Anaerobic Biodegradation of Haloaromatic Compounds. *Science,* **218**:1115-1117.

Vogel, T.M. 1988. Biotic and Abiotic Transformation of Halogenated Aliphatic Compounds. Ph.D. thesis, Department of Civil Engineering, Stanford University, Stanford, CA.

Vogel, T.M. and P.L. McCarty. 1985. Biotransformation of Tetrachloroethylene to Trichloroethylene, Dichloroethylene, Vinyl Chloride, and Carbon Dioxide under Methanogenic Conditions. *Appl. Environ. Microbiol.,* **49**:1080–1083.

Vogel, T.M. and P.L. McCarty. 1987. Abiotic and Biotic Transformations of 1,1,1-Trichloroethane under Methanogenic Conditions. *Environ. Sci. Technol.,* **21**:1208–1213.

Vogel, T.M., C.S. Criddle, and P.L. McCarty. 1987. Transformations of Halogenated Aliphatic Compounds. *Environ. Sci. Technol.,* **21**(8):722–736.

Wolf, C.R., L.K. King, and D.V. Park. 1978. The Anaerobic Dechlorination of Trichlorofluoromethane by Rat Liver Preparations In Vitro. *Chem.-Biol. Interactions,* **21**:277–288.

Wilson, B.H. et al. 1991. Reductive Dechlorination of Trichloroethylene in Anoxic Aquifer Material from Picatinny Aresenal, NJ. In U.S. Geological Survey, Water-Resources Investigations Report, 91-4034, Reston, VA, pp. 704–707.

Zinder, S., T. DiStefano, V. Tandoi, P. Bowser, and J. Gossett. 1992. Characterization of an Anaerobic Enrichment Culture Which Rapidly Converts Tetrachloroethene to Ethane. Abstract, ASM Conference, Anaerobic Dehalogenation and Its Environmental Implications, Athens, GA.

Chapter 27

Enhanced Biological Electron Acceptor H_2O_2

J.T. Wilson

INTRODUCTION

Remediation of fuel spills using biotic processes is an older technology within the developing field of bioremediation. The earlier work delivered oxygen to the contaminated aquifer by sparging air into standing water in a well (Raymond et al., 1978). Results were promising; however, the rate of remediation was limited by the low solubility of molecular oxygen in water (Brown et al., 1984; Lee et al., 1988). Groundwater equilibrated with air contains about 8 mg/L of oxygen. If equilibrated with pure oxygen gas at one atmosphere pressure, it contains up to 40 mg/L dissolved oxygen. In contrast, the oxygen demand to completely bioremediate aquifer material contaminated with a typical concentration of 5000 mg/kg total petroleum hydrocarbon (TPH) is equivalent to 100,000 mg/L dissolved oxygen in the groundwater in contact with hydrocarbon contamination.

To overcome the limitation imposed by oxygen solubility, hydrogen peroxide has been used to supply oxygen in a soluble form to the contaminated aquifer (Gibson et al., 1985; API, 1987; Hinchee et al., 1989; Hinchee et al., 1991; Frankenberger et al., 1989; Lee and Raymond, 1991). In groundwater, hydrogen peroxide decomposes into water and oxygen, either through enzymatic catalysis, or through interactions with subsurface minerals.

$$2H_2O_2 \rightarrow 2H_2O + O_2$$

The oxygen released from the decomposition of hydrogen peroxide supports aerobic microbial metabolism of organic contaminants. Subsurface microorganisms have tolerated hydrogen peroxide concentrations up to 1,000 mg/L, equivalent to 470 mg/L dissolved oxygen. The use of hydrogen peroxide can increase the oxygen supplying capacity of groundwater by an order of magnitude compared to water equilibrated with gaseous oxygen, or two orders of magnitude compared to water equilibrated with air.

THE PRIMARY APPLICATIONS OF *IN SITU* BIOREMEDIATION USING HYDROGEN PEROXIDE

In situ bioremediation using hydrogen peroxide has been widely used to remediate fuel spills from underground storage tanks or their associated plumbing, as well as spills from pipelines and fuel transfer terminals. These fuel storage and transfer facilities have been installed in almost every possible geological lithology. Most of the known leaks, however, have occurred in unconsolidated sandy material. There are several reasons for this. Many of our inland cities are built on floodplains or river terraces because they are flat and near water. Major portions of our coastal cities are built on old beaches or glacial outwash. Because these materials are transmissive, releases from underground storage tanks drain readily into the water table. Groundwater flow in these areas is usually rapid, resulting in plumes of contamination spreading over wide areas in a short period of time. Water table aquifers in sandy unconsolidated materials are commonly used for domestic water supply. When there is a high density of wells, detecting a release is much more likely.

The pattern of contamination from a release is complex (Figure 27.1). As the release drains through the unsaturated zone, a portion is left behind, trapped by capillary forces. If the released material is volatile, a plume of vapors fills the soil air in the vadose zone. If the released material is a light hydrocarbon, it drains down to the water table, and then spreads laterally. Groundwater moving through the aquifer comes in contact with the released materials and transports the water-soluble

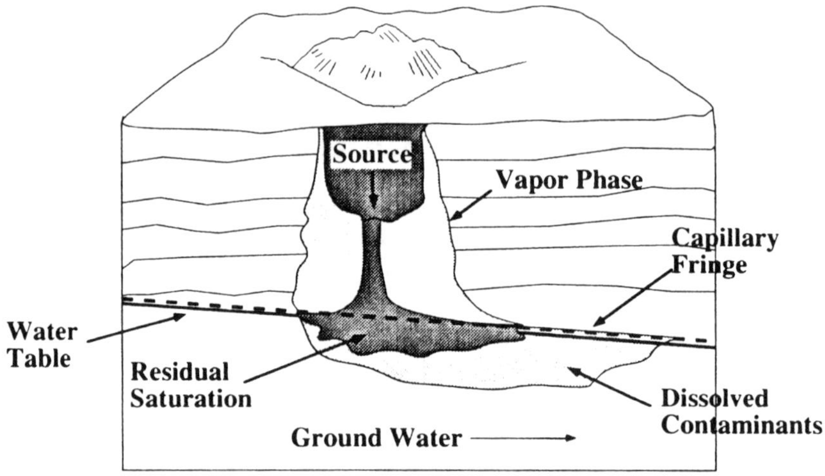

Figure 27.1. Distribution of a spill of petroleum hydrocarbons in the subsurface.

components. As a result, there are three distinct regions: a plume of vapors in the soil air; a groundwater solute plume; and the region that contains the oily phase material, which serves as a source area for both plumes. In practice, the source area is usually the object of remedial activities. There is little point in treating the plumes if the source area is left in place to continuously release contaminants.

ADVANTAGES OF *IN SITU* BIOREMEDIATION WITH HYDROGEN PEROXIDE

In situ bioremediation is frequently described as an inexpensive alternative to physical or chemical treatment, or excavation and removal. However, the real advantage of *in situ* bioremediation is its ability to clean groundwater at a spill in a few years, with minimum disruption to industrial or commercial activity on a site. Many remediations are driven by the desire to sell or redevelop the land and timeliness is important. Pump-and-treat rarely cleans groundwater to acceptable levels in a reasonable period of time. Excavation and incineration can be very expensive, while excavation and reburial does not remove the liability associated with the spill.

The technology is not appropriate for every spill. Depending on groundwater chemistry and the nutrient amendments also added to the contaminated material, mineral precipitates may form which can plug the formation (Wetzel et al., 1987). Hydrogen peroxide can also prematurely decompose and vent to the atmosphere (Hinchee et al., 1989; Hinchee et al., 1991; Spain et al., 1989; Huling et al., 1991).

The performance of hydrogen peroxide technology has been illustrated in well documented case studies.

CASE STUDY AT EGLIN AIR FORCE BASE, FLORIDA

Between 80,000 and 150,000 L of JP-4 jet fuel were spilled to a very shallow aquifer in fine-grained to coarse-grained sand. At the time of the study, hydrocarbon concentrations in the subsurface ranged from 1800 to 3600 mg/kg (Hinchee et al., 1991). The hydrocarbon was distributed from 0.6 to 2.1 m below land surface, and the water table was 1.0 m below land surface. The sands were examined for their capacity to decompose hydrogen peroxide. First-order rates of hydrogen peroxide decomposition in groundwater in contact with the sands varied from 1.3 to 2.7 per day.

Hydrogen peroxide decomposition catalyzed by inorganic mineral components of aquifer material is usually caused by iron and manganese minerals. Various phosphate salts can complex these mineral surfaces and reduce hydrogen peroxide decomposition. If the sands were treated in the laboratory with Restore® 375, a proprietary product containing 50% ammonium chloride, 20% disodium phosphate, 17.5% sodium tripolyphosphate, and 12.5% monosodium phosphate, the rate of hydrogen peroxide decomposition was reduced to 0.42 per day. An infiltration gallery was installed above the water table and the aquifer matrix adjacent to the gallery was treated with Restore® at a concentration of 1000 mg/L for two weeks prior to introduction of hydrogen peroxide. Then hydrogen peroxide was introduced into the infiltration gallery at a concentration of 300 mg/L and a flow of 7.6 L/min.

Over 105 days of operation, the hydrogen peroxide failed to move to the closest monitoring well, 0.6 m from the infiltration gallery. Oxygen concentrations in the injection gallery and the closest monitoring well averaged 20 mg/L, near the concentrations that would be expected from equilibration with gaseous oxygen at the groundwater temperatures of Florida. Apparently, hydrogen peroxide decomposed to oxygen very rapidly. A small portion dissolved in the injected water, while the rest bubbled away to the atmosphere. Gas samples collected above the infiltration gallery contained as much as 83% oxygen.

To measure the kinetics of hydrogen peroxide decomposition, and oxygen consumption, Hinchee et al. (1991) conducted *in situ* utilization tests. Flow to the galleries and from extraction wells was shut down, and concentrations of hydrogen peroxide and oxygen were measured in wells that sampled immediately below the gravel backfill of the infiltration gallery.

Figure 27.2 presents typical data. At initial hydrogen peroxide concentration of 230 mg/L rapidly decayed to oxygen, the first-order rate constant was 4.6 per day. As long as hydrogen peroxide was available, the concentration of dissolved oxygen was poised near 20 mg/L. When hydrogen peroxide was exhausted, the concentration declined at a first-order rate of 0.67 per day. Hinchee et al. (1991) fit a first-order rate to the decline in oxygen concentration. If a zero-order rate is fit to the data, the decline is 11 mg/L per day.

In followup laboratory work, Aggarwal et al. (1991) examined the efficacy of known hydrogen peroxide stabilizers, including trimetaphosphate, fluoride, boric acid, and hypophosphite. There was no environmentally significant improvement in peroxide stability. Spain et al. (1989) identified catalase activity associated with microbial biomass in the aquifer as an important contributor to the decomposition of hydrogen peroxide to oxygen. Hinchee et al. (1991) conceded that hydrogen peroxide decomposition in the field was probably due to catalase, which explains the failure of Restore® to stabilize the hydrogen peroxide.

CASE STUDY AT THE U.S. COAST GUARD AIR STATION AT TRAVERSE CITY, MICHIGAN

The U.S. EPA and the U.S. Coast Guard conducted a joint full-scale evaluation of *in situ* bioremediation using hydrogen peroxide at an aviation gasoline spill at the U.S. Coast Guard Air Station in East Bay Township, Michigan. A spill from a failed flange on an underground storage tank released aviation gasoline to the water table aquifer. Cores were acquired and analyzed to determine the quantity and location of aviation gasoline in the aquifer. Table 27.1 presents typical concentrations of hydrocarbon contaminants.

Hydraulic modeling was used to design an infiltration system that flooded all the fuel contaminated subsurface material (Wilson et al., 1993, see Figure 27.3). The spill was perfused with groundwater amended with 380 mg/L ammonium chloride, 190 mg/L disodium phosphate, and 190 mg/L potassium phosphate. Concentrations of hydrogen peroxide were raised slowly to a maximum of 750 to 1000 mg/L. The amended water was infiltrated at a rate of 38 L/min, in shallow wells screened across the depth interval contaminated with aviation gasoline. Unamended water was injected deep below the spill at 76 L/min to raise the water table high enough to flood the interval contaminated with gasoline.

Table 27.2 shows hydrogen peroxide electron acceptors (oxygen and nitrate) and contaminant concentrations at various distances from the injection area after seventeen months of system operation. Nitrate probably generated by nitrification of the added ammonium chloride can also serve as a microbial electron acceptor via denitrification. The concentration of benzene in groundwater in the most contaminated depth interval was less than 1 µg/L at distances as far as 110 ft from the infiltration wells. The other alkylbenzenes were also below federal drinking water standards at distances up to 110 ft from the infiltration wells.

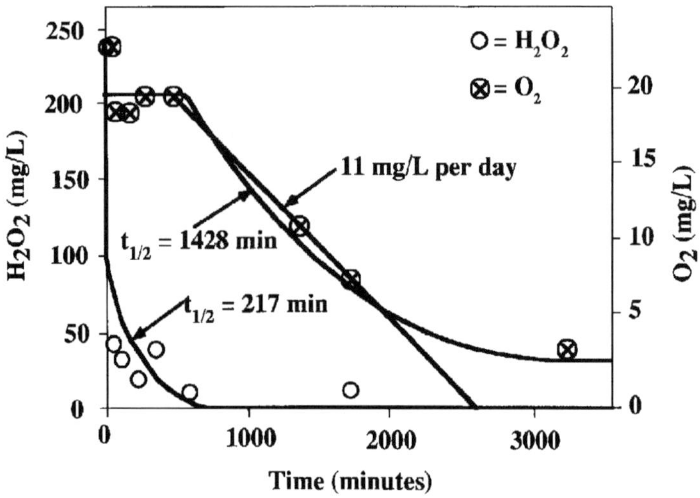

Figure 27.2. Decomposition of hydrogen peroxide and consumption of oxygen in aquifer material at Eglin AFB, Florida, after *in situ* bioremediation using hydrogen peroxide.

Table 27.1. Concentrations of Alkylbenzenes and Total Petroleum Hydrocarbons in Core Material Acquired Prior to Bioremediation of an Aquifer Contaminated with Aviation Gasoline.

Total Petroleum Hydrocarbons	Benzene	Toluene	Total Xylenes
	———(mg/kg wet sample)———		
5590	0.6	235	121
6500	0.3	544	48
2500	0.7	112	39

Cores were collected near the monitoring wells to ensure that the alkylbenzenes were actually being removed from the oily phase hydrocarbons in the aquifer (Table 27.3). The individual concentrations of the BTEX compounds were less than 0.07 mg/kg at distances up to 50 ft from the infiltration wells. However, the concentration of total petroleum hydrocarbons in the most contaminated interval varies from 992 to 10,800 mg/kg at the end of the demonstration.

The alkylbenzenes were extensively and selectively removed, while other components of the aviation gasoline, primarily branched chain alkanes, were more persistent. This result is difficult to explain based on the formulation of nutrients added to the groundwater used to perfuse the spill. However, the system acclimated to nitrify the ammonium ion in the nutrient mix. Up to 30 mg/L of nitrate was generated near the infiltration wells in the most contaminated flow path (Table 27.2). The concentration of nitrate declined in regions of the aquifer downgradient of the zone of active nitrification, indicating that nitrate was being consumed. Nitrate has been shown to support biodegradation of alkylbenzenes in groundwater (Hutchins et al., 1991). It is likely that the extensive and selective removal of alkylbenzenes was supported by nitrate biologically generated from ammonium chloride in the nutrient formulation.

Huling et al. (1991) sampled the concentration of oxygen in soil air above the demonstration to evaluate the importance of premature decomposition of the hydrogen peroxide. They measured the concentration of oxygen at 1.0 m, 2.0 m, and 3.0 m below the surface at three locations 2.0 m downgradient of the infiltration wells. The concentration gradient of oxygen at the three locations was

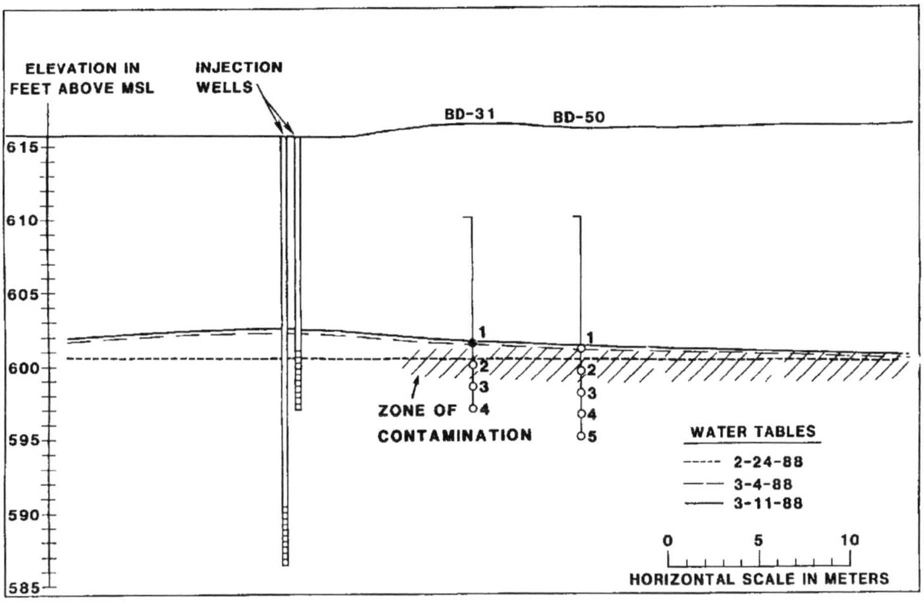

Figure 27.3. Location of infiltration wells and the zone of contamination at an aviation gasoline spill at the U.S. Coast Guard Air Station at Traverse City, Michigan. BD-31 and BD-50 are cluster monitoring wells. The circles represent the region sampled by individual wells.

Table 27.2. Projection of Electron Acceptors into an Aquifer Contaminated with Aviation Gasoline and the Corresponding Reduction in BTEX Contamination in the Groundwater.

Distance from the Injection Wells (m)	Travel Time of Injected Water (days)	Hydrogen Peroxide	Nitrate as Oxygen	Nitrogen	Benzene	BTEX
			(mg/L)		(μg/L)	
2.1	0.7	930	210	0.4	<0.1	<0.1
9.5	3.6	<0.1	8.2	30	<0.1	0.8
15.3	8.3	<0.1	1	22	<0.1	11.4
18.9	14.8	<0.1	0.7	27	<0.1	39.2
25.3	33	0.2	1.1	19	<0.1	379
32.9	>33	<0.1	0.4	4.3	<0.1	402

Hydrogen peroxide and oxygen data collected after 533 days of operation. Nitrate data were collected 07/13/89 after 499 days of operation.

10.5%/m, 8.2%/m, and 7.5%/m. At an average gradient of 8.8%/m, using a diffusion coefficient of 0.178 cm^2/sec, and assuming an air filled porosity of 0.3, this corresponds to a diffusive loss of 19 mg/m^2/sec.

From Table 27.2, the average velocity of the infiltrated water to a well 2.1 m downgradient was 3.0/day, and 2.6 m/day to a well 9.5 m downgradient. In the interval between the two samples, hydrogen peroxide disappeared. At a velocity of 2.6 m/day, a representative volume of water would transit under a square meter of surface area in 9.2 hr. In this time 22 g of oxygen would diffuse from a square meter of surface area. Two vertical meters of aquifer were infiltrated with groundwater containing at least 750 mg/L peroxide. Assuming a porosity of 0.3, the water under a square meter contained hydrogen peroxide equivalent to 211 g of oxygen. Approximately 10% of the hydrogen peroxide was lost to premature decomposition.

Table 27.3. Concentration of Alkylbenzenes and Total Petroleum Hydrocarbons in Core Material from the Most Contaminated Depth Interval, After Seventeen Months of Perfusion with Mineral Nutrients and Hydrogen Peroxide.

Distance from Infiltration Wells (m)	Total Petroleum Hydrocarbons	Benzene	Toluene	Ethylbenzene	*o*-Xylene	*m+p* Xylene
	(mg/kg wet sample)					
3.1	922	<0.07	<0.07	<0.07	<0.07	<0.07
9.8	2310	<0.07	<0.07	<0.07	<0.07	<0.07
16.1	10800	<0.07	<0.07	<0.07	<0.07	<0.07
19.5	1280	<0.07	<0.07	3.2	<0.07	0.14

CASE STUDY AT THE PUBLIC SERVICES FACILITY IN DENVER, COLORADO

A temporary holding tank under a garage leaked used crankcase oil, diesel fuel, gasoline, and other materials into a shallow water table aquifer. Groundwater Technology, Inc. remediated the spill (Nelson et al., 1993). Figure 27.4 shows the relationship between the garage, the work pit containing the leaking holding tank, and the approximate area of the spill.

Remediation involved removal of separate oily phases, *in situ* bioremediation with hydrogen peroxide and mineral nutrients, and bioventing. Groundwater flow under ambient conditions was to the north or northeast. The flow of water during the remediation paralleled the natural gradient. Water was produced from a recovery well on the northeast side of the spill. The flow from the well was split. Part of the flow was amended with hydrogen peroxide and nutrients and recharged to the aquifer in a nutrient recharge gallery on the south side of the spill (Figure 27.4). The remainder of the flow was delivered to a groundwater recharge gallery located to the south of the nutrient recharge gallery. From 3 to 6 gpm was delivered to the nutrient recharge gallery, and 4 to 8 gpm was delivered to the groundwater recharge gallery for a total flow of 9 to 11 gpm. The system was designed to sweep the groundwater containing hydrogen peroxide and mineral nutrients through the spill to the recovery well.

The system was operated from October 1989 to March 1992. At a flow of 10 gpm, from ten to fifteen pore volumes would have been exchanged in the area between the nutrient recharge gallery and the recovery well.

Table 27.4 compares the reduction in concentration of benzene and total BTEX compounds in groundwater that was achieved by *in situ* bioremediation at the site in Denver. Examine Figure 27.4 to locate the wells. Monitoring wells MW-1 and MW-8 are in areas with oily phase hydrocarbons. Well MW-2A is just outside the region with oily phase hydrocarbon. Well MW-3 is a significant distance from the region with oily phase hydrocarbon; it sampled the plume of contaminated groundwater that moved away from the spill.

Prior to remediation, concentrations in wells MW-1 and MW-8 were equivalent. Well MW-1 was closest to the nutrient gallery and the aquifer surrounding MW-1 was completely remediated; BTEX compounds were undetectable in groundwater. In well MW-8, immediately adjacent to the point of release, the concentration of benzene was reduced at least one order of magnitude and the concentrations of benzene and BTEX compounds in well MW-3 were also reduced an order of magnitude.

Water in the monitoring wells and the recirculation well contained low concentrations of contaminants by March 1992. Active remediation was terminated and the site entered a period of post-remediation monitoring. The staff of the Robert S. Kerr Environmental Research Lab (U.S. EPA) conducted a retrospective performance evaluation of remediation achieved at the site.

Core samples were taken from the aquifer to determine the extent of hydrocarbon contamination remaining and to determine whether a plume of contamination could return once active remediation ceased. The site was cored along a transect downgradient of the release. The transect extended laterally from clean material, through part of the spill, into clean material on the other side (Figure 27.4).

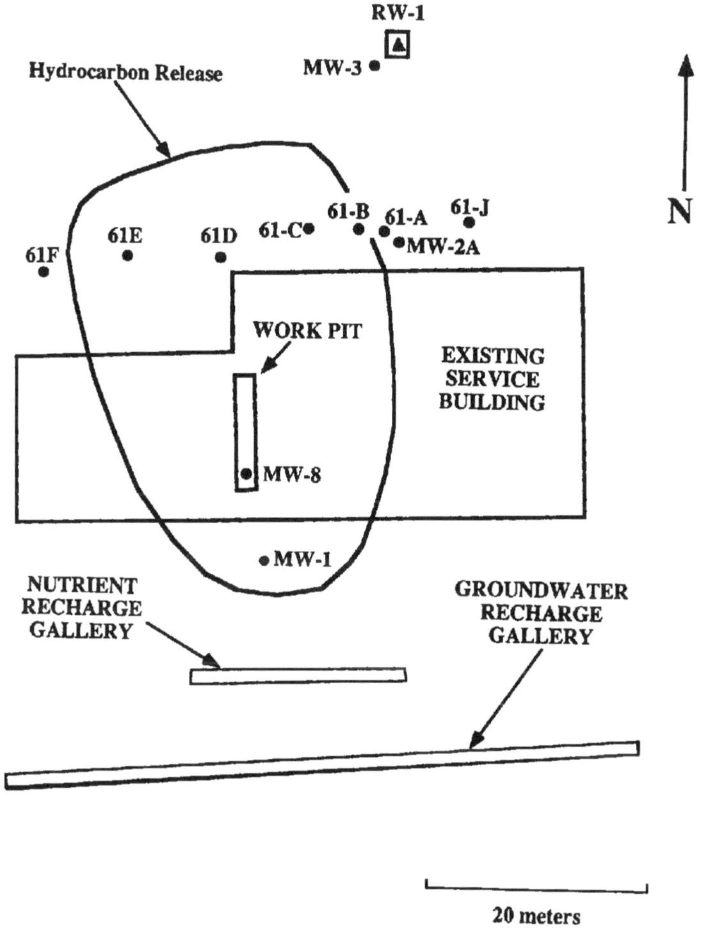

Figure 27.4. Infrastructure associated with *in situ* bioremediation and monitoring of a spill from a temporary holding tank under a garage in Denver, Colorado. RW-1 is the ground water capture and recirculation well. MW-1, MW-2A, MW-3, and MW-8 are monitoring wells, and 61-A through 61-J are bore holes for core samples.

Table 27.4. Reduction in Concentration of Hydrocarbon Contaminants in Groundwater Achieved by *In Situ* Bioremediation.

Well	Benzene Before	Benzene During	Benzene After	Total BTEX Before	Total BTEX During	Total BTEX After
			(µg/L)			
MW-1	220	<1	<1	2030	164	<6
MW-8	180	130	16	1800	331	34
MW-2A	?	11	0.8	?	1200	13
MW-3	11	5	2	1200	820	46
RW-1	<1	2	<1	<1	2	<1

In each borehole, continuous cores extended vertically from clean material above the spill, through the spill to clean material below.

Significant amounts of hydrocarbons remain within a narrow interval of fine-textured material approximately 2.0 ft thick, near the water table. The total saturated thickness of the aquifer was approximately 20 ft. At the time of sampling all the hydrocarbons were below the water table. The highest concentrations of hydrocarbons at the site in Denver were obtained in samples from borehole D, which was closest to the work pit (Table 27.5). The material in the interior of the spill had higher proportions of BTEX compounds.

Infiltration of hydrogen peroxide and mineral nutrients at the aviation gasoline spill in Michigan preferentially removed BTEX compounds from the oily phase gasoline, leaving a TPH residual low in aromatic hydrocarbons (Wilson et al., 1993). Apparently, at the Denver site, a cortex of material that has been physically and biologically weathered surrounds a central core of material that has not been depleted of BTEX compounds. Table 27.6 summarizes the performance of the combined treatment processes at the Denver site.

IMPLICATIONS FOR MONITORING AND CLOSURE

The usual expectation for *in situ* bioremediation is total removal of the contaminant from the subsurface environment. The extent of remediation achieved in this effort was removal of the contaminant from the circulated groundwater.

In situ bioremediation merely accelerates the natural physical and biological weathering processes that occur in the subsurface. The oily material in most intimate contact with the circulated groundwater is weathered to the greatest extent. After extensive remediation of the more transmissive regions, the release of contaminants to the circulated groundwater is controlled by physicochemical processes such as diffusion and slow advection from the subsurface material that still contains significant quantities of contaminants.

In such circumstances, the disposition of contamination in the groundwater can only be understood as a dynamic system. This release may best be described through the chemical engineering concept of a mass transfer coefficient. As the circulated water passes through the weathered spill, a certain quantity of hydrocarbon is transferred to the water; the amount transferred is directly proportional to the exposure time of the water in the contaminated area. If the circulated water contains enough nutrients and electron acceptors to meet the demands of the microorganisms degrading the contaminants transferred from the fine-textured material, the plume will be destroyed by biological activity as rapidly as it is produced.

When active remediation is stopped, the concentration of electron acceptor returns to ambient conditions in the aquifer and the hydraulic gradient returns to the normal condition. As a result, the residence time of water in the spill area is longer and the total amount of hydrocarbon transferred to the water is greater, although the supply of electron acceptor for biological destruction of the hydrocarbon is less.

The hydraulic conductivity in the depth interval containing the hydrocarbons at this site is approximately 8.5 m per day (personal communication, David Szlag, University of Colorado, Boulder). The distance from the nutrient recharge gallery to the recovery well is 45 m. Assume that the distance from the upgradient edge of the spill, through the spill, to monitoring wells downgradient at the property boundary is also 45 m. Assuming an effective porosity of 0.35, Darcy's law can be used to estimate the interstitial flow velocity of the groundwater and its residence time along the flow path. When active remediation was terminated, the residence time of water was 75 times longer and the supply of oxygen was 85 times less than conditions during active remediation (Table 27.7).

In June 1992, core material from the spill was assayed for the potential rate of oxygen consumption. Core material was de-watered by placing it in a Buchner funnel and applying a partial vacuum. Then the core material was sealed in a glass mason jar. After 24 hr, a sample of air in contact with the core material was passed through an oxygen indicating tube to estimate oxygen consumption potential. Acclimated material exerted oxygen demands of 6 to more than 36 mg/kg core material per day (Table 27.8). Forty to more than 240 mg/L of groundwater per day would need to contact the contaminated material to supply the oxygen required. At these rates, the oxygen supplied as hydrogen peroxide during active remediation would be consumed in two to twelve days.

In the assay, the microbial consumption of oxygen was faster than the rate of supply of oxygen during active remediation. If the microbes in the aquifer expressed the potential rate of oxygen con-

Table 27.5. Lateral Distribution of Total BTEX Compounds and Total Petroleum Hydrocarbons Along the Transect (Figure 27.4) at the Most Contaminated Depth Interval.

Bore Hole	TPH	BTEX (mg/kg)	Benzene
B	167	0.8	<0.2
C	156	3.5	<0.2
D	1176	260	4.3
E	156	3.5	0.06

Table 27.6. Performance of Free Product Recovery, Bioventing, and Groundwater Bioremediation Using Hydrogen Peroxide and Mineral Nutrients at a Site in Denver, Colorado (Nelson et al., 1993).

Initial estimate of hydrocarbon contaminant mass	12000 kg hydrocarbon
Initial estimate of aquifer volume contaminated	600 m^3
Oily phase material recovered	690 kg hydrocarbon
Directly volatilized mass of hydrocarbon contaminant	350 kg hydrocarbon
Biodegradation (CO_2 from bioventing, 40% converted)	15000 kg hydrocarbon
(this author's estimate, 60% converted to CO_2)	10000 kg hydrocarbon
Oily phase material remaining (this author's estimate)	60 kg hydrocarbon
Volume of groundwater recovered and reinjected	33000 m^3 (<15 pore volumes)
Time required	3 years
Cost	approximately $500,000 ($800 per m^3)

Table 27.7. Contrast in Conditions During Active *In Situ* Bioremediation and Conditions at the Termination of Remediation at a Spill from an Underground Storage Tank in Denver, Colorado.

Parameter	During Active Remediation	Ambient Conditions after Remediation
Introduced concentration of oxygen	470 mg/L	5.5 mg/L
Hydraulic gradient	0.097 m/m	0.0012 m/m
Interstitial flow velocity	2.4 m/day	0.03 m/day
Travel time of water across the spill	20 days to recovery well	1500 days to monitoring well
Maximum oxygen demand supported	20 mg/L day	0.004 mg/L day

sumption, oxygen would have been depleted before the recharge water moved across the spill. In the absence of oxygen, BTEX compounds would have partitioned to the groundwater and should have been detected in the monitoring wells. In fact, oxygen concentrations between 2 and 5 mg/L were always present in water produced by the recovery well and BTEX compounds were virtually absent. Oxygen consumption must have been limited by mass transfer of hydrocarbon to the groundwater circulated through the spill.

No established procedures exist for determining, under ambient conditions, if the mass transfer of hydrocarbons from oily residual material will exceed the supply of oxygen or other natural electron acceptors. As a result, it is impossible to predict if bioremediation with hydrogen peroxide will prevent the regeneration of a plume, or if a plume of contaminated groundwater will regenerate and at what concentrations. Groundwater moving under the natural gradient must be allowed to travel all the

Table 27.8. Relationship Between the Potential Oxygen Uptake Rate of Freshly Collected Core Material and the Location of the Core Material with Respect to the Interval Contaminated with Hydrocarbons.

Location in Borehole	Position of Borehole in the Transect (mg oxygen consumed per kg core material per day)					
	A	B	C	D	E	F
Just Above			<4		7.4	
Within	15.5	>30	>36		>34	23.5
Just Below	6	<3	5.7	7.3		21

way through the spill and then to the monitoring wells before it is possible to determine whether mass transfer effects will reestablish a plume.

COMPARISON OF THE RATE OF TREATMENT

Barenschee et al. (1991) present a useful conceptual model of the consumption of electron acceptors during bioremediation of hydrocarbons. They describe three phases of remediation as water moves through a contaminated flow path from the point of infiltration of electron acceptor to a monitoring point. The first phase has a zero-order rate of consumption of oxygen and hydrocarbons, with no breakthrough of oxygen (complete consumption along the flow path). The second stage is a short transitional phase between oxygen-limited and hydrocarbon-limited conditions. At this stage oxygen breaks through at the monitoring point. The third stage is limited by hydrocarbon degradation. It is also zero-order for oxygen and hydrocarbon, but the rate of degradation is slower than the first stage. Table 27.9 collects some estimates of the rate of oxygen consumption, and by inference, of hydrocarbon oxidation in geological materials contaminated with hydrocarbons. The rates vary more than an order of magnitude. Oxygen depletion in the laboratory studies was greater than those observed in the field studies, indicating that laboratory studies will provide conservative estimates of oxygen consumption in the field.

COST COMPARISONS

Three separate performance evaluations were performed on fuel spills on the U.S. Coast Guard Air Station at Traverse City, Michigan. The evaluations dealt with the same aquifer, were approximately the same size, and achieved the same regulatory endpoint. Benzene and BTEX compounds were removed from groundwater, but as previously stated, the reduction in total petroleum hydrocarbons was modest and did not meet TPH standards in place in many of the states of the USA. Construction, operation, and maintenance were carried out by the same contractor and subcontractor. This allows a comparison of the use of hydrogen peroxide to the use of nitrate as an electron acceptor, and to the use of bioventing and biosparging. The use of hydrogen peroxide is expensive compared to other *in situ* bioremediation technologies that have been developed subsequently. The costs of the use of hydrogen peroxide at Traverse City, Michigan ($641 per m^3, Table 27.10) are very similar to the costs at Denver, Colorado ($800 per m^3, Table 27.6).

SUMMARY

Bioremediation using hydrogen peroxide to supply oxygen as the terminal electron acceptor for microbial degradation of hydrocarbon contamination of groundwater has proven to be feasible. Its effectiveness, however, is primarily limited to degradation of compounds solubilized into the groundwater. Hydrogen peroxide treatment does not effectively enhance the degradation of contaminants that remain in a distinct oily phase. These criticisms also apply to other electron acceptor addition

Table 27.9. Rates of Oxygen Consumption in Contaminated Geological Material Subjected to *In Situ* Bioremediation Using Hydrogen Peroxide as the Electron Acceptor.

Location or Source	Reference	Basis of Estimate	Rate of Oxygen Depletion (mg/L)
Field Studies			
Traverse City, MI	Wilson et al. (1993)	Depletion in groundwater along a flow path	83
Traverse City, MI	Wilson et al. (1993)	Depletion of hydrocarbons	70
Eglin AFB, FL	Hinchee et al.(1991)	*In situ* die-away test	11
Denver, CO	Wilson (1993)	Mass balance	20
Laboratory Studies			
Traverse City, MI	Huling et al. (1991)	Oxygen depleted	114
The Netherlands	Verheul et al. (1988)	Oxygen depleted	170
Germany	Barenschee et al. (1991)	Oxygen depleted	250

Table 27.10. Comparison of Unit Costs to Implement Bioremediation Using Hydrogen Peroxide as an Electron Acceptor, to Bioremediation Using Nitrate or Atmospheric Oxygen as an Electron Acceptor, in Fuel Spills to a Shallow, Sandy, Water Table Aquifer at the U.S. Coast Guard Air Station at Traverse City, Michigan.

Cost Element	Cost ($ per cubic meter of aquifer material)		
	Hydrogen Peroxide	Nitrate	Bioventing
Construction	45	118	26
Labor/Monitoring	72	96	40
Chemicals	500	30	0.4
Electricity	24	12	6.8
Total	641	256	73

*Cost data are a personal communication from Robert H. Douglas, employed at the time of the demonstrations by The Traverse Group, Inc., subcontractor to Solar Universal Technologies, Inc., Traverse City, Michigan. Mr. Douglas is now employed by Global Environmental Engineering, Inc., Traverse City, Michigan.

treatments of contaminated groundwater. Hydrogen peroxide treatment is also fairly expensive, but the technology works well when applied in an appropriate manner.

NOTICE

This chapter has been reviewed in accordance with the U.S. Environmental Protection Agency's peer and administrative review policies and approved for presentation and publication.

REFERENCES

Aggarwal, P.K., J.L. Means, D.C. Downey, and R.E. Hinchee. 1991. Use of hydrogen peroxide as an oxygen source for *in situ* biodegradation: Part II. Laboratory Studies. *J. Haz. Mat.*, **27**:301–314.

American Petroleum Institute. 1987. Field study of enhanced subsurface biodegradation of hydrocarbons using hydrogen peroxide as an oxygen source. A.P.I. Pub. 4448, American Petroleum Institute, Washington, DC, 76 pp.

Barenschee, E.R., P. Bochem, O. Helmling, and P. Weppen. 1991. Effectiveness and kinetics of hydrogen peroxide and nitrate-enhanced biodegradation of hydrocarbons. *In* R. Hinchee and R. Olfenbuttel, eds., *In*

Situ Bioreclamation: Applications and Investigations for Hydrocarbon and Contaminated Site Remediation, Butterworth-Heinemann, pp. 103–124.

Brown, R.A., R.D. Norris, and R.L. Raymond. 1984. Oxygen transport in contaminated aquifers. Proceedings of the NWWA/API Conference on Petroleum Hydrocarbons and Organic Chemicals, NWWA, Dublin, OH. p. 441.

Frankenberger, W.T., K.D. Emerson, and D.W. Turner. 1989. *In situ* bioremediation of an underground diesel fuel: a case history. *Environ. Man.,* **13**(3):325–332.

Gibson, D.T., L.N. Britton, W.L. Wootan, and T. Voynick. 1985. Feasibility study of the use of hydrogen peroxide to enhance microbial degradation of gasoline. A.P.I. Publication 4389. American Petroleum Institute, Washington, DC.

Hinchee, R.E., D.C. Downey, and P.K. Aggarwal. 1991. Use of hydrogen peroxide as an oxygen source for *in situ* biodegradation: Part I. Field Studies, *J. Haz. Mat.* **27**:287–299.

Hinchee, R.E., D.C. Downey, J.K. Slaughter, D.A. Selby, M.S. Westray, and G.M. Long. 1989. Enhanced bioreclamation of jet fuels—a full scale test at Eglin AFB, FL. ESL-TR-88-78. Air Force Engineering & Services Center, Engineering & Services Laboratory, Tyndall AFB, FL.

Huling, S.G., B.E. Bledsoe, and M.V. White. 1991. The feasibility of utilizing hydrogen peroxide as a source of oxygen in bioremediation. *In* R.E. Hinchee and R.F. Olfenbuttel, eds., *In Situ* Bioreclamation: Applications and Investigations for Hydrocarbons and Contaminated Site Remediation. Butterworth-Heinemann, Boston. pp. 83–102.

Hutchins, S.R., W.C. Downs, J.T. Wilson, G.B. Smith, D.A. Kovacs, D.D. Fine, R.H. Douglas, and D.J. Hendrix. 1991. Effect of nitrate addition on biorestoration of fuel-contaminated aquifer: field demonstration. *Ground Water,* **29**(4):571–580.

Lee, M.D., J.M. Thomas, R.C. Borden, P.B. Bedient, C.H. Ward, and J.T. Wilson. 1988. Biorestoration of aquifers contaminated with organic compounds. *CRC Critical Reviews in Environmental Control.* 18(1):28–89.

Lee, M.D. and R.L. Raymond. 1991. Case history of the application of hydrogen peroxide as an oxygen source for *in situ* bioreclamation. *In* R.E. Hinchee and R.F. Olfenbuttel, eds. *In Situ* Bioreclamation: applications and investigations for hydrocarbons and contaminated site remediation. Butterworth-Heinemann, Boston. pp. 429–436.

Nelson, C., R.J. Hicks, and S.D. Andrews. 1993. *In situ* bioremediation: an integrated system approach. *In* P.E. Flathman, D.E. Jerger, and J.H. Exner, eds. Bioremediation: Field Experiences. Lewis Publishers, Boca Raton, FL.

Raymond, R.L., V.W. Jamison, J.O. Hudson, R.E. Mitchell, and V.E. Farmer. 1978. Field application of subsurface biodegradation of hydrocarbons in sand formation. Project No. 307-77. American Petroleum Institute, Washington, DC, 137 pp.

Spain, J.E., J.D. Milligan, D.C. Downey, and J.K. Slaughter. 1989. Excessive bacterial decomposition of hydrogen peroxide during enhanced biodegradation. *Ground Water,* **27**(2):163–167.

Verheul, J.H.A.M., R. van den Berg, and D.H. Eikelboom. 1988. *In situ* biorestoration of a subsoil contaminated with gasoline. *In* K. Wolf, W.J. van den Brink, and F.J. Colon, eds., *Contaminated Soil,* Kluwer Academic Publishers, Amsterdam, Netherlands, pp. 705–715.

Wetzel, R.S., C.M. Darst, D.H. Davidson, and D.J. Sarno. 1987. *In situ* biological treatment test at Kelly Air Force Base, Volume II Field test results and cost model. Technical Report 85-52, U.S. Air Force Engineering and Services Laboratory, Tyndall AFB, FL.

Wilson, J.T. 1993. Testing Bioremediation in the Field. In: *In situ* Bioremediation: When Does It Work? National Academy Press, Washington, DC.

Wilson, J.T., J.M. Armstrong, and H. Rafai. 1993. A full scale field demonstration on the use of hydrogen peroxide for *in situ* bioremediation of an aviation gasoline-contaminated aquifer. *In* P.E. Flathman, D.E. Jerger, and J.H. Exner, eds. Bioremediation: Field Experiences. Lewis Publishers, Chelsea, MI.

Section VII

Overview and Research Directions

Chapter 28

Research Needs for Subsurface Remediation

C.G. Enfield, S.C. Mravik, T.E. Short, and A.L. Wood

INTRODUCTION

There are many possible alternative methods to prioritize and evaluate subsurface remediation research needs. If one chooses to classify the needs according to specific contaminants, then problems arise due to the fact that different organizations and government agencies have different perspectives on environmental problems. For example, DoD might select energetics—including explosives and propellants—and chemical warfare agents as the most important contaminants related to their mission. DoE might select radionuclides and mixed (radioactive and hazardous) wastes as being the most significant contaminants related to their mission. EPA might select petroleum hydrocarbons, oils and lubricants (POLs), solvents and metals as being the most important "point" source pollutants because of their wide distribution and common release into the environment; and nitrates and pesticides as the most important "non-point" contaminants due to their frequency of occurrence and contamination of many public water supplies.

An alternative method of classifying research needs is to evaluate different remediation technologies which can be used for various classes of contaminants. An evaluation of the status of technology permits a determination of where research emphasis should be placed.

In addition to making new technologies available, site characterization research is required. The key to the application of any technology begins with a conceptual understanding of the site geohydrology in relation to the contaminant distribution. Also required is an understanding of the fundamental site processes controlling contaminant transport and transformation, and the processes involved in the available technologies. This conceptual and process knowledge is needed to implement an optimal remedial system.

Figure 28.1 illustrates, along the time axis, the steps normally followed in the development of a remedial technology. The number of projects required for any particular step is represented by the magnitude along the number of projects axis. During the conceptualization phase, the major systems operating in the subsurface are identified, and remediation technologies are proposed based on assumed controlling systems. Most proposed technologies will be discarded as impractical before they ever reach common practice. The nurturing of a conceptual idea into a proven/accepted technology is a tedious, time-consuming process with many false starts.

To effectively assign resources for the development of technologies, consideration must be given to those technologies that are currently being applied in the field. Figure 28.2 (Steimle, 1993) displays the percentages of innovative remediation technologies installed in the field for EPA. The figure contains three pie charts showing the percentage of innovative technologies implemented under three EPA programs: the Superfund program records of decision (RODs) for fiscal years '82–'91, the leaking Underground Storage Tank (UST) program for '92, and the Resource Conservation and Recovery Act (RCRA) corrective actions for '91. Over 90% of the sites being actively remediated utilize the only accepted technology for subsurface remediation—"pump-and-treat." As used here, pump-and-treat refers to using wells to extract contaminated water followed by aboveground treatment. Innovative technologies are anything other than pump-and-treat and, as can be seen, they are not commonly implemented.

Innovative remediation technologies can be classified into three major categories: containment, enhanced extraction, and *in situ* contaminant destruction. A normalized evaluation of selected innovative subsurface remediation technologies is presented in Table 28.1. Each technology is rated suggesting its state of progress along the technology pyramid in Figure 28.1. The length of the bar under each heading is the estimated fraction of the research and development completed in this area. If the

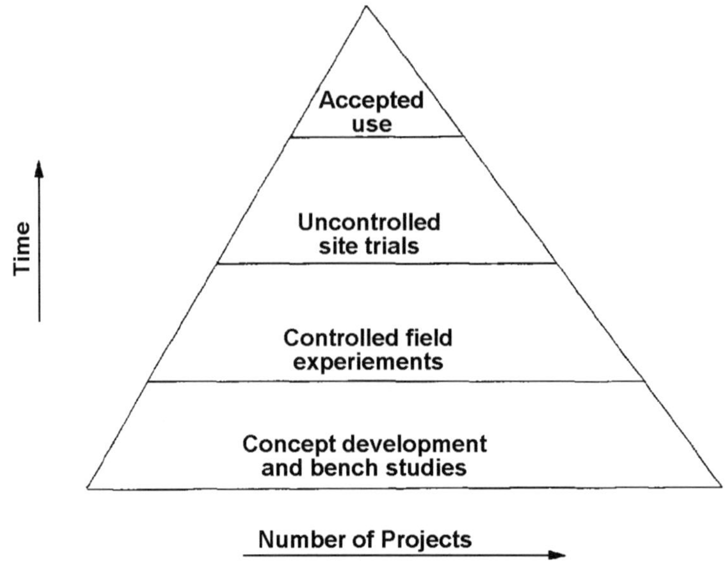

Figure 28.1. Technology development pyramid.

technology is mature and it is widely accepted, no research is needed and one would expect the bar to fill out the entire line. A significant amount of work has been completed on several of these technologies; however, design manuals appropriate for use by practicing engineers are not available. Lack of acceptance by engineers is not always a technical issue, but is based on business and attitudes of regulatory agencies of the city, state, and federal level. It should be noted that progress in technology development is not linear. Our understanding generally increases in steps with each new significant discovery, and the duration of a given plateau is difficult to estimate. There is always more that can be learned, but the cost required to make the next incremental step may be more than the value of the knowledge. Regardless of the technology, proven design models do not exist. Therefore, each remediation site should be considered a research facility and designed to collect the information that can be utilized to advance the installed technology.

SITE CHARACTERIZATION

Regardless of the remediation technology selected, there are common data requirements for design and implementation that describe the subsurface system. Four questions about the site and contaminant(s) must be asked:

1. Where is it? (Both macro and micro scale distributions!)
2. What is it? (Not only the contaminant of interest but also associated contaminants!)
3. How much is there?
4. How will it react in the contaminated hydrogeologic environment?

To answer these questions, one must obtain basic information concerning the specific site to be remediated—in other words, site characterization.

Site characterization, though, is not a one-time activity and should be an iterative process. The data needs and the data quality objectives vary as a function of the stage in the remediation process. Therefore, the detail of information required in the site description should change during each step in the remediation process.

An economic analysis, coupled with a good understanding of contaminant transport, is needed to define the data requirements and data quality objectives for each stage of the remedial investigation

Table 28.1. Pump-and-Treat Alternatives.

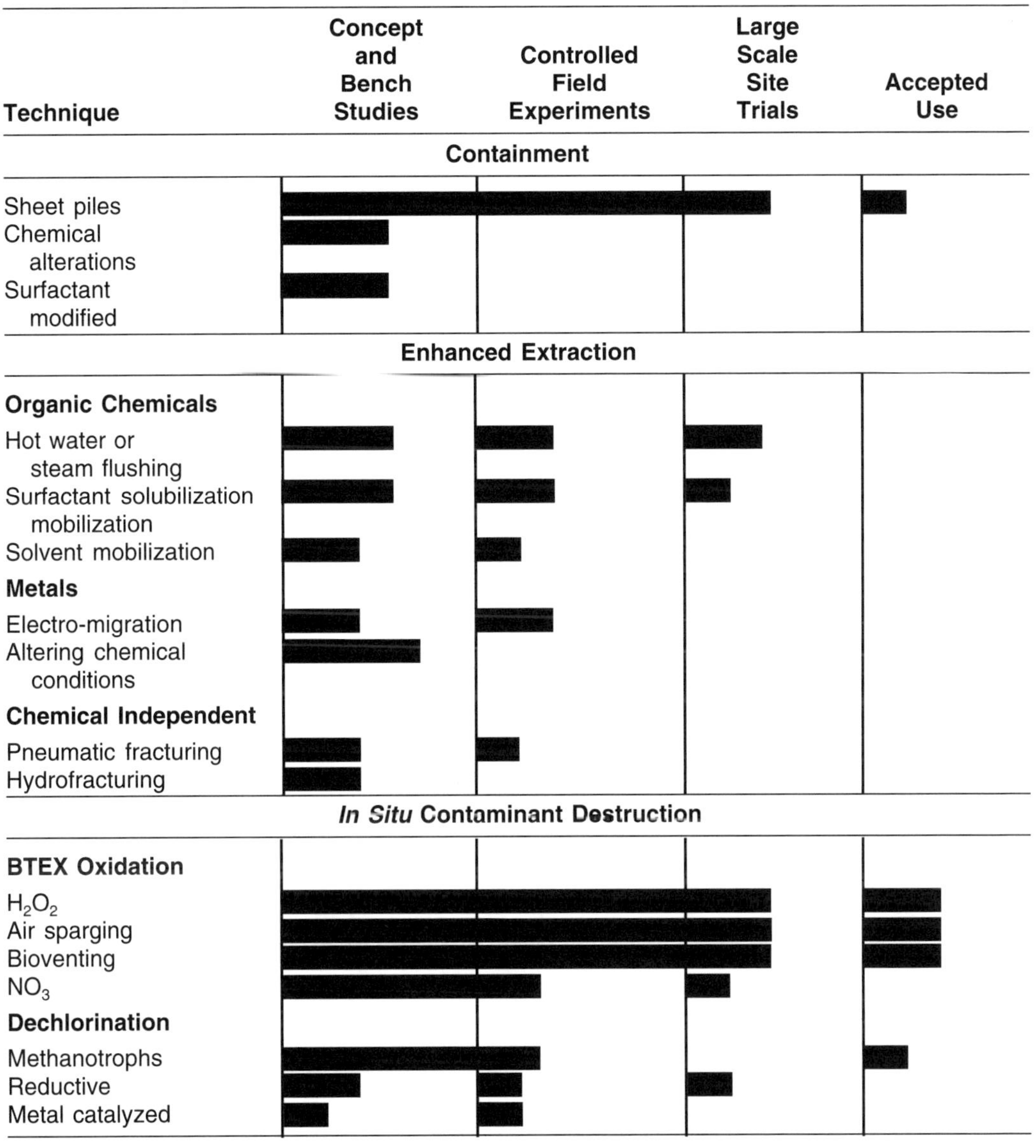

and remedial design. It is essential that techniques be developed that will optimize data collection strategies and ensure that the collected data is of adequate quality for the need. Sampling, analysis, or application of a technique should not be conducted merely because of an ability to make the measurement with a high degree of accuracy, but because the results are required to answer the proposed question(s).

DELINEATION—DEGREE OF CONTAMINATION

The first step in the remediation process is generally to delineate the spatial extent of groundwater contamination. Since the first indication of contamination is frequently the water supply, delineation of pollutants is often biased toward concentration in the groundwater and misses sediments and/or vadose (unsaturated) zone degradation. When wells are drilled to collect groundwater samples and obtain information on the extent of contamination, it is a good idea to collect soil samples and describe,

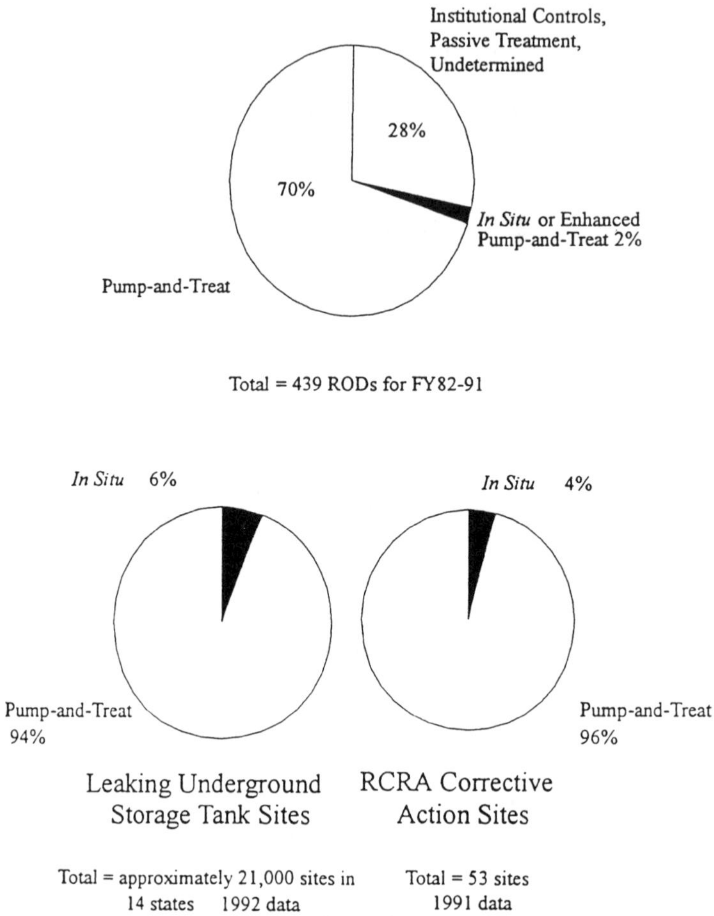

Figure 28.2. Current usage of innovative technologies by program.

hydrogeologically and chemically, the sediments during the well installation. This gives information regarding the hydrogeology of the system and can be used to estimate how fluids and contaminants will move. Obtaining hydrogeologic information adds little to the cost of the data collection. Ultimately, the additional information will add significantly to an ability to design a remediation system.

At this point in the investigation, data needs and objectives should be analyzed so that the level of information obtained is adequate to answer the question(s) posed. For example, if the objective is to delineate where the contaminant is present at concentrations above the maximum concentration level (MCL), then only a yes or no answer is required from the analyst; and there is no need for quantification in the aqueous phase to a high degree of accuracy. However, if the objective is to determine the mass of contaminant present, quantification in the aqueous phase would be required, as well as analysis of the stationary (sediment) and immiscible phases. Aqueous phase analysis and estimation techniques for the stationary phase using organic carbon contents may be adequate for the dissolved plume, but not in the source area where "free-product," nonaqueous phase may be present. Contaminant mass associated with the stationary and immiscible phases cannot presently be accurately evaluated from aqueous samples alone.

SAMPLE HANDLING PROBLEMS

Traditional handling of soil material which includes drying, sieving and mixing of soil samples to achieve homogeneous sub-samples can cause a significant loss of mass of the contaminant from site materials that contain volatile components. Losses of nonaqueous fluids can also occur during this process, which will significantly affect the measurement of partition coefficients. Consider the measurement of a partition coefficient for naphthalene at a site also containing JP-4 fuel. The standard procedure at RSKERL is to collect material from the depth of interest using a hollow-stem auger. The core is extruded in the field and the outside of the core is pared in a glovebox to minimize contamination from sediments above. The unconsolidated material is sealed in a glass jar and shipped back to the laboratory on ice to minimize the effect of biological activity on the sample. When the biological integrity of the sample is not critical, samples for chemical activity are extracted immediately in the field to minimize volatile losses. The material is taken directly from the sample container and repacked into a soil column where it is resaturated with deaerated water at a constant ionic strength, typical of groundwater at the site. Saturation is determined by measuring a series of breakthrough curves using $^{3}H_2O$ (tritium) and observing when the water-filled porosity no longer changes. A naphthalene pulse is then displaced through the saturated column and used to estimate the partition coefficient for the material. In Figure 28.3a, there is a series of breakthrough curves each of which can be used to obtain a partition coefficient for naphthalene. The first breakthrough curve in Figure 28.3a represents naphthalene displacement through a column that had been flushed previously with just enough water to resaturate the soil (labeled moist contaminated). The second curve represents a breakthrough curve that was obtained by first drying the soil material prior to packing it into the column and then performing the experiment as above. The third breakthrough curve is for a similar soil material but from an area that had not been contaminated (labeled pristine). The largest partition coefficient (the column requiring the greatest number of pore volumes of water to achieve breakthrough) was obtained with the soil that was packed at field moisture content with minimal handling. The column where the soil was dried yielded a lower partition coefficient. Similar results are observed in Figure 28.3b, demonstrating the effect of passing excess water through the column before the measurements are made. The more water that is flushed through the column, the lower the apparent partition coefficient. The discrepancies in these breakthrough curves are the result of changes in the nature of the sorbent, i.e., the sediment. The sorbent changed as a result of the soil handling and column preparation. Most likely, these changes reflect an alteration in the composition and quantity of the organic carbon in the system. Significantly different estimations of contaminant mass or cleanup times would be obtained from these results. Several conclusions can be made from these observations, as follows:

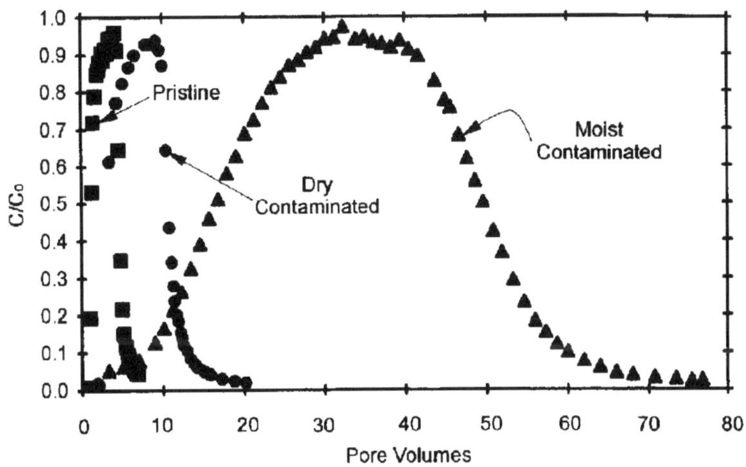

Figure 28.3a. Naphthalene displacement through soils with different sample handling (packed at field moisture content and dried prior to packing compared to a similar soil which was not contaminated).

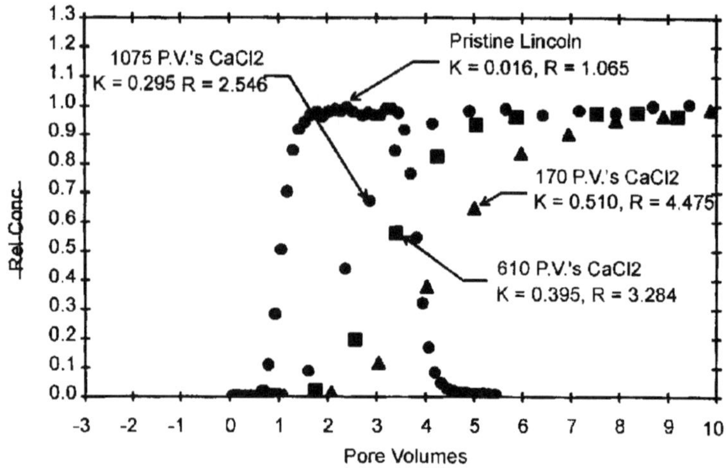

Figure 28.3b. Effect of passing excess water through soil column before taking measurements.

1. Sample handling is important in site characterization, and care must be taken to ensure the sample is representative of the site materials.
2. The true *in situ* value of partition coefficients is probably greater than any of the measurements made. Losses are likely to occur using current techniques which cause an underestimation of parameters. There is a need to develop better techniques for obtaining the correct *in situ* value. Alternatively, there is a need for techniques to estimate the magnitude and direction of the error generated using the current procedures.
3. The partition coefficients measured in these experiments can be used to estimate the time required for conventional pump-and-treat remediation. Models currently in use calculate remediation time assuming a constant partition coefficient. The partition coefficient does not have a constant value due to the heterogeneity in the original material; and as the remediation progresses, the partition coefficient is likely to decrease. Research is needed to determine if assuming a constant partition coefficient will adversely impact decisions regarding remediation strategies. A correct model for the source area would account for the contaminant of interest as well as the other relatively soluble contaminants using a relationship such as Raoult's law. Raoult's law applied to the source area predicts that the concentration of a contaminant in the mobile aqueous phase will be proportional to the mole fraction in the stationary phase times the water solubility of the contaminant. In a dynamic remediation system, the apparent partition coefficient will vary as the more soluble compounds are leached from the system.

PHYSICAL AND CHEMICAL HETEROGENEITIES

Physical and chemical heterogeneities may pose serious limitations to remediation alternatives. Traditional pump tests, which integrate the influence of several stratigraphic units, are inadequate when designing remediation systems if one objective is a knowledge of the time required to achieve the remedial objective. Most remediation technologies require the movement of mass into or out of a contaminated region to achieve remediation. In an effort to evaluate the importance of heterogeneity, a simplified approach is proposed which extends the conceptual model presented by Millington and Quirk (1961). They described the unsaturated hydraulic conductivity function of a porous medium by describing the flow of water through a series of capillary tubes with selected diameters which would mimic the pore size distribution of the soil. This approach accounts for both the imbibition and drainage hysteretic nature of the moisture characteristic curve and hydraulic conductivity.

Using a similar approach but employing a Darcy scale rather than a pore scale analogy and applying it to chemical transport rather than determining hydraulic properties was proposed. In the analogy shown conceptually in Figure 28.4, each of N Darcy scale units would have a characteristic hydraulic

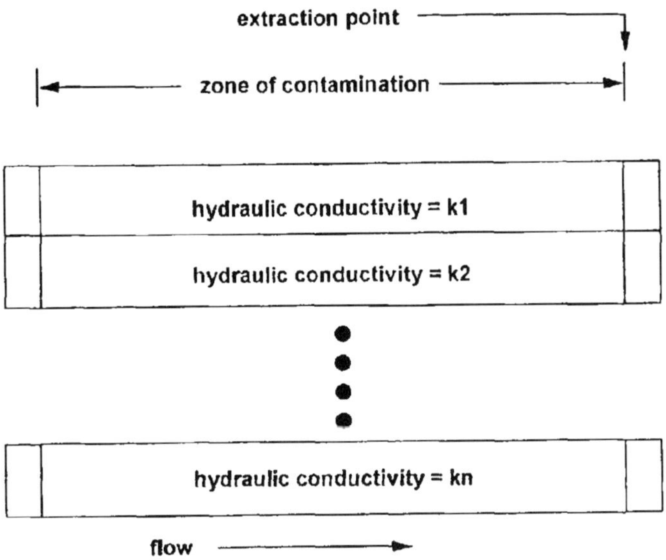

Figure 28.4. Conceptual model of soil profile with N soil layers. Each layer having characteristic hydraulic conductivity, dispersivity and retardation factor.

conductivity, characteristic dispersivity, and characteristic partition coefficient which could be measured using laboratory techniques. Each of the N units would represent an equal fraction of the area normal to the flow field. The hydraulic conductivity distribution is used to weight the flux through each of the N layers. Chemical transport is calculated using a one-dimensional, convective-dispersive solution; and the contaminant flux is recombined using the principal of superposition at the point of interest to describe the net contaminant flux for the system. A significant limitation of this approach is that it does not allow for diffusion from one flow path to another even though there is a concentration gradient. The reader is referred to the appendix to this chapter for a presentation of the equations used based on a convective dispersive equation in one dimension. Assuming a log-normal distribution of the hydraulic conductivity (e.g., Nielsen et al., 1973; Warrick et al., 1977; and Sisson and Wierenga, 1981) and a constant dispersivity (d = 0.1 L) (Freeze and Cherry, 1979), the effect of the variance (the square of the standard deviation) in hydraulic conductivity can be observed (Figures 28.5a and b). In Figure 28.5a and b, R = 1 (no partitioning of the compound to the stationary phase(s) was allowed) and a unit total flux was applied to the system. Therefore, Figure 28.5a and b displays the effect of variance in hydraulic conductivity on the remediation of a uniformly mixed, non-sorbing conservative contaminant such as ^3H_2O (neglecting radioactive decay). Figure 28.5a displays the results in a conventional manner, while Figure 28.5b presents the log of the relative concentration as a function of time. Figure 28.5b is more informative for evaluating remediation success since a three- to five-log reduction in concentration is frequently required to achieve cleanup goals. The figure shows that, if there is a significant amount of variance in the hydraulic conductivity, the amount of time required to reach cleanup goals is much greater than would be calculated using more traditional assumptions of homogeneous materials.

The log-normal distribution assumption was based on observations of spatial changes in hydraulic conductivity of surface soils. There is no particular reason to assume that the vertical distribution of hydraulic conductivity should be log-normally distributed. Vertical distributions are more likely related to depositional episodes, while the change within a given episode is described as log-normally distributed. Figure 28.6 demonstrates the effect of a two-layer system where one curve reproduces the homogeneous conditions displayed in Figure 28.5b. The second curve shows the effect of one layer having ten times the hydraulic conductivity of the second. The final curve shows the effect of having a higher partition coefficient in the finer-textured soil (where the hydraulic conductivity is lower). Both of the inhomogeneous curves look similar in shape to what is reported at pump-and-treat

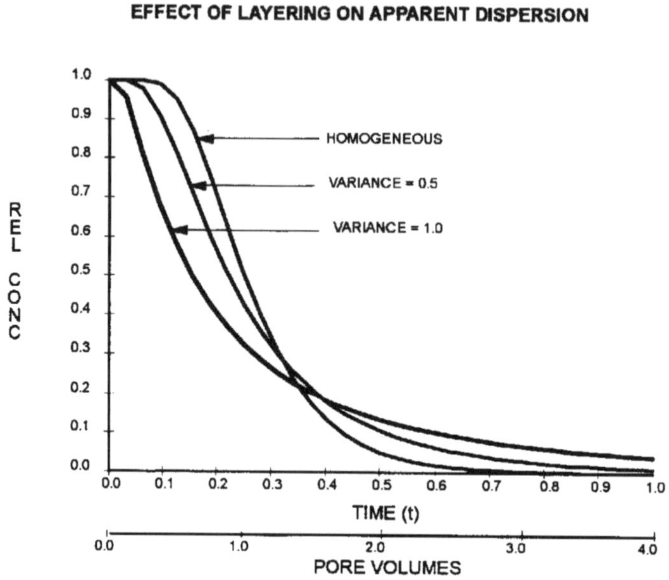

Figure 28.5a. Effect of layering on the remediation of a conservative non-reactive contaminant.

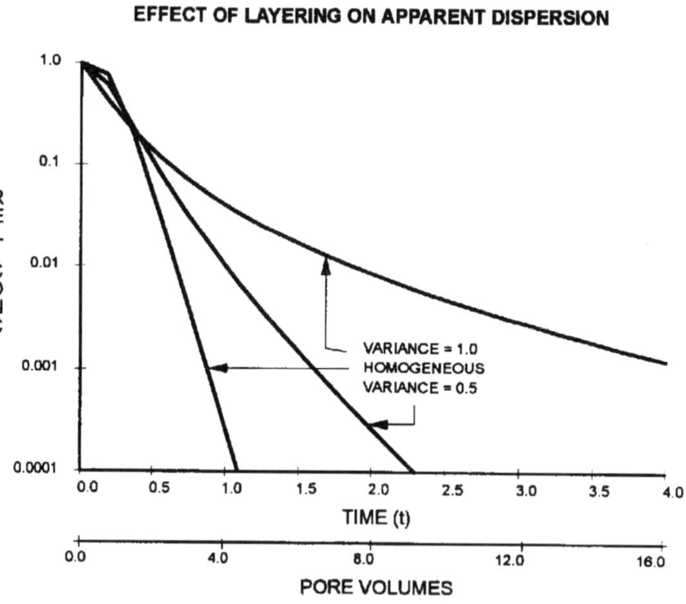

Figure 28.5b. Log-transformed relative concentration versus pore volumes demonstrating the effect of layering on the time required to achieve cleanup objectives for a conservative non-sorbing contaminant.

systems where the concentration is reported to start high and fall to some asymptotic value generally greater than the MCL.

In this discussion, a hypothetical example was used with a simplistic model to illustrate the importance of determining a detailed description of the hydrogeology when remediation is the objective.

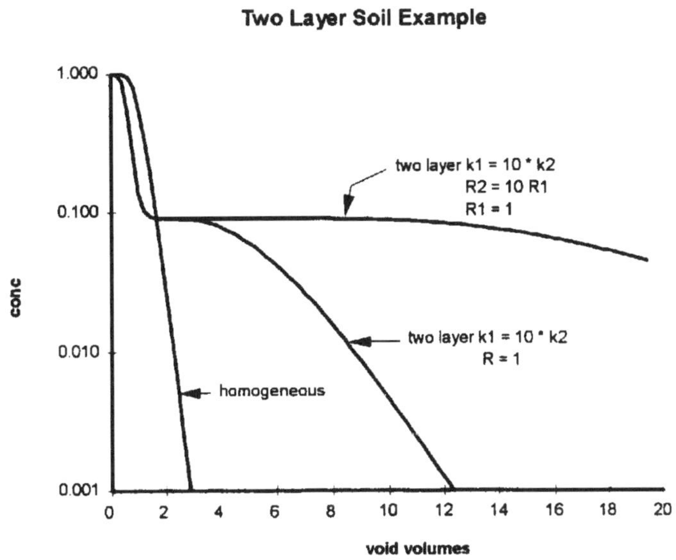

Figure 28.6. Estimated volume of water required to achieve cleanup when two soil layers with selected properties are present.

REMEDIATION TECHNOLOGY

Pump-and-treat is an accepted technology and the foundation for most of the innovative technologies. Pump-and-treat remediation has been receiving negative reviews in recent years because even after several years of operation, contaminant concentrations remain above target levels. The utility of any remedial technology should be evaluated with respect to stated remedial objectives. Before rejecting traditional pump-and-treat, consider the suitability of pump-and-treat related to containment and extraction. Pump and treat technology can prevent the migration of contamination in the dissolved phase from moving beyond property lines. For containment, a gallery of extraction wells near the property line with an aboveground treatment system is often used. Proper design and operation of a system will prevent contaminant movement beyond the capture zone while extracting a minimum amount of water. Excess pumping will result in the contamination of currently pristine groundwater by pulling it through contaminated soils or the mixing of contaminated waters with pristine waters. It is not reasonable to assume that excess pumping will significantly accelerate the rate of aquifer remediation.

Pump-and-treat is potentially effective for the objective of extracting a dissolved plume emanating from a source of contamination. If the actual source area can be removed or contained such that it does not continuously feed the plume, then pump-and-treat remediation should be effective in removing the dissolved plume. Plume removal using pump-and-treat is much more difficult than containment. The objective can be achieved by pumping just the amount of water that is flowing through the system. However, it is not possible to estimate the volume of water to be pumped without a detailed knowledge of the system's physical and chemical heterogeneities. Physical heterogeneities control the spatial variability of the hydraulic conductivity, while chemical heterogeneities cause nonuniform distribution of the contaminant in the system as well as variability in the partitioning or sorption of the contaminant. If heterogeneities could be characterized, it would be possible to optimally design a pumping system and estimate the rate at which the aquifer will be remediated.

The third application of pump-and-treat has an objective of extracting contaminants in the "source area" such that the entire aquifer could be returned to its previous unrestricted use. The "source area," as used here, refers to the volume of the subsurface where the contamination is present in excess of the amount that can be dissolved with respect to that in equilibrium with natural pristine soil or sediment. It is not limited to the immediate area where the contaminant was released. It is estimated that more than 90% of the organic contaminant mass is associated with either mobile or immobile nonaqueous phase fluids in the source area which are only sparingly soluble in water. It would be physically impossible for water to dissolve these nonaqueous fluids under natural gradients in reasonable time frames. Pump-and-treat, as currently practiced, will most likely be ineffective in remediating this area. It is within the source area, however, that there is a maximum potential for reducing long-term risk while minimizing the incremental cost of risk reduction. Detailed spatial data are required in the source area; however, the ability to obtain this information is currently inadequate.

IN SITU CONTAMINANT DESTRUCTION

Contaminant destruction strategies generally require two or more chemicals to be present at essentially the same place and time for a reaction to take place. A question arises as to what is the rate-limiting step—chemical reaction kinetics or the physical mixing? To evaluate this question, consider a one-dimensional analysis of the equation proposed by Yates and Enfield (1989). A solution to this equation (Equation 28.9 in the appendix to this chapter) can be used to examine the effect of mixing by dispersion, the effect of transformation rate, and the effect of mixing because one chemical moves at a different effective rate than the other due to a difference in partitioning. A qualitative conclusion can be reached based on which processes appear to be most important. If it is assumed that the entire system is initially at equilibrium with only Compound 1 equally distributed throughout the system, then the system can be observed as a pulse of Compound 2 is injected into the aquifer, as shown diagrammatically in Figure 28.7. The results are displayed as spatial distribution of Compound 1 and 2 at a point in time (t). Transformation can take place only when both compounds are present. The results of the analysis are presented in Figures 28.8 through 28.14, which show the distribution of both Compounds 1 and 2 at the same elapsed time (t) after the introduction of a pulse of Compound 2. First consider the effect of dispersion where Compounds 1 and 2 both move at the same rate (the retardation factor (R) is the same for both 1 and 2) and the reaction rate is fast (0.005). This situation might represent the remediation of a nitrate-contaminated aquifer by the injection of an alcohol (e.g., methanol) to act as an electron donor. There is no significant retardation of either nitrate or methanol. It is known from aboveground test cells that if anaerobic conditions exist, the denitrification process is relatively fast. Figures 28.8 and 28.9, then, display the solution to Equation 28.9 in the appendix to this chapter. The greater the dispersion, the greater the area containing both compounds and the greater the removal of both nitrate and alcohol. The effect of changing the transformation rate, as shown in Figures 28.8 and 28.10 through 28.12, indicates that there is some qualitative limit where an increase in the transformation rate no longer makes a significant difference in the amount of nitrate or alcohol removed because the reaction rates are large compared to the residence times of the compounds, resulting in the consumption of both compounds. On the other hand, if the reactions are slow, there is good mixing without effective removal (Figure 28.10). In the next example (Figures 28.8, 28.13 and 28.14), consider the effect of differences in the retardation factor. If, for example, a nitrate-contaminated aquifer is to be remediated by adding a hydrophobic electron donor, the nitrate would move faster in the aquifer than the electron donor. The greater the contrast in relative velocities, the better the ability to achieve remediation.

Given that the contaminant residence times in aquifers are generally large in relation to many transformation rates, research emphasis should be placed on the elucidation of mixing processes and transformation pathways. If environmental conditions are favorable to transformation, and the required constituents are present, effective remediation should be achieved.

ENHANCED EXTRACTION

Several extraction strategies involve the injection of nonaqueous phase fluids such as surfactants or co-solvents. When one of these remedial solutions passes through a contaminated region, it is expected that the concentration will increase in the mobile phase. As an example of this type of problem, consider the addition of a co-solvent to a contaminated system. One-dimensional equations to describe the process have been developed by Augustijn et al., 1994 (Appendix Equations A.15 and

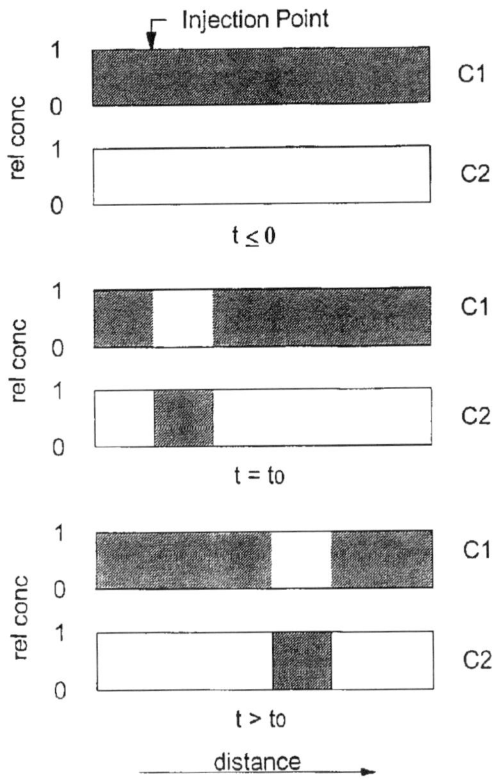

Figure 28.7. Conceptualization at three points in time of a system with two reactive components flowing. Both components have the same retardation factor and no dispersion. Compound 2 is injected for a period of time t_0. Shaded area indicates where component is present.

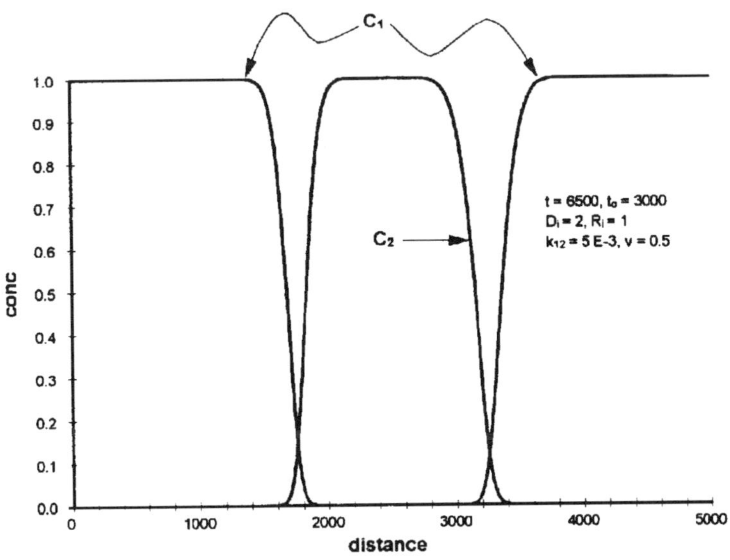

Figure 28.8. Spatial distribution of two compounds at time 6500 that react with each other. Neither compound is retarded by the soil. Both compounds have the same dispersion. Compound 2 was injected for 3000 units of time. The interstitial velocity of the fluid is 0.5.

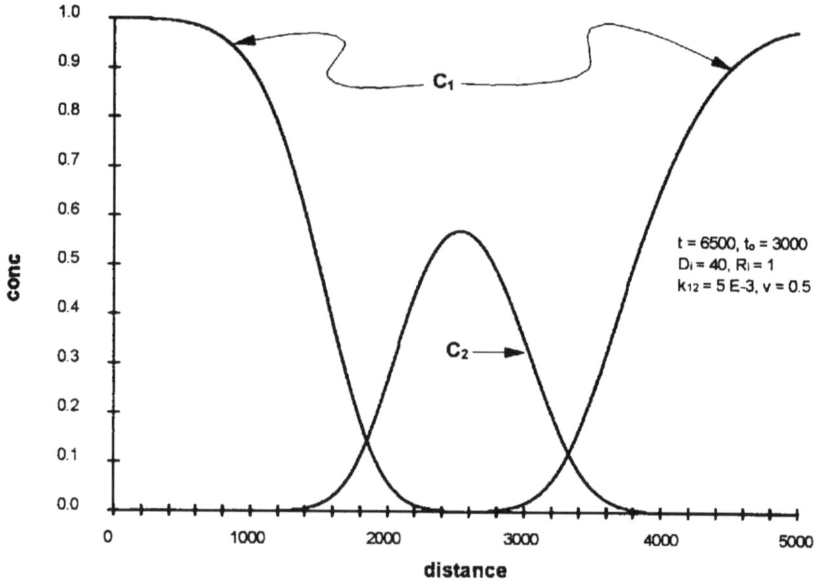

Figure 28.9. The spatial distribution of two compounds at time 6500 that react with each other. Neither compound reacts with the soil. Both compounds have the same dispersion coefficient 40. Compound 2 was injected for 3000 units of time. The interstitial velocity of the fluid is 0.5.

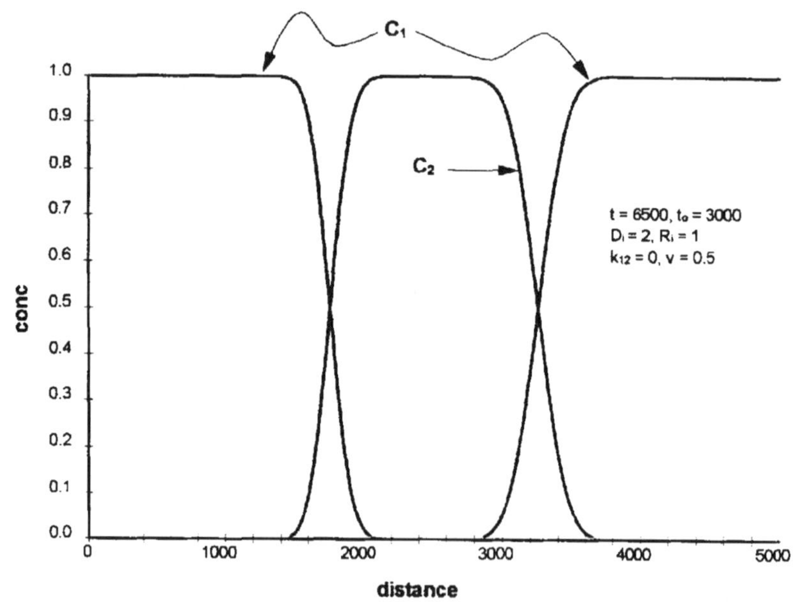

Figure 28.10. Spatial distribution of two compounds at time 6500 that do not react with each other. Neither compound is retarded by the soil. Both compounds have the same dispersion. Compound 2 was injected for 3000 units of time. The interstitial velocity of the fluid is 0.5.

A.16). Solutions to the equation yield results shown in Figure 28.15. Since the reaction can be projected in one dimension, it is often assumed that there is sufficient knowledge to perform remediation in the field. However, the problem of delivering the co-solvent to the contamination region and forcing the co-solvent through the contamination may be a serious limitation; and three-dimensional pore-scale variability or changes in hydraulic properties resulting from partial remediation are not

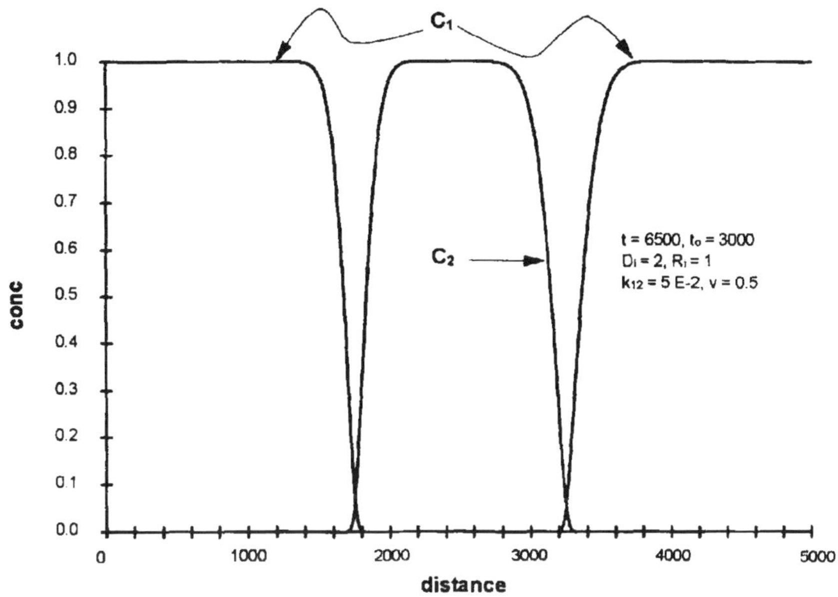

Figure 28.11. Spatial distribution of two compounds at time 6500 that react with each other at a rate of 0.05. Neither compound is retarded by the soil. Both compounds have the same dispersion. Compound 2 was injected for 3000 units of time. The interstitial velocity of the fluid is 0.5.

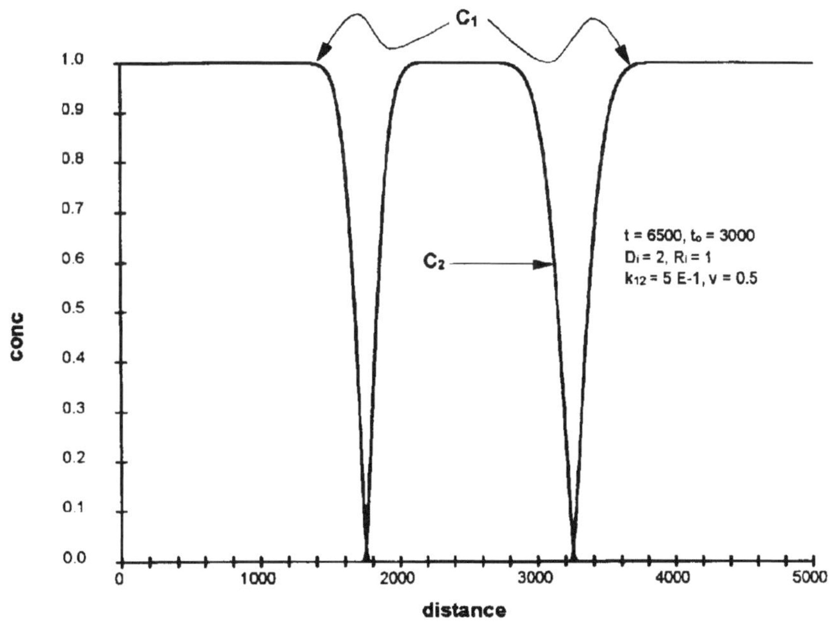

Figure 28.12. Spatial distribution of two compounds at time 6500 that react with each other at a rate of 0.5. Neither compound is retarded by the soil. Both compounds have the same dispersion. Compound 2 was injected for 3000 units of time. The interstitial velocity of the fluid is 0.5.

currently included in the transport equation. Design models which address field-scale complexities are needed, and these models should be evaluated using data from pilot- or field-scale tests of the remediation technologies.

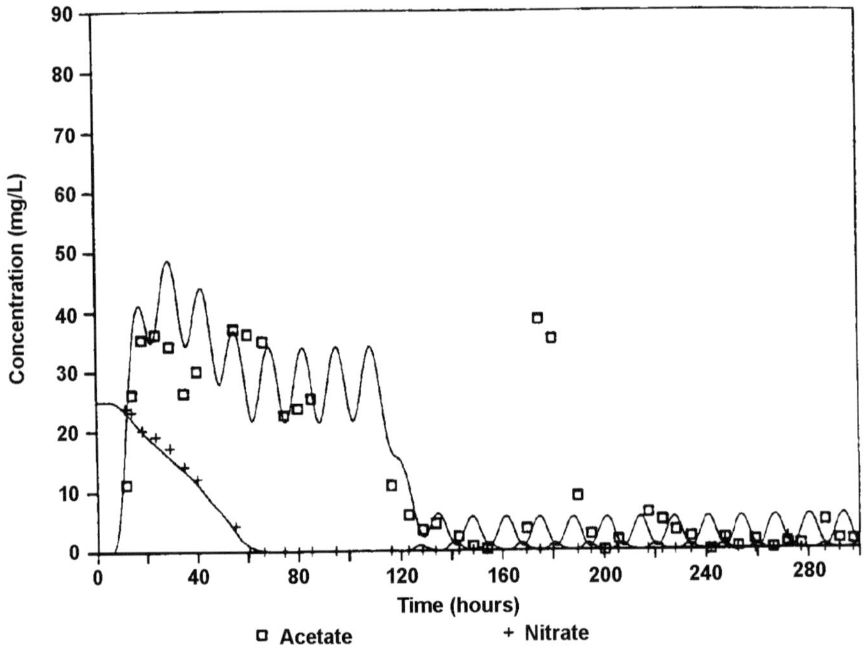

Figure 28.13. Spatial distribution of two compounds at time 6500 that react with each other at a rate of 0.005. Compound 1 is not retarded by the soil but Compound 2 has a retardation factor of 1.5 . Both compounds have the same dispersion. Compound 2 was injected for 3000 units of time. The interstitial velocity of the fluid is 0.5.

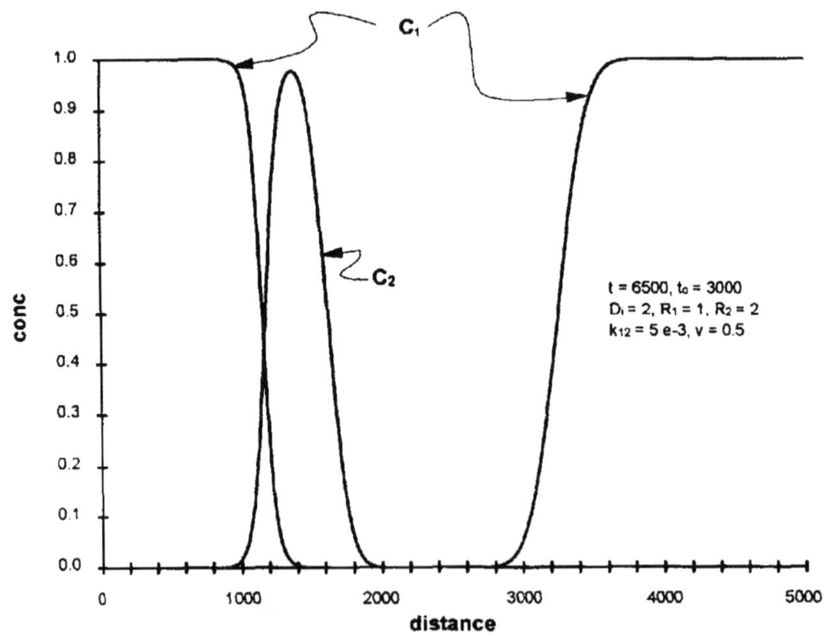

Figure 28.14. Spatial distribution of two compounds at time 6500 that react with each other at a rate of 0.005. Compound 1 is not retarded by the soil but Compound 2 has a retardation factor of 2.0. Both compounds have the same dispersion. Compound 2 was injected for 3000 units of time. The interstitial velocity of the fluid is 0.5.

DESIGN MODELS

To advance from the implementation of traditional pump-and-treat systems to the more innovative technologies, remedial design models for containment, extraction, and *in situ* destruction will need to be developed. There are numerous model codes in existence which describe the transport and transformation of contaminants in aqueous systems. Many of these codes, if they are implemented correctly, can be utilized to design hydraulic containment systems for dissolved plumes of contaminants. As one approaches the more innovative remediation technologies, the number and appropriateness of model codes are dramatically reduced. Remedial design models for extraction and destruction are significantly different from exposure models, which are generally limited to relatively dilute systems and do not need to consider how the fluid changes the system. Exposure models are aimed at assessing the risk of an activity. The actual location of the risk is frequently not critical. Design models, on the other hand, need to consider differences in fluid properties, how injected fluids and energy will react with resident fluids and hydrology, and particle transport.

A considerable emphasis of an *in situ* design model will be the delivery of the remedial fluid or energy to a selected target. The scale of resolution for a remedial design model will frequently be on the order of tens to hundreds of centimeters, while the scale for an exposure model will be on the order of tens to hundreds of meters. Numerical scaling problems become important with remediation models because the remedial fluid must be delivered to the particular points of contamination and not just to the general contaminated area. Significant numerical problems (e.g., numerical dispersion, stability) exist with most models being used. Biological and chemical reactions are often complex. For example, removal of one compound leads to the subsequent production of others, which in turn continue to be transformed. Some reactions produce compounds which may temporarily restrict flow in the aquifer. These changes and their effect on delivery systems are not yet being included in remedial design models.

Induced gas flow to enhance remediation is not yet fully understood, particularly below the water table. Few design models have been developed and none of the existing models have been adequately evaluated under anisotropic field conditions. Heat flow has been studied as much as any other process in the subsurface environment due to the efforts related to nuclear waste disposal. Codes to describe heat flow are reasonably accurate but must be coupled to mass transfer codes to be appropriate to remediation. Again, if innovative treatment technologies are to be implemented by technologists, remedial design models must be developed, evaluated, and distributed with adequate training.

CONCLUSIONS

Subsurface remediation technology is in its early stages of development and will need significant funding for the next several decades. With respect to subsurface remediation, there are four primary areas needing research: (1) the continued development of the basic transport and transformation pathways for both contaminants and remedial materials/energy; (2) techniques for delineating the physical and chemical heterogeneity of the subsurface environment to the required accuracy and precision, along with the implications of this heterogeneity on remedial technologies; (3) the development of methods to evaluate the effectiveness of remediation technologies; and (4) the development and verification of remedial design models for each of the *in situ* destruction, extraction or containment technologies. Each of these research areas presents significant challenges. The research efforts need to be conducted in parallel, integrating basic process information, site characterization methodologies, and design models for a particular remediation technology. The findings need to be utilized to develop design guidance manuals which include a comparison between an installed technology and a no-intervention alternative.

Teams need to be organized around existing potential technologies which can create the interdisciplinary environment required to develop a successful remediation technology. Field testing is an important aspect of remediation technology development and should be conducted through this interdisciplinary approach, while basic process research is the foundation of subsurface remediation technology and should be continued to further develop understanding of transport and transformation of chemicals in the environment.

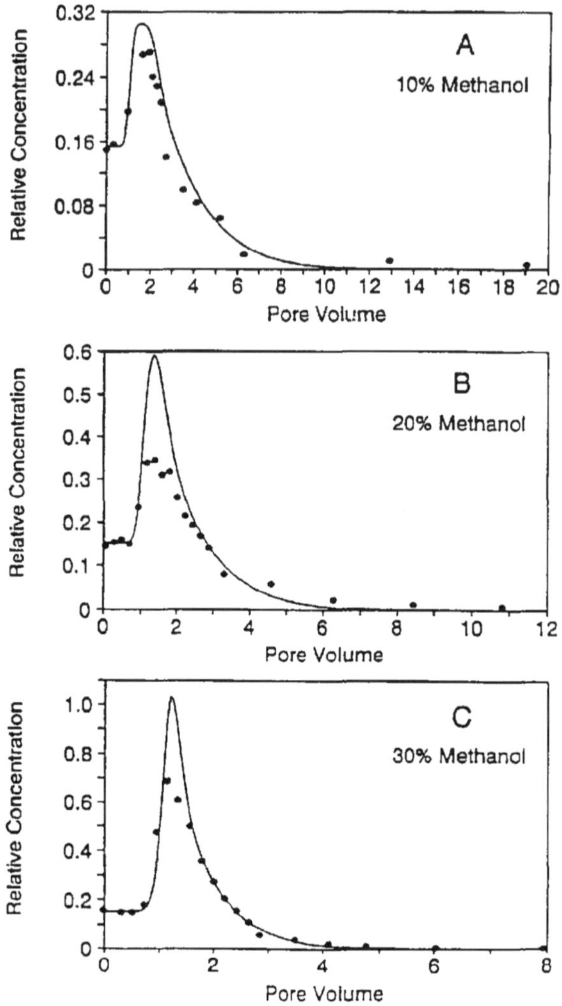

Figure 28.15. Experimental data (filled circles) and independent model predictions (solid lines) for naphthalene elution curves at three different cosolvent fractions.

APPENDIX

Conservative one-dimensional, convective-dispersive transport can be described by the equation

$$R \frac{\partial c}{\partial t} = D \frac{\partial^2 c}{\partial x^2} - v \frac{\partial c}{\partial x} \tag{A.1}$$

where:

R = retardation factor;
c = concentration of the contaminant (M/M);
D = hydrodynamic dispersion coefficient (L^2/T);
t = time (T);
x = distance along the flow path (L); and
v = interstitial fluid velocity (L/T).

Equation A.1 was solved with flux boundary conditions

$$c(x,0) = c_i \tag{A.2}$$

$$\left(-D \frac{\partial c}{\partial x} + vc\right)\Bigg|_{x=0} = \begin{cases} vc_0 & 0 < t < t_o \\ 0 & t > t_o \end{cases} \tag{A.3}$$

$$\frac{\partial c}{\partial x}(\infty, t) = \text{finite} \tag{A.4}$$

after van Genuchten and Alves (1982) as

$$c(x, t) = \begin{cases} c_i + (c_o - c_i) \quad A(x, t) & 0 < t < t_o \\ c_i + (c_o - c_i) \quad A(x, t) - c_o \quad A(x, t - t_o) & t > t_o \end{cases} \tag{A.5}$$

where

$$\begin{aligned} A(x, t) = \frac{1}{2} \operatorname{erfc}\left[\frac{Rx - vt}{2(DRt)^{1/2}}\right] + \left[\frac{v^2 t}{\pi DR}\right]^{1/2} \exp\left[-\frac{(Rx - vt)^2}{4DRt}\right] \\ - \frac{1}{2}\left[1 + \frac{vx}{D} + \frac{v^2 t}{DR}\right] \exp\left[\frac{vx}{D}\right] \operatorname{erfc}\left[\frac{Rx + vt}{2(DRt)^{1/2}}\right] \end{aligned} \tag{A.6}$$

Freeze and Cherry (1979) suggested that the dispersion coefficient can be described by the equation

$$D = D^* + dv \tag{A.7}$$

where

D^* = molecular diffusion of the solute in the porous medium (L^2/T); and
d = dynamic dispersivity (L).

Bresler (1973) suggested that D^* could be estimated by the equation:

$$D^* = D_o \, a \exp(b\theta) \tag{A.8}$$

where

D_o = molecular diffusion of the solute in the aqueous phase (L^2/T);
a = empirical constant with a value ranging from 0.001 to 0.005;
b = empirical constant with a value of approximately 10; and
θ = aqueous filled porosity (L^3/L^3).

One-dimensional convective-dispersive transport with second-order transformation can be described by the equation

$$R_i \frac{\partial C_i}{\partial t} = D_i \frac{\partial^2 C_i}{\partial x^2} - v_i \frac{\partial C_i}{\partial x} - k_{12} C_1 C_2 \qquad i = 1, 2 \tag{A.9}$$

where

k_{12} = the second order transformation rate for the reaction of Chemical 1 with Chemical 2.

Equation A.9 was solved numerically with the boundary and initial conditions:

$$C_1\ (x, 0) = 1 \tag{A.10}$$

$$C_2\ (x, 0) = 0 \tag{A.11}$$

$$C_1\ (0, t) = \begin{cases} 0 & 0 < t \leq t_o \\ 1 & t > t_o \end{cases} \tag{A.12}$$

$$C_2\ (0, t) = \begin{cases} 1 & 0 < t \leq t_o \\ 0 & t > t_o \end{cases} \tag{A.13}$$

$$\frac{\partial C_i}{\partial x}\ (L, t) = 0 \qquad i = 1, 2 \tag{A.14}$$

Augustijn et al. (1994) have described the extraction of a nonaqueous immobile fluid by a co-solvent with the equations:

$$\frac{\partial X*}{\partial p} + \frac{\partial F(R-1)C*}{\partial p} + \frac{\partial S*}{\partial p} = \frac{1}{P}\frac{\partial^2 C*}{\partial X^2} + \frac{\partial C*}{\partial X} \tag{A.15}$$

$$\frac{\partial S*}{\partial p} = \omega\big((1-F)(R-1)C* - S*\big) \tag{A.16}$$

Equations A.15 and A.16 were solved numerically with initial conditions:

$$S* = \frac{(1-F)(R-1)}{R} \qquad p = 0;\ 0 \leq X \leq 1 \tag{A.17}$$

$$C* = \frac{1}{R} \qquad p = 0;\ 0 \leq X \leq 1 \tag{A.18}$$

and boundary conditions:

$$C* - \frac{1}{P}\frac{\partial C*}{\partial X} = 0 \qquad p > 0;\ X = 0 \tag{A.19}$$

$$\frac{\partial C*}{\partial X} = 0 \qquad p > 0;\ X = 1 \tag{A.20}$$

with non-dimensional parameters:

$$C^* = \frac{\theta C}{M} \tag{A.21}$$

$$S^* = \frac{\rho S_2}{M} \tag{A.22}$$

$$p = \frac{vt}{L} \tag{A.23}$$

$$X = \frac{x}{L} \tag{A.24}$$

$$P = \frac{vL}{D} \tag{A.25}$$

$$R = 1 + \frac{\rho}{\theta} K_p \tag{A.26}$$

$$\omega = \frac{k_2 L}{v} \tag{A.27}$$

C = concentration in solution phase (M/L^3);
C* = relative concentration;
F = fraction of instantaneous sorption domain;
K_p = equilibrium sorption coefficient (L^3/M);
k_2 = reverse first-order sorption rate coefficient (T^{-1});
L = length of flow path (L);
M = total initial contaminant mass in the column (M/L^3) = θRC_o, where C_o = initial concentration (M/L^3);
P = Peclet number;
p = dimensionless time in pore volumes;
S_2 = sorbed concentration in the rate-limited domain (M/M);
S* = normalized sorbed concentration in the rate-limited domain;
X = relative distance in the column;
ρ = dry bulk density (M/L^3); and
ω = Damkohler number

DISCLAIMER

The information in this chapter has been funded by the United States Environmental Protection Agency. It has not been reviewed in accordance with the U.S. Environmental Protection Agency's peer and administrative review policies. Statements made do not constitute Agency endorsement or recommendation.

REFERENCES

Augustijn, D.C.M., R.E. Jessup, P.S.C. Rao, and A.L. Wood. 1994. Remediation of contaminated soils by solvent flushing. *J. Environ. Eng.,* **120**:42–57.

Bresler, E. 1973. Simultaneous transport of solutes and water under transient unsaturated flow conditions. *Water Resources Research,* **9**:975–986.

Freeze, R.A. and J.A. Cherry. 1979. *Groundwater.* Prentice-Hall. Englewood Cliffs, NJ.

Millington, R.J. and J.P. Quirk. 1961. Permeability of solids. *Trans. Faraday Soc.,* **57**:1200–1206.

Nielsen, D.R., J.W. Bigger, and K.T. Erh. 1973. Spatial variability of field-measured soil-water properties. *Hilgardia,* **42**:215–260.

Sisson, J.B. and P.J. Wierenga. 1981. Spatial variability of steady state infiltration rates as a stochastic process. *Soil Sci. Soc. Am. J.,* **45**:699–704.

Steimle R.R. 1993. An Inventory of Research and Field Demonstrations and Strategies for Improving Ground Water Remediation. EPA 500/K-93/001.

van Genuchten, M.Th. and W.J. Alves. 1982. Analytical Solutions of the One-Dimensional Convective-Dispersive Solute Transport Equation. United States Department of Agriculture. Technical Bulletin No. 1661.

Warrick A.W., G.J.Mullen, and D.R. Nielsen. 1977. Scaling field-measured soil hydraulic properties using a similar media concept. *Water Resources Research,* **13**:355–362.

Yates, S.R. and C.G. Enfield. 1989. Transport of dissolved substances with second-order reaction. *Water Resources Research,* **25**:1757–1762.

Index